Cooperative Communications for Improved Wireless Network Transmission:
Framework for Virtual Antenna Array Applications

Murat Uysal
University of Waterloo, Canada

INFORMATION SCIENCE REFERENCE

Information Science REFERENCE

Hershey · New York

Director of Editorial Content:	Kristin Klinger
Senior Managing Editor:	Jamie Snavely
Assistant Managing Editor:	Michael Brehm
Publishing Assistant:	Sean Woznicki
Typesetter:	Jamie Snavely
Cover Design:	Lisa Tosheff
Printed at:	Yurchak Printing Inc.

Published in the United States of America by
Information Science Reference (an imprint of IGI Global)
701 E. Chocolate Avenue
Hershey PA 17033
Tel: 717-533-8845
Fax: 717-533-8661
E-mail: cust@igi-global.com
Web site: http://www.igi-global.com/reference

Library of Congress Cataloging-in-Publication Data

Cooperative communications for improved wireless network transmission :
framework for virtual antenna array applications / Murat Uysal, editor.
      p. cm.
  Includes bibliographical references and index.
  Summary: "This book provides readers with a comprehensive understanding of
the fundamental principles of cooperative communications with a particular
emphasis on physical layer issues and further presents the latest advances and
open research problems in the wireless network field"--Provided by publisher.
  ISBN 978-1-60566-665-5 (hbk.) -- ISBN 978-1-60566-666-2 (ebook) 1.
Wireless communication systems. 2. Antenna arrays. I. Uysal, Murat, 1973-
TK5103.2.C664 2010
  621.384--dc22
                        2009007012

British Cataloguing in Publication Data
A Cataloguing in Publication record for this book is available from the British Library.

All work contributed to this book is new, previously-unpublished material. The views expressed in this book are those of the authors, but not necessarily of the publisher.

List of Reviewers

B. Sundar Rajan, *Indian Institute of Science, India*
Muhammad Ali Imran, *University of Surrey, UK*
Aydin Sezgin, *Stanford University, USA*
Ioannis Krikidis, *University of Edinburgh, UK*
Mustafa M. Matalgah, *University of Mississippi, USA*
Sami (Hakam) Muhaidat, *Simon Fraser University, Canada*
Arumugam Nallanathan, *King's College London, UK*
Il-Min Kim, *Queen's University, Canada*
John S. Thompson, *University of Edinburgh, UK*
Ibrahim Altunbas, *Istanbul Technical University, Turkey*
Gi-Hong Im, *Pohang University of Science and Technology (POSTECH), Republic of Korea*
Athina P. Petropulu, *Drexel University, USA*
Behrouz Maham, *University of Oslo, Norway*
Are Hjørungnes, *University of Oslo, Norway*
George N. Karystinos, *Technical University of Crete, Greece*
Ibrahim Körpeoğlu, *Bilkent University, Turkey*
Diego Piazza, *Politecnico di Milano, Italy*
Shih Yu Chang, *National Tsing Hua University, Taiwan*
Hsiao-Chun Wu, *Louisiana State University, USA*
Birsen Sirkeci-Mergen, *San Jose State University, USA*
Mahinthan Veluppillai, *University of Waterloo, Canada*
Samuel Cheng, *University of Oklahoma, USA*
Zhu Han, *University of Houston, USA*
Melda Yuksel, *TOBB Ekonomi ve Teknoloji Universitesi, Turkey*
Onur Kaya, *Işık University, Turkey*
Josephine Chu, *University of Toronto, Canada*
George K. Karagiannidis, *Aristotle University of Thessaloniki, Greece*
Zhiguo Ding, *Lancaster University, UK*
Ahmed Ibrahim, *University of Maryland, USA*

Table of Contents

Melda Yuksel, TOBB University of Economics and Technology, Turkey
Elza Erkip, Polytechnic Institute of New York University, USA

Ioannis Krikidis, University of Edinburgh, UK
John S. Thompson, University of Edinburgh, UK

Onur Kaya, Işık University, Turkey
Sennur Ulukus, University of Maryland, USA

Symeon Chatzinotas, University of Surrey, UK
Muhammad Ali Imran, University of Surrey, UK
Reza Hoshyar, University of Surrey, UK

Section 2
Practical Coding Schemes for Cooperative Communications

Section 3
Distributed Transmit and Receive Diversity Techniques for Cooperative Communications

Chapter 18

Birsen Sirkeci-Mergen, San Jose State University, USA
Anna Scaglione, University of California at Davis, USA
Michael Gastpar, University of California at Berkeley, USA

Section 6
An Industrial Perspective on Cooperative Communications

Chapter 19

Mischa Dohler, CTTC, Spain
Djamal-Eddine Meddour, Orange Labs, France
Sidi-Mohammed Senouci, Orange Labs, France
Hassnaa Moustafa, Orange Labs, France

Detailed Table of Contents

Section 1
Information Theoretical Results on Cooperative Communications

Chapter 1

Melda Yuksel, TOBB University of Economics and Technology, Turkey
Elza Erkip, Polytechnic Institute of New York University, USA

This chapter provides an overview of the information theoretic foundations of cooperative communications. Earlier information theoretic achievements, as well as the more recent developments, are discussed. The analysis accounts for full/half-duplex nodes and for multiple relays. Various channel models such as discrete memory less, additive white Gaussian noise (AWGN), and fading channels are considered. Cooperative communication protocols are investigated using capacity, diversity, and diversity-multiplexing tradeoff (DMT) as performance metrics. Overall, this chapter provides a comprehensive view on the foundations of and the state-of-the-art reached in the theory of cooperative communications.

Chapter 2

Ioannis Krikidis, University of Edinburgh, UK
John S. Thompson, University of Edinburgh, UK

Amplify-and-Forward (AF) is a simple cooperative strategy for ad-hoc networks with critical power constraints. It involves an amplification of the received signal in the analogue domain at the relays without further signal processing. This chapter gives an overview of the basic AF protocols in the literature and discusses recent research contributions in this area. Based on some well-defined AF-based cooperative configurations, it focuses on the behaviour of AF in block-fading channels, in power allocation problems, in relay selection and in cross-layer coordination. Mathematical models and outage probability simulations are used in order to show the enhancements of the presented AF techniques.

Chapter 3
Onur Kaya, Işık University, Turkey
Sennur Ulukus, University of Maryland, USA

In this chapter, we review the optimal power allocation policies for fading channels, in single user and multiple access scenarios. We present some background on cooperative communications, starting with the relay channel, and moving onto mutually cooperative systems. Then, we consider power control and user cooperation jointly, and for a fading Gaussian multiple access channel (MAC) with user cooperation, we present a channel adaptive encoding policy which relies on block Markov superposition coding. We obtain the power allocation policies that maximize the average rates achievable by this policy, subject to average power constraints. The optimal policies result in a coding scheme that is simpler than the one for a general multiple access channel with generalized feedback. This simpler coding scheme also leads to the possibility of formulating an otherwise non-concave optimization problem as a concave one. Using the perfect channel state information (CSI) available at the transmitters to adapt the powers, we demonstrate gains over the achievable rates for existing cooperative systems. We consider both backwards and window decoding, and show that, window decoding, which incurs less decoding delay, achieves the same sum rate as backwards decoding, when the powers are optimized.

Chapter 4
Symeon Chatzinotas, University of Surrey, UK
Muhammad Ali Imran, University of Surrey, UK
Reza Hoshyar, University of Surrey, UK

In the information-theoretic literature, it has been widely shown that multicell processing is able to provide high capacity gains in the context of cellular systems. What is more, it has been proved that the per-cell sum-rate capacity of multicell processing systems grows linearly with the number of base station (BS) receive antennas. However, the majority of results in this area has been produced assuming that the fading coefficients of the MIMO subchannels are completely uncorrelated. In this direction, this chapter investigates the ergodic per-cell sum-rate capacity of the Gaussian MIMO Cellular Channel under correlated fading and BS cooperation (multicell processing). More specifically, the current channel model considers Rayleigh fading, uniformly distributed user terminals (UTs) over a planar cellular system and power-law path loss. Furthermore, both BSs and UTs are equipped with correlated multiple antennas, which are modelled according to the Kronecker product correlation model. The per-cell sum-rate capacity is evaluated while varying the cell density of the system, as well as the level of receive and transmit correlation. In this context, it is shown that the capacity performance is compromised by correlation at the BS-side, whereas correlation at the UT-side has a negligible effect on the system's capacity.

Section 2
Practical Coding Schemes for Cooperative Communications

Chapter 5

John M. Shea, University of Florida, USA

Tan F. Wong, University of Florida, USA

Chan Wong Wong, University of Florida, USA

Byonghyok Choi, University of Florida, USA

This chapter provides a survey of practical cooperative coding schemes currently available in the literature, with focus on those schemes that achieve performance close to capacity or the best known achievable rates. To provide an insight into the construction of practical coding schemes for various cooperative communication scenarios, we first summarize the main design principles and tools that are used. We then present a survey of cooperative communication scenarios, and the progress on practical coding schemes for each of these scenarios is discussed in detail. Throughout the chapter, we demonstrate how the common design principles and tools are exploited to construct the existing practical coding schemes. We hope that the integrated view presented in this chapter can lead to further advances in this area.

Chapter 6

Meng Yu, Lehigh University, USA

Jing (Tiffany) Li, Lehigh University, USA

Haidong Wang, Thales Communications Inc., USA

We consider practical network coding, a useful generalization of routing, in multi-hop multicast wireless networks. The model of interest comprises a set of nodes transmitting data wirelessly to a set of destinations across an arbitrary, unreliable, and possibly time-varying network. This model is general and subsumes peer-to-peer, ad-hoc, sensory and mobile networks. It is first shown that, in the single-hop case, the idea of adaptively matching code-on-graph with network-on-graph, first developed in the adaptive-network-coded-cooperation (ANCC) protocol, provides a significant improvement over the conventional strategies. To generalize the idea to the multi-hop context, we propose to transform an arbitrarily connected network to a possibly time-varying "trellis network," such that routing design for the network becomes equivalent to path discovery in the trellis. Then, exploiting the distributed, real-time graph-matching technique in each stage of the trellis, a general network coding framework is developed. Depending on whether or not the intermediate relays choose to decode network codes, three practical network coding categories, progress network coding, concatenated network coding and hybrid network coding, are investigated. Analysis shows that the proposed framework can be as dissemination-efficient as those with random codes, but only more practical.

Lun Dong, Drexel University, USA
Athina P. Petropulu, Drexel University, USA
H. Vincent Poor, Princeton University, USA

Cooperative beamforming (CB) is a signal transmission technique that enables long-range communications in an energy efficient manner. CB relies on cooperation from a set of distributed network nodes, each carrying a single transmit antenna and acting as elements of a virtual antenna array. By appropriately weighting their transmissions, the cooperating nodes form one or more beams to cooperatively transmit one or more message signals to the desired destinations. In this chapter, a cross-layer framework is presented that brings the CB ideas closer to implementation in a wireless network setting. The process of sharing among the network nodes the information to be beamed is studied and evaluated in terms of its effect on the spectral efficiency of the overall system. Optimal or suboptimal beamforming weights are designed, and queuing analysis is provided to study delay characteristics of source messages.

Zhihang Yi, Queen's University, Canada
Il-Min Kim, Queen's University, Canada

This chapter focuses on distributed space-time codes (DSTCs) in cooperative networks. DSTCs can substantially improve the bandwidth efficiency of the cooperative network without requiring any feedback overheads from the destination to the relays. The first part of this chapter is dedicated to reviewing the existing works on DSTCs. In the second part of this chapter, distributed orthogonal space-time block codes (DOSTBCs) are presented in detail. It is shown that the DOSTBCs can achieve the single-symbol maximum likelihood decodability and full diversity order in an amplify-and-forward cooperative network. Then some special DOSTBCs, which generate uncorrelated noises at the destination, are introduced. Those codes are referred to as the row-monomial DOSTBCs. An upper bound of the data-rate of the row-monomial DOSTBC is derived and the codes achieving the upper bound are presented as well.

Elzbieta Beres, University of Toronto, Canada
Raviraj Adve, University of Toronto, Canada

Cooperative diversity has the potential of implementing spatial diversity and mitigating the adverse effects of channel fading without requiring multiple antennas at transmitters and receivers. Traditionally, cooperative diversity is implemented using maximal ratio combining (MRC), where all available nodes relay signals between the source and destination. It has recently been proposed, however, that for each source-destination transmission, only a single best node should be selected to act as a relay. The resulting scheme, referred to as selection cooperation or opportunistic relaying, outperforms MRC schemes and can be implemented in a distributed fashion with limited feedback. This result is not unexpected, as selection requires some (though very limited) information regarding instantaneous channel conditions, while MRC does not. When implemented in a distributed network, however, MRC does require feedback for the synchronization of nodes, rendering a comparison of the two schemes worthwhile and fair. In this chapter, we provide a detailed overview of selection. We begin with a single source-destination pair, and discuss its implementation and performance under various constraints and scenarios. We then discuss a less-common scenario, a multi-source network where nodes act both as sources and as relays.

Chapter 10

Zhong Zhou, University of Connecticut, USA
Jun-Hong Cui, University of Connecticut, USA
Shengli Zhou, University of Connecticut, USA
Shuguang Cui, Texas A&M University, USA

In this chapter, we focus on the energy efficient cooperative communication with random node cooperation for wireless networks. By "random," we mean that the cooperative nodes for each communication event are randomly selected based on the network and channel conditions. Different from the conventional deterministic cooperative communication where cooperative nodes are determined prior to the communication, here the number of cooperative nodes and the cooperation pattern may be random, which is more practical given the random nature of the channels among the source nodes, relay nodes, and destination nodes. In addition, it is more robust to the dynamic wireless network environment. Starting with a thorough literature survey, we then discuss the challenges for random cooperative communication systems. Afterwards, two examples are presented to illustrate the design methodologies. In the first example, we analyze a simple scheme for clustered wireless networks, where cooperative communication is deployed in the long-haul inter-cluster transmissions to improve the energy efficiency. We quantify the energy performance and emphasize its difference from the conventional deterministic ones. In the second example, we consider the cross-layer design between the physical layer and the medium access control (MAC) layer for the one-hop random single-relay networks. We unify the power control and the relay selection at the physical layer into the MAC signalling in a distributed fashion. This example clearly shows the strength of cross-layer design for energy-efficient cooperative systems with random node collaboration. Finally, we conclude with discussions over possible future research directions.

Chapter 11

Diomidis S. Michalopoulos, Aristotle University of Thessaloniki, Greece
George K. Karagiannidis, Aristotle University of Thessaloniki, Greece

A major advantage of cooperative communications is the potential for forming distributed antenna arrays, that is, arrays whose elements are not co-located but carried by independent relaying terminals.

This allows for a study and design of cooperative communications under a novel perspective, where the inherent end-to-end paths between the source and destination terminal constitute the multiple branches of a virtual, distributed diversity receiver. As a result, the well-known combining methods used in conventional diversity receivers can be implemented in a distributed fashion, resulting in novel relaying protocols and generally in new ways for exploiting the resources available in cooperative relaying setups. This chapter provides an overview of this distributed diversity concept, as well as a performance analysis of the corresponding distributed diversity schemes, with particular emphasis on the analysis of distributed switch-and-stay combining. Further insights regarding the potential of implementing the distributed diversity concept in practical applications are obtained.

Chapter 12

    Manav R. Bhatnagar, University of Oslo, Norway
    Are Hjørungnes, University of Oslo, Norway

In this chapter, we discuss single and double-differential coding for two-user cooperative communication system. The single-differential coding is important for the cooperative systems as the data at the destination/relaying node can be decoded without knowing the channel gains. The double-differential modulation is useful as it avoids the need of estimating the channel and carrier offsets for the decoding of the data. We explain single-differential coding for a cooperative system with one relay utilizing orthogonal transmissions with respect to the source. Next, we explain two single-differential relaying strategies; active user strategy (AUS) and passive users relaying strategy (PURS), which could be used by the base-station to transmit data of two users over downlink channels in the two-user cooperative communication network with decode-and-forward protocol. The AUS and PURS follow an improved time schedule in order to increase the data rate. A probability of error based approach is also discussed, which can be used to reduce the erroneous relaying of data by the regenerative relay. In addition, we also discuss how to implement double-differential (DD) modulation for decode-and-forward and amplify-and-forward based cooperative communication system with single source-destination pair and a single relay. The DD based systems work very well in the presence of random carrier offsets without any channel and carrier offset knowledge at the receivers, where the single differential cooperative scheme breaks down. It is further shown that optimized power distributions can be used to improve the performance of the DD system.

Chapter 13

    J. Harshan, Indian Institute of Science, India
    G. Susinder Rajan, Indian Institute of Science, India
    B. Sundar Rajan, Indian Institute of Science, India

Cooperative communication in a wireless network can be based on the relay channel model where a set of users act as relays to assist a source terminal in transmitting information to a destination terminal. Recently, the idea of space-time coding (STC) has been applied to wireless networks wherein the relay nodes cooperate to process the received signal from the source and forward them to the destination such that the signal received at the destination appears like a space-time block code (STBC). Such STBCs (referred as distributed space time block codes [DSTBCs]) when appropriately designed are known to offer spatial diversity. It is known that separate classes of DSTBCs can be designed based on the destination's knowledge of various fading channels in the network. DSTBCs designed for the scenario when the destination has either the knowledge of only a proper subset of the channels or no knowledge of any of the channels are called non-coherent DSTBCs. This chapter addresses the problems and results associated with the design, code construction, and performance analysis (in terms of pairwise error probability [PEP]) of various non-coherent DSTBCs.

Section 4
Broadband Cooperative Communications

Chapter 14

Ibrahim Y. Abualhaol, Broadcom Corporation, USA
Mustafa M. Matalgah, University of Mississippi, USA

In this chapter, a cooperative broadband relay-based resource allocation technique is proposed for adaptive bit and power loading multiple-input-multiple-output/orthogonal frequency division multiplexing (MIMO-OFDM) system. In this technique, sub-channels allocation, M-QAM modulation order, and power distribution among different sub-channels in the relay-based MIMO-OFDM system are jointly optimized according to the channel state information (CSI) of the relay and the direct link. The transmitted stream of bits is divided into two parts according to a suggested cooperative protocol that is based on sub-channel-division. In this protocol, the first part is sent directly from the source to the destination, and the second part is relayed to the destination through an indirect link. Such a cooperative relay-based system enables us to exploit the inherent system diversities in frequency, space and time to maximize the system power efficiency. The BER performance using this cooperative sub-channel-division protocol with adaptive sub-channel assignment and adaptive bit/power loading are presented and compared with a non-cooperative ones. The use of cooperation in a broadband relay-based MIMO-OFDM system showed high performance improvement in terms of BER.

Chapter 15

Tae-Won Yune, POSTECH, Republic of Korea
Dae-Young Seol, POSTECH, Republic of Korea
Dongsik Kim, POSTECH, Republic of Korea
Gi-Hong Im, POSTECH, Republic of Korea

Cooperative diversity is an effective technique to combat the fading phenomena in wireless communications without additional complexity of multiple antennas. Multiple terminals in the network form a virtual antenna array in a distributed fashion. Even though each of them is equipped with only one antenna, spatial diversity gain can be achieved through cooperation. In this chapter, we discuss relay-assisted single carrier transmissions extending conventional transmit diversity schemes. We focus on distributed space-frequency block coded single carrier transmission, in order to operate over fast fading channels. A pilot design technique is also discussed for channel estimation of this single carrier cooperative system, which shows better channel tracking performance than conventional block-type channel estimations. In addition, spectral efficient cooperative diversity protocols are presented, where the users access a relay simultaneously or transmit superposed data blocks. Interference from the other user is effectively removed by using an iterative detection technique.

Section 5
Mathematical Tools for the Analysis and Design of Cooperative Networks

This chapter discusses important aspects in cooperative communications such as power allocation and node distributions using majorization theory, spanning both theoretical foundations and practical issues. Majorization theory provides a large amount of tools and techniques which can be used in order to accelerate the pace of developments in this fascinating research area of cooperative communications. The aim of the chapter is to build good intuition and insight into this important field of cooperative communications and how majorization theory can be used in order to solve quite complex problems in a very efficient and elegant way. Although we focus on some specific applications, the tools can be also applied to other setups and processing techniques.

Chapter 17

This chapter considers the problem of data gathering in correlated wireless sensor networks with distributed source coding (DSC), and virtual multiple input and multiple output (MIMO) based cooperative transmission. Using the concepts of super and sub modularity on a lattice, we analytically quantify as how the optimal constellation size, and optimal number of cooperating nodes, vary with respect to the correlation coefficient. In particular, we show that the optimal constellation size is an increasing function of the correlation coefficient. For the virtual MIMO transmission case, the optimal number of cooperating nodes is a decreasing function of the correlation coefficient. We also prove that in a virtual MIMO based transmission scheme, the optimal constellation size adopted by each cooperating node is a decreasing function of number of cooperating nodes. Also, it is shown that the optimal number of cooperating nodes is a decreasing function of the constellation size adopted by each cooperating node.

We also study numerically that for short distance ranges, SISO transmission achieves better energy-mutual information (MI) tradeoff. However, for medium and large distance ranges, the virtual MIMO transmission achieves better energy-MI tradeoff.

Birsen Sirkeci-Mergen, San Jose State University, USA
Anna Scaglione, University of California at Davis, USA
Michael Gastpar, University of California at Berkeley, USA

This chapter studies the cooperative broadcasting in wireless networks. We especially focus on multistage cooperative broadcasting in which the message from a source node is relayed by multiple groups of cooperating nodes. Interestingly, group transmissions become beneficial in the case of broadcasting as opposed to the case in traditional networks where receptions from different transmitters are considered as collision and disregarded. Different aspects of multistage cooperative broadcasting are analyzed in the chapter: (i) coverage behavior; (ii) power efficiency; (ii) error propagation; (iv) maximum communication rate. Whenever possible, performance is compared with multi-hop broadcasting where transmissions are relayed by a single node at each hop. We consider a large-scale network with many nodes distributed randomly in a given area. In order to analyze such networks, an important methodology, the continuum limit, is introduced. In the continuum limit, random networks are approximated by their dense limits under sum relay power constraint. This method allows us to obtain analytical results for the analysis of cooperative multistage broadcasting.

Section 6
An Industrial Perspective on Cooperative Communications

Mischa Dohler, CTTC, Spain
Djamal-Eddine Meddour, Orange Labs, France
Sidi-Mohammed Senouci, Orange Labs, France
Hassnaa Moustafa, Orange Labs, France

An ever-growing demand for higher data-rates has facilitated the growth of wireless networks in the past decades. These networks, however, are known to exhibit capacity and coverage problems, hence jeopardizing the promised quality of service towards the end-user. To overcome these problems, prohibitive investment costs in terms of base station or access point rollouts would be required if traditional non-scalable cell-splitting and micro-cell capacity dimension procedures were applied. The prime aim of current R&D initiatives is hence to develop innovative network solutions that decrease the cost per bit/s/Hz over the wireless link. To this end, cooperative networks have emerged as an efficient and promising solution. We discuss in this chapter some key research and deployment issues, with emphasis on cooperative architectures, networking and security solutions. We expose some motivations to use such networks, as well as latest state-of-the-art developments, open research challenges and business models.

Preface

The increasing demands for wireless multimedia and interactive Internet services, along with the rapid proliferation of a multitude of communications and computational gadgets, are fuelling intensive research efforts on the design of novel wireless communication system architectures for high-speed, reliable, and cost-effective transmission solutions. Within the last decade, a notable development in the area of communication theory has been the introduction of MIMO (multiple-input multiple-output) communications which have led to practical schemes such as space-time coding and spatial multiplexing. Such systems provide significant improvement in link reliability and spectral efficiency through the use of multiple antennas at the transmitter and/or receiver side. Multiple-antenna techniques are very attractive for deployment in cellular applications at base stations and have already been included in the 3rd generation cellular standards. Variations of MIMO techniques are also now a part of many existing and emerging wireless standards, such as IEEE 802.11 (WiFi), IEEE 802.16 (WiMax), and IEEE 802.20 (MBWA).

Deployment of MIMO techniques, however, is not always feasible mainly due to size and power constraints such as in cellular mobile devices, as well as in wireless sensor and ad-hoc networks which are gaining popularity in recent years. An innovative approach to harness the spatial diversity without deploying multiple antennas is *cooperative diversity*, also known as *cooperative communications*. The concept of cooperative communication stands out as a fundamental shift from the conventional network design based on point-to-point communications. Instead of a classical network with isolated communicating pairs, cooperative communication builds upon a network architecture in which nodes help each other in relaying information to realize spatial diversity advantages, thereby improving their own performance and that of the whole network. Cooperative communication techniques take advantage of the broadcast nature of wireless transmission, effectively creating a *virtual antenna array*[1] through cooperating nodes. This new transmission paradigm promises significant performance gains in terms of link reliability, spectral efficiency, system capacity, and transmission range.

Cooperative communication can be applied to both *infrastructure-based* networks, such as cellular systems, WLANs (wireless local area networks), WMANs (wireless metropolitan area networks), and *infrastructure-less* networks, such as MANETs (mobile ad-hoc networks), VANETs (vehicular ad-hoc area networks), and WSNs (wireless sensor networks). With large existing target markets and indisputable advantages, cooperative communication is one of the best opportunities today for high impact research in the area of wireless communications. As such, this emerging research area has spurred tremendous excitement within the academia and industry circles and resulted in a surge of research papers over the last few years.

This book aims to provide readers a comprehensive understanding of the fundamental principles of cooperative communications with a particular emphasis on physical layer issues and further present the latest advances and open research problems in this rapidly-evolving field. It would serve as a valuable reference for graduate students, scientists, faculty members who are conducting or wishing to conduct research in this area, as well as for engineers and research strategists in the relevant industries.

For the convenience of the reader, this book is organized in five intertwined thematic areas, namely as *Information Theoretical Results on Cooperative Communications* (Chapters 1-4), *Practical Coding Schemes for Cooperative Communications* (Chapters 5-6), *Distributed Transmit and Receive Diversity Techniques for Cooperative Communications* (Chapters 7-13), *Broadband Cooperative Communications* (Chapters 14-15), and *Mathematical Tools for the Analysis and Design of Cooperative Networks* (Chapters 16-18), followed by a chapter providing an industrial perspective on cooperative communications. The contributions of individual chapters are outlined in the following.

In Chapter 1, M. Yuksel and E. Erkip provide a comprehensive overview of the information theoretic foundations of cooperative communications. Adopting capacity, diversity, and diversity-multiplexing tradeoffs as performance metrics, they outline the performance limits of major cooperation protocols for Gaussian, fading, and multiple-access channels considering both full-duplex and half-duplex relay nodes. While Chapter 1 places a particular emphasis on decode-and-forward (DF) and compress-and-forward (CF) relaying, the focus of Chapter 2 by I. Krikidis and J. S. Thompson shifts to amplify-and-forward (AF) relaying, which is particularly attractive with its low complexity. This chapter first summarizes information theoretic performance limits of AF relaying, then discusses relay selection, cross-layer coordination, and power allocation for AF cooperative networks. As briefly discussed in Chapter 2, optimum power control is a key technique to realize the full potentials of cooperative transmission. In Chapter 3, O. Kaya and S. Ulukus present an information theoretic approach to power control with a particular focus on a two-user fading multiple access channel. They derive optimal power control methods which maximize the long term average rates achievable by user cooperation, and the associated improved rate regions. Their results lead to the characterization of jointly optimum encoding, medium access and routing policies, yielding the importance of a cross-layer approach to wireless network design. In Chapter 4, S. Chatzinotas, M. A. Imran, and R. Hoshyar focus on a particular application of cooperative communications and investigate the information theoretical performance limits of a cellular system with multi-cell processing. Specifically, they study the ergodic per-cell sum-rate capacity of the MIMO cellular channel under the assumption of correlated Rayleigh fading and uniformly distributed user terminals over a planar cellular system. Their results demonstrate that sum-rate capacity is compromised by spatial correlation at the base station while it grows linearly with the number of receive antennas under independent fading sub-channels.

While information theoretic results lay the ultimate performance limits for cooperative communications, an important practical question is how to achieve performance close to capacity or the best known achievable rates. To address this question, J. M. Shea, T. F. Wong, C. W. Wong, and B. Choi provide a framework in Chapter 5 for multi-user communications including relay-assisted transmission and cooperative multiple-access as special cases. Under this integrated overview, they summarize the main design principles and tools for code construction and present a detailed survey of research progress on practical coding schemes. In Chapter 6, M. Yu, J. Li, and H. Wang consider a multi-hop relay scenario and investigate practical and efficient coding schemes for multicast data transmission based on the principles of networking coding.

In multi-relay cooperative networks, a crucial design aspect is to decide on the method how the relay nodes will participate in the cooperation. Although repetition-based cooperation schemes (in which only one relay is allowed to transmit the signals for each time slot) are able to provide the full diversity, they suffer from poor bandwidth efficiency. Cooperative beamforming, distributed space-time coding, and relay selection have been proposed as spectral-efficient methods to coordinate the transmissions from multiple relays which are, respectively, addressed in Chapters 7, 8, and 9.

In Chapter 7, L. Dong, A. P. Petropulu, and H. V. Poor discuss the concept of cooperative beamforming in which a proper weight coefficient is assigned to each relay in order to make the signals from different relay nodes add up constructively at the destination. Particularly, they present two cross-layer cooperative beamforming techniques both of which involve the MAC (medium access control) layer during the information-sharing stage and the physical layer during the beamforming stage. These techniques, however, differ from each other on how the message signals are shared between source and cooperating nodes, and on how they are weighted and forwarded. The performance of the two techniques is compared in terms of spectral efficiency. Under the envisioned cross-layer design, a queuing analysis is further presented to provide insight into how packet delays and the stability regions of traffic rate are affected when source and relay nodes are subject to a queue.

Distributed space-time coding, which is the focus of Chapter 8, provides an open-loop alternative to cooperative beamforming avoiding the feedback requirement for weight coefficients. It is relatively easy to integrate conventional space-time codes (i.e., designed for co-located antennas) in a DF cooperative system. Particularly, orthogonal space-time block codes are attractive with their robustness in the presence of node failures. Since these codes have orthogonal structure, node failure corresponds to deletion of a column in the code matrix, but the other columns remain orthogonal. This lets the distributed scheme still exploit the residual diversity benefits from the remaining nodes. The application of space-time codes to AF cooperative systems is, however, much more challenging. In Chapter 8, Z. Yi and I.-M. Kim address the design of single-symbol decodable distributed orthogonal space-time block codes (DOSTBCs), which are particularly desirable from the viewpoint of low-complexity receiver design. A major problem in the construction of the DOSTBCs for AF relaying comes from the fact that the noise covariance matrix at the destination is not diagonal in general due to the presence of terms related to forwarded signals. To overcome this problem, Yi and Kim present row-monomial DOSTBCs which guarantee uncorrelated noise terms at the destination. They present an upper bound on the data-rate of the row-monomial DOSTBCs and show that these codes can achieve approximately twice higher bandwidth efficiency than the repetition-based cooperative schemes.

Unlike cooperative beamforming and distributed space-time coding, which require the participation of all (active) relay nodes in relaying phase, relay selection schemes avoid multiple relay transmissions by having only a single relay to forward the information from the source. The "best" relay with favourable channel conditions is selected based on a performance metric such as end-to-end signal-to-noise ratio or mutual information. In Chapter 9, E. Beres and R. Adve outline relay selection methods for AF and DF relaying and further discuss various practical implementation aspects including feedback overhead and system requirements. Chapter 10, by Z. Zhou, J.-H. Cui, S. Zhou, and S. Cui, continues the discussion on relay selection. Different from the conventional deterministic cooperative communication where relay node(s) are determined prior to the communication, this chapter investigates "random" cooperation where the number of cooperative nodes and the cooperation pattern changes based on the random nature of the channels among source, relay, and destination nodes. The authors demonstrate that such a random cooperative scheme is more robust to dynamic wireless network environments and further discuss the challenges that lie ahead in designing general random cooperative schemes.

In Chapter 11, D. S. Michalopoulos and G. K. Karagiannidis consider the multi-relay cooperative system design from the receiver design point of view. They treat inherent end-to-end paths between source and destination nodes as the multiple branches of a virtual diversity receiver and address conventional diversity combining techniques in the context of cooperative communications. Specifically, they investigate the performance of distributed implementation of MRC (Maximal Ratio Combining), SSC (Switch and Stay Combining), and SC (Selection Combining) methods along with discussions on practical implementation aspects.

A common assumption in the existing literature on cooperative communications is the availability of perfect channel state information at relay and destination nodes. Although knowledge on fading channel coefficients is required for coherent communication systems, this can be avoided through differential techniques as discussed in Chapter 12. In this chapter, for a two-user cooperative communication system, M. R. Bhatnagar and A. Hjørungnes present single-differential and double-differential coding, the latter of which also avoids the need of estimating carrier offsets besides the fading channel coefficients. Relay selection and power allocation are further discussed for potential performance improvements. In Chapter 13, J. Harshan, G. S. Rajan, and B. S. Rajan consider a multi-relay scenario and address the design, construction, and performance analysis of differential distributed space-time block codes. They demonstrate that differential codes from Clifford algebras are the best among the existing designs in terms of the encoding-decoding complexity and also in terms of error rate performance.

The existing literature on cooperative communication has heavily focused on the frequency-flat fading channel model. Such a model fails to provide an accurate modeling for broadband wireless channels which exhibit frequency-selectivity and result in intersymbol interference (ISI). An efficient approach to mitigate ISI is OFDM (orthogonal frequency division multiplexing); a multicarrier transmission system where the high-rate data stream is demultiplexed and transmitted over a number of frequency subcarriers. In Chapter 14, I. Y. Abualhaol and M. M. Matalgah consider a cooperative broadband MIMO-OFDM system with DF relaying and present adaptive bit and power loading algorithms to optimize the error rate performance. An alternative low-complexity approach to ISI mitigation is the single-carrier frequency-domain equalization (SC-FDE). Historically shadowed by OFDM, SC-FDE provides a powerful alternative air interface architecture with a similar implementation complexity and has started receiving significant attention recently. In Chapter 15, T.-W. Yune, D.-Y. Seol, D. Kim, and G.-H. Im investigate this promising technique in the context of cooperative communications. They focus on distributed space-frequency block coded (DSFBC) single carrier transmission which is particularly useful over rapidly time varying channels. They devise a pilot-assisted channel estimation technique for DSFBC SC-FDE under consideration and present spectrally efficient cooperation protocols based on the principles of iterative multiuser detection.

A common feature of Chapters 16-18 is how they smartly take advantage of some sophisticated mathematical tools for the analysis and design of cooperative networks. In Chapter 16, A. Sezgin and E. A. Jorswieck discuss how *majorization theory* can be applied to solve some typical design problems encountered in cooperative communications. After they provide an overview on majorization and order preserving functions (namely Schur-convex and Schur-concave functions), they use these mathematical tools to design power allocation schemes and determine optimal node distribution along with a capacity analysis. In Chapter 17, L. S. Pillutla and V. Krishnamurthy study the problem of data gathering in correlated wireless sensor networks with distributed source coding. Using *monotone comparative statics* borrowed from the economics literature, they analytically study the effect of data correlation on physical layer design variables such as constellation size and number of cooperating nodes. In Chapter 18, B. Sirkeci-Mergen, A. Scaglione, and M. Gastpar study cooperative broadcasting in large-area wireless networks with randomly distributed nodes. To analyze the network behaviour, they adopt the so-called *continuum limit* methodology which approximates random networks by their dense limits under sum relay power constraint. They derive several analytical results including coverage behaviour, power efficiency, error propagation, and maximum communication rate. Although the focus of this chapter is mainly on cooperative broadcasting, the deployed mathematical tools can be useful in the analysis of other cooperative systems as well.

The last chapter of our book presents an industrial R&D perspective on cooperative communications. In this chapter, M. Dohler, D.-E. Meddour, S.-M. Senouci, and H. Moustafa first provide an overview of state-of-the-art developments from both academia and industry, then address possible deployment architectures for cooperative multi-hop cellular networks as a case study. They expose some motivations, including business models, to use such networks and present a comprehensive discussion on technical challenges for real-world deployment scenarios, such as routing, appropriate QoS metrics, authentication, and authorization to services' access.

With over fifty contributors from the field, this comprehensive book would introduce students and practitioners to the diverse research on cooperative communications and expose the underlying fundamental issues in the analysis, design, and optimization of cooperative wireless communication systems. Considering the scarcity of books in this new research area, I hope it would fill a void in the current literature as a valuable reference guide.

Murat Uysal
University of Waterloo, Canada
2009

ENDNOTE

[1]   The phrase of "virtual antenna array" was coined by Dr. Mischa Dohler during his Ph.D. study at the King's College London.

Acknowledgment

I would like to thank all contributing authors for their time and efforts which made this book possible. I feel fortunate to be able to bring together such a distinguished body of eminent researchers for this project. Their high-quality contributions and insights on the topic make this book worthwhile.

It is also my pleasure to thank staff members at IGI Global for their support throughout the project. Particularly, I thank Kristin M. Klinger, Managing Acquisitions Editor, for her invitation and encouragement to initiate this book project and Julia Mosemann, Editorial Assistant, for her help in the development process of the manuscript.

Murat Uysal
University of Waterloo, Canada

Section 1
Information Theoretical Results on Cooperative Communications

Chapter 1
Information Theoretical Limits on Cooperative Communications

Melda Yuksel
TOBB University of Economics and Technology, Turkey

Elza Erkip
Polytechnic Institute of New York University, USA

ABSTRACT

This chapter provides an overview of the information theoretic foundations of cooperative communications. Earlier information theoretic achievements, as well as the more recent developments, are discussed. The analysis accounts for full/half-duplex node, and for multiple relays. Various channel models such as discrete memoryless, additive white Gaussian noise (AWGN), and fading channels are considered. Cooperative communication protocols are investigated using capacity, diversity, and diversity-multiplexing tradeoff (DMT) as performance metrics. Overall, this chapter provides a comprehensive view on the foundations of and the state-of-the-art reached in the theory of cooperative communications.

INTRODUCTION

In traditional communication networks data transmission directly occurs between the transmitter and the receiver. No user solicits the assistance of another one. However, in a general communication network, there are many intermediate nodes that are available to help. For example in wireless networks, when one node broadcasts its messages, all nearby nodes overhear this transmission. Processing and forwarding these messages to the intended destination, system performance, whether it be throughput, lifetime, or coverage area, can be improved. To understand how much performance improvement can ideally be possible by this "cooperative" network, we need an information theoretical study. Such a study also elucidates how cooperation should take place and helps construct the backbone for future cooperative communication applications.

DOI: 10.4018/978-1-60566-665-5.ch001

Figure 1. The three-terminal communication channel

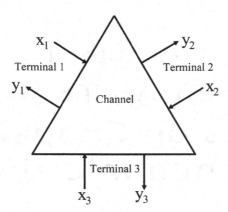

In information theory the idea of cooperation was first presented by van der Meulen (1971), which established the foundations of the *relay channel*. The relay channel is a three-terminal network, in which Terminal 1 (source) aims to transmit to Terminal 3 (destination) with the help of Terminal 2 (relay) as in Figure 1. The aim is to attain the highest communication rate from Terminal 1 to Terminal 3. In general the intermediate relay enhances communication rates of the direct link from Terminal 1 to 3.

To elaborate the idea of enhanced communication rates via relaying, here we introduce an example from (van der Meulen, 1971). Table 1 gives the channel output probabilities $P(y_2, y_3|x_1, x_2)$ conditioned on input pairs (x_1, x_2) when the destination is assumed to be silent $(x_3=0)$. Observe that if $x_2 = 1$, both y_2 and y_3 are equal to 1 no matter what x_1 is. Neither the relay, nor the destination can distinguish the correct x_1 and hence, communication rates between the source and the relay, and the source and the destination are equal to zero given $x_2 = 1$. When $x_2 = 0$, the channels between the source and the relay, and the source and the destination both become equivalent to a binary symmetric channel with crossover probability 1/2, whose capacity is also equal to zero. Hence, no direct communication is possible from the source to the relay, or from the source to the destination. However, non-zero communication rates from the source to the destination can be achieved with the help of the relay. Observe that if $x_2 = 0$ and the destination knows the channel output y_2, then the source can send 1 bit noiselessly to the destination. On the other hand, the destination can learn about y_2, if the relay helps the source and transmits its output signal after its own observation is complete. This information can be sent at rate 0.32 bits/channel use when x_1 is set to 0. By appropriate time division between these two strategies, (van der Meulen, 1971) proves that the capacity of this example channel is equal to 0.243 bits/channel use with vanishing error probability. As this example clearly shows, relaying can substantially increase achievable rates with respect to direct transmission.

The relay channel is essential to both wired and wireless networks. In wired networks many source and destination pairs are connected via intermediate relay nodes. In wireless networks, due to the *wireless broadcast advantage* idle nodes overhear nearby transmissions. These nodes can relay the information and contribute to achieving higher rates.

The capacity of the relay channel was calculated in (van der Meulen, 1971) only for some particular example channels. Guidelines for communication/relaying principles in a general relay channel were

Table 1. A relaying example (Adapted from van der Meulen, 1971)

		y_2y_3			
		00	**10**	**01**	**11**
x_1x_2	00	½	0	0	½
	10	0	½	½	0
	01	0	0	0	1
	11	0	0	0	1

developed later in the landmark paper by Cover and El Gamal (1979). In (Cover & Gamal, 1979), the memoryless relay channel is investigated and two fundamental relaying techniques; decode-and-forward (DF) and compress-and-forward (CF), are presented. The paper also includes capacity theorems for both degraded and reversely degraded[1] relay channels and the relay channel with feedback.

Despite the substantial advancement (Cover & Gamal, 1979) provided, the capacity of the general relay channel has been unsolved for over thirty years. The recent activity on cooperative communications mostly stems for the potential wireless applications and is spurred by recent papers (Sendonaris et al., 2003a; Sendonaris et al., 2003b) and (Laneman et al., 2004). In (Sendonaris et al., 2003a) and (Sendonaris et al., 2003b), motivated by Willems' multiple-access channel with generalized feedback (GMAC) model (Willems et al., 1983), the authors considered the benefits of mutual cooperation in fading environments and suggested how cooperation can be carried out in code-division multipleaccess (CDMA) systems. In (Laneman et al., 2004) the authors suggested *simple* yet high performance relaying protocols. Promising significant gains and having vast application areas, the ideas presented in these papers triggered an extensive literature on cooperative communication systems.

Later, Kramer, Gastpar and Gupta generalized the DF and CF protocols suggested in (Cover & Gamal, 1979) to arbitrary number of relay nodes (Kramer et al., 2005). Even though the capacity of the relay channel is unknown, this paper proves that the DF and CF protocols can be capacity achieving depending on the location of the relay node(s). Recent results on multiple relay terminal networks suggested strategies that are a fixed number of bits away from the capacity (Avestimehr et al., 2007; Avestimehr et al., 2008).

An important constraint in the relay channel is the processing capability of the relay node. The relay can either be full or half-duplex. If the relay is full-duplex it can transmit and receive simultaneously in the same frequency band. In half-duplex systems, transmission and reception takes place in orthogonal channels. Although the full-duplex assumption is not practically feasible, it helps us understand the fundamental characteristics of the relay channel. On the other hand, the half-duplex assumption is required to study practical aspects. Half-duplex operation in the relay channel was first considered (Laneman et al., 2004). Later (Khojastepour et al., 2003a) and (Kramer, 2004) investigated the capacity of the relay channel under the half-duplex assumption.

The above mentioned papers consider probability of error and achievable rates to measure the level of reliability and the throughput of relaying/cooperation schemes. Another performance measure is the diversity-multiplexing tradeoff (DMT) (Zheng and Tse, 2003), which unifies the reliability and rate perspectives. The DMT is a high SNR analysis suitable for fading channels and establishes the fundamental tradeoff between diversity and multiplexing for multiple-input multiple-output (MIMO) systems. DMT is also a useful performance measure for cooperative/relay systems. While the capacity of the relay channel is not known in general, it is possible to find relaying schemes that are optimal from the DMT perspective. In the literature (Azarian et al., 2005; Bletsas et al., 2006; Laneman et al., 2004; Prasad & Varanasi, 2004) were the first to investigate the relay channel DMT. In these works, either the source or the destination (or both) has a single antenna. The multiple-antenna multiple-relay channel was first studied in (Yuksel & Erkip, 2007). In (Yuksel & Erkip, 2007) DMT upper bounds are found, and a relaying strategy, which achieves the bound for both full and half-duplex relays, is suggested.

This chapter surveys the above mentioned models for cooperation as well as how cooperation is useful under achievable rate, diversity and DMT metrics. It is important to note that there has been a large body of information theoretical work on cooperative communications in the past few years (ITs, 2007; Kramer et al., 2006) and this chapter only covers some of the fundamentals. For example, we consider neither the ergodic capacity, nor the resource allocation problems. Similarly, scaling laws in large cooperative networks and cooperative channel coding are out of this chapter's scope.

In the next section, we first introduce the DF and CF protocols and introduce the generalized multiple access channel model. In the third section, we extend these basic models to multiple relay/ cooperating nodes. In the fourth section, we explain the effect of relay processing constraints on achievable rates. In the fifth section, we examine the relay channel when there is fading. Next, in the sixth section, we provide the DMT analysis for the relay channel. Finally, in the seventh section, we conclude.

The Basic Models for Cooperation

In this section we introduce the full-duplex relay channel to capture unilateral cooperation and describe the two fundamental relaying protocols DF and CF. We then explain the G-MAC model to explain two-way cooperation.

A. The Relay Channel

The basic relay channel model is illustrated in Figure 2, where $p(y_R, y_D | x_S, x_R)$ indicates the discrete memoryless channel, W denotes the message, X_S and X_R are the signals the source and relay transmit, Y_R and Y_D are the received signals at the relay and at the destination, and $\widehat{W}$ is the destination's estimate of W.

If the channel is real additive white Gaussian noise (AWGN), then we can write

$$Y_R = a_{SR} X_S + Z_R \tag{2.1}$$

Figure 2. The relay channel

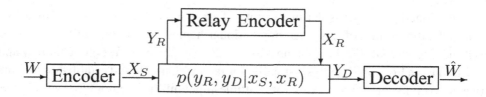

$$Y_D = a_{SD}X_S + a_{RD}X_R + Z_D \tag{2.2}$$

where a_{SR}, a_{SD} and a_{RD} ($\in \mathbb{R}$) are respectively the channel gains between the source and the relay, the source and the destination, and the relay and the destination. The AWGN at the relay and destination are respectively denoted by Z_R and Z_D, which are assumed to be zero mean and unit variance. Both the source and the relay have individual power constraints P_S and P_R. The objective is to find the capacity, the maximum achievable rate beyond which reliable communication is not possible. Note that the cut-set upper bound (Cover & Thomas, 1991) suggests that if R is an the achievable rate, then

$$R \leq \max_{p(x_S, x_R)} \min\{I(X_S; Y_R, Y_D); I(X_S, X_R; Y_D)\},$$

which becomes

$$R \leq \max_{\rho \in [0,1]} \min \left\{ \frac{1}{2} \log\left(1 + \rho a_{SR}^2 P_S + \rho a_{SD}^2 P_S\right), \right.$$
$$\left. \frac{1}{2} \log\left(1 + a_{SD}^2 P_S + a_{RD}^2 P_R + 2\sqrt{(1-\rho)a_{SD}^2 P_S a_{RD}^2 P_R}\right)\right\} \tag{2.3}$$

for the Gaussian case.

DF and CF are two of the relaying schemes proposed in (Cover & El Gamal, 1979) for the general relay channel. In DF, the relay decodes its received message, re-encodes it, and forwards it to the destination. In CF, the relay first compresses its received signal and then forwards the compressed signal through the relay-destination channel. The compression is Wyner-Ziv type (Wyner & Ziv, 1976); i.e. the relay compresses its received signal taking into account that the destination has side information available directly from the source. Next, we introduce the ideas behind these two protocols. The formal proofs can be found in (Cover & El Gamal, 1979).

1. **Decode-and-forward:** In the DF protocol, the source, and the relay perform block Markov superposition coding. The destination can do sliding window or backward decoding (Kramer et al., 2005; Willems & van der Meulen, 1985; Xie & Kumar, 2004; Zeng et al., 1989). We will briefly describe the backward decoding here. The encoding structure is depicted in Figure Figure 3. Transmission takes place in B blocks, where each block consists of N symbols, with N and B both large. The source transmits a length-N codeword $x_S^N(w_{b-1}, w_b)$ in each block b, for which $w_0 = w_B = 1$, where w_b denotes the new message transmitted in block b.

The source rate is chosen so that the relay can reliably decode. To satisfy this constraint the source rate $R^{(DF)}$ should satisfy

$$R^{(DF)} \leq I(X_S; Y_R \mid X_R) \tag{2.4}$$

Figure 3. The decode-and-forward block structure

Block 1	Block 2	Block 3		Block B
$x_S^N(1, w_1)$	$x_S^N(w_1, w_2)$	$x_S^N(w_2, w_3)$	...	$x_S^N(w_{B-1}, 1)$
$x_{R_1}^N(1)$	$x_{R_1}^N(w_1')$	$x_{R_1}^N(w_2')$		$x_{R_1}^N(w_{B-1}')$

for a fixed input distribution. Under this condition, the relay finds an estimate w_b' for w_b at the end of block b and sends $x_R^N(w_b')$ in block $b+1$. In the first block the relay sends a predetermined codeword $x_R^N(1)$. With very high probability $w_b' = w_b$ and the relay decodes the source message reliably. This way, the relay removes the effects of the channel from its received signal, and thus obtains a *clean* copy of the original source message. As the source repeats w_b in block $b+1$, the relay can then act collectively with the source. Note that the full-duplex assumption is critical to realize the block Markov encoding structure, which requires the relay to transmit and receive simultaneously.

The destination starts decoding after all B blocks are received and moves backwards (Kramer et al., 2005; Willems & van der Meulen, 1985; Zeng et al., 1989). At block B, no fresh information is sent and the destination is interested in decoding w_{B-1}. The correct message index can be identified with high probability if

$$R^{(DF)} \le I(X_S, X_R; Y_D).$$ (2.5)

Once the destination decodes w_{B-1}, it can move backwards to decode $w_{B-2}, ..., w_1$ in a similar manner. Combining both constraints (2.4) and (2.5), we conclude that for a fixed input distribution $p(x_S, x_R)$

$$R^{(DF)} \le \min\{I(X_S; Y_R \mid X_R), I(X_S, X_R; Y_D)\}$$ (2.6)

is achievable. Maximizing over all input distributions, we can write the maximum rate the DF protocol achieves is

$$R_{max}^{(DF)} = \max_{p(x_S, x_R)} \min\{I(X_S; Y_R \mid X_R), I(X_S, X_R; Y_D)\}.$$

For the Gaussian relay channel using Gaussian code books $R_{max}^{(DF)}$ becomes

$$R_{max}^{(DF)} = \max_{\rho \in [0,1]} \min\left\{\frac{1}{2}\log\left(1 + \rho a_{SR}^2 P_S\right),\right.$$
$$\left.\frac{1}{2}\log\left(1 + a_{SD}^2 P_S + a_{RD}^2 P_R + 2\sqrt{(1 - \rho)a_{SD}^2 P_S a_{RD}^2 P_R}\right)\right\}.$$ (2.7)

The first term in the minimization accounts for the rate the relay can reliably decode. The second term is the rate, which the destination can decode, when the source and the relay transmit together. The parameter

ρ denotes the correlation between source and relay signals and is related to the coherent combining gain at the destination. As the source and the relay are synchronized, their signals add coherently at the destination. This leads to the additional gain in the received signal-to-noise ratio $2\sqrt{(1-\rho)a_{SD}^2 P_S a_{RD}^2 P_R}$. Changing ρ the source node divides its power among sending *fresh* information and repeating *old* information to coherently add with the relay's signal.

It is interesting to note that the DF protocol achieves the capacity, when the relay channel is physically degraded (Cover & El Gamal, 1979). Physical degradedness ensures that decoding at the relay does not impose an additional constraint on the system, and DF becomes optimal.

When the relay channel is not physically degraded, then decoding at the relay may impose strict constraints; i.e. the source-relay channel can limit the achievable rates. In such cases partial decode-and-forward (PDF) can be used instead of DF. As its name suggests, in PDF the relay is required to decode part of the source message only, and the remaining part is directly sent to the destination without the relay's help. This strategy improves upon DF achievable rates (Cover & El Gamal, 1979). We will discuss the PDF strategy more in the second section, B..

2. **Compress-and-forward:** In the DF protocol the relay performs a *hard decision* about the source message. Forcing the relay to make a decision can incur losses in achievable transmission rates. In some cases compression based protocols, which forward relay's *soft* signal are desirable.

A simple *B*-block compression protocol, which we call simple compress-and-forward (SCF), is as follows: The relay compresses its received signal Y_R in block b to form $\hat{Y}_R$, maps this to the channel codeword X_R and sends X_R to the destination in block $b+1$. As the relay is full duplex, the relay can listen and compress the message in the current block while transmitting the compressed signal of the previous block. The received signal Y_D in block b is a function of both the source and the relay signals X_S and X_R transmitted in the same block.

The destination starts decoding after all B blocks are transmitted. It first decodes the relay signal X_R, treating the source signal as noise and recovers $\hat{Y}_R$. The reliability of these steps are ensured, if the relay's compression rate is below the relay-to-destination achievable rate considering the source signal as noise. In other words, the condition

$$I(\hat{Y}_R; Y_R \mid X_R) \le I(X_R; Y_D) \tag{2.8}$$

has to be satisfied for a fixed input distribution $p(x_S)p(x_R)p(\hat{y}_R|x_R,y_R)p(y_R,y_D|x_S,x_R)$. After the destination decodes $\hat{Y}_R$, the destination uses both $\hat{Y}_R$ and Y_D from the previous block to determine the original source signal X_S. This can be done reliably if

$$R^{(SCF)} < I(X_S; \hat{Y}_R, Y_D \mid X_R). \tag{2.9}$$

The performance of the above scheme could be improved if Wyner-Ziv type compression (Wyner & Ziv, 1976) is used instead of simple compression. The Wyner-Ziv technique allows for lower compression rates in the presence of correlated side information at the decoder. This is indeed the case in the relay channel, in which the direct signal from the source received at the destination in the previous block can be thought of as the side information. This side information lowers the compression rate from $I(\hat{Y}_R ; Y_R | X_R)$ to $I(\hat{Y}_R ; Y_R | X_R, Y_D)$ leading to the condition

$$I(\hat{Y}_R; Y_R \mid X_R, Y_D) \leq I(X_R; Y_D). \qquad (2.10)$$

Other decoding steps remain the same and the overall CF achievable rate is

$$R^{(CF)} = I(X_S; \hat{Y}_R Y_D \mid X_R) \qquad (2.11)$$

subject to (2.10) where the joint probability distribution is $p(x_S)p(x_R)p(\hat{y}_R|x_R,y_R)p(y_R,y_D|x_S,x_R)$. Since (2.10) is looser than (2.8), achievable rates are potentially higher when Wyner-Ziv type compression is employed. For the Gaussian case, $R^{(CF)}$ becomes equal to

$$R^{(CF)} = \frac{1}{2} \log \left(1 + a_{SD}^2 P_S + \frac{a_{SR}^2 P_S}{1 + \hat{N}_R} \right) \qquad (2.12)$$

subject to

$$\hat{N}_R \geq \frac{1 + a_{SD}^2 P_S + a_{SR}^2 P_S}{a_{RD}^2 P_R}. \qquad (2.13)$$

Note that two terms, $a_{SD}^2 P_S$ and $a_{SR}^2 P_S / (1 + \hat{N}_R)$, contribute to the received SNR at the destination. The former is the received SNR due the direct source-destination link. The latter is the received SNR at the relay except the additional $\hat{N}_R$ in the denominator. This shows that the destination receives the relay's observation with an additional compression noise.

Kramer et al. (2005) provide an extensive analysis of the achievable rates of DF and CF for general channel models. One of the models considered is the path loss model, which follows the Gaussian channel in (2.1)-(2.2) with $a_{ij} = 1 / \sqrt{d_{ij}^\alpha}$, $i,j = S, R, D$, $i \neq j$, where α indicates the path loss exponent, and d_{ij} indicates the inter-node distances.

In Figure 4, we plot the DF and CF rates of (2.7) and (2.12) for $\alpha = 2$, and $P_S = P_R = 10$. The source and the destination are respectively located at *0* and *1*. The relay's location varies on the line joining the source and the destination. The DF and CF rates are also compared to the upper bound of (2.3) as well as the direct transmission rates. When d_{SR} distance is small, the first terms in (2.3) and (2.7) become very large. The second term dominates both equations and DF becomes optimal. Similarly, when d_{RD} distance is small, the compression noise in (2.13) approaches zero and the CF rate in (2.7) becomes equal to the upper bound in (2.3). The figure clearly confirms that the DF protocol achieves the capacity when the relay is close to the source, and the CF protocol achieves the capacity when the relay is close to the destination.

Figure 4. DF, CF and SCF achievable rates in comparison to the upper bound and direct transmission (Adapted from Kramer et al., 2005)

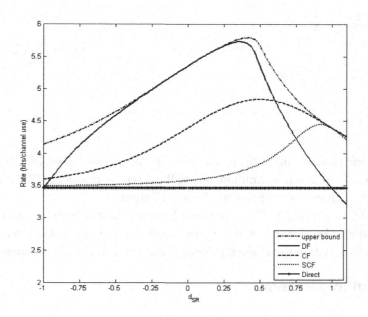

B. Multiple-Access Channel with Generalized Feedback

The relay channel model describes the basic form of cooperation. In the relay channel, the relay node does not have its own information to send; it only assists the source node to enhance achievable communication rates. On the other hand, the full cooperation model captured in the multiple access channel with generalized feedback considers two or more sources, which mutually help each other to attain better performance.

The two-user discrete-memoryless GMAC is shown in Figure 5, where W_1 and W_2 denote the messages of the first and second sources, X_{S_1} and X_{S_2} are the source signals, $p\left(y_{S_1}, y_{S_2}, y_D \mid x_{S_1}, x_{S_2}\right)$ is the channel, Y_{S_1}, Y_{S_2} and Y_D are respectively the received signals at S_1, S_2 and the destination, and $\hat{W}_1$ and $\hat{W}_2$ are the destination's estimate of W_1 and W_2. This generalized feedback model allows for both users to *overhear* each other and thus mutual cooperation can take place. In the GMAC, the relay is another source node, which has its own information to send. The achievable rate region is the collection of all achievable rate pairs. This model reduces to the relay channel model, if $W_2 = \varnothing$ and $Y_{S_1} = \varnothing$.

The achievable rate region suggested in (Willems et al., 1983) for the GMAC is PDF based. In PDF some part of the message is directly sent to the destination without the help of the other. Similar to the DF in the second section, A.1 communication lasts for B blocks. In each block, each source's messages are considered to be composed of three parts. For the first part neither of the sources seek the other's assistance. This part of the message is sent directly to the destination. The second part of S_1's message in block b is aimed to be decoded only at S_2 at the end of block b. Decoding this part of the message, S_2 forms the third part of its message in block $b+1$. Similarly, at the end of block b, S_1 decodes the second

Figure 5. The multiple access channel with generalized feedback

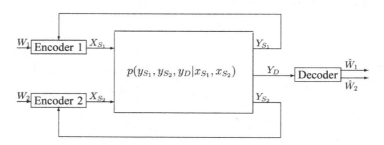

part of S_2's message, and re-encoding it forms the third part of its own message in block $b+1$. This way, each source node can help each other simultaneously. We refer the readers to (Willems et al., 1983) and (Willems, 1982) for the achievable rate region characterization.

Although the GMAC results existed for more than 20 years, their interpretation as a means for mutual cooperation was not until (Sendonaris et al., 2003a; Sendonaris et al., 2003b). We will mention these results in the fifth section, within the context of cooperation in wireless communications.

Extensions to Multiple Relays

In the last section, we considered the relay channel and GMAC, which consist of only two transmitters. In a general multiple-terminal network more than one relay is available to help or multiple users form coalitions. Communication strategies and achievable rates for the multiple relay channel is another important question, which was considered in (Xie & Kumar, 2004; Kramer et al., 2005). The paper (Xie & Kumar, 2004) generalizes the DF protocol, and (Kramer et al., 2005) generalizes the CF protocol to multiple relays, which we explain next.

When we described the DF protocol in the second section, A. 1, we observed that $R^{(DF)}$ is the minimum of two rates, the source-relay communication rate, and the rate the source and relay collectively transmit to the destination. When there are multiple relays and all relays perform DF, this idea can be generalized. Suppose there are two relays. Then the DF achievable rate is the minimum of three rates: the rate source can reliably send to the first relay, the rate source and the first relay can reliably send to the second relay, and the rate the source and both relays altogether can send to the destination. Of course, this depends on the relay assignment; i.e. which relay will be the first and which relay will be the second. Therefore, the minimum of the above mentioned rates can be maximized over all relay permutations. If there are M relays, then the DF achievable rate is the minimum of $M+1$ rates maximized over all relay permutations.

Generalizing the CF protocol to multiple relays is more intricate. When there are multiple relays, the relays observe correlated signals. Forwarding by the relays becomes equivalent to sending arbitrarily correlated sources over a multiple access channel, whose complete solution is not known (Cover et al., 1980). Moreover, the destination has side information available in addition to the signals it receives from the relays.

In the case of multiple relays, all relays do not have to perform the same relaying strategy. When some relays do DF and the rest do CF an achievable rate region is given in (Kramer et al., 2005). Here, we explain this scenario for the two-relay case, where the first relay does DF and the second does CF after listening to the first relay. Then the achievable rate becomes

$$R^{(DCF)} = \min\left\{I(X_S; Y_{R_1} \mid X_{R_1}), I(X_S, X_{R_1}; \hat{Y}_{R_2}, Y_D \mid X_{R_2})\right\}$$

$$\text{(3.1)}$$

subject to

$$I(\hat{Y}_{R_2}; Y_{R_2} \mid X_{R_2}, Y_D) \leq I(X_{R_2}; Y_D)$$

$$\text{(3.2)}$$

for the joint distribution $p(x_S, x_{R_1})p(x_{R_2})p(\hat{y}_{R_2} \mid x_{R_2}, y_{R_2})p(y_{R_1}, y_{R_2}, y_D \mid x_S, x_{R_1}, x_{R_2})$. The first mutual information expression in (3.1) indicates the rate the first relay can decode. If the first relay can decode, then the source and the first relay together act as a joint transmitter. Via the second relay the destination obtains a second observation $\hat{Y}_{R_2}$ in addition to its own Y_D. Therefore, the second relay and the destination mimic a joint receiver. Overall, the second term in (3.1) mimics the 2 x 2 multiple-antenna system. In a path loss model as in the second section- A,, this mixed strategy is capacity achieving when the first relay is in close proximity with the source and the second relay is in close proximity with the destination (Kramer et al., 2005).

The multiple relay channel problem has also been recently considered in (Avestimehr et al., 2008). In this work, inspired from a deterministic model (Avestimehr et al., 2007), the authors find a lower bound on the capacity of Gaussian relay networks that is within a constant number of bits away from the cut-set upper bound. This constant depends on the topology of the network but not on the channel parameters. This result provides a good approximation of the capacity, which becomes tight as signal-to-noise ratio increases.

Half-Duplex

In the second and third sections, we had the idealized assumption that the relay is full-duplex. However, full-duplex operation is not possible in practical applications. Wireless transceivers cannot transmit and receive at the same time in the same band. In this section, we consider a half-duplex relay and study the impact of half-duplex operation on relaying protocols and achievable rates.

We can model half-duplex operation using the state variable Q that controls the relay operation. Q takes the value q_1 if the relay is listening, and q_2 if the relay is transmitting. A more general state configuration assumes that the relay can be in sleep, listen or talk states (Kramer, 2004), the fundamental idea in both configurations being the same. In this chapter we consider the former for simplicity. We also consider fixed protocols, in which the relay listens for a fixed time interval fraction (t, with $0 \leq t \leq 1$) and then transmits in the remaining portion ($1-t$), Figure 6. The relay does not aim to pass along additional information by breaking its transmission and reception intervals into smaller blocks and controlling its state variable (Yuksel & Erkip, 2007).

The cut-set upper bound on achievable rates for the half-duplex relay channel is found in (Khojaste-pour et al., 2003b). We will state this upper bound for the multiple-antenna half-duplex relay channel in the sixth section, B.

For the half-duplex relay channel, the DF achievable rate extends from (2.6), if the half-duplex constraint is taken into account. When the relay is half-duplex and employs DF

Figure 6. Fixed relaying: The relay listens for t fraction of the time and transmits in the remaining 1-t fraction

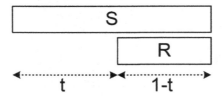

$$R^{(DF)} = \min\{tI(X_S;Y_R \mid q_1) + (1-t)I(X_S;Y_D \mid X_R,q_2),$$
$$tI(X_S;Y_D \mid q_1) + (1-t)I(X_S,X_R;Y_D \mid q_2)\} \tag{4.1}$$

is achievable for a fixed input distribution and a fixed t, can be maximized over all input distributions and $t \in [0,1]$. The first term in (4.1) is the sum of two mutual information expressions. The first one indicates the amount of information the relay can decode in t fraction of the time. The second is the mutual information the destination collects from the source during the time the relay is transmitting. The second term in (4.1) is also a sum of two terms, the first of which is the mutual information at the destination while the relay is silent, and the second is the rate the source and the relay can together send in *1-t* fraction of the time. Similar to the full-duplex case, if the half-duplex relay channel is physically degraded, then the above rate is capacity achieving.

When the relay employs CF, the Wyner-Ziv type compression rate is such that the compressed signal at the relay can reach the destination error-free in the remaining *(1-t)* fraction of time, in which the relay transmits. Then, for a fixed t the instantaneous mutual information at the destination is

$$R^{(CF)} = tI(X_S;\hat{Y}_R,Y_D \mid q_1) + (1-t)I(X_S;Y_D \mid X_R,q_2)$$

subject to

$$tI(\hat{Y}_R;Y_R \mid Y_D,q_1) \leq (1-t)I(X_R;Y_D \mid q_2). \tag{4.2}$$

Note that the above equations incorporate the half-duplex constraint into (2.11) and (2.10) and can be maximized over $t \in [0,1]$.

The real Gaussian noise half-duplex relay channel is very similar to the full-duplex case of (2.1)-(2.2). However, in half-duplex case the relay cannot transmit and receive at the same time. Then we can write

$$Y_{R,1} = a_{SR}X_{S,1} + Z_{R,1}$$
$$Y_{D,1} = a_{SD}X_{S,1} + Z_{D,1}$$

for state q_1, and

$$Y_{D,2} = a_{SD}X_{S,2} + a_{RD}X_{R,2} + Z_{D,2}$$

for state q_2. The channel gains a_{SR}, a_{SD} and a_{RD} are defined similarly as in the second section.

For the Gaussian half-duplex relay channel the DF achievable rate becomes (Khojastepour et al., 2003b)

$$R_{max}^{(DF)} = \max_{\rho, t \in [0,1]} \min \left\{ \frac{t}{2} \log \left(1 + a_{SR}^2 P_{S,1}\right) + \frac{1-t}{2} \log \left(1 + \rho a_{SD}^2 P_{S,2}\right), \right.$$
$$\left. \frac{t}{2} \log \left(1 + a_{SD}^2 P_{S,1}\right) + \frac{1-t}{2} \log \left(1 + a_{SD}^2 P_{S,2} + a_{RD}^2 P_R + 2\sqrt{(1-\rho)a_{SD}^2 P_{S,2} a_{RD}^2 P_R}\right) \right\}.$$

Similarly, the half-duplex CF achievable rates are (Lai et al., 2006)

$$R^{(CF)} = \max_{t \in [0,1], P_{S,1}, P_{S,2}} \frac{t}{2} \log \left(1 + a_{SD}^2 P_{S,1} + \frac{a_{SR}^2 P_{S,1}}{1 + \hat{N}_R}\right) + \frac{1-t}{2} \log \left(1 + a_{SD}^2 P_{S,2}\right),$$

where

$$\hat{N}_R = \frac{1 + (a_{SD}^2 + a_{SR}^2)P}{(1 + a_{SD}^2 P)\left[\left(1 + \frac{a_{RD}^2 P_R}{1 + a_{SD}^2 P_{S,2}}\right)^{\frac{1-t}{t}} - 1\right]}.$$

In the above expressions $P_{S,1}$ and $P_{S,2}$ are the average source power constraints in states q_1 and q_2 respectively with $P_{S,1} + P_{S,2} = P$. Both rates $R^{(DF)}$ and $R^{(CF)}$ can be optimized over $P_{S,1}$ and $P_{S,2}$ as well.

In addition to DF and CF based protocols, amplify-forward (AF) (Laneman et al., 2004) and non-orthogonal amplify-and-forward (NAF) (Nabar et al., 2004; Azarian et al., 2005) are two linear relaying protocols that are important for the Gaussian half-duplex relay channel. In AF the source and the relay share the time equally, and the source node remains silent while the relay transmits in the second half of the time. In AF the relay simply scales $Y_{R,1}$ according to its own power constraint and forwards $X_R = \beta Y_R$ to the destination, where $\beta \leq \sqrt{P_R / (a_{SR}^2 P_S + 1)}$. Assuming β is equal to its upper bound, the AF protocol achieves the rate $R^{(AF)} = \frac{1}{4} \log \left(1 + a_{SD}^2 P_S + \frac{a_{SR}^2 P_S a_{RD}^2 P_R}{1 + a_{SR}^2 P_S + a_{RD}^2 P_R}\right).$

The NAF protocol is an extension of AF, where the source can simultaneously transmit with the relay in state q_2.

WIRELESS APPLICATIONS: COOPERATIVE DIVERSITY

In the previous sections we have considered cooperation in discrete memoryless and Gaussian channels. One of the most important advancements in cooperative communication has been the unearthing of its potential in wireless networks.

Wireless channels experience fading, which degrades the system performance when signal components that are received over different propagation paths add destructively (Proakis, 2000). Most of the error correcting codes can recover from even very deep fade levels, if the fading coherence time is on the order of a symbol time. But if the fading coherence time is much longer than the symbol duration, then a deep fade affects many symbols consecutively. Retransmission or forward error correction is too costly in this case and no recovery is possible. These slowly fading channels do not guarantee reliable communication for any transmission rate and have zero capacity. For these channels, error probability is due to two main causes: deep fade levels, and channel noise. When the wireless channel experiences a deep fade and the channel cannot support the transmission rate, it is said that the channel is in outage (Tse and Viswanath, 2005). The probability of error can then be written as

$$P(\text{error}) = P(\text{error} \mid \text{outage})P(\text{outage}) + P(\text{error} \mid \text{no outage})P(\text{no outage}).$$

If the channel is in outage, then the error probability is almost equal to 1. Similarly, if there is no outage, good channel codes ensure arbitrarily small error probability. Thus the outage event dominates the error event and probability of error is approximately equal to probability of outage.

To mitigate the adverse effects of fading, time, frequency or spatial diversity techniques are employed. For example when there are multiple antennas at the receiver, each antenna observes an independent copy of the transmitted message. It is less likely that all observations are bad, and thus reliability is increased. Similarly, when there are multiple transmit antennas, each antenna can suitably repeat the same message to increase reliability. In addition to these traditional diversity techniques, cooperation/relaying can also be used to provide diversity (Laneman et al., 2004; Sendonaris et al., 2003a; Sendonaris et al., 2003b).

In this section, we first assume the relay is full-duplex. Under fading, the received signals at the relay and at the destination are given by (2.1) and (2.2), but now the channel inputs, outputs and the channel noise are complex valued. Assuming Rayleigh fading, $a_{ij} = h_{ij}$, $i,j = S,R,D$, $i \neq j$, are independent, identically distributed (i.i.d.) zero mean complex Gaussian random variables with zero mean and unit variance. We consider slow fading, which stays constant for the duration of a channel block. There is no channel state information at the source. We assume there are pilot signals to measure receiver channel gains, and thus the relay and the destination know their incoming fading levels. In addition, the relay knows its outgoing gain and the source-destination channel gain, which can happen at a negligible cost with proper feedback. This information becomes necessary to implement the CF protocol and will become clear when we explain CF later in this section and in the fourth section, A.

As the source node does not have channel side information, and the application is delay-limited and constant rate, the source node transmits at a fixed target data rate $R^{(T)}$. An outage occurs if the mutual information at the destination is not large enough to support this fixed target rate, and we can write the outage probability as

$$P(\text{outage}) = P(I < R^{(T)}),$$

where I is the mutual information at the destination when receiver side channel state information is available.

Under fading the DF protocol is slightly different than the one described in the second section, A.1 for non-fading channels. When there is fading, the mutual information at the destination depends on whether the relay decodes the source message or not (Laneman et al., 2004). The relay can reliably decode the source message and repeat it to the destination if the mutual information it collects $I(X_S; Y_R) = I_R$

$$I_R = \log(1+ \mid h_{SR} \mid^2 P_S)$$

is larger than the target data rate. If not, we assume the relay remains silent. Let

$$I_{SD} = \log(1+ \mid h_{SD} \mid^2 P_S)$$
$$I_{SR,D} = \log(1+ \mid h_{SD} \mid^2 P_S + \mid h_{RD} \mid^2 P_R).$$

Then the mutual information at the destination is equal to

$$I^{(DF)} = \begin{cases} I_{SD} & \text{if the relay cannot decode} \\ I_{SR,D} & \text{if the relay decodes} \end{cases}.$$

^{2} Note that unlike the Gaussian case, there is no coherent combining gain between the source and the relay, when they transmit together. This is because the source node does not have channel state information and cannot synchronize its phase with the relay.

Then we can write the probability of outage at the destination as

$$P(\text{outage at D}) = P(\text{outage} \mid \text{relay decodes})P(\text{relay decodes})$$
$$+P(\text{outage} \mid \text{relay cannot decode})P(\text{relay cannot decode}), \tag{5.1}$$

which becomes equal to

$$P(\text{outage at D}) = P(I_{SR,D} < R^{(T)} \mid I_R > R^{(T)})P(I_R > R^{(T)})$$
$$+P(I_{SD} < R^{(T)} \mid I_R < R^{(T)})P(I_R < R^{(T)}). \tag{5.2}$$

When the relay employs CF then the mutual information at the destination is same as in (2.12) if a_{ij}^2 are replaced with $|h_{ij}|^2$, and the 1/2 is removed. Note that the relay has to know all to channel gains in the system to determine the compression noise in (2.13). The probability of outage for CF is then $P(I^{(CF)} < R^{(T)})$.

Figure 7 shows the outage probability of DF and CF protocols as a function of the target data rate $R^{(T)}$ for $P_1 = P_2 = 10$. Direct transmission and 1 x 2 MIMO outage probabilities are also included in the figure for comparison. We observe that cooperative protocols significantly improve upon direct transmission.

Figure 7. Probability of outage as a function of rate. $P_1 = P_2 = 10$.

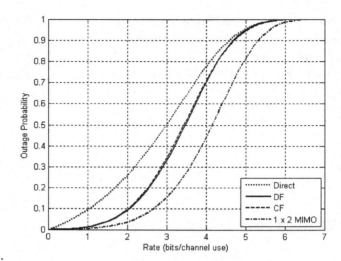

The above analysis shows the probability of outage as a function of the target data rate for a fixed P_S and P_R. It is also important to understand how probability of outage behaves as a function of P_S and P_R for a fixed target data rate. In particular the probability of error decay rate is of interest, as fast decay implies superior reliability. This decay rate is called the diversity gain and is defined as

$$d = - \lim_{\text{SNR} \to \infty} \frac{\log P_e(\text{SNR})}{\log \text{SNR}}. \tag{5.3}$$

Here SNR denotes the average received signal-to-noise ratio. For high SNR values (5.3) can be restated to write

$$P_e(\text{SNR}) \doteq \frac{1}{\text{SNR}^d}.$$

Diversity gains do not change with a constant scaling in the transmit power levels P_S and P_R. In the relay channel, we thus assume $P_S = P_R$ without any loss of generality. Note that the average transmit power is equal to the average SNR as we assumed the noise variance is equal to 1.

Now, we find the diversity gains for DF and CF protocols. As probability of error dominated by outage, we simply investigate the outage probability decay rate.

For the DF protocol, we first look into the relay outage probability, which is

$$P(I_R < R^{(T)}) = P\left(|h_{SR}|^2 < \frac{2^{R^{(T)}} - 1}{P_S}\right)$$

$$= 1 - \exp\left(-\frac{2^{R^{(T)}} - 1}{P_S}\right)$$

$$\doteq \frac{1}{\text{SNR}}.$$

This also implies that $P(I_R > R^{(T)}) \doteq 1$. Note that direct transmission outage probability also results in the same limit as the channel gains are assumed to be identically distributed. Similarly, one can show that $P(\text{outage} \mid \text{relay cannot decode}) \doteq \text{SNR}^{-1}$ and $P(\text{outage} \mid \text{relay can decode}) \doteq \text{SNR}^{-2}$. Overall,

$$P(\text{outage at D}) \doteq \frac{1}{\text{SNR}^2}$$

and the DF protocol diversity gain is equal to 2.

Similar to the DF protocol, the CF protocol also achieves 2 levels of diversity gain (Yuksel & Erkip, 2007). Intuitively, from a diversity perspective, the CF protocol makes the relay channel equivalent to a single-antenna source, two-antenna destination point-to-point system, while DF results in a two-antenna source, a single-antenna destination system. In the point-to-point system each receiver antenna suffers from additive noise. In the CF protocol it is as if one of the receiver antennas experiences additional noise due to compression. However, the relay can adjust this compression noise to enable full diversity. We omit the details and refer the reader to (Yuksel & Erkip, 2007) for a more complicated discussion.

Figure 8 displays DF and CF outage probabilities as a function of per user SNR $P_S = P_R$ for $R^{(T)} = 5$ bits/ channel use. The figure clearly shows that DF and CF have 2 levels of diversity as in 1 x 2 MIMO, whereas direct transmission has only 1 levels of diversity. This result implies that DF and CF protocols can be used to form virtual antenna arrays and to provide diversity gains similar to spacial diversity techniques.

Figure 8. Probability of outage as a function of SNR. $R^{(T)} = 5$ bits/ channel use

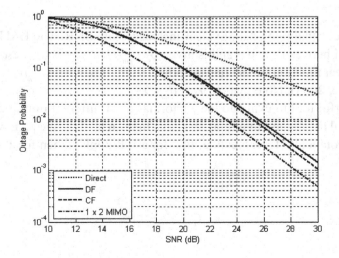

It is important to note that half-duplex DF and CF and the AF and NAF protocols introduced in the fourth section, also achieve 2 levels of diversity (Laneman et al., 2004; Nabar et al., 2004). This is because the source-destination and the relay-destination links observe independent channel gains, and the diversity gain mainly depends on the number of independent observations at the destination even when half-duplex constraint is imposed.

Another protocol, which achieves 2 levels of diversity, is the dynamic decode-and-forward (DDF) protocol suggested in (Azarian et al., 2005). Unlike DF in DDF the relay listens to the source until it is able to decode reliably. When this happens, the relay re-encodes the source message and sends it in the remaining portion of the frame. In DDF the fraction of the time the relay listens, *t*, depends on the source-relay link quality. If this channel is of good quality, the relay can decode quickly, and transmit for a longer time interval. If this channel is bad, the relay has a short time left for transmission.

When there are multiple relays in the network, protocols other than DF, CF, AF, NAF or DDF can be used to obtain diversity gains. One method is to employ the relays to form a distributed space time code (Laneman & Wornell, 2003). The other method is to choose the best relay and to use one relay at a time (Bletsas et al., 2006). Both strategies result in full-diversity gains.

Diversity-Multiplexing Tradeoff of Cooperative Communications

In the previous section we investigated the diversity gains of cooperation for single antenna terminals. We showed that cooperative protocols enable the formation of *virtual* antenna arrays to increase reliability. In a general network, nodes can have multiple antennas, increasing the system's degrees of freedom. It is important to understand how these additional degrees of freedom affect diversity gains, and how other resources, such as rate, coverage area, or network lifetime, need to be sacrificed for higher reliability. In this section we provide such an analysis for the single relay case under Rayleigh fading.

The tradeoff between reliability and rate, known as diversity-multiplexing tradeoff (DMT), is established for point-to-point multiple antenna slow-fading channels in (Zheng & Tse, 2003). In DMT the reliability measure is the diversity gain as defined in (5.3). The transmission rate $R^{(T)}$ is measured by the multiplexing gain, which is

$$r = \lim_{\text{SNR}\to\infty} \frac{R^{(T)}(\text{SNR})}{\log \text{SNR}}.$$

It shows how fast the actual rate of the system increases with SNR. The DMT of an *m* x *n* MIMO is defined as the tradeoff between *r* and *d* and is given by $d_{mn}(r)$. $d_{mn}(r)$ is the best achievable diversity, which is a piecewise-linear function connecting the points $(k, d_{mn}(k))$, where $d_{mn}(k) = (m-k)(n-k)$, *k*=0,1,..., min {*m,n*} (Zheng & Tse, 2003). Note that $d_{mn}(r) = d_{nm}(r)$.

The DMT is a powerful tool to evaluate the performance of different multiple antenna schemes at high SNR; it is also useful for cooperative/relay systems. In the rest of the chapter, we study the multiple-antenna relay channel from a DMT perspective, first for a full-duplex, then for a half-duplex relay.

A. Full-Duplex Relay

We assume the source, the destination and the relay have m, n and k antennas respectively. Note that our model is general enough to account for the already existing spatial diversity in the form of multiple antennas. The received signals at the relay and at the destination are

$$\mathbf{Y}_R = \mathbf{H}_{SR}\mathbf{X}_S + \mathbf{Z}_R \qquad (6.1)$$

$$\mathbf{Y}_D = \mathbf{H}_{SD}\mathbf{X}_S + \mathbf{H}_{RD}\mathbf{X}_R + \mathbf{Z}_D, \qquad (6.2)$$

where $\mathbf{Z}_R$ and $\mathbf{Z}_D$ are the independent complex Gaussian noise vectors at the corresponding node. $\mathbf{H}_{SR}$, $\mathbf{H}_{SD}$ and $\mathbf{H}_{RD}$ are the $k \times m$, $n \times m$ and $n \times k$ channel gain matrices between the source and the relay, the source and the destination, and the relay and the destination respectively. These channel gain matrices have i.i.d. complex Gaussian entries, with real and imaginary parts zero mean and variance 1/2 each. Next, we first find an upper bound on DMT, then study the DMT the DF and CF protocols achieve.

An upper bound on DMT can be derived using the well-known cut-set upper bound (Cover & Thomas, 1991; Yuksel & Erkip, 2007). In the relay channel there are two cut-sets of interest: the cut-set around the source and the cut-set around the destination. These cut-set mutual information expressions are respectively

$$I_{C_S} = I(\mathbf{X}_S; \mathbf{Y}_R, \mathbf{Y}_D \mid \mathbf{X}_R) \qquad (6.3)$$

$$I_{C_D} = I(\mathbf{X}_S, \mathbf{X}_R; \mathbf{Y}_D). \qquad (6.4)$$

To find the best DMT upper bound, we need to find the maximum of these mutual information expressions. We can further upper bound the mutual information expressions as

$$I_{C_S} \le I'_{C_S} = \log K_{S,RD} \qquad (6.5)$$

$$I_{C_D} \le I'_{C_D} = \log K_{SR,D}, \qquad (6.6)$$

where

$$K_{S,RD} \triangleq \left| \mathbf{I}_{k+n} + \mathbf{H}_{S,RD}\mathbf{H}_{S,RD}^{\dagger} P_S \right|, \qquad (6.7)$$

$$K_{SR,D} \triangleq \left| \mathbf{I}_n + \mathbf{H}_{SR,D}\mathbf{H}_{SR,D}^{\dagger}(P_S + P_R) \right|, \qquad (6.8)$$

and

$$\mathbf{H}_{S,RD} = \begin{bmatrix} \mathbf{H}_{SR} \\ \mathbf{H}_{SD} \end{bmatrix}, \ \mathbf{H}_{SR,D} = \begin{bmatrix} \mathbf{H}_{SD} & \mathbf{H}_{RD} \end{bmatrix}. \qquad (6.9)$$

Note that $P(I'_{C_i} < R^{(T)}) \doteq \text{SNR}^{-d'_{C_i}(r)}$, $i = S, D$, with the target data rate $R^{(T)} = r \log \text{SNR}$, $d'_{C_S}(r) = d_{m(n+k)}(r)$ and $d'_{C_D}(r) = d_{(m+k)n}(r)$. Then for transmission rate $R^{(T)}$, one can easily upper bound the system DMT by

$$d(r) \le \min\{d_{m(n+k)}(r), d_{(m+k)n}(r)\}.$$

This DMT upper bound is quite intuitive. The cut-set around the source node idealizes the relay-destination link and assumes it to be perfect. The system becomes equivalent to an *m* x *(n+k)* MIMO, whose DMT is $d_{m(n+k)}(r)$. Similarly, the cut-set around the destination assumes that the source-relay link is perfect leading to the DMT $d_{(m+k)n}(r)$.

We first study the CF protocol and argue that it is DMT optimal. In the multiple antenna CF protocol the relay performs vector Wyner-Ziv compression with side information taken as the destination's received signal. The CF protocol achieves the rate in (2.11) subject to (2.10) if the random variables X_S, X_R, Y_R, Y_D and $\hat{Y}_R$ are respectively replaced with their vector counterparts $\mathbf{X}_S$, $\mathbf{X}_R$, $\mathbf{Y}_R$, $\mathbf{Y}_D$ and $\hat{\mathbf{Y}}_R$. Note that the relay needs to know all the channel gains in the system to ensure that the compression rate constraint in (2.10) is satisfied. Then CF achieves the rate (Yuksel & Erkip, 2007)

$$R^{(CF)} = \log \frac{L_{S,RD}}{\left(\sqrt[k]{\frac{L_{S,RD}}{L_{SR,D}}} + 1 \right)^k}, \qquad (6.10)$$

subject to

$$\log \frac{L_{S,RD}}{L_{S,D} \hat{N}_R^k} \le \log \frac{L_{SR,D}}{L_{S,D}},$$

where

$$L_{S,D} \triangleq \left| \mathbf{H}_{SD} \mathbf{H}_{SD}^\dagger \frac{P_S}{m} + \mathbf{I}_n \right|$$

$$L_{SR,D} \triangleq \left| \mathbf{H}_{SD} \mathbf{H}_{SD}^\dagger \frac{P_S}{m} + \mathbf{H}_{RD} \mathbf{H}_{RD}^\dagger \frac{P_R}{k} + \mathbf{I}_n \right|,$$

$$L_{S,RD} \triangleq \left| \mathbf{H}_{S,RD} \mathbf{H}_{S,RD}^\dagger \frac{P_S}{m} + \begin{bmatrix} (\hat{N}_R + 1)\mathbf{I}_k & \mathbf{0} \\ \mathbf{0} & \mathbf{I}_n \end{bmatrix} \right|,$$

$$L'_{S,RD} \triangleq \left| \mathbf{H}_{S,RD} \mathbf{H}_{S,RD}^\dagger \frac{P_S}{m} + \mathbf{I}_{k+n} \right|.$$

Then it can be shown that CF achieves the DMT

$$d_{CF}(r) = \min\{d_{m(n+k)}(r), d_{(m+k)n}(r)\}.$$

We avoid the mathematical details of the proof in this chapter and refer the interested readers to (Yuksel & Erkip, 2007).

In the CF protocol as the relay knows all the channel gains the network, it adjusts its compression rate such that the destination can always reliably decode the relay signal. The destination then obtains a perfect copy of the compressed version ($\hat{\mathbf{Y}}_R$) of the relay's observation ($\mathbf{Y}_R$). The compression noise at the relay is never too large and does not hinder diversity gains. The CF protocol adapts well to the current channel conditions, the relay and destination work in accordance and imitate *m* x *(n+k)* MIMO behavior. It is worth mentioning that the SCF protocol in the second section, A. 2 has a suboptimal performance and the CF adaptivity is required for DMT optimal behavior.

As an alternative to CF, the relay can use the DF protocol. When the source, the destination and the relay all have a single antenna each, following the analysis in the second section, A..1 we can show that the DF protocol also achieves the DMT upper bound, which is equal to $d_{12}(r)$. But this optimality does not generalize to arbitrary *m, n* and *k*. The DF protocol achieves the DMT

$$d_{DF}(r) = \begin{cases} \min\{d_{(m+k)n}(r), d_{mn}(r) + d_{mk}(r)\} & \text{if } 0 \leq r \leq \min\{m,n,k\} \\ d_{mn}(r) & \text{if } \min\{m,n,k\} < r \leq \min\{m,n\} \end{cases}$$

If *m* or *n* (or both) is equal to 1, we find that DF meets the DMT upper bound. Similarly we can show that for cases such as *(m,n,k)*=(3,2,2) or *(m,n,k)*=(4,2,3), as $d_{(m+k)n}(r) < d_{mn}(r) + d_{mk}(r)$ for all *r*, DF is optimal. A general necessary condition for DF to be optimal for all multiplexing gains is $m \geq n$. If *m<n*, then $d_{mn}(r) + d_{mk}(r) \leq d_{m(n+k)}(r) < d_{(m+k)n}(r)$, and DF is suboptimal. Whenever $\min\{m,n,k\} = k$, the degrees of freedom in the direct link is larger than the degrees of freedom in the source to relay link, that is $\min\{m,n\} \geq \min\{m,k\}$. For multiplexing gains in the range $\min\{m,n,k\} < r \leq \min\{m,n\}$, the relay can never help and the system has the direct link DMT $d_{mn}(r)$. Therefore, DF loses its optimality.

Intuitively the suboptimality of DF is because of the hard decision at the relay. The point-to-point MIMO DMT, which is the piecewise linear function connecting the points *(k,(m-k)(n-k))*, *k* = 1,..., min *{m,n}*, suggests that each additional unit of multiplexing gain uses an antenna both from the transmitter and the receiver, while the remaining antennas provide the diversity gain. Similarly, in the DF protocol each additional multiplexing gain effectively uses an antenna at every node (source, relay and destination) due to hard decision. So the cost of each additional multiplexing gain costs is, in terms of antennas, is one more than the ideal MIMO behavior.

In Figure 9 we compare CF and DF DMT for *(m,n,k)* =(2,2,1). We see that the CF protocol is always DMT optimal, but the DF protocol is significantly worse. Note that any minor difference in the DMT curve is quite important because diversity gain is the exponent of 1/SNR at high SNR. The suboptimal behavior of DF arises because the outage event when the relay cannot decode can dominate for general *m, n* and *k*. In addition to this, for multiplexing gains larger than min*{m,n,k}*, the relay never participates in the communication because it is degrees of freedom limited and cannot decode large multiplexing gain signals. For this region, we observe the direct link behavior. For DF, the available relay CSI does not improve the DMT performance compared to the case where the relay only knows the source-relay channel gains. With this CSI the relay can at best perform beamforming, which brings in power gains but no improvement on DMT. We conclude that soft information transmission, as in the CF protocol, is necessary at the relay not to lose diversity or multiplexing gains.

In the second section, we showed that there is a symmetry between DF and CF. DF achieves capacity when the relay is close to the source, and CF achieves capacity if the relay is close to the destination. However, the above analysis reveals that CF and DF protocols do not always behave similar, unlike the single antenna relay system. The degrees of freedom available also has an effect on relaying strategies.

B. Half-Duplex Relay

In the previous section, we studied the relay channel when the relay is full-duplex. Although this is an ideal assumption about the relay's physical capabilities, it helps us understand the fundamental differences between the DF and CF protocols. In this section we assume a half-duplex relay to study how this affects the DMT behavior of the relay channel. In this section we investigate *static* half-duplex channels, in which the relay state is not controlled based on channel realizations. The relay can also perform *dynamic* behavior as in the DDF protocol. In dynamic protocols, the relay determines the fraction of time it listens and transmits depending on the channel gains.

In the previous subsection we found an upper bound on DMT, when the relay is full-duplex. As this bound may not be tight when the relay is half-duplex, next we consider a half-duplex DMT upper bound. The cut-set mutual information expressions for a half-duplex relay are respectively (Khojastepour et al., 2003a)

$$I_{C_s}(t) = tI(\mathbf{X}_S; \mathbf{Y}_R, \mathbf{Y}_D \mid q_1) + (1-t)I(\mathbf{X}_S; \mathbf{Y}_D \mid \mathbf{X}_R, q_2) \qquad (6.11)$$

Figure 9. The source has 2, the destination has 2, and the relay has 1 antenna, (m,n,k)=(2,2,1).

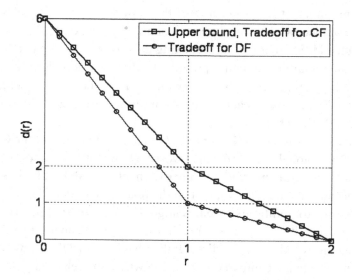

$$I_{C_D}(t) = tI(\mathbf{X}_S; \mathbf{Y}_D \mid q_1) + (1-t)I(\mathbf{X}_S, \mathbf{X}_R; \mathbf{Y}_D \mid q_2). \tag{6.12}$$

Note that these expressions depend on t, the amount of time the relay listens. Then an upper bound on these expressions are

$$I_{C_S}(t) \le I'_{C_S}(t) = t \log K_{S,RD} + (1-t) \log K_{S,D} \tag{6.13}$$

$$I_{C_D}(t) \le I'_{C_D}(t) = t \log K_{S,D} + (1-t) \log K_{SR,D} \tag{6.14}$$

where $K_{S,RD}$ and $K_{SR,D}$ are respectively defined in (6.7) and (6.8), and

$$K_{S,D} \triangleq \left| \mathbf{H}_{SD} \mathbf{H}_{SD}^\dagger P_S + \mathbf{I}_n \right|. \tag{6.15}$$

For a target data rate $R^{(T)} = r \log \text{SNR}$, and for a fixed t, if $P(I'_{C_i}(t) < R^{(T)}) \doteq \text{SNR}^{-d'_{C_i}(r,t)}$, $i = S, D$, then $d(r,t)$, the best achievable diversity for the half-duplex relay channel for fixed t, satisfies

$$d(r,t) \le \min\{d'_{C_S}(r,t), d'_{C_D}(r,t)\}. \tag{6.16}$$

Optimizing over t we find an upper bound on the static multiple antenna half-duplex relay channel DMT as

$$d(r) \le \max_{t \in [0,1]} \min\{d'_{C_S}(r,t), d'_{C_D}(r,t)\}. \tag{6.17}$$

In general it is hard to compute the exact DMT of (6.17). In particular for static protocols, to find $d'_{C_S}(r,t)$ and $d'_{C_D}(r,t)$ for general m, n and k we need to calculate the joint eigenvalue distribution of two correlated Hermitian matrices, $\mathbf{H}_{SD}\mathbf{H}_{SD}^\dagger$ and $\mathbf{H}_{S,RD}\mathbf{H}_{S,RD}^\dagger$ or $\mathbf{H}_{SD}\mathbf{H}_{SD}^\dagger$ and $\mathbf{H}_{SR,D}\mathbf{H}_{SR,D}^\dagger$. However, when $m=1$, both $\mathbf{H}_{SD}\mathbf{H}_{SD}^\dagger$ and $\mathbf{H}_{S,RD}\mathbf{H}_{S,RD}^\dagger$ reduce to vectors and $d'_{C_S}(r,t)$ is easier to compute. Similarly, when $n=1$, $\mathbf{H}_{SD}\mathbf{H}_{SD}^\dagger$ and $\mathbf{H}_{SR,D}\mathbf{H}_{SR,D}^\dagger$ are vectors, and $d'_{C_D}(r,t)$ can be found. An explicit form for $d'_{C_S}(r,t)$ for $m=1$ is given as

$$d'_{C_S}(r,t) = \begin{cases} n + k - k\dfrac{r}{t} & \text{if} \quad r \le t, \text{and} \, t \le \dfrac{k}{n+k} \\[3mm] n\left(\dfrac{1-r}{1-t}\right) & \text{if} \quad r \ge t, \text{and} \, t \le \dfrac{k}{n+k} \\[3mm] (n+k)(1-r) & \text{if} \qquad t \ge \dfrac{k}{n+k} \end{cases}.$$

For $n = 1$ and for arbitrary m and k, $d'_{C_D}(r,t)$ has the same expression as $d'_{C_S}(r,t)$ if n and t are replaced with m and $(1-t)$ in the above expressions (Yuksel & Erkip, 2007).

Although we do not have an explicit expression for $d'_{c_S}(r,t)$ or $d'_{c_D}(r,t)$ for general (m,n,k), we can comment on some special cases and get insights about multiple antenna, half-duplex behavior. First we observe that $d'_{c_S}(r,t)$ and $d'_{c_D}(r,t)$ depend on the choice of t, and the upper bound of (6.17) is not always equal to the full-duplex bound. As an example consider $(m,n,k) = (1,1,2)$, for which $d'_{c_S}(r,t)$ is shown in Figure 10. To achieve the full-duplex bound for all r, $d'_{c_S}(r,t)$ needs to have $t \geq 2/3$, whereas $d'_{c_D}(r,t)$ needs $t \leq 1/3$. As both cannot be satisfied simultaneously, $d(r,t)$ is less than the full-duplex bound for all t.

On the other hand, to maximize the half-duplex DMT it is optimal to choose $t = 1/2$ whenever $m=n$. To see this, we compare (6.13) with (6.14), and note that both $K'_{S,RD} \geq K_{S,D}$ and $K_{SR,D} \geq K_{S,D}$ for $m=n$. Furthermore, for $m=n$ $d'_{c_S}(r,t) = d'_{c_D}(r,1-t)$, and $d'_{c_S}(r,t)$ is a non-decreasing function in t. Therefore $\min\{d'_{c_S}(r,t), d'_{c_D}(r,t)\}$ must reach its maximum at $t=1/2$.

Similar to the full-duplex case, in the multiple-antenna half-duplex relay channel (Yuksel & Erkip, 2007) the CF protocol is DMT optimal and achieves the half-duplex cut-set bound DMT

$$d_{CF}(r) = \max_{t \in [0,1]} \min\{d'_{c_S}(r,t), d'_{c_D}(r,t)\}.$$

Note that if the relay is dynamic, CF can also behave dynamically, choosing t as a function of CSI available at the relay, and still achieve the DMT upper bound for dynamic protocols at high SNR.

Figure 11 shows the CF DMT in comparison to DDF and NAF DMT for $(m,n,k) = (1,1,1)$. For this single antenna case the half-duplex DMT bound is equal to the full-duplex case. We observe that DDF does not meet the upper bound for $1/2 \leq r \leq 1$, as in this range, the relay does not have enough time to transmit the high rate information it received (Azarian et al., 2005). NAF protocol is also suboptimal. However, CF is DMT optimal for $(m,n,k) = (1,1,1)$ as well as arbitrary antenna configurations.

Figure 10. DMT upper bound for the cut-set around the source, C_S. The source has 1, the destination has 1, and the relay has 2 antennas, $(m,n,k) = (1,1,2)$. Note that as $m=n$, $d'_{c_S}(r,t) = d'_{c_D}(r,1-t)$. The upper bound in (6.17) reaches its maximum for $t=1/2$. The solid line in the figure is also equal to the full-duplex bound.

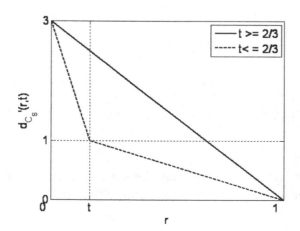

Figure 11. The DMT for CF, DDF, and NAF protocols for single antenna nodes

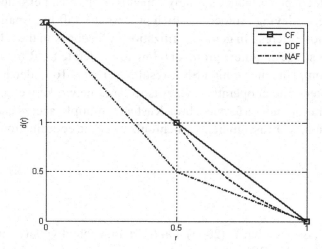

We showed that for a multiple antenna full-duplex relay system, the probability that the relay cannot decode is dominant and the DF protocol becomes suboptimal. Therefore, we do not expect any relay decoding based protocol to achieve the DMT upper bound in the multiple antenna half-duplex system either.

CONCLUSION

In this chapter we outline some of the major information theoretical advances in the cooperative communications literature. For discrete memoryless and Gaussian channels, we show that the relay has a large potential to improve achievable rates with respect to direct transmission. We explain two fundamental strategies DF and CF for the relay channel. We also provide an

overview of mutual cooperation among two users under the generalized multiple access channel model. For multiple relay systems we give an example on how multiple relays can be coordinated.

In addition to non-fading channels, cooperation offers substantial gains under fading. We show that via cooperative protocols *virtual* antenna arrays can be formed and spatial diversity gains similar to physical antenna arrays can be achieved. To provide a complete characterization, we also study the diversity gains in the DMT context. We investigate the effects of increased degrees of freedom and relay processing capability on cooperative DMT. We show that although full-duplex DF and CF have the same DMT when each node has a single antenna, this is not the case for multiple-antenna relay systems. DF can be suboptimal, yet CF is DMT optimal for arbitrary antenna configurations. For the half-duplex relay CF remains to be DMT optimal. Although it is hard to find the DMT upper bound explicitly for arbitrary number of antennas at the nodes, we have solutions for special cases. We also show that the static half-duplex DMT bound is tighter than the full-duplex bound in general.

Despite the recent significant advances we outlined in this chapter, there are many open problems related to the relay channel in particular and to cooperative networks in general. For example there are no practical channel codes designed for CF, which hinders the implementation of this robust protocol. On the other hand, CF protocol performance strongly depends on the channel state information available at the relay. Suggesting new relaying protocols, which process the soft information, but do not require this information are of utmost interest. In general, utilization of feedback in the relay channel is not fully understood. In addition to suggesting new protocols, low to moderate SNR studies need to be carried for the relay channel to complement the high SNR results of DMT. To understand larger cooperative networks better and to extend the cooperation benefits to large networks we also need to understand the role of relaying/cooperation in interference, broadcast and multiple-access channels, and to explore network gains of cooperation, relay assignment, synchronization and coordination issues, and incentives to cooperate.

REFERENCES

Azarian, K., El-Gamal, H., & Schniter, P. (2005). On the achievable diversity-multiplexing tradeoff in half-duplex cooperative channels. *IEEE Transactions on Information Theory, 51*(12), 4152. doi:10.1109/TIT.2005.858920

Bletsas, A., Khisti, A., Reed, D. P., & Lippman, A. (2006). A simple cooperative diversity method based on network path selection. *IEEE Journal on Selected Areas in Communications, 24*(3), 659. doi:10.1109/JSAC.2005.862417

Cover, T. M., & Gamal, A. E. (1979). Capacity theorems for the relay channel. *IEEE Transactions on Information Theory, 25*(5), 572. doi:10.1109/TIT.1979.1056084

Cover, T. M., Gamal, A. E., & Salehi, M. (1980). Multiple access channels with arbitrarily correlated sources. *IEEE Transactions on Information Theory, 26,* 648. doi:10.1109/TIT.1980.1056273

Cover, T. M., & Thomas, J. A. (1991). *Elements of information theory.* John-Wiley & Sons, Inc.

ITs. (2007). *IEEE Transactions on Information Theory.* Special Issue on Models, Theory, and Codes for Relaying and Cooperation in Communication Networks. Avestimehr, S., Diggavi, S., & Tse, D. (2007). A deterministic approach to wireless relay networks. In *Proceedings of Allerton Conference on Communication, Control, and Computing.*

Avestimehr, S., Diggavi, S., & Tse, D. (2008). Approximate capacity of Gaussian relay networks. In *Proceedings of IEEE International Symposium on Information Theory.*

Khojastepour, M., Sabharwal, A., & Aazhang, B. (2003a). On capacity of Gaussian 'cheap' relay channel. In *Proceedings of IEEE Global Telecommunications Conference* (p. 1776).

Khojastepour, M., Sabharwal, A., & Aazhang, B. (2003b). On the capacity of 'cheap' relay networks. In *Proceedings of 37th Conference on Information Sciences and Systems.*

Kramer, G. (2004). Models and theory for relay channels with receive constraints. In *Proceedings of 42nd Allerton Conference on Communication, Control and Computing.*

Kramer, G., Gastpar, M., & Gupta, P. (2005). Cooperative strategies and capacity theorems for relay networks. *IEEE Transactions on Information Theory, 51*(9), 3037. doi:10.1109/TIT.2005.853304

Kramer, G., Maric, I., & Yates, R. D. (2006). Cooperative communications. *Foundations and Trends in Networking, 1*(3-4). NOW Publishers Inc.

Lai, L., Liu, K., & Gamal, H. E. (2006). The three node wireless network: Achievable rates and cooperation strategies. *IEEE Transactions on Information Theory, 52*(3), 805. doi:10.1109/TIT.2005.864421

Laneman, J. N., Tse, D. N. C., & Wornell, G. W. (2004). Cooperative diversity in wireless networks: Efficient protocols and outage behavior. *IEEE Transactions on Information Theory, 50*(12), 3062. doi:10.1109/TIT.2004.838089

Laneman, J. N., & Wornell, G. W. (2003). Distributed space-time coded protocols for exploiting cooperative diversity in wireless networks. *IEEE Transactions on Information Theory, 49*(10), 2415. doi:10.1109/TIT.2003.817829

Nabar, R. U., Bolcskei, H., & Kneubuhler, F. W. (2004). Fading relay channels: Performance limits and space-time signal design. *IEEE Journal on Selected Areas in Communications, 22*(6), 1099. doi:10.1109/JSAC.2004.830922

Prasad, N., & Varanasi, M. K. (2004). Diversity and multiplexing tradeoff bounds for cooperative diversity protocols. In *Proceedings of IEEE International Symposium on Information Theory* (p. 268).

Proakis, J. G. (2000). *Digital Communications,* fourth ed. New York: McGraw-Hill, Inc.

Sendonaris, A., Erkip, E., & Aazhang, B. (2003a). User cooperation diversity-part I: System description. *IEEE Transactions on Communications, 51*(11), 1927. doi:10.1109/TCOMM.2003.818096

Sendonaris, A., Erkip, E., & Aazhang, B. (2003b). User cooperation diversity-part II: Implementation aspects and performance analysis. *IEEE Transactions on Communications, 51*(11), 1939. doi:10.1109/TCOMM.2003.819238

Tse, D., & Viswanath, P. (2005). *Fundamentals of wireless communication.* Cambridge University Press.

van der Meulen, E. C. (1971). Three-terminal communication channels. *Advances in Applied Probability, 3*, 121.

Willems, F. M. J. (1982). *Informationtheoretical results for the discrete memoryless multiple access channel.* Unpublished doctoral thesis, Katholieke Universiteit, Leuven, Belgium.

Willems, F. M. J., & van der Meulen, E. C. (1985). The discrete memoryless multiple-access channel with cribbing encoders. *IEEE Transactions on Information Theory, 31*, 313. doi:10.1109/TIT.1985.1057042

Willems, F. M. J., van der Meulen, E. C., & Schalkwijk, J. P. M. (1983). Achievable rate region for the multiple access channel with generalized feedback. In *Proc. Annu. Allerton Conf. Communication, Control, Computing* (p. 284).

Wyner, A., & Ziv, J. (1976). The rate-distortion function for source coding with side information at the decoder. *IEEE Transactions on Information Theory, 22*, 1. doi:10.1109/TIT.1976.1055508

Xie, L., & Kumar, P. R. (2004). A network information theory for wireless communication: Scaling laws and optimal operation. *IEEE Transactions on Information Theory*, *50*(5), 748. doi:10.1109/TIT.2004.826631

Yuksel, M., & Erkip, E. (2007). Multiple-antenna cooperative wireless systems: A diversity-multiplexing tradeoff perspective. *IEEE Transactions on Information Theory, Special Issue on Models, Theory, and Codes for Relaying and Cooperation in Communication Networks*, *53*(10), 3371.

Zeng, C., Kuhlmann, F., & Buzo, A. (1989). Achievability proof of some multiuser channel coding theorems using backward decoding. *IEEE Transactions on Information Theory*, *35*(6), 1160. doi:10.1109/18.45272

Zheng, L., & Tse, D. N. C. (2003). Diversity and multiplexing: A fundamental tradeoff in multiple-antenna channels. *IEEE Transactions on Information Theory*, *49*, 1073. doi:10.1109/TIT.2003.810646

ENDNOTES

[1]   In the physically degraded relay channel the destination's observation is a physically degraded version of the relay's observation. It is the opposite for the reversely degraded relay channel, where the relay's signal is physically degraded with respect to the destination's.

[2]   Note that this DF protocol is different than the one defined in (Laneman et al., 2004), in which the signal received from the source is completely ignored if the relay cannot decode the message.

Chapter 2
Overview of Amplify-and-Forward Relaying

Ioannis Krikidis
University of Edinburgh, UK

John S. Thompson
University of Edinburgh, UK

ABSTRACT

Amplify-and-Forward (AF) is a simple cooperative strategy for ad-hoc networks with critical power constraints. It involves an amplification of the received signal in the analogue domain at the relays without further signal processing. This chapter gives an overview of the basic AF protocols in the literature and discuss recent research contributions in this area. Based on some well-defined AF-based cooperative configurations, it focuses on the behaviour of AF in block-fading channels, in power allocation problems, in relay selection, and in cross-layer coordination. Mathematical models and outage probability simulations are used in order to show the enhancements of the presented AF techniques.

INTRODUCTION

Cooperative diversity has emerged as a promising technique to combat fading in wireless communications (Sendonaris et. al., 2003a; Sendonaris et. al., 2003b). It is based on the broadcast nature of the wireless medium and enables single-antenna users to "enjoy" space diversity benefits by sharing their physical resources through a virtual transmit and/or receive antenna array. The basic relay channel model is comprised of three terminals: a source that transmits information, a destination that receives information, and a relay that both receives and transmits information in order to enhance communication between the source and the destination (Lai et al., 2006). Models with multiple relays have been also examined which can be regarded as an extension of this basic configuration (Azarian et. al., 2005; Kramer et. al., 2005; Ribeiro et. al., 2005; Yang & Belfiore, 2007; Fan et. al., 2007). Since the work of Sendonaris et al. (2003a, 2003b) that introduced the notion of cooperative diversity, a number of

DOI: 10.4018/978-1-60566-665-5.ch002

Figure 1. (a) The cooperative concept; (i) Decode-and-Forward, (ii) Amplify-and-Forward.(b) AF protocols-slot structure; (i) OAF, (ii) NAF, (iii) BFNAF.

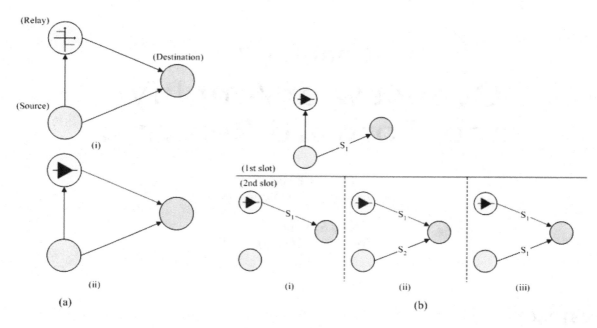

relaying protocols have been proposed in the literature (Laneman et. al., 2004; Larsson & Vojcic, 2005; Azarian et. al., 2005). These schemes can be grouped into two basic classes: Decode-and-Forward (DF) and Amplify-and-Forward (AF). In the DF schemes, the relay decodes the received source message, re-encodes it, and forwards the resulting signal to the destination. On the other hand, in AF schemes, the relays simply amplify the received signal and resend it without any further signal processing in the analogue domain. The amplification process can be regarded as a multiplication with an amplification factor (G) which normalizes the received power. Between these two essential cooperative schemes, AF seems to be a low-complexity solution for practical ad-hoc networks with critical power constraints as it does not require decoding process at the relays. In addition to complexity benefits, it has been shown in (Laneman et al., 2004; Nabar et al., 2004) that AF asymptotically approaches the DF scheme as far as the diversity performance is concerned. Furthermore, in some cases avoiding decoding the signal at the relay nodes actually prevents propagation of decoding errors at the relay (Yang & Belfiore, 2007). Figure 1a schematically presents the two basic cooperative schemes for a three-node configuration.

Although in the original work of Sendonaris et al. (2003a, 2003b) the nodes can transmit and receive simultaneously, in practice the nodes are limited by the half-duplex constraint so they must receive data before retransmitting it later. The first AF cooperative protocol which obeys this constraint was proposed by Laneman et al. (2002, 2003, 2004) where time division multiplexing separated the source and relay transmissions. However, this approach results in a loss of data rate (bandwidth) as the source is inactive for half of the time. In order to recover the rate loss, Nabar et. al. (2004) has proposed the Non-orthogonal AF scheme (NAF) where the source continues to transmits new data during the relaying process. Finally, for the case of block-fading channels where the channel changes during the cooperative frame, Krikidis et. al. (2008c) has proposed the Block-Fading NAF (BFNAF) scheme, where the source retransmits the same data in order to increase the diversity gain instead.

Because cooperative communications is inherently a network problem, issues of protocol layering and cross-layer architectures naturally arise. Cross-layer techniques enable joint protocol design by actively exploiting the dependence between protocol layers to obtain performance gains. The coordination of scheduling and routing with the physical layer as well as the relay selection in multi-node environments are the basic cross-layer problems in the cooperative literature (Bletsas et. al., 2006; Krikidis & Belfiore, 2007b; Krikidis et al., 2008a). Proposed approaches take into account the instantaneous channel conditions and optimize jointly performance (capacity, error probability), complexity (feedback overhead) and fairness.

The scope of this chapter is to discuss all the above issues. More specifically, it introduces the AF cooperative strategy and summarizes the basic approaches of the literature. Based on an appropriate power allocation, a comparison in terms of outage probabilities is also provided. Furthermore, it deals with the problem of cross-layer design for some well-defined AF configurations. The proposed schemes incorporate different layers of the protocol stack with the AF cooperative concept and are analyzed according to three essential quality criteria which are performance, complexity and fairness.

The remainder of this chapter is organized as follows. Section 1 summarizes the basic AF schemes and introduces the BFNAF protocol. The power allocation problem based on their outage probabilities is also provided. Section 2 deals with the problem of AF relay selection in two basic system configurations. The first one focuses on a relay selection based on a partial channel knowledge and the second one assumes an interference-limited environment. Section 3 considers the problem of cross-layer optimization and presents two AF schemes which incorporate scheduling and routing with the instantaneous physical channel. Finally, Section 4 concludes the chapter with some discussion and future directions for research and development.

AF TECHNIQUES OVER BLOCK-FADING CHANNELS

In this Section, we study cooperative protocols for classical quasi-static block-fading environments. Almost all of the ``cooperative literature'' to date assumes that the channel spans several slots and it is constant during the whole cooperative frame. Although this assumption simplifies the analysis, it cannot always be realistic, especially for the case where there is node mobility. Here, we relax this constraint and study cooperative protocols for conventional quasi-static channels with a coherence time smaller than the duration of the complete cooperative frame. This new assumption offers an additional diversity degree and allows the source of the conventional NAF protocols to retransmit the same data during the cooperative slot in order to increase the diversity gain at the receiver. The resulting new NAF-based protocol is termed Block-Fading Non-Orthogonal Amplify-and-Forward (BFNAF) and is suitable for low spectral efficiency scenarios. Another issue that is addressed in this paper is the optimal power allocation of the proposed scheme. Based on the asymptotic expressions of the outage probabilities and using some appropriate average capacity bounds, we propose an appropriate power allocation policy for all the considered cooperative schemes. The considered power constraint deals with a total amount of transmit power over the entire cooperative frame and the optimization criterion is the minimum outage probability. Moreover, the avoidance of the required instantaneous feedback (related with the power allocation) makes the proposed optimizations suitable for practical ad-hoc and sensor applications with crucial power constraints.

System Model

Consider a three node cooperative structure consisting of one source S, one relay R and one destination D (similar with the system model of Figure 1a). All nodes are half duplex and thus they cannot transmit and receive simultaneously. The physical links between terminals are, subject to block-fading, modeled as zero-mean, independent, circularly-symmetric complex Gaussian random variables with unit variance. In contrast with previous NAF-related work (Nabar et al., 2004; Yang & Belfiore, 2007) which assumed that the channel gains do not change during the transmission of a cooperation frame (i.e. two data slots) here we suppose a classical quasi-static channel which remains constant during the transmission of one data slot but changes independently from one slot to another. This system assumption, which corresponds to a coherence time smaller than the frame interval, characterizes some ad-hoc applications with mobility and is popular in the literature (Chen & Laneman, 2006; Hunter & Nosratinia, 2006). For the NAF-based protocols, this assumption offers an additional diversity degree which will be efficiently used in the proposed protocols. Furthermore, additive noise is represented as a zero-mean mutually independent, circularly-symmetric, complex Gaussian random sequence with unit variance. Although our study focuses on a symmetric-link configuration, the proposed techniques can also be applied for the more general non-symmetric case. A cooperation frame is composed of 2 slots of l symbols, denoted by $x_i \in C^{l \times 1}$ for $i = 1, 2$. Without loss of generality we focus on two sequential time slots which correspond to a basic cooperative frame. The system model can be described as

$$
\begin{aligned}
y_D^{(1)} &= \sqrt{P_S^{(1)}} f_{S,D}^{(1)} x_1 + n_{S,D}^{(1)}, \\
y_R^{(1)} &= \sqrt{P_S^{(1)}} f_{S,R}^{(1)} x_1 + n_{S,R}^{(1)}, \\
y_D^{(2)} &= \sqrt{P_S^{(2)}} f_{S,D}^{(2)} x_2 + \sqrt{P_R} f_{R,D}^{(2)} \left(G y_R^{(1)} \right) + n_{S,D}^{(2)},
\end{aligned}
\tag{1}
$$

where $y_D^{(t)}, y_R^{(t)} \in C^{l \times 1}$ denote the received symbols at the destination and the relay, respectively, for the t-th slot ($t = 1, 2$), $f_{k,l}^{(t)}$ is the channel coefficient for the physical link $k \to l$ at the t-th slot, $n_{k,l}^{(t)}$ is the corresponding noise term, $P_S^{(t)}$ and P_R are the transmitted power from the source at the t-th slot and the relay respectively. Finally $P = P_S^{(1)} + P_S^{(2)} + P_R$ is the total power constraint and $G = \sqrt{1 / P_S^{(1)} \left| f_{S,R}^{(1)} \right|^2 + 1}$ denotes the amplification factor at the relay. The power constraint used here refers to the total energy which is available or the transmission of a data packet. In the case of cooperation, this power has to be appropriately distributed between the different slots and links in order to optimize the performance. The optimal power distribution for the cooperative protocols under consideration will now be discussed. The system can be represented by an equivalent channel matrix. Therefore,

$$
\begin{bmatrix} y_D^{(1)} \\ y_D^{(2)} \end{bmatrix} = \begin{bmatrix} \sqrt{P_S^{(1)}} f_{S,D}^{(1)} & 0 \\ \dfrac{\sqrt{P_S^{(1)} P_R}}{\sqrt{1 + P_R G^2 \left| f_{R,D}^{(2)} \right|^2}} G f_{S,R}^{(1)} f_{R,D}^{(2)} & \dfrac{\sqrt{P_S^{(2)}} f_{S,D}^{(2)}}{\sqrt{1 + P_R G^2 \left| f_{R,D}^{(2)} \right|^2}} \end{bmatrix} \cdot \begin{bmatrix} x_1 \\ x_2 \end{bmatrix} + Z \Rightarrow Y = H X + Z,
\tag{2}
$$

where $X \in C^{2 \times l}$ is the codeword vector; $H \in C^{2 \times 2}$ is the equivalent channel matrix; $Z \in C^{2 \times l} \sim CN\left(0, I\right)$ is the AWGN at the destination.

Direct Transmission

In direct transmission there is no cooperation and the system behaves as a classical single-input single-output (SISO) system. The transmission is performed in only one slot and all the available power resources (P) are allocated to the link $S \rightarrow D$. The classical case is included to act as a baseline reference for a comparison of the advantages and disadvantages offered by cooperative diversity schemes. The corresponding capacity of the direct transmission is given as (Cover & Thomas, 1991)

$$C_D = \log_2 \left(1 + P \left| f_{S,D}^{(1)} \right|^2 \right). \tag{3}$$

Orthogonal Amplify-and-Forward (OAF)

This is the first AF protocol which has been proposed in order to obey the half-duplex constraint (Laneman, 2002). Its basic idea is depicted in Figure 1b(i). In the Orthogonal Amplify-and-Forward (OAF) protocol, data transmission is performed in two slots, where in the second slot the relay scales and retransmits a noisy version of its received signal from the source in time slot 1. However, in the second slot the source is inactive ($P_S^{(2)} = 0$) and the relay is the only node which transmits. The capacity of the system can be expressed as (Laneman et al., 2003; Laneman et. al., 2004)

$$C_{OAF} = \frac{1}{2} \log_2 \left(1 + \underbrace{P_S^{(1)} \left| f_{S,D}^{(1)} \right|^2}_{1st\ slot} + \underbrace{\frac{P_S^{(1)} P_R \left| f_{S,D}^{(1)} \right|^2 \left| f_{R,D}^{(2)} \right|^2}{P_S^{(1)} \left| f_{S,R}^{(1)} \right|^2 + P_R \left| f_{R,D}^{(2)} \right|^2 + 1}}_{2nd\ slot} \right), \tag{4}$$

where the factor of a $1 / 2$ accounts for the fact that information is conveyed to the destination terminal over two time slots.

Non-Orthogonal Amplify-and-Forward (NAF)

The NAF protocol was proposed by Nabar et al. (2004) and has been proved to be the optimal AF scheme for a half-duplex single-relay channel by Azarian et al. (2005) in terms of diversity-multiplexing-tradeoff (DMT). It is an optimization of the OAF scheme proposed by Laneman et al. (2004) where orthogonal channels (i.e. time slots) separate the source and the relay transmissions. More specifically, in order to overcome the data rate loss which arises from the inactivity of the source during the cooperative channel in OAF, the NAF protocol allows the source to be active during the relaying transmission. Its superiority comes from the fact that the source keeps transmitting new data (x_2) while the relay forwards previously "overheard" data (x_1) with $x_2 \neq x_1$. The structure of the NAF protocol is schematically presented in Figure 1b(ii). The NAF protocol improves multiplexing gain as the source continuously transmit new data to the destination. However, for a single-relay setting, it provides relaying in half of time and therefore data are not always "protected" by diversity. The generalization of the NAF protocol for multi-relay cases (Azarian et al., 2005; Yang & Belfiore, 2007) can overcome this limitation but these configurations are beyond the scope of this chapter. For the NAF protocol, the corresponding capacity can be written by using the equivalent channel model in Eq. (2). Therefore, we have

$$C_{NAF} = \frac{1}{2} \log_2 \det \left(\mathrm{I} + \mathrm{H} \, \mathrm{H}^H \right). \tag{5}$$

where I denotes the identity matrix. The complexity of the NAF capacity expression is a bottleneck in the analytical computation of the outage probability which is the basic tool for the power allocation algorithm. However, according to (Krikidis et. al., 2008c) we can establish an efficient capacity upper-bound which simplifies the outage computations and results in a reliable power allocation.

Block-Fading Non-Orthogonal AF (BFNAF)

The protocol proposed here is similar to the classical NAF except that the source retransmits the same packet during the cooperative slot (Krikidis et al., 2008c). The principal motivation behind this protocol is the additional diversity degree which is offered by the slot-basic channel model. The physical link between the source and the destination changes during the cooperative slot $\left(f_{S,D}^{(2)} \neq f_{S,D}^{(1)} \right)$ due to the slot-based block-fading nature of the channel. The source retransmission of the same data via another independent channel increases the receiver diversity and therefore can improve the system reliability. As we will show in the following discussion, this new NAF behavior is interesting for low data rates where the diversity gain is more important than the multiplexing-gain. Using the previous formulation of the conventional NAF scheme, the BFNAF protocol is characterized by the property $x_2 = x_1$ for the t-th cooperative frame. Figure 1b(iii) schematically presents the BFNAF scheme.

Furthermore, in order to increase the performance of BFNAF scheme, a cophase operation of the simultaneous transmissions is assumed (Zhao & Valenti, 2005). This operation results in an addition in phase of the copies of the transmitted signal at the destination and can be regarded as phase-only beamforming. In the following discussion, all the references to the BFNAF protocol concern its co-phase version, which requires the relay to know the phase shift between $S \rightarrow D$ and $S \rightarrow R \rightarrow D$ links (reasonable for a time division duplex scenario). The behaviour of the BFNAF protocol during the cooperative slot can be represented by an equivalent one-dimensional channel which is the algebraic sum of the two individual channels ($S \rightarrow D$, $R \rightarrow D$). In this case, the corresponding capacity can be written as (Krikidis et. al., 2008c)

$$C_{BFNAF} = \frac{1}{2} \log_2 \left(1 + P_S^{(1)} \left| f_{S,D}^{(1)} \right|^2 + \frac{P_S^{(1)} P_R}{1 + P_R G^2 \left| f_{R,D}^{(2)} \right|^2} G \left| f_{R,D}^{(2)} \right|^2 \left| f_{S,R}^{(1)} \right|^2 + \underbrace{\frac{P_S^{(2)}}{1 + P_R G^2 \left| f_{R,D}^{(2)} \right|^2} \left| f_{S,D}^{(2)} \right|^2 + \frac{2\sqrt{P_S^{(1)} P_S^{(2)} P_R}}{1 + P_R G^2 \left| f_{R,D}^{(2)} \right|^2} G \left| f_{R,D}^{(2)} \right| \left| f_{S,R}^{(1)} \right| \left| f_{S,D}^{(2)} \right|}_{\equiv \Phi} \right). \tag{6}$$

An Upper Bound

Given the complexity of Eq. (6), the establishment of a useful capacity upper-bound which can simplify our power allocation optimization is required (Lang, 2004). The proposed bound supposes the two simultaneous transmissions of the second slot are orthogonal (i.e. different frequencies, CDMA codes). This assumption efficiently exploits all the diversity degrees of the system and gives a capacity expression similar to conventional multiple-input single-output (MISO) systems (the factor $1/2$ corresponds to two time slots). We note that the basic motivation for the establishment of this upper-bound is the loss of the third diversity degree due to the corresponding normalization factor (first term in Φ of Eq. (6). This capacity upper-bound is expressed as

$$
C_{BFNAF}^{(Upper)} = \frac{1}{2} \log_2 \left(1 + \underbrace{P_S^{(1)} \left| f_{S,D}^{(1)} \right|^2}_{1st \ slot} + \underbrace{P_S^{(2)} \left| f_{S,D}^{(2)} \right|^2}_{1st \ channel} + \underbrace{\underbrace{\frac{P_S^{(1)} P_R \left| f_{S,R}^{(1)} \right|^2 \left| f_{R,D}^{(2)} \right|^2}{P_S^{(1)} \left| f_{S,R}^{(1)} \right|^2 + P_R \left| f_{R,D}^{(2)} \right|^2 + 1}}_{2nd \ channel}}_{2nd \ slot} \right).
$$

(7)

The robustness of the proposed capacity upper-bound is related with the outage probabilities and concerns the high signal-to-noise ratios (SNRs) where a full diversity gain is achieved ($SNR \to \infty$). Therefore, the efficiency of the proposed capacity upper-bound requires the convergence to its asymptotic outage expression. For the low SNR regions where this convergence is not achieved, the proposed upper-bound is not reliable (Krikidis et al., 2008c).

Power Allocation

Power allocation is another efficient technique which can improve performance of cooperative protocols. The appropriate distribution of the available transmitted power between the links optimizes performance and resource utilization (Hasna & Alouini, 2004; Zhang et al., 2004; Yao et al., 2005; Deng & Haimovich, 2005; Hong et al., 2007; Mahinthan et al., 2008). In this study, we are interested in designing power allocation algorithms that do not depend on the instantaneous channel conditions. Waterfilling techniques which maximize the instantaneous data rates and depend on continuous feedback are beyond the scope of our work. Therefore, the selected optimization criterion is the minimization of the outage probability subject to a total transmit power constraint over the two time-slots. This constraint corresponds to the maximum power that a given packet is allowed to consume throughout its propagation from source to destination.

The outage probability is an important tool for the analysis of ergodic (quasi-static) systems. It is defined as the probability that the instantaneous capacity is lower than the required spectral efficiency. Moreover, it has been shown in (Zheng & Tse, 2003) that the error probability is dominated by the outage probability in the high SNR regime for long codewords. Mathematically speaking, the outage probability is given as

$$P_{out} = \Pr\left\{C\left(\text{H}\right) < R_0\right\},\qquad(8)$$

where R_0 (bits/s/Hz) denotes the desired spectral efficiency of the system. If the channel is in outage ($C\left(\text{H}\right) < R_0$) the reliability of the transmission is not guaranteed.

Consequently, the problem is formulated as

$$\min\ P_{out}$$

$$subject\quad to\quad \left\{\begin{array}{c} P_S^{(1)} + P_S^{(2)} + P_R = P \\ P_S^{(1)}, P_S^{(2)}, P_R \neq 0\ \left(ex.\,P_S^{(2)} = 0\ for\ OAF\right) \end{array}\right\} \Rightarrow \left\{\begin{array}{c} P_S^{(1)} = \alpha P, P_S^{(2)} = \beta P, P_R = \gamma P \\ \alpha + \beta + \gamma = 1 \\ 0 < \alpha, \beta, \gamma < 1\ \left(ex.\,\beta = 0\ for\ OAF\right) \end{array}\right\}.$$

$$(9)$$

It should be noted here that the optimization problem under consideration is a convex problem, which means that it has a global solution (Boyd & Vandenberghe, 2004). This is to be expected because the optimization function is an outage probability function, which is convex, and the considered constraint is linear (Hasna & Alouini, 2004). In the following subsections, we present the application of this optimization problem to the three considered AF protocols. We note that to the best of our knowledge, this is the first time that the power allocation problem is applied to NAF-based protocols.

OAF Case

For the conventional OAF protocol (Krikidis et al., 2008c), the optimal power allocation is equal to:

$$P_S^{(1)} = \frac{2}{3}P,\qquad\qquad P_R = \frac{1}{3}P.\qquad(10)$$

Therefore, for the conventional application of the OAF protocol, the allocation of double the power in the source node in comparison with the relay node will optimize the performance.

BFNAF Case

Given the complexity of the initial expression of BFNAF capacity, the analytical computation of the outage probability is not possible. However, the proposed BFNAF capacity-upper bound ($C_{BFNAF}^{(Upper)}$) simplifies the optimization problem and allows analytical expressions for the outage probabilities. Simulation results in the next subsection show that this solution approaches the optimal solution of the BFNAF case. The optimal power allocation policy for the BFNAF case is given by (Krikidis et. al., 2008c)

$$P_S^{(1)} = \frac{4}{9}P,\qquad\qquad P_S^{(2)} = \frac{1}{3}P,\qquad\qquad P_R = \frac{2}{9}P.\qquad(11)$$

An interesting observation is that the optimal power allocation solution for the BFNAF case also requires the allocation of double the power to the first direct link in comparison with the relaying one.

NAF Case

As has been proven in (Krikidis et. al., 2008c), the optimal power allocation for the NAF case is equivalent to the BFNAF protocol in (11). As with the BFNAF protocol, the corresponding outage minimization is based on the related NAF upper-bound in order to simplify computation.

NUMERICAL RESULTS

In this subsection, we investigate the numerical results obtained by Monte-Carlo simulations. By default, we consider a three-node symmetric network, where all the channels coefficients are independent and identically-distributed (i.i.d.) Rayleigh distributed with unit variance. The selected performance metric is the outage probabilities (Zheng & Tse, 2003), and information rate is measured in bits per channel use (BPCU).

In Figure 2, we compare direct transmission, OAF, NAF, BFNAF and BFNAF-bound with an optimal power allocation policy for three indicative spectral efficiencies $R_0 = 1$, 4 and 7 BPCU, respectively. As can be seen, NAF protocols significantly outperform the conventional OAF scheme. The multiplexing loss of the OAF scheme results in a poor performance for all cases. Furthermore, the comparison between the BFNAF and the related BFNAF-bound shows that the proposed bound is an efficient capacity bound for the BFNAF case. This bound is used for the solution of the power allocation problem. Finally, the most important observation comes from the comparison between the two NAF-based schemes. As we can see, the BFNAF protocol outperforms classical NAF at low spectral efficiencies (1 BPCU) but is outperformed by NAF at the high spectral efficiencies (4 and 7 BPCU). For example at 10^{-3} outage probability, the gain of BFNAF is equal to 1 dB for $R_0 = 1$ BPCU and for $R_0 = 4$ BPCU, the gain of NAF is equal to 3 dB. This result shows that the proposed BFNAF scheme is useful for the low spectral efficiencies where the diversity gain is more important than the multiplexing gain. This important observation is the principal motivation for the switching between NAF and BFNAF in the proposed hybrid scheme in (Krikidis et al., 2008c).

Relay Selection

Scaling cooperation to more than one relay is still an open area of research, despite the recent interest in cooperative communication. One possible approach is the repetition coding approach (Laneman, 2002; Laneman et. al., 2003), where each relay forwards the data in a dedicated orthogonal channel (time slot). However, as with the basic three-node configuration, this approach results in a high data rate loss which increases as the number of relays increases. Another approach for multi-relay environments is the use of distributed space-time coding (DSTC) among the participating nodes (Laneman et. al., 2003). This approach allows a simultaneous transmission of all the relays under a well-defined power constraint and therefore does not suffer from data rate loss. However, in practice, such code design has a high degree of complexity due to the distributed and ad-hoc nature of cooperative links (the number of useful anten-

Figure 2. Outage probabilities for conventional protocols; non-cooperation, OAF, NAF, BFNAF and BFNAF-bound with optimal power allocation. Considered information rate: $R_0 = 1,4,7$ BPCU; the brackets include the power allocation for the links $\{S \rightarrow D \ (slot\text{-}1), \ R \rightarrow D, \ S \rightarrow D \ (slot\text{-}2)\}$, respectively (normalization to unity). (@2008, IEEE. Used with permission.).

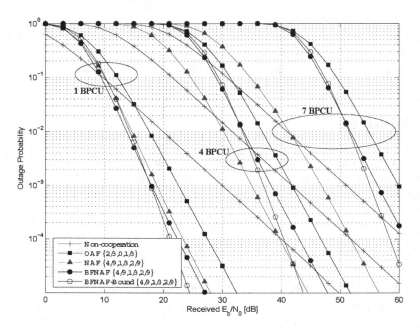

nas is generally unknown and time-variant) and therefore hardware implementations are quite difficult (Laneman, 2002; Yang & Belfiore, 2007).

Currently, there is a lot of interest in AF cooperative schemes, where the relay nodes are selected according to some well-defined system parameters (Hammerström et al., 2004; Bletsas et al., 2006; Krikidis et al., 2008b; Krikidis et. al, 2008d). In these approaches, selecting one relay based on the best instantaneous channel conditions offers the same diversity benefits with a lower complexity than DSTC (Scutari & Barbarossa, 2005; Bletsas et. al., 2007) which require simultaneous relay transmissions. In the following discussion, we present two novel relay selection policies which are related to two different ad hoc problems. The first one selects the relay based on partial instantaneous channel knowledge and is suitable for applications with critical power constraints, and the second one is appropriate for applications with interference during the AF relaying process.

Partial Relay Selection

Bletsas et. al. (2006; 2007) proposed a distributed relay selection for a two-hop AF system, where the selected criterion is the best instantaneous SNR composed of the SNR across the two-hops. For protocols where the relay selection takes into account all relay links, the statistical behaviour of the SNR can be expressed by a simple application of order statistics. However, a distributed relay selection is time sensitive and requires perfect time synchronization among the nodes which is a crucial issue for practical systems. On the other hand, centralized approaches require continuous channel feedback from all the links of the network and therefore result in high power consumption. The related complexity is increased as the number of nodes and hops increases. In resource-constrained wireless systems such as sensor networks (Karaki & Kamal, 2004) by monitoring the connectivity among the nodes locally (one-hop) rather than globally, one can prolong the network lifetime. In this section, we deal with a scheme where a cluster of relays is selected based on average SNR and the instantaneous channel-information is limited to the source's neighbourhood for a two-hop AF system. Our basic scope is to present the statistical analysis of a basic AF relaying link, when the relay node is selected based only on source to relay links.

System Model

A simple ad-hoc configuration which is composed of one source S, one destination D and a set $S_{relay} = \{1, 2, ..., K\}$ of potential K relays ($k \in S_{relay}$) is assumed. The source has no direct link with the destination and the transmission is performed only via relays. For convenience, we assume that the K relays are clustered relatively closely together (location-based clustering) and have been selected by a long-term routing process for establishing communication between source-destination (Karaki & Kamal, 2004). Figure 3 schematically presents the system model. The related routing scheme can track variations in path-loss and shadowing but not those due to small scale fading. This clustered (layered) structure, where a cluster of relays is selected based on average SNR, is a common model in the literature and ensures equivalent average channels for the relays (the same average SNRs for $S \rightarrow k$, $k \rightarrow D$ links) (Yang & Belfiore, 2008). The cooperation is performed via the OAF protocol in order to minimize the relay complexity. The source continuously monitors the quality of its connectivity with the relays via the transmission of local (one-hop) feedback. In contrast with previous work, where the relay selection is based on the best instantaneous relaying link (Bletsas et al., 2006), it is supposed here that the available instantaneous knowledge is limited to the one-hop neighbourhood of the source. Due to this partial information the relay selection follows the best link between the source and the relays ($S \rightarrow k$). For each link, the channel is assumed to be block flat-fading (quasi-static) and modelled as a zero-mean, independent, circularly-symmetric complex Gaussian random variable with unit variance. The additive noise is represented as a zero-mean mutually independent, circularly-symmetric, complex Gaussian random sequence which, without loss of generality, has variance α for the links $S \rightarrow k$ and β for the links $k \rightarrow D$ which correspond to an average SNR equal to $\overline{\gamma}_{S,k} = 1 / \alpha$ and $\overline{\gamma}_{k,D} = 1 / \beta$, respectively. For clarity of exposition we suppose that channel asymmetries (different path-loss and shadowing values between source-relays and destination-relays) are lumped into the difference of the average SNRs. The system model can be represented by the following equations:

$$y_k = f_{S,k} \cdot x + n_{S,k},$$
$$y_D = f_{k^*,D} \cdot \left(G \cdot y_{k^*} \right) + n_{k^*,D}, \tag{12}$$

where y_k denotes the received signal at the k-th relay, y_D is the received signal at the destination, x is the transmitted signal with $E\left(xx^* \right) = 1$, $f_{A,B}$ and $n_{A,B}$ denote the channel coefficient and the noise term, respectively, for the link $A \rightarrow B$, $G = \sqrt{1 / \left| f_{S,k^*} \right|^2 + \alpha}$ is the amplification factor (unity normalization) and $k^* = \arg_k \max \left\{ \left| f_{S,k} \right|^2 / E\left(\left| n_{S,k} \right|^2 \right) \right\}$ denotes the selected relay. The problem under consideration

Figure 3. The system model for the partial relay selection.

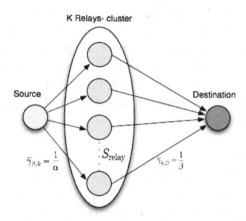

is the statistical description of the SNR for the considered relaying link with partial relay selection, as well as the estimation of its performance.

STATISTICAL EXPRESSIONS

The cumulative distribution function (CDF) of the SNR for the relaying link under consideration can be approximated as

$$P(\gamma) = 1 - \sum_{k=1}^{K} \binom{K}{k} \cdot 2\sqrt{k\alpha\beta}\gamma \cdot e^{-(\beta+k\alpha)\gamma} \cdot K_1\left(2\sqrt{k\alpha\beta}\gamma\right) \cdot (-1)^{k+1}, \tag{13}$$

where γ denotes the instantaneous SNR and $K_1(\cdot)$ is the first order modified Bessel function of the second kind (Abramowitz, 1970, Eq. 9.6.22). The corresponding probability density function (PDF) derivation can be found in (Krikidis et. al., 2008b). The above statistical expression generalizes the results presented in (Hasna & Alouini, 2003).

For the high SNR region, the previous analytic approximation of the SNR yields a simple exponential-based expression which is useful for estimating the performance without complicated computations. More specifically, for high average SNRs per link ($\alpha, \beta \to 0$) the first order modified Bessel function of the second kind can be approximated by the equation $K_1(x) = 1 / x$ (Abramowitz, 1970, Eq. 9.6.9). By substituting this approximation into Eq. (13), the CDF of the considered link can be simplified as follows

$$P'(\gamma) = 1 - \sum_{k=1}^{K} \binom{K}{k} \cdot e^{-(\beta+k\alpha)\gamma} \cdot (-1)^{k+1}, \tag{14}$$

which corresponds to a linear combination of exponential functions.

Error Probability and Diversity Analysis

Based on the above simplified expression, the error probability can be expressed as a linear combination of the Q-function and therefore its analytical computation is well-known in the literature (Proakis, 2001; Tse, 2004). Furthermore, as has been proved in (Krikidis et. al., 2008b), for $K > 1$ relays, the considered relaying link $S \rightarrow k^* \rightarrow D$ asymptotically approaches the behaviour of the single $k \rightarrow D$ link and thus does not result in diversity gain for the end-to-end relay link. In addition to this observation, the asymptotic performance of the considered system does not depend on the source-relay links and thus it is unchanged for any choice of α.

NUMERICAL RESULTS

Computer simulations were carried out in order to validate the proposed analytical expressions. Figure 4 presents the Bit Error Rate (BER) performance of the considered relaying link for different numbers of relays. More specifically, the simulation environment supposes a Binary Phase Shift Keying (BPSK) modulation (straightforwardly it can be extended to other modulations according to Alouini et. al., 1999), $K = 1, 2, 7$ relays and $\alpha = \beta$ (all links have the same average SNR). The proposed exponential-based approximation (Eq. 13) and the performance of the link $k \rightarrow D$ (with average SNR equal to $1 / \beta$) are given also as reference curves. From this figure it can be seen that the considered partial-based relay selection improves the performance of the conventional relaying link ($K = 1$) and the resulting gain is equal to 3 dB at high SNRs. Moreover, an interesting observation is that from two relays ($K \geq 2$) the relaying link converges to the performance of the relay-destination link for the high SNR region. Therefore, finding two potential relays and selecting between them is enough to overcome the performance degradation which results from the AF relaying process. This result, in combination with the partial nature of the considered "feedback", minimizes the required complexity and signalling overhead. On the other hand for low SNRs we can see that as the number of relays increases the performance approaches that of the relay-destination link. Finally, as far as the proposed asymptotic exponential-based approximation in Eq. (13) is concerned, it can be seen that it provides accurate performance prediction for the high SNR region in all cases. Furthermore, this result provides a reliable lower-bound on BER performance for the case of low SNRs.

Relay Selection for Interference Limited Environments

To the best of our knowledge, the behaviour of the AF scheme for ad-hoc scenarios with multi-user interference has not yet been reported in the literature. In this Section, the basic OAF protocol (Laneman et all., 2004) is considered when an external structured interference affects the cooperation process. This assumption represents practical ad-hoc systems where neighbouring clusters are not perfectly isolated and inter-cluster interference results in performance degradation (Krikidis et. al., 2008d). It is shown that interference modifies the conventional AF statistic expression of the "harmonic mean" (Hasna & Alouini, 2003) and the related system capacity converges to a well-defined static point. Based on this modified asymptotic expression, two novel relay selection criteria which are optimal for interference-limited systems with many relays are investigated. These criteria appropriately take into account the interference component and are suitable for distributed implementations.

Figure 4. BER performance of a relaying link with partial relay selection versus the SNR; $\alpha = \beta$, $K = 1, 2, 7$ relays and BPSK modulation is used. Results are shown for both Monte Carlo simulation and the asymptotic exponential approximation (Eq. 13). (@2008, IEEE. Used with permission.).

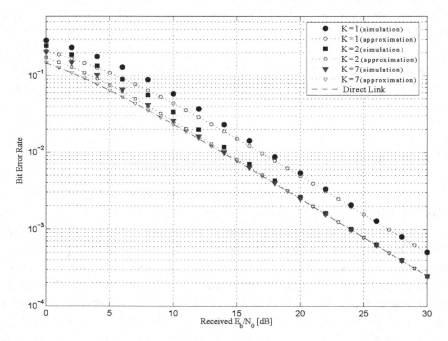

System Model

As the target of this work is the analysis of the inter-cluster interference, a simple ad-hoc configuration which consists of two neighbouring clusters (C, C') is assumed, the cluster of interest being C without loss of generality. Figure 5 schematically presents the considered channel model. The basic structure of the cluster C follows the system model of the previous section. As far the neighbouring cluster is concerned, it is assumed that it is based on the direct transmission and consists of one source S' and one destination D' (could be a relay or the final destination). More complicated interference structures (i.e. relaying at both clusters or asynchronous sources) are beyond the scope of this study and will be covered in future work. Furthermore, the interference (inter-cluster interference) is assumed to degrade the $S \rightarrow k$ links but is negligible at the destination. This assumption isolates the impact of the structural interference on the AF process and simplifies our analysis. The interference considered can be regarded as a direct link between the neighbouring source S' and the relays S_{relay} ($S' \rightarrow k$). In this case, the system model of the previous subsection can be updated as follows

$$y_k = \underbrace{f_{S,k} \cdot x}_{\text{Desired signal}} + \underbrace{f_{S',k} \cdot x'}_{\text{Interference signal}} + n_k,$$

$$y_D = f_{k^*,D} \cdot \left(G \cdot y_{k^*} \right) + n_D. \tag{15}$$

Figure 5. The general system model; C : cluster of interest (communication via cooperative link), S : source, D : destination, S_{relay} : relay cluster, C' : neighbouring cluster (interference via direct link), D' : destination (final destination or intermediate relay), I_{INF_k} : interference signal for the k-th relay ($S' \rightarrow k$).@2008, IEEE, Used with permission.

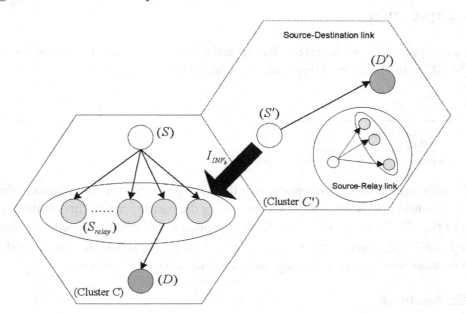

Furthermore, for the clarity of exposition, it is supposed that the average signal-to-interference power ratio is defined as $L = E\left(\gamma_{S,k}\right) / E\left(\gamma_{INF_k}\right) = \sigma_{S,k}^2 / \sigma_{S',k}^2$, $\gamma_{S,k}$ is the SNR of the link $S \rightarrow k$ and $\gamma_{INF_k} \equiv \gamma_{S',k}$ denotes the interference-to-noise ratio (INR) for the k-th relay. However, our analysis can be extended straightforwardly to completely asymmetric configurations ($\sigma_{S,k}^2 \neq \sigma_{k,D}^2$). Note that the parameter L controls the level of interference in the cluster of interest and differentiates the problem under consideration from previously reported studies.

Amplification Factor

In the AF case, the relay nodes can not differentiate between source and interference signals. The amplification process is performed in the analogue domain and thus consists of a simple normalization of the total received power without further processing. Therefore, dirty paper coding theory (Costa, 1983) and interference mitigation techniques cannot easily be performed at the relays. The amplification factor can be expressed as

$$G = \sqrt{\frac{P_S}{\left|f_{S,k^*}\right|^2 P_S + \left|f_{S',k^*}\right|^2 P_{S'} + N_0}},$$

(16)

where G is the AF amplification factor and P_s ($P_{s'}$) denotes the source's transmitted power for the cluster of interest (neighboring cluster). For our system model, it is assumed that the source powers are normalized to unity and therefore power control issues are not discussed.

Statistical Expression

Assuming coherent detection with perfect channel estimation, the resulting signal-to-interference and noise ratio (SINR) of the decision variable can be obtained from Eq. (15) as

$$\gamma_D = \frac{\gamma_{S,k^*} \cdot \gamma_{k^*,D}}{\underbrace{\gamma_{INF_k} \cdot \left(\gamma_{k^*,D} + 1\right)}_{\equiv \Theta \ (\text{interference-based term})} + \underbrace{\gamma_{S,k^*} + \gamma_{k^*,D} + 1}_{\equiv \Phi \ (\text{conventional-based term})}} . \tag{17}$$

Therefore, the impact of the interference during the AF process modifies the statistical description of conventional AF scheme (Ribeiro et al., 2005) by adding the parameter Θ. This additional term in the denominator of the SINR significantly complicates the process of finding a closed form expression for the probability density function of γ_D. However, in the following paragraphs, some useful approximations and asymptotic bounds which simplify the related analysis are proposed.

Asymptotic Analysis

For high SNRs ($SNR \rightarrow \infty$), the denominator terms in Eq. (17) reduce to one term. More specifically, the second order term (included in Θ) becomes significantly higher than the linear terms (i.e. $\gamma_{INF_k^*} \cdot \gamma_{k^*,D} >> \gamma_{INF_k^*} + \gamma_{S,k^*} + \gamma_{k^*,D}$) and thus dominates the denominator. Therefore, the SINR expression is now in a simplified form which can be written as

$$\gamma_D \approx \frac{\gamma_{S,k^*} \cdot \gamma_{k^*,D}}{\gamma_{INF_k} \cdot \gamma_{k^*,D}} = \frac{\gamma_{S,k^*}}{\gamma_{INF_k}}, \tag{18}$$

which is the ratio between the SNR of the first hop and the INR of the interference. Therefore, at the high SNRs, the statistical description of the system is independent of the second hop ($k \rightarrow D$). In this case, the PDF is given in closed form and corresponds to an F-distribution (ratio between two exponential random variables). More specifically, the asymptotic description of the system is given as (Nadarajah & Kotz, 2006)

$$p_\Gamma\left(\gamma\right) = \frac{L}{\left(L + \gamma\right)^2}, \qquad P_\Gamma\left(\gamma\right) = \frac{\gamma}{L + \gamma}, \tag{19}$$

where $p_\Gamma\left(\cdot\right)$ and $P_\Gamma\left(\cdot\right)$ denote the PDF and the CDF of the SINR, respectively.

RELAY SELECTION WITH INTERFERENCE

Conventional Selection

First consider the conventional proactive relay selection policy which is used in no-interference configurations. The policy considered here was proposed in (Bletsas et al., 2006; Bletsas et al., 2007) and is suitable for distributed implementation. It requires the instantaneous signal strength (SNR) between the links $S \rightarrow k$ and $k \rightarrow D$ ($k \in S_{relay}$) and the selected relay is chosen to maximize the minimum signal strength. The conventional relay selection policy can be expressed as

$$k_{Conv}^* = \arg_{k \in S_{relay}} \max \min \left\{ \gamma_{S,k}, \gamma_{k,D} \right\}. \tag{20}$$

This policy ensures that the relay with the "best" end-to-end path between the source and the destination is used and provides diversity gain on the order of the number of relays (Bletsas et al., 2006). However, this selection criterion has been designed for environments without interference and thus does not take into account its effects. In the work presented here, the investigated relay selection criteria can be regarded as an extension of this basic selection scheme. Note that although the above selection scheme seems to be more natural for DF cooperative schemes and the consideration of the harmonic mean more appropriate for AF protocols, it has been proven in (Bletsas et al., 2006) that both selection metrics are efficient for the AF case. The *max-min* criterion is a simple one and as we prove in the following discussion easily can be generalized for interference environments.

Asymptotic Selection

The first proposed relay selection criterion is motivated by the simplified expression of the system at high SNRs. As has been seen in Eq. (18) the asymptotic behaviour of the system converges to the ratio between $S \rightarrow k$ and interference links. Therefore, from an asymptotic point of view, an appropriate relay selection is to choose the relay which gives the maximum value of this ratio. The asymptotic relay selection criterion can be expressed as

$$k_{Asym}^* = \arg_{k \in S_{relay}} \max \left\{ \frac{\gamma_{S,k}}{\gamma_{INF_k}} \right\}. \tag{21}$$

Outage Performance

Outage probability is an important metric for the analysis of ergodic (quasi-static) systems. Based on (Krikidis et al., 2008d), the outage probability for the asymptotic relay selection can be expressed as

$$\Pr\left\{ C_{Asym} < R_0 \right\} = \left[\frac{2^{2R_0} - 1}{L + 2^{2R_0} - 1} \right]^K, \tag{22}$$

where C_{Asym} denotes the instantaneous capacity for the asymptotic relay selection and R_0 denotes the spectral efficiency.

Distributed Implementation

The proposed asymptotic selection criterion is suitable for distributed implementation. It requires only the SNRs of the $S \to k$ links and the INRs of the interference links which can be estimated by the relay nodes during the first-slot transmission. This estimation can be performed locally (each relay estimates its corresponding ratio from the received signal) without the need for dedicated channel feedback. More specifically, the potential relays which are assumed to be perfectly synchronized, set-up local timers based on the estimated instantaneous signal-to-interference ratio in order to access the channel, following (Bletsas et al., 2006; 2007). The timer of the relay with the best signal-to-interference ratio expires first, and therefore the corresponding relay is selected for retransmission. The remainder of the relays, whose timers have not expired yet, overhear the transmission of the selected relay and back off. Note that the absence of the second-hop link ($k \to D$) from the selection function gives the proposed asymptotic selection a lower complexity than the one in (Bletsas et al., 2006), which requires channel knowledge of the $k \to D$ links during the feedback-based set-up period.

Semi-Asymptotic Selection

The semi-asymptotic selection criterion combines the previous two criteria in Eq.'s (20) and (21). It takes into account the conventional and the asymptotic system behavior and it is suitable for "intermediate" SNR regions which experience a mixture of conventional and asymptotic SNR characteristics. The semi-asymptotic selection policy is given as

$$k_{Semi}^{*} = \arg_{k \in S_{relay}} \max \left\{ \frac{\min \left\{ \gamma_{S,k}, \gamma_{k,D} \right\}}{\gamma_{INF_k}} \right\}. \tag{23}$$

Outage Performance

According to (Krikidis et al., 2008d) the outage probability of semi-asymptotic relay selection is bounded by

$$\Pr \left\{ C_{Semi} < R_0 \right\} = \frac{1}{2} \cdot \left[\frac{2^{2R_0} - 1}{L + 2^{2R_0} - 1} \right]^K + \frac{1}{2} \cdot \left[\frac{2 \left(2^{2R_0} - 1 \right)}{L + 2 \left(2^{2R_0} - 1 \right)} \right]^K, \tag{24}$$

where C_{Semi} denotes the instantaneous capacity for the semi-asymptotic relay selection.

Switching Algorithm for Relay Selection Criteria

The selection criteria considered takes into account different "expressions" of the general statistic in Eq. (17) and thus are suitable for different SNR regions where these simplified expressions apply. In (Krikidis et. al., 2008d), the key regimes for the above relay selection techniques have been identified, and a theoretical framework for switching among them has been provided. More specifically, it has been proved that for the low SNRs $(-\infty \quad \bar{\gamma}_{LOW}]$ the system is dominated by the AWGN noise and therefore the conventional criterion is the most appropriate. For the intermediate SNRs $(\bar{\gamma}_{LOW} \quad \bar{\gamma}_{HIGH}]$, the statistical behaviour of the system is a mixture of the conventional and the asymptotic statistic and thus the semi-asymptotic selection criterion can be used. The third SNR region $[\bar{\gamma}_{HIGH} \quad \infty)$ is related to the high SNRs and corresponds to the asymptotic selection criterion. The proposed switching scheme defines the bounds of these regions $\bar{\gamma}_{LOW}, \bar{\gamma}_{HIGH}$ which are dependent on an average knowledge of the channel statistics. This behaviour of the proposed switching algorithm maintains system stability and minimizes signalling overhead.

Numerical Results

Computer simulations have been carried out in order to validate the performance of the proposed relay selection schemes. The first simulation results deal with the low SNR regions and focus on the conventional relay selection scheme. Figure 6 compares the BER performance versus the SNR ($\bar{\gamma}$) of different selection schemes for the case of $K = 2$ relays and signal-to interference power ratio $L = 11$. An uncoded system with perfect channel estimation and BPSK modulation is assumed without loss of generality. The conventional selection scheme is compared with the semi-asymptotic and the asymptotic selection policies. The random relay selection (non-selection), the selection of the relay with the minimum inter-ference link as well as the relay selection with the best $S \rightarrow k$ link are used as reference BER curves. Firstly, it can be seen that the structural interference under consideration bounds the diversity gain of the AF cooperation and yields a convergence region for all of the curves. However, the "selection" process significantly improves the system performance. The comparison of the selection schemes with the non-selection policy shows that selection is a useful tool in order to decrease interference effects and thus increase the capacity bound. The second important observation is that the three selection criteria considered are the most efficient. Their comparison with the reference selection criteria (minimum interference and best $S \rightarrow k$) show that they are the most appropriate ones as they have been designed according to the statistical behaviour of the considered interference-limited AF system. Furthermore, it can be seen that the conventional relay selection criterion outperforms the semi-asymptotic and asymptotic schemes at low SNRs. In this region, additive white Gaussian noise (AWGN) dominates the system degradation and therefore the conventional criterion is the most efficient selection scheme. As far as the upper-limit ($\bar{\gamma}_{LOW}$) of the corresponding SNR region is concerned, the proposed limit is an efficient estimate.

The following simulation results focus on the semi-asymptotic and asymptotic selection criteria and the efficiency of their switching algorithm. Figure 7 presents the outage probability performance versus the SNR ($\bar{\gamma}$) of the semi-asymptotic and asymptotic selection policies as well as the proposed theoretical approximations for different interference levels ($L = 10, 100$). The considered spectral efficiency is $R_0 = 2$ BPCU. The plotted curves validate that the semi-asymptotic criterion outperforms the asymptotic selection at the intermediate SNR regions but it is outperformed by the asymptotic selection

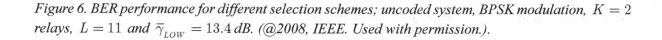

Figure 6. BER performance for different selection schemes; uncoded system, BPSK modulation, $K = 2$ relays, $L = 11$ and $\bar{\gamma}_{LOW} = 13.4\,dB$. (@2008, IEEE. Used with permission.).

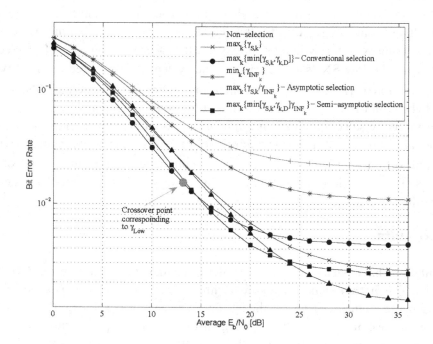

at the high SNRs. The corresponding gain is increased as the interference factor is decreased (L is increased). For example, the gain of the semi-asymptotic selection against the asymptotic one is 5 dB for outage probability 10^{-2}. On the other hand, the asymptotic selection is used when the semi-asymptotic selection has already converged and therefore the related gain is not meaningful. As for the switching upper-limit between the two proposed asymptotic criteria, it can be seen that the proposed method approaches in terms of the corresponding average SNR the true crossover point of the semi-asymptotic and asymptotic curves.

SCHEDULING AND CROSS-LAYER ISSUES FOR AF SYSTEMS

In all the previous examples, we assumed that the relay nodes are particular structures of a network which have as their main task to support the transmission of a source. However, in real systems, source, relay and destination nodes may all be peer devices. Optimal performance in peer-to-peer networks involves a high dependence of the cooperative diversity concept with other layers of the protocol stack and requires a cross-layer design. The following examples deal with the cross-layer nature of the cooperative diversity and shows the relation of the AF with the problem of routing and scheduling.

Figure 7. Outage probabilities for asymptotic-based relay selection schemes; $K = 4$ relays, $L = 10, 100$, $R_0 = 2 \, BPCU$ and $\overline{\gamma}_{HIGH} = \begin{bmatrix} 21.5 & 40.7 \end{bmatrix} dB.$ (@2008, IEEE. Used with permission.).

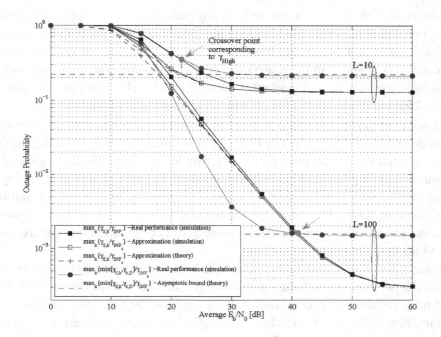

A Cross-Layer Problem for AF Networks

In this subsection, we study the joint combination of cooperative diversity with channel-based scheduling and routing under a two-hop AF and DF framework. Based on a simple well-selected ad-hoc configuration, we introduce an optimization problem which requires an appropriate assignment of three cross-layered roles. The proposed approach selects the relay nodes, the intermediate router and the final destination according to the instantaneous channel conditions. This cross-layer Multiple Access Control (MAC) decision which incorporates information from the Physical (PHY) and Network (NET) layers can optimize the system performance. Furthermore, we focus on centralized approaches which despite the extra signalling are more robust and less sensitive to timing synchronization than the distributed schemes. However, in order to overcome the problem of feedback and decrease the required complexity, a partial and quantized feedback is proposed. This feedback is limited to a qualitative description of a well-selected sub-set of links. The resulting performance and complexity trade-off is analyzed by theoretical and numerical studies and we show that the proposed sub-optimal solution gives reliable results. The final issue we address is the way that the available resources are shared between the nodes. The proposed algorithm distributes the roles among the nodes with the same probability and thus supports the fairness concept (Wang et al., 2004; Krikidis et al., 2008a). Although the presented analysis refers to a particular ad-hoc configuration, it is also useful for general network structures. The considered system is the simplest one in order to combine the considered cross-layer issues and can be regarded as the basis of more complex topologies.

System Model

Consider a 2-hop ad-hoc structure consisting of one source S, two routers R_1, R_2 and two destinations D_1, D_2 (Krikidis et. al., 2008a). Figure 8 schematically presents the cross-layer problem under consideration. It is assumed that the source continuously has data to transmit for both destinations and that all the nodes are half-duplex and thus cannot transmit and receive simultaneously. The source has no direct link with the destinations, and both DF routers can be used as intermediate nodes for both destinations. This long-term routing information (2 available routers) is provided by multipath-based routing protocols (Marina & Das, 2001) and is available for the role assignment under consideration.

The considered configuration is organized in two clusters C_1, C_2, where each cluster corresponds to a basic three-node cooperative system. The first cluster includes the source and the two routers and the second one the selected intermediate router and the two destinations. The basic goal of this particular system configuration is to present the considered cross-layer problem in a simple way. The selected topology (2 hops with 2 nodes) is the simplest one in order to combine routing, scheduling and fairness issues under a combined AF/DF cooperative framework. It is the basis of real systems and thus the proposed solution can be useful for practical peer-to-peer configurations which are characterized by many nodes and hops.

In each cluster, the cooperative diversity scheme is based on the conventional OAF protocol, where each transmission is performed in a dedicated channel (i.e. different frequencies). Each destination node (final receiver, intermediate router) combines multiple copies of the transmitted packet by using a Maximum Ratio Combining (MRC) and demodulates (decodes) the received signal. The modulation scheme is BPSK, but generalization to other modulation formats is straightforward. The statistical description of the channel model (channel statistics, additive noise etc) follows this one of the previous Sections. However, the system here is characterized by asymmetric average SNRs for the different links in order to introduce the fairness problem.

The considered cross-layer problem is to decide, for each transmission, the destination node (scheduling) and the inter-mediate router (routing) based on the instantaneous channel conditions. Our routing

Figure 8. The system model for the cross layer problem.

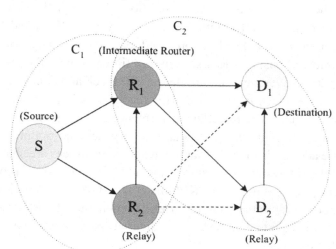

optimization focuses on a short-term definition of the next-hop router among the available ones and can be regarded as a "complementary routing process" to a long-term routing scheme. More specifically, we assume that an existing long-term routing algorithm provides a set of available routers (2 routers in our case) in order to establish communication between source and destinations. The short-term routing decision under consideration is to select the most appropriate router among the available ones based on an instantaneous channel feedback. In addition to this cross-layer concept, the basic targets of our design are to optimize the aggregate performance, to minimize complexity as well as to support a fairness concept. Here, the fairness parameter is defined as an equal distribution on average of the considered roles (Krikidis & Belfiore, 2007b) among all nodes. Therefore, each one-hop node becomes a router (or relay) with the same probability and each two-hop node becomes a destination (or relay) with the same probability.

The Proposed Solution

In this section we present the algorithm which solves the above optimization problem. The proposed cross-layer approach incorporates opportunistic scheduling with channel-based routing and achieves fairness and optimized performance. The performance of the proposed algorithm is validated by theoretical studies and an estimation of its computational complexity is also given.

A Round Robin (RR) Approach

First of all, we present briefly the classical RR approach which is used as a reference throughout this section. For the considered problem, RR corresponds to a simple scheduling and routing approach which ensures the fairness concept. It does not require any feedback from the wireless medium, and distributes the roles (intermediate router, final destination, diversity-relays) periodically among the nodes. This resource allocation corresponds to a fair distribution of the roles. However, despite the implementation simplicity and the support of the fairness concept, not using the channel parameter gives RR a poor performance.

Performance Optimization

In order to optimize the performance, the scheduling and routing decisions have to be adapted to the instantaneous channel conditions. However, an optimal real-time adaptation requires a high degree of signalling and searching which corresponds to a *NP*-complete problem for multi-hop configurations. Here, we propose a sub-optimal solution which adapts the scheduling and routing decisions to partial feedback information. More specifically, instead of deciding the appropriate router and destination based on global feedback from all the possible links to the source, we take the decisions hop by hop.

Therefore, firstly the algorithm selects the best available DF router which will be used as an intermediate node in order to communicate with the destinations. This selection is based on the instantaneous channel conditions between the source and the two routers. Therefore, the router which has the best direct link is selected as an intermediate node and the other one as the cooperative AF relay for the first cluster. This decision can be regarded as a dynamic routing strategy with cooperative diversity. On the other hand, the decision of the final destination is based on the instantaneous channel conditions between the selected router and the two destinations. Therefore, the destination which has the best direct link with the selected router is the final destination and the other one becomes the cooperative AF relay of the second cluster. This decision can be regarded as an opportunistic scheduling with cooperative diversity. Based on the above formalization, the solution which optimizes the performance can be presented as

$$
\begin{aligned}
I_{opt} &= \arg\max_{i\in\{1,2\}} \left\{ \gamma_{S,R_i} \right\}, \\
J_{opt} &= \arg\max_{j\in\{1,2\}} \left\{ \gamma_{R_{I_{opt}},D_j} \right\},
\end{aligned}
\tag{25}
$$

where $\gamma_{A,B}$ denotes the instantaneous SNR for the link $A \rightarrow B$.

Fairness Concept

The fairness concept is measured from a long term point of view and with respect to the resource utilization (channel access, power consumption) (Krikidis & Belfiore, 2007b; Wang et al., 2004). More specifically, the support of fairness is translated to an equivalent distribution of the roles to each category of nodes. Therefore, from the router point of view the fairness concept ensures that each router can be the intermediate router (or C_1 relay) with the same probability. This consideration ensures the same power consumption for the two routers. More specifically, the intermediate router has to perform DF relaying and this results in higher power consumption than the AF relay-router. The fairness concept enforces the same power consumption for both routers on average.

From the destination's point of view, the fairness concept ensures that each node will communicate with the source with the same probability. This consideration avoids monopolistic use of the wireless medium (final destination) and also continuous and unnecessary power consumption (C_2 relay). When the channel statistics of the different links are i.i.d. random variables, each link has the best instantaneous conditions (i.e. SNR) with the same probability and thus the fairness concept automatically is supported. However, in the case where the links have different statistics, the router and the destination with the higher average channels will be selected with a higher probability than the other ones. In order to overcome this problem and support the fairness concept for all the cases, we use the normalized instantaneous SNR as a criterion of the instantaneous quality of a link (Krikidis & Belfiore, 2007b). More specifically, instead of comparing the absolute SNR value of the direct links, we compare the difference of each instantaneous link from its average value. The considered instantaneous channel quality criterion can be expressed as

$$
\hat{\gamma}_{A,B} = \frac{\gamma_{A,B}}{E\left(\gamma_{A,B}\right)}.
\tag{26}
$$

Complexity Discussion

The computational complexity of the proposed algorithm is related to the required overhead. In contrast with the full-feedback centralized approaches which require feedback from all the links (6 links) to the source, the proposed sub-optimal scheme requires a limited number of feedback packets. More specifically, the links which are involved in the decision process are the direct links between source and routers and selected-router and destinations (4 links). Moreover, in contrast with the full-feedback approaches which require probability computations of analytical performance for all the possible combinations in order to find to optimal solution, the proposed algorithm bases its decision on simple SNR comparisons.

Furthermore, another factor which is related with the feedback complexity is its numerical (digital) representation of feedback. Almost all the literature to date supposes a perfect feedback representation which is not possible for practical applications. In order to reduce the feedback overhead, we propose a quantized form of feedback which is limited to N bits. In this case the feedback does not represent the absolute value of the link quality, but a qualitative description based on 2^N quality levels. In order to have a probabilistically equivalent division of the SNR region, the switching SNR levels are given as follows

$$\mathrm{Pr}\,ob\left\{\hat{\gamma} < \Gamma_\nu\right\} = \frac{n}{2^N} \Rightarrow P_{\hat{\gamma}}\left(\Gamma_\nu\right) = \frac{n}{2^N} \Rightarrow 1 - e^{-\Gamma_\nu} = \frac{n}{2^N} \Rightarrow \Gamma_\nu = \ln\frac{2^N}{2^N - 1}, \tag{27}$$

where $P_{\hat{\gamma}}\left(\cdot\right)$ denotes the CDF of the normalized SNR (the SNR is exponentially distributed with unit parameter for Rayleigh fading channels) and $\nu = 1 \ldots 2^N - 1$. The quantized approach and the limited number of the required feedback bits significantly decrease the computational complexity in comparison with the conventional full-feedback centralized schemes. It is obvious that the complexity gain is more important for complex ad-hoc configurations which contain many nodes and hops.

Performance Analysis

Based on the previous sub-sections, we can now describe the proposed cross-layer approach. In the first phase of the proposed protocol, the source receives quantized (N bits) feedback from the two routers which indicate the quality of their links with the source. If the two SNR values belong to different subregions, the link with the higher SNR level defines the intermediate router. In the case that the two feedback values are equivalent, the selection is based on a simple RR policy which periodically selects a router as the intermediate node. We note that RR ensures the support of fairness for the case of equivalent links. In the second phase of the protocol, the source receives (quantized) feedback from the two destinations which describes the quality of the links from the selected router. The destination decision process is the same as for the intermediate router.

The average error probability of the proposed protocol can be expressed as

$$P_{A,B,C}(e) = \frac{2^N - 1}{2^N + 1} \cdot P_{A,B,C}^{(opt)}(e) + \frac{2}{2^N + 1} \cdot P_{A,B,C}^{(conv)}(e),$$

$$P^{D_J}(e) = \frac{1}{2} \cdot \sum_{I=1}^{2}\left(P_{S,R_{-I},R_I}(e) + P_{I,D_{-J},D_J}(e)\right), \tag{28}$$

where $P^{D_J}(e)$ is the error probability for the D_J destination, $P_{A,B,C}(e)$ is the error probability for the cooperative links $A \rightarrow C$ and $A \rightarrow B \rightarrow C$, $P_{A,B,C}^{(opt)}(e)$ denotes the error probability for the case where the direct link has the best instantaneous SNR, $P_{A,B,C}^{(conv)}(e)$ is the error probability for the conventional case and the $-I$ denotes the complement of I. Their analytical expressions and the related weight factors can be found in (Krikidis et al., 2008a).

Numerical Results

Computer simulations were carried out in order to evaluate the performance of the proposed cross-layer design. The different links are not symmetric and their relation can be represented by the equation: $E\left(\gamma_{S,R_1}\right) = E\left(\gamma_{R_1,D_1}\right) = \sigma^2 + 4$ dB, $E\left(\gamma_{R_2,D_1}\right) = \sigma^2 + 2$ dB where σ^2 is the received power (in dB) of all the other links. Figure 9 presents the BER performance for the destination D_1 and for different configurations of the proposed algorithm. The performance for a RR non-cooperative scheme and a channel-based scheme with absolute partial feedback are given as reference results. Moreover, the performance of a full feedback scheme which bases the scheduling and routing decision on an analytical estimation of the performance for all the possible combinations is also shown for comparison. As it can be shown, the gain of the perfect-feedback approach is equal to 2 dB and for quantized-feedback it is equal to 0.5 dB for each used bit (1 dB for 2 bits). An important observation is that a quantized feedback with a low number of bits ($N = 2$) gives a satisfactory performance-complexity trade-off. The comparison between analytical and simulation results validate the presented theoretical analysis. Furthermore, we can see that the perfect feedback normalized solution has a similar performance with the perfect feedback absolute SNR solution which suffers from a lack of fairness. Therefore, the normalized approach jointly optimizes the performance and supports the fairness concept. The comparison of the proposed algorithm with the full feedback scheme shows a difference of about 1.5 dB at high SNR's. This difference has been expected as the proposed algorithm only uses partial channel information. However, the decrease in complexity by considering partial feedback in the scheduling and routing decision is more important than the resulting performance degradation.

Figure 9. BER performance for the D_1 destination and for different levels of feedback versus σ^2 ; Dotted lines: analytical results, Points: simulation results. (@2008, IEEE. Used with permission.).

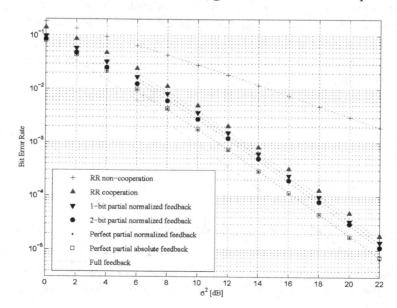

A DISTRIBUTED SCHEDULING SCHEME

According to the previous cross-layer problem, a scheduling decision that takes into account the instantaneous channel conditions requires a continuous feedback from the channel. Here, we introduce a distributed scheduling scheme for cooperative systems, which efficiently elaborates the instantaneous channel conditions without the need of an instantaneous channel feedback. We note that previously reported distributed implementations are based on channel-depended local timers at the relays and require an instantaneous signalling from the channel (Bletsas et al, 2006; Bletsas et al., 2007).

The system model is similar to the cluster C_2 of the figure 8 and thus consists of one source S and two destinations D_1, D_2, which can also be used as relays. In order to focus our study on the distributed issues, a symmetric channel is assumed to ensure fairness. The source has two data flows to transmit, one for each destination, grouped in the same frame that is broadcasted by the source (Krikidis & Belfiore, 2007a). The transmission is carried out into two orthogonal channels of different lengths in order to support the cooperation. In the first slot (source slot), each destination receives its corresponding data flow. The second slot (relay slot) is used for the cooperation, that is, a destination scales and forwards the data flow of the other destination. The main problem here is to determine in a distributed fashion (without feedback) which destination will be used as a relay for the other one (Srivasan et al., 2003).

A Centralized Algorithm

A centralized scheme, which bases the scheduling decision on the quality of the direct links, is used as a reference protocol. More specifically, this algorithm supposes a centralized mechanism which provides the source with the best instantaneous source-destination link. Based on the related instantaneous feedback, the destination with the best direct link is used as a relay. For the case of an i.i.d. channel model, this algorithm can be viewed as a form of RR, which uses instead of one direct link, the best of the two direct links. This statistical behaviour (PDF) can be expressed with the help of the order statistics and thus can be expressed as

$$p_s(\gamma) = 2 \cdot p_d(\gamma) \cdot P_d(\gamma),$$ (29)

where $p_s(\cdot)$ is the PDF of the selection diversity and $p_d(\gamma) = \beta e^{-\beta\gamma}$, $P_d(\gamma) = 1 - e^{-\beta\cdot\gamma}$ are the PDF and the CDF the direct link, respectively. Based on this expression, the average error probability is given by

$$P_1^{D_j}(e) = \frac{1}{2} \cdot \int_0^\infty P(e\,|\,\gamma) \cdot p_s(\gamma) d\gamma + \frac{1}{2} \cdot \int_0^\infty P(e\,|\,\gamma) \cdot p_c(\gamma) d\gamma.$$ (30)

where $P(e\,|\,\gamma)$ denotes the instantaneous error probability (Alouini et. Al., 1999) and $p_c(\cdot)$ is the PDF for the conventional OAF cooperation (Hasna & Alouini, 2003).

The centralized algorithm is the most efficient one, as the scheduling decision is based on the instantaneous channel conditions of the two direct links. It is also a fair solution because all the channels change symmetrically (i.i.d.), and thus each direct link can be the best one with the same probability. However, the knowledge of the instantaneous channel quality requires a high degree of signalling which makes it very hard to implement.

A Distributed Algorithm

In order to reduce the implementation complexity, the proposed distributed algorithm does not require a channel feedback and the scheduling decision is taken by each destination. More specifically, each destination estimates the quality of the direct link (using the received signal) and decides if it will relay or if it will be relayed to. The decision is based on a comparison of its instantaneous SNR with a threshold (T). If the SNR is higher than T, the destination behaves as a relay and thus broadcasts again the signal of the other destination. If the SNR is lower than the threshold, it continues to behave as a destination and waits for a possible relaying link. In the case where both destinations decide to relay, a collision is generated and the cooperation is considered as unsuccessful. Moreover, in the case where both destinations fail to exceed the predetermined threshold the cooperation cannot be performed (outage scheduling). The average error probability can be calculated as

$$P_2^{D_j}(e,T) = \mu \cdot \underbrace{\int_0^\infty P(e\,|\,\gamma)\, p_c(\gamma)\,d\gamma + (1-\mu) \cdot \int_0^\infty P(e\,|\,\gamma)\, p_d(\gamma)\,d\gamma}_{P^{s \to D_j \leq T}(e)} + \underbrace{\mu \cdot \int_0^\infty P(e\,|\,\gamma)\, p_d(\gamma)\,d\gamma + (1-\mu) \cdot \int_0^\infty P(e\,|\,\gamma)\, p_s(\gamma)\,d\gamma}_{P^{s \to D_j > T}(e)},$$

(31)

where $\mu = e^{-\beta T}$ is the probability that the instantaneous SNR is higher than a threshold T ($\Pr\{\gamma > T\} = 1 - P_d(T)$). The weight factors correspond to the behaviour of the other direct link ($S \to D_{-j}$).

The optimal threshold is the one which minimizes expression (31),

$$T_{opt} = \arg\min_T \left\{ P_2(e,T) \right\}.$$

(32)

The optimal threshold depends on the average SNR and can be calculated numerically by using different iterative optimization approaches (Conjugate Gradient, Quasi-Newton etc.). The proposed distributed algorithm is a fair solution as it is a particular version of the centralized algorithm. Moreover, it has a low complexity and thus is an attractive solution for practical implementations.

Numerical Results

Computer simulations were carried out in order to evaluate the performance of the scheduling algorithms. The tested modulation is BPSK and but a generalization to another modulation format is straightforward (Alouini et. al., 1999). Figure 10 compares the performance using as a measure the BER. The compared algorithms are RR, the distributed algorithm with a fixed threshold ($T = 3$), the distributed algorithm with the optimal threshold, and the centralized algorithm. Analytical curves are also given. From this figure we can see that the distributed scheduling significantly outperforms RR. Moreover, the distributed

Figure 10. The BER performance of different scheduling algorithms; (points) simulation results, (dotted lines) theoretical results. (@2008, IEEE. Used with permission.).

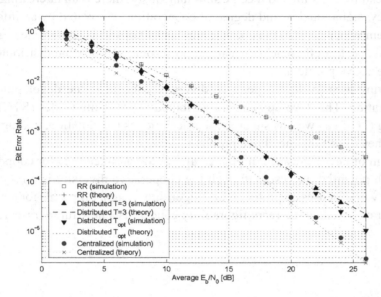

algorithm with the optimal threshold outperforms its fixed-threshold version. We note that the selected fixed-threshold is equal to the optimal threshold for E_b/N_0=16.5 dB. Therefore, the difference between the two distributed approaches (optimal, fixed) is negligible in the SNR region which is close to this value. The difference of the two approaches is more important at high SNR, where the optimal threshold significantly exceeds the fixed-threshold ($T_{opt} >> 3$). In comparison with the centralized algorithm, we can see that the distributed algorithm is 3 dB worse at high SNR. This difference is a result of the possible outage scheduling and collision. Finally, the difference between the analytical and simulated curves of the centralized scheduling is due to the fact that the instantaneous SNR of the simulation is estimated for the total frame (group of symbols), while the theoretical curve refers to each symbol.

FUTURE TRENDS

The basic ideas of this chapter can be extended to more practical configurations and problems that could be considered for future research work. The Multi-access relay channel (MARC) is an essential cooperative channel model which allows many sources to share a common relay in order to communicate with a destination. It corresponds to the uplink connection of practical systems (ad hoc networks, cellular systems) which use relays in order to increase transmission coverage. The design of new cooperative protocols (i.e. BFNAF), the solution of the power allocation problem as well as the investigation of new relay selection metrics are some open research issues for this particular cooperative structure. The shared nature of relays and the multi-user interference are two basic system parameters that have to be taken into account in order to efficiently solve the above questions.

Cross-layer design and cooperative diversity is another research area with practical future interest. The basic problem here is to design distributed algorithms which jointly optimize routing and scheduling in multi-hop ad-hoc networks with relaying cooperation. Due to the half-duplex constraint of the relay nodes (nodes cannot transmit and receive simultaneously) there is an interesting trade-off between bandwidth efficiency, signal quality and degree of cooperation. Motivated by an information-theoretic analysis of the system which defines the optimal number of hops in a regular linear network in order to support the required spectral efficiency, we can propose some sub-optimal distributed schemes which efficiently solve the above problem.

Another interesting area in the cooperative diversity concept is the development of real coding schemes for cooperation. The development of distributed Space-Time Codes (STC), consideration of Dirty-Paper Coding (DPC) as well as superposition techniques (modulation, coding) seem to be some interesting approaches to solving the above problem.

Finally the design of new Hardware implementations and architectures for cooperative networks is a research area with potential interest. Although the cooperative diversity is a hot research topic, the majority of existing work focuses on protocols and algorithm design. Proposals for new reprogrammable hardware architectures based on the reconfigurability concept is a new future research direction.

CONCLUSION

This chapter has dealt with the amplify-and-forward relaying strategy in ad-hoc networks. Due to the absence of a decoding process at the relays, AF seems to be interesting for practical applications with critical computational constraints. An analysis of the basic AF protocols in the literature has been presented and their comparison from an information theory point of view has been discussed. Furthermore, the interplay of the AF cooperative diversity concept with higher layers of the protocol stack has been considered. It has been shown that cooperative diversity is a multi-layer concept and a cross-layer design is required for a global system optimization. The problem of AF relay selection as well as its coordination with scheduling and routing has been analyzed based on some well-defined system configurations. Optimization criteria such as performance, complexity and fairness have been used in order to qualify the proposed solutions. Mathematical models and simulation results have been used in order to support the presented approaches.

REFERENCES

Abramowitz, M., & Stegun, I. A. (1970). *Handbook of mathematical functions with formulas, graphs, and mathematical tables*, 9th ed. New York: Dover.

Alouini, M.-S., Tang, X., & Goldsmith, A. J. (1999). An adaptive modulation scheme for simultaneous voice and data transmission over fading channels. *IEEE Journal on Selected Areas in Communications*, 5(17), 837–850. doi:10.1109/49.768199

Azarian, K., El Gamal, H., & Schniter, P. (2005). On the achievable diversity-multiplexing tradeoff in half-duplex cooperative channels. *IEEE Transactions on Information Theory*, 12(51), 4152–4172. doi:10.1109/TIT.2005.858920

Bletsas, A., Khisti, A., Reed, D. P., & Lippman, A. (2006). A simple cooperative diversity method based on network path selection. *IEEE Journal on Selected Areas in Communications*, *3*(24), 659–672. doi:10.1109/JSAC.2005.862417

Bletsas, A., Shin, H., & Win, M. Z. (2007). Cooperative communications with outage-optimal opportunistic relayins. *IEEE Transactions on Wireless Communications*, *6*(9), 3450–3460. doi:10.1109/TWC.2007.06020050

Boyd, S., & Vandenberghe, L. (2004). *Convex optimization*. Cambridge University Press.

Chen, D., & Laneman, J. N. (2006). Modulation and demodulation for cooperative diversity in wireless systems. *IEEE Transactions on Wireless Communications*, *7*(5), 1785–1794. doi:10.1109/TWC.2006.1673090

Costa, M. (1983). Writing on dirty paper. *IEEE Transactions on Information Theory*, *3*(29), 439–441. doi:10.1109/TIT.1983.1056659

Cover, T., & Thomas, J. A. (1991). *Elements of information theory*. New York: Wiley.

Deng, X., & Haimovich, A. M. (2005). Power allocation for cooperative relaying in wireless networks. *IEEE Communications Letters*, *11*(9), 994–996. doi:10.1109/LCOMM.2005.11012

Fan, Y., Wang, C., Thompson, J., & Poor, H. V. (2007). Recovering multiplexing loss through successive relaying using repetition coding. *IEEE Transactions on Wireless Communications*, *6*(12), 4484–4493. doi:10.1109/TWC.2007.060339

Hammerström, I., Kuhn, M., & Wittneben, A. (2004, September). Channel adaptive scheduling for cooperative relay networks. *Proc. IEEE Vehic. Tech. Conf.*, Los Angeles, CA (pp. 2784-2788).

Hasna, M. O., & Alouini, M.-S. (2003). End-to-end performance of transmission systems with relays over rayleigh-fading channels. *IEEE Transactions on Wireless Communications*, *6*(2), 1126–1131. doi:10.1109/TWC.2003.819030

Hasna, M. O., & Alouini, M.-S. (2004). Optimal power allocation for relayed transmissions over rayleigh-fading channels. *IEEE Transactions on Wireless Communications*, *6*(3), 1999–2004. doi:10.1109/TWC.2004.833447

Hong, Y.-W., Huang, W.-J., Chiu, F.-H., & Kuo, C.-C. J. (2007). Cooperative communications in resource-constrained wireless networks. *IEEE Signal Processing Magazine*, *3*(24), 47–57. doi:10.1109/MSP.2007.361601

Hunter, T. E., & Nosratinia, A. (2006). Diversity through coded cooperation. *IEEE Transactions on Wireless Communications*, *2*(5), 283–289. doi:10.1109/TWC.2006.1611050

Karaki, J. N. A., & Kamal, A. E. (2004). Routing techniques in wireless sensor networks: A survey. *IEEE Wirel. Commun.*, *6*(11), 6–28. doi:10.1109/MWC.2004.1368893

Kramer, G., Gastpar, M., & Gupta, P. (2005). Cooperative strategies and capacity theorems for relay networks. *IEEE Transactions on Information Theory*, *51*(9), 3037–3063. doi:10.1109/TIT.2005.853304

Krikidis, I., & Belfiore, J.-C. (2007a). Three scheduling schemes for amplify-and-forward relay environments. *IEEE Communications Letters*, *5*(11), 414–416. doi:10.1109/LCOMM.2007.070077

Krikidis, I., & Belfiore, J.-C. (2007b). Scheduling for amplify-and-forward cooperative networks. *IEEE Trans. Vehic. Tech.*, *6*(56), 3780–379. doi:10.1109/TVT.2007.901062

Krikidis, I., Thompson, J., & Goertz, N. (2008a). A cross-layer approach for cooperative networks. *IEEE Trans. Vehic. Tech.*, *57*(5), 3257–3263. doi:10.1109/TVT.2008.915508

Krikidis, I., Thompson, J., McLaughlin, S., & Goertz, N. (2008b). Amplify-and-forward with partial relay selection. *IEEE Communications Letters*, *4*(12), 235–237. doi:10.1109/LCOMM.2008.071987

Krikidis, I., Thompson, J., McLaughlin, S., & Goertz, N. (2008c). Optimization issues for cooperative amplify-and-forward systems over block-fading channels. *IEEE Trans. Vehic. Tech.*, *57*(5), 2868–2884. doi:10.1109/TVT.2008.917235

Krikidis, I., Thompson, J., McLaughlin, S., & Goertz, N. (2008d). Relay selection in interference-limited amplify-and-forward systems. *IEEE Trans. Wirel. Commun.*, submitted for publication.

Lai, L., Liu, K., & El Gamal, H. (2006). The three-node wireless network: Achievable rates and cooperation strategies. *IEEE Transactions on Information Theory*, *52*(3), 805–828. doi:10.1109/TIT.2005.864421

Laneman, J. N. (2002). *Cooperative diversity in wireless networks: Algorithms and architectures*. Unpublished doctoral thesis, Massachusetts Institute of Technology, Cambridge, MA.

Laneman, J. N., Tse, D. N. C., & Wornell, G. W. (2004). Cooperative diversity in wireless networks: Efficient protocols and outage behaviour. *IEEE Transactions on Information Theory*, *12*(50), 3062–3080. doi:10.1109/TIT.2004.838089

Laneman, J. N., & Wornell, G. W. (2003). Distributed space-time-coded protocols for exploiting cooperative diversity in wireless networks. *IEEE Transactions on Information Theory*, *10*(49), 2415–2425. doi:10.1109/TIT.2003.817829

Lange, K. (2004). *Optimization (Springer texts in statistics)*. Springer-Verlag.

Larsson, E. G., & Vojcic, B. R. (2005). Cooperative transmit diversity based on superposition modulation. *IEEE Communications Letters*, *9*(9), 778–780. doi:10.1109/LCOMM.2005.1506700

Mahinthan V., Cai L., Mark J. W., & Shen X. (S.) (2008). Partner selection based on optimal power allocation in cooperative-diversity systems. *IEEE Trans. Vehic. Tech.*, *1*(57), 511-520.

Marina, M. K., & Das, S. R. (2001). On-demand multipath distance vector routing in ad hoc networks. In *Proc. IEEE Int. Conf. Netw. Protoc.*, Mission Inn, CA (pp. 14-23).

Nabar, R. U., Bölcskei, & Kneubuhler, H. F. W. (2004). Fading relay channels: Performance limits and space-time signal design. *IEEE J. Select. Areas Commun.* *6*(22), 1099-1109.

Nadarajah, S., & Kotz, S. (2006). On the product and ratio of gamma and weibull random variables. *Econometric theory-Cambridge University Press*, 22, 338-344.

Proakis, J. G. (20001). *Digital communications,* 4th edition. McGraw Hill.

Ribeiro, A., Cai, X., & Giannakis, G. B. (2005). Symbol error probabilities for general cooperative links. *IEEE Transactions on Wireless Communications, 3*(4), 1264–1273. doi:10.1109/TWC.2005.846989

Scutari, G., & Barbarossa, S. (2005). Distributed space-time coding for regenerative relay newtroks. *IEEE Transactions on Wireless Communications, 5*(4), 2387–2399. doi:10.1109/TWC.2005.853883

Sendonaris, A., Erkip, E., & Aazhang, B. (2003a). User cooperation diversity part I-system description. *IEEE Transactions on Communications, 11*(51), 1927–1938. doi:10.1109/TCOMM.2003.818096

Sendonaris, A., Erkip, E., & Aazhang, B. (2003b). User cooperation diversity part II- implementation aspects and performance analysis. *IEEE Transactions on Communications, 11*(51), 1939–1948. doi:10.1109/TCOMM.2003.819238

Srinivasan, V., Nuggehalli, P., Chiasserini, C. F., & Rao, R. R. (2003, March). Cooperation in wireless ad hoc networks. In *Proc. IEEE INFOCOM*, San Francisco, CA (pp. 808-817).

Tse, D., & Viswanath, P. (2004). *Fundamentals of wireless communications*. Cambridge University Press.

Wang, L., Kwok, Y.-K., Lau, W.-C., & Lau, V. K. N. (2004). Efficient packet scheduling using channel adaptive fair queueing in distributed mobile computing systems. *ACM Mobile Netw. & Appl., 9*, 297–309. doi:10.1023/B:MONE.0000031589.32967.0f

Yang, S., & Belfiore, J.-C. (2007). Towards the optimal amplify-and-forward cooperative diversity scheme. *IEEE Transactions on Information Theory, 9*(53), 3114–3126. doi:10.1109/TIT.2007.903133

Yang, S., & Belfiore, J.-C. (2008). Diversity of MIMO multi-hop relay channels. *IEEE Trans. Inf. Theory*. Submitted for publication.

Yao, Y., Cai, X., & Giannakis, G. B. (2005). On energy efficiency and optimum resource allocation of relay transmissions in the low-power regime. *IEEE Transactions on Wireless Communications, 6*(4), 2917–2927.

Zhang, J., Zhang, Q., Shao, C., Wang, Y., Zhang, P., & Zhang, Z. (2004). Adaptive optimal transmit power allocation for two-hop non-regenerative wireless relay system. In *Proc. IEEE Vehicular Technology Conf.*, Milan, Italy (pp. 1213-1217).

Zhao, B., & Valenti, M. C. (2005). Practical relay networks: A generalization of hybrid ARQ. *IEEE Journal on Selected Areas in Communications, 1*(23), 7–18. doi:10.1109/JSAC.2004.837352

Zheng, L., & Tse, D. N. (2003). Diversity and multiplexing: A fundamental tradeoff in multiple-antenna channels. *IEEE Transactions on Information Theory, 5*(49), 1073–1096. doi:10.1109/TIT.2003.810646

Chapter 3
Power Allocation for Cooperative Communications

Onur Kaya
Işık University, Turkey

Sennur Ulukus
University of Maryland, USA

ABSTRACT

In this chapter, we review the optimal power allocation policies for fading channels in single user and multiple access scenarios. We provide some background on cooperative communications, starting with the relay channel, and moving onto mutually cooperative systems. Then, we consider power control and user cooperation jointly, and for a fading Gaussian multiple access channel (MAC) with user cooperation, we present a channel adaptive encoding policy, which relies on block Markov superposition coding. We obtain the power allocation policies that maximize the average rates achievable by block Markov coding, subject to average power constraints. The optimal policies result in a coding scheme that is simpler than the one for a general multiple access channel with generalized feedback. This simpler coding scheme also leads to the possibility of formulating an otherwise non-concave optimization problem as a concave one. Using the perfect channel state information (CSI) available at the transmitters to adapt the powers, we demonstrate significant gains over the achievable rates for existing cooperative systems. We consider both backwards and window decoding, and show that, window decoding, which incurs less decoding delay, achieves the same sum rate as backwards decoding, when the powers are optimized.

INTRODUCTION

The wireless medium brings along its unique challenges such as fading and multiuser interference, which complicate the design and analysis of the communication systems. On the other hand, these very same challenges are what give rise to the concept of diversity, which can be carefully exploited to the advantage of the network capacity.

DOI: 10.4018/978-1-60566-665-5.ch003

Fading may become a major limiting factor on the performance of wireless networks, unless appropriate measures are taken against it. Perhaps the most prominent measure to combat fading is power control, which relies on some sort of channel state information to adapt the transmit strategies to the evolving channel conditions. When adaptive transmission is an option, the presence of fading creates diversity across time, which is called temporal diversity, and across users, which is called multiuser diversity. Power control can be cleverly used to maximize the system performance, by making use of these forms of diversity to adjust the signals transmitted by the users.

Another distinguishing feature arising from the physical nature of the wireless channel is what is traditionally thought of as interference in networks: overheard information. The fact that signals from all sources in the network are superposed in the transmit medium can be taken advantage of in the design of wireless networks, by allowing cooperation between the nodes in the network, yielding cooperative diversity.

Although user cooperation and power control are often seen as two separate ways to combat fading, they are inherently coupled approaches. Specifically, in cooperative communication, a user has to decide on how much of its available power should be allocated for cooperation, and how much for only its own transmission. Going a step further, it may come down to deciding whether to cooperate or not in the first place, during a given channel condition.

How power control needs to be performed for a specific wireless network strongly depends on the underlying application. The initial wireless networks that have been deployed in practice were invariably intended to carry voice traffic, and therefore had to make sure that a desired quality of service is attained, so that the delay sensitive voice application is not disrupted. Such networks need to find ways to combat the possible deep attenuations caused by the fading at given time instants, and the associated power control policies therefore have to compensate for worse channel states by boosting the transmit power.

Recently, the growing volume of data traffic (email, files, etc.) and the emergence of wireless networks as a medium for communicating higher rate but less delay sensitive traffic have made it essential to investigate resource allocation policies that are more efficient and opportunistic. In this case, it is possible to exploit the time diversity within a long block of transmission: if the transmitter faces a deep fade level, knowing that the better channel states are to be realized in the future, it can save its available resources, for example transmit power, to the upcoming favorable states. This type of approach leads to worse instantaneous performance at some channel states, but is expected to improve the average performance metrics, such as the long term average Shannon capacity.

In this chapter, we provide an information theoretic approach to power control in cooperative wireless networks. We will focus on a two user fading multiple access channel, where both users mutually cooperate using a decode and forward type of strategy, which relies on block Markov superposition coding, and backwards or window decoding. The fading will be modeled as a stationary and ergodic random process, whose statistics, as well as the instantaneous realization are known to the communicating parties. The main goals of this chapter are:

- to provide a general background on ways to approach user cooperation and power control,
- to present a channel adaptive encoding mechanism which integrates power control to block Markov superposition coding,
- to obtain the optimal power control policies which maximize the long term average rates achievable by user cooperation, and the associated improved rate regions,

- to show some key properties of the optimal power allocation policy; in particular, how the information theoretic approach to power control dictates the optimal cooperation and routing strategies, thereby yielding a cross-layer optimization, and
- to extend the optimal power control to the more practical case of window decoding.

Throughout this chapter, we use boldface letters **x** to denote vector variables, as opposed to scalar variables, which are simply denoted by an italicized *x*. Capital letters *X* are used to distinguish random variables from their realizations *x*, where necessary.

BACKGROUND

Increasing demand for higher rates in wireless communication systems continuously triggers major research which aims to characterize their capacities. In this section, we present an evolution of capacity results for fading wireless channels under various types of communication paradigms. Most part of this section will be devoted to multiuser systems, and we will consider settings with a single ultimate receiver. We will present some recent and not-so-recent results about multiple access channels and relay channels, with different degrees of cooperation.

This section is organized as follows. First, we will present some ergodic capacity results on several types of fading channels, which will constitute the foundation for the power control approach in our treatment of cooperative communications. Next, we will review some of the literature on user cooperation, specifically in multiple access and relay channels.

Power Control for Fading Channels

We will limit our review of power control in fading channels to multiplicative fading in the presence of additive white Gaussian noise, and consider only the ergodic capacity of such systems. An excellent comprehensive survey of fading channels and an extensive set of references on fading can be found in (Biglieri, Proakis, & Shamai (Shitz), 1998). We would like to point out that power control for wireless networks is a very broad topic in its own right, and what follows is just a very concise summary of only a few ideas which will be used in our treatment of cooperative channels.

The capacity of fading channels has attracted interest mostly after the pioneering work by Goldsmith and Varaiya (1997), where the ergodic (long term average) capacity of a single transmitter single receiver system described by

$$r = b\sqrt{p(h)h} + n,$$

(1)

was obtained. Here, n denotes the zero-mean real-valued additive white Gaussian noise with variance σ^2, h denotes the multiplicative fading, b denotes the transmitted symbol, and $p(h)$ is the transmit power, which can be chosen as a function of the channel fading state h. The fading is modeled as a stationary and ergodic random process, whose statistics are known to the transmitter as well as the receiver. The fading is further assumed to be slow enough so that, its realization h remains constant over at least one symbol interval, and it can be tracked by the receiver and fed back to the transmitter, leading to perfect channel state information. Some insight on how the fading is modeled as a random process and why these assumptions, however stringent they may seem, are in fact practical, can be gained by referring to (Gallager, 1994). Goldsmith and Varaiya (1997) showed that the ergodic capacity is achieved by modulating the transmitted symbols by channel adaptive powers $p(h)$, subject to an average power constraint. The resulting single user capacity for the fading additive white Gaussian noise channel, assuming a real channel model is,

$$
\begin{aligned}
C = \max_{\{p(h)\}} \quad & E_h\left[\frac{1}{2}\log\left(1 + \sigma^{-2}p(h)h\right)\right] \\
\text{s.t.} \quad & E_h\left[p(h)\right] \le \overline{p}, \quad p(h) \ge 0.
\end{aligned}
\tag{2}
$$

The optimal power allocation that achieves this capacity is characterized as "waterfilling in time," which exploits the knowledge of the channel state h by allocating more power to better channel states and less power to worse channel states, i.e.,

$$
p(h) = \left(\frac{1}{\lambda} - \frac{\sigma^2}{h}\right)^+,
\tag{3}
$$

where, λ denotes the Lagrange multiplier associated with the constrained optimization problem, whose inverse, which appears in (3), can be viewed as the "level of the water surface."

When we move to systems with multiple users sharing the same channel, i.e., a multiple access channel, the rates at which reliable communication can be carried out constitute a "capacity region," rather than a single capacity value. A remarkably important generalization of the optimal power control strategy of a single user channel to a K-user MAC modeled by

$$
r = \sum_{i=1}^{K} \sqrt{p_i(\mathbf{h})h_i}\, b_i + n,
\tag{4}
$$

was obtained by Knopp and Humblet (1995). Here, $\mathbf{h}$ denotes the vector of instantaneous channel states of all users, and the index i represents different users. Knopp and Humblet considered a specific point on the capacity region boundary, i.e., the ergodic sum capacity, as their objective function. The sum capacity is shown to be the solution to the optimization:

$$
\begin{aligned}
\max_{\{p_i(\mathbf{h})\}_{i=1}^{K}} \quad & E_{\mathbf{h}}\left[\frac{1}{2}\log\left(1 + \sigma^{-2}\sum_{i=1}^{K}h_i p_i(\mathbf{h})\right)\right] \\
\text{s.t.} \quad & E_{\mathbf{h}}\left[p_i(\mathbf{h})\right] \le \overline{p}_i, \quad p_i(\mathbf{h}) \ge 0, \quad i = 1,\cdots,K,
\end{aligned}
\tag{5}
$$

and the optimum power control yields a TDMA like transmit policy, where, for each given time instant, only the user with the best channel state at that time instant transmits. In this setting, the solution to the power optimization problem still points to a single user waterfilling type of strategy, but over disjoint sets of channel states:

$$
p_k(\mathbf{h}) = \begin{cases} \left(\dfrac{1}{\lambda_k} - \dfrac{\sigma^2}{h_k} \right)^+, & \text{if} \quad h_k / \lambda_k > h_j / \lambda_j, \quad \forall j \neq k, \\ \\ 0 \quad, & \text{otherwise.} \end{cases}
\tag{6}
$$

where, λ_k is the Lagrange multiplier associated with the power constraint of user k.

While Knopp and Humblet (1995) focused only on the sum capacity of the fading MAC (4), Hanly and Tse (1998) characterized its entire ergodic capacity region, i.e.,

$$
\bigcup_{\{\mathbf{p}(\mathbf{h}):\, E_{\mathbf{h}}[p_i(\mathbf{h})] \leq \bar{p}_i, \, \forall i\}} \left\{ \mathbf{R} : \sum_{i \in \Gamma} R_i \leq E_{\mathbf{h}} \left[\frac{1}{2} \log \left(1 + \sigma^{-2} \sum_{i \in \Gamma} h_i p_i(\mathbf{h}) \right) \right], \quad \forall \Gamma \subset \{1, \cdots, K\} \right\},
\tag{7}
$$

where R_i denotes the long term average transmission rate of user i. The structural characteristics of the capacity region (7) were established, and an algorithm to compute the optimal power allocation policies that achieve the rate points on the capacity region boundary was proposed in (Hanly & Tse 1998). A sample capacity region is illustrated in Figure 1. The capacity region for each fixed power control policy turns out to be a polymatroid (a pentagon in the two user scenario), whose corners can be achieved by successive decoding. Moreover, every point on the capacity region boundary is in fact a corner of one of the underlying polymatroids, and hence, time sharing is not required to achieve any rate tuple on the capacity region. Therefore, given a point on the capacity region boundary, the associated optimum power control policy can be uniquely determined. The rate region corresponding to the power allocation policy that maximizes the sum capacity is the rectangle in Figure 1.

The optimal power allocation for the MAC, for points other than the sum capacity point cannot be expressed in closed form as in (6), but is rather described algorithmically. In fact, the optimum power allocation which achieves any point on the capacity region boundary can also be found by maximizing a weighted sum of rates (see Figure 1), using a one-user-at-a-time iterative algorithm called "generalized iterative waterfilling" (Kaya & Ulukus 2006). This algorithm is also applicable to waveform (or vector) channel models, such as CDMA.

Let us pause for a moment and stress what is common to all three problems discussed so far. Clearly, once the channel fading can be tracked by the transmitters, the transmitters can adapt their coding strategies accordingly to achieve higher rates. In the case of scalar single user and multiple access channels, optimally utilizing the channel state information at the transmitters is simply equivalent to performing power control: in order to achieve the capacity, it is sufficient to create a fixed codebook only once, and just scale it with an instantaneous transmit power value, which is a function of the channel state. More importantly, the results given in equations (3) and (6) have a very fundamental practical implication: optimum power control policies may also dictate the optimum medium access strategies. Therefore, power control strategies designed to optimize an information theoretic objective also give us an idea about how to jointly design the physical and the medium access control layers of the communication systems.

Figure 1. Sample capacity region for the fading MAC, and optimization of rates on its boundary.

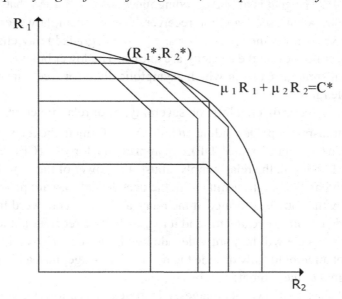

User Cooperation in Wireless Networks

The nature of the wireless medium allows all transmissions to be heard by all communicating parties. Although the overheard information is often not intended to be decoded by the overhearing transmitters, systems where the transmitters get involved in each others' transmission and cooperate while transmitting to the ultimate receiver(s) are likely to achieve higher rates. The capacity region for the MAC stated in the previous subsection was obtained, assuming that the messages of the users are independently generated, and that the users do not decode the messages of each other. In this subsection, we will summarize some of the techniques used to exploit the additional information arising from the users being able to listen to each other and cooperate. In particular, we will review systems with one or more users trying to communicate to a single receiver, in additive white Gaussian noise.

The simplest form of user cooperation is observed in a relay channel, where the sole task of the relay is to assist the transmission of the source node to its prescribed destination. A model for the relay channel was developed by van der Meulen (1968; 1971). In their seminal paper on relay channels, Cover and El Gamal (1979) proposed several encoding and decoding strategies, and obtained the corresponding achievable rate/capacity results for a single relay network. Since the focus of this chapter is mainly on Gaussian channels, without loss of generality of the results, we will limit our review to the Gaussian relay channel:

$$Y_i = X_{1i} + X_{2i} + N_i,$$
$$Y_{1i} = X_{1i} + N_{1i}, \quad i = 1, \cdots, n. \tag{8}$$

Here, user 1 transmits its message w_1 by sending a length n codeword $\mathbf{X}_1(w_1)$, while user 2, based on its previous observations Y_{1i}, creates a codeword $\mathbf{X}_2$ using a relay function $X_{2i} = f_i(Y_{11}, \cdots, Y_{1i-1})$. The signal received at the receiver is Y_i. The goal is to find encoding functions $\mathbf{X}_1$ and $\mathbf{X}_2$ that yield the maximum reliable transmission rate between user 1 and the receiver. This problem, however easy it may seem, has not yet been solved except for some special cases, e.g., the degraded relay channel. On the other hand, some upper and lower bounds on the capacity are known. The lower bounds refer to rates, which can be achieved by known strategies, and at which asymptotically error free transmission is possible, hence the name achievable rates.

Before we go any further, let us dwell a little on several types of relay functions that can be used in assisting the source in its transmission: Depending on how X_{2i} is formed, the relaying strategy is called amplify and forward (AF), decode and forward (DF), or compress and forward (CF). As its name suggests, in the amplify and forward protocol, the relay simply adjusts the power of the signal it receives to meet a transmit power constraint, and then retransmits it. In the decode and forward protocol, an estimate of the message transmitted by the source is formed at the relay, and a new codeword from possibly a new codebook is used to encode this message and to send it to the ultimate receiver in the next block, where the relay's version of the message will be jointly decoded with the sender's version. In compress and forward, the relay does not attempt to fully decode the received message; instead it sends a compressed version of its received signal to the ultimate receiver.

Each of these protocols has their own advantages: AF strategy has a lower complexity, as the relay does not re-encode the message, and the receiver employs relatively simpler decoding techniques. However, since it also amplifies the noise, it often achieves poorer performance than decode and forward and compress and forward. DF strategy performs best if the relay is close to the source, since in that case the cost of decoding the source signal at the relay is very low, and the system becomes one with two virtual transmit antennas with the same information. On the other hand, if the relay is far away from the source, the overall achievable rate is limited to the achievable rate on the link from the source to the relay. This is the case where the advantage of using CF strategy kicks in: the CF strategy is known to perform best when the relay is very close to the destination, since the receiver's copy of the relay signal will be very high quality, and this may be viewed as a setting in which the receiver has two versions of the transmitted signal, one of which is less noisy, thanks to the compression of the noise. The CF strategy does not suffer from the bottleneck effect of DF on the source-relay link, since the relay does not attempt to decode the signal in CF. These heuristic arguments are quantified in a recent paper by Kramer, Gastpar and Gupta (2005), which also includes an excellent review of relay and other cooperative channel models.

For reasons which will become clearer when we start discussing the achievable rates for the cooperative MAC, we restrict our attention to decode and forward type of strategies, which are more widely studied than the others in the context of user cooperation. An achievable rate for the arbitrary relay channel, which uses DF relaying is given by (Cover & El Gamal, 1979),

$$\max_{p(x_1,x_2)} \min \left\{ I(X_1;Y_1 \mid X_2), I(X_1,X_2;Y) \right\}, \tag{9}$$

where $I(.)$ denotes the mutual information. This rate is achieved by the so called block Markov superposition encoding, where instead of a single block of transmission, B blocks, each consisting of length n codewords are considered. In each block i, user 1 sends a codeword $\mathbf{X}_1$, which depends on the message w_i it wants to transmit in block i, as well as the previous message w_{i-1}. This allows the relay, who has already decoded w_{i-1}, to cooperate with the transmitter in the ith slot to communicate message w_{i-1}, thereby reducing the receiver's ambiguity about w_{i-1}. For the Gaussian relay channel in question here, this coding scheme can be realized by forming the codewords as follows:

$$\mathbf{X}_1 = \sqrt{\bar{\alpha}P_1 / P_2}\,\mathbf{X}_2 + \mathbf{X}_{10}, \tag{10}$$

$$\mathbf{X}_{10} \quad \text{i.i.d} \sim N_n(0, \alpha P_1 \mathbf{I}_n), \tag{11}$$

$$\mathbf{X}_2 \quad \text{i.i.d} \sim N_n(0, P_2 \mathbf{I}_n). \tag{12}$$

where $\bar{\alpha} = 1 - \alpha$. It can be verified that power constraints P_1, P_2 are satisfied, and (9) becomes

$$C = \max_{0 \leq \alpha \leq 1} \min \left\{ \frac{1}{2}\log\left(1 + \frac{P_1 + P_2 + 2\sqrt{\bar{\alpha}P_1 P_2}}{\sigma^2}\right), \frac{1}{2}\log\left(1 + \frac{\alpha P_1}{\sigma_1^2}\right) \right\}. \tag{13}$$

An upper bound for the capacity is obtained by the max-flow-min-cut theorem (Cover & Thomas, 1991)

$$\max_{p(x_1, x_2)} \min \left\{ I(X_1; Y, Y_1 \mid X_2), I(X_1, X_2; Y) \right\}. \tag{14}$$

For some special cases of relay channels, i.e., when the relay channel is degraded, or there is feedback from the outputs Y, Y_1 to X_1 and X_2, the upper and lower bounds are tight, and in fact give the capacity (Cover & El Gamal, 1979).

The achievable rates (9) have also been obtained by different methods by several authors. Willems (1982) uses block Markov superposition encoding, where the codebook sizes are identical, together with backward decoding. Xie and Kumar (2004) also use the same block Markov encoding, together with window decoding, using a window length of two blocks.

An interesting problem that also relates to power control discussed in the previous subsection is the characterization of long term achievable rates for relay channels that are subject to fading. Since there are no capacity expressions in the general case, one cannot characterize the ergodic capacity with the current knowledge of relay channels. However, if the channel states are available at the transmitter and the relay, upper and lower bounds for the capacity can be maximized as a function of valid power control policies. This problem was solved in part for a set of bounds in (Host-Madsen & Zhang, 2005).

Power control for relay channels is a topic which attracted a lot of attention, and there exists a very wide range of approaches to the problem of power control in this context. The variations to this problem include, but certainly are not limited to the permutations of: i) a choice of the encoding/decoding policy among amplify and forward (AF), decode and forward (DF), compress and forward (CF) strategies; ii) a choice of the cost function among probability of error, outage probability, achievable rates, throughput; iii) a choice between half duplex or full duplex communication protocols, as well as orthogonal or non-orthogonal scenarios. Obviously, it is not feasible to delve into all these variations in this chapter; instead, we will look at the less studied and seemingly more interesting problem of power control for cooperative multiple access channels, where cooperation is mutual as opposed to the relay channels where the sole task of the relay is to assist a source node in its transmission to a destination. For the reader interested in references on various resource allocation strategies on relay channels, including non-information theoretical approaches, a thorough survey on resource allocation for relay networks can be found in Hong et. al (2007).

The efforts in obtaining achievable rates for multi-terminal cooperative communication systems are certainly not limited to one transmitter - one relay scenarios. One major setting involving multiple terminals is the multiple relay channel, where many relays, which do not have their own messages, help a single transmitter to send its message to an intended receiver. This type of channel has recently attracted considerable attention, and resulted in many interesting results on the capacity of such systems. In (Xie & Kumar, 2004), where information theoretic results for several wireless networks were derived, an achievable rate for a multiple relay Gaussian channel was also obtained. This achievability result was then generalized to arbitrary relay networks by Kramer, Gastpar and Gupta (2005), using a multi-hopping strategy. The capacity of parallel Gaussian relay networks was first studied by Schein and Gallager (2000) for the case of two relays, where upper and lower bounds for the capacity were obtained. Parallel Gaussian relay problem was then attacked for the limiting case of infinite number of relays by Gastpar and Vetterli (2002), and upper and lower bounds on the asymptotic capacity were obtained. It was shown that the difference of upper and lower bounds is bounded in the limit as number of relays goes to infinity, and an example where this difference goes to zero was given.

A second major setting in multi-terminal cooperative networks is the case of multiple access relay channel, which is composed of an M-user MAC, and one additional relay whose sole task is to assist the MAC users in their transmissions (Kramer & van Wijngaarden, 2000; Sankaranarayanan, Kramer & Mandayam, 2004). The list of references on multi-terminal relay networks, and the associated capacity results could be further exhausted; the interested reader may find a more complete set of references on several relay networks in (Kramer, Gastpar & Gupta, 2005). However, relay networks all operate based on the principle that the terminals participating in the transmission are either dedicated sources or dedicated relays. In this chapter, we are interested in the practically more interesting scenario where all participating terminals have their own messages to be transmitted. In wireless systems, this "mutual" cooperation can be achieved by simply manipulating the overheard information at each transmitter.

The foundations for the idea of mutual user cooperation in wireless networks were laid in the early 1980s, with the consecutive solutions of a series of toy problems on multiple access channels. First, Cover and Leung (1981) considered a two user multiple access channel where both users have access to the channel output, and obtained an achievable rate region for this channel model. Later, it was demonstrated by Willems and van der Meulen (1983) that the channel output need not be known at both of the cooperating parties to achieve Cover and Leung's rate region; a feedback link from the channel output to only one of the transmitters is sufficient.

Further relaxations on the amount of available side information at the cooperating encoders are possible. One such scenario is the MAC with partially cooperating encoders, the capacity of which was obtained by Willems (1983). In this setting, cooperation among the encoders is established through finite capacity communication links connecting them to each other. A limiting case for this setting arises if the encoders are allowed to "crib" from each other, that is, they learn each others' codewords before the next transmission (Willems & van der Meulen, 1985). Several scenarios regarding which encoder(s) crib, and how much of the codewords the encoders learn, are treated and the capacity region for each case is obtained in (Willems & van der Meulen, 1985). Note that, these problems involving cribbing might seem impractical because of the assumption of free exchange of information among the encoders

Figure 2. MAC with generalized feedback

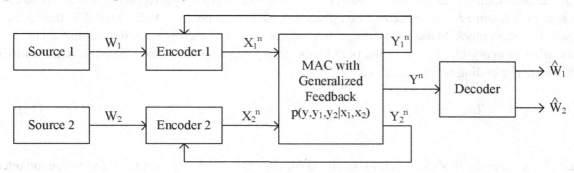

without the need for allocating resources to any such exchange. However, it is still instructive to obtain their capacities, as they provide an upper bound to the rates achievable by more practical cooperative schemes.

Among the channel models that allow some sort of cooperation among the users, the MAC with generalized feedback, which is shown in Figure 2, is worth special attention. This channel accurately models the over-heard information by the transmitters in a wireless network, since each of the transmitters gets a version of the channel output, which in turn depends on the other's transmission. In particular, the two user MAC with generalized feedback is described by $(X_1 \times X_2, P(y, y_1, y_2 \mid x_1, x_2), Y \times Y_1 \times Y_2)$, where user 1 has access to channel output Y_1 and user 2 has access to channel output Y_2. For this model, a region of achievable rates was first obtained by Carleial (1982), followed by a simpler and larger rate region, obtained by Willems, van der Meulen and Schalkwijk (1983). Carleial and Willems et al. used the different versions of what is so called superposition block Markov encoding, as well as different decoding policies, namely sliding window versus backwards decoding, to obtain the corresponding rate regions. The latter of these regions obtained using backward decoding, where the receiver waits to receive all B blocks of codewords before starting to decode, is in general larger, and simpler to characterize. On the other hand, the delay performance of such a decoding policy is considerably worse.

Recently, these achievability results for the rather general MAC with generalized feedback were adapted to a fading Gaussian MAC by Sendonaris, Erkip and Aazhang (2003), leading to a new form of diversity called user cooperation diversity in wireless channels. In this setting, both the receiver and the transmitters receive noisy versions of the transmitted messages. Hence, the transmitters form their codewords not only based on their own information, but also on the information they have received from each other, in previous blocks of transmission. The idea of encoding builds on a modification of the basic Gaussian relay channel. The system model for the two user cooperative multiple access channel is

$$Y_0 = K_{10}X_1 + K_{20}X_2 + Z_0,$$
$$Y_1 = K_{21}X_2 + Z_1,$$
$$Y_2 = K_{12}X_1 + Z_2,$$

(15)

where X_i is the symbol transmitted by node i, Y_i is the symbol received at node i, and the receiver is denoted by $i = 0$. Z_i is the zero mean additive white Gaussian noise at node i, having variance σ_i^2. K_{ij}

are the random fading coefficients, assumed known at their corresponding receiver j, but nowhere else. The system is assumed to be coherent, i.e., phase information is present at all nodes. The transmitters use superposition block Markov encoding; they allocate some of their powers to establish a common information in every block, and in the next block, they coherently combine part of their transmitted codewords. The coding scheme is given by

$$X_i = X_{i0} + X_{ij} + U_i, \quad i,j \in \{1,2\}, \quad i \neq j, \tag{16}$$

where X_{i0} carries the fresh information intended for the receiver, X_{ij} carries the information intended for transmitter j for cooperation in the next block, and U_i is the common information sent by both transmitters for resolution of the remaining uncertainty from the previous block. Labeling the associated power levels accordingly, and imposing the power constraints

$$P_i = P_{i0} + P_{ij} + P_{Ui}, \tag{17}$$

the achievable rate region is given by the convex hull of all rate pairs satisfying

$$
\begin{aligned}
R_{12} &< E\left[\frac{1}{2}\log\left(1 + \frac{K_{12}^2 P_{12}}{K_{12}^2 P_{10} + \sigma_2^2}\right)\right] \\
R_{21} &< E\left[\frac{1}{2}\log\left(1 + \frac{K_{21}^2 P_{21}}{K_{21}^2 P_{20} + \sigma_1^2}\right)\right] \\
R_{10} &< E\left[\frac{1}{2}\log\left(1 + \frac{K_{10}^2 P_{10}}{\sigma_0^2}\right)\right] \\
R_{20} &< E\left[\frac{1}{2}\log\left(1 + \frac{K_{20}^2 P_{20}}{\sigma_0^2}\right)\right] \\
R_{10} + R_{20} &< E\left[\frac{1}{2}\log\left(1 + \frac{K_{10}^2 P_{10} + K_{20}^2 P_{20}}{\sigma_0^2}\right)\right] \\
R_{10} + R_{20} + R_{12} + R_{21} &< E\left[\frac{1}{2}\log\left(1 + \frac{K_{10}^2 P_1 + K_{20}^2 P_2 + 2K_{10}K_{20}\sqrt{P_{U1}P_{U2}}}{\sigma_0^2}\right)\right],
\end{aligned}
\tag{18}
$$

for some power assignment (17). This achievable rate region is shown to significantly exceed the rates promised by traditional MAC capacity region without cooperation, especially when the channel between the two users is relatively good. This also serves as a justification of why decode and forward is a legitimate choice as a cooperation protocol, as far as transmit cooperation is concerned: in practice, the users in a network are more likely to cooperate with other nearby users, and the inter-user links will therefore be much stronger than the direct links to the receiver, otherwise cooperation is not meaningful.

Note that, the rates obeying (18) constitute the ergodic achievable rate region in the Shannon sense, i.e., rates achievable with probability of error as close to zero as desired. In delay sensitive scenarios, where the transmitters do not have channel state information, it is more meaningful to consider outage probability as the performance metric, so that a quality of service requirement can be imposed. Although this chapter mostly deals with ergodic rates, and hence high data rate delay insensitive traffic, there is also considerable amount of work which focuses on outage analysis. Interested reader may find a detailed treatment of the outage behavior of cooperative communication systems in (Laneman, Tse & Wornell, 2004), where a system, in which the users are allowed to cooperate only in half-duplex mode, without channel state information at the transmitters is studied. In that setting, based on the availability of a low rate feedback link and an error detection mechanism at the receiver, practical strategies such as incremental relaying are also proposed.

In the remainder of this chapter, we consider a two user fading cooperative Gaussian MAC with complete channel state information at the transmitters and the receiver, and average power constraints on transmit powers. Note that, this requires only a small quantity of additional feedback, namely the amplitude information on the forward links, over the systems requiring coherent combining (Sendonaris, Erkip & Aazhang, 2003). In this case, the transmitters can adapt their coding strategies as a function of the channel states, by adjusting their transmit powers as in (Goldsmith & Varaiya, 1997; Knopp & Humblet, 1995; Hanly & Tse, 1998). A relaxation of the mutually cooperative fading Gaussian MAC; i.e., a user cooperation system with finite capacity cooperation links, was considered in (Erkip, 2000), where a suboptimal solution to the problem of maximizing the sum rate in the presence of full channel state information was proposed. There it was also noted that the resulting optimization problem is non-concave. In this chapter, we will demonstrate that, with the powers being allocated optimally, the non-concavity of the rate maximization problem is relegated.

In particular, we start by characterizing the optimal power allocation policies which maximize the set of ergodic rates achievable by block Markov superposition coding. Noting that some of the transmit power levels are essentially zero at every channel state, we prove that the seemingly non-concave optimization problem of maximizing the achievable rates can be reduced to a concave problem. The dimensionality of the optimization problem is also reduced, as the number of power variables is considerably less. By this, we show that the usage of the channel state information to adapt the transmission policies, unexpectedly simplifies the block Markov superposition coding strategy proposed in (Willems, van der Meulen & Schalkwijk, 1983) and employed in (Sendonaris, Erkip & Aazhang, 2003) for a fading Gaussian channel. Because of the non-differentiable nature of the objective function, the optimization problem first needs to be replaced by an equivalent differentiable one, in order to find the structure of the optimal power allocation policies via convex optimization techniques. We numerically obtain the optimal power distributions that maximize the achievable rates using sub-gradient methods. For various fading distributions, we compute the resulting regions of achievable rates, and demonstrate that controlling the transmit powers in conjunction with user cooperation provides significant gains over the existing rate regions for cooperative systems. Finally, we show that the more practical window decoding achieves the same sum rate as backwards decoding in cooperative MAC employing block Markov coding. The results on power control and achievable rates for the fading cooperative MAC, given in Section 3 of this chapter are compiled from (Kaya & Ulukus, © 2007 IEEE) and (Kaya, © 2006 IEEE).

Figure 3. Two user fading cooperative MAC

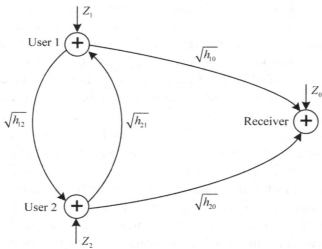

POWER CONTROL AND ACHIEVABLE RATES FOR FADING COOPERATIVE MULTIPLE ACCESS CHANNELS

System Model

We consider a two user fading Gaussian MAC, where both the receiver and the transmitters receive noisy versions of the transmitted messages. Our model, illustrated in Figure 3, is equivalent to the model given in (15), and is repeated here for the sake of completeness:

$$
\begin{aligned}
Y_0 &= \sqrt{h_{10}}\,X_1 + \sqrt{h_{20}}\,X_2 + Z_0, \\
Y_1 &= \sqrt{h_{21}}\,X_2 + Z_1, \\
Y_2 &= \sqrt{h_{12}}\,X_1 + Z_2.
\end{aligned}
\tag{19}
$$

To simplify the notation, we defined the power attenuation coefficients $h_{ij} = \left| K_{ij} \right|^2$, which by a small abuse of terminology will simply be called the fading coefficients, or channel states in the sequel. Since the phase information is available at each node in the network and any phase shift can therefore be corrected, we may as well replace each link gain by the real coefficients $\sqrt{h_{ij}}$ multiplying the transmitted signals. Unlike the assumption in (Sendonaris, Erkip & Aazhang, 2003) about perfect CSI at the receivers only, we further assume that the instantaneous realizations of all channel states are known by both transmitters and the receiver. We assume that the channel variation is slow enough so that the fading parameters can be tracked accurately at the transmitters, yet fast enough to ensure that the long term ergodic properties of the channel are observed within the blocks of transmission (Biglieri, Proakis & Shamai (Shitz), 1998).

Encoding and Decoding Strategy

For the channel model depicted in Figure 3, it is assumed that the transmitters are capable of making decoding decisions based on the signals they receive. Hence, their transmitted codewords are formed not only based on their own information, but also based on the information they have received from each other, in the previous blocks. We further assume that all nodes operate in full-duplex mode. This channel model is a special case of the MAC with generalized feedback (Willems, van der Meulen & Schalkwijk, 1983), whose capacity region is not known to this date. However, there exist cooperative schemes, which provide achievable rates well beyond the capacity region of the traditional MAC with no cooperation. For example, an achievable rate region may be obtained by using what is so called superposition block Markov encoding scheme, followed by backwards decoding, or alternatively, window decoding.

In this chapter, we employ a channel adaptive version of the block Markov encoding scheme, where the powers with which the cooperation information and the fresh information are transmitted are varied as functions of the channel state. As far as decoding is concerned, we will first stick with backwards decoding, where the receiver waits to receive all B blocks of codewords before starting to decode. Note that, an alternative is to use window decoding, which starts the decoding process only after a delay of one block. However, although window decoding has recently been shown to achieve the same rates as backwards decoding under several relay channel models (Xie & Kumar, 2004; Sankaranarayanan, Kramer & Mandayam, 2004), it is in general inferior to backwards decoding in terms of the achievable rate region for a MAC with generalized feedback, when block Markov encoding is used (Laneman & Kramer, 2004). Nevertheless, the properties of the sum rate optimum power allocation policy for a system which employs window decoding will also be discussed towards the end of this chapter, and it will be shown that the achievable sum rates for systems using window and backwards decoding are in fact the same.

The codebook generation, the power controlled encoding strategy, and the backwards decoding technique are summarized below.

Codebook Generation

The codebook is generated according to the following order, and revealed to both transmitters and the receiver before transmission:

1. Independently generate $2^{n(R_{12}+R_{21})}$ length n sequences $\mathbf{U}$, with entries from independent and identically distributed (i.i.d) unit Gaussian distributions. Assign each realization of these sequences to a distinct set of messages $\left\{w'_{12}, w'_{21}\right\} \in \left[1, \cdots, 2^{nR_{12}}\right] \times \left[1, \cdots, 2^{nR_{21}}\right]$, thereby forming $\mathbf{u}(w'_{12}, w'_{21})$.

2. For every $\mathbf{u}(w'_{12}, w'_{21})$, independently generate $2^{nR_{12}}$ sequences $\mathbf{X}_{12}$ from an i.i.d unit Gaussian distribution and label each realization as $\mathbf{x}_{12}\left(w_{12}, \mathbf{u}(w'_{12}, w'_{21})\right)$, for $w_{12} \in \left[1, \cdots, 2^{nR_{12}}\right]$.

3. For every $\mathbf{u}(w'_{12}, w'_{21})$, independently generate $2^{nR_{21}}$ sequences $\mathbf{X}_{21}$ from an i.i.d unit Gaussian distribution and label each realization as $\mathbf{x}_{21}\left(w_{21}, \mathbf{u}(w'_{12}, w'_{21})\right)$, for $w_{21} \in \left[1, \cdots, 2^{nR_{21}}\right]$.

4. For every pair $\mathbf{u}(w'_{12}, w'_{21})$ and $\mathbf{x}_{12}\left(w_{12}, \mathbf{u}(w'_{12}, w'_{21})\right)$, independently generate $2^{nR_{10}}$ sequences $\mathbf{X}_{10}$ from an i.i.d unit Gaussian distribution, and label each realization as $\mathbf{x}_{10}\left(w_{10}, w_{12}, \mathbf{u}(w'_{12}, w'_{21})\right)$, for $w_{10} \in \left[1, \cdots, 2^{nR_{10}}\right]$.

5.   For every pair $\mathbf{u}(w'_{12}, w'_{21})$ and $\mathbf{x}_{21}\left(w_{21}, \mathbf{u}(w'_{12}, w'_{21})\right)$, independently generate $2^{nR_{20}}$ sequences $\mathbf{X}_{20}$ from an i.i.d unit Gaussian distribution, and label each realization as $\mathbf{x}_{20}\left(w_{20}, w_{21}, \mathbf{u}^{n}(w_{12}', w_{21}')\right)$, for $w_{20} \in \left[1, \cdots, 2^{nR_{20}}\right]$.

Note that, the codebook generation is identical to those in (Willems, van der Meulen & Schalkwijk, 1983; Sendonaris, Erkip & Aazhang, 2003). However, the encoding is done differently: the generated codewords are scaled on a symbol by symbol basis by variable power levels, thus adjusting the variance of each transmitted component.

Encoding

The transmission of $B-1$ consecutive messages for each user is completed in B blocks. The messages are denoted by $w_1[b] = (w_{10}[b], w_{12}[b])$ and $w_2[b] = (w_{20}[b], w_{21}[b])$, $b = 1, \cdots, B-1$, respectively. In the presence of channel state information, the encoding in each block b is performed by

$$
\begin{aligned}
x_i^{(k)} = {} & \sqrt{p_{i0}(\mathbf{h}^{(k)})} x_{i0}^{(k)}\left(w_{i0}[b], w_{ij}[b], \mathbf{u}\left(w_{12}[b-1], w_{21}[b-1]\right)\right) \\
& + \sqrt{p_{ij}(\mathbf{h}^{(k)})} x_{ij}^{(k)}\left(w_{ij}[b], \mathbf{u}\left(w_{12}[b-1], w_{21}[b-1]\right)\right) + \sqrt{p_{u_i}(\mathbf{h}^{(k)})} u^{(k)}\left(w_{12}[b-1], w_{21}[b-1]\right),
\end{aligned}
$$

$$(20)$$

for $i, j \in \{1, 2\}$, $i \neq j$, and $k = 1, \cdots, n$, where $x_i^{(k)}$ denotes the kth entry of vector $\mathbf{x}_i$, and $\mathbf{h}^{(k)}$ is the channel realization during the transmission of the kth symbol. Here x_{i0} carries the fresh information intended for the receiver, x_{ij} carries the information intended for transmitter j for cooperation in the next block, and u is the common information sent by both transmitters for the resolution of the remaining uncertainty from the previous block. Observe that, the transmitters allocate part of their available powers, i.e., $p_{ij}(\mathbf{h}^{(k)})$, to establish common information in each block, so that they can obtain coherent combining gain in the next block through the powers $p_{u_i}(\mathbf{h}^{(k)})$, assigned to the common cooperative codeword. Assuming that the fading process is stationary and ergodic, we can replace the samples $\mathbf{h}^{(k)}$ of the fading process with a random variable $\mathbf{h}$ that obeys the stationary distribution, and drop the time dependence. This way, time averages are replaced by statistical averages over the distribution of $\mathbf{h}$. Therefore, the transmit power levels for the codeword components, associated with each joint fading state $\mathbf{h}$ can simply be denoted by $p_{i0}(\mathbf{h})$, $p_{ij}(\mathbf{h})$ and $p_{U_i}(\mathbf{h})$, which are required to satisfy the average power constraints,

$$
E[p_{i0}(\mathbf{h}) + p_{ij}(\mathbf{h}) + p_{U_i}(\mathbf{h})] = E[p_i(\mathbf{h})] \leq \overline{p}_i, \quad i = 1, 2.
$$

$$(21)$$

Note that, this channel adaptive encoding argument roots from the capacity achieving encoding policy for the traditional multiple access channels (Hanly & Tse, 1998) discussed previously. The optimum encoding for the MAC was simply achieved by generating a fixed codebook, and adjusting the transmitted symbol by instantaneous amplitude values which are functions of the channel state. Finally, since the transmitted messages are unknown to the corresponding receivers, we need to replace the realizations of the transmitted codewords by their random equivalents and model each sample from the transmitted codeword as

$$X_i = \sqrt{p_{i0}(\mathbf{h})}X_{i0} + \sqrt{p_{ij}(\mathbf{h})}X_{ij} + \sqrt{p_{U_i}(\mathbf{h})}U, \tag{22}$$

for $i,j \in \{1,2\}$, $i \neq j$. The received signals at each terminal can now be obtained by plugging (22) into (19).

Decoding

Following block Markov encoding, decoding may be done in several ways. Perhaps the most widely studied decoding policy in the literature for multiple access channels with generalized feedback is backwards decoding, which waits for all blocks to be received, and then decodes starting from the cooperation information in the last block. Since no fresh information is transmitted in the last block B, that is, $w_1(B)$ and $w_2(B)$ are just known messages, the receiver can decode the cooperation signal $\mathbf{u}(w_{12}(B-1), w_{21}(B-1))$ without any interference, and then use the signal received in the previous block to decode $w_{10}(B-1)$ and $w_{20}(B-1)$. Meanwhile, the cooperation information $\mathbf{u}(w_{12}(B-2), w_{12}(B-2))$ is also decoded, thereby aiding the decoding of the codewords in block $B-2$, and so on. The apparent disadvantage of this scheme is that it requires the receiver to wait for all B blocks to be received before starting the decoding process, causing a significant delay. We will discuss an alternative to this encoding scheme, namely window decoding, later in this chapter.

Region of Achievable Rates with Backwards Decoding

In this section, we characterize the region of achievable rates for the fading Gaussian MAC with cooperating encoders, for which we employ the power controlled encoding scheme (22), followed by backwards decoding. The inequalities that need to be satisfied by the rates for reliable communication can be obtained by appropriately modifying the results in (Willems, van der Meulen & Schalkwijk, 1983; Sendonaris, Erkip & Aazhang, 2003; Laneman & Kramer 2004). The achievable rate region is given by the convex hull of R_1 and R_2 that satisfy $R_1 = R_{10} + R_{12}$ and $R_2 = R_{20} + R_{21}$, where

$$R_{12} \leq I(x_{12}; y_2 \mid x_2, u, \mathbf{h})$$
$$R_{21} \leq I(x_{21}; y_1 \mid x_1, u, \mathbf{h})$$
$$R_{10} \leq I(x_1; y_0 \mid x_2, x_{12}, x_{21}, u, \mathbf{h})$$
$$R_{20} \leq I(x_2; y_0 \mid x_1, x_{12}, x_{21}, u, \mathbf{h})$$
$$R_{10} + R_{20} \leq I(x_1, x_2; y_0 \mid x_{12}, x_{21}, u, \mathbf{h})$$
$$R_{10} + R_{20} + R_{12} + R_{21} \leq I(x_1, x_2; y_0, \mathbf{h}). \tag{23}$$

Using standard properties of mutual information, it can be shown that the achievable rate region, which we denote by $\mathcal{R}$, can be alternatively expressed in terms of the powers as the convex hull of all rate pairs satisfying (24)-(26) below, where the convex hull is taken over all power allocation policies that satisfy

(21). Note that, (24)-(26) are a more compact version of (18) for the power controlled scenario.

$$R_1 < E\left[\log\left(1 + \frac{h_{12}p_{12}(\mathbf{h})}{h_{12}p_{10}(\mathbf{h}) + \sigma_2^2}\right) + \log\left(1 + \frac{h_{10}p_{10}(\mathbf{h})}{\sigma_0^2}\right)\right] \tag{24}$$

$$R_2 < E\left[\log\left(1 + \frac{h_{21}p_{21}(\mathbf{h})}{h_{21}p_{20}(\mathbf{h}) + \sigma_1^2}\right) + \log\left(1 + \frac{h_{20}p_{20}(\mathbf{h})}{\sigma_0^2}\right)\right] \tag{25}$$

$$R_1 + R_2 < \min\left\{E\left[\log\left(1 + \frac{h_{10}p_1(\mathbf{h}) + h_{20}p_2(\mathbf{h}) + 2\sqrt{h_{10}h_{20}p_{U_1}(\mathbf{h})p_{U_2}(\mathbf{h})}}{\sigma_0^2}\right)\right],\right.$$
$$E\left[\log\left(1 + \frac{h_{12}p_{12}(\mathbf{h})}{h_{12}p_{10}(\mathbf{h}) + \sigma_2^2}\right) + \log\left(1 + \frac{h_{21}p_{21}(\mathbf{h})}{h_{21}p_{20}(\mathbf{h}) + \sigma_1^2}\right)\right]$$
$$\left. + E\left[\log\left(1 + \frac{h_{10}p_{10}(\mathbf{h}) + h_{20}p_{20}(\mathbf{h})}{\sigma_0^2}\right)\right]\right\}. \tag{26}$$

For a given valid power allocation $\mathbf{p}(\mathbf{h})$, let us denote the rate region in (24)-(26) by $\mathcal{R}\,\mathbf{p}(\mathbf{h})$. Note that, although this region is usually a pentagon, the sum rate constraint in (26) may dominate the individual rate constraints, and unlike the traditional MAC, this might yield a triangular shape (this happens if the user-destination links are consistently very bad), hence the need for the minimum operation.

The overall achievable rate region $\mathcal{R}$ can then be represented as the convex hull of the union of all such regions, i.e.,

$$\mathcal{R} = conv\left(\bigcup_{\{E[p_i(\mathbf{h})] \le \bar{p}_i, i=1,2\}} \mathcal{R}(\mathbf{p}(\mathbf{h}))\right). \tag{27}$$

In the next two sections, we describe methods which can be used to obtain the optimal power allocation policies that maximize the rate tuples on the rate region boundary, and characterize the properties of the optimal policy.

STRUCTURE OF THE OPTIMAL POLICIES, AND THE RATE REGION

Sum Rate Optimal Policies

Let us first consider the problem of optimizing the sum rate of the system, the solution to which will be instructive for the optimization of arbitrary points on the rate region boundary.

The sum rate (26) is clearly not a concave function of the vector of variables $\mathbf{p}(\mathbf{h}) = [p_{10}(\mathbf{h})\ p_{12}(\mathbf{h})\ p_{U_1}(\mathbf{h})$ $p_{20}(\mathbf{h})\ p_{21}(\mathbf{h})\ p_{U_2}(\mathbf{h})]$, as we have some of the power variables appearing in the denominators. However, the following theorem states that, for the sum rate to be maximized, at least two of the four components of $[p_{10}(\mathbf{h})\ p_{12}(\mathbf{h})\ p_{20}(\mathbf{h})\ p_{21}(\mathbf{h})]$ should be equal to zero for any given channel state $\mathbf{h}$. This result not only yields a concave optimization problem, but it also reduces the size of the variable space.

Theorem 1: Let the effective channel gains normalized by the noise powers be defined as $s_{ij} = h_{ij} / \sigma_j^2$. Then, for the power control policy $\mathbf{p}^*(\mathbf{h})$ that maximizes (26), we need

1) $p_{10}^*(\mathbf{h}) = p_{20}^*(\mathbf{h}) = 0$, if $s_{12} > s_{10}$ and $s_{21} > s_{20}$,
2) $p_{10}^*(\mathbf{h}) = p_{21}^*(\mathbf{h}) = 0$, if $s_{12} > s_{10}$ and $s_{21} \le s_{20}$,
3) $p_{12}^*(\mathbf{h}) = p_{20}^*(\mathbf{h}) = 0$, if $s_{12} \le s_{10}$ and $s_{21} > s_{20}$,

$$\left.\begin{array}{c} p_{12}^*(\mathbf{h}) = p_{21}^*(\mathbf{h}) = 0 \\ \text{or} \\ p_{10}^*(\mathbf{h}) = p_{21}^*(\mathbf{h}) = 0 \\ \text{or} \\ p_{12}^*(\mathbf{h}) = p_{20}^*(\mathbf{h}) = 0 \end{array}\right\} \text{if } s_{12} \le s_{10} \text{ and } s_{21} \le s_{20}$$

4) $p_{12}^*(\mathbf{h}) = p_{20}^*(\mathbf{h}) = 0$

Proof: In order to simplify the notation, the dependence of the powers on the channel states will be dropped, whenever it is obvious from the context. Let $p_i = p_{i0} + p_{ij} + p_{U_i}$ be the total power allocated to a given channel state. Let us define

$$A = 1 + s_{10}p_1 + s_{20}p_2 + 2\sqrt{s_{10}s_{20}p_{U_1}p_{U_2}}, \tag{28}$$

$$B = \frac{1 + s_{10}p_{10} + s_{20}p_{20}}{(1 + s_{12}p_{10})(1 + s_{21}p_{20})}, \tag{29}$$

$$C = \left(1 + s_{12}(p_{10} + p_{12})\right)\left(1 + s_{21}(p_{20} + p_{21})\right). \tag{30}$$

Then, the sum rate in (26) can equivalently be expressed as,

$$R_{sum} = \min\left\{E\left[\log(A)\right], E\left[\log(BC)\right]\right\}. \tag{31}$$

Now, for a given channel state, let us fix arbitrary total power levels, say p_i, and cooperative power levels, $p_{U_i} \le p_i$, for each user. For each such allocation, the variables A and C appearing in the sum rate expression are fixed, regardless of how we allocate the remaining available powers $p_i - p_{U_i}$ among p_{i0} and p_{ij}. Moreover, the way $p_i - p_{U_i}$ is divided among p_{i0} and p_{ij} does not alter the total power consumption at the given state. Hence, we may limit our attention to the maximization,

$$\begin{aligned} \max_{\{p_{10}, p_{20}\}} \quad & B\left(p_{10}, p_{20}\right) \\ s.t. \quad & p_{10} + p_{12} = p_1 - p_{U_1}, \\ & p_{20} + p_{21} = p_2 - p_{U_2}. \end{aligned} \tag{32}$$

The partial derivatives of B with respect to p_{10} and p_{20} can be obtained as

$$\frac{\partial B}{\partial p_{10}} = \frac{s_{10} - s_{12}(1 + s_{20}p_{20})}{(1 + s_{12}p_{10})^2(1 + s_{21}p_{20})}, \tag{33}$$

$$\frac{\partial B}{\partial p_{20}} = \frac{s_{20} - s_{21}(1 + s_{10}p_{10})}{(1 + s_{21}p_{20})^2(1 + s_{12}p_{10})}. \tag{34}$$

Here, a crucial observation is that the sign of the partial derivatives vary with the relative qualities of the effective channel gains. Therefore, we should consider the following four cases:

1) $s_{12} > s_{10}$, $s_{21} > s_{20}$. Then, $\frac{\partial B}{\partial p_{10}} < 0$ and $\frac{\partial B}{\partial p_{20}} < 0$, i.e., $B(p_{10}, p_{20})$ is monotonically decreasing in both p_{10} and p_{20}, therefore the sum rate is maximized at $p_{10} = p_{20} = 0$.

2) $s_{12} > s_{10}$, $s_{21} \leq s_{20}$. Then, $\frac{\partial B}{\partial p_{10}} < 0$, and the function is maximized at $p_{10} = 0$ for any p_{20}. But this gives $\left.\frac{\partial B}{\partial p_{20}}\right|_{p_{10}=0} > 0$, meaning p_{20} should take its maximum possible value, i.e., $p_{21} = 0$.

3) $s_{12} \leq s_{10}$, $s_{21} > s_{20}$. This is the same as case 2 with roles of users 1 and 2 reversed.

4) $s_{12} \leq s_{10}$, $s_{21} \leq s_{20}$. In this case, both of the partial derivatives of B can be made equal to zero within the constraint set, yielding a critical point. However, second order derivative tests show that the determinant of the Hessian matrix is always negative, and therefore this solution corresponds to a saddle point, meaning that B is again maximized at one of the boundaries, $(p_{10}, p_{20}) \in \left\{0, p_1 - p_{U_1}\right\} \times \left\{0, p_2 - p_{U_2}\right\}$, i.e., two of the components of the optimal power vector are again guaranteed to be zero. Inspection of the gradient on these boundary points yields one of the three corner points $\left\{(p_1 - p_{U_1}, 0), (0, p_2 - p_{U_2}), (p_1 - p_{U_1}, p_2 - p_{U_2})\right\}$ as candidates for optimal solution in case 4. □

At this point, a small discussion is in order: Theorem 1 guarantees that, for each case, two of the components of the power vector are zero. Except for case 4, it also states precisely which ones will be zero. In case 4, which two of the powers will become zero actually depends on the p_i and p_{U_i} that we fix. Hence, in case 4, the solution is not completely specified independently of p_i and p_{U_i}. However, one should note that the settings of practical interest are those where the channels between the cooperating users are on average much better than their direct links; since it is in these settings when cooperative diversity yields high capacity gains (Sendonaris, Erkip & Aazhang, 2003). In such scenarios, the probability of both users' direct link gains exceeding their corresponding cooperation link gains (case 4) is very low. Therefore, in practice the average achievable rate performance should be virtually invariant to the choice among the three possible operating points, and we can safely set the power allocation policy to be either one of them, and carry on with our optimization problem, as a function of the remaining variables. Although this argument is likely to cause some sub-optimality in our scheme, numerical examples still demonstrate a significant gain in achievable rates.

The significance of Theorem 1 is two-fold. Firstly, it greatly simplifies the well known block Markov encoding, in a very intuitive way: given a channel state, if the direct links of both users are inferior to their cooperation links, the users cease direct transmission of fresh information towards the receiver, and they simply use each other as relays. If either one of the users' direct channel is stronger than its cooperation channel, while the other user is in the opposite situation, then the user with the strong direct channel chooses to transmit directly to the receiver, while the weaker direct channel user still chooses to relay its information over its partner. Note that, in all cases the cooperative signal **u** is kept out of this discussion, and may be sent regardless of the channel ordering. Also note that, although it may seem useless to transmit the cooperative message **u**, when the common information is not being established in a given transmission block (i.e., p_{ij} is equal to 0), since we consider ergodic fading, there will always be common information established in the previous blocks, hence this situation is perfectly valid. Second important implication of Theorem 1 is that it simplifies the constraint set on the variables, and more importantly, as shown in Corollary 1 below, this makes the sum rate a concave function over the reduced set of constraints and variables. Therefore, the problem of obtaining the optimal power allocation policy becomes more tractable.

Corollary 1: The sum rate R_{sum} given by (26) is a concave function of $\mathbf{p}(\mathbf{h})$, over the reduced constraint set described by Theorem 1.

Proof: The proof of this result follows by direct substitution of the zero power components into the sum rate expression in (31) at each channel state. Note that in each of the four cases, the second function in the minimization, i.e., $\log(BC)$ takes either the form $\log(1+a) + \log(1+b)$, or $\log(1+a+b)$, both of which are clearly jointly concave in a and b. Also, $\log(A)$ is clearly a concave function of $\mathbf{p}(\mathbf{h})$ since it is a composition of a concave function with the concave and increasing logarithm. The desired result is obtained by noting that the minimum of two concave functions is concave. □

Properties of the Achievable Rate Region

So far, we have focused our attention only on the sum rate, discussed its structure, and some key properties of the optimal power allocation that maximizes it. We now turn back to the problem of maximizing an arbitrary rate point on the rate region boundary.

The following result establishes the applicability of the power control policies described by Theorem 1 to arbitrary rates.

Corollary 2: The power allocation strategy in Theorem 1 enlarges the entire achievable rate region in all directions, compared to any other valid power allocation policy, provided $p_{12} = 0$, $p_{21} = 0$ is chosen, and is sum-rate optimal, in case 4.

Proof: First consider maximizing the bound on R_1, given by (24). It is easy to check that, the right hand side of the constraint (24) is a monotone increasing function of p_{10}, when $s_{10} > s_{12}$, and it is a monotone decreasing function of the same variable, if $s_{12} > s_{10}$. Therefore, the power control policy that maximizes the constraint in (24) should allocate all the available power $p_1 - p_{U_1}$ to p_{10} in the former case, and to p_{12} in the latter. Due to symmetry, a similar argument applies to the maximization of R_2 given in (25), yielding a transmit-over-the-better-link type of policy. This shows that, the properties that need to be satisfied by the power control policies that maximize the individual rates and the sum rate are identical, and are as given in Theorem 1, as long as we choose the operating point in case 4 to be $p_{12} = 0$, $p_{21} = 0$, which proves the desired result. □

In case the sum-rate optimal policy for case 4 is not $p_{12} = 0$, $p_{21} = 0$, setting these powers to zero will cause a slight loss in the sum rate, although the region is still enlarged in the directions of individual rate constraints. However, this low probability event can simply be ignored.

Corollary 2 helps us reduce the dimensionality of our variable space considerably, as the set of valid power allocation policies which are candidates for optimality has now shrunk in size. In particular, if we define Γ as the set of valid power control policies $\mathbf{p}(\mathbf{h})$ which satisfy Theorem 1, with $p_{12} = 0$, $p_{21} = 0$ being chosen in case 4, i.e.,

$$\Gamma = \{\mathbf{p}(\mathbf{h}) : \{E[p_i(\mathbf{h})] \le \overline{p}_i\} \cap \{p_{ij} = 0 \text{ if } s_{ik} > s_{ij},\} \quad i = 1, 2, j \ne k\}, \tag{35}$$

then, as a direct consequence of Corollary 2, the union required to obtain the achievable rate region $\mathcal{R}$ given by (27) can be taken just over Γ, thus replacing the equivalent set of rates $\mathcal{R}$ by

$$\mathcal{R}_\Gamma = conv\left(\bigcup_{\mathbf{p}(\mathbf{h}) \in \Gamma} \mathcal{R}(\mathbf{p}(\mathbf{h}))\right), \tag{36}$$

barring the possible minor sub-optimality that may be caused by fixing $p_{12} = 0$, $p_{21} = 0$ in case 4. Therefore, from now on it is sufficient to focus on the achievable rates given in (36).

The presence of the convex hull operation in the definition of the achievable rate region poses a problem, if one wants to obtain the optimal power allocation policy that corresponds to an arbitrary rate point on the boundary. The reason is that we need to determine whether the point comes directly from a valid power control policy or it requires time sharing between two power control policies. Therefore, in obtaining the power control policies that achieve arbitrary points on the rate region boundary, the following convexity result is very useful.

Lemma 1. The achievable rate region $\mathcal{R}_\Gamma$ defined in (36) is equivalent to

$$\mathcal{R}_\Gamma = \bigcup_{\mathbf{p}(\mathbf{h}) \in \Gamma} \mathcal{R}(\mathbf{p}(\mathbf{h})). \tag{37}$$

Proof: The proof of this lemma closely follows the proof of a similar result by Hanly and Tse (1998), for the capacity region of traditional MAC. First, consider any two valid power allocation policies $\mathbf{p}^1(\mathbf{h})$, $\mathbf{p}^2(\mathbf{h})$ from the set Γ; and their respective sets of achievable rates $\mathcal{R}(\mathbf{p}^1(\mathbf{h})), \mathcal{R}(\mathbf{p}^2(\mathbf{h}))$. Let us define a new power control policy $\alpha\mathbf{p}^1(\mathbf{h}) + (1-\alpha)\mathbf{p}^2(\mathbf{h})$, for $0 \le \alpha \le 1$, which also clearly satisfies the average power constraints and the conditions of Theorem 1. Since the power allocation policies satisfy the conditions in Theorem 1 by assumption, the upper bounds in (24)-(26) are concave. Then, using (24)-(26), we get

$$\alpha\mathcal{R}(\mathbf{p}^1(\mathbf{h})) + (1-\alpha)\mathcal{R}(\mathbf{p}^2(\mathbf{h})) \subseteq \mathcal{R}(\alpha\mathbf{p}^1(\mathbf{h}) + (1-\alpha)(\mathbf{p}^2(\mathbf{h}))), \tag{38}$$

which states that any linear combination of two achievable rate tuples can be achieved by another valid power allocation policy, and is therefore in the achievable rate region, thereby proving the convexity. $\square$

This lemma simply states that the union of achievable rate regions, over all valid candidate power allocation policies for optimality, produces a convex region, and the convex hull operation in (36) is no longer required and can be removed. Therefore, it should be possible to obtain the optimal power allocation policies that achieve any given point on the rate region boundary, through a maximization of weighted sum of rates, corresponding to a tangent line to $\mathcal{R}$, as in Figure 1. However, for this approach to work, we need one final modification to the convexity result: there should be no linear region on the rate region boundary, or else some combinations of weights will not correspond to optimum policies. This is proved in the following lemma.

Lemma 2. The rate region $\mathcal{R}$ is strictly convex, i.e., it has a curved boundary, except for the portions parallel to R_1 and R_2 axes.

Proof: Strict concavity of the logarithm dictates that the subset relation in (38) is satisfied with strict inequality, provided $\mathbf{p}^{(1)}(\mathbf{h}) \neq \mathbf{p}^{(2)}(\mathbf{h})$ over a set of channel states with nonzero probability. Therefore, for any two corner points $\{R_1^{(1)}, R_2^{(1)}\}$, $\{R_1^{(2)}, R_2^{(2)}\}$ on the boundary of $\mathcal{R}$, resulting from distinct power control policies, there exists a rate tuple strictly outside the line connecting these two points. Hence, if there is to be a linear region on $\mathcal{R}$, it should result from a single power control policy, which in turn can only be the sum rate maximizing power control policy. On the other hand, it can be shown that the sum rate maximizing policy always yields a rectangular rate region, and therefore has only a single rate point on the boundary of $\mathcal{R}$. This is because, the sum rate is maximized when the arguments of the minimum operation in (26) are equal, i.e., when the sum of the right hand sides of (24) and (25) is identical to the right hand side of (26). Hence, it is easy to see that each of the points on the boundary of the rate region $\mathcal{R}$ is due to a distinct power allocation policy, and the boundary is strictly convex. $\square$

As a result of Lemmas 1 and 2, it can now be claimed that the achievable rate region boundary can be obtained by maximizing a weighted sum of rates, say $R_{\mu} = \mu_1 R_1 + \mu_2 R_2$, for varying weight (priority) values. Note that, the function R_{μ} carries the same concavity properties of the sum rate, since, for $\mu_i > \mu_j$, the weighted sum of rates can be written as a linear combination of the concave functions R_{sum} and R_i, i.e., $R_{\mu} = \mu_j R_{sum} + (\mu_i - \mu_j)R_i$. In the next section, we deal with the problem of maximizing R_{μ} to obtain the optimum power control policies that achieve the points on the achievable rate region boundary.

Rate Maximization using Convex Optimization

The weighted sum rate maximization approach proposed in Section 3.4 can be carried out by expressing the rate expressions in terms of the powers, while taking into account each of the cases in Theorem 1, and setting the corresponding power components to zero. Let us assume, without loss of generality, that $\mu_1 \geq \mu_2$, and state the optimization problem explicitly:

$$\max_{\mathbf{p}(\mathbf{h})} (\mu_1 - \mu_2)\left\{E_{1,2}\left[\log(1 + p_{12}(\mathbf{h})s_{12})\right]\right.$$

$$\left. + E_{3,4}\left[\log(1 + p_{10}(\mathbf{h})s_{10})\right]\right\} + \mu_2 \min\left\{E[\log(A)],\right.$$

$$+E_1\left[\log(1+p_{12}(\mathbf{h})s_{12})+\log(1+p_{21}(\mathbf{h})s_{21})\right]$$

$$+E_2\left[\log(1+p_{12}(\mathbf{h})s_{12})+\log(1+p_{20}(\mathbf{h})s_{20})\right]$$

$$+E_3\left[\log(1+p_{10}(\mathbf{h})s_{10})+\log(1+p_{21}(\mathbf{h})s_{21})\right]$$

$$+E_4\left[\log(1+p_{10}(\mathbf{h})s_{10}+p_{20}(\mathbf{h})s_{20})\right]\Big\}$$

$$s.t. \quad E_{3,4}\left[p_{10}(\mathbf{h})\right]+E_{1,2}\left[p_{12}(\mathbf{h})\right]+E\left[p_{U_1}(\mathbf{h})\right]\le \bar{p}_1,$$

$$E_{2,4}\left[p_{20}(\mathbf{h})\right]+E_{1,3}\left[p_{21}(\mathbf{h})\right]+E\left[p_{U_2}(\mathbf{h})\right]\le \bar{p}_2,$$

$$p_{10}(\mathbf{h}),p_{12}(\mathbf{h}),p_{U_1}(\mathbf{h}),p_{20}(\mathbf{h}),p_{21}(\mathbf{h}),p_{U_2}(\mathbf{h})\ge 0,\quad \forall\mathbf{h}. \tag{39}$$

Here, E_S, $S\subset\{1,2,3,4\}$ denotes the expectation, taken over the set of channel states satisfying case(s) S from Theorem 1, and A is given by (28). The optimization problem given in equation (39) has a special and desirable form: it is a maximization of a concave objective function over a convex constraint set, for which any local optimum is a global optimum. However, although it is known that a concave function is differentiable almost everywhere, the maximum value of R_{V_4} is attained along the discontinuity of its gradient, when the two arguments of the minimum operation in (39) are equal. In order to prove this, it is sufficient to observe that when the arguments of the minimum operation are not equal, it is possible to increase the smaller argument at the expense of the larger, by reallocating the total transmit power at any channel state $\mathbf{h}$, among the components $p_{U_i}(\mathbf{h})$ and $p_{ij}(\mathbf{h})$. Such reallocation of powers simply transfers some power from one component to the other; but it keeps their sum constant, so that the power constraint is still satisfied.

The problem of discontinuity of the gradient, which complicates the process of carrying out the maximization, can be resolved by introducing a new auxiliary variable r, replacing the minimum operation with two inequality constraints, and considering the equivalent problem

$$\max_{\mathbf{p}(\mathbf{h})} (\mu_1-\mu_2)\Big\{E_{1,2}\left[\log(1+p_{12}(\mathbf{h})s_{12})\right]$$

$$+E_{3,4}\left[\log(1+p_{10}(\mathbf{h})s_{10})\right]\Big\}+\mu_2 r$$

$$s.t. \quad r\le E[\log(A)],$$

$$r\le E_1\left[\log(1+p_{12}(\mathbf{h})s_{12})+\log(1+p_{21}(\mathbf{h})s_{21})\right]$$

$$+E_2\left[\log(1+p_{12}(\mathbf{h})s_{12})+\log(1+p_{20}(\mathbf{h})s_{20})\right]$$

$$+E_3\left[\log(1 + p_{10}(\mathbf{h})s_{10}) + \log(1 + p_{21}(\mathbf{h})s_{21})\right]$$

$$+E_4\left[\log(1 + p_{10}(\mathbf{h})s_{10} + p_{20}(\mathbf{h})s_{20})\right],$$

$$E_{3,4}\left[p_{10}(\mathbf{h})\right] + E_{1,2}\left[p_{12}(\mathbf{h})\right] + E\left[p_{U_1}(\mathbf{h})\right] \leq \overline{p}_1,$$

$$E_{2,4}\left[p_{20}(\mathbf{h})\right] + E_{1,3}\left[p_{21}(\mathbf{h})\right] + E\left[p_{U_2}(\mathbf{h})\right] \leq \overline{p}_2,$$

$$p_{10}(\mathbf{h}), p_{12}(\mathbf{h}), p_{U_1}(\mathbf{h}), p_{20}(\mathbf{h}), p_{21}(\mathbf{h}), p_{U_2}(\mathbf{h}) \geq 0, \quad \forall \mathbf{h}. \tag{40}$$

Clearly, the objective function of this new problem is still concave in $\mathbf{p}(\mathbf{h})$. Moreover, it is easy to see that, all of the constraints can be written in the form $f_i(\mathbf{p}) \leq 0$, where $f_i(\mathbf{p})$ are convex functions. Therefore, Slater's conditions for constraint qualification hold, and the optimal solution to the constrained maximization problem given in (40) can be obtained by using Karush-Kuhn-Tucker (KKT) conditions (Boyd & Vandenberghe, 2004).

In what follows, we illustrate the use of KKT conditions and provide some insight about the structure of the optimum solution. Without loss of generality, we focus our attention to the special case of maximizing the sum rate, i.e., $\mu_1 = \mu_2$. Note that, our treatment may as well be carried out by deriving the KKT conditions for the general problem (40) with arbitrary priorities, but the derivations in that case are rather lengthy, and the resulting expressions are less intuitive.

Analytical Derivation of Sum-Rate Optimal Power Allocation

In practice, the gains from cooperative communications are more pronounced when the cooperating transmitters are closely located with relatively small number of obstructions and scatterers along the signal paths between them, as opposed to their paths to the intended receiver, which are in general much weaker. This physical scenario corresponds to case 1 in Theorem 1, for which $s_{12} > s_{10}$ and $s_{21} > s_{20}$. Therefore, in our analysis we shall assume that the fading realizations are such that they always satisfy $s_{12} > s_{10}$ and $s_{21} > s_{20}$, an assumption which will also simplify the derivations considerably. Since we are interested in maximizing the sum rate, according to Theorem 1, we should pick $p_{10}(\mathbf{h}) = p_{20}(\mathbf{h}) = 0$ for all $\mathbf{h}$. Then, letting $\mu_1 = \mu_2$ in (40), the sum rate maximization problem reduces to

$$\max_{\mathbf{p}(\mathbf{h})} \quad r \tag{41}$$

$$s.t. \quad r \leq E\left[\log\left(1 + s_{12}p_{12}(\mathbf{h})\right) + \log\left(1 + s_{21}p_{21}(\mathbf{h})\right)\right], \tag{42}$$

$$r \leq E\left[\log(D)\right], \tag{43}$$

$$E\left[p_{12}(\mathbf{h}) + p_{U_1}(\mathbf{h})\right] \leq \overline{p}_1, \tag{44}$$

$$E\left[p_{21}(\mathbf{h}) + p_{U_2}(\mathbf{h})\right] \leq \overline{p}_2, \tag{45}$$

$$p_{12}(\mathbf{h}), p_{U_1}(\mathbf{h}), p_{21}(\mathbf{h}), p_{U_2}(\mathbf{h}) \geq 0, \quad \forall \mathbf{h}, \tag{46}$$

where

$$D = 1 + s_{10}\left(p_{12}(\mathbf{h}) + p_{U_1}(\mathbf{h})\right) + s_{20}\left(p_{21}(\mathbf{h}) + p_{U_2}(\mathbf{h})\right) + 2\sqrt{s_{10}s_{20}p_{U_1}(\mathbf{h})p_{U_2}(\mathbf{h})}. \tag{47}$$

The KKT conditions, which are necessary and sufficient for optimality, are obtained by assigning the Lagrange multipliers $\gamma_1, \gamma_2 > 0$ to the inequality constraints (42), (43), $\lambda_1, \lambda_2 > 0$ to the power constraints (44), (45), and $\xi_i(\mathbf{h}) \geq 0$, $i = 1, \cdots, 4$ to the non-negativity constraints (46); and noting that the power constraints need to be satisfied by equality. Clearly, for any given $\mathbf{h}$, the optimal components $p_{U_1}(\mathbf{h})$ and $p_{U_2}(\mathbf{h})$ should be either both positive or both zero, or else the available power will be wasted. Therefore, it is sufficient to analyze the following two cases separately:

1. First let $p_{U_1}(\mathbf{h}), p_{U_2}(\mathbf{h}) > 0$. Then, the KKT conditions reduce to

$$\gamma_1 \frac{s_{12}}{1 + s_{12}p_{12}(\mathbf{h})} + \gamma_2 \frac{s_{10}}{D} \leq \lambda_1, \tag{48}$$

$$\gamma_1 \frac{s_{21}}{1 + s_{21}p_{21}(\mathbf{h})} + \gamma_2 \frac{s_{20}}{D} \leq \lambda_2, \tag{49}$$

$$\gamma_2 \frac{\sqrt{s_{10}s_{20}p_{U_2}(\mathbf{h})} + s_{10}\sqrt{p_{U_1}(\mathbf{h})}}{D\sqrt{p_{U_1}(\mathbf{h})}} = \lambda_1, \tag{50}$$

$$\gamma_2 \frac{\sqrt{s_{10}s_{20}p_{U_1}(\mathbf{h})} + s_{20}\sqrt{p_{U_2}(\mathbf{h})}}{D\sqrt{p_{U_2}(\mathbf{h})}} = \lambda_2, \tag{51}$$

where $\gamma_1 + \gamma_2 = 1$. Note that, dividing (50) by (51), it can be shown that the optimal $p_{U_1}(\mathbf{h})$ varies linearly with optimal $p_{U_2}(\mathbf{h})$, i.e.,

$$\lambda_1^2 s_{20} p_{U_1}(\mathbf{h}) = \lambda_2^2 s_{10} p_{U_2}(\mathbf{h}), \tag{52}$$

2. Next, let $p_{U_1}(\mathbf{h}) = p_{U_2}(\mathbf{h}) = 0$. Since this point also lies on the line (52), we conclude that the optimal solution is guaranteed to lie on that line regardless of whether the cooperative powers are zero or non-zero. Therefore, although the partial derivatives $\dfrac{\partial D}{\partial p_{U_i}(\mathbf{h})}$, $i = 1, 2$ are not well defined when

$p_{U_i}(\mathbf{h})$ are both zero, it is sufficient to consider the directional derivatives, and evaluate the optimality conditions along the line (52) for $p_{U_1}(\mathbf{h}) = p_{U_2}(\mathbf{h}) = 0$, as well as for $p_{U_1}(\mathbf{h}), p_{U_2}(\mathbf{h}) > 0$.

Now, we can combine these two cases by substituting (50) into (48), (51) into (49), using (52), and extending (50) and (51) to include the case $p_{U_1}(\mathbf{h}), p_{U_2}(\mathbf{h}) = (0,0)$. The last extension is simply performed by replacing the equalities in (50) and (51) by inequalities, thereby allowing the complementary slackness constraints to be active when required. The resulting final set of conditions for optimality of the power allocation policy is,

$$\frac{s_{12}}{1 + s_{12}p_{12}(\mathbf{h})} \leq \frac{1}{\gamma_1} \frac{\lambda_1^2 s_{20}}{\lambda_2 s_{10} + \lambda_1 s_{20}}, \tag{53}$$

$$\frac{s_{21}}{1 + s_{21}p_{21}(\mathbf{h})} \leq \frac{1}{\gamma_1} \frac{\lambda_2^2 s_{10}}{\lambda_2 s_{10} + \lambda_1 s_{20}}, \tag{54}$$

$$\frac{1 - \gamma_1}{D}\left(s_{10} + \frac{\lambda_1}{\lambda_2} s_{20}\right) \leq \lambda_1, \tag{55}$$

$$\frac{1 - \gamma_1}{D}\left(s_{20} + \frac{\lambda_2}{\lambda_1} s_{10}\right) \leq \lambda_2, \tag{56}$$

where equalities in each of (53)-(56) hold if $p_{12}(\mathbf{h}) > 0$, $p_{21}(\mathbf{h}) > 0$, $p_{U_1}(\mathbf{h}) > 0$, $p_{U_2}(\mathbf{h}) > 0$, respectively.

Based on these conditions, we make the following observation about the structure of the optimal power allocation policy: with the transmitter-receiver channel gains s_{10}, s_{20} being fixed, the optimal power levels associated with the components $p_{12}(\mathbf{h})$ and $p_{21}(\mathbf{h})$, intended for establishing common information among the transmitters have the form,

$$p_{12}(\mathbf{h}) = \left(\frac{1}{\nu_1} - \frac{1}{s_{12}}\right)^+, \tag{57}$$

$$p_{21}(\mathbf{h}) = \left(\frac{1}{\nu_2} - \frac{1}{s_{21}}\right)^+, \tag{58}$$

where $(\cdot)^+$ denotes $\max(0, \cdot)$. Equations (57) and (58) are obtained by denoting the right hand sides of (53) and (54) by the constants ν_1 and ν_2 for fixed s_{10} and s_{20}, and solving for $p_{12}(\mathbf{h})$ and $p_{21}(\mathbf{h})$. Note that, the equalities are satisfied only when the power levels are positive, otherwise the powers are set to zero. The resulting power levels $p_{12}(\mathbf{h})$ and $p_{21}(\mathbf{h})$ can be computed independently of each other, and obey a *single user waterfilling* type of solution, over the individual link gains s_{12} and s_{21}, respectively. The direct link gains s_{10} and s_{20} are used to determine the water level. After $p_{12}(\mathbf{h})$ and $p_{21}(\mathbf{h})$ are calculated, they can be substituted into (55) and (56) with D being defined as in (47), to obtain $p_{U_i}(\mathbf{h})$.

The downside of the above procedure for obtaining the optimal power allocation is that it requires solving for the Lagrange multipliers λ_i and γ_i that satisfy the power constraints, which, in general would involve a multi-dimensional search. Besides, the more general problem of maximizing the weighted sum of rates results in more complicated optimality conditions, which may be even harder to solve analytically. Therefore, although it is useful to obtain the analytical solutions, as they shed some light on the structure of the optimum transmit strategy, we still need to devise some numerical methods, to compute the achievable rates and the associated power allocation. In the next section, we demonstrate one such method, based on projected subgradients, to obtain the optimal power allocation policy numerically. Note that, interesting alternatives to the results presented in this section could be the development of efficient algorithms which directly obtain the Lagrange multipliers, thereby solving the KKT conditions, or the solution of the convex and differentiable version of the problem given in (40) using the interior point method (Boyd & Vandenberghe, 2004).

Numerical Optimization of Powers and Sample Rate Regions

In this section we adopt the method of subgradients from non-differentiable optimization theory (Shor, 1985; Bertsekas, 1995) to provide some numerical examples which illustrate the performance of the proposed joint power allocation and cooperation scheme. The properties of the resulting optimum power allocation policies will then be compared to those predicted by the analytical solution.

The subgradient methods are very similar to gradient ascent methods; in fact, when the objective function is differentiable, which in our case is true almost everywhere, the subgradient is exactly equal to the gradient. Unlike gradient ascent methods however, subgradient methods do not necessarily produce monotonically non-decreasing objective function values.

A subgradient for the concave objective function $R_{\gamma_A}(\mathbf{p})$ given in (39) is any vector $\mathbf{g}$ that satisfies (Bertsekas, 1995),

$$R_{\gamma_A}(\mathbf{p'}) \leq R_{\gamma_A}(\mathbf{p}) + (\mathbf{p'} - \mathbf{p})^{\mathrm{T}}\mathbf{g}, \tag{59}$$

where $\mathbf{p'}$ and $\mathbf{p}$ are two arbitrary but distinct power vectors. The projected subgradient method for constrained maximization uses the update

$$\mathbf{p}(k+1) = [\mathbf{p}(k) + \alpha_k \mathbf{g}_k]^{\dagger}, \tag{60}$$

where $[\cdot]^{\dagger}$ denotes the Euclidean projection onto the constraint set, and α_k is the step size parameter at iteration k. The step size α_k may be chosen in a variety of ways, to guarantee convergence to the global optimum. In our simulations, we choose the diminishing step size, normalized by the norm of the subgradient to ensure the convergence (Shor, 1985)

$$\alpha_k = \frac{a}{b + \sqrt{k}} \frac{1}{\|\mathbf{g}_k\|}. \tag{61}$$

We first consider a rather hypothetical system, with uniformly distributed fading coefficients $h_{i0} \in \{0.025, 0.050, \cdots, 0.25\}$ and $h_{ij} \in \{0.26, 0.27, \cdots, 0.35\}$, where each fading value is observed with probability 1/10. The reason for these particularly odd choices of channel distributions is that, with such a selection, the cooperation (inter-user) links are always better than the direct (user-receiver) links. Therefore, the system operates only in case 1 of Theorem 1. Consequently, in this particular setting, our power allocation scheme is indeed the optimal power allocation policy for the block Markov superposition encoding, and the achievable rates should not suffer from the potential suboptimality caused by case 4.

Clearly, the inter-user links are statistically symmetric, and so are the direct links. The transmit power, and noise variance are chosen such that $\bar{p}_i = \sigma_i^2 = 1$ (i.e., 0dB SNR without fading). Therefore, we expect to obtain a symmetric achievable rate region. Figure 4 illustrates the achievable rate region we obtain for the jointly optimized power control and cooperation system, using the described parameter set in uniform fading, along with several benchmark systems. The region for joint power control and cooperation is generated by multiple uses of the subgradient method, each for a different set of priorities μ_i of the users. As the priorities are varied, we trace the points on the rate region boundary, and then finally a convex hull operation over these points is performed. Note that, had we been able to run the algorithm for all possible priority pairs (a continuous set), there would be no need for the convex hull operation, but this is clearly infeasible in practice. In the subgradient algorithm, the step size parameters were fixed at $a = 50$ and $b = 5$.

In Figure 4, we observe that for rate pairs close to the sum rate, performing power control alone yields higher rates than a cooperative system with no power control. This may be attributed to the more efficient utilization of the direct link by the power control policy. Note that, power control itself yields a form of diversity, which is called multiuser diversity. As far as the sum rate is concerned, multiuser diversity is more pronounced than cooperative diversity, as the optimum transmit strategy for a non-cooperative MAC is a greedy one-user-at-a-time policy, where only the best user transmits, which allows the other users to save their powers for future transmissions. Yet, the joint user cooperation and power control scheme does significantly improve on all other schemes, as it takes advantage of both cooperation diversity and time (or multiuser) diversity in the system. In fact, this joint utilization of diversity, which owes to the joint adaptation of coding, medium accessing and routing, yields a cross-layer approach for the design of the wireless communication system.

Next, we consider the more practical scenario of Rayleigh fading, where the power gains introduced by the channel become exponential random variables. We choose the means of the exponential power gains as $E[h_{10}] = E[h_{20}] = 0.3$ and $E[h_{12}] = E[h_{21}] = 0.6$, so that the inter-user links are stronger, on the average. The transmit powers and noise powers are not changed. The resulting achievable rate regions are given in Figure 5. Unlike the uniform fading considered above, in this setting, all four cases in Theorem 1 are realized, and there is potentially some loss over the optimally achievable rates. However, the set of rate regions, and the way they compare to each other, are very similar to those given in Figure 4 for the uniform fading case. This indicates that the loss, if any, is very small thanks to both of the direct links outperforming the cooperation link being a very low probability event.

The careful reader might have realized that in both examples, we have chosen relatively low signal to noise ratios. This is because; the rate gains from power control are especially significant at low SNR values (Goldsmith & Varaiya, 1997). Therefore, joint user cooperation and power control is more appropriate for the most challenging low SNR scenarios. To quantify the rate gain achieved by joint use

Figure 4. Rates achievable by joint power control and user cooperation for uniform fading

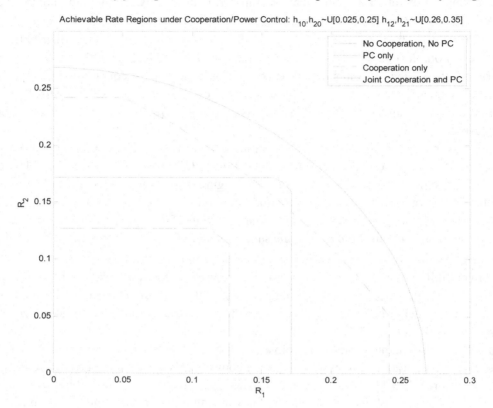

Achievable Rate Regions under Cooperation/Power Control: $h_{10}, h_{20} \sim U[0.025, 0.25]$ $h_{12}, h_{21} \sim U[0.26, 0.35]$

of cooperation and power control, let us compare the rate regions in Figure 5. With average transmit power and noise variance set to 1, and the average of the exponentially distributed direct link fading value set to 0.3, the average received SNR can be computed as -5.2 dB. In this case, it is observed that the individual rate of either user is more than doubled when compared to a system with no power control and user cooperation. Similarly, there is a 1.8 dB improvement in the sum rate, which is increased from 0.48 symbols/transmission to 0.73 symbols/transmission.

A final observation, common to Figures 4 and 5, is the fact that, near the sum rate point, joint cooperation and power control improves relatively less over only power control. This situation parallels the earlier observation about the effect of power control being more pronounced near the sum rate, compared to cooperation. Note that, for the traditional MAC, the sum rate is achieved by time division among the users, which does not allow for coherent combining gain (Knopp & Humblet, 1995). Therefore, it is not surprising that in order to attain cooperative diversity gain, users may have to sacrifice some of the gain they obtain from exploiting the time diversity through power control.

The convergence of the subgradient method is illustrated in Figure 6. The priorities in the objective function $R_{\frac{1}{4}}$ are selected as $\mu_1 = 2$ and $\mu_2 = 1$. We compare the convergence behavior for three sets of step size parameters, all used to determine the diminishing step size given by equation (61). With larger step size parameters, the non-monotonic behavior of the subgradient algorithm becomes more apparent; on the other hand, the convergence is significantly faster than the convergence for smaller step sizes. This owes to the algorithm being more likely to approach the optimal value of the function in the initial

Figure 5. Rates achievable by joint power control and user cooperation for Rayleigh fading

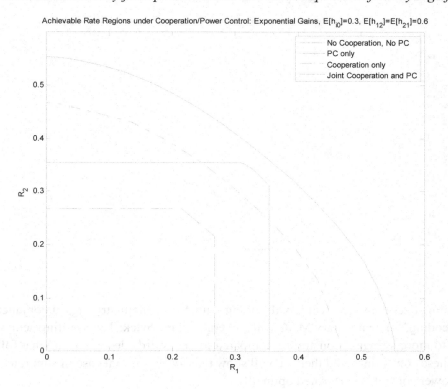

Achievable Rate Regions under Cooperation/Power Control: Exponential Gains, $E[h_{i0}]=0.3$, $E[h_{12}]=E[h_{21}]=0.6$

iterations, for large step sizes. The solid curve nearly converges after about 500 iterations. Note that, although the three curves seem to be converging to different values, if we let the subgradient algorithm run for a sufficiently large number of iterations, they shall indeed converge to the same limit.

The sum rate maximizing power allocation obtained by the projected subgradient iterations, for the uniform fading case of Figure 4, is illustrated in Figures 7(a) and 7(b). These figures demonstrate the waterfilling structure of the optimal power levels $p_{12}(\mathbf{h})$, and $p_{21}(\mathbf{h})$, as derived in Section 3.5.1. The power allocation obtained after 1000 iterations nearly satisfies the KKT conditions derived earlier. The power distributions are actually functions of all four channel states, and Figures 7(a) and 7(b) correspond only to the distribution of the powers over a 2-D "slice" of the 4-D channel state space. The powers are sketched with the direct link gains set at $s_{10} = 0.2$ and $s_{20} = 0.15$. As dictated by (57) and (58), the "water level" for p_{12} is higher, and the threshold beyond which user 1 starts allocating its power for common information build-up is lower.

Region of Achievable Rates Using Window Decoding

The decoding delay can be significantly reduced if backwards decoding is replaced by window decoding introduced by Carleial (1982). Due to its improved delay performance, window decoding has recently attracted increasing attention, but mostly in the context of relay channels, where the relays do not have their own messages to transmit. For relay channels, it is known that window decoding achieves the same rate as backwards decoding. However, this is not in general true for MAC with generalized feedback. Laneman and Kramer (2004) obtained the achievable rate regions for a system which employs two

Figure 6. Convergence of $R_{1/4}$ using subgradient method for different step size parameters

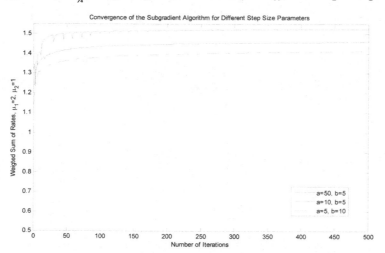

variations of window decoding, i.e., window decoding with and without striping, in conjunction with block Markov encoding, for an arbitrary MAC with generalized feedback. The resulting achievable rate region is subject to more stringent constraints compared to backwards decoding. In what follows, we will first review these constraints, and then we will show that for the special case of sum rate, they will not be active if power control is performed optimally.

In window decoding, the decoding process is carried out using a sliding window of two blocks, and therefore can begin after only a one block delay. The messages $w_{12}(b-1)$, $w_{21}(b-1)$, $w_{10}(b-1)$ and $w_{20}(b-1)$ are decoded using the signals received in blocks $b-1$ and b, at the end of block b. If the cooperative codewords, conveying $w_{12}(b-1)$ and $w_{21}(b-1)$, are decoded first, subtracted from the received signal, and fresh information is decoded afterwards, the decoding technique is called window decoding with stripping. If no specific decoding order is imposed, and the messages are simultaneously decoded using the information from both blocks, the resulting scheme is called window decoding without stripping (Laneman & Kramer, 2004).

When backwards decoding is used, the decoding starts from the last block, where no fresh information is transmitted, and the cooperative signal does not face any interference. However, in window decoding, the cooperative codeword $\mathbf{u}$ is decoded while fresh information is still being transmitted, and therefore needs to be decoded in the presence of interference from other codeword components. Hence, the rates achievable for the cooperative messages w_{12} and w_{21} are in general potentially lower. For window decoding with stripping, the following extra constraints need to be imposed for reliable decoding of the cooperative part of the message, which are required in addition to the constraints for backwards decoding (23):

$$
\begin{aligned}
R_{12} &\leq I(u; y_0) + I(x_{12}; y_0 \mid x_{21}, u, \mathbf{h}), \\
R_{21} &\leq I(u; y_0) + I(x_{21}; y_0 \mid x_{12}, u, \mathbf{h}), \\
R_{12} + R_{21} &\leq I(u, x_{12}, x_{21}; y_0, \mid \mathbf{h}).
\end{aligned}
\tag{62}
$$

Figure 7. Power distributions obtained using the projected subgradient method

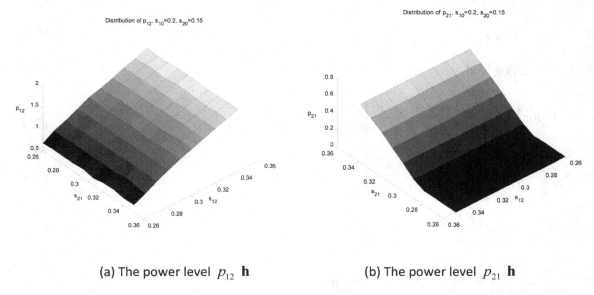

(a) The power level p_{12} **h**   (b) The power level p_{21} **h**

If window decoding without stripping is used instead, the cooperative and direct messages are decoded jointly, while the fresh information is treated as interference, leading to the additional rate constraints

$$R_{12} + R_{10} \leq I(u; y_0) + I(x_1; y_0 \mid x_2, x_{21}, u, \mathbf{h}),$$
$$R_{21} + R_{20} \leq I(u; y_0) + I(x_2; y_0 \mid x_1, x_{12}, u, \mathbf{h}),$$
$$R_{12} + R_{10} + R_{20} \leq I(u; y_0) + I(u, x_1, x_2; y_0, \mid x_{21}, \mathbf{h}),$$
$$R_{21} + R_{10} + R_{20} \leq I(u; y_0) + I(u, x_1, x_2; y_0, \mid x_{12}, \mathbf{h}). \tag{63}$$

Now, let us express the overall rate constraints (23), (62) and (63) in terms of the transmit powers and channel states, for our Gaussian MAC with generalized feedback, shown in Figure 3. It can be shown by standard properties of mutual information, and results regarding the capacity of fading channels (Hanly & Tse, 1998), that the achievable rate region is given by the convex hull of rates $R_1 = R_{10} + R_{12}$ and $R_2 = R_{20} + R_{21}$, which obey equations (64) through (72) for window decoding with stripping; and equations (64)-(69), (73)-(76) for window decoding without stripping:

$$R_{12} < E\left[\log\left(1 + \frac{s_{12}p_{12}(\mathbf{h})}{s_{12}p_{10}(\mathbf{h}) + 1}\right)\right] \tag{64}$$

$$R_{21} < E\left[\log\left(1 + \frac{s_{21}p_{21}(\mathbf{h})}{s_{21}p_{20}(\mathbf{h}) + 1}\right)\right] \tag{65}$$

$$R_{10} < E\left[\log\left(1 + s_{10}p_{10}(\mathbf{h})\right)\right] \tag{66}$$

$$R_{20} < E\left[\log\left(1 + s_{20}p_{20}(\mathbf{h})\right)\right] \tag{67}$$

$$R_{10} + R_{20} < E\left[\log\left(H\right)\right] \tag{68}$$

$$R_1 + R_2 < E\left[\log\left(F\right)\right] \tag{69}$$

$$R_{12} < E\left[\log\left(\frac{F}{G}\right) + \log\left(\frac{H + s_{10}p_{12}(\mathbf{h})}{H}\right)\right] \tag{70}$$

$$R_{21} < E\left[\log\left(\frac{F}{G}\right) + \log\left(\frac{H + s_{20}p_{21}(\mathbf{h})}{H}\right)\right] \tag{71}$$

$$R_{12} + R_{21} < E\left[\log\left(\frac{F}{H}\right)\right] \tag{72}$$

$$R_1 < E\left[\log\left(\frac{F}{G}\right) + \log\left(1 + s_{10}(p_{10}(\mathbf{h}) + p_{12}(\mathbf{h}))\right)\right] \tag{73}$$

$$R_2 < E\left[\log\left(\frac{F}{G}\right) + \log\left(1 + s_{20}(p_{20}(\mathbf{h}) + p_{21}(\mathbf{h}))\right)\right] \tag{74}$$

$$R_1 + R_{20} < E\left[\log\left(\frac{F}{G}\right) + \log\left(C + s_{10}p_{12}(\mathbf{h})\right)\right] \tag{75}$$

$$R_2 + R_{10} < E\left[\log\left(\frac{F}{G}\right) + \log\left(C + s_{20}p_{21}(\mathbf{h})\right)\right] \tag{76}$$

where

$$F = 1 + s_{10}p_1(\mathbf{h}) + s_{20}p_2(\mathbf{h}) + 2\sqrt{s_{10}s_{20}p_{u_1}(\mathbf{h})p_{u_2}(\mathbf{h})}$$
$$G = 1 + s_{10}(p_{10}(\mathbf{h}) + p_{12}(\mathbf{h})) + s_{20}(p_{21}(\mathbf{h}) + p_{20}(\mathbf{h}))$$
$$H = 1 + s_{10}p_{10}(\mathbf{h}) + s_{20}p_{20}(\mathbf{h}).$$

In Section 3.4.1, we have obtained some key properties of the power control policy which maximizes the sum rate, for backwards decoding. In what follows, we will investigate the maximum sum rate achievable by our proposed power controlled encoding scheme in (22), and the structure of the sum rate optimal policies, when window decoding is used.

Sum Rate Achievable by Window Decoding

The sum rate achievable by power controlled block Markov encoding in (22), followed by backwards decoding was given in equation (26). We have shown that, the optimal power allocation policy that maximizes the sum rate (26) has to satisfy Theorem 1. We will now use this result to further show that, when power control is employed, the sum rate achieved by window decoding is the same as that achieved by backwards decoding, and the corresponding optimal power control policies are identical.

Theorem 2. For a two user fading Gaussian multiple access channel where the encoders employ the channel adaptive block Markov superposition encoding given by (22), let the maximum sum rates achievable by window decoding and backwards decoding be denoted by R_w^*, and R_b^*, respectively. Then, $R_w^* = R_b^*$. Furthermore, the maximum sum rates achievable by window decoding with and without stripping are identical.

Proof (Theorem 2): Let us denote by $\mathbf{p}^*(\mathbf{h})$ the power control policy that maximizes the sum rate R_b achievable by backwards decoding, given in (26). The following two lemmas are sufficient to prove Theorem 2.

Lemma 3. $R_b(\mathbf{p}^*(\mathbf{h})) \geq R_w(\mathbf{p}^*(\mathbf{h}))$.

Proof: Since the rate constraints for backwards decoding are a subset of the constraints for window decoding, for any valid power allocation policy $\mathbf{p}(\mathbf{h})$, $R_b(\mathbf{p}(\mathbf{h})) \geq R_w(\mathbf{p}(\mathbf{h}))$. The result follows directly by choosing $\mathbf{p}(\mathbf{h}) = \mathbf{p}(\mathbf{h})^*$. $\square$

Lemma 4. $R_b(\mathbf{p}^*(\mathbf{h})) \leq R_w(\mathbf{p}^*(\mathbf{h}))$.

Proof: Note that, window decoding with stripping is a special case of window decoding without stripping, with a specific decoding order. Therefore, it has a potentially smaller achievable rate region, and its sum rate is less than or equal to that for window decoding without stripping. Hence, it is sufficient to prove the lemma for window decoding with stripping; and the result will automatically follow for window decoding without stripping.

Listed below are all possible combinations of rate constraints (64)-(72) which yield a bound on the sum rate R_w:

1. (69),
2. (64)+(65)+(68),
3. (68)+(72),
4. (68)+(70)+(71),
5. (64)+(68)+(71),
6. (65)+(68)+(70).

Note that, the constraints (66) and (67) never show up in the above list, as they are always dominated by (68), as far as sum rate is concerned.

Lemma 4 can be proved by showing that neither of the above bounds is tighter than the backwards decoding bound given in (26), as long as the optimal power allocation policy is used. Below, we establish this desired result by treating each of the cases 1-6 separately.

1. The bound (69), which is common to both backwards and window decoding, automatically satisfies the lemma, with equality.
2. The bound (64)+(65)+(68) is also common to both backwards and window decoding, and therefore automatically satisfies the lemma, with equality.
3. The right hand sides of (68) and (72) combine to yield the right hand side of (69), thereby reducing to case 1.
4. Adding the right hand sides of the equations (68), (70) and (71), we get

$$E\left[2\log\left(\frac{F}{G}\right) + \log\left(\frac{H + s_{10}p_{12}(\mathbf{h})}{H}\right) + \log\left(\frac{H + s_{20}p_{21}(\mathbf{h})}{H}\right) + \log(H)\right]$$

$$\overset{(a)}{\geq} E\left[\log\left(\frac{FH}{G}\right) + \log\left(\frac{H + s_{10}p_{12}(\mathbf{h}) + s_{20}p_{21}(\mathbf{h})}{H}\right)\right] \overset{(b)}{=} E\left[\log(F)\right], \tag{77}$$

where, (77)(a) follows from the concavity of the logarithm, and the fact that we threw away a positive term, and (77) (b) follows since numerator of the second term on the right hand side of (77) is G. But then, the bound on the sum rate obtained in this case is always looser than $\log(F)$, which is the bound in case 1; hence the statement of the lemma is satisfied.

5. Note that, the first four cases hold irrespective of the power allocation policy being used. However, for cases 5 and 6, we will need to assume that the power control possesses the properties required by Theorem 1.

Adding the right hand sides of the equations (64), (68) and (71), we obtain

$$E\left[\log\left(\frac{F}{G}\right) + \log\left(1 + \frac{s_{12}p_{12}(\mathbf{h})}{s_{12}p_{10}(\mathbf{h}) + 1}\right) + \log\left(\frac{H + s_{20}p_{21}(\mathbf{h})}{H}\right) + \log(H)\right]. \tag{78}$$

This bound is active only if it gives a value which is lower than the minimum of the bounds in the previous cases, namely, $E[\log(F)]$. Simplifying and rearranging the terms, a sufficient condition for the bound being active is

$$E\left[\log\left(1 + \frac{s_{12}p_{12}(\mathbf{h})}{s_{12}p_{10}(\mathbf{h}) + 1}\right)\right] < E\left[\log\left(\frac{G}{H + s_{20}p_{21}(\mathbf{h})}\right)\right]. \tag{79}$$

Let us now use Theorem 1, which states that some components of $\mathbf{p}^*(\mathbf{h})$ are always zero depending on the relative channel qualities, to check the validity of condition (79). In the following, E_i, $i = 1, 2, 3, 4$ denotes the conditional expectation given that case i from Theorem 1 occurs. We will show that neither of the conditional expectations satisfies the ordering in (79), therefore, the average over all cases cannot satisfy (79).

- Let $s_{10} < s_{12}$ and $s_{20} < s_{21}$, which requires $p_{10}^*(\mathbf{h}) = p_{20}^*(\mathbf{h}) = 0$. Then, evaluating (79) at $\mathbf{p}^*(\mathbf{h})$, we get

$$
\begin{aligned}
E_1\left[\log\left(1 + s_{12}p_{12}(\mathbf{h})\right)\right] &< E_1\left[\log\left(1 + \frac{s_{10}p_{12}(\mathbf{h})}{1 + s_{20}p_{21}(\mathbf{h})}\right)\right] \\
&\leq E_1\left[\log\left(1 + s_{10}p_{12}(\mathbf{h})\right)\right] \\
&\leq E_1\left[\log\left(1 + s_{12}p_{12}(\mathbf{h})\right)\right],
\end{aligned}
\tag{80}
$$

which is a contradiction, hence (79) is not satisfied, conditioned on this particular channel ordering, and the sum rate is bounded by $E_1[\log(F)]$.

- Let $s_{10} \geq s_{12}$ and $s_{20} < s_{21}$, which requires $p_{12}^*(\mathbf{h}) = p_{20}^*(\mathbf{h}) = 0$. Then, it is easy to check that both sides of (79) are equal to 0, the strict inequality is not satisfied, and the sum rate is bounded by $E_2[\log(F)]$.

- Let $s_{10} < s_{12}$ and $s_{20} \geq s_{21}$, which requires $p_{10}^*(\mathbf{h}) = p_{21}^*(\mathbf{h}) = 0$. Then, for the bound to be active, we need,

$$
E_3\left[\log\left(1 + s_{12}p_{12}(\mathbf{h})\right)\right] < E_3\left[\log\left(1 + \frac{s_{10}p_{12}(\mathbf{h})}{1 + s_{20}p_{20}(\mathbf{h})}\right)\right],
\tag{81}
$$

which is never possible since $s_{10} < s_{12}$, and the sum rate is bounded by $E_3[\log(F)]$.

- Let $s_{10} \geq s_{12}$ and $s_{20} \geq s_{21}$ in which case $p_{12}^*(\mathbf{h}) = p_{21}^*(\mathbf{h}) = 0$ or $p_{12}^*(\mathbf{h}) = p_{20}^*(\mathbf{h}) = 0$ or $p_{10}^*(\mathbf{h}) = p_{21}^*(\mathbf{h}) = 0$. It is easy to check that the first two possible choices of powers give both sides of (79) equal to 0, thereby yielding no tighter bounds. The last choice requires a little more work. In this case, the bound is clearly *inactive* if

$$
\log\left(1 + \frac{s_{12}p_{12}(\mathbf{h})}{s_{12}p_{10}(\mathbf{h}) + 1}\right) > \log\left(\frac{G}{H + s_{20}p_{21}(\mathbf{h})}\right),
\tag{82}
$$

which is equivalent to

$$
1 + s_{20}p_{20}(\mathbf{h}) > \frac{s_{10}}{s_{12}}.
\tag{83}
$$

However, this condition is always satisfied for the optimal $p_{10}^*(\mathbf{h})$, since from (33), it is equivalent to the partial derivative of the sum rate with respect to p_{10} being always negative, which is required if $p_{10}^*(\mathbf{h}) = 0$ is to be optimal. Hence, in this final case, the sum rate is bounded by $E_4[\log(F)]$.

Therefore, we have shown that, the left hand side of (79) is always greater than or equal to the right hand side, and the bound (64)+(68)+(71) is always inactive, and is dominated by the sum rate constraint for backwards decoding, $E[\log(F)]$.

6.   Finally, due to the symmetry between cases 5 and 6, the window decoding bound on the sum rate in case 6 is looser or at best equivalent to that for backwards decoding.

Since all the extra constraints imposed by window decoding yield either identical or looser bounds when compared to backwards decoding, we have established $R_b(\mathbf{p}^*(\mathbf{h})) \leq R_w(\mathbf{p}^*(\mathbf{h}))$. □

The proof of Theorem 2 immediately follows by combining Lemmas 3 and 4. □

CONCLUSION

In this chapter, we considered two major means of exploiting the diversity presented by time varying nature of the wireless channels: power control and user cooperation. This chapter was intended to provide a background on both power control and user cooperation, and then bridge the gap between these two seemingly independently researched subjects, and demonstrate how instructive it is to study the problem of joint power allocation and user cooperation, in the design of wireless networks.

Focusing on fading channels, we first presented an information theoretical approach to the power control problem, which led to the following crucial results. Firstly, in the presence of channel state information, power control is all we have to do to adapt our transmit strategies to the changing channel conditions: the underlying codebook generation and encoding strategies do not need to be updated. Secondly, the power control problem dictates the optimum medium access policies, by telling us which users are going to transmit, and when they are going to transmit.

Next, we reviewed some major results on cooperative communication, starting from the simplest form of cooperation, i.e., relay channels, and moving towards the more beneficial and fairer scheme of mutual cooperation. We focused on a decode and forward type cooperation protocol, and employed block Markov superposition encoding policy, together with both backwards, and window decoding. When the channel state information is not used to adapt the transmission strategies, block Markov encoding policy is composed of three transmit signal components: one which creates common information among the transmitters, one which injects fresh information to be decoded at the receiver, and one which is used for cooperatively conveying the common information built in the earlier blocks of communication.

Combining power control with user cooperation diversity, we have addressed the problem of optimal power allocation for a fading cooperative MAC, where the transmitters and the receiver have channel state information, and are therefore able to adapt their coding and decoding strategies by allocating their resources. We have obtained the power control policies that maximize the rates achievable by block Markov superposition coding, and proved that, when channel state information is available, the coding strategy is significantly simplified: given any channel state, for each of the users, either the signal intended directly for the receiver, or for the other transmitter, is transmitted, but not both. Using this result, we were also able to formulate the otherwise non-concave problem of maximizing the achievable rates as a concave maximization (or equivalently convex minimization) problem. Using techniques from convex optimization, we obtained the analytical structure of the optimal power allocation policies. The power control policies, which are jointly optimal with block Markov coding, were then obtained through simulations, using subgradient methods. The resulting achievable rate regions for joint power control and cooperation were shown to improve significantly over those for both cooperative systems without power control, and power controlled systems without cooperation, since our joint approach makes use of both cooperative diversity and time diversity. As a result, the power optimization problem led to the characterization of jointly optimum encoding, medium access and routing policies, yielding a cross layer approach to wireless network design.

Finally, we have compared the sum rates achievable by backwards decoding and window decoding, for the fading Gaussian multiple access channel in consideration. Window decoding is a more practical choice than backwards decoding, owing to its much improved delay performance. However, rate-wise it is in general known to be inferior to backwards decoding. Nevertheless, we have shown that, window decoding in fact achieves the same sum rate as backwards decoding, when the powers are allocated optimally. This result is extremely important, since using window decoding, the decoding delay can be reduced from the B block delay of the backwards decoding, down to only a single block of transmission. Therefore, using window decoding is always preferable as far as the total information rate in the system is concerned. We have also argued that the sum-rate optimal power allocation policies for backwards and window decoding are identical, since the expressions for the objective function are the same. This means, for window decoding, our findings regarding the optimal power control are still valid. That is, at any given channel state, two of the transmitted signal components need to be assigned zero powers, thereby simplifying the block Markov encoding strategy significantly. We further showed that since window decoding with stripping achieves the same maximum sum rate as backwards decoding, its more general version, window decoding without stripping, also achieves the same sum rate, therefore, under optimal power control, the sum rates achievable by window decoding with and without stripping are identical.

All of the results presented in this chapter point to the same striking fact: both user cooperation, and power control are crucial means of benefiting from diversity in communication systems, however, neither one is complete without considering the other. Joint consideration of power allocation and user cooperation not only improves the performance limits of the wireless networks, but also brings more insight into the problem of jointly designing various aspects of the communication protocols, and may even unexpectedly simplify or alleviate seemingly complex design issues.

REFERENCES

Bertsekas, D. P. (1995). *Nonlinear programming*. Belmont, MA: Athena Scientific.

Biglieri, E., Proakis, J., & Shamai (Shitz), S. (1998). Fading channels: Information-theoretic and communications aspects. *IEEE Transactions on Information Theory, 44*(6), 2619–2692. doi:10.1109/18.720551

Boyd, S., & Vandenberghe, L. (2004). *Convex optimization*. Cambridge, UK: Cambridge University Press.

Carleial, A. B. (1982). Multiple access channels with different generalized feedback signals. *IEEE Transactions on Information Theory, 28*(6), 841–850. doi:10.1109/TIT.1982.1056587

Cover, T. M., & El Gamal, A. A. (1979). Capacity theorems for the relay channel. *IEEE Transactions on Information Theory, 25*(5), 572–584. doi:10.1109/TIT.1979.1056084

Cover, T. M., & Leung, C. S. K. (1981). An achievable rate region for the multiple access channel with feedback. *IEEE Transactions on Information Theory, 27*(3), 292–298. doi:10.1109/TIT.1981.1056357

Cover, T. M., & Thomas, J. A. (1991). *Elements of information theory*. Wiley Interscience.

Erkip, E. (2000). Capacity and power control for spatial diversity. In *Proceedings of the Conference on Information Sciences and Systems,* Princeton, NJ (pp. WA4-28/WA4-31).

Gallager, R. G. (1994). An inequality on the capacity region of multiaccess multipath channels. In Blahut, Costello, Maurer & Mittleholzer (Eds.), *Communications & cryptography, two sides of one tapestry* (pp. 129-140). Norwell, MA: Kluwer Academic Publishers.

Gastpar, M., & Vetterli, M. (2002). On the asymptotic capacity of Gaussian relay networks. In *Proceedings of the IEEE International Symposium on Information Theory*, Lausanne, Switzerland (p. 195).

Goldsmith, A. J., & Varaiya, P. P. (1997). Capacity of fading channels with channel side information. *IEEE Transactions on Information Theory, 43*(6), 1986–1992. doi:10.1109/18.641562

Hanly, S., & Tse, D. N. C. (1998). Multiaccess fading channels-part I: Polymatroid structure, optimal resource allocation and throughput capacities. *IEEE Transactions on Information Theory, 44*(7), 2796–2815. doi:10.1109/18.737514

Hong, Y.-W., Huang, W.-J., Chiu, F.-H., & Kuo, C.-C. J. (2007). Cooperative communications in resource-constrained wireless networks. *IEEE Signal Processing Magazine, 24*, 47–57. doi:10.1109/MSP.2007.361601

Host-Madsen, A., & Zhang, J. (2005). Capacity bounds and power allocation for wireless relay channels. *IEEE Transactions on Information Theory, 51*(6), 2020–2040. doi:10.1109/TIT.2005.847703

Kaya, O. (2006). Window and backwards decoding achieve the same sum rate for the fading cooperative Gaussian multiple access channel. In *Proceedings of the IEEE Global Communications Conference,* San Francisco, CA (pp. 1-5, CTH06-2).

Kaya, O., & Ulukus, S. (2006). Achieving the capacity region boundary of fading CDMA channels via generalized iterative waterfilling. *IEEE Transactions on Wireless Communications, 5*(11), 3215–3223. doi:10.1109/TWC.2006.04785

Kaya, O., & Ulukus, S. (2007). Power control for fading cooperative multiple access channels. *IEEE Transactions on Wireless Communications, 6*(8), 2915–2923. doi:10.1109/TWC.2007.05858

Knopp, R., & Humblet, P. A. (1995). Information capacity and power control in single-cell multiuser communications. In *Proceedings of the IEEE International Conference on Communications*, Seattle, WA (pp. 331–335).

Kramer, G., Gastpar, M., & Gupta, P. (2003). Capacity theorems for wireless relay channels. In *Proceedings of the Allerton Conference*, Monticello, IL (pp. 1074-1083).

Kramer, G., Gastpar, M., & Gupta, P. (2005). Cooperative strategies and capacity theorems for relay networks. *IEEE Transactions on Information Theory, 51*(9), 3037–3063. doi:10.1109/TIT.2005.853304

Kramer, G., & van Wijngaarden, A. J. (2000). On the white Gaussian multiple-access relay channel. In *Proceedings of the IEEE Inernational Symposium on Information Theory (ISIT)*, Sorrento, Italy (p. 40).

Laneman, J. N., & Kramer, G. (2004). Window decoding for the multiple access channel with generalized feedback. In *Proceedings of the IEEE International Symposium on Information Theory*, Chicago, IL (p. 279).

Laneman, J. N., Tse, D. N. C., & Wornell, G. W. (2004). Cooperative diversity in wireless networks: Efficient protocols and outage behavior. *IEEE Transactions on Information Theory*, *50*(12), 3062–3080. doi:10.1109/TIT.2004.838089

Laneman, J. N., & Wornell, G. W. (2003). Distributed space-time-coded protocols for exploiting cooperative diversity in wireless networks. *IEEE Transactions on Information Theory*, *49*(10), 2415–2425. doi:10.1109/TIT.2003.817829

Sankaranarayanan, L., Kramer, G., & Mandayam, N. B. (2004). Capacity theorems for the multiple-access relay channel. In *Proceedings of the 42nd Annual Allerton Conference on Communication, Control, and Computing,* Montocello, IL (pp. 1782-1791).

Schein, B., & Gallager, R. (2000). The parallel Gaussian relay network. In *Proceedings of the IEEE International Symposium on Information Theory*, Sorrento, Italy (p. 22).

Sendonaris, A., Erkip, E., & Aazhang, B. (2003). User cooperation diversity-part I: System description. *IEEE Transactions on Communications*, *51*(11), 1927–1938. doi:10.1109/TCOMM.2003.818096

Shor, N. Z. (1985). *Minimization methods for non-differentiable functions.* Berlin, Germany: Springer-Verlag.

Van der Meulen, E. C. (1968). *Transmission of information in a t-terminal discrete memoryless channel.* Unpublished doctoral dissertation, Univ. of California, Berkeley, CA.

Van der Meulen, E. C. (1971). Three-terminal communication channels. *Advances in Applied Probability*, *3*, 120–154. doi:10.2307/1426331

Willems, F. M. J. (1982). *Informationtheoretical results for the discrete memoryless multiple access channel.* Unpublished doctoral dissertation, Katholieke Universiteit Leuven, Leuven, Belgium.

Willems, F. M. J. (1983). The discrete memoryless multiple access channel with partially cooperating encoders. *IEEE Transactions on Information Theory*, *29*(3), 441–445. doi:10.1109/TIT.1983.1056660

Willems, F. M. J., & van der Meulen, E. C. (1983). Partial feedback for the discrete memoryless multiple access channel. *IEEE Transactions on Information Theory*, *29*(2), 287–290. doi:10.1109/TIT.1983.1056646

Willems, F. M. J., & van der Meulen, E. C. (1985). The discrete memoryless multiple access channel with cribbing encoders. *IEEE Transactions on Information Theory*, *31*(3), 313–327. doi:10.1109/TIT.1985.1057042

Willems, F. M. J., van der Meulen, E. C., & Schalkwijk, J. P. M. (1983). An achievable rate region for the multiple access channel with generalized feedback. In *Proceedings of the Allerton Conference,* Monticello, IL (pp. 284-292).

Xie, L. L., & Kumar, P. R. (2004). A network information theory for wireless communication: Scaling laws and optimal operation. *IEEE Transactions on Information Theory*, *50*(5), 748–767. doi:10.1109/TIT.2004.826631

Chapter 4
Capacity Limits of Base Station Cooperation in Cellular Networks

Symeon Chatzinotas
University of Surrey, UK

Muhammad Ali Imran
University of Surrey, UK

Reza Hoshyar
University of Surrey, UK

ABSTRACT

In the information-theoretic literature, it has been widely shown that multicell processing is able to provide high capacity gains in the context of cellular systems. What is more, it has been proved that the per-cell sum-rate capacity of multicell processing systems grows linearly with the number of base station (BS) receive antennas. However, the majority of results in this area have been produced assuming that the fading coefficients of the MIMO subchannels are completely uncorrelated. In this direction, this chapter investigates the ergodic per-cell sum-rate capacity of the Gaussian MIMO cellular channel under correlated fading and BS cooperation (multicell processing). More specifically, the current channel model considers Rayleigh fading, uniformly distributed user terminals (UTs) over a planar cellular system, and power-law path loss. Furthermore, both BSs and UTs are equipped with correlated multiple antennas, which are modelled according to the Kronecker product correlation model. The per-cell sum-rate capacity is evaluated while varying the cell density of the system, as well as the level of receive and transmit correlation. In this context, it is shown that the capacity performance is compromised by correlation at the BS-side, whereas correlation at the UT-side has a negligible effect on the system's capacity.

DOI: 10.4018/978-1-60566-665-5.ch004

INTRODUCTION

Since their conceivement, wireless communications have been greatly evolved while finding applications in every aspect of the contemporary life. The ubiquity of wireless systems has raised the demand for cost-efficient high-rate wireless services increases and thus the network operators have been in search for new transmission techniques which will allow them to reach the channel capacity limits. However, the improvement margin of the traditional cellular paradigm is getting more and more limited, as the increase in the system complexity becomes disproportional with respect to the provided capacity gain. In this direction, academia and industry have started investigating alternative cellular architectures which have the ability to provide high spectral efficiencies. In this direction, cooperative wireless cellular architectures are gaining momentum as a dominant candidate for an alternative approach in wireless cellular networks.

Cooperation in wireless networks can take many forms, such as User Terminal (UT) cooperation, Base Station (BS) cooperation and Relaying. UT cooperation is theoretically possible but practically it involves many complications, since the UTs have to communicate either on a separate wireless frequency band or through the BS in order to exchange cooperative information. This fact results in a waste of bandwidth and energy, which is very important in terms of battery life in mobile devices. Relaying can be beneficial but it either consumes the resources of relaying UTs or requires the installation of additional transponders by the network operator.

Based on the previous discussion, the approach of BS cooperation or Multicell Processing (Wyner, 1994; Somekh & Shamai, 2000; Letzepis, 2005; Shamai, Somekh, & Zaidel, 2004) is investigated, focusing on the *information-theoretic capacity limits*. In the literature, it has been widely shown that multicell processing is able to provide high capacity gains (Telatar, 1999; Foschini & Gans, 1998) in the context of cellular systems. What is more, it has been proved that the per-cell sum-rate capacity of multicell processing systems grows linearly with the number of Base Station (BS) antennas (Chatzinotas, Imran, & Tzaras, 2008e). However, the majority of results in this area has been produced based on the unrealistic assumption that the fading coefficients of the MIMO subchannels are completely uncorrelated. In general, MIMO point-to-point channels may appear correlated due to inadequate antenna separation and/or poor local scattering (Shiu, Foschini, & Kahn, 2000). More specifically, if the components of an antenna array are separated by a distance less than half of the communication wavelength, then the fading coefficients appear correlated. Furthermore, if the number of local scatterers is insufficient, then the regularities between the multiple paths can lead to correlated fading. In a typical macrocellular scenario, the inadequate antenna separation mainly affects the correlation at the UT side due to their size limitations, whereas poor local scattering affects the correlation at BS side due to their elevated position.

Taking into account the previous discussion, the purpose of this book chapter is to investigate the performance of the cooperative cellular systems which employ multi-cell joint processing under correlated fading conditions. In this direction, this chapter investigates the ergodic per-cell sum-rate capacity of the Gaussian MIMO Cellular Channel in both uplink and downlink. The employed channel model considers flat fading, uniformly distributed User Terminals (UTs) over a planar cellular system and power-law path loss. Furthermore, both BSs and UTs are equipped with multiple antennas, which are correlated according to the Kronecker product model.

More specifically, the main objectives of this book chapter are to:

- review the various approaches towards determining the capacity of the point-to-point MIMO correlated channel
- review the evolution of the information-theoretic cellular channel models
- define a channel model which combines the characteristics of both cellular channels and MIMO correlated channels
- evaluate the information-theoretic capacity limits of this model for both uplink and downlink
- interpret the capacity results based on realistic path loss and correlation parameters in the context of a typical macrocellular scenario

Outline

The *first* section of the book chapter is dedicated on introducing the main concepts of cooperation in wireless cellular system.

The *second* section focuses on multicell processing and provides a brief account on the channel coding schemes, which achieve the capacity limit in the context of the considered cooperative cellular system. More specifically, the operation of superposition coding is analyzed for the uplink cellular channel, whereas for the downlink cellular channel the dirty paper coding scheme (coding with known interference cancellation) is investigated.

The *third* section provides a detailed review of the point-to-point MIMO correlation channel models, focusing on the generic correlation model (Gesbert, Bolcskei, Gore, & Paulraj, 2002), the Kronecker correlation model (Kermoal, Schumacher, Pedersen, Mogensen, & Frederiksen, 2002) and the exponential correlation model (Loyka, 2001). Subsequently, a wide range of approaches for determining the capacity of the MIMO correlated channel are reviewed, including techniques from Random Matrix Theory, Free Probability and Replica methods. Based on these approaches, capacity results are overviewed for various settings and regimes, such as single/double-side Kronecker correlation and single/double-side asymptotics of the channel matrix dimensions.

The *fourth* section presents the evolution of the multicell processing models. Starting from single-tier collocated interference (Wyner, 1994), these models have evolved to multi-tier collocated interference (Letzepis, 2005) so as to reach the - more realistic - multi-tier distributed interference (Chatzinotas, Imran, & Tzaras, 2008a).

The *fifth* section defines the considered channel model, which combines the cellular channel model with the MIMO correlation model. Furthermore, the capacity limits for both uplink and downlink are evaluated in the context of the employed model by varying the cell density and the correlation level. Finally, practical capacity results are obtained for a typical macrocellular scenario by considering realistic path loss and correlation parameters.

The *sixth* section offers a set of conclusive observations and touches on some open issues in the area of multicell processing systems with correlated multiple antennas.

Notation

Throughout the formulations of this chapter, R is the cell radius, N is the number of BSs, K is the number of UTs per cell and η is the power-law path loss exponent. Additionally, n_{BS} and n_{UT} are the number of multiple antennas at each Base Station (BS) $\alpha\nu\delta$ $\epsilon\alpha\chi\eta$ $\Upsilon\sigma\epsilon\rho$ $\mathrm{T}\epsilon\rho\mu\iota\nu\alpha\lambda$ (UT) respectively.

$\mathbb{E}[\cdot]$ denotes the expectation, $(\cdot)^*$ denotes the complex conjugate, $(\cdot)^\dagger$ denotes the conjugate transpose matrix, $\circ$ denotes the Hadamard product, $\otimes$ denotes the Kronecker product and $\cong$ denotes asymptotic equivalence of the eigenvalue distributions. The norm of a complex scalar is denoted by $|\cdot|$, whereas the Frobenius norm of a matrix or vector is denoted by $\|\cdot\|$. The inequality $\mathbf{A} \geq \mathbf{B}$, where $\mathbf{A}, \mathbf{B}$ are positive semidefinite matrices, denotes that $\mathbf{A} - \mathbf{B}$ is also positive semidefinite.

BACKGROUND

Cooperative Cellular Systems

Over the last few decades, the main thrust in the area of communications has been the mobile wireless communications. The main characteristic of the wireless communication is that the medium, namely the available radiofrequencies, has to be shared by multiple users. This phenomenon necessitates a "cooperative" approach which optimally exploits the shared medium in order to jointly accomodate all the users. Three types of cooperation scenarios emerge in the context of a cellular wireless system: cooperation amongst the UTs, cooperation amongst the geographically spaced BSs and relaying over intermediate transceivers (RSs - Relay Stations) (Figure 1). A consice comparison amongst these three types of cooperation in terms of feasibility can be seen in Table 1.

BS Cooperation

It is well known that the wireless signal is attenuated while travelling along a distance. Therefore, a single BS is able to provide coverage to a limited area, depending on the spatial characteristics of the terrain. To overcome this problem, the cellular architecture was devised, where multiple BSs are deployed in order to efficiently cover the system area. In traditional cellular systems, the UT constantly monitors the BS channel gain levels and associates with the BS which has the strongest channel gain. As a result, the signals transmitted on a BS-UT wireless link act as interference to neighboring BS and UTs. However, BS cooperation can convert this interference to useful information-bearing signal which can be utilized while coding or decoding the UT signals. The idea of cooperation between multiple receivers to jointly decode multiuser signals is called multiuser detection (Verdu, 1986, 1998) and it has the potential of turning the unwanted received signal into useful received signal by means of cooperation between the receivers. The concept of BS cooperation at the uplink of a cellular system was first introduced by (Hanley & Whiting, 1993) and also independently proposed by the seminal work of (Wyner, 1994), further extended in (Somekh & Shamai, 1998, 2000) and discussed in (Shamai & Wyner, 1997). In this context, the term "Hyper-Receiver" was proposed for a powerful processor which is able to jointly decode the uplink signals received by multiple BSs. Similarly, the cooperating BSs can pre-cancel the interference by jointly constructing the transmitted signals ("Hyper-Transmitter"), using a pre-coding technique termed as Dirty Paper Coding (Costa, 1983). It should be noted that a fundamental assumption of multicell processing is the availability of Channel State Information (CSI) at the cooperating BSs.

Figure 1. Cooperation in a cellular system

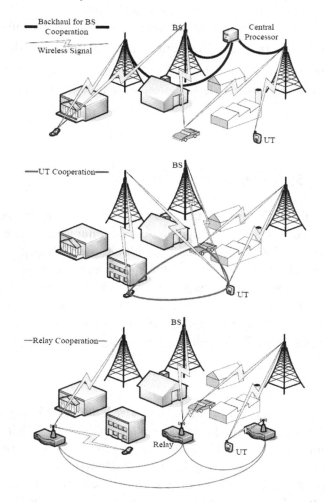

Practical Considerations

In a BS cooperation system, the requisite infrastructure comprises the central processor and the backhaul that interconnects the BSs. First, we focus on the complexity of the multiuser signal processing on the central processor. Since it has to take into account all the UTs of the cellular system, one would expect that the computational load increases with the system size. However, as shown in (Aktas, Evans, & Hanly, 2008) distributed multiuser detection based on the belief propagation algorithm results in complexity and delay per base station which do not grow with the system size.

Furthermore, based on the concept of clustering (also known as distributed antenna system) we can allow cooperation amongst smaller clusters of BSs without greatly compromising the capacity enhancements achieved with global BS cooperation (Bacha, Evans, & Hanly, 2006; Ng, Evans, Hanly, & Grant, 2004; Grant, Hanly, Evans, & Muller, 2004). This concept relies on the fact that the wireless signal decays fast with distance and therefore it suffices to consider only a limited number of cooperating cell tiers. This approach has also the advantage of decreasing the cost of infrastructure investment, since the

Table 1. Comparison of various cooperation scenarios

Cooperation Type	Advantages	Disadvantages
UT	• No need for additional infrastructure • Coverage extension	• Transceiver complexity moved at UT-side • Battery life penalty and signaling overhead
RS *fixed transponders	• No UT cooperation needed • Battery life saving • Coverage extension	• New sites needed • Increased complexity for BS and RS signal processing • Careful interference management needed
BS	• No UT cooperation needed • Use of the existing infrastructure • No harmful inter-cell interference	• BS interconnection needed (backhaul cabling) • Increased complexity for BS signal processing

cabling and the processing can take place locally (within the clusters) instead of globally. However, it should be noted that clustering results in a hybrid of BS cooperation and interference-limited systems, as the inter-cluster interference still affects the performance of the system. As a result, the cluster-edge UTs will be inadequately served in comparison to the cluster-centre UTs. In general, the effect of clustering increases as the cluster size decreases. This effect could be mitigated by designing overlapping clusters and/or incorporating an additional layer of inter-cluster processors. Coordination amongst clusters can recover the major part of this loss by engaging effective interference control and management techniques through shared control information. For example potential inter-cluster interfering UTs could be dynamically orthogonalized through inter-cluster coordinated radio resource allocation mechanisms (Choi & Andrews, 2008). Additionaly, the soft-handover technique (Lee, 1990; Viterbi, Viterbi, Gilhousen, & Zehavi, 1994; Wong & Lim, 1997) introduced in 3G CDMA systems may be considered as a form of 2-BS clustering. In this scheme diversity gain is achieved by associating a UT to two BS antennas simultaneously. Hence, the instantaneous chance of having a strong channel to at least one of the two antennas is increased. A cellular system with an extension to Wyner's model that closely represent this soft handover scenario is investigated in (Liang, Yoo, & Goldsmith, 2006; Somekh, Zaidel, & Shamai, 2007).

Another important limiting factor is the rate and delay limitations of the backhaul system that interconnects the BSs. Since it is not always possible to deploy optic fibres due to financial or geographical reasons, the backhaul can comprise less reliable connections, such as copper cables or microwave links. The imperfectly-connected BSs can be considered as relays and thus this setting resembles the case of relaying in combination to BS cooperation. The recent survey in (Shamai et al., 2007) summarises the effects of limited-capacity backhaul for non-fading uplink and downlink channels for simplified cellular system models. In addition, the authors in (Marsch & Fettweis, 2008) consider the multi-cell signal processing paradigm for cellular systems with a strongly constrained backhaul and they study the occuring performance-complexity trade-offs.

In this point, it is worth noting that multiple cooperation strategies can be deployed simultaneously, although initial studies indicate that the sum-rate capacity gain derived from this combination is not considerable. More specifically, it was shown (Simeone, Somekh, Bar-Ness, & Spagnolini, 2006, 2007; Somekh, Simeone, Poor, & Shamai, 2007) that the capacity gain due to relaying is not significant, when BS cooperation is already in place. On the other hand, according to recent findings (Chatzinotas, Imran, & Tzaras, 2008a, 2008b, 2008c) multi-cell joint processing can increase the order of magnitude of the spectral efficiency in a typical macrocellular scenario. Based on the above discussion, the rest of this chapter is dedicated on investigating the effect of fading correlation on a multicell processing system.

CELLULAR UPLINK CHANNEL

A conventional cellular uplink channel is classified as the interference channel in information theoretic terms. Introducing the joint processing at the end of the multiple receivers (BSs), transforms the set-up into a multiple input multiple output (MIMO) multiple access channel (MAC). The main difference from the conventional MIMO-MAC is the presence of receive-antennas at large geographical separation. This requires high capacity backhaul connections between the distributed receive antennas for delay-less pooling of the received signals in order to perform the joint decoding. In this context, the seminal works (Hanley & Whiting, 1993; Wyner, 1994; Somekh & Shamai, 1998) assume unlimited backhaul and perfect cooperation between the antennas with simplified assumptions for channel propagation to gain insights into the problem.

Joint Decoding of Signals The problem of jointly decoding the multiple signals can be intuitively explained by considering multiple signals with different received powers. The receiver can start the decoding process by attempting to decode the strongest signal while tolerating all other weaker signals as unwanted interference. Once the signal is decoded (assuming correct detection) the receiver can strip-off the signal from the aggregate received signal and then attempt to decode the next strongest signal. This scheme (termed successive interference cancellation - SIC) combined with minimum mean square error detection at each stage, provides optimal capacity (Varanasi & Guess, 1997).

This idealised scheme of interference cancellation has some practical limitations. One limitation is the latency involved in performing the decoding process serially for a large number of signals. There are several approaches to perform interference cancellation in practice. Parallel interference cancellation increases the required computational power by introducing several decoders operating in parallel but reduces the latency involved in decoding the signals one after the other (Andrews, 2005). A comprise scheme with affordable computational power and tolerable latency can be obtained by using the multi-stage decoding (Varanasi, 1995; Wijk, Janssen, & Prasad, 1995). In this scheme subsets of users are decoded in parallel and after their reliable detection the group of users is cancelled from the aggregate signal. In practical channel coded systems, channel decoding and interference cancellation can exchange the soft decisions iteratively to improve the overall performance of the detection of transmitted message (Wang & Host-Madsen, 1999; Alexander & Grant, 2000). Main limitations of these interference cancellation schemes are the performance degradation owing to error propagartion from the earlier to the later stages of decoding process and the high complexity of the decoder.

CELLULAR DOWNLINK CHANNEL

Similar to the case of cellular uplink channel, the cooperation of the spatially distributed BS transmitters renders the MIMO interference channel to a MIMO broadcast channel. The base station transmitters cooperate to agree on a joint transmission strategy that preempts the interference experienced at each reciever to minimise or cancel it by adjusting the multiple transmissions. This preemptive transmission strategy is termed as Dirty Paper Coding (DPC).

Dirty Paper Coding (DPC) Dirty paper coding (Costa, 1983) can achieve the capacity of the cellular downlink channel (Weingarten, Steinberg, & Shamai, 2006; Vishwanath, Jindal, & Goldsmith, 2003; Viswanath & Tse, 2003; Yu, 2006a). The main idea behind DPC can be intuitively understood using the example of a single link whose receiver is experiencing some interference that is known to the transmitter. As the transmitter knows the interference signal a priori it can preemptively transmit a signal that will combat the interference and facilitate the correct detection of the desired message (Peel, 2003). It shall be noted that the receiver does not need to be aware of the interference in this case. In multiple transmission scenario, the signals encoded first will not benefit from the pre-cancellation of any interference as there are no other signals which are known to the encoders. The signals encoded subsequently will have the advantage of pre-cancelling all the signals which are now encoded and are known a priori. The last encoded signal will benefit the most as all interfering signals can be appropriately dealt at the transmitter end.

Practical coding schemes, such as Tomlinson-Harashima precoding (Harashima & Miyakawa, 1972; Tomlinson, 1971) and the vector perturbation technique (Hochwald, Peel, & Swindlehurst, 2005) have been based on the DPC principles.

MIMO Correlation Channel Models

This section reviews the physical channel models which have been used in the literature for studying MIMO correlated point-to-point channels. According to the *generic correlation model* in (Gesbert et al., 2002), the channel matrix of a point-to-point MIMO link can be written as:

$$\mathbf{H} = \mathbf{R}_R^{1/2}\mathbf{G}_R\mathbf{R}_H^{1/2}\mathbf{G}_T\mathbf{R}_T^{1/2} \tag{1}$$

where $\mathbf{G}_R$ and $\mathbf{G}_T$ are Gaussian matrices with independent identically distributed (i.i.d.) complex circularly symmetric (c.c.s.) elements, whereas $\mathbf{R}_R$, $\mathbf{R}_H$ and $\mathbf{R}_T$ are deterministic matrices. The matrices $\mathbf{R}_R$ and $\mathbf{R}_T$, also known as the receive and transmit correlation matrix, depend on the angle spread, the antenna beamwidth and the antenna spacing at the receive and the transmit end respectively. The matrix $\mathbf{R}_H$ introduces the notion of the keyhole or pinhole channel, which appears when $\mathbf{R}_H$ is a low-rank matrix. This kind of channel can be created when the ensemble of the multipaths travels through a thin tunnel (Figure 2).

In cases where there is adequate scattering to prevent the keyhole effects ($\mathbf{R}_H = \mathbf{I}$), the channel matrix can be written as:

$$\mathbf{H} = \mathbf{R}_R^{1/2}\mathbf{G}_R\mathbf{G}_T\mathbf{R}_T^{1/2} = \mathbf{R}_R^{1/2}\mathbf{G}\mathbf{R}_T^{1/2} \tag{2}$$

where $\mathbf{G}$ is the product of two Gaussian matrices, also known as Rayleigh product channel (Yang & Belfiore, 2006). However, in the limiting case where the matrix dimensions tend to infinity, $\mathbf{G}$ converges to a Gaussian matrix derived by applying the Central Limit Theorem (CLT) to the product $\mathbf{G}_R\mathbf{G}_T$. This channel matrix represents the *Kronecker correlation model* (Kermoal et al., 2002), since the covariance

Figure 2. Graphical representation of the multipath rays in a typical cellular scenario. BS antennas appear correlated due to poor local scattering. Keyhole effects may rise when rays are guided through a thin tunnel (Adapted from (Jorswieck & Boche, 2006a)).

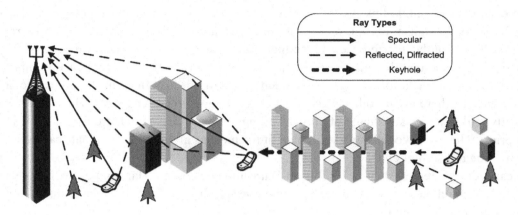

of the vectorized channel matrix can be written as the Kronecker product of the receive and transmit correlation matrix, namely:

$$\operatorname{cov}\left(\operatorname{vec}\left(\mathbf{H}\right)\right) = \mathbf{R}_R \otimes \mathbf{R}_T \tag{3}$$

or equivalently

$$\mathrm{E}\left[\left(\mathbf{H}\right)_{pq}\left(\mathbf{H}\right)_{rs}^{*}\right] = \left(\mathbf{R}_R\right)_{pr}\left(\mathbf{R}_T\right)_{qs} \tag{4}$$

where $\left(\cdot\right)_{ij}$ is the $\left(i,j\right)$th element of a matrix. According to the Kronecker correlation model, the transmit correlation is independent of the receive antenna and vice versa. Furthermore, the correlation between two subchannels equals to the product of the corresponding transmit and receive correlation. From a physical point-of-view, the Kronecker model appears when the antennas are arranged in regular arrays and the correlation vanishes fast with distance (Chuah, Tse, Kahn, & Valenzuela, 2002). In this point, it is worth mentioning that according to (Chizhik, Foschini, & Valenzuela, 2000; Chizhik, Foschini, Gans, & Valenzuela, 2002) a MIMO channel with a large number of keyholes converges to the Kronecker MIMO model on the basis of the CLT.

It has been well established across the literature (Chuah et al., 2002; Hachem, Khorunzhiy, Loubaton, Najim, & Pastur, 2006) that the Kronecker model can be transformed into a *separable correlation model*, while studying the eigenvalue distribution of $\mathbf{H}\mathbf{H}^{\dagger}$. More specifically, if $\mathbf{R}_R = \mathbf{U}\mathbf{D}_R\mathbf{U}^{\dagger}$ and $\mathbf{R}_T = \mathbf{V}\mathbf{D}_R\mathbf{V}^{\dagger}$ are the eigenvalue decompositions of the receive and transmit correlation matrices respectively, then the eigenvalue distribution of $\mathbf{H}\mathbf{H}^{\dagger} = \mathbf{R}_R^{1/2}\mathbf{G}\mathbf{R}_T\mathbf{G}^{\dagger}\mathbf{R}_R^{1/2}$ is equivalent to the one of $\mathbf{D}_R^{1/2}\mathbf{G}\mathbf{D}_T\mathbf{G}^{\dagger}\mathbf{D}_R^{1/2}$. In this direction, the equivalent MIMO channel matrix can be written as:

$$\mathbf{H} \cong \mathbf{D}_R^{1/2} \mathbf{G} \mathbf{D}_T^{1/2} = \left(\mathbf{d_R}^\dagger \mathbf{d_T}\right)^{\frac{1}{2}} \circ \mathbf{G} \tag{5}$$

where $\mathbf{d_R}$ and $\mathbf{d_T}$ are row vectors containing the eigenvalues of $\mathbf{R_R}$ and $\mathbf{R_T}$ respectively. The term "separable" is due to the structure of the right-hand side of equation (5), according to which the channel matrix can be expressed as the Hadamard product of a separable variance profile and a Gaussian matrix.

In this point, it is worth overviewing the exponential correlation model (Martin & Ottersten, 2004; Mestre, Fonollosa, & Pages-Zamora, 2003; Skupch, Seethaler, & Hlawatsch, 2005) which has been often used in order to effectively quantify the level of spatial correlation. More specifically, it is a model for constructing the receive/transmit correlation matrix utilizing a single coefficient $\rho_e \in \mathbf{C}$ with $\left|\rho_e\right| \leq 1$ as follows:

$$R_{ij} = \begin{cases} \left(\rho_e\right)^{\mathrm{abs}(j-i)}, & i \leq j \\ \left[\left(\rho_e\right)^{\mathrm{abs}(j-i)}\right]^*, & i > j \end{cases} \tag{6}$$

where $\mathrm{abs}(\cdot)$ denotes the absolute value. It has been shown that the exponential model can approximate the correlation in a uniform linear array under rich scattering conditions (Loyka, 2001). Similar correlation models, such as the square exponential and the tridiagonal model can be found in (Shin, Win, & Chiani, 2008).

Finally, a number of analytical correlation models which are based on propagation theory can also be found in the literature. For example, (Shiu et al., 2000) employs the one-ring correlation model which is derived from ray tracing principles and is appropriate for fixed wireless links where the BS is elevated and seldom obstructed. In the same direction, (Muller, 2002) considers a correlation model which incorporates both spatial and temporal correlation for Non Line-Of-Sight (NLOS) links under poor scattering conditions. In this point, it should be noted that the Kronecker and separable correlation models are not able to efficiently describe the correlation structure of every wireless channel. For example, the authors in (Raghavan, Kotecha, & Sayeed, 2008) study the capacity mismatch between sparse non-regular channels described by their canonical coordinates and the Kronecker-correlated channel. Nevertheless, the Kronecker and separable correlation models can be extremely useful in terms of capacity analysis due to their tractable mathematical structure.

ADVANCES IN POINT-TO-POINT MIMO CHANNEL CAPACITY LIMITS

The performance of multi-antenna channels was originally investigated in (Telatar, 1999; Foschini & Gans, 1998) and it was shown that the capacity grows linearly with $\min\left(n_r, n_t\right)$ where n_r and n_t are the number of receive and transmit antennas respectively. However, the correlated fading amongst the multiple antennas compromises the capacity performance with respect to the independent fading case. This phenomenon is widely established in various regimes and settings, as described in the rest of this section.

The already existing results for the point-to-point MIMO channel can be classified in two main categories: exact analysis and asymptotic analysis. In the exact analysis, the probability distributions of finite-dimension matrices are investigated, resulting in closed forms which can produce exact results. On the other hand, in the asymptotic analysis a single or both dimensions of the random channel matrix grow infinitely large in order to allow approximations and simplifications due to the law of large numbers. Although the asymptotic analysis may seem less accurate, it has been widely shown that asymptotic closed forms are able to produce accurate results even for finite dimensions (Tulino & Verdu, 2004). What is more, the asymptotic analysis is ideal for studying cases where the system size is of no importance, since it reveals the effect of normalized parameters and provides insights into the system's performance (Muller, 2002).

In the category of exact analysis, authors in (Marques & Abrantes, 2008) present a self-contained analysis for the case of receive correlation by employing principles from the theory of group representations. In the same direction, the authors in (Chiani, Win, & Zanella, 2003) present a closed form for the non-asymptotic capacity of the single-side correlation model by utilizing joint eigenvalue distribution expressions for Wishart matrices. In the category of asymptotic analysis, the single-side Kronecker correlation model is employed in (Mestre et al., 2003; Skupch et al., 2005), i.e. correlation can affect either the receiving or the transmitting end. Based on the principles of free probability (Voiculescu, 1983), the authors establish the limiting eigenvalue distribution of the single-side exponential correlation model, although they have to resort to numerical evaluations in some parts of the derivation. Furthermore, in (Martin & Ottersten, 2004) the single-side asymptotics of the eigenvalue distribution are considered, meaning that the antennas of the one side approach infinity, whereas the antennas of the other side are kept fixed. Based on eigenvalue distributions of sample covariance matrices, the authors derive a closed form for the channel capacity.

Although the aforementioned asymptotic approaches have provided some useful insights on the capacity of the correlated MIMO channel, the majority of the availabe results focus on the generic setting where correlation affects both transmit and receive end and the numbers of both transmit and receive antennas grow large together while preserving a fixed ratio. Although the asymptotic eigenvalue distribution analysis comprises an approximation for matrices of finite dimensions, it is often employed in order to isolate the effect of specific physical parameters and to produce analytical closed forms. This setting is also suitable for studying the channel capacity of multicell processing cellular systems, since the ratio of transmit and receive antennas is a constant proportional to the per-cell number of UTs K. In this regime, the capacity of the Kronecker correlated (also known as doubly correlated) MIMO channel is expressed as a fixed-point equation based on the Steltjes' transform (Chuah et al., 2002) of the limiting eigenvalue distribution of $\mathbf{HH}^{\dagger}$. In the same direction, authors in (Tulino, Lozano, & Verdu, 2005) study the capacity of the Kronecker correlated MIMO channel based on the principles of Random Matrix Theory (Tulino & Verdu, 2004). The derivation results in a fixed-point equation including functionals of the *SINR* and *MMSE*. In (Sengupta & Mitra, 2006) and (Hachem et al., 2006), the expectation and the variance of the capacity are evaluated using closed forms based on the solution of 2×2 equation systems. More specifically, the results in (Sengupta & Mitra, 2006) are based on the saddle point in integration (Sengupta & Mitra, 1999), whereas the results in (Hachem et al., 2006) are derived using the Poincare-Nash inequality and integration by parts. In (Jorswieck & Boche, 2006b), the principles of majorization theory (Jorswieck & Boche, 2006a) are applied in order to show that the average mutual information is a Schur-concave function with respect to the ordered eigenvalue vector of the correlation matrix. In addition, the doubly correlated MIMO channel for Toeplitz correlation matrices is analyzed in (Shin et al., 2008) based on the concept of linear spectral statistics. Finally, in (Moustakas & Simon, 2007; Taricco, 2007) the performance of Kronecker correlated MIMO channels is studied using the replica method, which originates in theoretical physics.

It should be noted that the aforementioned results specifically focus on the point-to-point correlated MIMO channel. In the following paragraph, we describe the channel matrix characteristics of a multiple-access channel which is the information-theoretic basis for modelling the cellular uplink channel.

Cellular Channel Models

The main characteristic of the cellular channel model is that the Rayleigh fading coefficients do not have a uniform variance. Thus, the channel model of the multicell processing is affected by an additional variance profile which is dictated by the considered path loss model. The evolution of the path-loss variance profile in the context of multicell processing literature can be seen graphically in Figure 3. Initially, (Wyner, 1994) introduced the concept of interference factor a, which quantifies the amount of intercell interference with the first interfering tier. Subsequently, (Somekh & Shamai, 2000), introduced the i.i.d. fading coefficients b,c,d. Multiple tiers of interference were introduced by (Letzepis, 2005), which has considered variable interference factors depending on the distance from the interfering tier and the power-law path loss. Finally, (Chatzinotas, Imran, & Tzaras, 2008a) have alleviated the assumption of collocated UTs by introducing user distribution. The last subfigure of Figure 3 depicts a comparison of the interference factors used in each model for a linear cellular array.

In the current model, we adopt the configuration in (Chatzinotas, Imran, & Tzaras, 2008a). Assuming power-law path loss, Rayleigh flat fading and uniformly distributed users, the received signal at cell n, at time index i, is given by:

$$y^n[i] = \sum_{m=1}^{N}\sum_{k=1}^{K} \varsigma_k^{nm} g_k^{nm}[i] x_k^m[i] + z^n[i] \tag{7}$$

where $x_k^m[i]$ is the ith complex channel symbol transmitted by the kth UT of the mth cell and $\{g_k^{nm}\}$ are independent, strictly stationary and ergodic complex random processes in the time index i, which represent the flat fading processes experienced in the transmission path between the nth BS and the kth UT in the mth cell. The fading coefficients are assumed to have unit power, i.e. $\mathbb{E}[|g_k^{nm}[i]|^2] = 1$ for all (n, m, k) and all UTs are subject to an average power constraint, i.e. $\mathbb{E}[|x_k^m[i]|^2] \leq P$ for all (m, k). The parameter $\gamma = P / \sigma^2$ is defined as the UT transmit power normalized by the receiver noise power. The interference factors ς_k^{nm} in the transmission path between the mth BS and the kth UT in the nth cell are calculated according to the "modified" power-law path loss model (Letzepis, 2005; Ong & Motani, 2007):

$$\varsigma_k^{nm} = (1 + d_k^{nm})^{-\eta/2}. \tag{8}$$

Dropping the time index i, the aforementioned model can be more compactly expressed as a vector memoryless channel of the form:

Figure 3. Evolution of information-theoretic cellular models

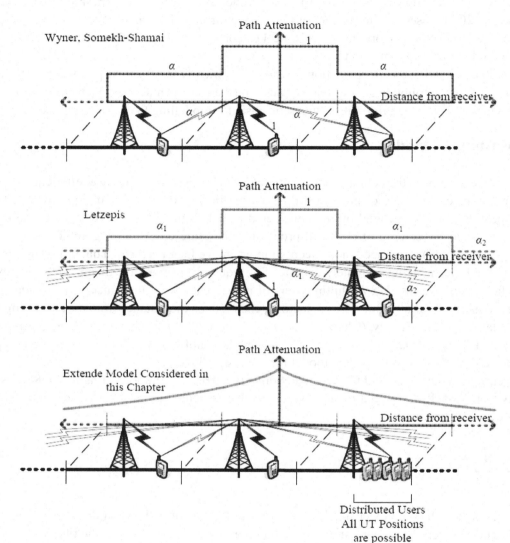

$$\mathbf{y} = \mathbf{H}\mathbf{x} + \mathbf{z}. \tag{9}$$

The channel matrix $\mathbf{H}$ can be written as,

$$\mathbf{H} = \boldsymbol{\Sigma} \circ \mathbf{G} \tag{10}$$

where $\boldsymbol{\Sigma}$ is a $N \times KN$ deterministic matrix and $\mathbf{G}$ is a Gaussian $N \times KN$ matrix with c.c.s. i.i.d. elements of unit variance, comprising the corresponding Rayleigh fading coefficients. The entries of the $\boldsymbol{\Sigma}$ matrix are defined by the variance profile function

$$\varsigma(u,v) = \left(1 + d\left(u,v\right)\right)^{-\eta/2} \tag{11}$$

where $u \in [0,1]$ and $v \in [0,K]$ are the normalized indexes for the BSs and the UTs respectively and $d(u,v)$ is the normalized distance between BS u and user v. In the case of multiple UT and/or BS antennas, the channel matrix $\mathbf{H}$ can be written as,

$$\mathbf{H} = \mathbf{\Sigma}_M \circ \mathbf{G}_M \qquad (12)$$

where $\mathbf{G}_M$ is a standard complex Gaussian $Nn_{BS} \times KNn_{UT}$ matrix with elements of variance 1, comprising the Rayleigh fading coefficients between the KNn_{UT} transmit and the Nn_{BS} receive antennas. Similarly, $\mathbf{\Sigma}_M$ is a $Nn_{BS} \times KNn_{UT}$ deterministic matrix, comprising the path loss coefficients between the KNn_{UT} transmit and the Nn_{BS} receive antennas. Since the multiple antennas of each UT / BS are collocated, $\mathbf{\Sigma}_M$ can be written as a block matrix based on the variance profile matrix $\mathbf{\Sigma}$ of equation (10)

$$\mathbf{\pounds}_M = \mathbf{\pounds} \otimes \mathbf{J} \qquad (13)$$

where $\mathbf{J}$ is a $n_{BS} \times n_{UT}$ matrix of ones.

According to (Tulino & Verdu, 2004), the asymptotic sum-rate capacity C_{opt} for the uncorrelated model assuming a very large number of cells, is given by

$$C_{\text{opt}} = \lim_{N \to \infty} \frac{1}{N} \mathrm{I}\left(\mathbf{x}; \mathbf{y} \mid \mathbf{H}\right)$$

$$= \lim_{N \to \infty} \mathbb{E}\left[\frac{1}{N} \sum_{i=1}^{Nn_{BS}} \log\left(1 + \frac{\tilde{\gamma}}{Kn_{UT}} \lambda_i\left(\frac{1}{N}\mathbf{H}\mathbf{H}^\dagger\right)\right)\right]$$

$$= n_{BS} \int_0^\infty \log(1 + \frac{\tilde{\gamma}}{Kn_{UT}} x) d\mathrm{F}_{\frac{1}{N}\mathbf{H}\mathbf{H}^\dagger}(x)$$

$$= n_{BS} \mathrm{V}_{\frac{1}{N}\mathbf{H}\mathbf{H}^\dagger}\left(\tilde{\gamma} / Kn_{UT}\right) = n_{BS} Kn_{UT} \mathrm{V}_{\frac{1}{N}\mathbf{H}^\dagger\mathbf{H}}\left(\tilde{\gamma} / Kn_{UT}\right) \qquad (14)$$

where $\tilde{\gamma} = KN\gamma$ and $\gamma = P / \sigma^2$ are the system- and UT-transmit power normalized by the receiver noise power respectively, $\lambda_i\left(\mathbf{X}\right)$ denotes the eigenvalues of matrix $\mathbf{X}$ and

$$\mathrm{V}_\mathbf{X}(y) = \mathrm{E}[\log(1 + y\lambda_i\left(\mathbf{X}\right))] = \int_0^\infty \log\left(1 + y\lambda_i\left(\mathbf{X}\right)\right) d\mathrm{F}_\mathbf{X}(x) \qquad (15)$$

is the Shannon transform (Tulino & Verdu, 2004) of a random square Hermitian matrix $\mathbf{X}$, whose limiting eigenvalue distribution has a cumulative function denoted by $\mathrm{F}_\mathbf{X}(x)$. Assuming that the channel matrix is just a Gaussian matrix $\mathbf{G} : \mathrm{CN}\left(\mathbf{0}, \mathbf{I}\right)$, the empirical eigenvalue distribution of $(1 / N)\mathbf{G}^\dagger\mathbf{G}$

converges almost surely (a.s.) to the nonrandom limiting eigenvalue distribution of the Marcenko-Pastur law (Marcenko & Pastur, 1967), whose Shannon transform is given by

$$
V_{\frac{1}{N}\mathbf{G}^\dagger\mathbf{G}}(y) \overset{a.s.}{\to} V_{MP}(y, K) \tag{16}
$$

where

$$
V_{MP}(y, K) = \log\left(1 + y - \frac{1}{4}\varphi(y, K)\right) + \frac{1}{K}\log\left(1 + yK - \frac{1}{4}\varphi(y, K)\right) - \frac{1}{4Ky}\varphi(y, K) \tag{17}
$$

and $\quad \varphi(y, K) = \left(\sqrt{y\left(1 + \sqrt{K}\right)^2 + 1} - \sqrt{y\left(1 - \sqrt{K}\right)^2 + 1}\right)^2.$ $\tag{18}$

The analysis for the multiple-antenna uncorrelated model (Chatzinotas, Imran, & Tzaras, 2008e) is based on the row-regularity (Tulino & Verdu, 2004) of the channel matrix and the Free Pobability approach initially presented in (Letzepis, 2005). According to this result, the probability density function (p.d.f.) of the limiting eigenvalue distribution of $(1 / N)\mathbf{H}^\dagger\mathbf{H}$ follows a scaled version of the Marcenko-Pastur law and hence the Shannon transform of the limiting eigenvalue distribution of $(1 / N)\mathbf{H}^\dagger\mathbf{H}$ can be approximated by

$$
V_{\frac{1}{N}\mathbf{H}^\dagger\mathbf{H}}\left(\frac{\tilde{\gamma}}{Kn_{UT}}\right) \approx V_{MP}\left(q(\mathbf{\Sigma})\frac{\tilde{\gamma}}{Kn_{UT}}, Kn_{UT}\right) \tag{19}
$$

where

$$
q(\mathbf{\Sigma}) = \frac{\|\mathbf{\Sigma}_M\|^2}{Nn_{BS} \cdot KNn_{UT}} = \frac{\|\mathbf{\Sigma}\|^2}{N \cdot KN} = \frac{1}{K}\int_0^K \varsigma^2(u, v)dv. \tag{20}
$$

As a result, the per-cell capacity is given by

$$
C_{opt}(\gamma, N, n_{BS}, K, n_{UT}) \approx n_{BS} \cdot Kn_{UT}V_{MP}\left(q(\mathbf{\Sigma})\frac{\tilde{\gamma}}{Kn_{UT}}, Kn_{UT}\right) \tag{21}
$$

CORRELATED MIMO CELLULAR CHANNEL MODEL & ANALYSIS

As described in the previous section, the correlated channel matrix of the point-to-point MIMO channel can be expressed in terms of the separable variance profile, which depends on the eigenvalues of the correlation matrices. Similarly, the channel matrix of a cellular channel can be expressed in terms of the path-loss variance profile, which depends on the considered UT distribution and path loss exponent. These two distinct problems have been tackled individually by exploiting the separability and the row-regularity of the variance profile respectively. However, the channel matrix produced in the context of a correlated cellular channel is expressed as a combination (Hadamard product) of a separable and a row-regular variance profile. As a result, the variance profile of the produced channel matrix is neither separable nor row-regular and hence it cannot be tackled by any of the aforementioned approaches. In this chapter, we present an asymptotic approach, which relies on free-probabilistic arguments, in order to tackle the problem of the combined variance profile and derive a closed form for the per-cell sum-rate capacity. The derived closed forms provide insights into the effect of correlation on a multicell processing cellular system and in addition they can be utilized for capacity calculations for large systems where Monte Carlo simulations become computationally intensive.

In this direction, let us provide the basic assumptions of the channel model. Firstly, it is considered that the signals are transmitted on a bandwidth W over which the channel's frequency response is assumed to be flat. In addition, the channel is assumed to be quasi-static and thus the channel gains can be considered fixed over a time duration, adequately long for Gaussian codes to be employed. Now assume that the K UTs are uniformly distributed in each cell of a planar cellular system comprising N base stations and that each BS and each UT are equipped with n_{BS} and n_{UT} antennas respectively. Under conditions of correlated flat fading, the received signal at cell n, at time index i, will be given by:

$$\mathbf{y}^n[i] = \sum_{m=1}^{N}\sum_{k=1}^{K} \varsigma_k^{nm} \left(\mathbf{R}_{\mathbf{R}k}^{nm}\right)^{\frac{1}{2}} \mathbf{G}_k^{nm}[i] \left(\mathbf{R}_{\mathbf{T}k}^{nm}\right)^{\frac{1}{2}} \mathbf{x}_k^m[i] + \mathbf{z}^n[i] \tag{22}$$

where $\mathbf{x}_k^m[i]$ is the ith complex channel symbol vector $n_{UT} \times 1$ transmitted by the kth UT of the m th cell and $\{\mathbf{G}_k^{nm}\}$ is a $n_{UT} \times n_{UT}$ random matrix with independent, strictly stationary and ergodic complex random elements in the time index i. According to the Kronecker correlation model, $\mathbf{R}_{\mathbf{T}k}^{nm}$ and $\mathbf{R}_{\mathbf{R}k}^{nm}$ are deterministic transmit and receive correlation matrices of dimensions $n_{UT} \times n_{UT}$ and $n_{BS} \times n_{BS}$ respectively. In this context, the following normalizations are considered in order to ensure that the correlation matrices do not affect the path loss gain of the BS-UT links: $\mathrm{Tr}\left(\mathbf{R}_{\mathbf{T}k}^{nm}\right) \le n_{UT}$ and $\mathrm{Tr}\left(\mathbf{R}_{\mathbf{R}k}^{nm}\right) \le n_{BS}$ for all (n,m,k). The matrix product $\mathbf{R}_{\mathbf{R}k}^{nm}\mathbf{G}_k^{nm}[i]\mathbf{R}_{\mathbf{T}k}^{nm}$ represents the multiple-antenna correlated flat fading processes experienced in the transmission path between the n_{BS} receive antennas of the n th BS and the n_{UT} transmit antennas of the k th UT in the m th cell. The fading coefficients are assumed to have unit power, i.e. $\mathbb{E}_i[\mathbf{G}_k^{nm}[i]\mathbf{G}_k^{nm}[i]^\dagger] = \mathbf{I}$ for all (n,m,k) and all UTs are subject to a power constraint P, i.e. $\mathrm{E}_i[\mathbf{x}_k^m[i]\mathbf{x}_k^m[i]^\dagger] \le \dfrac{P}{n_{UT}}\mathbf{I}_{n_{UT}}$ for all (m,k). The variance coefficients ς_k^{nm} in the transmission path between the m th BS and the k th UT in the n th cell are calculated according to the "modified" power-law path loss model (cf. (8)).

Dropping the time index i, the aforementioned model can be more compactly expressed as a vector memoryless channel of the form

$$\mathbf{Y} = \mathbf{H}\mathbf{X} + \mathbf{Z} \tag{23}$$

where the vector $\mathbf{Y} = [\mathbf{y}^{(1)} \dots \mathbf{y}^{(N)}]^T$ with $\mathbf{y}^{(n)} = [y^1 \dots y^{n_{BS}}]$ represents received signals by the BSs, the vector $\mathbf{X} = [\mathbf{x}^{(1)}_{(1)} \dots \mathbf{x}^{(N)}_{(K)}]^T$ with $\mathbf{x}^{(n)}_{(k)} = [x^1 \dots y^{n_{UT}}]$ represents transmit signals by all the UTs of the cellular system and the components of vector $\mathbf{z} = [\mathbf{z}^{(1)} \dots \mathbf{z}^{(N)}]^T$ with $\mathbf{z}^{(n)} = [z^1 \dots z^{n_{BS}}]$ are i.i.d c.c.s. random variables representing AWGN with $\mathbb{E}[\mathbf{z}^n] = \mathbf{0}, \mathbb{E}[|\mathbf{z}^n|^2] = \sigma^2 \mathbf{I}$. The channel matrix $\mathbf{H}$ can be written as

$$\mathbf{H} = \boldsymbol{\Sigma}_{\mathbf{M}} \circ \left(\left(\mathbf{I}_N \otimes \mathbf{R}_{\mathbf{R}}^{\frac{1}{2}} \right) \mathbf{G}_{\mathbf{M}} \left(\mathbf{I}_{KN} \otimes \mathbf{R}_{\mathbf{T}}^{\frac{1}{2}} \right) \right) \tag{24}$$

where $\mathbf{G}_{\mathbf{M}}$ is a $Nn_{BS} \times KNn_{UT}$ Gaussian matrix with i.i.d. c.s.s. elements of unit variance. As explained before, the Kronecker correlation model is equivalent to a separable variance profile model in terms of its eigenvalue distribution. Based on this equivalence, the channel matrix can be rewritten as follows:

$$\begin{aligned}\mathbf{H} &= \boldsymbol{\Sigma}_{\mathbf{M}} \circ \left(\left(\mathbf{I}_N \otimes \mathbf{R}_{\mathbf{R}}^{\frac{1}{2}} \right) \mathbf{G}_{\mathbf{M}} \left(\mathbf{I}_{KN} \otimes \mathbf{R}_{\mathbf{T}}^{\frac{1}{2}} \right) \right) \\ &\cong \boldsymbol{\Sigma}_{\mathbf{M}} \circ \left(\tilde{\mathbf{D}}_{\mathbf{R}}^{\frac{1}{2}} \mathbf{G}_{\mathbf{M}} \tilde{\mathbf{D}}_{\mathbf{T}}^{\frac{1}{2}} \right) \\ &= \boldsymbol{\Sigma}_{\mathbf{M}} \circ \left(\tilde{\mathbf{d}}_{\mathbf{R}}^{\dagger} \tilde{\mathbf{d}}_{\mathbf{T}} \right)^{\frac{1}{2}} \circ \mathbf{G}_{\mathbf{M}} \end{aligned} \tag{25}$$

where $\tilde{\mathbf{D}}_{\mathbf{R}}$ and $\tilde{\mathbf{D}}_{\mathbf{T}}$ are the diagonal eigenvalue matrices of $\mathbf{I}_{N \times N} \otimes \mathbf{R}_{\mathbf{R}}$ and $\mathbf{I}_{N \times N} \otimes \mathbf{R}_{\mathbf{T}}$ respectively and $\tilde{\mathbf{d}}_{\mathbf{R}}$ and $\tilde{\mathbf{d}}_{\mathbf{T}}$ are row vectors containing the diagonal elements of $\tilde{\mathbf{D}}_{\mathbf{R}}$ and $\tilde{\mathbf{D}}_{\mathbf{T}}$ respectively. As it can be seen, the MIMO correlation model has been transformed into an uncorrelated model with a variance profile $\boldsymbol{\Omega} = \boldsymbol{\Sigma}_{\mathbf{M}} \circ \left(\tilde{\mathbf{d}}_{\mathbf{R}}^{\dagger} \tilde{\mathbf{d}}_{\mathbf{T}} \right)^{\frac{1}{2}}$ which is neither row regular nor separable. Taking into account the simplifying assumption that all the BS-UT links are equally correlated, then the channel matrix could be further rewritten as:

$$\mathbf{H} = \left(\boldsymbol{\Sigma} \otimes \left(\mathbf{d}_{\mathbf{R}}^{\dagger} \mathbf{d}_{\mathbf{T}} \right)^{\frac{1}{2}} \right) \circ \mathbf{G}_{\mathbf{M}} \tag{26}$$

where $\mathbf{d}_{\mathbf{R}}$ and $\mathbf{d}_{\mathbf{T}}$ are row vectors containing the eigenvalues of $\mathbf{R}_{\mathbf{R}}$ and $\mathbf{R}_{\mathbf{T}}$ respectively.

Cellular Uplink Channel Analysis

This section describes a free probability approach which can be utilized to find the asymptotic eigenvalue distribution in the case of single-side and double-side Kronecker correlation. Free probability has been also used in (Mestre et al., 2003; Skupch et al., 2005) to investigate the case of point-to-point MIMO channels correlated on a single side according to the exponential model. The findings are organized in form of propositions which highlight the derived equations and the insights gained from the analysis. The proofs of the propositions can be skipped unless the reader is interested in the underlying principles of the free probability approach.

UT-Side Correlation

Proposition 1: Considering only UT-side correlation, the per-cell capacity is given by (21) which coincides with the case of uncorrelated multiple antennas at the UT side. Therefore, for large values of K ($K >> n_{UT}$) UT-side correlation has no effect on the system's performance. This ascertainment is expected, since the capacity scaling is dictated by the rank of the channel matrix $\mathbf{H}$, which depends only on the number of BS antennas in a cellular scenario.

Proof: In this case, we assume that $\mathbf{R}_T$ has the same eigenvalues for all UTs. The following analysis can be easily generalized to encompass the case of different transmit correlation matrix at each UT, but we keep this assumption to simplify the notations. Assuming that there is no receive correlation at the BS side i.e $\mathbf{R}_R = \mathbf{I}$, the channel matrix of equation (24) can be rewritten as follows:

$$\mathbf{H} = \left(\mathbf{W} \left(\mathbf{I}_{KN} \otimes \mathbf{R}_T^{\frac{1}{2}} \right) \right) \cong \left(\mathbf{W} \left(\mathbf{I}_{KN} \otimes \mathbf{D}_T^{\frac{1}{2}} \right) \right) \tag{27}$$

where $\mathbf{W} = \dfrac{1}{\sqrt{N}} \boldsymbol{\Sigma}_M \circ \mathbf{G}_M$ and therefore

$$\frac{1}{N} \mathbf{H}^\dagger \mathbf{H} = \sum_{i=1}^{Nn_{BS}} \mathbf{h}_i^\dagger \mathbf{h} \cong \sum_{i=1}^{Nn_{BS}} \left(\mathbf{I}_{KN} \otimes \mathbf{D}_T^{\frac{1}{2}} \right) \mathbf{w}_i^\dagger \mathbf{w}_i \left(\mathbf{I}_{KN} \otimes \mathbf{D}_T^{\frac{1}{2}} \right)$$

$$= \sum_{i=1}^{Nn_{BS}} \left(\left(\mathbf{1}_{KN} \otimes \boldsymbol{\lambda}_T^{\frac{1}{2}} \right) \circ \mathbf{w}_i \right)^\dagger \left(\left(\mathbf{1}_{KN} \otimes \boldsymbol{\lambda}_T^{\frac{1}{2}} \right) \circ \mathbf{w}_i \right) \tag{28}$$

where $\mathbf{w}_i$ denotes the ith $1 \times KNn_{UT}$ row vector of $\mathbf{W}$, $\mathbf{1}_{KN}$ is a $1 \times KN$ row vector of ones and $\boldsymbol{\lambda}_T$ is a row vector containing the eigenvalues of $\mathbf{R}_T$. Hence, the R-transform can be written as

$$\lim_{N \to \infty} R_{\frac{1}{N} \mathbf{H}^\dagger \mathbf{H}}(w) = \lim_{N \to \infty} \sum_{i=1}^{Nn_{BS}} R_{\mathbf{h}_i^\dagger \mathbf{h}_i}(w) = \lim_{N \to \infty} \frac{1}{Kn_{UT}N} \sum_{i=1}^{Nn_{BS}} \frac{\left\| \mathbf{h}_i \right\|^2}{1 - w \left\| \mathbf{h}_i \right\|^2}$$

$$= \frac{q(\Omega)}{1 - \dfrac{Kn_{UT}}{n_{BS}} w q(\Omega)} = R_{q(\Omega)\frac{1}{N} \mathbf{G}_M^\dagger \mathbf{G}_M}(\omega) \tag{29}$$

where

$$q(\Omega) = \frac{\left\| \mathbf{h}_i \right\|^2}{KNn_{UT}} = \frac{\left\| \left(\mathbf{1}_{KN} \otimes \boldsymbol{\lambda}_T^{\frac{1}{2}} \right) \mathbf{w}_i \right\|^2}{KNn_{UT}} = \frac{1}{n_{UT}} \sum_{j=1}^{n_{UT}} \boldsymbol{\lambda}_T(j) \cdot \frac{1}{K} \int_0^K \varsigma^2(u,v) dv$$

$$= \frac{1}{K}\int_0^K \varsigma^2(u,v)dv = q(\mathbf{\Sigma}) \tag{30}$$

It can be seen that the scaling of the Marcenko-Pastur law is identical with the case of uncorrelated UT-side antennas (equation (20)). □

BS-Side Correlation

Proposition 2: Considering only BS-side correlation, the per-cell capacity is given by

$$C_{\text{opt}}(\gamma, N, n_{BS}, K, n_{UT}) = n_{BS}\int_0^\infty \log\left(1 + \frac{\tilde{\gamma}}{Kn_{UT}}x\right)f_{\frac{1}{N}\mathbf{HH}^\dagger}^\infty(x)dx$$

$$= Kn_{UT}\int_0^\infty \log\left(1 + \frac{\tilde{\gamma}}{Kn_{UT}}x\right)f_{\frac{1}{N}\mathbf{H}^\dagger\mathbf{H}}^\infty(x)dx \tag{31}$$

while each point of the asymptotic eigenvalue pdf (AEPDF) of $(1/N)\mathbf{H}^\dagger\mathbf{H}$ can be calculated analytically by determining the imaginary part (Im{}) of the Cauchy transform G for real arguments

$$f^\infty(x) = \lim_{y\to 0^+}\frac{1}{\pi}Im\{G(x+jy)\} \tag{32}$$

considering that the Cauchy transform is derived from the R-transform (Raj-Rao & Edelman, 2007) as follows

$$G^{-1}(w) = R(-w) - \frac{1}{w} \tag{33}$$

and the R-transform of $(1/N)\mathbf{H}^\dagger\mathbf{H}$ is given by

$$R_{\mathbf{H}^\dagger\mathbf{H}}(w) = \sum_{j=1}^{n_{BS}}\lambda_\mathbf{R}(j)R_\mathbf{A}(\lambda_\mathbf{R}(j)w) \tag{34}$$

The effect of BS-side correlation cannot be straightforwardly deduced by the above equations and thus it is discussed in detail in the "Uplink Numerical Results" section.

Proof: In this case, we assume that $\mathbf{R_R}$ has the same eigenvalues for all BSs. The following analysis can be easily generalized to encompass the case of receive different correlation matrix at each BS, but we keep this assumption to simplify the notations. Assuming that there is no transmit correlation at the UT side i.e. $\mathbf{R_T} = \mathbf{I}$, the channel matrix of equation (24) can be rewritten as follows:

$$\mathbf{H} = \left(\left(\mathbf{I}_N \otimes \mathbf{R_R}^{\frac{1}{2}} \right) \mathbf{W} \right) \cong \left(\left(\mathbf{I}_N \otimes \mathbf{D_R}^{\frac{1}{2}} \right) \mathbf{W} \right) \tag{35}$$

and therefore

$$\frac{1}{N} \mathbf{H}^\dagger \mathbf{H} = \frac{1}{N} \sum_{i=1}^{N} \mathbf{H}_i^\dagger \mathbf{H}_i = \frac{1}{N} \sum_{i=1}^{N} \mathbf{W}_i^\dagger \mathbf{D_R} \mathbf{W}_i = \sum_{j=1}^{n_{BS}} \lambda_{\mathbf{R}}(j) \sum_{i=1}^{N} \mathbf{w}_i^\dagger \mathbf{w}_i \tag{36}$$

where $\mathbf{H}_i$ and $\mathbf{W}_i$ are submatrices of $\mathbf{H}$ and $\mathbf{W}$ respectively with dimensions $n_{BS} \times KNn_{UT}$ and $\lambda_{\mathbf{R}}$ is a row vector containing the eigenvalues of $\mathbf{R_R}$. Based on the previous analysis, the asymptotic eigenvalue distribution of $\mathbf{A} = \sum_{i=1}^{N} \mathbf{w}_i^\dagger \mathbf{w}_i$ follows a scaled version of the Marcenko-Pastur law. Hence, the R-transform of $\mathbf{A}$ can be written as

$$R_{\mathbf{A}}(w) \approx R_{q(\Sigma)\frac{1}{N}\tilde{\mathbf{G}}^\dagger \tilde{\mathbf{G}}}(w) = \frac{q(\Sigma)}{1 - Kn_{UT} w q(\Sigma)} \tag{37}$$

where $\tilde{\mathbf{G}}$ is a $N \times KNn_{UT}$ matrix distributed as $CN\left(\mathbf{0}, \mathbf{I}\right)$ and

$$q(\Sigma) = \frac{\|\mathbf{w}_i\|^2}{KNn_{UT}} = \frac{1}{K} \int_0^K \varsigma^2(u,v) dv \tag{38}$$

Based on (Tulino & Verdu, 2004, Theorems 2.31 and 2.64), the R-transform of $(1 / N)\mathbf{H}^\dagger \mathbf{H}$ is given by equation (34). □

Double-Side Correlation

Proposition 3: By combining the two previous cases, it can be easily seen that the AEPDF for the double-side Kronecker correlation model and therefore the resulting capacity coincides with the BS-side correlation case, since transmit correlation has no effect on the asymptotic eigenvalue distribution of $(1 / N)\mathbf{H}\mathbf{H}^\dagger$.

Uplink Numerical Results

The analytical results have been verified by running Monte Carlo simulations over 100 random instances of the system and by averaging the produced results. More specifically, for each system instance the complex matrix $\left(\tilde{\mathbf{d}}_{\mathbf{R}}^\dagger \tilde{\mathbf{d}}_{\mathbf{T}} \right)^{\frac{1}{2}} \circ \mathbf{G}$ is constructed by randomly generating correlated fading coefficients according to the exponential model with ρ_R and ρ_T being the correlation coefficient at the BS-side and the UT-side respectively. Subsequently, the variance profile matrix Σ is constructed by randomly placing the UTs according to the considered distribution in the coverage area of each cell and by calculating the variance profile coefficients using equation (8). In the context of the mathematical analysis, the distance d_k^{nm} can be calculated assuming that the UTs are positioned on a uniform planar grid. After constructing the channel matrix $\mathbf{H}$, the sum-rate capacity is calculated by evaluating the formula in (Telatar, 1999)

$$C_{\text{opt}} = \frac{1}{N} \mathbb{E}\left[\text{logdet}\left(\mathbf{I}_{Nn_{BS}} + \frac{\gamma}{n_{UT}} \mathbf{H}\mathbf{H}^{\dagger}\right)\right]$$

Figure 4 depicts the per-cell sum-rate capacity vs. the normalized cell radius R varying the level of receive correlation $\rho_R = [0, 0.9, 0.99, 1]$ in a planar cellular system with uniformly distributed UTs. As it can be seen, the receive correlation decreases the degrees of freedom due to the multiple receive antennas and therefore compromises the capacity performance of the system. In the no-correlation extreme $\rho_R = 0$, the capacity curves are identical to the curves derived in (Chatzinotas, Imran, & Tzaras, 2008e) for multicell processing cellular systems with multiple antennas. In the full-correlation extreme $\rho_R = 1$, the capacity curve degrades to the single-antenna capacity, since no diversity gain is achieved by the multiple BS antennas (Chatzinotas, Imran, & Tzaras, 2008a). It should be noted that correlation does not affect the linear capacity scaling with the number of receive antennas n_{BS}. Nevertheless, correlation results in a decreased growth rate with respect to the uncorrelated case.

Cellular Downlink Channel Analysis

In this section, we study the effect of fading correlation on the sum-rate capacity of the downlink cellular channel. The cellular downlink channel is defined in accordance with the uplink channel, as described in the previous section. In this context, the dual downlink model can be compactly expressed as a vector memoryless channel of the form

$$\mathbf{Y} = \mathbf{H}_{DL}\mathbf{X} + \mathbf{Z} \tag{39}$$

where the vector $\mathbf{X} = [\mathbf{x}^{(1)} \dots \mathbf{x}^{(N)}]^T$ with $\mathbf{x}^{(n)} = [x^1 \dots x^{n_{BS}}]$ represents transmitted signals by the BSs, the vector $\mathbf{Y} = [\mathbf{y}_{(1)}^{(1)} \dots \mathbf{y}_{(K)}^{(N)}]^T$ with $\mathbf{y}_{(k)}^{(n)} = [y^1 \dots y^{n_{UT}}]$ represents transmit received by all the UTs of the cellular system and the components of vector $\mathbf{Z} = [\mathbf{z}_{(1)}^{(1)} \dots \mathbf{z}_{(K)}^{(N)}]^T$ with $\mathbf{z}_{(k)}^{(n)} = [z^1 \dots z^{n_{UT}}]$ are i.i.d c.c.s. random variables representing AWGN with $\mathbb{E}[\mathbf{z}^n] = \mathbf{0}, \mathbb{E}[|\mathbf{z}^n|^2] = \sigma^2\mathbf{I}$. The channel matrix $\mathbf{H}$ can be written as

$$\mathbf{H}_{DL} = \mathbf{H}^{\dagger} = \mathbf{\Sigma_M}^{\dagger} \circ \left(\left(\mathbf{I}_{KN} \otimes \mathbf{R_T}^{\frac{1}{2}}\right)\mathbf{G_M}^{\dagger}\left(\mathbf{I}_N \otimes \mathbf{R_R}^{\frac{1}{2}}\right)\right) \tag{40}$$

The capacity region for the downlink cellular channel is given by employing Dirty Paper Coding along with the optimal power allocation (Weingarten, Steinberg, & Shamai, 2006):

Figure 4. Per-cell uplink sum-rate capacity vs. the normalized cell Radius R varying the level of BS-side correlation $\rho_R = [0, 0.9, 0.99, 1]$ *in a planar cellular system with uniformly distributed UTs. No UT-side correlation is considered. Parameters:* $N = 7, K = 4, \gamma = 10, n_{BS} = 2, n_{UT} = 2, \eta = 2.$

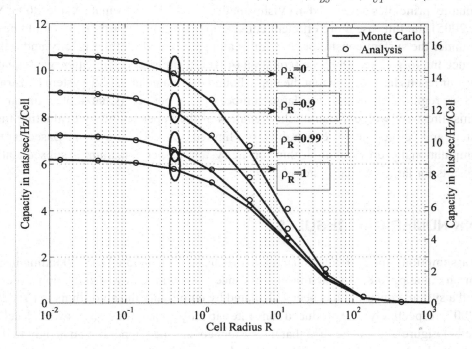

$$C_{opt} = \frac{1}{N} E \left[\max_{\mathbf{Q}} \sum_{i=1}^{KN} \log \frac{\det(\mathbf{I} + \mathbf{H}_i^\dagger \sum_{j=1}^{i} \mathbf{Q}_j(\mathbf{H})\mathbf{H}_i)}{\det(\mathbf{I} + \mathbf{H}_i^\dagger \sum_{j=1}^{i-1} \mathbf{Q}_j(\mathbf{H})\mathbf{H}_i)} \right] \qquad (41)$$

where $\mathbf{H}_i$ is the ith $Nn_{BS} \times n_{UT}$ column matrix of $\mathbf{H}_{DL}$ and $\mathbf{Q}_j$ are the $Nn_{BS} \times Nn_{BS}$ downlink input covariance matrices with $\mathbf{Q}^{(n)}_i$ being the nth $n_{BS} \times n_{BS}$ block matrix across the diagonal of $\mathbf{Q}_i$. The per-cell power constraint can be expressed as:

$$Tr(\mathbf{x}^{(n)}\mathbf{x}^{(n)\dagger}) = \sum_{i=1}^{KN} Tr(\mathbf{Q}_i^{(n)}) \leq P_{tot} / N, \forall n \qquad (42)$$

whereas the system power constraint as

$$Tr(\mathbf{X}\mathbf{X}^\dagger) = \sum_{i=1}^{KN} Tr(\mathbf{Q}_i) \leq P_{tot} \qquad (43)$$

In practical systems, there is a power constraint per BS imposed by the amplifier saturation limit (per-cell constraint). The case of per-cell power constraint has been studied in (Yu & Lan, 2007), but the proposed interior-point algorithm is too computational intensive, especially for our setting which includes simulating a large number of cells and averaging over a large number of fading instances. However, by consider a power constraint which is imposed over all the BSs of the system (system constraint), we can employ the duality principles developed in (Vishwanath et al., 2003; Viswanath & Tse, 2003; Yu, 2006b; Jindal et al., 2005) and thus use more efficient convex optimization algorithms. It should be noted that the capacity under the system constraint constitutes an upper bound on the capacity under the per-cell constraint, since the former search space (equation (43)) is a superset of the latter one (equation (42)). Interestingly, our results have shown that the upper bound calculated under the system constraint is tight in the high-SNR regime, when there is a large number of UTs per cell (Chatzinotas, Imran, & Tzaras, 2008d). In this direction, we demonstrate the tightness of the upper bound by evaluating the capacity for both system and per-cell power constraints in a limited-size cellular system. Based on this result, the more computationally-efficient algorithm for the system power constraint is subsequently utilized to produce results for large cellular systems.

DOWNLINK NUMERICAL RESULTS

Under these assumptions, the correlated channel matrix has been generated for 100 fading instances and for a correlation value ρ_R=0.8. For each fading instance, the optimal power allocation under a per-cell constraint and a system constraint has been calculated by employing the algorithms in (Yu, 2006b) and (Yu & Lan, 2007) respectively. The produced sum-rate capacity for a limited-size downlink cellular channel is depicted in Figure 5. It can be seen that the system constrained upper bound is tight to the realistic per-cell constrained capacity curve. Based on this result, we employ the more efficient system constrained optimization algorithms to calculate the large-system capacity in the section "Practical Results".

Practical Results

In order to apply the aforementioned results to real-world cellular systems, a reference distance d_0 is required to interconnect the actual distance $\hat{d}$ and the normalized distance $d = \hat{d} / d_0$. If the power loss at the reference distance d_0 is L_0, the scaled variance profile function is given by

$$\varsigma = \sqrt{L_0(1 + \hat{d} / d_0)^{-\eta}} \tag{44}$$

The values of L_0 and η have been fitted to the three scenarios presented in (ETSI, 2007) as follows: a) Suburban Macro L_0=31.5dB, η=3.5, b) Urban Macro L_0=34.5dB, η=3.5 and Urban Micro L_0=34.53dB, η=3.8. The value of the correlation parameter ρ_R has been selected based on the specifications in (ETSI, 2007). The selected scenario considers an antenna spacing of 4λ, which corresponds to 60cm in an operation frequency of 2 GHz. Assuming 2 degrees angle spread and 50 degrees angle of arrival, the correlation should be ρ_R=0.8624. The presented results are plotted versus a variable cell radius R using the following parameters:

- Cell Radius R = 0.1-3 Km
- Reference Distance d_0 = 1 m

Figure 5. Per-cell downlink sum-rate capacity vs. the normalized cell radius R in a planar cellular system with uniformly distributed UTs under per-cell and system power constraint. No UT-side correlation is considered. Parameters: $N=2$, $K=4$, $\gamma_{SYS}=30dB$, $n_{BS}=2$, $n_{UT}=1$, $\eta=2$, $\rho_R=0.8$.

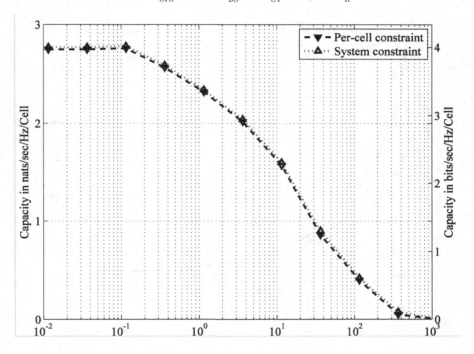

- UTs per cell $K = 4$
- UT Transmit Power $P_T = 200mW$
- BS Transmit Power $P_{BS} = 50$ W
- Thermal Noise Density $N_0 = -169$ dBm/Hz
- Channel Bandwidth $W = 5$ Mhz

As it can be seen (Figure 6 and Figure 7), in practical systems the capacity gap due to fading correlation at the BS side is evident, but fortunately it does not grow linearly with the level of correlation. As a result, even for high correlation levels the capacity gain due to multiple antennas at the BS is not greatly compromised.

FUTURE TRENDS

- Uplink Input optimization using Statistical CSI

The current study has considered that in the cellular uplink channel UTs are completely unaware of the channel conditions (CSI), while the BSs have perfect CSI meaning that they are aware of the instantaneous channel gains of all UTs. A more practical scenario which is worth investigating is when both UT and

Figure 6. Per-cell uplink sum-rate capacity vs. cell radius R varying the level of BS-side correlation $\rho_R=[0,0.8624,1]$ in a planar cellular system with uniformly distributed UTs. No UT-side correlation is considered.

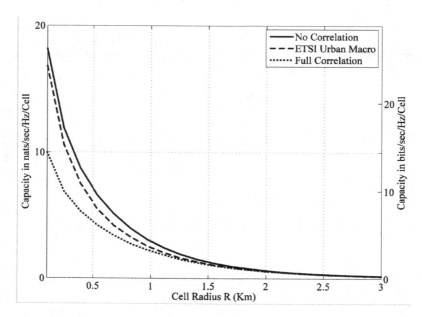

Figure 7. Per-cell downlink sum-rate capacity vs. cell radius R varying the level of BS-side correlation $\rho_R=[0,0.8624,1]$ in a planar cellular system with uniformly distributed UTs. No UT-side correlation is considered.

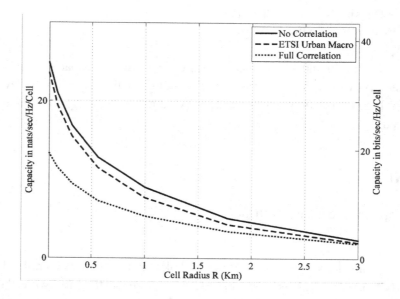

BS have statistical channel state information. This entails that the statistics of the fading process e.g. the correlation matrices are available to the transmitters, but the actual instantaneous fading realization is unknown. From an implementation point-of-view, the statistical CSI can be more effectively measured at the UT and fed back to the BS, since it remains fixed for longer channel blocks. For more details, the reader is referred to (Tulino, Lozano & Verdu, 2006).

- Multiobjective resource allocation

In the information-theoretic literature, the majority of results have been focused on the sum-rate capacity or the weighted sum-rate capacity as a metric for the performance of cellular systems. However, according to (Chatzinotas, Imran, & Tzaras, 2008c) maximizing the sum-rate capacity may not be always beneficial for the rate fairness of the system. Therefore, more sophisticated power control and allocation algorithms are needed in order to optimize the system performance with respect to multiple objectives e.g. sum-rate capacity maximization, fair user rate distribution and spatial coverage. In this context, the combination of multicell processing and relaying techniques would be a case of interest. For more details the reader is referred to (Simeone, Somekh, Bar-Ness & Spagnolini, 2007).

CONCLUSION

In this chapter, we have considered a multicell processing system with MIMO links and distributed UTs. In this context, we have investigated the effect of antenna correlation on the capacity performance of the system. The presented results has been derived considering that the variances of the Gaussian channel gains are scaled by a generic variance profile which incorporates both path loss and antenna correlation. In this direction, we have presented an asymptotic Free Probability approach in order to derive closed-forms for the uplink channel and we have ininvestigated the downlink channel performance using optimal power allocation algorithms. The main findings can be summarized as follows: antenna correlation degrades the capacity performance of the system, especially if it appears on the BS side. In practical systems, the capacity gain due to multiple antennas at the BS is not greatly compromised even for high correlation levels, since the capacity gap due to fading correlation does not grow linearly with the level of correlation.

ACKNOWLEDGMENT

The work reported in this chapter has formed part of the "Fundamental Limits to Wireless Network Capacity" Elective Research Programme of the Virtual Centre of Excellence in Mobile & Personal Communications, Mobile VCE, www.mobilevce.com. This research has been funded by the following Industrial Companies who are Members of Mobile VCE - BBC, BT, Huawei, Nokia, Nokia Siemens Networks, Nortel, Vodafone. Fully detailed technical reports on this research are available to staff from these Industrial Members of Mobile VCE. The authors would like to thank Prof. G. Caire and Prof. D. Tse for the useful discussions.

REFERENCES

Aktas, E., Evans, J., & Hanly, S. (2008). Distributed decoding in a cellular multiple-access channel. *IEEE Transactions on Wireless Communications*, *7*(1), 241–250. doi:10.1109/TWC.2008.060469

Alexander, P., & Grant, A. (2000). Iterative channel and information sequence estimation in CDMA. In *IEEE Sixth International Symposium on Spread Spectrum Techniques and Applications,* (Vol. 2, pp. 593–597).

Andrews, J. (2005). Interference cancellation for cellular systems: A contemporary overview. *IEEE Transactions on Wireless Communications*, *12*(2), 19–29. doi:10.1109/MWC.2005.1421925

Bacha, M., Evans, J., & Hanly, S. (2006). On the capacity of MIMO cellular networks with macrodiversity. In *7th Australian Communications Theory Workshop Proceedings*, (pp. 105–109).

Chatzinotas, S., Imran, M., & Tzaras, C. (2008a). Optimal information theoretic capacity of planar cellular uplink channel. In *9th IEEE International Workshop on Signal Processing Advances in Wireless Communications*, Recife, Brazil (pp. 196-200).

Chatzinotas, S., Imran, M., & Tzaras, C. (2008d). On the multicell processing rates of the cellular downlink fading channel. Submitted to IET Communications.

Chatzinotas, S., Imran, M., & Tzaras, C. (2008e). Uplink capacity of MIMO cellular systems with multicell processing. In *IEEE International Symposium on Wireless Communication Systems,* Reykjavik, Iceland.

Chatzinotas, S., Imran, M. A., & Tzaras, C. (2008b). On the capacity of variable density cellular systems under multicell decoding. *IEEE Communications Letters*, *12*(7), 496–498. doi:10.1109/LCOMM.2008.080439

Chatzinotas, S., Imran, M. A., & Tzaras, C. (2008c). Spectral efficiency of variable density cellular systems. In *IEEE International Symposium on Personal, Indoor and Mobile Radio Communications, International Workshop on Efficiency*, Cannes, France.

Chiani, M., Win, M., & Zanella, A. (2003). On the capacity of spatially correlated MIMO Rayleigh-fading channels. *IEEE Transactions on Information Theory*, *49*(10), 2363–2371. doi:10.1109/TIT.2003.817437

Chizhik, D., Foschini, G., Gans, M., & Valenzuela, R. (2002). Keyholes, correlations, and capacities of multielement transmit and receive antennas. *IEEE Transactions on Wireless Communications*, *1*(2), 361–368. doi:10.1109/7693.994830

Chizhik, D., Foschini, G., & Valenzuela, R. (2000). Capacities of multi-element transmit and receive antennas: Correlations and keyholes. *Electronics Letters*, *36*(13), 1099–1100. doi:10.1049/el:20000828

Choi, W., & Andrews, J. G. (2008). The capacity gain from intercell scheduling in multi-antenna systems. *IEEE Transactions on Wireless Communications*, *7*(2), 714–725. doi:10.1109/TWC.2008.060615

Chuah, C.-N., Tse, D., Kahn, J., & Valenzuela, R. (2002). Capacity scaling in MIMO wireless systems under correlated fading. *IEEE Transactions on Information Theory*, *48*(3), 637–650. doi:10.1109/18.985982

Costa, M. (1983). Writing on dirty paper. *IEEE Transactions on Information Theory, IT-29,* 439–441. doi:10.1109/TIT.1983.1056659

ETSI TR 125 996 V7.0.0. (2007). Universal mobile telecommunications system (UMTS): Spatial channel model for multiple input multiple output (MIMO) simulations (3GPP tr 25.996 version 7.0.0 release 7).

Foschini, G. J., & Gans, M. J. (1998). On limits of wireless communications in a fading environment when using multiple antennas. *Wireless Personal Communications, 6*(3), 311–335. doi:10.1023/A:1008889222784

Gesbert, D., Bolcskei, H., Gore, D., & Paulraj, A. (2002). Outdoor MIMO wireless channels: Models and performance prediction. *IEEE Transactions on Communications, 50*(12), 1926–1934. doi:10.1109/TCOMM.2002.806555

Grant, A., Hanly, S., Evans, J., & Muller, R. (2004). Distributed decoding for Wyner cellular systems. In *5th Australian Communications Theory Workshop* (pp. 77–81).

Hachem, W., Khorunzhiy, O., Loubaton, P., Najim, J., & Pastur, L. (2006). A new approach for capacity analysis of large dimensional multi-antenna channels. Submitted to *IEEE Transactions on Information Theory.*

Hanley, S., & Whiting, P. (1993). Information-theoretic capacity of multi-receiver networks. *Telecommunication Systems, 1,* 1–42. doi:10.1007/BF02136153

Harashima, H., & Miyakawa, H. (1972). Matched-transmission technique for channels with intersymbol interference. *IEEE Transactions on Communications, 20,* 774–780. doi:10.1109/TCOM.1972.1091221

Hochwald, B., Peel, C., & Swindlehurst, A. (2005). A vector-perturbation technique for near-capacity multiantenna multiuser communication-part II: Perturbation. *IEEE Transactions on Communications, 53,* 537–544. doi:10.1109/TCOMM.2004.841997

Jindal, N., Rhee, W., Vishwanath, S., Jafar, S., & Goldsmith, A. (2005). Sum power iterative water-filling for multi-antenna Gaussian broadcast channels. *IEEE Transactions on Information Theory, 51*(4), 1570–1580. doi:10.1109/TIT.2005.844082

Jorswieck, E., & Boche, H. (2006a). Majorization and matrix-monotone functions in wireless communications. *Foundations and Trends in Communications and Information Theory, 3*(6), 553–701. doi:10.1561/0100000026

Jorswieck, E., & Boche, H. (2006b). Performance analysis of MIMO systems in spatially correlated fading using matrix-monotone functions. *IEICE Transactions, 89-A*(5), 1454–1472.

Kermoal, J., Schumacher, L., Pedersen, K., Mogensen, P., & Frederiksen, F. (2002). A stochastic MIMO radio channel model with experimental validation. *IEEE Journal on Selected Areas in Communications, 20*(6), 1211–1226. doi:10.1109/JSAC.2002.801223

Lee, W. (1990). *Mobile cellular telecommunications systems.* New York: McGraw-Hill, Inc.

Letzepis, N. (2005). *Gaussian cellular multiple access channels*. Unpublished doctoral dissertation, Institute for Telecommunications Research, University of South Australia.

Liang, Y., Yoo, T., & Goldsmith, A. (2006). Coverage spectral efficiency of cellular systems with cooperative base stations. In *IEEE Global Telecommunications Conference*, San Francisco, CA.

Loyka, S. (2001). Channel capacity of MIMO architecture using the exponential correlation matrix. *IEEE Communications Letters, 5*(9), 369–371. doi:10.1109/4234.951380

Marcenko, V., & Pastur, L. (1967). Distributions of eigenvalues of some sets of random matrices. *Math. USSR-Sb., 1*, 507–536.

Marques, P., & Abrantes, S. (2008). On the derivation of the exact, closed-form capacity formulas for receiver-sided correlated MIMO channels. *IEEE Transactions on Information Theory, 54*(3), 1139–1161. doi:10.1109/TIT.2007.915692

Marsch, P., & Fettweis, G. (2008). On multicell cooperative transmission in backhaul-constrained cellular systems. *Annales des Télécommunications, 63*(5-6). doi:10.1007/s12243-008-0028-3

Martin, C., & Ottersten, B. (2004). Asymptotic eigenvalue distributions and capacity for MIMO channels under correlated fading. *IEEE Transactions on Wireless Communications, 3*(4), 1350–1359. doi:10.1109/TWC.2004.830856

Mestre, X., Fonollosa, J., & Pages-Zamora, A. (2003). Capacity of MIMO channels: Asymptotic evaluation under correlated fading. *IEEE Journal on Selected Areas in Communications, 21*(5), 829–838. doi:10.1109/JSAC.2003.810352

Moustakas, A., & Simon, S. (2007). On the outage capacity of correlated multiple-path MIMO channels. *IEEE Transactions on Information Theory, 53*(11), 3887–3903. doi:10.1109/TIT.2007.907468

Muller, R. (2002). A random matrix model of communication via antenna arrays. *IEEE Transactions on Information Theory, 48*(9), 2495–2506. doi:10.1109/TIT.2002.801467

Ng, B., Evans, J., Hanly, S., & Grant, A. (2004). Distributed linear multiuser detection in cellular networks. In *5th Australian Communications Theory Workshop* (pp. 127–132).

Ong, L., & Motani, M. (2007). On the capacity of the single source multiple relay single destination mesh network. *Ad Hoc Networks, 5*(6), 786–800. doi:10.1016/j.adhoc.2006.12.006

Peel, C. (2003). On "dirty-paper coding." *IEEE Signal Processing Magazine, 20*(3), 112–113. doi:10.1109/MSP.2003.1203214

Raghavan, V., Kotecha, J. H., & Sayeed, A. M. (2008). Why does a Kronecker model result in misleading capacity estimates? Submitted to *IEEE Transactions on Information Theory*.

Raj Rao, N., & Edelman, A. (2007). The polynomial method for random matrices. In *Foundations of Computational Mathematics*.

Sengupta, A., & Mitra, P. (1999). Distributions of singular values for some random matrices. *Physical Review E: Statistical Physics, Plasmas, Fluids, and Related Interdisciplinary Topics, 60*(3), 3389–3392. doi:10.1103/PhysRevE.60.3389

Sengupta, A., & Mitra, P. (2006). Capacity of multivariate channels with multiplicative noise: Random matrix techniques and large-n expansions (2). *Journal of Statistical Physics, 125*(5), 1223–1242. doi:10.1007/s10955-006-9076-0

Shamai, S., Somekh, O., Simeone, O., Sanderovich, A., Zaidel, B., & Poor, H. (2007). Cooperative multicell networks: Impact of limited-capacity backhaul and interusers links. In *Joint Workshop on Coding and Communications*, Austria.

Shamai, S., Somekh, O., & Zaidel, M. (2004). Multicell communications: An information theoretic perspective. In *Joint Workshop on Communications and Coding*, Donnini (Florence), Italy.

Shamai (Shitz), S., & Wyner, A. (1997). Information theoretic considerations for symmetric, cellular multiple access fading channels part I and part II. *IEEE Transactions on Information Theory, 43*, 1877–1911. doi:10.1109/18.641553

Shin, H., Win, M., & Chiani, M. (2008). Asymptotic statistics of mutual information for doubly correlated MIMO channels. *IEEE Transactions on Wireless Communications, 7*(2), 562–573. doi:10.1109/TWC.2008.060271

Shiu, D.-S., Foschini, G., & Kahn, M. G. J. (2000). Fading correlation and its effect on the capacity of multielement antenna systems. *IEEE Transactions on Communications, 48*(3), 502–513. doi:10.1109/26.837052

Simeone, O., Somekh, O., Bar-Ness, Y., & Spagnolini, U. (2006). Low-SNR analysis of cellular systems with cooperative base stations and mobiles. In *Fortieth Asilomar Conference on Signals, Systems, and Computers* (pp. 626–630).

Simeone, O., Somekh, O., Bar-Ness, Y., & Spagnolini, U. (2007). Uplink throughput of TDMA cellular systems with multicell processing and amplify-and-forward cooperation between mobiles. *IEEE Transactions on Wireless Communications, 6*(8), 2942–2951. doi:10.1109/TWC.2007.051026

Skupch, A., Seethaler, D., & Hlawatsch, F. (2005). Free probability based capacity calculation for MIMO channels with transmit or receive correlation. In *International Conference on Wireless Networks, Communications and Mobile Computing* (Vol.2, pp. 1041–1046).

Somekh, O., & Shamai, S. (1998). A Shannon-theoretic view of Wyner's multiple-access cellular channel model in the presence of fading. In *IEEE International Symposium on Information Theory* (p. 393).

Somekh, O., & Shamai, S. (2000). Shannon-theoretic approach to a Gaussian cellular multiple-access channel with fading. *IEEE Transactions on Information Theory, 46*(4), 1401–1425. doi:10.1109/18.850679

Somekh, O., Simeone, O., Poor, H. V., & Shamai, S. (2007). Cellular systems with full-duplex amplify-and-forward relaying and cooperative base-stations. In *IEEE International Symposium on Information Theory*, Nice, France.

Somekh, O., Zaidel, B., & Shamai, S. (2007). Sum rate characterization of joint multiple cell-site processing. *IEEE Transactions on Information Theory, 53*(12), 4473–4497. doi:10.1109/TIT.2007.909170

Taricco, G. (2008). Asymptotic mutual information statistics of separately-correlated Rician fading MIMO channels. *IEEE Transactions on Information Theory*, *54*(8), 3490–3504. doi:10.1109/TIT.2008.926415

Telatar, I. E. (1999). Capacity of multi-antenna Gaussian channels. *European Transactions on Telecommunications*, *10*(6), 585–595. doi:10.1002/ett.4460100604

Tomlinson, M. (1971). New automatic equalizer employing modulo arithmetic. *Electronics Letters*, *7*, 138–139. doi:10.1049/el:19710089

Tulino, A., Lozano, A., & Verdu, S. (2005). Impact of antenna correlation on the capacity of multiantenna channels. *IEEE Transactions on Information Theory*, *51*(7), 2491–2509. doi:10.1109/TIT.2005.850094

Tulino, A. M., Lozano, A., & Verdu, S. (2006). Capacity-achieving input covariance for single-user multi-antenna channels. *IEEE Transactions on Wireless Communications*, *5*(3), 662–671. doi:10.1109/TWC.2006.1611096

Tulino, A. M., & Verdu, S. (2004). Random matrix theory and wireless communications. *Communications and Information Theory*, *1*(1), 1–182. doi:10.1561/0100000001

Varanasi, M. (1995). Group detection for synchronous Gaussian code-division multiple-access channels. *IEEE Transactions on Information Theory*, *41*(4), 1083–1096. doi:10.1109/18.391251

Varanasi, M. K., & Guess, T. (1997). Optimum decision feedback multiuser equalization with successive decoding achieves the total capacity of the Gaussian multiple-access channel. In *Thirty-First Asilomar Conference on Signals, Systems, & Computers,* (Vol.2, pp.1405-1409).

Verdu, S. (1986). Minimum probability of error for asynchronous Gaussian multiple-access channels. *IEEE Transactions on Information Theory*, *32*(1), 85–96. doi:10.1109/TIT.1986.1057121

Verdu, S. (1998). *Multiuser detection*. Cambridge: University Press.

Vishwanath, S., Jindal, N., & Goldsmith, A. (2003). Duality, achievable rates, and sum-rate capacity of Gaussian MIMO broadcast channels. *IEEE Transactions on Information Theory*, *49*(10), 2658–2668. doi:10.1109/TIT.2003.817421

Viswanath, P., & Tse, D. (2003). Sum capacity of the vector Gaussian broadcast channel and uplink-downlink duality. *IEEE Transactions on Information Theory*, *49*(8), 1912–1921. doi:10.1109/TIT.2003.814483

Viterbi, A., Viterbi, A., Gilhousen, K., & Zehavi, E. (1994). Soft handoff extends CDMA cell coverage and increases reverse link capacity. *IEEE Journal on Selected Areas in Communications*, *12*(8), 1281–1288. doi:10.1109/49.329346

Voiculescu, D. (1983). Asymptotically commuting finite rank unitary operators without commuting approximants. *Acta Sci. Math.*, *45*, 429–431.

Wang, X., & Host-Madsen, A. (1999). Group-blind multiuser detection for uplink CDMA. *IEEE Journal on Selected Areas in Communications*, *17*(11), 1971–1984. doi:10.1109/49.806826

Weingarten, H., Steinberg, Y., & Shamai, S. (2006). The capacity region of the Gaussian multiple-input multiple-output broadcast channel. *IEEE Transactions on Information Theory, 52*(9), 3936–3964. doi:10.1109/TIT.2006.880064

Wijk, F., Janssen, G., & Prasad, R. (1995). Groupwise successive interference cancellation in a DS/CDMA system. In *IEEE International Symposium on Personal, Indoor and Mobile Radio Communications,* (Vol. 2, pp. 742–746), Toronto, Canada.

Wong, D., & Lim, T. (1997). Soft handoffs in CDMA mobile systems. *IEEE Wireless Communications, 4*(6), 6–17.

Wyner, A. (1994). Shannon-theoretic approach to a Gaussian cellular multiple-access channel. *IEEE Transactions on Information Theory, 40*(6), 1713–1727. doi:10.1109/18.340450

Yang, S., & Belfiore, J-C. (2006). Diversity-multiplexing tradeoff of double scattering MIMO channels. Submitted to *IEEE Transactions on Information Theory.*

Yu, W. (2006a). Uplink-downlink duality via minimax duality. *IEEE Transactions on Information Theory, 52*(2), 361–374. doi:10.1109/TIT.2005.862102

Yu, W. (2006b). Sum-capacity computation for the Gaussian vector broadcast channel via dual decomposition. *IEEE Transactions on Information Theory, 52*(2), 754–759. doi:10.1109/TIT.2005.862106

Yu, W., & Lan, T. (2007). Transmitter optimization for the multi-antenna downlink with per-antenna power constraints. *IEEE Transactions on Signal Processing, 55*(6), 2646–2660. doi:10.1109/TSP.2006.890905

Section 2
Practical Code Schemes for Cooperative Communications

Chapter 5
Source and Channel Coding Techniques for Cooperative Communications

John M. Shea
University of Florida, USA

Tan F. Wong
University of Florida, USA

Chan Wong Wong
University of Florida, USA

Byonghyok Choi
University of Florida, USA

ABSTRACT

This chapter provides a survey of practical cooperative coding schemes currently available in the literature, with focus on those schemes that achieve performance close to capacity or the best known achievable rates. To provide an insight into the construction of practical coding schemes for various cooperative communication scenarios, we first summarize the main design principles and tools that are used. We then present a survey of cooperative communication scenarios, and the progress on practical coding schemes for each of these scenarios is discussed in detail. Throughout the chapter, we demonstrate how the common design principles and tools are exploited to construct the existing practical coding schemes. We hope that the integrated view presented in this chapter can lead to further advances in this area.

DOI: 10.4018/978-1-60566-665-5.ch005

I. INTRODUCTION

In a wireless communication channel, the signals transmitted by the terminals suffer from a variety of degradations, such as additive white Gaussian noise (AWGN), shadowing, and multi-path fading. Until recently, most of the efforts on mitigating the effects of these degradations have focused on signal processing and coding design in the physical layer for each communication link between a transmitter and receiver. Many wireless communication systems contain multiple terminals that share a common medium, and information theory indicates that in efficient communication schemes, the terminals jointly use the medium as a shared resource instead of one that is divided orthogonally among pairs of terminals.

Recently, cooperative communication techniques have been proposed that provide practical designs for wireless networks to exploit the shared medium. This resurgence of interest in multi-terminal communications was motivated by the investigation of practical approaches to user cooperation for cellular communications in (Sendonaris, Erkip, & Aazhang, 2003a, 2003b) and related work on the relay channel in (Laneman, Tse, & Wornell, 2004; Laneman, & Wornell, 2003; Laneman, Wornell, & Tse, 2001). Much of the following work has focused on the relay channels, and the term "cooperative communications" is sometimes used to refer to relay channels. However, the resurgence in interest in multi-terminal communications has also included other types of cooperation among radios, and we use the term cooperative communications in this broader sense.

The focus of this chapter is to discuss practical coding schemes for cooperative communications. In particular, we are interested in schemes that can achieve performance close to capacity or the best known achievable rates. The practical cooperative coding schemes that we discuss use some common principles and tools, three of which we discuss in Section II. The first is the discovery of practical capacity-approaching error-control coding schemes starting in the 1990s, which has been a key enabler for efficient cooperative communication schemes. In Section II-A, we provide a brief overview of two classes of such codes: turbo codes (Berrou, Glavieux, & Thitimajshima, 1993) and low-density parity-check (LDPC) codes (Gallager, 1962, 1963; MacKay & Neal, 1997) and the related irregular repeat-accumulate (RA) codes (Jin, Khandekar, & McEliece, 2000). The introduction of the turbo decoder also brought new perspectives on iterative signal processing of communication signals, which is the topic of Section II-B. Finally, in Section II-C, we discuss practical distributed source coding schemes, also known as **Slepian-Wolf** (Slepian & Wolf, 1973b) or **Wyner-Ziv** (Wyner & Ziv, 1976) coding.

The majority of this chapter is devoted to introducing various cooperative communication scenarios and progress on practical coding schemes for these scenarios. The cooperative communication schemes that we consider are shown in the network illustrated in Figure 1. The numbered circles illustrate these different cooperative communication scenarios:

1) Nodes A and B have information to send to node C. This is a ***multiple-access channel (MAC)***, which is discussed in Section III and Section VI.

2) Node C has information to send to both nodes D and E. This is a ***broadcast channel***, which is considered in Section IV.

3) Nodes F and G have correlated sensor measurements of some phenomena (here F and G observe a jeep from different locations). The nodes wish to compress their measurements and send to node D. Since the measurements are correlated, F and G can use ***distributed source coding*** to compress their measurements separately, while decoding is performed jointly at D. Distributed source coding is discussed in Section V.

Figure 1. Cooperative communication

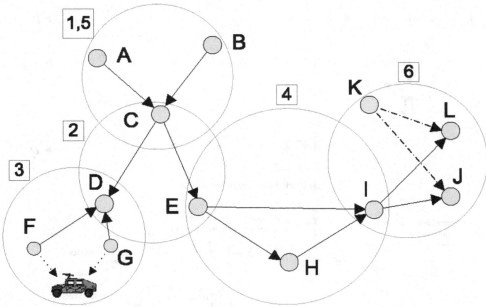

4) Node E has information to send to node I. Node H helps to relay the information to I. This is an example of the ***relay channel***, which is discussed in Section VI.

5) Nodes A and B can listen to each other's transmission and help forward each other's information to node C. This is an example of the ***cooperative multiple-access channel (CMAC)***, which is discussed in Section VII.

6) Nodes I and K have separate information to send to nodes J and L, respectively. If I and K transmit simultaneously, then I's transmission causes interference at L, and K's transmission causes interference at J. This is an example of the *interference channel*, which is discussed in Section VIII.

II. COMMON DESIGN PRINCIPLES AND TOOLS

A. Capacity-Approaching Codes

One of the keys to practical cooperative communication schemes has been the development of capacity-approaching codes with encoders and decoders that are practical to implement. The existing capacity-approaching codes can be broadly classified as either turbo codes (Berrou et al., 1993) or low-density parity-check (LDPC) codes (Gallager, 1962; MacKay & Neal, 1997). In this section, we focus on the structure of the codes. In Section II-B, we discuss the iterative decoding of these codes, which is also one of the key developments that make these codes practical.

Figure 2. Block diagrams of turbo encoder and iterative decoder

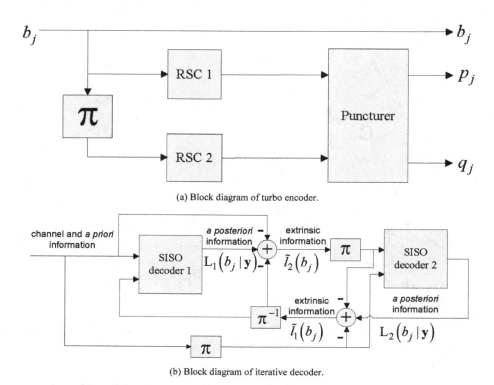

(a) Block diagram of turbo encoder.

(b) Block diagram of iterative decoder.

1. Turbo Codes

Turbo codes, first presented in (Berrou et al., 1993), represent one of the most important developments in error-control coding, as these were the first codes that were found to achieve performance close to the capacity of the AWGN channel with reasonable decoder complexity. A turbo encoder consists of a parallel[1] concatenation of two (or more) convolutional encoders as shown in Figure 2(a). In conventional serial concatenated codes (Lin & Costello, 1983), an outer and inner encoder are connected with a rectangular interleaver in between. The purpose of the interleaver is to distribute any errors at the output of the inner decoder across the input block for the outer decoder. In contrast, turbo codes use a pseudo-random interleaver on the input to one of the codes, which serves to make any low-weight error event for one code be, with high probability, a high-weight error event at the other code. This form of concatenation results in the overall code having a distance spectrum similar to random codes. While the conventional approach to code design generally focuses on achieving the largest possible minimum distance, turbo codes have small numbers of codewords at the minimum distance. Moreover, as the size of the interleaver increases, the proportion of codewords at small distances decreases. This property of the turbo codes is known as spectral thinning (Perez, Seghers, & Costello, 1996).

Another key to the performance of turbo codes is that the constituent codes are recursive, systematic convolutional (RSC) codes. As discussed further in Section II-B.1, convolutional codes have an efficient *a posteriori* probability (APP) decoder, known as the **Bahl-Cocke-Jelinek-Raviv (BCJR) algorithm** (Bahl, Cocke, Jelinek, & Raviv, 1974). Moreover, the choice of RSC codes has important implications in terms of distance properties (Divsalar & Pollara, 1995) and decoder performance (Divsalar, Dolinar, & Pollara, 2001; ten Brink, 1999, 2001), respectively. As shown in Figure 2(a), a puncturer can be used to easily achieve higher code rates by deleting selected parity bits at the outputs of the encoders.

2. Low-Density Parity-Check and Irregular Repeat Accumulate Codes

LDPC codes were proposed by Gallager (1962, 1963). However, the full potential of these codes was not realized until almost 35 years later when they were "rediscovered" by McKay and Neal (MacKay & Neal, 1997). The primary reason that these codes were forgotten by the coding community is that at the time of their development by Gallager, these codes could not be used in any practical communication scheme because of insufficient computational power. LDPC codes are characterized by their parity-check matrix, which is a non-systematic, sparse matrix. Gallager proposed a class of LDPC codes that are now referred to as *regular* LDPC codes because they have an equal number of 1s in each row and column of their parity-check matrices. An (n, k, j, ℓ) regular LDPC code has a parity-check matrix with n columns, $n - k$ rows, j 1's per column, and ℓ 1's per row. A useful observation is that LDPC codes can be represented as bipartite graphs (Tanner, 1981), often called Tanner graphs, between a set of *variable nodes* and *check nodes*. The variable nodes correspond to the code symbols, and the check nodes correspond to the parity-check constraints from the parity-check matrix. For regular LDPC codes, each type of node has the same number of connections to the other type of node. The number of connections is called the *degree* of the node. Since the parity-check matrix has low density, the degree of each type of node is small.

The performance of LDPC codes was further improved by their generalization to *irregular* LDPC codes that have varying numbers of 1's in the rows and columns of their parity-check matrices. This is equivalent to allowing different nodes in the Tanner graph to have different degree distributions. The early work on LDPC codes with irregular degree distributions focused on the design of codes for the erasure channel that have good performance and low encoding and decoding complexity (Luby, Mitzenmacher, Shokrollahi, & Spielman, 1998, 2001; Luby, Mitzenmacher, Shokrollahi, Spielman, & Stemann, 1997). Rather than finding specific codes, however, the techniques in (Luby et al., 1998, 2001; Luby et al., 1997) give ways to find degree distributions for ensembles of codes that offer good average performance. This approach was extended in (Richardson, Shokrollahi, & Urbanke, 2001; Richardson & Urbanke, 2001) to many other channels, including the binary-input AWGN channel. By optimizing the degree distribution, irregular LDPC codes can achieve performance extremely close to the channel capacity. For example, irregular LDPC codes have been designed that can achieve performance within 0.0045 dB of the capacity of the binary-input AWGN channel (Chung, Forney, Jr., Richardson, & Urbanke, 2001).

LDPC codes can be decoded in linear time, as discussed in Section II-B.2. However, a major disadvantage of LDPC codes, both regular and irregular LDPC, is the encoding complexity, which is quadratic in the block length for the natural encoding algorithm (Jin et al., 2000). One way to overcome this approach is to combine simple codes to form a code that has good distance properties, an efficient decoding algorithm, and an efficient encoding algorithm. This approach was suggested by Tanner in (Tanner, 1981), and also forms the basis for turbo codes. One class of codes that has the characteristics of both turbo codes and LDPC codes is the repeat-accumulate (RA) codes, proposed in (Divsalar, Jin, & McEliece, 1998). The RA codes, whose encoder includes repetition, permutation and accumulation units, have been shown to be capable of operating near capacity limits with a linear-time encoding complexity; however, they have the weakness that they are naturally low rate (lower than $1 / 2$). The encoder for these codes

Figure 3. Tanner graph and encoder for IRA codes

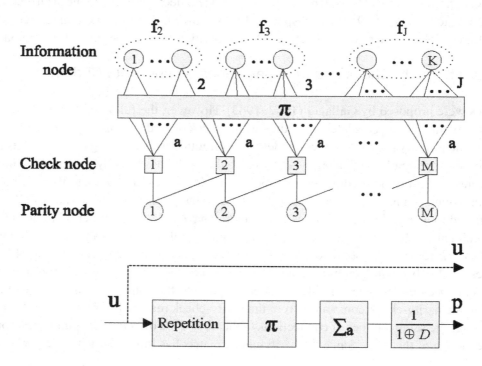

is illustrated in Figure 3. Here π represents a permutation, which is a pseudo-random reordering of its input. For regular RA codes, each input bit is repeated the same number of times.

Irregular RA codes (IRA), which were introduced in (Jin et al., 2000), also have an encoder of the same form as in Figure 3. However, in IRA codes, different bits are repeated different numbers of times, and the distribution of the repetitions can be optimized to improve the decoding performance in a way similar to irregular LDPC codes (Jin et al.; Roumy, Guemghar, Caire, & Verdu, 2004). The IRA codes offer performance close to that of the irregular LDPC codes, achieving a decoding threshold within 0.059 dB of capacity for a rate 1/2 code (Roumy et al.). The IRA codes have the advantages of linear-time encoding complexity, simple encoder structure, and higher code rates than the RA codes. This combination of performance and low complexity is the reason why IRA codes also are being used in practical coding schemes for multi-terminal communications.

The IRA code in Figure 3 can be represented by a Tanner graph with N variable nodes and M check nodes. The N variable nodes consist of K information nodes and M parity nodes. Note that there are as many check nodes as parity nodes. Each check node connects to a information nodes. On the other hand, the information nodes can connect to different numbers of check nodes. Let f_i denote the fraction of all edges connected to information nodes with degree i. Then clearly $\sum_i f_i = 1$. The $a \cdot M$ connections between check nodes and information nodes are determined by the pseudo-random permuter. In addition, the check nodes are connected to parity nodes in a zigzag pattern (by the accumulator). An IRA code can thus be described by the parameters $(\mathbf{f}; a) = (f_2, f_3, \cdots, f_J; a)$. Decoding of the IRA and LDPC codes can be performed by passing messages on the Tanner graph and is discussed in Section II-B.2.

B. Iterative Processing

Iterative processing has been used for many years in communications systems, e.g., in adaptive equalization (Lucky, 1965) and successive interference cancellation (Kohno, Imai, Hatori, & Pasupathy, 1990). However, it was the introduction of the turbo decoder in (Berrou et al., 1993) that has revolutionized how iterative signal processing is applied in communication systems. The development of the turbo decoder revealed a technique for building a suboptimal detection system with good performance from relatively simple subsystems that exchange *a posteriori* probability (APP) information about common pieces of information. The name "turbo" in fact refers to the action of the decoder, which like a turbocharger in an engine, uses feedback to enhance performance. Since APP information can be generated in many detection and decoding processes, turbo processing has since been applied to joint equalization and decoding (Douillard et al., 1995; Tuchler, Koetter, & Singer, 2002), joint multi-user detection and decoding (Liang & Stark, 2001; Wang & Poor, 1999), and joint demodulation and decoding (Li & Ritcey, 1999a, 1999b).

Iterative processing can be used to approximate joint detection or decoding using separate detection or decoding blocks that are connected together. Each block that is included in the iterative process must have two features. First, it must have the ability to generate soft outputs, usually in the form of APPs. Second, each block should be able to accept soft information generated by other blocks. Thus, these blocks are often called soft-input soft-output (SISO) blocks. Maximum-likelihood (ML) and maximum *a posteriori* probability (MAP) detectors/decoders can often be used as SISO blocks. An important discovery in (Berrou et al., 1993) is that only the additional (**extrinsic**) information generated by a block should be forwarded to the next block in the iterative process. This is usually accomplished by removing the contribution of the input soft information from the soft outputs generated by a block before passing them along. In other words, each block should only output the information that is not shared by other blocks in the iterative processing to avoid having the output of a block feeding back to its own input.

1. Turbo Decoding for Turbo Codes

Turbo decoding is the classic example of modern **iterative processing**. Figure 2 shows the schematic of the encoder and decoder for a **turbo code** constructed from two rate-$1/2$ RSC constituent codes. As indicated in Figure 2(a), the binary input sequence $\mathbf{b} = \left\{ b_j \middle| j = 1, \cdots, N \right\}$ is encoded into codeword $\mathbf{d} = \left\{ d_t \middle| t = 1, \cdots, 3N \right\} = \left\{ b_1, p_1, q_1, b_2, p_2, q_2, \cdots, b_N, p_N, q_N \right\}$ if no puncturing is performed. Here, we use the notation and terminology that j is the *index* of an input information bit, and t is the *time* of a transmitted symbol. Clearly $\left\lfloor j = \frac{t}{3} \right\rfloor$ in this example. The received word is represented by $\mathbf{y} = \mathbf{d} + \mathbf{n}$, where $\mathbf{n}$ represents the noise sequence. We also denote subsequences of $\mathbf{y}$ by $\mathbf{y}_k^\ell = \left\{ u_k, v_k, w_k, u_{k+1}, v_{k+1}, w_{k+1}, \cdots, u_\ell, v_\ell, w_\ell \right\}$ for $k \leq \ell$, where u_k, v_k, w_k are noisy versions of the systematic bit b_k and parity bits p_k and q_k, respectively.

The turbo decoder iteratively estimates the APPs $\Pr \left\{ b_j \middle| \mathbf{y} \right\}$ by exchanging information between the SISO decoders for the two RSC constituent codes. To provide some insight into this process, we first describe how the APPs can be generated in each component SISO decoder through the well known **BCJR algorithm** (Bahl et al., 1974). As previously mentioned, the BCJR algorithm is a MAP decoder that computes the APP log-likelihood ratios (LLRs)

$$L\left(b_j \mid \mathbf{y}\right) = \log\left(\frac{\Pr\left\{b_j = +1 \mid \mathbf{y}\right\}}{\Pr\left\{b_j = -1 \mid \mathbf{y}\right\}}\right) \qquad (1)$$

for $i = 1, 2, \ldots, N$. The MAP decision for b_j is then simply

$$\hat{b}_j = \operatorname{sgn}\left\{L\left(b_j \mid \mathbf{y}\right)\right\}. \qquad (2)$$

By using the structure of the trellis for the RSC, (1) can be rewritten as

$$L\left(b_j \mid \mathbf{y}\right) = \log\left(\frac{\sum_{U^+} p\left(S_j = s, S_{j-1} = s', \mathbf{y}\right)}{\sum_{U^-} p\left(S_j = s, S_{j-1} = s', \mathbf{y}\right)}\right), \qquad (3)$$

where S_j is the encoder state at index j. Here, U^+ is the set of branches (denoted by the previous and next state pair (s', s)), which correspond to input value $b_j = +1$, and U^- is the set of branches corresponding to $b_j = -1$.

It is shown that $p\left(S_j = s, S_{j-1} = s', \mathbf{y}\right)$ can be factored as (Bahl et al., 1974)

$$p\left(S_j = s, S_{j-1} = s', \mathbf{y}\right) = \alpha_{j-1}\left(s'\right)\gamma_j\left(s', s\right)\beta_j\left(s\right), \qquad (4)$$

where

- $\alpha_{j-1}\left(s'\right) = p\left(S_{j-1} = s', \mathbf{y}_1^{j-1}\right)$ is the forward-looking state probability, which is found using forward recursion from index 0 to time $j-1$,
- $\beta_j\left(s\right) = p\left(\mathbf{y}_j^N \mid S_j = s\right)$ is the backward-looking state probability, which is calculated using reverse recursion from index N to index j, and
- $\gamma_j\left(s', s\right) = p\left(\mathbf{y}_j^j, S_j = s \mid S_{j-1} = s'\right)$ is the state-transition probability, which can be calculated directly from the channel likelihoods for the received symbols at time j.

All of the α, β, and γ values can be computed using one forward and one reverse pass through the code trellis (Bahl et al., 1974).

The turbo decoder in Figure 2(b) utilizes two **BCJR** MAP decoders as the constituent decoders. The overall turbo decoder iterates between the constituent SISO decoders as described below. The input to the turbo decoder consists of the channel LLRs

$$L\left(u_j\right) = \log\left(\frac{p\left\{u_j \mid b_j = +1\right\}}{p\left\{u_j \mid b_j = -1\right\}}\right)$$

and the *a priori* LLRs,

$$l(b_j) = \log\left(\frac{\Pr\{b_j = +1\}}{\Pr\{b_j = -1\}}\right).$$

(5)

Each constituent decoder takes as input the channel information and the *a priori* probabilities for the bits, as well as extrinsic information generated from the output of the other MAP decoder. The constituent decoders use the extrinsic information as if it were an independent source of *a priori* information. Each decoder outputs the APP LLRs for the information bits b_j. The output of decoder i for bit j is denoted by $L_i(b_j \mid \mathbf{y})$. Let $\tilde{l}_i(b_j)$ denote the extrinsic information for bit j at the output of decoder i. As previously mentioned, the extrinsic information for a bit represents new information derived in the decoding process about the bit that was not present in the channel observation for the bit, the *a priori* information for the bit, or the extrinsic information from the other decoder for that bit. Thus

$$\tilde{l}_i(b_j) = L_i(b_j \mid \mathbf{y}) - L(u_j) - l(b_j) - \tilde{l}_{\bar{i}}(b_j),$$

(6)

where $\bar{1} = 2$ and $\bar{2} = 1$.

To further explain how the iterative decoder works, we briefly discuss the decoding iterations. Consider the first iteration of turbo decoding. The decoders have not yet produced any extrinsic (new) information, so $\tilde{l}_1(b_j) = \tilde{l}_2(b_j) = 0$, $\forall j \in \{1, \cdots, N\}$. SISO decoder 1 executes first and generates *a posteriori* information $L_1(b_j \mid \mathbf{y})$ from the received symbols $\mathbf{y}$ and *a priori* information $l(b_j)$. The extrinsic information $\tilde{l}_1(b_j)$ is then calculated according to (6), with $\tilde{l}_2(b_j) = 0$. The channel LLRs for the systematic bits $\{L(u_j)\}$, the *a priori* information $\{l(b_j)\}$, and the extrinsic information from decoder 1 $\{\tilde{l}_1(b_j)\}$ are then all interleaved to match the order of the symbols in the second constituent code. SISO decoder 2 then executes and generates *a posteriori* information $L_2(b_j \mid \mathbf{y})$ from the received symbols $\mathbf{y}$, the *a priori* information $l(b_j)$ and the extrinsic information $\tilde{l}_1(b_j)$ that was produced by decoder 1. The extrinsic information $\tilde{l}_2(b_j)$ at the output of decoder 2 can then be computed from (6). These values are in the order of the systematic bits in code 2, so they must be deinterleaved before they are input to SISO decoder 1. At this point, the first iteration of iterative decoding is complete. As the iterations continue, the two SISO decoders exchange **extrinsic information** to refine their estimates of the information sequence. The iteration process continues until some stopping criterion is satisfied.

2. Iterative Decoding for LDPC and IRA Codes

LDPC codes, including RA and IRA codes, can be decoded using **message passing algorithms** (MPAs), which can be visualized as computing and exchanging soft-information iteratively among the variable and check nodes in the Tanner graph. Typically, the sum-product (SP) algorithm (cf. (Kschischang, Frey, & Loeliger, 2001)), a form of belief propagation, is used for the MPA. As before, let $\mathbf{d} = (d_1, \cdots, d_n)$ be the transmitted codeword, and $\mathbf{y} = \mathbf{d} + \mathbf{n}$ be the received word. The SP algorithm estimates the *a posteriori* LLRs for the coded bits,

$$L\left(d_i\right) = \log\left(\frac{\Pr\left\{d_i = +1|\mathbf{y}\right\}}{\Pr\left\{d_i = -1 \mid \mathbf{y}\right\}}\right), \tag{7}$$

for $i = 1, \cdots, n$. Note that unlike turbo codes, the LDPC codes are typically nonsystematic codes, and the SP algorithm estimates the values for the coded bits, not the message bits. The coded bits can be recovered from the estimated codeword through matrix operations.

In MPAs, computation is performed at each vertex of the graph, and messages are exchanged along the edges. For the LDPC codes, the vertices are either check nodes or variable nodes. Although many different message-passing schedules are possible, it is convenient to discuss the MPA as an iterative process in which each iteration consists of two steps. In the first step, the check nodes perform computations on messages received from the variable nodes. In the second step, the variable nodes perform computation on messages received from the check nodes. MPAs are usually performed under the assumption that the messages involved in the algorithm are independent. Although this is true for certain types of graphs, such as trees, it is not true for most codes of interest, including the LDPC and (I)RA codes. Thus, the resulting algorithm is an approximation to the MAP decoder, even if the computations performed at the variable and check nodes are done according to the MAP rule.

We now briefly overview the sum-product MPA for decoding LDPC codes. The variable nodes input messages consisting of the channel LLRs $L\left(y_j\right)$, and extrinsic information from the check nodes. Let $\tilde{l}_k(d_j)$ be the extrinsic information from the kth check node about coded bit j, and let $\bar{l}_i(d_j)$ be the sum of the channel LLR and extrinsic information about code bit j to the ith check node. Then by applying the independence assumption, $\bar{l}_i(d_j)$ is the sum of the LLRs received on all of the edges into the variable node j, except for the LLR received on the edge from check node i. That is,

Figure 4. Message passing algorithm for the first and the second half iteration

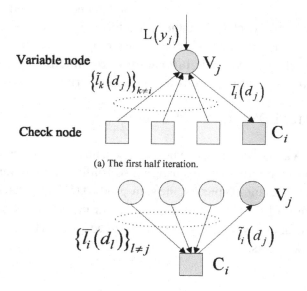

(a) The first half iteration.

(b) The second half iteration.

$$\bar{l}_i(d_j) = L(y_j) + \sum_{k \neq i} \tilde{l}_k(d_j).$$

This processing is illustrated in Figure 4(a). Note that at the beginning of the first iteration, the variable nodes have not received any messages yet, so the variable node j has only the LLR of the channel observation $L(y_j)$. Each variable node passes a message equal to the channel LLR $L(y_j)$ on the vertices to each of the check nodes to which it is connected.

Each check node enforces a parity check equation, either from the low density parity-check matrix of the LDPC codes or the code structure for the (I)RA codes. The check nodes use the messages from the variable nodes to compute extrinsic information to pass back to the variable nodes. The extrinsic information about bit d_j from check node i, $\tilde{l}_i(d_j)$, is given by (Hagenauer, Offer, & Papke, 1996)

$$\tilde{l}_i(d_j) := 2 \tanh^{-1} \left\{ \prod_{\ell \neq j} \tanh \left[\frac{\bar{l}_i(d_\ell)}{2} \right] \right\} \tag{8}$$

and illustrated in Figure 4(b). After some stopping criterion has been met, the decoder computes the LLR and makes a decision on the bits d_j according to

$$\hat{d}_j = sgn \left\{ L(y_j) + \sum_i \tilde{l}_i(d_j) \right\}.$$

C. Slepian-Wolf/Wyner-Ziv Coding

Consider a system in which two sensors make dependent observations of some phenomena. Let the signals at sensor 1 and 2 be denoted by X and Y, respectively. Furthermore, assume that the sensors do not know each other's observations but that they do have information about how the signals are dependent. Now consider how the observations of the signals at the two sensors can be compressed for communication to a common sink. Obviously, the two sensors could independently compress their observations. Thus, a compression rate as low as the sum of the entropies of X and Y, i.e., $H(X) + H(Y)$, is possible. Encoding and decoding can be performed separately on X and Y. If X and Y were observed at a single sensor, then a compression rate equal to the joint entropy $H(X,Y) \leq H(X) + H(Y)$ is possible. In this case, encoding and decoding are done jointly on X and Y. For the case of two separate sensors but a common sink, encoding must be done separately, but decoding can be done jointly. This is called **distributed source coding** (DSC). This raises the question of what compression rates are possible in this case. Surprisingly, Slepian and Wolf (1973b) proved that all rates R_1 and R_2 are possible such that

$$R_1 \geq H(X \mid Y), \quad R_2 \geq H(Y \mid X), \text{ and}$$

$$R_1 + R_2 \geq H(X,Y). \tag{9}$$

The rate $H(X,Y)$ that is achieved by joint encoding can also be achieved if encoding is performed separately, provided joint decoding is performed.

In this section, we first motivate the idea behind **Slepian-Wolf coding**. Then we discuss the fundamental practical approaches that have been applied for Slepian-Wolf and Wyner-Ziv coding. Applications of these approaches and their performances on specific channels are discussed in Section IV. We note that the literature on this topic is large and rapidly expanding, and we have not attempted to include every reference. (The reader is suggested to consult the long version of (Xiong, Liveris, & Cheng, 2004) for a detailed reference list up to mid-2004.) **Wyner-Ziv coding** is also applied in the compress-forward variant of the relay channel, which is discussed in Section VI-C.

1. Motivating Example

To provide some insight into Slepian-Wolf coding, we present a simple but interesting illustrative example from (Pradhan & Ramchandran, 2003). In this scenario, it is assumed that Y is compressed and sent to the sink. The problem then is to compress X given that the sink knows Y, which is usually referred as side information for X. This is asymmetric DSC. By the Slepian-Wolf coding theorem (Cover & Thomas, 1991), it is possible to compress X at the rate $H(X \mid Y)$, thus achieving the same compression rate as joint source coding $H(X,Y) = H(X \mid Y) + H(Y)$.

To illustrate this, let X and Y be binary three-tuples. It is convenient to consider X and Y to be codewords from the codebook $C = \{000, 001, 010, 011, 100, 101, 110, 111\}$. We assume that each codeword is equiprobable. The correlation structure between X and Y is that they differ in at most one position.

We consider three different compression scenarios:

- **Separate Encoding and Separate Decoding:** If the encoder does not know Y and the decoder does not use its knowledge of Y, then three bits are needed to compress X.

- **Joint Encoding and Joint Decoding:** Since the encoder and decoder both know Y, it is sufficient to compress the difference between X and Y. Since X and Y differ by at most one bit, the difference is in the set $\{000, 001, 010, 011\}$. Thus, in this scenario, only two bits are needed to convey X.

- **Separate Encoding and Joint Decoding:** In this case, Y is known only to the decoder but not to the encoder. Note that C is a group and $S = \{000, 111\}$ is subgroup of C (In fact, S is a normal subgroup of C since C is abelian). Given Y, the minimum Hamming distance can be used to determine X if $X \in S$ since all three positions are different between the elements of S. Then S can be used to partition C into disjoint cosets, where each coset is of the form $cS = \{c \oplus s \mid s \in S\}$ for some $c \in C$. The cosets of C thus formed are $\{000, 111\}$, $\{100, 011\}$, $\{010, 101\}$, and $\{001, 110\}$. Note that the members of each coset are separated by a Hamming distance of 3. Thus, since Y is know at the decoder, it is sufficient just send the index of the coset in which X lies. The decoder then computes the Hamming distance between Y and the members of the specified coset. The correct value of X will be within Hamming distance 1 of Y, whereas the other member of the coset will be at a Hamming distance of either 2 or 3. Thus, the decoder can recover the correct value of X by using Y as side information. As in the case of joint encoding and decoding, only two bits are needed, even though encoding was done separately (without knowledge of Y).

2. Practical Slepian-Wolf Coding Schemes

Although the result of Slepian and Wolf is promising, there were virtually no developments of practical Slepian-Wolf coding schemes before 1999. The key observation that has enabled the development of many practical Slepian-Wolf coding schemes is that Slepian-Wolf coding can be addressed using error-correction coding techniques. This observation was originally made by Wyner (1974), but the first practical Slepian-Wolf schemes were reported by Pradhan and Ramchandran (1999). Later work has built on capacity-approaching codes, such as those discussed in Section II-A, to provide performance close to the compression limits given in (9).

Currently, practical coding schemes for efficient DSC can be classified into two different approaches. The first approach (Chou, Pradhan, & Ramchandran, 2003; Liu, Cheng, Liveris, & Xiong, 2004; Liveris, Xiong, & Georghiades, 2002a; Pradhan & Ramchandran, 1999, 2000b, 2003; Servetto, 2000, 2007; Stankovic, Liveris, Xiong, & Georghiades, 2004; Yang, Cheng, Xiong, & Zhao, 2003), which is based on explicit random binning techniques (Cover & Thomas, 1991; Wyner, 1974), partitions a source codebook onto cosets of a channel code, such that the minimum distance among the codewords in a coset is as large as possible. As in the motivating example of Section II-C.1, the source sends the index of the coset, and the side information is used to select the correct codeword within the coset. In the second approach (Aaron & Girod, 2002; Bajcsy & Mitran, 2001; Garcia-Frias, 2001a; Zhao & Garcia-Frias, 2005), binning is implicitly used. In this approach, the data is encoded using capacity-approaching codes, which are then punctured to very high rates to achieve compression.

Pradhan and Ramchandran (1999) first outlined a method of constructing Slepian-Wolf codes based on practical syndrome and coset techniques. Most of the results in (Pradhan & Ramchandran, 1999) and the journal version (Pradhan & Ramchandran, 2003) actually focus on the Wyner-Ziv problem of lossy coding of a continuous source, but most of the Wyner-Ziv coding constructions can be interpreted as quantization followed by practical Slepian-Wolf coding. The original method proposed in (Pradhan & Ramchandran, 1999, 2003) is for asymmetric coding, where one of the sources is known at the decoder. However, the approach was extended in (Pradhan & Ramchandran, 2000b) to the symmetric coding case, and it is well known that over many transmissions, time-sharing can be used with asymmetric coding to operate at any achievable rate.

In (Pradhan & Ramchandran, 1999, 2003), two schemes are proposed for decomposing a (possibly quantized) source alphabet ∇ into cosets such that the minimum distance among the members of the coset is as large as possible. The first scheme is memoryless coset construction, in which ∇ is directly partitioned into subsets to maximize the distance among the members of the subset. This is generally straightforward. For example, if $\nabla = \{r_0, r_1, \cdots, r_{V-1}\}$, where $r_i < r_{i+1}$ for $0 \le i < V - 1$, then if two cosets are desired, the cosets are taken to be $\{r_0, r_2, \cdots, r_{V-2}\}$ and $\{r_1, r_3, \cdots, r_{V-1}\}$. The cosets are assigned indexes, and the index of the coset in which the source symbol lies is transmitted. For this example, a rate of 1 bit per source symbols is required. Equivalently, one of the cosets can be selected as the channel code, C, and the syndrome s of the source symbol is transmitted. The decoder selects the codeword that is closest to the side information in the coset selected by the syndrome.

Partitioning ∇ into subsets with large minimum distances can be improved by adding memory to the process. The key observation is the view of the partitioning process as the decomposition of ∇ into cosets using a channel code. When memory is introduced, ∇^L is decomposed into cosets. The second scheme in (Pradhan & Ramchandran, 1999) uses trellis-coded modulation (TCM) to construct the cosets.

Again, the source alphabet ∇ is decomposed into subsets, but now one subset is used as the input to a TCM encoder with output alphabet ∇. The set of all sequences of the TCM encoder is then used to partition the source vectors ∇^L into cosets that give the same syndrome s. The syndrome of a source vector is transmitted. At each stage of the decoder, the syndrome can be used to select among a set of alternative trellis labelings in the Viterbi algorithm (cf. (Pradhan & Ramchandran, 1999, 2003)).

Another approach to explicit binning uses nested **lattice codes** (Servetto, 2000, 2007) (which were known to be useful for the Wyner-Ziv problem (Zamir & Shamai, 1998), but no good nested coding schemes were known at that time). LDPC codes were used for binning in (Liveris et al., 2002a). Stankovic et al. (2004) provide a detailed design based on systematic IRA and turbo codes so that powerful capacity approaching codes can be employed for efficient DSC.

The second approach to Slepian-Wolf coding is based on the observation in (Pradhan & Ramchandran, 1999) that in many cases the side information at the decoder, Y, is a noisy version of the information to be compressed as the channel input, X. From another perspective, Y is the output of a *virtual channel* (also known as the *dependency channel* (Aaron & Girod, 2002)) with X as the input. Thus, to recover X, additional parity information from a good error-correcting code should be sufficient to allow the "errors" in Y to be corrected. This approach was proposed by Garcia-Frias in (Garcia-Frias, 2001b) in which turbo codes are used for joint source-channel coding of correlated binary sequences. In this approach, symmetric coding is used, where one source sequence is directly turbo coded, and the second source sequence is interleaved and turbo coded. The decoder essentially decodes across the combination of the two turbo codes, which has four constituent decoders. A simpler approach that used turbo coding just for Slepian-Wolf coding was given in (Bajcsy & Mitran, 2001) and (Garcia-Frias, 2001a). The approach in (Garcia-Frias, 2001a) is a symmetric coding approach, in which both encoders operate at the same rate. By contrast, the approach in (Bajcsy & Mitran, 2001) is an asymmetric approach, which has the advantage of being able to achieve all possible rates via time-sharing, as previously discussed.

To illustrate the basic ideas in using an error-control code for the virtual channel induced by the correlation between two sources, we further discuss the details of the scheme in (Garcia-Frias, 2001a). Let the two sources be X and Y, which are assumed to differ according to some binary symmetric channel. Let $p = \mathrm{Pr}\{X \neq Y\}$ be the crossover probability between the two sources. For optimal operation, the decoder should know p, although it is not required.

The scheme proposed in (Garcia-Frias, 2001a) is illustrated in Figure 5. In this scheme, the outputs of the sources are collected into equal-length vectors, $\mathbf{x}$ and $\mathbf{y}$. Each of these vectors is independently encoded using a turbo code, although $\mathbf{y}$ is interleaved before encoding. The outputs of the turbo codes are then punctured so that only one-half of the systematic bits are transmitted from each of the turbo encoders and the remaining parity bits are punctured to achieve the desired compression rate, although no details of the puncturing scheme are presented in (Garcia-Frias, 2001a). Let $\mathbf{d}^{(x)}$ and $\mathbf{d}^{(y)}$ be the encoded sequences at the outputs of the punctures for $\mathbf{x}$ and $\mathbf{y}$, respectively. It should be noted that the encoders do not have to know the crossover probability p, although the coding rate necessary for successful decoding at the sink will depend on p.

Joint decoding is performed on $\mathbf{d}^{(x)}$ and $\mathbf{d}^{(y)}$, as shown in Figure 5. For each of the turbo codes, there is an associated turbo decoder whose decoding proceeds in the standard fashion over each iteration. The two turbo decoders exploit the correlation between $\mathbf{x}$ and $\mathbf{y}$ by exchanging extrinsic information between them. Let $\mathrm{Pr}\left\{x_j = 1 \mid \mathbf{d}^{(x)}\right\}$, $\mathrm{Pr}\left\{x_j = 0 \mid \mathbf{d}^{(x)}\right\}$ denote the APPs for bit j generated by the turbo decoder for $\mathbf{x}$, and let $\mathrm{Pr}\left\{y_j = 1 \mid \mathbf{d}^{(y)}\right\}$, $\mathrm{Pr}\left\{y_j = 0 \mid \mathbf{d}^{(y)}\right\}$ denote the APPs for bit j generated by the

Figure 5. Encoder and decoder structure for compressing two binary correlated sources X and Y

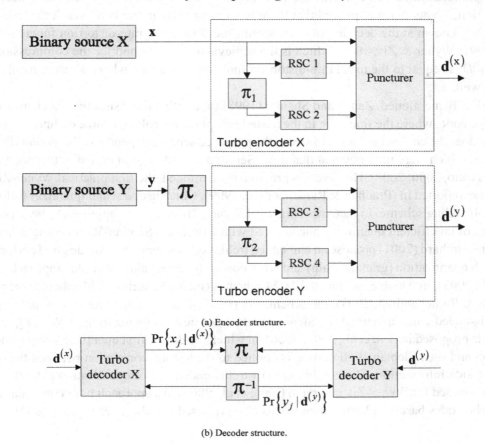

(a) Encoder structure.

(b) Decoder structure.

turbo decoder for $\mathbf{y}$. For convenience of notation, we ignore the presence of the interleaver at the input to the encoder for $\mathbf{y}$ in this discussion. Then at the decoder for $\mathbf{y}$, the APPs from the decoder for $\mathbf{x}$ are combined and used as *a priori* probabilities for the j th bit according to

$$
\Pr\{y_j = 1\} = (1 - p)\Pr\{x_j = 1 \mid \mathbf{d}^{(x)}\} + p\Pr\{x_j = 0 \mid \mathbf{d}^{(x)}\}
$$
$$
\Pr\{y_j = 0\} = (1 - p)\Pr\{x_j = 0 \mid \mathbf{d}^{(x)}\} + p\Pr\{x_j = 1 \mid \mathbf{d}^{(x)}\}. \tag{10}
$$

The APPs from $\mathbf{y}$ are combined and used as *a priori* information in the decoding of $\mathbf{x}$ in a similar way.

The performance of this technique and extensions to other source models is discussed further in Section V.

3. Practical Wyner-Ziv Coding Schemes

Wyner and Ziv (1976) extended the work of Slepian and Wolf to the case of lossy compression, in which the distortion (Cover & Thomas, 1991) between X and the reconstruction of X is required to be less than some predetermined value D. In the case of Gaussian memoryless sources and mean-squared error distortion, Wyner and Ziv proved that there is no compression rate loss even if the side information Y is only known at the decoder. In other words, the Wyner-Ziv rate-distortion function (Cover & Thomas, 1991; Wyner & Ziv, 1976) which is the achievable lower bound for the compression rate for a distortion D, is equal to the usual rate-distortion function in the case where Y were available at the encoder as well.

As previously mentioned, Zamir and Shamai (1998) first outlined constructive mechanisms using a nested lattice code, where the fine code in the nested pair plays the role of source coding and each coset coarse code does the channel coding. However, the proposed scheme depends on the availability of good nested codes, which were not known at that time. Servetto provided good nested lattice code constructions in (Servetto, 2000, 2007). However, as previously mentioned, the first practical Wyner-Ziv coding schemes were reported in (Pradhan & Ramchandran, 1999), which uses a scalar quantizer followed by a Slepian-Wolf coding scheme. In fact, most practical Wyner-Ziv coding techniques can be classified as a combination of an efficient quantizer concatenated with a practical Slepian-Wolf coding scheme.

Wang and Orchard (2001) propose an embedded trellis coded scheme to provide good performance in terms of both quantization (granular gain) and error-control (coding gain). A similar approach is taken in (Chou et al., 2003), in which concatenated TCM schemes (include a turbo TCM scheme) are proposed. In (Pradhan & Ramchandran, 2003), the schemes in (Pradhan & Ramchandran, 1999) are extended to utilize trellis-coded scalar quantization. Slepian-Wolf coded nested quantization (SWC-NQ) for Wyner-Ziv coding is proposed in (Liu et al., 2004, 2006), in which the data is first quantized using nested lattice quantization and then Slepian-Wolf coding is applied to exploit the correlation between the quantized data and the side information. The combination of trellis-coded vector quantization (TCQ) and LDPC codes was proposed for Wyner-Ziv coding in (Yang et al., 2003). An approach based on scalar quantization and turbo codes based on Latin Squares was also proposed in (Mitran & Bajcsy, 2002).

III. MULTIPLE ACCESS CHANNEL

The **multiple access channel (MAC)** consists of multiple transmitters sending information simultaneously to a common receiver. A typical two-user MAC is shown in Figure 1; namely terminals A and B transmit two independent signals to terminal C. The signals from the transmitters collide, causing what is known as multiple access interference (MAI) at the receiver. The MAC capacity region was first found by Ahleswede (1971) and Liao (1972). The capacity (or best known rate) regions for many variations of the MAC, such as the MAC with common information (Slepian & Wolf, 1973a), MAC with feedback (Cover & Leung, 1981; Gaarder & Wolf, 1975; Ozarow, 1984; Willems, 1982), and MAC with correlated sources (Ahlswede & Han, 1983; Cover, El Gamal, & Salehi, 1980) have since been studied.

The most studied MAC is the Gaussian MAC (Cover & Thomas, 1991) in which the signal seen by the receiver is the superposition of all the transmitted signals and AWGN. The practical coding schemes that we describe in this section are all designed for the Gaussian MAC. The receiver needs to recover the original transmitted information from all users in the presence of MAI. Practically, spreading (Viterbi, 1995) is often applied to signals before they are transmitted as a means to separate the transmitted signals and hence partially control MAI. The transmission of spread signals over a MAC is often referred to as code-division multiple-access (CDMA).

Figure 6. Iterative detection/decoding in a turbo-coded CDMA system

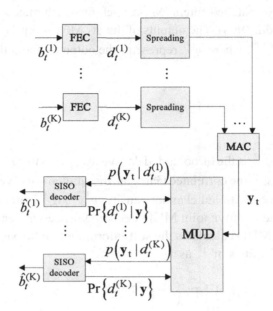

Due to strong commercial interest in CDMA, practical signal processing and coding techniques for the MAC have been extensively studied in the last two decades (Verdú, 1998). Different types of detectors that suppress the effect of MAI have been proposed. Among these detectors, **multiuser detectors** (MUDs) provide the best performance. A MUD jointly demodulates signals from all users to overcome the problem of MAI. Different MUDs may process the received signal using a number of different techniques including interference cancellation, linear equalization, decision-aided equalization and maximum-likelihood sequence estimation (Verdú).

Forward error correction (FEC) is usually applied to improve performance over the MAC. With FEC, optimal detection can be performed by joint MUD-decoders such as the one suggested in (Giallorenzi & Wilson, 1996). However joint MUD-decoder schemes are too computationally complex for practical implementation. **Iterative processing** is often employed to solve this computational complexity problem. An iterative detection/decoding architecture for the CDMA system has first been proposed in (Hagenauer, 1996). The FEC-CDMA system is viewed as a serially concatenated coded system as shown in Figure 6. Soft information is exchanged between a SISO MUD and the individual SISO decoders for the K users (transmitters). Unfortunately, no practical implementation or performance results are reported in (Hagenauer).

Following the architecture proposed in (Hagenauer, 1996), a turbo-coded CDMA system was implemented in (Reed, Schlegel, Alexander, & Asenstorfer, 1998). The data of each user is turbo-coded and then spread for transmission. In the receiver, the MAP-based MUD and the turbo decoders for each user exchange information to form a joint iterative detection-decoding structure. Both the MAP-based MUD and the turbo decoders produce soft outputs to enable the exchange of **extrinsic information** in the iterative detection-decoding process. Figure 6 shows the overall block diagram of this turbo-coded CDMA system, with the various information generated and exchanged in the iterative detection-decoding process indicated.

Suppose that there are K users in the turbo-coded CDMA system. The received signal is first despread at the receiver by a set of K matched filters (MFs), each of which matches the spreading code used by the corresponding user (Verdú, 1998). The outputs of the K MFs at time t are grouped together into the vector $\mathbf{y}_t = [y_t^{(1)}, y_t^{(2)}, \cdots, y_t^{(K)}]^T$, where $y_t^{(k)}$ represents the output of the kth MF for $k = 1, 2, \cdots, K$. It is easy to obtain that

$$\mathbf{y}_t = \mathbf{H}_t \mathbf{d}_t + \mathbf{z}_t \tag{11}$$

where $\mathbf{d}_t = [d_t^{(1)}, d_t^{(2)}, \cdots, d_t^{(K)}]^T$ is the turbo-coded data vector, $\mathbf{H}_t$ is the $K \times K$ cross-correlation matrix of the spreading codes, and $\mathbf{z}_t$ is the correlated Gaussian distributed noise vector (Reed et al., 1998). Since the receiver knows $\mathbf{H}_t$, the conditional channel density $p\left(\mathbf{y}_t \mid \mathbf{d}_t\right)$ is multivariate Gaussian and hence can be easily calculated. The iterative joint MUD-decoding processing can be described as follows.

In the first iteration, the MUD calculates the soft information to be passed to kth turbo decoder, for each $k = 1, 2, \ldots, K$ and all values of t, as

$$p\left(\mathbf{y}_t \mid d_t^{(k)} = d\right) = \sum_{\mathbf{d}_t : d_t^{(k)} = d} p\left(\mathbf{y}_t \mid \mathbf{d}_t\right) \prod_{i \neq k} \Pr\left\{d_t^{(i)}\right\} \tag{12}$$

for each $d \in \left\{+1, -1\right\}$. Since there is no *a priori* information about the coded symbols at this point, we set $\Pr\left\{d_t^{(i)} = +1\right\} = \Pr\left\{d_t^{(i)} = -1\right\} = \frac{1}{2}$. We note that the product in (12) does not include the *a priori* information of $d_t^{(k)}$ (i.e. $\Pr\left\{d_t^{(k)}\right\}$). This is because only the extrinsic information about $d_t^{(k)}$ should be passed to the kth turbo decoder as discussed in Section II-B. For $k = 1, 2, \cdots, K$, the kth turbo decoder then uses the soft information in (12) provided by the MUD as the channel likelihoods (*cf.* Section II-B) to calculate the state-transition probability in the BCJR algorithm,

$$\gamma_j\left(s', s\right) = \Pr\left\{S_j = s \mid S_{j-1} = s'\right\} \prod_t p\left(\mathbf{y}_t \mid d_t^{(k)}\right). \tag{13}$$

Here j is the index of the current information bit, S_j is the state of the kth user's RSC decoder trellis at index j, and the product is over all the $d_t^{(k)}$ values that are associated with the transition from state s to state s' in the trellis. Iterative decoding, as discussed in Section II-B, can hence be performed by the kth turbo decoder to obtain the APP $\Pr\left\{d_t^{(k)} \mid \mathbf{y}\right\}$ for all bits (both systematic and parity), where the vector $\mathbf{y}$ denotes the concatenation of the MF output vectors $\mathbf{y}_t$ over the whole transmission time. The kth turbo decoder then feeds this APP back to the MUD.

In the next iteration, the MUD uses the APP $\Pr\left\{d_t^{(k)} \mid \mathbf{y}\right\}$ as *a priori* information for the coded bit $d_t^{(k)}$ to recalculate the channel likelihood $p\left(\mathbf{y}_t \mid d_t^{(k)}\right)$ by replacing $\Pr\left\{d_t^{(k)}\right\}$ in (12) with $\Pr\left\{d_t^{(k)} \mid \mathbf{y}\right\}$. The new channel likelihoods are then fed to the turbo decoders for them to redo decoding. The iteration between the MUD and turbo decoders goes on until some stopping criterion is satisfied. Finally, for $k = 1, 2, \cdots, K$, the kth turbo decoder outputs the bit decision $b_j^{(k)}$ for the jth information bit of the kth user.

Simulations results in (Reed et al., 1998) show that the performance obtained by the iterative joint MUD-decoder in a 5-user system is within 0.3 dB of the single-user performance. The computational complexity of this iterative joint MUD-decoder, however, is still prohibitive for systems with a medium to large number of users. A convolutional coded CDMA system is suggested in (Wang & Poor, 1999) with a SISO MUD detector that works based on the minimum mean-squared error (MMSE) principle. The objective of employing the MMSE-based MUD and the convolutional code is to reduce the complexity of the joint MUD-decoding process.

Serial concatenated convolutional codes (SCCCs) have also been suggested as the FEC component of a coded CDMA system in (Shi & Schlegel, 2001). With SCCC, the FEC-CDMA system can be viewed as one with three serial concatenated components. As before, extrinsic information is exchanged between these three components, forming two loops with the inner code of SCCC participates in both loops. This design differs from the one in (Reed et al., 1998) in that the inner code of SCCC works as a bridge between the MUD and the outer code of SCCC. In (Shi & Schlegel), a variance transfer function is used to show that an unconventional iteration schedule with several iterations inside each loop is superior to other schedules.

Interleave-division multiple-access (IDMA) (Ping, Liu, & Leung, 2003; Ping, Liu, Wu, & Leung, 2004) is another form of signal design that has been suggested for the **MAC**. Instead of separating the users by different spreading codes, IDMA separates users by the use of different interleavers. In IDMA, the data of each user is first coded using a low-rate code. The main reason for the use of low-rate codes is to exploit their high coding gains. Different interleavers are then used to separate the coded signals of different users. Figure 7 shows the transmitter and receiver structures proposed in IDMA.

As shown in the figure, the coded symbol sequence $\mathbf{d}^{(k)} = \left\{ d_t^k, \forall t \right\}$ is the output of a low-rate code, whose input is the data sequence $\mathbf{b}^{(k)} = \left\{ b_j^k, \forall j \right\}$ of user k. An interleaver π_k is used to produce the interleaved sequence $\mathbf{x}^{(k)} = \left\{ x_t^k, \forall t \right\}$ from $\mathbf{d}^{(k)}$. That is, $\mathbf{x}^{(k)} = \pi_k \left\{ \mathbf{d}^{(k)} \right\}$. The receiver, which consists of an elementary signal estimator (ESE) and K SISO decoders (DECs), operates under the iterative detection-decoding principle developed in (Reed et al., 1998). The ESE is a SISO MUD that operates

Figure 7. Transmitter and receiver of an IDMA system

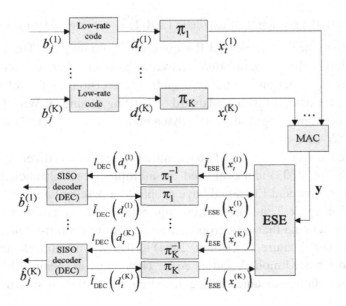

under the MMSE principle similar to the one suggested in (Wang & Poor, 1999).

To illustrate that the iterative MUD-decoding process in (Reed et al., 1998) can also be performed in the log domain, we describe the log-domain iterative joint MUD-decoding of the IDMA system in Figure 7.

In each iteration, for each time t, the ESE employs the channel output y_t and its *a priori* LLRs $\left[\mathbf{l}_{ESE}\right]_t = \left\{l_{ESE}(x_t^{(k)}), \forall k\right\}$ to generate extrinsic information about the symbol $x_t^{(k)}$ as

$$\tilde{l}_{ESE}\left[x_t^{(k)}\right] = \log\left(\frac{\Pr\left\{x_t^{(k)} = +1 \mid y_t, \left[\mathbf{l}_{ESE}\right]_t\right\}}{\Pr\left\{x_t^{(k)} = -1 \mid y_t, \left[\mathbf{l}_{ESE}\right]_t\right\}}\right) - l_{ESE}\left[x_t^{(k)}\right] \tag{14}$$

for all $k = 1, 2, \cdots, K$. Once again, we point out that the *a priori* LLR is subtracted from the *a posteriori* LLR in (14) to obtain the extrinsic information as discussed in Section II-B before. As mentioned above, the ESE proposed in (Ping et al., 2004) works under the MMSE principle. Hence it produces only approximations to the LLRs in (14) above. If the number of users is small, one can also implement a SISO MUD based on the MAP principle in place of the ESE to give the exact LLRs. Moreover, for the first iteration, since the ESE has no *a priori* information, $l_{ESE}\left[x_t^{(k)}\right]$ is set to 0 for all $k = 1, 2, \cdots, K$ and all time t in (14) above. Then the set of extrinsic information generated by the ESE for user k is deinterleaved to be used as *a priori* LLRs for its SISO decoder. That is, $\mathbf{l}_{DEC}^{(k)} = \left\{l_{DEC}(d_t^{(k)}), \forall t\right\} = \pi_k^{-1}\left\{\tilde{\mathbf{l}}_{ESE}^{(k)}\right\}$.

For each $k = 1, 2, \cdots, K$, the decoder of user k then takes the sequence $\mathbf{l}_{DEC}^{(k)}$ as its *a priori* LLRs input to generate the extrinsic information about $d_t^{(k)}$ by either the BCJR or **MPA** algorithm, depending on the code used, as described in Section II-B:

$$\tilde{l}_{DEC}\left[d_t^{(k)}\right] = \log\left(\frac{\Pr\left\{d_t^{(k)} = +1 \mid \mathbf{l}_{DEC}^{(k)}\right\}}{\Pr\left\{d_t^{(k)} = -1 \mid \mathbf{l}_{DEC}^{(k)}\right\}}\right) - l_{DEC}\left[d_t^{(k)}\right] \tag{15}$$

for all time t. Afterward, the sequence of extrinsic LLR $\tilde{\mathbf{l}}_{DEC}^{(k)}$ generated by the decoder of user k is re-interleaved to be used to as the new *a priori* LLR sequence $\mathbf{l}_{ESE}^{(k)}$ for the ESE. That is, $\mathbf{l}_{ESE}^{(k)} = \pi_k\left\{\tilde{\mathbf{l}}_{DEC}^{(k)}\right\}$. For the following iterations, the above information exchange process between the ESE and the decoders repeats until some stopping criterion is satisfied. Finally the decisions on the information bits are determined by the decoders. By using a turbo-Hadamard code (Ping, Leung, & Wu, 2003) at an approximate rate of $1/69$, it is reported in (Ping et al., 2004) that a performance of 1.4 dB from the MAC capacity limit is achieved in a 35-user system.

On fading channels, the performance is often measured in terms of the diversity-multiplexing tradeoff proposed in (Zheng & Tse, 2003) and evaluated for multiple-access channels in (Tse, Viswanath, & Zheng, 2004). Recently, Nam and El Gamal have proposed a practical approach to transmission over the MAC called LAttice Space-Time (LAST) coding (Nam & El Gamal, in press). Their scheme extends the lattice coding scheme that was proposed for single-user multi-input, multi-output (MIMO) communication in (El Gamal, Caire, & Damen, 2004) to the case of multiple-access by extending the mod-Λ lattice code to a nested mod-Λ lattice code. They show in (Nam & El Gamal, in press) that under joint minimum-Euclidean-distance lattice decoding, the proposed scheme achieves the optimal diversity-multiplexing tradeoff, provided the block length is sufficiently large. However, the authors point out that although the LAST scheme provides the optimal diversity-multiplexing tradeoff, their preliminary results indicate that it does not achieve the optimal outage probability. They also observe that joint decoding may be essential to achieve the optimal diversity-multiplexing tradeoff in outage-limited multiple-access channels.

IV. BROADCAST CHANNEL

The **broadcast channel** consists of a single source sending common and disparate information simultaneously to several receivers. Figure 1 contains an example of the broadcast channel with a single source (terminal C) and two receivers (terminal D and E). Cover (1972) showed that superposition coding achieves rates strictly exceeding those achieved by simple time sharing in a broadcast channel. Later, the capacity region of the degraded broadcast channel (including the Gaussian broadcast channel) was established by Bergmans (1973). The converse showing the optimality of Bergmans' rate region was established a year later in (Bergmans, 1974; Gallager, 1974).

There are two disparate bodies of current work on coding for the broadcast channel. The first, which focuses on broadcast channels with random channel noise (especially AWGN channels) is primarily based on **dirty paper coding** strategies. Coding techniques for this channel are discussed in Section IV-A. The second body of work considers deterministic broadcast channels in which the channel output is a deterministic function of its input. Coding for these channels is discussed in Section IV-B.

A. Coding for Broadcast Channel with Random Channel Noise

The existing work on coding for the broadcast channel with random channel noise is based on practical coding schemes for dirty paper coding (DPC). DPC has been shown to achieve the capacity region for the Gaussian multi-input multi-output (MIMO) broadcast channel (Weingarten, Steinberg, & Shamai, 2006). When DPC is used for broadcasting, the signals to different users are iteratively encoded, and at each stage the current user is encoded using DPC with all of the previous users' signals treated as known interference. Most of the existing work on practical DPC schemes are based on transmitter precoding strategies that are equivalent to Tomlinson-Harashima precoding (THP) (Miyakawa & Harashima, 1969; Tomlinson, 1971). In this section, we focus on THP schemes for broadcasting with single antennas at the transmitter and receivers (Erez & Brink, 2005; Erez, Shamai, & Zamir, 2000; Ginis & Cioffi, 2000; Sun, Liveris, Stankovic, & Xiong, 2005; Yu & Cioffi, 2001; Yu, Varodayan, & Cioffi, 2005). THP schemes for MIMO channels are given in (Airy, Forenza, Heath, Jr., & Shakkottai, 2004; Fischer, Stierstorfer, & Windpassinger, 2003; Fischer, Windpassinger, Lampe, & Huber, 2002a, 2002b; Kusume, Joham, Utschick, & Bauch, 2005; Windpassinger, Fischer, Vencel, & Huber, 2004).

THP was originally developed as an alternative to equalization for the ISI channel (Miyakawa & Harashima, 1969; Tomlinson, 1971). The idea behind THP is that if power were no constraint, then filtering for channel equalization could be performed before transmission instead of after reception; when applied at the transmitter, it is known as precoding. Precoding avoids several problems associated with equalization at the receiver, including error propagation for decision-feedback equalizers and noise amplification in other equalizers. The reason that these problems are avoided is that the precoder acts on the correct, noiseless symbols (which are only available at the transmitter) rather than their noisy or reconstructed versions that are available at the receiver. However, the problem with directly moving the equalizer filter from the receiver to the transmitter is that the symbols precoded in this way may require significantly higher power than in the absence of the precoder. Tomlinson and Harashima proposed the use of a modulo operation after the precoding filter to limit the transmit power to a specified constraint. They showed that by implementing the same modulo operation at the receiver and then appropriately demodulating the signal, the performance can be close to implementing the precoder without the power constraint.

When used for equalization, THP essentially takes advantage of *a priori* knowledge of a single user's symbols that interfere with each other through the ISI channel. In (Yu & Cioffi, 2001), it is observed that when multiple users' signals are transmitted from a source, such as at a cellular base station or the access point of a wireless local area network, the transmitter has *a priori* knowledge of signals that may cause multi-user interference if the channel is not completely orthogonalized through some multiplexing or multiple-access technique. Thus, in (Airy et al., 2004; Fischer et al., 2002a, 2002b; Ginis & Cioffi, 2000; Joham, Brehmer, & Utschick, 2002 ; Kusume et al., 2005; Windpassinger et al., 2004; Yu & Cioffi, 2001, 2002; Yu et al., 2005) THP techniques are proposed for the presubtraction of multi-user interference (MUI) at the transmitter.

The first work on THP for DPC was for single antennas at each of the communicators (Ginis & Cioffi, 2000; Yu & Cioffi, 2001). This approach was later extended to MIMO channels (Fischer et al., 2002a, 2002b) and multi-user MIMO channels (Windpassinger et al., 2004). These initial works exploit zero-forcing techniques for frequency flat channels. THP based on the MMSE criterion was introduced in (Joham et al., 2002) for frequency-selective channels. Suboptimal MMSE THP algorithms that have performance close to the optimal while maintaining complexity close to linear precoding for flat fading channels are studied in (Kusume et al., 2005).

To understand the basic operation of THP algorithms, consider the application of DPC to transmission of a signal U in the presence of a strong interfering signal S, where S is know at the transmitter for U. Since S is known at the transmitter, a precoder may send $U - S$ instead of U to cancel out the interference during the transmission through the channel. However, when the power of S is large, the power of $U - S$ may be exceedingly large. Therefore, instead of sending $U - S$, the precoder sends $X = (U - S) \bmod \Delta$, where the modulo operator is designed to make the transmitted signal X constrained within the finite interval $[-\Delta / 2, \Delta / 2)$. The decoder implements the same modulo-Δ operation to the received signal in order to recover the intended signal U. In the absence of noise, this equivalence relation allows U be completely recovered at the receiver.

The straightforward application of THP suffers some non-negligible precoding losses. In (Yu et al., 2005), three types of precoding losses are categorized:

- *Shaping Loss*: For the AWGN channel, it is well known that the channel input should have a Gaussian distribution to achieve Shannon's channel capacity. However, when U is amplitude-limited and S is large, X is uniformly distributed in $[-\Delta / 2, \Delta / 2)$. The use of the uniform shape causes a performance loss that is known as "shaping loss". Shaping losses for communication over AWGN channels with no interference are negligible at low SNR; however, this is not the case for the broadcast channel, in which shaping losses can be even more important at low SNRs. It is shown in (Erez et al., 2000) that in the high SNR regime, THP with scalar quantization suffers a 1.53 dB shaping loss on the Gaussian broadcast channel and that this shaping loss can not be eliminated if the transmitter uses only causal knowledge of the interference signal.

- *Power Loss*: Consider an input U that is uniformly distributed over discrete points in $[-\Delta/2, \Delta/2)$. If after S is presubtracted from U and the Δ-modulo operation is applied, X is uniformly distributed over the continuous region $[-\Delta/2, \Delta/2)$, then X may have a higher power than U, resulting in *power loss*. This is most significant for small signal constellations (and hence low SNRs, where such constellations would be used).

- *Modulo Loss*: At the receiver, the modulo operation results in a "folding" of the received signal into the region $[-\Delta/2, \Delta]$. Thus, all signals Δ away are folded on to the same value after the modulo operation. In the presence of noise, this can result in signals that are near the edge of the signal constellation being wrapped around to the other edge of the signal constellation. Again, this effect is most pronounced at low SNRs, where small signal constellations are used, because a greater proportion of the points in the signal constellation are near the edges of the signal constellation.

The shaping loss can be overcome if the precoder better uses its knowledge of the entire interference sequence. Consider again the modulo operation used in the THP scheme described above. The modulo operation $X = (U - S) \bmod \Delta$ is equivalent to first applying scalar quantization to $U - S$, where the quantization is to points on a lattice with spacing Δ, and then letting X be the quantization error. To overcome the shaping loss, a natural extension is to take a vector input $\mathbf{U}'$ and apply vector quantization and transmit the quantization error $\mathbf{S}'$. Whereas the Voronoi region for a vector of scalar-quantized symbols is a hypercube, the Voronoi region for high-dimensional vector-quantized symbols is a hypersphere. Optimal Gaussian coding also results in a hypersphere shape, so vector quantization can provide the desired shaping gain.

Trellis-shaping codes were proposed by Forney (Forney, 1992) to recover the shaping gain for trellis-coded communication over AWGN channels. In the approach described in (Forney), a binary convolutional code is used to control an allowable set of quantization points, and the Viterbi algorithm is used to select the quantization point that minimizes the energy of the error. The error signal is then the resulting shaped signal. This approach is used in (Yu & Cioffi, 2001) to improve the performance of THP.

Erez and ten Brink (2005) use a concatenation of a repeat-irregular nonsystematic RA code and a lattice-based trellis shaping code. The RA decoder is designed for iterative quantization detection and decoding based on EXIT charts. The overall system model is illustrated in Figure 8. At the receiver, an iterative detection scheme is used. The BCJR algorithm is used on an appropriately defined trellis structure to compute the APP values. Then extrinsic information for the RA code symbols are forwarded to the message-passing decoder for the RA code. The RA decoder computes extrinsic information for the coded symbols, which are fed back to the BCJR algorithm for the vector quantizer. After the desired number of iterations, the *a posteriori* estimates of the information bits are output from the RA decoder.

Sun et al. (2005) use a trellis-coded quantization (TCQ) (Marcellin & Fischer, 1990) for trellis shaping (Forney, 1992) and IRA codes for channel coding. Their approach is similar to that in (Erez & Brink, 2005), but the use of TCQ give improved vector quantization, and they use EXIT charts (ten Brink, 1999, 2001) to optimize the degree distributions of the IRA codes for the channel induced by the combination of vector quantization and channel noise. The iterative decoding structure is essentially the same as discussed above.

As previously mentioned, power loss and modulo loss are the most significant precoding losses at low SNRs. The performance can be improved by only partially cancelling the interference to minimize the combined effect of noise and residual interference. This approach is called partial interference pre-

Figure 8. Dirty paper coding based on lattice-based trellis shaping

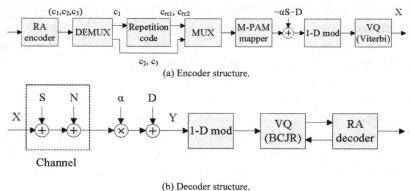

(a) Encoder structure.

(b) Decoder structure.

subtraction (PIP) (Yu et al., 2005). (An equivalent approach was proposed earlier in (Erez et al., 2000) (see also (Erez, Shamai, & Zamir, 2005)) based on quantizing to an "inflated" lattice, but the PIP approach provides a more intuitive explanation.) In PIP, αS, where $0 < \alpha \leq 1$, is subtracted from U at the transmitter instead of S and at the receiver end, the receiver multiplies the received signal by α and then implements the same modulo operation to decode the intended signal U as

$$
\begin{aligned}
\hat{U} &= \left(\alpha X + \alpha S + \alpha N \right) \bmod \Delta \\
&= \left(\alpha X + (U - X) + \alpha N \right) \bmod \Delta \\
&= \left(U + \left(1 - \alpha \right)(-X) + \alpha N \right) \bmod \Delta
\end{aligned}
$$

The effective noise is then the modulo operation of the weighted sum of two independent random variables, $\left((1 - \alpha)(-X) + \alpha N \right) \bmod \Delta$. The optimal value of α maximizes $I(\hat{U}; U)$; however, the value of α that minimizes the power of $(1 - \alpha)X + \alpha N$ works well in practice. In (Yu et al., 2005), it is shown that the gain due to PIP is more than 2 dB at less than or equal to 0.5 bits per transmission for THP transmissions using M-PAM.

Erez and ten Brink's broadcasting scheme based on **lattice codes**, MMSE scaling, and vector quantization performs within 1.3 dB of capacity at 0.5 bits/s/Hz spectral efficiency. Sun et al.'s (2005) scheme, which is based on trellis-coded quantization (TCQ) and optimized irregular repeat accumulate (IRA) codes, achieves performance that is approximately 0.883 dB away from capacity at 0.5 bits/s/ Hz. However, the TCQ scheme in (Sun et al.) uses a vector quantizer with 1024 states thus requiring substantially higher complexity at then encoder and decoder than the scheme in (Erez & Brink, 2005), which uses a 64-state quantizer.

B. Deterministic Broadcast Channels

One special case of the **broadcast channel** is the discrete memoryless deterministic broadcast channel (DBC). In the DBC, the M channel outputs are deterministic functions of the channel input X as $Y^{(t)} = f_t(X), t = \left\{ 1, 2, \cdots, M \right\}$. The capacity region of the DBC is known, and the achievability is shown in general by random binning (Marton, 1979). Several approaches to develop capacity-approaching codes for DBCs have been proposed (Coleman, Effros, Martinian, & Medard, 2005; Coleman, Martinian, Effros, & Medard, 2005; Haim & Zamir, 2005; Yu & Aleksic, 2005). Among them, Coleman, Martinian et al. claim to have developed the first practical, non-trivial and capacity-achieving code construction for the DBC.

A **LDPC**-like code has been proposed in (Yu & Aleksic, 2005) for practical implementation of binning schemes at rates close to the capacity region of the Blackwell channel. The Blackwell channel, which is a particular kind of DBC, is given by

$$(Y^{(1)}, Y^{(2)}) = \begin{cases} (0,1), X = 0 \\ (0,0), X = 1 \\ (1,0), X = 2 \end{cases}$$

The main coding idea for the Blackwell channel proposed in (Yu & Aleksic, 2005) is binning. Let $\mathbf{x} = \left[x_1, x_2, \cdots, x_n \right]$ be the length n channel input and $\mathbf{y}^{(t)} = \left[y_1^{(t)}, y_2^{(t)}, \cdots, y_n^{(t)} \right], t = \{1, 2\}$ be the channel output. The code construction involves the generation of $2^{nH(Y^{(1)})}$ $\mathbf{y}^{(1)}$ sequences and $2^{nH(Y^{(2)})}$ $\mathbf{y}^{(2)}$ sequences and a random assignment of these sequences into 2^{nR_1} and 2^{nR_2} bins, where $\left[R_1, R_2 \right]$ is an arbitrary achievable rate pair. The information to be encoded is the bin indices (i, j), and the encoder starts by looking for a pair of $\mathbf{y}^{(1)}$ in bin i and $\mathbf{y}^{(2)}$ in bin j such that they are jointly typical (Cover & Thomas, 1991). Then the encoder selects the channel input $\mathbf{x}$ to produce this pair at the channel output.

The graphical coding structure for $\left[R_1, R_2 \right]$ is shown in Figure 9, in which binning is accomplished via parity-check codes. In this figure, circles are used to denote the variable nodes (i.e., $\mathbf{y}^{(1)}$ and $\mathbf{y}^{(2)}$), and squares are used to denote constraint nodes. The set of $\mathbf{y}^{(1)}$s in bin i is defined as the set of n-sequences that satisfy a particular set of parity-check constraints, which are shown in the bottom of Figure 9 and connected to the variable nodes at random. The values of these bottom parities are the bin indices i and j. The other constraint nodes at the top of Figure 9 are NAND operators that enforce the constraint $(y_k^{(1)}, y_k^{(2)}) \neq (1, 1), \forall k = 1, 2, \cdots n$, which is not a possible output of the Blackwell channel.

The encoding and decoding processes proposed in (Yu & Aleksic, 2005) are as follows. The information bits i and j are first placed at the parity nodes, which are shown at the bottom of Figure 9. The encoding task is to find the set of variable assignments $\left[\mathbf{y}^{(1)}, \mathbf{y}^{(2)} \right]$ that satisfy the parity-check and the NAND constraints. Then the corresponding channel input that produces $\left[\mathbf{y}^{(1)}, \mathbf{y}^{(2)} \right]$ is sent through the Blackwell channel. The decoder obtains $\left[\mathbf{y}^{(1)}, \mathbf{y}^{(2)} \right]$ and recovers the information i and j by forming the parity checks. The only error events occur exclusively in the encoding process when the encoder fails to find $\left[\mathbf{y}^{(1)}, \mathbf{y}^{(2)} \right]$ such that all constraints are satisfied. On the other hand, the decoding process is error-free.

A survey propagation algorithm, which is similar to the belief propagation used in LDPC decoding, is proposed in (Yu & Aleksic, 2005). Survey propagation is used to find $\mathbf{y}^{(1)}$ and $\mathbf{y}^{(2)}$ that are consistent with the parity-check and NAND constraints. Survey propagation differs from belief propagation in that a "don't care" state is allowed for every variable node, apart from the traditional 0 and 1. Since the NAND constraints have the property that if one of the variables connecting to it is 0, the other variable can be either 0 and 1, any variable node can then be assigned the "don't care" state. The detailed survey propagation equations for the NAND and parity-check constraints, which characterize the message

Figure 9. Coding for the Blackwell channel

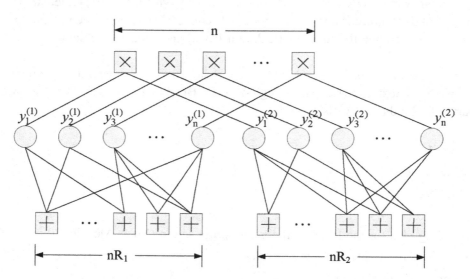

passed between the constraint nodes and variables nodes, can be founded in (Yu & Aleksic). In (Yu & Aleksic), simulations results show that the survey propagation works well at rates close to the Blackwell channel capacity region (about 95% of the sum rate).

V. DISTRIBUTED SOURCE CODING

As previously introduced in Section II-C, **distributed source coding** (DSC) can be used when two or more sensors make dependent observations of some phenomena. This scenario is illustrated in Figure 1, in which terminals F and G have correlated observations that need to be conveyed to terminal D. By applying practical **Slepian-Wolf** or **Wyner-Ziv coding** techniques, as described in Section II-C.2, compression can be performed individually at the sensors but still achieve overall compression equal to joint compression provided that the decoding at the sink is performed jointly. The first work showing that this was possible is due to Slepian and Wolf (1973b). An early set of upper and lower bounds on performance were found by Berger (1977) and Tung (1978) with many refinements presented in later works (Berger & Yeung, 1989; Gastpar, 2004; Kaspi & Berger, 1982; Oohama, 1997, 1998, 2005; Wagner & Anantharam, 2008; Zamir & Berger, 1999).

A. Lossless Distributed Source Coding

Lossless DSC is performed via Slepian-Wolf coding, as discussed in Section II-C.2. The earliest work on practical Slepian-Wolf coding schemes was in the context of Wyner-Ziv coding (Pradhan & Ramchandran, 1999) and joint source-channel coding (Garcia-Frias, 2001b), both of which we discuss separately later. The first pure Slepian-Wolf coding technique was the symmetric turbo-code based technique proposed in (Garcia-Frias, 2001a). The channel studied therein is the most popular one for investigating Slepian-Wolf coding, in which the sources are binary and differ with some probability p. For block lengths of

16384 bits, the lossless[2] compression rate was within 0.17 of the achievable rate for values of p from 0.025 to 0.2. The compression rates in this range were from 1.31 to 1.89. In (Bajcsy & Mitran, 2001), an asymmetric turbo coding scheme was applied for the same binary channel, and performance with 0.18 of the achievable rate for asymmetric compression was achieved. Another asymmetric Slepian-Wolf coding scheme based on turbo codes with rate $n / (n-1)$ constituent codes was investigated in (Aaron & Girod, 2002) and applied to the binary correlated channel. Using 16-state, rate 4/5 constituent codes and block length 10^6, rates within 0.036 bits/sample and 0.154 bits/sample of the Slepian-Wolf limit were achieved for $p \approx 0.0112$ and $p \approx 0.065$, respectively. Schemes based on the syndrome approach (cf. Section II-C) and irregular LDPC codes were proposed in (Liveris et al, 2002a). Using LDPC codes with length 16384, the authors were able to improve the performance over those in (Aaron & Girod) by up to 0.12 bits/sample (operating at rate 0.5 compression on a channel with conditional entropy 0.466 bits/sample vs. 0.346 bits/sample for the scheme in (Aaron & Girod, 2002). The best result reported in (Liveris et al., 2002a) is within 0.034 bits/sample of the Slepian-Wolf limit. In (Stankovic et al., 2004), a channel code partitioning technique is proposed that can operate at general rate combinations (from asymmetric to symmetric). Then using turbo and IRA codes, performance within 0.039 bits/sample of the rate limit are achieved for a binary correlated channel with crossover probability 0.11.

Practical lossless DSC schemes have been proposed for a variety of other sources. In (Liveris, Xiong, & Georghiades, 2002b), coding schemes based on turbo codes are applied to correlated images. In (Garcia-Frias & Zhong, 2003), binary correlated sources are considered in which the crossover probability varies according to a hidden Markov model (HMM). Using LDPC coding as in (Liveris et al., 2002a), performance within 0.14 bits/sample of the rate limit is achieved for a variety of different two-state HMMs. In (Li, Tu, & Blum, 2004), turbo coding-based DSC schemes are developed for non-uniform sources. The performance is optimized by designing the constituent codes to match the nonuni-form source distribution and utilizing variable-length syndrome sequences. By doing so, performance within 1.53 dB of the rate limit is achieved for source distributions that emit a zero with probabilities from 0.7 to 0.9. In (Johnson, Ishwar, Prabhakaran, Schonberg, & Ramchandran, 2004), it is shown that Slepian-Wolf and Wyner-Ziv coding techniques can be used to compress encrypted data (in contrast to the usual approach of encrypting compressed data). Estimation and coding schemes for nonbinary sources with unknown correlations are proposed in (Zhao & Garcia-Frias, 2005). The performance is shown to be within 0.9 bits/sample of the limit (which varies from 3.35 bits/sample to 4.65 bits/sample) over a variety of correlations.

B. Lossy Distributed Source Coding

Much of the work on DSC focuses on the lossy case, or DSC with a fidelity criterion. The rate-distortion limits for coding with side information at the decoder were given by Wyner and Ziv (1976). The DISCUS scheme proposed in (Pradhan & Ramchandran, 1999) are asymmetric DSC schemes based on scalar quantization and practical Slepian-Wolf coding based on trellis codes. These schemes have some sig-nificant losses with respect to capacity. The sources of the losses of the DISCUS schemes in (Pradhan & Ramchandran, 1999) and (Pradhan & Ramchandran, 2003) are quantified in (Xiong, Liveris, & Cheng, 2004). For example, the scalar quantization schemes in (Pradhan & Ramchandran, 1999) will suffer at least 1.53 dB in granular gain (this corresponds to shaping gain in other contexts; see, for example, Section IV-A). To overcome these losses, trellis-coded quantization is used in some of the DISCUS schemes in

(Pradhan & Ramchandran, 2003). The other source of loss is due to the weakness of the error-correction codes that are used in (Pradhan & Ramchandran, 1999, 2003). For example, it is shown in [176] that even lattice codes of dimension 250 perform more than 1 dB away from capacity. The DISCUS schemes in (Pradhan & Ramchandran, 2003) perform within 3.2 dB to 5.2 dB of the Wyner-Ziv bound.

Symmetric lossy DSC schemes were developed in (Pradhan & Ramchandran, 2000a, 2000b). In (Wang & Orchard, 2001) an embedded trellis coding structure is used for both quantization and binning. Gains of 1.16 dB granular (quantization) gain and over 1 dB in coding gain are reported via this approach. Aaron and Girod (2002) present a Wyner-Ziv coding scheme based on scalar quantization and Slepian-Wolf coding using turbo codes with implicit binning (the second of the approaches discussed in Section II-C.2). They show that by using a capacity-approaching code, gains of up to 3 dB are possible if a residual error rate of 10^{-4} is tolerable. However, the use of scalar quantization still limits the overall performance. In (Yang et al., 2003), a scheme is proposed that uses both efficient vector quantization and capacity-approaching codes. TCQ or trellis-coded vector quantization (TCVQ) is used for quantization, and irregular LDPC codes are used for binning. The overall schemes are called Slepian-Wolf coding with TCQ (SWC-TCQ) schemes. With two-dimensional TCVQ and irregular LDPC codes of block length 10^6, performance within 0.47 dB to 0.67 dB of the Wyner-Ziv limit is reported for rates from 3.3 bits/sample down to 1.0 bits/sample. Good results with turbo and trellis-coded constructions are also reported in (Chou et al., 2003).

C. Distributed Joint Source-Channel Coding

In distributed joint source-channel (DJSC) coding, the source and channel code are designed jointly, encoding is performed separately, but decoding is performed jointly. DJSC schemes can be separated into three scenarios (Garcia-Frias & Xiong, 2005). In the first scenario, the asymmetric case, the outputs of one of the sources is assumed to be available at the decoder as side information for the decoding of the other source. In the second scenario, the independent channels case, each source is separately coded and transmitted through independent noisy channels. In the final scenario, the MAC case, the correlated sources are separately coded but transmitted through a multiple access channel.

Practical implementations for the asymmetric case have been proposed in (Aaron & Girod, 2002; Liveris, Xiong, & Georghiades, 2002c; Mitran & Bajcsy, 2002). By using systematic IRA codes as source-channel codes for the transmission of an equiprobable memoryless binary source with side information at the decoder, the design of (Liveris et al., 2002c) achieves performance that is 0.7–0.8 dB away from the theoretical limit for AWGN channels. Designs for the independent channels case and the MAC case have been proposed in (Aaron & Girod, 2002; Garcia-Frias, Zhong, & Zhao, 2002; Murugan, Gopala, & El Gamal, 2004; Zhong, Zhao, & Garcia-Frias, 2003). Linear codes with low density generator matrices (LDGMs) have been applied to the transmission of correlated sources over independent channels in (Garcia-Frias et al., 2002). By using iterative decoding between two received sequences, the performance gap is around 1–2 dB from the theoretical limit. Deterministic and statistical cooperative source channel coding techniques has been studied in (Murugan et al. 2004) and are shown to outperform the scheme in (Garcia-Frias et al., 2002) for highly correlated sources.

VI. RELAY CHANNEL

The **relay channel**, first introduced in (van der Meulen, 1971), consists of a source and a destination, along with a relay that can aid the source in transmitting messages to the destination. A typical three-node relay channel can be found in Figure 1 with E, H, and I acting as the source, relay, and destination, respectively. The capacities of the degraded and inversely degraded relay channels and relay channel with feedback were given in (Cover & El Gamal, 1979). For an arbitrary relay channel, lower and upper bounds on the capacity were also provided. Recently, much effort has been reported on capacity bounds, coding strategies and protocols, and multiplexing-diversity tradeoff analysis of relay channels with fading and multiple antennas (Azarian, El Gamal, & Schniter, 2005; Gastpar & Vetterli, 2005; Høst-Madsen, 2002; Høst-Madsen & Zhang, 2005; Khojastepour, Sabharwal, & Aazhang, 2003; Kramer, Gastpar, & Gupta, 2005; Lai, Liu, & El Gamal, 2006; Laneman et al., 2004; Lo, Vishwanath, & Heath, 2005; Nabar & Bolcskei, 2003; Reznik, Kulkarni, & Verdu, 2004; Tang, Chae, Heath, Jr., & Cho, 2006; Wang, Zhang, & Høst-Madsen, 2005; Xie & Kumar, 2005). In addition to these theoretical works, there has also been significant development on practical coding schemes (Anghel, Leus, & Kaveh, 2003; Hu & Duman, 2007; Hu & Li, 2005; Khojastepour, Ahmed, & Aazhang, 2004; Liu, Stankovic, & Xiong, 2005; Yang & Belfiore, 2007; Yu & Li, 2005; Zhang & Duman, 2002, 2007; Zhao & Valenti, 2003) for the relay channel based on the amplify-and-forward (AF), decode-and-forward (DF) (Laneman et al., 2004) and compress-and-forward (CF) (Høst-Madsen & Zhang, 2005) approaches, which are covered in Sections VI-A, VI-B amd VI-C, respectively.

A. Coding for Decode-and-Forward

In the decode-and-forward (DF) approach, the relay decodes the information that it receives from the source, and then re-encodes the decoded information before forwarding to the destination. Intuitively, DF is applicable when the channel between the source and the relay is good, as the relay needs to decode the information first. Here we focus on the more practical situation in which the relay operates in the half-duplex mode; i.e., it can either transmit or receive, but not both, at any time. Several practical DF implementations based on different half-duplex DF protocols and codes have been proposed (Hu & Duman, 2007; Khojastepour et al., 2004; Zhang & Duman, 2002, 2007; Zhao & Valenti, 2003). Here we focus first on the implementations based on turbo codes.

The idea of distributed **turbo codes** was suggested in (Zhao & Valenti, 2003) for the orthogonal relay channel in which the source and relay do not transmit simultaneously. Consider the RSC component code of a rate-$1/3$ turbo code as described in Section II-A. The source and relay transmit in successive time slots. In the first time slot, the source sends the RSC-coded codeword to both the relay and destination. The relay decodes the received codeword from the source, interleaves the decoded data, and re-encodes the data using the RSC code. Then it transmits the parity symbols to the destination in the second time slot. As a result, the transmissions from the source and relay combine to form a full codeword of the rate-$1/3$ turbo code at the destination.

The restriction of orthogonal transmissions between the relay and source is removed in (Zhang & Duman, 2002, 2007). Like before, only the source transmits in the first time slot. In the second time slot, the relay still transmits to the destination after decoding and re-encoding. But the source also transmits in this time slot. Thus, the destination observes the superposition of the signals sent by the source and relay. Finally, the destination performs joint decoding on all its observations in the two time slots. Optimization of the length of the first time slot (the listening period for the relay) plays an important role in approaching the best achievable DF rate of the half-duplex relay channel (Zhang & Duman, 2007). The details of the code design and implementation in (Zhang & Duman, 2007) are outlined below.

The source encodes a block of N information bits with a rate-R systematic turbo code, which is obtained by puncturing a lower rate standard "mother" turbo code. For example, for the case of $R = 1/2$, a rate-$1/2$ turbo code is obtained by puncturing the parity bits of the component RSC codes of the rate-$1/3$ systematic turbo code described before. Let $0 < \alpha \leq 1$. In the first time slot, the source sends $\alpha N / R$ coded bits[3] to both the relay and destination. Generally, we should choose $\alpha \geq R$ so that all the systematic bits are transmitted in the first time slot. For example, if the above rate-$1/2$ turbo code obtained from the rate-$1/3$ mother code is used, we should choose $\alpha \geq 1/2$. The source sends the remaining $(1-\alpha)N / R$ coded bits to the destination in the second time slot.

The relay treats the transmission from the source in the first time slot as the codeword of a turbo code of rate R / α obtained by puncturing the source's rate-R turbo code. It decodes the received codeword and re-encodes the decoded N information bits using the mother turbo code employed by the source. It further punctures the output of its encoder to obtain $(1-\alpha)N / R$ coded bits. In the puncturing process, coded bits are removed in order according to their priority (bits of lower priority are removed first) until the number of remaining bits reaches $(1-\alpha)N / R$. The parity bits of the mother turbo code that are punctured to obtain the rate-R code employed by the source have the highest priority. The systematic bits have the next highest priority. The remaining parity bits have the lowest priority. The relay sends the output of the puncturer in the second time slot to the destination. Note that the signals transmitted by the source and the relay superimpose in the second time slot.

Let

$$R_e = \frac{N}{(1-\alpha)N / R + N / R} = \frac{R}{2 - \alpha}.$$

If R_e is larger than or equal to the rate of the mother turbo code, then all the coded bits received by the destination in the two time slots form a received codeword of the rate-R_e turbo code obtained by puncturing the mother code. On the other hand, if R_e is smaller than the rate of the mother turbo code, then we instead obtain a codeword of the mother turbo code by repeating some of the coded bits. The iterative turbo decoder for the corresponding turbo code is employed at the destination.

Since the received signal seen by the destination in the second time slot is the superposition of the signals transmitted by the source and relay, we need to employ a SISO MAP-**MUD** as described in Section III to obtain the soft-inputs of the respective coded bits mentioned above for the turbo decoder. In general, a joint MUD-decoding structure similar to the one in Section III is employed. As no spreading is suggested in (Zhang & Duman, 2007), the MAP-MUD is particularly simple in this case. Let $d_t^{(S)}$ and $d_t^{(R)}$ be the respective coded bits transmitted simultaneously by the source and relay at time t over the second time slot. Assuming that binary phase shift keying (BPSK) is employed, let $l(d_t^{(S)}) = \log \frac{\Pr\{d_t^{(S)}=+1\}}{\Pr\{d_t^{(S)}=-1\}}$ and $l(d_t^{(R)}) = \log \frac{\Pr\{d_t^{(R)}=+1\}}{\Pr\{d_t^{(R)}=-1\}}$ be the *a priori* LLR of the coded bits $d_t^{(S)}$ and $d_t^{(R)}$, respectively. Then the *a posteriori* LLR of the coded bit $d_t^{(S)}$ is given by

$$
\begin{aligned}
L(d_t^{(S)} \mid y_t) &= \log \frac{\Pr\left\{d_t^{(S)} = +1 \mid y_t\right\}}{\Pr\left\{d_t^{(S)} = -1 \mid y_t\right\}} \\
&= \log \frac{p(y_t \mid d_t^{(S)} = +1, d_t^{(R)} = -1) + p(y_t \mid d_t^{(S)} = +1, d_t^{(R)} = +1)\exp l(d_t^{(R)})}{p(y_t \mid d_t^{(S)} = -1, d_t^{(R)} = -1) + p(y_t \mid d_t^{(S)} = -1, d_t^{(R)} = +1)\exp l(d_t^{(R)})} \\
&\quad + l(d_t^{(S)}),
\end{aligned}
\tag{16}
$$

where y_t is the destination's received channel symbol that contains the superposition of $d_t^{(S)}$ and $d_t^{(R)}$. That is $y_t = h_t^{(S)} d_t^{(S)} + h_t^{(R)} d_t^{(R)} + n_t$, where $h_t^{(S)}$ and $h_t^{(R)}$ are the respective channel gains of the links from the source to the relay and destination, and n_t is the Gaussian noise sample. A similar expression can be obtained for the *a posteriori* LLR $L(d_t^{(R)} \mid y_t)$ of the coded bit $d_t^{(R)}$.

The iterative joint MUD-decoding process is described as follows. In the first iteration, the MUD executes first by setting $l(d_t^{(i)}), i \in \{S, R\}$ to 0 since the turbo decoder has not yet produced any outputs. The turbo decoder then executes by using the extrinsic information generated by the MUD as the channel LLR for the correspondent bit. The extrinsic information for $d_t^{(i)}, i \in \{S, R\}$ is given by

$$
\tilde{l}(d_t^{(i)}) = L(d_t^{(i)} \mid y_t) - l(d_t^{(i)}).
\tag{17}
$$

At this point, the first iteration is finished. For the second iteration, the extrinsic information is first subtracted from the output of turbo decoder before being fed into the MUD as the updated *a priori* LLR. The MUD then takes the newly updated *a priori* LLRs to generate $L(d_t^{(i)} \mid y_t)$ and $\tilde{l}(d_t^{(i)}), i \in \{S, R\}$ using (16) and (17), respectively. The second iteration is completed by executing the turbo decoder, and the iteration process continues until some stopping criterion is satisfied.

It is clear that a smaller α provides more diversity gain, as the relay will be able to transmit more coded bits in the second time slot. However, as mentioned before, α should also be large enough so that the relay can correctly decode the codeword that it receives from the source in the first time slot. This requires the effective code rate R / α of this transmission be smaller than the capacity of the link from the source to relay by a certain (small) margin. Hence the diversity gain achievable by the code design mentioned above is limited by this requirement on α.

A partial encoding approach is introduced in (Zhang & Duman, 2007) to overcome this limitation. Consider the encoding and decoding structures shown in Figure 10. The source first breaks the N information bits into two blocks of lengths N_1 and N_2, where $N_1 + N_2 = N$. These two blocks of information bits will be encoded separately. To simplify the discussion, let us assume that a rate-$1/3$ systematic turbo code is employed. Recall that for this turbo code the information bits are interleaved before entering the encoder of the second RSC component codes (cf. Figure 2(a)). To encode the two blocks of information bits, we employ two distinct random interleavers π_1 and π_2 of lengths N_1 and N_2 in two rate-$1/3$ turbo codes, respectively. The parity bits are then punctured to get code rates R_{s1} and R_{s2} for the first and second information blocks, respectively. In the first time slot, the source sends the N_1 / R_{s1} coded bits of the rate-R_{s1} systematic turbo code to the relay and destination. In the second time slot, the source sends the N_2 / R_{s2} coded bits of the rate-R_{s2} systematic turbo code to the destination. Thus the overall code rate of the combination of the two turbo codewords sent by the source is

$$R = \frac{N}{N_1 / R_{s1} + N_2 / R_{s2}}.$$

Now, in order for the relay to successfully decode the codeword sent by the source in the first time slot, $R_{s1} = \frac{N_1}{N} \cdot \frac{R}{\alpha}$ (instead of R / α) needs to be smaller than the capacity of the link from the source to relay by a certain margin. Hence α can be made smaller by reducing N_1. After decoding the N_1 information bits, the relay re-encodes them using the rate-$1 / 3$ turbo code with interleaver π_1. If the number of coded bits ($3N_1$) is larger than $(1 - \alpha)N / R$, then the coded bits are punctured according to the same priority order mentioned before. The punctured codeword is sent by the relay in the second time slot. Otherwise, all the $3N_1$ coded bits will be sent together with repetition of first the systematic bits, then the parity bits sent by the source in the first time slots, and finally the parity bits not sent by the source in the first time slot.

After receiving all the transmitted bits from the source and relay, the destination separately decodes the two blocks of information bits using two turbo decoders. In addition, the SISO MAP MUD described above is employed to generate the soft inputs to both turbo decoders for the respective coded bits received in the second time slot as shown in Figure 10. The same iterative joint MUD-decoding technique discussed before then applies between the MUD and the two turbo decoders. Note that there is no diversity advantage for the second block of information bits, since all the corresponding coded bits are transmitted directly by the source to the destination in the second time slot. To make sure that they can be correctly decoded, we may need to choose a smaller R_{s2}. In general, the optimal choice of the various parameters depends on the channel gains of the three links in the relay channel (Zhang & Duman, 2007). With a proper choice of the parameters, it is reported in (Zhang & Duman, 2007) that the partial encoding design can give performance that is 1.2 dB away from the DF rate of the relay channel.

Although (Zhang & Duman, 2007) describes the necessary extensions to utilize the approach with higher-order signal constellations or multi-antenna/MIMO communications, only limited results for these scenarios are provided. In (Murugan, Azarian, & El Gamal, 2007), a practical implementation of the dynamic decode and forward (DDF) protocol, which was originally proposed in (Azarian et al, 2005), is developed. In (Murugan et al., 2007), the DDF was reported to provide the best diversity-multiplexing tradeoff of the known DF protocols. As in (Zhang & Duman, 2002, 2007), non-orthogonal transmission is used by the source and relay, although the proposed approach is quite different.

Figure 10. Block diagram of partial encoding turbo codes for the half-duplex relay channel

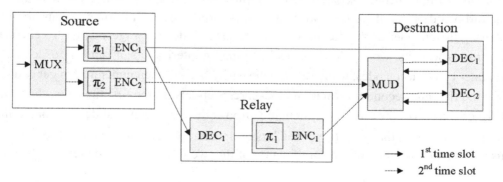

In the modified DDF protocol proposed in (Murugan et al., 2007), the source divides its codeword into sub-blocks. The relay listens for a sufficient number of sub-blocks to achieve correct decoding (but at least 1/2 the total number of sub-blocks, denoted by M). The relay then transmits symbols so the transmissions from the source and the relay together form an Alamouti constellation (Alamouti, 1998). After linear processing in the receiver, the MIMO channel is translated into a time-selective single-input single-output channel, which can reduce the decoding complexity at the destination. Lattice coding is used for error-correction, and the minimum mean-square error decision feedback equalizer (MMSE-DFE) Fano decoder of (Murugan, El Gamal, Damen, & Caire, 2006) is used to provide practical decoding complexity at the destination. Frame error rate results for two and three bits per channel use are provided, and the performance of the modified DDF protocol is shown to strictly dominate that of the non-orthogonal amplify-and-forward (NAF) protocol, which is discussed in the next section.

B. Coding for Amplify-and-Forward

In the amplify-and-forward (AF) approach, the source sends to the relay and destination in the first time slot. The relay does not attempt to decode the information sent by the source; instead it just amplifies and retransmits its received signal to the destination in the second time slot. For the relay, AF is of much lower complexity than DF. The destination attempts to decode the source information based on all the signals that it receives from the source and relay. Unlike DF, the relay does not need to decode the signal sent by the source, and hence the performance of AF is not limited by the link from the source to the relay. On the other hand, since the destination receives two copies of information from the source and relay that experience independent fading, diversity gain can be achieved. However, since the relay may amplify the noise in its received signal when forwarding it to the destination, AF may not work well in the low SNR regime.

A non-orthogonal AF (NAF) approach is proposed in (Azarian et al., 2005; Nabar & Bolcskei, 2003) allowing both the source and relay transmit in the second time slot. In fact, it is shown in (Azarian et al.) that the transmissions from the source and relay in two time slots can be considered as a multi-input multi-output (MIMO) channel to the destination. The diversity-multiplexing tradeoff achievable with NAF for a channel with $N-1$ relays is shown to be

$$d(r) = (1-r) + (N-1)(1-2r)^+,$$

where d denotes the diversity gain, r denotes the multiplexing gain (cf. (Tse, Viswanath, & Zheng, 2004; Zheng & Tse, 2003)), and x^+ denotes $\max\{x, 0\}$. With the MIMO interpretation, many standard MIMO coding designs can be modified for the AF approach. For instance, a slightly modified version of the Alamouti code was suggested in (Anghel et al., 2003) for the relay channel with AF, although dual frequency bands are assumed to allow for full-duplex operation at the relays. For the case of a single relay, the equivalent MIMO channel model for NAF is of the form (cf. (Azarian et al.; Murugan et al., 2007; Yang & Belfiore, 2007))

$$\mathbf{y}_k = \mathbf{H}\mathbf{x}_k + \mathbf{z}_k$$

for the kth cooperation frame (which consists of two consecutive symbols), where $\mathbf{x}_k = [x_k^{(S)} x_k^{(R)}]$ denotes the channel input from the source and relay, $\mathbf{y}_k$ denotes the received symbol at the destination, and $\mathbf{H}$ is the equivalent MIMO channel gain matrix, and $\mathbf{z}_k$ is Gaussian noise.

In (Yang & Belfiore, 2007), schemes to achieve the diversity multiplexing tradeoff of NAF are proposed using algebraic space-time codes based on the non-vanishing determinant criterion (Elia, Kumar, Pawar, Kumar, & Lu, 2006). These schemes are generalized in (Murugan et al., 2007) to provide better performance by coding over more than two cooperation frames using a lattice coding/decoding framework. For example, the authors suggest using the Golden constellation design of (Yang & Belfiore) as an inner code with an outer lattice code constructed using a convolutional code over a finite integer ring. The MMSE-DFE Fano decoder of (Murugan et al., 2006) is again used to provide for practical decoding complexity.

C. Coding for Compress/Estimate/Quantize-and-Forward

Cover and El Gamal propose a third approach for relay communication that is now implemented in various ways and known by various names, including compress-and-forward (CF), estimate-and-forward (EF), or quantize-and-forward (QF). These approaches are similar to the AF scheme in that correct decoding is not required at the relay. However, they differ from the AF scheme in that they can use digital modulation for transmission from the relay and the relay can transmit a different number of symbols than it receives. In the EF and QF schemes, the relay typically quantizes and encodes the received message (Chakrabarti, de Baynast, Sabharwal, & Aazhang, 2006; Djeumou, Lasaulce, & Klein, 2007). The CF schemes offer the most flexibility, as the relay quantizes the received message and then applies source and channel coding before transmission to the destination. Thus, CF schemes are the focus of this section.

Two research groups implemented the first practical CF schemes at nearly the same time (Hu & Li, 2005; Li & Hu, 2005; Liu et al., 2005). Hu and Li (2005) (also see (Li & Hu)) proposed schemes based on practical Slepian-Wolf coding schemes. The key observation made in (Hu & Li, 2005; Li & Hu) is that when decoding fails at the relay, the three-node relay channel is very similar to a DSC system. When decoding at the relay fails, the relay has a noisy version of the information at the source. Thus, **Slepian-Wolf** or **Wyner-Ziv coding** can be used to perform DSC on the two sets of information for joint decoding at the destination.

The simplest Slepian-Wolf compress-forward scheme of (Hu & Li, 2005; Li & Hu, 2005) takes place in three phases. In the first phase, the source broadcasts the (possibly coded) information X to the destination and relay. Let Y be the noisy version of X received at the relay, and Z be the noisy version of X received at the relay. For the schemes in (Hu & Li, 2005; Li & Hu), it is assumed that Y is the output of the hard-decision demodulator for X. Then, in the second phases, the relay transmits Y to the destination. Finally, in the third interval, the source uses asymmetric Slepian-Wolf coding at rate $H(X \mid Y)$ to allow for joint decoding of Y and Z at the destination. The three-phase transmission scheme can easily be modified to reduce it to two phases (Hu & Li, 2005; Li & Hu).

The schemes in (Hu & Li, 2005; Li & Hu, 2005) have several limitations. First, these schemes use hard-decision demodulation at the relay, which enables the use of Slepian-Wolf coding schemes but reduces the performance because useful soft information is discarded. Secondly, these schemes assume orthogonal transmissions from the relay and source. The first limitation can be overcome through the use of Wyner-Ziv coding of the soft information at the relay, which was originally proposed in (Liu et al.,

Figure 11. Block diagram of a practical CF coding scheme based on Wyner-Ziv coding

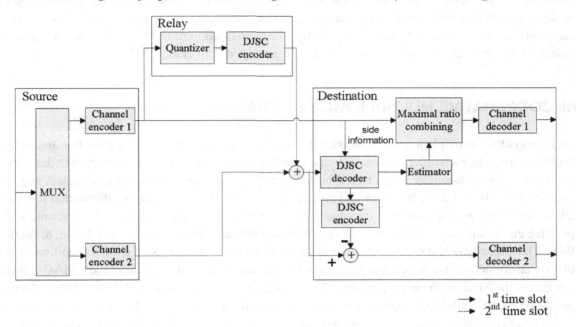

2005) and also investigated in (Hu & Li, 2006). The scheme proposed in (Liu et al., 2005) also allows simultaneous transmission by the source and relay, and is the only scheme that has been demonstrated to achieve performance close to the CF limits on the half-duplex relay channel. Thus, the remainder of this section is dedicated to describing this scheme.

The practical CF coding scheme in (Liu et al., 2005) is based on the practical joint source-channel coding scheme for a binary channel with side information in (Liveris et al., 2002c). The structure of this CF coding scheme, which is shown in Figure 11, is similar to the partial encoding approach described in Section VI-A. The source separates its information bits into two blocks. The first block is sent to the relay and the destination. Unlike the DF approach, the relay does not decode the transmission from the source. Instead, it first quantizes the signal that it receives from the source, and then encodes the quantized signal using the systematic distributed joint source and channel (DJSC) encoder in (Liveris et al., 2002c). The DJSC encoder jointly implements Wyner-Ziv compression as well as error protection. In the second time slot, the relay sends the parity bits given out by the DJSC encoder to the destination. During the same time slot, the source also sends out the second block of its information bits.

The destination performs DJSC decoding (Liverise et al., 2002c) based on the same idea as discussed in Section II-C, treating its received signal in the first time slot as the side information for the quantized received signal of the relay. Note that since the source also transmits in the second time slot, the "corruption" to the relay's quantized received signal includes both noise and the signal sent by the source. The DJSC decoder provides an estimate of the relay's quantized received signal, which is maximal ratio combined with the signal received in the first time slot by the destination for decoding of the first block of source information. This step makes sure that the proposed CF coding always gives better performance than direct transmission from the source to destination. For decoding of the second block of source information, the estimated relay's quantized received signal is re-DJSC-encoded to give the parity bits transmitted by the relay in the second time slot. Then the effect of these parity bits is subtracted from the destination's received signal in the second time slot, and the resulting signal is decoded to give the second block of source information. This practical implementation of the CF coding scheme comes as close as 0.76 dB to the theoretical limit of CF as reported in (Liu et al., 2005).

VII. COOPERATIVE MULTIPLE ACCESS CHANNEL

The **cooperative multiple access channel** (CMAC) is the same as the MAC, except that the users are allowed to listen to each other's transmission and help forward each other's information to the common destination. For the two-user MAC example shown in Figure 1, this means that terminals A and B can listen to each other's transmit signals and send their own as well as the other's information to terminal C. For simplicity, we will assume a two-user CMAC in the following discussion. The first practical cooperative communication scheme was in fact proposed for the CMAC (Sendonaris, Erkip, & Aazhang, 1998). The capacity of the CMAC channel is yet to be found, but it has been shown (Sendonaris et al., 2003a) that cooperation between the users gives a larger achievable rate region for the CMAC compared with that of the MAC without cooperation. Other works (Kaya, 2006; Kaya & Ulukus, 2007; Mesbah & Davidson, 2006a, 2006b) that consider power allocation for the fading CMAC also confirm this fact.

For practical coding schemes, orthogonalization of the users' access to the CMAC channel by time-, frequency-, or code-division multiplexing is often employed to simplify the design. For example in (Sendonaris et al., 2003b), a CDMA implementation is considered, wherein different spreading codes are used to obtain different channels for simultaneous transmission and reception by the users. With this user orthogonalization, the CMAC reduces to two orthogonal **relay channels**, in each of which a user acts as relay for the other user's transmission to the destination. Thus the distributed turbo coding approach discussed in Section VI applies directly. For example, such an approach is suggested in (Hunter & Nosratinia, 2002) (which terms this approach as coded cooperation), (Stefanov & Erkip, 2004) (which uses a convolutional code instead of a turbo code) and other related works (Bao & Li, 2005; Cao & Vojcic, 2005; Ho-Van & Le-Ngoc, 2007; Liu, Spasojevic, & Soljanin, 2003; Vardhe & Reynolds, 2007). Space-time cooperation (Janani, Hedayat, Hunter, & Nosratinia, 2004; Wei, Zhang, & Li, 2006) (also see (Laneman & Wornell, 2003) extends this idea by employing distributed space-time codes to allow the users to transmit simultaneously on each channel in the second time slot. The use of distributed space-time coding requires the users to phase-synchronize their transmission, but can help reduce the rate loss due to user orthogonalization (Laneman & Wornell).

In general, with suitable user orthogonalization, the CMAC reduces to a pair of relay channels, and hence most of the practical coding schemes for the relay channel discussed in Section VI apply with slight modifications. To give another example, the Slepian-Wolf cooperation scheme for the orthogonal relay channel, which is proposed in (Hu & Li, 2005) and discussed in Section VI-C, applies to the orthogonalized CMAC. A cooperative transmission scheme based on superposition modulation was proposed in (Ding, Ratnarajah, & Cowan, 2007; Larsson & Vojcic, 2005). This scheme uses a time division approach in which a user simultaneously transmits its own information and the other user's information that it receives in the previous time slot by using superposition modulation. A user decodes the other user's information from the other user's transmission in the previous time slot by making use of the known information that it sends two time slots before. The destination demodulates two consecutive time slots to get the channel LLRs for decoding the information from a user. This scheme is demonstrated (Ding et al.) to achieve the optimal diversity-multiplexing tradeoff for the symmetric rate requirement scenario. It is also clear that the performance of this approach can be improved if the capacity-approaching practical broadcast channel coding schemes discussed in Section IV are used in place of the suboptimal superposition modulation.

The non-orthogonal AF (NAF) approach used in (Azarian et al., 2005) can also be extended to the orthogonalized CMAC. In NAF, each user transmits only once in each cooperation frame consisting of two symbols. The first user sends its symbol in the first symbol interval of a cooperative frame. In the second symbol interval, the second user transmits a linear combination of its own symbol and the signal that it observes during the previous symbol in the frame. The role of the two users may swap in the next frame if needed. As in the case of the NAF for the relay channel, a MIMO channel results in each cooperation frame, and it is proved in (Azarian et al.) that this NAF approach achieves the optimal diversity-multiplexing tradeoff with symmetric data rate requirement for the two users. In (Murugan et al., 2007), a practical implementation of the CMA-NAF protocol is proposed based on lattice coding/decoding. Again, the channel is shown to be equivalent to MIMO channel. For the case of two cooperating uses, the channel is (cf. (Murugan et al., 2007), Section V) of the form

$$\mathbf{y} = \mathbf{Hx} + \mathbf{Bw} + \mathbf{v},$$

where $\mathbf{x}$ denotes the vector formed by multiplexing the codewords from the two sources, $\mathbf{H}$ is the equivalent MIMO channel gain matrix, $\mathbf{w}$ and $\mathbf{v}$ represent white Gaussian noise vectors at the sources and destination, respectively, and $\mathbf{B}$ is a matrix related to $\mathbf{H}$ that causes the noise at the destination to be colored. After noise whitening, the equivalent channel is of the form

$$\mathbf{y}' = \mathbf{H}'\mathbf{x} + \mathbf{z},$$

where $\mathbf{z}$ is white noise with unit variance. Lattice coding is applied with *joint* lattice decoding of the information from the two sources. Results are presented to show significant gains from using CMA-NAF over relaying NAF and that the proposed CMA-NAF lattice coding/decoding schemes offer good complexity/performance tradeoffs.

VIII. CONCLUSIONS

A recurring theme that emerges in cooperative coding schemes is their use of a common set of basic tools including turbo, LDPC, RA, and lattice codes; iterative decoding and signal-processing techniques; and practical Slepian-Wolf/Wyner-Ziv coding techniques. Moreover the coding schemes constructed from these tools for the "simpler" cooperative channels (e.g., the MAC, broadcast channel, and DSC) are used to construct more complicated coding schemes for the "more complicated" cooperative channels (e.g., the relay channel and CMAC). These observations lead to the extrapolation that these tools and basic coding schemes may also be used for some "even more complicated" cooperative channels for which few practical coding schemes have currently been proposed. This approach of employing the coding schemes of the "simpler" cooperative channels as building blocks is further supported by our recent works (Chatterjee & Wong, in press;Chatterjee, Wong, & Lok, 2008; Tam, Lok, & Wong, in press; Wong, Lok, & Shea, in press) showing that good performance can be obtained by simple flow designs that employ only the MAC and broadcast channel components for the relay, CMAC, and other similar channels. In addition, the designs based on the deterministic channel models proposed in (7, Avestimehr, Diggavi, & Tse, 2007) for wireless relay networks also suggest the use of the MAC and broadcast components.

For fading channels, many of the recently proposed schemes rely on evaluating the performance in terms of the diversity-multiplexing tradeoff (Azarian et al., 2005; Murugan et al., 2007; Yang & Belfiore, 2007) rather than the outage capacity. Indeed, the capacity is still not know for some channels, such as the cooperative multiple access channel, and thus the performance cannot be compared to capacity. Thus, additional theoretical work is required for this channel.

Another channel for which the capacity is not known and for which almost no practical schemes have been developed is the interference channel. In the interference channel, several senders want to simultaneously communicate with their corresponding receivers over a common channel. The best known achievable region is from the results of Han and Kobayashi (1981). If the interference is very strong, its effect is known to be negligible (Costa, 972). There has been very little work on practical coding schemes for the interference channel. A trellis-coded modulation was proposed in (Li, Murata, & Yoshida, 1998) to derive a uniquely decodable code over a two-user interference channel. For this TCM, a TCM pair is designed to ensure that the sum of the two users' signals with arbitrary phase rotations will be uniquely decodable in the absence of noise using the Viterbi algorithm. Recently, Etkin et al. (2007) showed that a simplified Han-Kobayashi type scheme can achieve rates that are within a single bit from the capacity under all interference regimes. Since this scheme can also be obtained using the deterministic channel models proposed in (Avestimehr et al., 2007), it is conceivable that a practical coding scheme can be constructed using the basic coding blocks described in this chapter. An interesting approach to dealing with interference via cooperation is given in (Yang & Høst-Madsen, 2007). In this scheme, cooperation is used to increase the interference experienced by one of the nodes in an interference channel, thereby making it easier for that node to detect the desired signal in the presence of the strong interference.

Thus, cooperative communications represents an area in which many contributions have been made in the past decade toward practical schemes for a variety of applications. Yet, there are still interesting open problems to be addressed, such as practical schemes for the interference channel and better understanding of the limits of the interference channel. We expect that the tools and techniques described in this chapter will lead to practical coding schemes for this and other open problems.

RERFERENCES

Aaron, A., & Girod, B. (2002). Compression with side information using turbo codes. *Proceedings of IEEE Data Compression Conference,* Snowbird, UT (pp. 252–261).

Ahlswede, R. (1971). Multiway communication channels. In *Proceedings of IEEE International Symposium on Information Theory, Tsahkadsor, Armenian S.S.R.*, (pp. 23–52).

Ahlswede, R., & Han, T. (1983). On source coding with side information via a multiple-access channel and related problems in multi-user information theory. *IEEE Transactions on Information Theory, 29,* 396–412. doi:10.1109/TIT.1983.1056669

Airy, M., Forenza, A., Heath, R. W. Jr, & Shakkottai, S. (2004). Practical Costa precoding for the multiple antenna broadcast channel. In . *Proceedings of IEEE Global Communications Conference, 6,* 3942–3946.

Alamouti, S. (1998). A simple transmit diversity technique for wireless communications. *IEEE Journal on Selected Areas in Communications, 16*(8), 1451–1458. doi:10.1109/49.730453

Anghel, P., Leus, G., & Kaveh, M. (2003). Multi-user space-time coding in cooperative networks. In . *Proceedings of IEEE International Conference on Acoustic, Speech, and Signal Processing, 4,* 73–76.

Avestimehr, A. S., Diggavi, S. N., & Tse, D. N. C. (2007). A deterministic approach to wireless relay networks. In *Proceedings of Allerton Conference on Communications, Control, and Computing,* Allerton, IL.

Azarian, K., El Gamal, H., & Schniter, P. (2005). On the achievable diversity-multiplexing tradeoff in half-duplex cooperative channels. *IEEE Transactions on Information Theory, 51*(12), 4152–4172. doi:10.1109/TIT.2005.858920

Bahl, L. R., Cocke, J., Jelinek, F., & Raviv, J. (1974). Optimal decoding of linear codes for minimizing symbol error rates. *IEEE Transactions on Information Theory, 20*(2), 284–287. doi:10.1109/TIT.1974.1055186

Bajcsy, J., & Mitran, P. (2001). Coding for the Slepian-Wolf problem using turbo codes. In *Proceedings of IEEE Global Communications Conference,* San Antonio, TX (pp. 1400–1404).

Bao, X., & Li, J. (2005). Decode-amplify-forward (DAF): A new class of forwarding strategy for wireless relay channels. In *Proceedings of IEEE Workshop on Signal Processing Advances Wireless Communications* (pp. 816–820).

Benedetto, S., Divsalar, D., Montorsi, G., & Pollara, F. (1998). Serial concatenation of interleaved codes: Performance analysis, design and iterative decoding. *IEEE Transactions on Information Theory, 44,* 909–926. doi:10.1109/18.669119

Berger, T. (1977, July). Multiterminal source coding. *Lectures Presented at CISM Summer School on the Information Theory Approach to Communication.*

Berger, T., & Yeung, R. (1989). Multiterminal source encoding with one distortion criterion. *IEEE Transactions on Information Theory, 35*(2), 228–236. doi:10.1109/18.32119

Bergmans, P. P. (1973). Random coding theorem for broadcast channels with degraded components. *IEEE Transactions on Information Theory, 19*(2), 197–207. doi:10.1109/TIT.1973.1054980

Bergmans, P. P. (1974). A simple converse for broadcast channels with additive white Gaussian noise. *IEEE Transactions on Information Theory, 20*(2), 279–280. doi:10.1109/TIT.1974.1055184

Berrou, C., Glavieux, A., & Thitimajshima, P. (1993). Near Shannon limit error-correcting coding and decoding. In *Proceedings of IEEE International Conference Communications, 2,* Geneva, Switzerland (pp. 1064–1070).

Cao, Y., & Vojcic, B. (2005). Cooperative coding using serial concatenated convolutional codes. In . *Proceedings of IEEE Wireless Communications and Networking Conference, 2*, 1001–1006. doi:10.1109/WCNC.2005.1424645

Chakrabarti, A., de Baynast, A., Sabharwal, A., & Aazhang, B. (2006). Half-duplex estimate-and-forward relaying: Bounds and code design. In *Proceedings of IEEE International Symposium on Information Theory* (pp. 1239–1243).

Chatterjee, D., & Wong, T. F. (in press). Active user cooperation in fading multiple-access channels. In . *Proceedings of IEEE Military Communications Conference*.

Chatterjee, D., Wong, T. F., & Lok, T. M. (2008). Cooperative transmission in a wireless cluster based on flow management. In *Proceedings of IEEE Wireless Communications Networking Conference*, Las Vegas, NV (pp. 30–35).

Chou, J., Pradhan, S., & Ramchandran, K. (2003). Turbo and trellis-based constructions for source coding with side information. In *Proceedings of IEEE Data Compression Conference*, Snowbird, UT (pp. 33–42).

Chung, S., Forney, G. D. Jr, Richardson, T. J., & Urbanke, R. (2001). On the design of low-density parity-check codes within 0.0045 dB of the Shannon limit. *IEEE Communications Letters, 5*(2), 58–60. doi:10.1109/4234.905935

Coleman, T., Effros, M., Martinian, E., & Medard, M. (2005). Rate-splitting for the deterministic broadcast channel. In *Proceedings of IEEE International Symposium on Information Theory* (pp. 2189–2192).

Coleman, T., Martinian, E., Effros, M., & Medard, M. (2005). Interference management via capacity-achieving codes for the deterministic broadcast channel. In *Proceeding of IEEE Information Theory Workshop, 5*.

Costa, M. (1972). On the Gaussian interference channel. *IEEE Transactions on Information Theory, 31*(5), 607–615. doi:10.1109/TIT.1985.1057085

Cover, T. M. (1972). Broadcast channels. *IEEE Transactions on Information Theory, 18*(1), 2–14. doi:10.1109/TIT.1972.1054727

Cover, T. M., El Gamal, A., & Salehi, M. (1980). Multiple access channels with arbitrarily correlated sources. *IEEE Transactions on Information Theory, 26*(6), 648–657. doi:10.1109/TIT.1980.1056273

Cover, T. M., & El Gamal, A. A. (1979). Capacity theorems for the relay channel. *IEEE Transactions on Information Theory, 25*(5), 572–584. doi:10.1109/TIT.1979.1056084

Cover, T. M., & Leung, C. (1981). An achievable rate region for the multiple-access channel with feedback. *IEEE Transactions on Information Theory, 27*(3), 292–298. doi:10.1109/TIT.1981.1056357

Cover, T. M., & Thomas, J. A. (1991). *Elements of information theory*. NY: John Wiley & Sons.

Ding, Z., Ratnarajah, T., & Cowan, C. C. F. (2007). On the diversity-multiplexing tradeoff for wireless cooperative multiple access systems. *IEEE Transactions on Signal Processing, 55*(9), 4627–4638. doi:10.1109/TSP.2007.896276

Divsalar, D., Dolinar, S., & Pollara, F. (2001). Iterative turbo decoder analysis based on density evolution. *IEEE Journal on Selected Areas in Communications, 19*(5), 891–907. doi:10.1109/49.924873

Divsalar, D., Jin, H., & McEliece, R. (1998). Coding theorems for 'turbo-like' codes. In *Proceedings of Allerton Conference on Communications, Control, and Computing* (pp. 201–210).

Divsalar, D., & Pollara, F. (1995, November). *On the design of turbo codes* (Tech. Rep.). *NASA Jet Propulsion Laboratory. TDA Progress Report 42-123*.

Djeumou, B., Lasaulce, S., & Klein, A. G. (2007). Practical quantize-and-forward schemes for the frequency division relay channel. *EURASIP J. Wireless Communications Networking, 4*, 1–11. doi:10.1155/2007/20258

Douillard, C., Jezequel, M., Berrou, C., Picart, A., Didier, P., & Glavieux, A. (1995). Iterative correction of intersymbol interference: Turbo-equalization. *European Transactions on Telecommunications, 6*(5), 507–511. doi:10.1002/ett.4460060506

El Gamal, H., Caire, G., & Damen, M. O. (2004). Lattice coding and decoding achieve the optimal diversity-multiplexing-delay tradeoff of mimo channels. *IEEE Transactions on Information Theory, 50*(6), 968–985. doi:10.1109/TIT.2004.828067

Elia, P., Kumar, K. R., Pawar, S. A., Kumar, P. V., & Lu, H. F. (2006). Explicit space-time codes achieving the diversity-multiplexing gain tradeoff. *IEEE Transactions on Information Theory, 52*(9), 3869–3884. doi:10.1109/TIT.2006.880037

Erez, U., & Brink, S. (2005). A close-to-capacity dirty paper coding scheme. *IEEE Transactions on Information Theory, 51*(10), 3417–3432. doi:10.1109/TIT.2005.855586

Erez, U., Shamai, S., & Zamir, R. (2000). Capacity and lattice strategies for canceling known interference. In *Proceedings of International Symposium on Information Theory and Its Application,* Honolulu, HI (pp. 681–684).

Erez, U., Shamai, S., & Zamir, R. (2005). Capacity & lattice strategies for canceling known interference. *IEEE Transactions on Information Theory, 51*(11), 3820–3833. doi:10.1109/TIT.2005.856935

Etkin, R., Tse, D. N. C., & Wang, H. (2007). Gaussian interference channel capacity to within one bit. Retrieved in 2007, from http://www.citebase.org/abstract?id=oai:arXiv.org:cs/0702045

Fischer, R., Stierstorfer, C., & Windpassinger, C. (2003). Precoding and signal shaping for transmission over MIMO channels. In *Proceedings of Canadian Workshop Information Theory,* Waterloo, ON, Canada (pp. 83–87).

Fischer, R., Windpassinger, C., Lampe, A., & Huber, J. (2002a). MIMO precoding for decentralized receivers. In *Proceedings of IEEE International Symposium on Information Theory* (p. 496).

Fischer, R., Windpassinger, C., Lampe, A., & Huber, J. (2002b). Space-time transmission using Tomlinson-Harashima precoding. In *Proceedings of ITG Conference on Source and Channel Coding* (pp. 139–147).

Forney, G. D. Jr. (1992). Trellis shaping. *IEEE Transactions on Information Theory, 38*(2), 281–300. doi:10.1109/18.119687

Gaarder, N., & Wolf, J. (1975). The capacity region of a multiple-access discrete memoryless channel can increase with feedback. *IEEE Transactions on Information Theory, 21*(1), 100–102. doi:10.1109/TIT.1975.1055312

Gallager, R. G. (1962). Low-density parity-check codes. *IEEE Transactions on Information Theory, 8*(1), 21–28. doi:10.1109/TIT.1962.1057683

Gallager, R. G. (1963). *Low-density parity-check codes.*Cambridge, MA: MIT Press.

Gallager, R. G. (1974). Capacity and coding for degraded broadcast channels. *Problemv Peredaci Informotsii, 10*(3), 3–14.

Garcia-Frias, J. (2001a). Compression of correlated binary sources using turbo codes. *IEEE Communications Letters, 5*(10), 417–419. doi:10.1109/4234.957380

Garcia-Frias, J. (2001b). Joint source-channel decoding of correlated sources over noisy channels. In *Proceedings of IEEE Data Compression Conference,* Snowbird, UT (pp. 283–292).

Garcia-Frias, J., & Xiong, Z. (2005). Distributed source and joint source-channel coding: From theory to practice. In *. Proceedings of IEEE International Conference on Acoustics, Speech, and Signal Processing, 5,* 1093–1096. doi:10.1109/ICASSP.2005.1416498

Garcia-Frias, J., & Zhong, W. (2003). LDPC codes for compression of multi-terminal sources with hidden Markov correlation. *IEEE Communications Letters, 7*(3), 115–117. doi:10.1109/LCOMM.2003.810001

Garcia-Frias, J., Zhong, W., & Zhao, Y. (2002). Iterative decoding schemes for source and joint source-channel coding of correlated sources. In *Proceedings of Asilomar Conference on Signals, Systems and Computers* (Vol. 1, pp. 250–256).

Gastpar, M. (2004). The Wyner-Ziv problem with multiple sources. *IEEE Transactions on Information Theory, 50*(11), 2762–2768. doi:10.1109/TIT.2004.836707

Gastpar, M., & Vetterli, M. (2005). On the capacity of large Gaussian relay networks. *IEEE Transactions on Information Theory, 51*(3), 765–779. doi:10.1109/TIT.2004.842566

Giallorenzi, T., & Wilson, S. (1996). Suboptimum multiuser receivers for convolutionally coded asynchronous DS-CDMA systems. *IEEE Transactions on Communications, 44*(9), 1183–1196. doi:10.1109/26.536924

Ginis, G., & Cioffi, J. M. (2000). A multi-user precoding scheme achieving crosstalk cancellation with application to DSL systems. In *Proceedings of Asilomar Conference on Signals, Systems and Computers* (Vol. 2, pp. 1627–1631).

Hagenauer, J. (1996). Forward error correcting for CDMA systems. In *. Proceedings of IEEE International Symposium on Spread Spectrum Techniques and Applications, 2,* 566–569. doi:10.1109/ISSSTA.1996.563190

Hagenauer, J., Offer, E., & Papke, L. (1996). Iterative decoding of binary block and convolutional codes. *IEEE Transactions on Information Theory, 42*(2), 429–445. doi:10.1109/18.485714

Haim, E., & Zamir, R. (2005). Quantization with variable resolution and coding for deterministic broadcast channels. In *Proceedings of Allerton Conference on Communications, Control, and Computing.*

Han, T., & Kobayashi, K. (1981). A new achievable rate region for the interference channel. *IEEE Transactions on Information Theory, 27*(1), 49–60. doi:10.1109/TIT.1981.1056307

Ho-Van, K., & Le-Ngoc, T. (2007). A bandwidth-efficient coded user-cooperation scheme for flat block fading channels. In *Proceedings of International Symposium on Wireless Communications Systems* (pp. 421–425).

Høst-Madsen, A. (2002). On the capacity of wireless relaying. In . *Proceedings of IEEE Vehicular Technology Conference, 3*, 1333–1337.

Høst-Madsen, A., & Zhang, J. (2005). Capacity bounds and power allocation for wireless relay channels. *IEEE Transactions on Information Theory, 51*(6), 2020–2040. doi:10.1109/TIT.2005.847703

Hu, J., & Duman, T. (2007). Low density parity check codes over wireless relay channels. *IEEE Transactions on Wireless Communications, 6*(9), 3384–3394. doi:10.1109/TWC.2007.06083

Hu, R., & Li, J. (2005). Exploiting Slepian-Wolf codes in wireless user cooperation. In *Proceedings of IEEE Workshop on Signal Processing Advances in Wireless Communications* (pp. 275–279).

Hu, R., & Li, J. (2006). Practical compress-forward in user cooperation: Wyner-Ziv cooperation. In *Proceedings of IEEE International Symposium on Information Theory* (pp. 489–493).

Hunter, T., & Nosratinia, A. (2002). Cooperation diversity through coding. In *Proceedings of IEEE International Symposium on Information Theory* (p. 220).

Janani, M., Hedayat, A., Hunter, T., & Nosratinia, A. (2004). Coded cooperation in wireless communications: Space-time transmission and iterative decoding. *IEEE Transactions on Signal Processing, 52*(2), 362–371. doi:10.1109/TSP.2003.821100

Jin, H., Khandekar, A., & McEliece, R. (2000, September). Irregular repeat-accumulate codes. In *Proceedings of International Symposium on Turbo Codes Related Topics,* Brest, France (pp. 1–8).

Joham, M., Brehmer, J., & Utschick, W. (2002). MMSE approaches to multiuser spatio-temporal Tomlinson-Harashima precoding. In *Proceedings of ITG Conference on Source and Channel Coding* (pp. 387–394).

Johnson, M., Ishwar, P., Prabhakaran, V., Schonberg, D., & Ramchandran, K. (2004). On compressing encrypted data. *IEEE Transactions on Signal Processing, 52*(10), 2992–3006. doi:10.1109/TSP.2004.833860

Kaspi, A., & Berger, T. (1982). Rate-distortion for correlated sources with partially separated encoders. *IEEE Transactions on Information Theory, 28*(6), 828–840. doi:10.1109/TIT.1982.1056586

Kaya, O. (2006). Window and backwards decoding achieve the same sum rate for the fading cooperative Gaussian multiple access channel. In *Proceedings of IEEE Global Communications Conference, San Francisco, CA* (pp. 1–5).

Kaya, O., & Ulukus, S. (2007). Power control for fading cooperative multiple access channels. *IEEE Transactions on Communications, 6*, 2915–2923. doi:10.1109/TWC.2007.05858

Khojastepour, M., Ahmed, N., & Aazhang, B. (2004). Code design for the relay channel and factor graph decoding. In *Proceedings of Asilomar Conference on Signals, Systems and Computers* (Vol. 2, pp. 2000–2004).

Khojastepour, M., Sabharwal, A., & Aazhang, B. (2003). On capacity of Gaussian 'cheap' relay channel. In . *Proceedings of IEEE Global Communications Conference, 3*, 1776–1780.

Kohno, R., Imai, H., Hatori, M., & Pasupathy, S. (1990). Combinations of an adaptive array antenna and a canceller of interference for direct-sequence spread-spectrum multiple-access system. *IEEE Journal on Selected Areas in Communications, 8*(4), 675–682. doi:10.1109/49.54463

Kramer, G., Gastpar, M., & Gupta, P. (2005). Cooperative strategies and capacity theorems for relay networks. *IEEE Transactions on Information Theory, 51*(9), 3037–3063. doi:10.1109/TIT.2005.853304

Kschischang, F., Frey, B., & Loeliger, H.-A. (2001). Factor graphs and the sum-product algorithm. *IEEE Transactions on Information Theory, 47*(2), 498–519. doi:10.1109/18.910572

Kusume, K., Joham, M., Utschick, W., & Bauch, G. (2005). Efficient Tomlinson-Harashima precoding for spatial multiplexing on flat MIMO channel. In . *Proceedings of IEEE International Conference Communications, 3*, 2021–2025.

Lai, L., Liu, K., & El Gamal, H. (2006). The three-node wireless network: Achievable rates and cooperation strategies. *IEEE Transactions on Information Theory, 52*(3), 805–828. doi:10.1109/TIT.2005.864421

Laneman, J. N., Tse, D. N. C., & Wornell, G. (2004). Cooperative diversity in wireless networks: Efficient protocols and outage behavior. *IEEE Transactions on Information Theory, 50*(12), 3062–3080. doi:10.1109/TIT.2004.838089

Laneman, J. N., & Wornell, G. W. (2003). Distributed space-time coded protocols for exploiting cooperative diversity in wireless network. *IEEE Transactions on Information Theory, 49*, 2415–2425. doi:10.1109/TIT.2003.817829

Laneman, J. N., Wornell, G. W., & Tse, D. N. C. (2001). An efficient protocol for realizing cooperative diversity in wireless networks. In *Proceedings of IEEE International Symposium on Information Theory* (p. 294).

Larsson, E. G., & Vojcic, B. R. (2005). Cooperative transmit diversity based on superposition coding. *IEEE Communications Letters, 9*(9), 778–780. doi:10.1109/LCOMM.2005.1506700

Li, J., & Hu, R. (2005). Slepian-Wolf cooperation: A practical and efficient compress-and-forward relay scheme. In *Proceedings of Allerton Conference on Communications, Control, and Computing, Champaign, IL*.

Li, J., Tu, Z., & Blum, R. (2004). Slepian-Wolf coding for nonuniform sources using turbo codes. In *Proceedings of IEEE Data Compression Conference,* Snowbird, UT (pp. 312–321).

Li, X., & Ritcey, J. A. (1999a). Bit-interleaved coded modulation with iterative decoding. In *Proceedings of IEEE International Conference on Communications* (Vol. 2, pp. 858–863).

Li, X., & Ritcey, J. A. (1999b). Trellis-coded modulation with bit interleaving and iterative decoding. *IEEE Journal on Selected Areas in Communications, 17*(4), 715–724. doi:10.1109/49.761047

Li, Y., Murata, H., & Yoshida, S. (1998). Coding for multi-user detection in interference channel. In *Proceedings of IEEE Global Communications Conference* (Vol. 6, pp. 3596–3601).

Liang, P.-P., & Stark, W. (2001). Iterative multiuser detection for turbo-coded FHMA communications. *IEEE Journal on Selected Areas in Communications, 19*(9), 1810–1819. doi:10.1109/49.947045

Liao, H. (1972). A coding theorem for multiple access communications. In *Proceedings of IEEE International Symposium on Information Theory,* Asilomar, CA.

Lin, S., & Costello, D. J. (1983). *Error control coding: Fundamentals and applications*. NJ: Prentice-Hall.

Liu, R., Spasojevic, P., & Soljanin, E. (2003). User cooperation with punctured turbo codes. In *Proceedings of Allerton Conference on Communications, Control, and Computing,* Monticello, IL (pp. 201–210).

Liu, Z., Cheng, S., Liveris, A., & Xiong, Z. (2004). Slepian-Wolf coded nested quantization (SWC-NQ) for Wyner-Ziv coding: Performance analysis and code design. In *Proceedings of IEEE Data Compression Conference,* Snowbird, UT (pp. 322–331).

Liu, Z., Cheng, S., Liveris, A., & Xiong, Z. (2006). Slepian-Wolf coded nested lattice quantization for Wyner-Ziv coding: High-rate performance analysis and code design. *IEEE Transactions on Information Theory, 52*(10), 4358–4379. doi:10.1109/TIT.2006.881708

Liu, Z., Stankovic, V., & Xiong, Z. (2005). Wyner-Ziv coding for the half-duplex relay channel. In *Proceedings of IEEE International Conference on Acoustic, Speech, and Signal Processing* (Vol. 5, pp. 1113–1116).

Liveris, A., Xiong, Z., & Georghiades, C. (2002a). Compression of binary sources with side information at the decoder using LDPC codes. *IEEE Communications Letters, 6*(10), 440–442. doi:10.1109/LCOMM.2002.804244

Liveris, A., Xiong, Z., & Georghiades, C. (2002b). A distributed source coding technique for correlated images using turbo-codes. *IEEE Communications Letters, 6*(9), 379–381. doi:10.1109/LCOMM.2002.803479

Liveris, A., Xiong, Z., & Georghiades, C. (2002c). Joint source-channel coding of binary sources with side information at the decoder using IRA codes. In *Proceedings of IEEE Multimedia Signal Processing Workshop* (pp. 53–56).

Lo, C., Vishwanath, S., & Heath, R. W., Jr. (2005). Rate bounds for MIMO relay channels using precoding. In *Proceedings of IEEE Global Communications Conference* (Vol. 3, pp. 1172–1176).

Luby, M. G., Mitzenmacher, M., Shokrollahi, M. A., & Spielman, D. A. (1998). Analysis of low density codes and improved designs using irregular graphs. In *Proceedings of ACM Symposium on Theory Computing,* Dallas, TX (pp. 249–258).

Luby, M. G., Mitzenmacher, M., Shokrollahi, M. A., & Spielman, D. A. (2001). Efficient erasure correcting codes. *IEEE Transactions on Information Theory, 47*(2), 569–584. doi:10.1109/18.910575

Luby, M. G., Mitzenmacher, M., Shokrollahi, M. A., Spielman, D. A., & Stemann, V. (1997). Practical loss-resilient codes. In *Proceedings of ACM Symposium on Theory Computing,* El Paso, TX (pp. 50–159).

Lucky, R. W. (1965). Automatic equalization for digital communication. *The Bell System Technical Journal, 44*(4), 547–588.

MacKay, D., & Neal, R. (1997). Near Shannon limit performance of low density parity check codes. *IEE Electronics Letters, 33*(6), 457–458. doi:10.1049/el:19970362

Marcellin, M., & Fischer, T. (1990). Trellis coded quantization of memoryless and Gauss-Markov sources. *IEEE Transactions on Communications, 38*(1), 82–93. doi:10.1109/26.46532

Marton, K. (1979, May). A coding theorem for the discrete memoryless broadcast channel. *IEEE Transactions on Information Theory, 25*(3), 306–311. doi:10.1109/TIT.1979.1056046

Mesbah, W., & Davidson, T. N. (2006a). Optimal power allocation for full-duplex cooperative multiple access. In *Proceedings of IEEE International Conference on Acoustic, Speech, and Signal Processing,* Toulouse, France (Vol. *4,* pp. 689–692).

Mesbah, W., & Davidson, T. N. (2006b). Optimal power and resource allocation for half-duplex cooperative multiple access. In *Proceedings of IEEE International Conference Communications,* Istanbul, Turkey (Vol. *10,* pp. 4469–4473).

Mitran, P., & Bajcsy, J. (2002). Coding for the Wyner-Ziv problem with turbo-like codes. In *Proceedings of IEEE International Symposium on Information Theory,* Sorrento, Italy (p. 91).

Mitran, P., & Bajcsy, J. (2002). Turbo source coding: A noise-robust approach to data compression. In *Proceedings of IEEE Data Compression Conference,* Snowbird, UT (p. 465).

Miyakawa, M., & Harashima, H. (1969). A method of code conversion for a digital communication channel with intersymbol interference. *Transactions of the Institute of Electronics and Communications Engineers of Japan, 52-A,* 272–273.

Murugan, A., Azarian, K., & El Gamal, H. (2007). Cooperative lattice coding and decoding in half-duplex channels. *IEEE Journal on Selected Areas in Communications, 25*(2), 268–279. doi:10.1109/JSAC.2007.070204

Murugan, A., El Gamal, H., Damen, M. O., & Caire, G. (2006). A unified framework for tree search decoding: Rediscovering the sequential decoder. *IEEE Transactions on Information Theory, 52*(3), 933–953. doi:10.1109/TIT.2005.864418

Murugan, A., Gopala, P., & El Gamal, H. (2004). Correlated sources over wireless channels: Cooperative source-channel coding. *IEEE Journal on Selected Areas in Communications, 22*(6), 988–998. doi:10.1109/JSAC.2004.830889

Nabar, R., & Bolcskei, H. (2003). Space-time signal design for fading relay channels. In *Proceedings of IEEE Global Communications Conference* (Vol. *4*, pp. 1952–1956).

Nam, Y.-H., & El Gamal, H. (in press). Joint lattice decoding achieves the optimal diversity-multiplexing tradeoff of multiple access channels. Retrieved from http://www.ece.osu.edu/ helgamal/MAC_TITR1_H.pdf

Oohama, Y. (1997). Gaussian multiterminal source coding. *IEEE Transactions on Information Theory, 43*(6), 1912–1923. doi:10.1109/18.641555

Oohama, Y. (1998). The rate-distortion function for the quadratic Gaussian CEO problem. *IEEE Transactions on Information Theory, 44*(3), 1057–1070. doi:10.1109/18.669162

Oohama, Y. (2005). Rate-distortion theory for Gaussian multiterminal source coding systems with several side informations at the decoder. *IEEE Transactions on Information Theory, 51*(7), 2577–2593. doi:10.1109/TIT.2005.850110

Ozarow, L. (1984). The capacity of the white Gaussian multiple access channel with feedback. *IEEE Transactions on Information Theory, 30*(4), 623–629. doi:10.1109/TIT.1984.1056935

Perez, L. C., Seghers, J., & Costello, D. J. Jr. (1996). A distance spectrum interpretation of turbo codes. *IEEE Transactions on Information Theory, 42*(6), 1698–1709. doi:10.1109/18.556666

Ping, L., Leung, W. K., & Wu, K. Y. (2003). Low-rate turbo-Hadamard codes. *IEEE Transactions on Information Theory, 49*(12), 3213–3224. doi:10.1109/TIT.2003.820018

Ping, L., Liu, L., & Leung, W. K. (2003). A simple approach to near-optimal multiuser detection: Interleave-division multiple-access. In *Proceedings of IEEE Wireless Communications and Networking Conference* (pp. 391–396).

Ping, L., Liu, L., Wu, K. Y., & Leung, W. K. (2004). Approaching the capacity of multiple access channels using interleaved low-rate codes. *IEEE Communications Letters, 8*(1), 4–6. doi:10.1109/LCOMM.2003.822534

Pradhan, S. S., & Ramchandran, K. (1999). Distributed source coding using syndromes (DISCUS): Design and construction. In *Proceedings of IEEE Data Compression Conference,* Snowbird, UT (pp. 158–167).

Pradhan, S. S., & Ramchandran, K. (2000a). A constructive approach to distributed source coding with symmetric rates. In *Proceedings of IEEE International Symposium on Information Theory* (p. 178).

Pradhan, S. S., & Ramchandran, K. (2000b). Distributed source coding: Symmetric rates and applications to sensor networks. In *Proceedings of IEEE Data Compression Conference,* Snowbird, UT (pp. 363–372).

Pradhan, S. S., & Ramchandran, K. (2003). Distributed source coding using syndromes (DISCUS): Design and construction. *IEEE Transactions on Information Theory, 49*(3), 626–643. doi:10.1109/TIT.2002.808103

Reed, M., Schlegel, C., Alexander, P., & Asenstorfer, J. (1998). Iterative multiuser detection for CDMA with FEC: Near-single-user performance. *IEEE Transactions on Communications, 46*(12), 1693–1699. doi:10.1109/26.737408

Reznik, A., Kulkarni, S., & Verdu, S. (2004). Degraded Gaussian multirelay channel: Capacity and optimal power allocation. *IEEE Transactions on Information Theory, 50*(12), 3037–3046. doi:10.1109/TIT.2004.838373

Richardson, T., Shokrollahi, M., & Urbanke, R. (2001). Design of capacity-approaching irregular low-density parity-check codes. *IEEE Transactions on Information Theory, 47*(2), 619–637. doi:10.1109/18.910578

Richardson, T., & Urbanke, R. (2001). The capacity of low-density parity-check codes under message-passing decoding. *IEEE Transactions on Information Theory, 47*(2), 599–618. doi:10.1109/18.910577

Roumy, A., Guemghar, S., Caire, G., & Verdu, S. (2004). Design methods for irregular repeat-accumulate codes. *IEEE Transactions on Information Theory, 50*(8), 1711–1727. doi:10.1109/TIT.2004.831778

Sendonaris, A., Erkip, E., & Aazhang, B. (1998). Increasing uplink capacity via user cooperation diversity. In *Proceedings of IEEE International Symposium on Information Theory,* Boston, MA (p. 156).

Sendonaris, A., Erkip, E., & Aazhang, B. (2003a). User cooperation diversity–part I: System description. *IEEE Transactions on Communications, 51,* 1927–1938. doi:10.1109/TCOMM.2003.818096

Sendonaris, A., Erkip, E., & Aazhang, B. (2003b). User cooperation diversity–part II: Implementation aspects and performance analysis. *IEEE Transactions on Communications, 51*(11), 1939–1948. doi:10.1109/TCOMM.2003.819238

Servetto, S. D. (2000). Lattice quantization with side information. In *Proceedings of IEEE Data Compression Conference,* Snowbird, UT (pp. 510–519).

Servetto, S. D. (2007). Lattice quantization with side information: Codes, asymptotics, and applications in sensor networks. *IEEE Transactions on Information Theory, 53*(2), 714–731. doi:10.1109/TIT.2006.889697

Shi, Z., & Schlegel, C. (2001). Joint iterative decoding of serially concatenated error control coded CDMA. *IEEE Journal on Selected Areas in Communications, 19*(8), 1646–1653. doi:10.1109/49.942525

Slepian, D., & Wolf, J. (1973a). A coding theorem for multiple access channels with correlated sources. *The Bell System Technical Journal, 52,* 1037–1076.

Slepian, D., & Wolf, J. (1973b). Noiseless coding of correlated information sources. *IEEE Transactions on Information Theory, 19*(4), 471–480. doi:10.1109/TIT.1973.1055037

Stankovic, V., Liveris, A., Xiong, Z., & Georghiades, C. (2004). Design of Slepian-Wolf codes by channel code partitioning. In *Proceedings of IEEE Data Compression Conference,* Snowbird, UT (pp. 302–311).

Stefanov, A., & Erkip, E. (2004). Cooperative coding for wireless networks. *IEEE Transactions on Communications, 52*(9), 1470–1476. doi:10.1109/TCOMM.2004.833070

Sun, Y., Liveris, A., Stankovic, V., & Xiong, Z. (2005). Near-capacity dirty-paper code designs based on TCQ and IRA codes. In *Proceedings of IEEE International Symposium on Information Theory* (pp. 184–188).

Tam, W. P., Lok, T. M., & Wong, T. F. (in press). Flow optimization in parallel relay networks with cooperative relaying. *IEEE Transactions on Wireless Communications.*

Tang, T., Chae, C. B., Heath, R. W., Jr., & Cho, S. (2006). On achievable sum rates of a multiuser MIMO relay channel. In *Proceedings of IEEE International Symposium on Information Theory* (pp. 1026–1030).

Tanner, R. (1981). A recursive approach to low complexity codes. *IEEE Transactions on Information Theory, 27*(5), 533–547. doi:10.1109/TIT.1981.1056404

ten Brink, S. (1999). Convergence of iterative decoding. *IEE Electronics Letters, 35*(13), 1117–1119.

ten Brink, S. (2001). Convergence behavior of iteratively decoded parallel concatenated codes. *IEEE Transactions on Communications, 49*(10), 1727–1737. doi:10.1109/26.957394

Tomlinson, M. (1971). New automatic equaliser employing modulo arithmetic. *IEE Electronics Letters, 7*(5), 138–139. doi:10.1049/el:19710089

Tse, D. N. C., Viswanath, P., & Zheng, L. (2004). Diversity-multiplexing tradeoff in multiple-access channels. *IEEE Transactions on Information Theory, 50*(9), 1859–1874. doi:10.1109/TIT.2004.833347

Tuchler, M., Koetter, R., & Singer, A. (2002). Turbo equalization: Principles and new results. *IEEE Transactions on Communications, 50*(5), 754–767. doi:10.1109/TCOMM.2002.1006557

Tung, S.-Y. (1978). *Multiterminal source coding.* Unpublished doctoral dissertation, Cornell University, Ithaca, NY.

van der Meulen, E. C. (1971). Three-terminal communication channels. *Advances in Applied Probability, 3*, 120–154. doi:10.2307/1426331

Vardhe, K., & Reynolds, D. (2007, July). Fast communication: User cooperation in an asynchronous cellular uplink. *Signal Processing, 87*(7), 1799–1807. doi:10.1016/j.sigpro.2007.01.002

Verdú, S. (1998). *Multiuser detection.* Cambridge University Press.

Viterbi, A. J. (1995). *CDMA: Principles of spread spectrum communication.* MA: Addison-Wesley.

Wagner, A., & Anantharam, V. (2008). An improved outer bound for multiterminal source coding. *IEEE Transactions on Information Theory, 54*(5), 1919–1937. doi:10.1109/TIT.2008.920249

Wang, B., Zhang, J., & Høst-Madsen, A. (2005). On the capacity of MIMO relay channels. *IEEE Transactions on Information Theory, 51*(1), 29–43. doi:10.1109/TIT.2004.839487

Wang, X., & Orchard, M. (2001). Design of trellis codes for source coding with side information at the decoder. In *Proceedings of IEEE Data Compression Conference,* Snowbird, UT (pp. 361–370).

Wang, X., & Poor, H. (1999). Iterative (turbo) soft interference cancellation and decoding for coded CDMA. *IEEE Transactions on Communications, 47*(7), 1046–1061. doi:10.1109/26.774855

Wei, N., Zhang, Z., & Li, S. (2006). An adaptive space-time coded cooperation scheme in wireless communication. *IEICE Transactions on Communications . E (Norwalk, Conn.), 89-B*(11), 2973–2981.

Weingarten, H., Steinberg, Y., & Shamai, S. (2006). The capacity region of the Gaussian multiple-input multiple-output broadcast channel. *IEEE Transactions on Information Theory, 52*(9), 3936–3964. doi:10.1109/TIT.2006.880064

Willems, F. (1982). The feedback capacity region of a class of discrete memoryless multiple access channels. *IEEE Transactions on Information Theory, 28*(1), 93–95. doi:10.1109/TIT.1982.1056437

Windpassinger, C., Fischer, R., Vencel, T., & Huber, J. (2004). Precoding in multiantenna and multiuser communications. *IEEE Transactions on Wireless Communications, 3*(4), 1305–1316. doi:10.1109/TWC.2004.830852

Wong, T. F., Lok, T. M., & Shea, J. M. (in press). Flow-optimized cooperative transmission for the relay channel. *IEEE Transactions on Information Theory,* Retrieved from http://arxiv.org/PScache/cs/pdf/0701/0701019v3.pdf

Wyner, A. (1974). Recent results in the Shannon theory. *IEEE Transactions on Information Theory, 20*(1), 2–10. doi:10.1109/TIT.1974.1055171

Wyner, A., & Ziv, J. (1976). The rate-distortion function for source coding with side information at the decoder. *IEEE Transactions on Information Theory, 22*(1), 1–10. doi:10.1109/TIT.1976.1055508

Xie, L.-L., & Kumar, P. (2005). An achievable rate for the multiple-level relay channel. *IEEE Transactions on Information Theory, 51*(4), 1348–1358. doi:10.1109/TIT.2005.844066

Xiong, Z., Liveris, A. D., & Cheng, S. (2004, September). Distributed source coding for sensor networks. [Long version retrieved from http://lena.tamu.edu/papers/DSC.pdf]. *IEEE Signal Processing Magazine, 21*(5), 80–94. doi:10.1109/MSP.2004.1328091

Yang, S., & Belfiore, J.-C. (2007). Optimal space-time codes for the MIMO amplify-and-forward cooperative channel. *IEEE Transactions on Information Theory, 53*(2), 647–663. doi:10.1109/TIT.2006.888998

Yang, Y., Cheng, S., Xiong, Z., & Zhao, W. (2003). Wyner-Ziv coding based on TCQ and LDPC codes. In *Proceedings of Asilomar Conference on Signals, Systems and Computers* (Vol. *1*, pp. 825–829).

Yang, Z., & Høst-Madsen, A. (2007). Cooperation through interference amplification. *IEEE Communications Letters, 11*(5), 369–371. doi:10.1109/LCOMM.2007.070028

Yu, M., & Li, J. (2005). Is amplify-and-forward practically better than decode-and-forward or vice versa? In *Proceedings of IEEE International Conference on Acoustic, Speech, and Signal Processing* (Vol. *3*, pp. 365–368).

Yu, W., & Aleksic, M. (2005). Coding for the Blackwell channel: A survey propagation approach. In *Proceedings of IEEE International Symposium on Information Theory* (pp. 1583–1587).

Yu, W., & Cioffi, J. (2001). Trellis precoding for the broadcast channel. In *Proceedings of IEEE Global Communications Conference* (Vol. *2*, pp. 1344–1348).

Yu, W., & Cioffi, J. (2002). Sum capacity of a Gaussian vector broadcast channel. In *Proceedings of IEEE International Symposium on Information Theory* (p. 498).

Yu, W., Varodayan, D., & Cioffi, J. (2005). Trellis and convolutional precoding for transmitter-based interference presubtraction. *IEEE Transactions on Communications*, *53*(7), 1220–1230. doi:10.1109/TCOMM.2005.851605

Zamir, R., & Berger, T. (1999). Multiterminal source coding with high resolution. *IEEE Transactions on Information Theory*, *45*(1), 106–117. doi:10.1109/18.746775

Zamir, R., & Shamai, S. (1998). Nested linear/lattice codes for Wyner-Ziv encoding. In *Proceedings of Information Theory Workshop* (pp. 92–93).

Zhang, Z., & Duman, T. (2002). Capacity approaching turbo coding for half duplex relaying. In *Proceedings of IEEE International Symposium on Information Theory* (pp. 1888–1892).

Zhang, Z., & Duman, T. (2007). Capacity-approaching turbo coding for half-duplex relaying. *IEEE Transactions on Communications*, *55*(10), 1895–1906. doi:10.1109/TCOMM.2007.906404

Zhao, B., & Valenti, M. C. (2003). Distributed turbo coded diversity for relay channel. *IEEE Electronics Letters*, *39*(10), 786–787. doi:10.1049/el:20030526

Zhao, Y., & Garcia-Frias, J. (2005). Joint estimation and compression of correlated nonbinary sources using punctured turbo codes. *IEEE Transactions on Communications*, *53*(3), 385–390. doi:10.1109/TCOMM.2005.843414

Zheng, L., & Tse, D. N. C. (2003). Diversity and multiplexing: A fundamental tradeoff in multiple-antenna channels. *IEEE Transactions on Information Theory*, *49*(5), 1073–1096. doi:10.1109/TIT.2003.810646

Zhong, W., Zhao, Y., & Garcia-Frias, J. (2003). Turbo-like codes for distributed joint source-channel coding of correlated senders in multiple access channels. In *Proceedings of Asilomar Conference on Signals, Systems, and Computers* (Vol. *1*, pp. 840–844).

ENDNOTES

[1] We focus on parallel concatenated codes in this chapter. However, the two convolutional encoders can also be concatenated in series. Interested readers can refer to (Benedetto, Divsalar, Montorsi, & Pollara, 1998).

2   Although no errors were observed in 2000 blocks, it is possible that some errors could occur.

3   For simplicity of notation, let us assume that $\dfrac{\alpha N}{R}$ is an integer.

Chapter 6
Network Coding for Multi–Hop Wireless Networks

Meng Yu
Lehigh University, USA

Jing (Tiffany) Li
Lehigh University, USA

Haidong Wang
Thales Communications Inc., USA

ABSTRACT

We consider practical network coding, a useful generalization of routing, in multi-hop multicast wireless networks. The model of interest comprises a set of nodes transmitting data wirelessly to a set of destinations across an arbitrary, unreliable, and possibly time-varying network. This model is general and subsumes peer-to-peer, ad-hoc, sensory, and mobile networks. It is first shown that, in the single-hop case, the idea of adaptively matching code-on-graph with network-on-graph, first developed in the adaptive-network-coded-cooperation (ANCC) protocol, provides a significant improvement over the conventional strategies. To generalize the idea to the multi-hop context, we propose to transform an arbitrarily connected network to a possibly time-varying "trellis network," such that routing design for the network becomes equivalent to path discovery in the trellis. Then, exploiting the distributed, real-time graph-matching technique in each stage of the trellis, a general network coding framework is developed. Depending on whether or not the intermediate relays choose to decode network codes, three practical network coding categories, progress network coding, concatenated network coding and hybrid network coding, are investigated. Analysis shows that the proposed framework can be as dissemination-efficient as those with random codes, but only more practical.

DOI: 10.4018/978-1-60566-665-5.ch006

INTRODUCTION

We consider multi-hop multicast data transmission in a general wireless network. The model of interest here comprises a set of nodes transmitting data wirelessly to a set of common destinations across an arbitrary, unreliable and possibly time-varying network. Our model is very general, and subsumes many specialized networks including peer-to-peer, ad-hoc, sensory and mobile networks.

Essential to network operation is *user cooperation*, where multiple nodes share resources to collaboratively accomplish a communication or computation task. User cooperation, arranged possibly in the physical, MAC, network, and application layer, or across these layers, is particularly important to the wireless scenario: While a single wireless channel may be useless due to fading or shadowing, combined together a set of channels may become useful again. The fundamental question we ask is: How do users cooperate in an efficient and trust-worthy manner? Because of the dynamic and unreliable nature of the underlying wireless network, an algorithm must be de-centralized and self-adaptive to be practical, scalable and robust.

In attacking this challenge, we resort to the recently-developed technology of *network coding* (Ahlswede, Cai, Li & Yeung, 2000; Li, Yeung & Cai, 2003; Zhang, Liew & Lan, 2006; Katti et al 2006), which is a form of user cooperation by nature. Traditional store-and-forward routing fails to exploit network capacity. By allowing intermediate relaying nodes to perform simple coding functions across different packets coming from different up-streams, network coding provides new capabilities to routing, and opens the possibility to achieve optimal network throughput. For example, in the renowned example of the butterfly network (Figure 7 in (Ahlswede, Cai, Li & Yeung, 2000)), the two receivers will each receive 1 symbol per cycle if the traditional store-and-forward routing is used, 1.5 symbols per cycle if time-sharing is used in combination with routing, but 2 symbols per cycle (which is also the max-flow min-cut bound of this network) if network coding is used. Hence, network coding is also known as *generalization of routing}*, or, *intelligent routing*.

Network coding was initially introduced to maximize the end-to-end throughput in lossless networks. Here, by lossless, we mean that the links have transmission rates limited by some positive number called the link capacity, but are otherwise free from noise and outage. Recent work has come to look at more practical networks such as the Internet and wireless networks.

From the practical point of view, network coding is more appealing to wireless networks than wireline networks. This is because the coding operation performed at each intermediate router, although presumably simple and fast, could amount to an unacceptably large complexity and delay for today's ultra-fast networks that typically operate at gigabits per second or faster. Comparatively, wireless networks are much slower and far less reliable, and can therefore take advantage of the diversity gain enabled by network coding without sacrificing other performance aspects. Furthermore, the wireless media actually makes network coding cheaper, since broadcasting (almost always needed for network coding) is achieved at no additional cost over unicast.

On the other hand, the wireless media also imposes two constraints on the design and operation of network codes:

(1) For wireline networks, it is in general beneficial for each out-edge of a network node to carry its own specific function of the in-edges symbols, namely, the coding function is edge-dependent or in the point-to-point mode (Appuswamy, Franceschetti & Zeger, 2006). For wireless networks, due to the broadcast nature of wireless media, every out-edge symbol carries the same function of the

in-edges symbols, forcing the coding function to be node-dependent or in the broadcast-mode.

(2) Due to the possibility of random fading, link outage and node mobility, the topology of a wireless network may be continually changing. What this implies on network coding design is the need for a *distributed* and *adaptive* algorithm.

However, the second aspect appears to have been largely ignored. The existing work on practical network codes almost exclusively used *fixed, pre-defined* coding schemes, whose applicability heavily relies on the assumptions that at least some wireless links are always available and that the network topology hardly changes. One notable exception is (Bao & Li, 2005), which proposed a real-time adaptive scheme termed *adaptive network-coded cooperation* (ANCC). The central idea therein is to match code-on-graph, i.e. low-density parity-check (LDPC) codes and LDPC-like codes, with network-on-graph, i.e. instantaneous network topology represented in graph, in a distributed and real-time manner. This idea has since been adapted in a number of other cooperative strategies (Kaewprapha, Puttarak, Wang & Li, 2008; Peng, Zhang, Zhao & Yao, 2007; Hausl & Dupraz, 2006; Wang & Giannakis, 2007), all of which fuel the practicality of network coding with practical codes of linear encoding and decoding time.

While these papers have exemplified the virtue of the graph-matching idea, the algorithm has only been carried out on simple network models where the source(s) and the destination(s) are one-hop reachable. How about a more sophisticated network having an arbitrary topology?

The objective of this chapter is to study practical and efficient coding strategies for efficient data transmission over a general wirelesss networks that may be cyclic or acyclic. The main result is the development of a distributed and adaptive network coding framework that can be viewed as a generalization of ANCC to networks with multiple hops and even loops. We show that an arbitrarily connected network can be uniquely transformed to a ``trellis network'', such that routing design for the network is equivalent to path discovery in the trellis. For a wireline network, the resulting trellis is invariant from stage to stage, but for a wireless network with changing link state and topology, the trellis is time-varying. We next discuss in detail the proposed network coding framework which exploits the graph-matching idea developed in ANCC in each stage of the trellis, in a distributed manner that is close in spirit to *gossip* (Mosk-Aoyama & Shah, 2006; Deb & Medard, 2004) or *rumor-spreading* (Karp, Schindelhauer, Shenker & Vocking, 2000). Depending on whether or not the intermediate relays decode network codes, three practical network coding strategies, progress network coding -- where the relays will decode before re-encoding data (possibly using a different code), concatenated network coding -- where the relays do not decode but encode on top the, and hybrid network coding, are investigated. Analysis is also performed on the message dissemination efficiency.

The remainder of the chapter is organized as follows. We start by discussing the key idea of the real-time, distributed graph-matching strategy. We next propose a discrete trellis representation for an arbitrary network graph, followed by a detailed discussion of the proposed network coding framework. Three possible coding strategies, theoretical analysis and simulation results will be provided. Finally we conclude the chapter.

GRAPH-MATCHING NETWORK CODING

This section motivates and explains the application of network coding, and particularly distributed and topology-adaptive network coding, in wireless networks, through simple but generalizable network models.

Figure 1. A simple network model consisting of two senders and one common destination. (A) Plain routing; (B) Network coding using a single-parity check code.

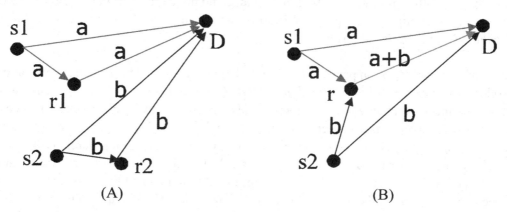

(A)   (B)

Following the convention of the network coding literature, in the sequel, we will use "symbol" and "packet" inter-changeably. A symbol, or, a packet, can be viewed as a vector consisting of many bit elements, and a coding operation between two symbols will be performed between the corresponding bit elements in these vectors. For simplicity, all the examples and the simulation results shown here use coding functions in the binary field, but the discussion of the algorithm is made general to the finite field $GF(q)$.

Recall that the fundamental technology to ensure reliable transmission through wireless channels is *diversity*, which refers to duplicate transmissions through *independent* channels, such that the probability of packet loss/corruption can be reduced from p for a single transmission to p^m for m duplicate transmissions. As an important physical-layer technology, diversity can be implemented in several different ways, either through time, frequency, space, or any combination of them. Here we focus on *spatial diversity*, which, in the context of routing, translates to redundant routing though different paths.

Example 1: (Parity Check Network Code) Consider two wireless users, s_1 and s_2, send binary symbols, a and b, respectively to a common destination. The easiest way to achieve redundant routing is to have two relay nodes, r_1 and r_2, which have overheard a and b respectively, forward a second copy to the destination. This simple scenario, depicted in Figure 1(A), promises a diversity order of 2 for each symbol. However, a more bandwidth- and power-efficient method is to let one node simultaneously help the two users by relaying the check-sum symbol, $a \oplus b$, instead of individual source-symbols a and b; see Figure 1(B) ($\oplus$ denotes the binary addition). It is easy to see that, using only three times slots instead of the original four time slots, each source continues to enjoy a diversity order of 2, and retrieving any two of the three transmitted symbols at the destination will recover both sources.

From the network coding perspective, the gain of the latter strategy results from the *coding gain* of a single-parity check code, $[a, b, a \oplus b]$, over a repetition code, $[a, b, a, b]$. Put another way, the conventional redundant routing can be viewed as network coding using simple but weak *repetition codes*, and network coding using non-trivial codes is bound to outperform it.

Example 2: ((7,4) Hamming Network Code) Here is another simple example: When four users have data to transmit, instead of involving four additional nodes to relay one-for-one to achieve a diversity

Figure 2. A simple network model consisting of four senders and one common destination. (A) Phase I: Senders broadcast; (B) Phase II: Relays perform network coding using a (7,4) Hamming code.

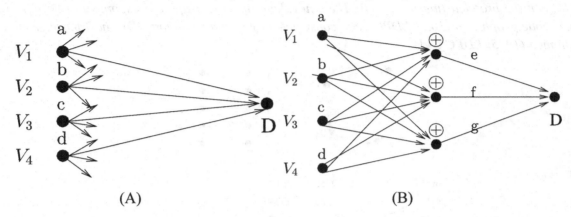

(A)   (B)

order of 2 for each source symbol, it suffices to take only three relays but to achieve a diversity order of 3 using a (7,4) *Hamming code*. This network coding scheme is illustrated in Figure 2, where the relay nodes perform coding functions $a \oplus b \oplus c$, $a \oplus c \oplus d$, *and* $a \oplus b \oplus d$, respectively. Again, it can be quickly verified (or use the known fact that a Hamming code has a minimum Hamming distance of 3) that missing any two symbols from the seven transmitted symbols will still enable the successful deduction of the four source-symbols.

Comments: It can be proven that, for channels with independent fading, a binary code promises a diversity that equals to the minimum Hamming distance d_m of that code. For a repetition code, d_m equals the number of times a source-symbol is repeated. If a total of n transmissions are offered to $k <$ n source-symbols, then the *(n,k)* repetition code would result in $d_m = \lfloor n/k \rfloor$. The minimum distance of a non-trivial *(n, k)* code is code-dependent, but almost always greater than n/k, and can be as large as $n-k+1$. (In the coding literature, the code whose minimum distance satisfies $d_m = n-k+1$ is called a maximum distance separable (MDS) *code*.)

For larger networks, more sophisticated linear codes can be exploited. This includes, for example, the classical BCH codes, Reed-Solomon code and convolutional codes, and the capacity-approaching turbo codes and low-density parity-check (LDPC) codes. The application of these codes follows straightly from the above discussion, except for the practicality that wireless channels are subject to random fading. A wireless channel will usually experience a certain percentage of *deep* fading, known as *outage*, such that transmission (using any finite amount of power) at those times will fail. Hence, even though two nodes may sit within each other's wireless transmit range, their communication channels will break from time to time. If we further account for node mobility or the possibility for a node to fail or power off at any time, then little can be assumed about the availability of a wireless link in the next moment.

Such wireless dynamics bring operational problems to the *fixed* network coding scheme we just discussed: Since a relay may fail to overhear an upstream symbol (say, symbol **a** in the examples) which is needed to perform its intended coding function, or a designated relay node may simply become unavailable, the pre-defined network code will break. This calls for an immediate adaptation, such that the system can generate a new (and good) network code to match to the new network topology.

Figure 3. Topology-adaptive network coding for a simple network consisting of five senders, one common destination (not shown), and randomly faded wireless channels. (A) The instantaneous network topology at the broadcasting phase; (B) The equivalent bipartite graph corresponds to (A); (C) The Tanner code graph of a (10,5) LDPC code constructed by graph matching; (D) The Tanner graph of the thinned (10, 5) LDPC code.

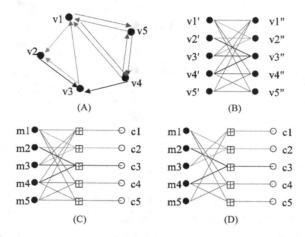

Example 3: (Topology-Adaptive LDPC Network Codes) (Bao & Li, 2005) To see how graph-matching may be performed in a distributed and real-time manner, consider a wireless network where 5 users v_1, v_2, v_3, v_4 and v_5 communicate to a common destination that sits one hop away. Initially, each user v_i broadcasts its message m_i using orthogonal channels, where orthogonality can be realized through either time-division multiple-access (TDMA), frequency-division multiple-access (FDMA), or code-division multiple-access (CDMA). For convenience, here we assume a time-division system, i.e., each user takes turns to transmit its message. Because of possible link outage, the destination may not receive all the messages, and because of wireless broadcasting, each user may overhear some of the messages. Suppose that for the time being, the connectivity of the *inter-user* channels is as shown in Figure 3(A) (for simplicity, the destination is not shown), where a directed edge indicates a successful transmission. Now this network graph can be easily transformed to a bipartite graph by:

(1)   Splitting each node v_i in two: v'_i incident to only the outbound edges and v''_i to only the inbound edges, and

(2)   Adding additional edges to connect v'_i to v''_i for all i.

The resulting bipartite graph, shown in Figure 3(B), provides clear indication on what symbols are available at each user before and after the initial phase of transmission. It also gives rise to a natural network coding scheme where each node v''_i can compute and relay the binary check sum, or, the *parity-symbol*, denoted as c_i, of the source symbols it has collected. From the coding theory, one realizes that the 5 source-symbols transmitted at the first phase, $m_1, \cdots, m_5$, and the 5 parity-symbols transmitted at the relay phase, $c_1, \cdots, c_5$ form a (10, 5) systematic LDPC-like code (MacKay, 1999). (For simplicity, we will not differentiate between LDPC and LDPC-like codes.) The destination can then decode this network code using a practical belief-propagation algorithm, to recover any source-symbol m_i that may have gotten lost during the transmission. The Tanner code graph of this (10, 5) LDPC code is shown in Figure 3(C), where black circles represent the source-symbols, white circles represent the parity-symbols, and boxes represent parity check constraints that all the symbols incident to a box have a binary sum of 0.

Comments: First, this coding strategy, referred to as adaptive network-coded cooperation, or, ANCC, in (Bao & Li, 2005), can match with the instantaneous network topology on-the-fly. The encoding process is performed by all the participating nodes in a distributed manner. Each node does not need to know the complete network topology; it simply computes one parity-symbol using what's available locally. Hence, not only are the parity-symbols computed at different nodes different, but they will also change with time, in accordance to the changing link state and network topology.

Second, to make the belief-propagation decoding (which is performed at the common destination) effective, instead of performing the check sum on *all* the symbols it has, each relay node can (randomly) select only *a few* symbols, thus "thinning" the code graph and eliminating the chances for short cycles. For example, in Figure 3(C), user v_1 may de-select the source-symbols from user v_4 and itself, m_4 and m_1, when computing the check sum c_1; and users v_3 and v_5 may each de-select the source-symbol from itself, m_3 and m_5, respectively. The new code graph, which now happens to be free of length-4 cycles, is shown in Figure 3(D). Since the iterative message-passing algorithm, with which LDPC codes are decoded, tend to converge faster and perform better on code graphs with a larger girth, code thinning thus helps improve the performance, as well as reduce the complexity (MacKay, 1999; Bao & Li, 2005).

Third, when each node relays the parity-symbol, it will also include a small bit-map field, so that the destination knows how the checks were formed and can correspondingly replicate the Tanner code graph and perform belief-propagation decoding. This requires an adaptive decoder architecture at the destination, which can be implemented, for example, using software-defined radio (SDR).

Forth, since a different LDPC code is constructed with each instantaneous network graph, the system performance is the average performance of an *ensemble* of LDPC codes rather than any single code. (Bao & Li, 2005) compared the efficiency of several practical LDPC code ensembles, including low-density generator-matrix (LDGM) codes of degrees from 7 to 12, and lower-triangular LDPC (LT-LDPC) codes of degrees from 4 to 8.

Fifth, in the example, we assumed that all the users participated in the relay phase. In practice, depending on the individual power budget and the respective user-destination channel quality, one, some or all of the nodes may choose to relay; and one node may choose to relay more than once, each time using a different parity check function. The exploitation of *user diversity* (i.e. nodes experiencing poor channel quality give way to nodes enjoying good channel quality), and possibly power allocation, will further improve the system performance.

Sixth, this network coding strategy can be applied to one-hop wireless networks of any size, and it will *always* outperform plain routing. Although the code construction appears rather crude, the performance is surprisingly good, especially for large networks, as verified by computer simulations and theoretical analysis (Bao & Li, 2008).

Finally, the coding function each relay performs need not be limited in the binary field, and LDPC codes in the Galois Field **GF(q)** can also be exploited. A non-binary LDPC code usually outperforms a binary LDPC code at the cost of a (considerably) higher complexity.

TRELLIS MODEL FOR A GENERAL NETWORK

Equipped with the idea of adaptive graph-matching in single-hop networks, we proceed to the general case of multi-hop networks. We propose a discrete-time trellis model for a general network. Time-invariant trellises for static networks were first introduced in (Ahlswede, Cai, Li & Yeung, 2000) in the name of a time-parameterized graph, and subsequently studied in (Li, Yeung & Cai, 2003) and (Wu, 2006). Here we generalize the model to include time-varying cases. For convenience, consider dividing time into units and performing any one-hop transmission in one unit time. This can be regarded as a slotted synchronous system.

Definition 1: (Discrete Time-Varying Network Graph) A discrete time-varying network is modeled by a time-varying directed graph $\mathbf{G}^{[N]} = (V, \mathbf{E}^{[N]})$, with $|V|$ nodes and sets of directed edges $\mathbf{E}^{[N]} = \{E^{[1]}, E^{[2]}, \cdots \}$, indicating the state of communication channels over the time series $n = 1, 2, \cdots$.

Comment: In general, an edge has associated to a link capacity. For notational convenience, the above definition assumes that all edges have unit capacity, and multiple parallel edges are allowed to represent possible difference in link capacity. Further, the instantaneous graph at time n, $(V, E^{[n]})$, can be either cyclic or acyclic, connected or partitioned, and the communication channels are not necessarily reciprocal. The latter usually appears in a wireless environment where, for example, node v_i reaches node v_j by using a high transmit power, but v_j is unwilling or unable to afford the same high transmit power and hence cannot reach back.

We demonstrated in the previous section that by splitting each node v_i to a "disseminating replicate" v_i' and a "collecting replicate" v_i'', a directed network graph (V, E) can be transformed to a bipartite graph with $|V'| = |V|$ left nodes, $|V''| = |V|$ right nodes, and $|E|+|V|$ edges connecting them. Such a bipartite graph can be regarded as a description of "network-flow" during one particular time unit. A discrete-time trellis thus results from concatenating these descriptions in consecutive time units.

Algorithm 1: (Discrete Time-Varying Trellis) Given a discrete time-varying directed graph $\mathbf{G}^{[N]} = (V, \mathbf{E}^{[N]})$, the associated discrete time-varying trellis $\mathbf{T}^{[N]}$ is constructed as follows:

1) For each node $v_i \in V$ and time index $n \in \{0, ..., N\}$, the trellis $\mathbf{T}^{[N]}$ includes a vertex $v_i^{[n]}$, such a vertex $v_i^{[n]}$ in the trellis corresponds to the original node v_i at time n.

2) For each edge $e^{[n]} \in E^{[n]}$ going from v_i to v_j and $n \in \{1, ..., N\}$, the trellis $\mathbf{T}^{[N]}$ includes an edge $e^{[n]}$ going from $v_i^{[n-1]}$ to $v_j^{[n-1]}$; such an edge, called a *transmission edge*, corresponds to the transmission from v_i to v_j during the nth time unit.

3) For each $v_i \in V$ and each $n \in \{1, ...N\}$, the trellis $\mathbf{T}^{[N]}$ also includes an edge (with capacity ∞, or ∞ unit-capacity edges) from $v_i^{[n-1]}$ to $v_j^{[n-1]}$; such an edge, called a *memory edge*, corresponds to the accumulation of information at node v_i up to time n.

Example 4: (Trellis Representation) Consider a 5-node wireless network starts at time $n=1$ with a topology as shown in Figure 3(A). Suppose that the channel from v_4 to v_1 experiences an outage at time 2 and does not come back until time 4, and that the channel from v_1 to v_5 is also temporarily down at time 4. The discrete, time-varying trellis of this network is presented in Figure 4.

Figure 4. A discrete time-varying trellis

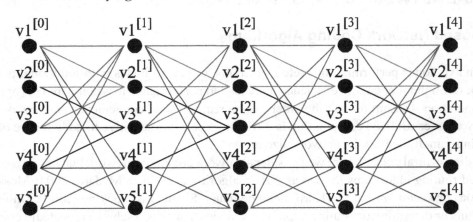

This trellis model resolves the cycle complication in a network, captures the time-varying nature of a wireless network, and provides a comprehensible platform to perform (and track) network coding form one time unit to the next. When the network graph is time-invariant, i.e. $\mathbf{G}^{[N]}=G=(V, E)$, the single-source multicast capacity has a clear definition: The min-cut value between the source s and the set of destinations D, where $D \subset V$. The following theorem due to (Ahlswede, Cai, Li & Yeung, 2000; Li, Yeung & Cai, 2003) shows that the multicast capacity is asymptotically achievable by the discrete (time-invariant) trellis.

Theorem 1: (Connectivity in Discrete, Time-Invariant Trellis) Consider a discrete time-invariant trellis $\mathbf{T}^{[N]}$ that conforms to a network graph $G = (V, E)$. Let $C(G)$ and $C(\mathbf{T}^{[N]})$ be the multicast capacity from source S to destinations D in G and from source $S^{[0]}$ to destinations $D^{[n]}$ in $\mathbf{T}^{[N]}$, respectively. At any time unit n,

$$(n- | V | +2)C(\mathrm{G}) \leq C(\mathbf{T}^{[N]}) \leq nC(\mathrm{G}),$$

which leads to

$$\lim_{n \to +\infty} \frac{C(\mathbf{T}^{[N]})}{n} = C(\mathrm{G}). \tag{1}$$

The multicast capacity of a time-varying graph is not well defined, but it is easy to verify that the trellis representation is unique and has one-to-one correspondence with the original network graph. Conventional routing in the original network graph then boils down to path search, either a single path or a set of (possibly edge-joint) paths, in the corresponding trellis; and network coding introduces an additional dimension by allowing intermediate nodes to perform intelligent coding on in-edge symbols before sending them to the out-edges.

A GENERAL NETWORK CODING FRAMEWORK

The Proposed Network Coding Algorithm

In general, naive nodes performing discrete algorithms, or selfish nodes with individual performance objectives of throughput or energy efficiency, may lead to performance loss compared to (optimal) centralized cooperation – assuming that the latter can be realized with convenience and low overhead. In practice, however, centralized computation is difficult, expensive, sensitive to node failure or mobility, and increasingly inefficient as the network size grows.

Disregarding central control, here we consider a network coding framework that has its root in a class of distributed routing algorithms known as *gossip* algorithms (Deb & Medard, 2004; Mosk-Aoyama & Shah, 2006) or *rumor spreading* (Karp, Schindelhauer, Shenker & Vocking, 2000). Unlike centralized algorithms or deterministic routing approaches, nodes in a gossip scheme repeatedly communicate with randomly selected neighbors, either using plain routing or network coding, to achieve the goal of distributing the computational burden among all the nodes and disseminating information across the network. The distributed graph-matching coding we discussed in Section II can be viewed, in a sense, as network coding-assisted gossip. The key idea of the proposed network coding framework is to regard the wireless network as a discrete time-varying trellis, and to perform graph-matching network coding at each trellis stage.

Algorithm 2: (Topology-Adaptive Network Coding for General Networks) Consider an arbitrary time-varying graph $\mathbf{G}^{[N]} = (V, \mathbf{E}^{[N]})$. Let $S \subset V$ be the set of source nodes, and $R \subset V$ be the set of sink nodes. S and V may or may not overlap. Each node $v_i \in S$ starts communication with a unique source-symbol m_i, and the goal is to efficiently spread all of these messages to all the nodes $u_j \in R$. (In fact, it is not necessary to assume that each source has only one source-symbol. It suffices to say that in each communication session, a set of sources S have a set of independent messages $M = \{m_i\}$ to disseminate to a set of destinations R.) A network-coding based data dissemination algorithm consists of three components: a gossip mechanism, which determines how messages flow from nodes to nodes when a communication is initiated, a practical linear coding algorithm, which specifies what message to transmit by a node to its out-edge neighbors during a communication, and a decoding algorithm, which governs how decoding is performed at a receiver node (destination).

Gossip Mechanism: In any time slot, each eligible node broadcasts some message to its neighbors (through orthogonal channels). An eligible node is defined as a node that is either a source node, or a relay node having accumulated information about the source-symbols (i.e. m_i's or functions of m_i's). A neighbor is defined as a node that locates within the transmit range of the sender. The message transmitted by node v_l at time n will reach its neighbor v_k only if edge $(lk) \in \mathbf{E}^{[n]}$. Disregarding possible node failure and mobility (a less common event that may be ignored if one is evaluating a rather short section of time), a wireless channel (lk) experiences outage with probability P_{lk} independently of all other communication events. In the stochastic sense, if v_k remains a neighbor of v_l, then with probability P_{lk} it hears v_l. This randomness, introduced by the wireless media, actually reflects the "gossip spirit" in a lossless graph, where node v_l *randomly* selects neighbors to communicate to, and with probability P_{lk} it selects v_k, independent of other communications events.

LDPC-Like Network Coding: For practical purpose, we restrict our attention to linear codes that can be encoded and decoded with manageable complexity. In the context of the gossip mechanism, the choice for codes appears to be much limited to the *general class of LDPC-like codes*, which include, for example, LDGM codes, lower-triangular LDPC codes, concatenated LDPC codes, zig-zag codes, and Luby-Transform (LT) codes (Luby, 2002).

In general, each message or symbol is a vector over a finite field **GF(q)**. Let r be the length of the vector (i.e. the size of the data packet). Let $M = \{ m_1, \cdots, m_{|M|} \}$ denote the set of source-symbols, where $m_i \in (\mathbf{GF(q)})^r$. We assume that all the messages in M are linearly independent. Consider at a certain instant n, an eligible node v_l has accumulated a set of messages $M_l(n) = \{ f_1, f_2, \cdots, f_{|Ml(n)|} \}$, some of which may be the original source-symbols while others may be linear functions (i.e. check-sums or coded messages) of the source-symbols. By definition of linear codes,

$$\forall f_i \in M_l(n), 1 \le i \le |M_l(n)|, f_i \in \sum_{j=1}^{|M|} \alpha_{ij} m_j, f_i \in (F_q)^r, \alpha_{ij} \in F_q, \tag{2}$$

where the addition and multiplication operations conform to the rules in $(\mathbf{GF(q)})^r$. As will be clear shortly, the protocol ensures that most of the coefficients where the addition and multiplication operations conform to the rules in $(\mathbf{GF(q)})^r$. As will be clear shortly, the protocol ensures that most of the coefficients α_{ij} are zero, and v_l knows all the (non-zero) α_{ij}, for all the symbols $f_i \in M_l(n)$.

The coding function an eligible node v_l performs, or the parity-symbol g it prepares to put on the out-edges, is some *sparse random linear combination* of the source-symbols. With regard to whether or not v_l decodes f_i's before (re-)encoding them, there come three different forms.

1)   In the progressive network coding scheme, an intermediate nodes is asked to decode **GF(q)** to extract as many source-symbols m_i's as possible (the decoding process will be discussed later). Node v_l then computes the out-edge symbol g as a sparse random linear combination of the source-symbols it has successfully deduced:

$$\sum_{i=1}^{|M|} \alpha'_j m_j, \alpha'_j \in F_q. \tag{3}$$

If m_i is not available to v_l, then the coefficient $\alpha'_j = 0$; otherwise, the coefficient α'_j can be either zero or a non-zero element randomly selected from **GF(q)**. During the transmission, node v_l piggybacks on g the non-zero coefficients α'_j, which are the control overhead necessary for the protocol. In the (rare) case where not a single source-symbol is available, node v_l refrains from transmission, waits for another time unit to receive more symbols, and tries belief-propagation decoding again.

2)   In the concatenated network coding scheme, intermediate nodes (other than the sink nodes) are exempt from mid-stage decoding. A relay node v_l in this case computes the out-edge symbol g directly from f_i's, but the check-sum coefficients that accompany g should reflect the direct relation between g and m_i's. Suppose v_l has used a set of sparse, random coefficients $\beta'_j \in \mathbf{GF(q)}$, to produce the out-edge symbol g from f_i's. Since

$$g = \sum_{i=1}^{|M_l(n)|} \beta_i f_i = \sum_{i=1}^{|M_l(n)|} \beta_i \sum_{j=1}^{|M|} \alpha_{ij} m_j, \tag{4}$$

$$= \sum_{j=1}^{|M|} \left(\sum_{i=1}^{|M_l(n)|} \beta_i \alpha_{ij} \right) m_j = \sum_{j=1}^{|M|} \alpha'_j m_j, \tag{5}$$

3) In the hybrid network coding scheme, an intermediate node makes its independent decision on whether or not to perform mid-stage decoding. The decision may be made based on, for example, the number of symbols and the particular set of symbols a node has accumulated thus far, the energy and the computational power provisioned to that node, and the current and immediate work load it expects.

A more detailed discussion of the three coding strategies, together with examples and simulation results, will be presented in the next subsection.

Decoding Algorithm: The decoding algorithm is performed by a relay node in a progressive or hybrid coding scheme, as well as by a sink node in all the schemes at any time n. The process is similar to the decoding process we discussed in Example 3: node v_l constructs a bipartite Tanner graph that specifies all the check-sum constraints available from the information it has $M_l(n) = \{f_i\}$, and performs the standard belief-propagation decoding on the Tanner graph (MacKay, 1999). If the Tanner graph bears a layering structure (e.g. resulted from a concatenated coding scheme), then multi-stage belief-propagation may be performed to reduce the decoding complexity.

Progressive, Concatenated and Hybrid Network Coding Schemes

The LDPC network code as seen by a sink node is in essence a set of linear constraints on a set of source-symbols $M = \{m_j\}$. From the pure mathematics point of view, two LDPC codes will provide comparable performance, if the vector spaces expanded by their individual set of constraints have similar *distance separation* properties. When the number of source-symbols $|M|$ is large and the network size $|V|$ is large, the *concentration theorem* (Chung, Richardson & Urbanke, 2001) guarantees that the performance of a random LDPC code will converge in probability to the average performance of the code ensemble. This should hold, ideally, independent of the context of progressive coding, concatenated coding or hybrid coding.

However, when network coding is operated as a network protocol in real systems, how encoding and decoding is performed along the message dissemination becomes a central technical issue, which directly affects the system performance and operational cost. For example, the concatenated coding scheme obviates the need for the relay nodes to perform mid-stage decoding, thus greatly reducing their complexity to not much higher than plain routing. This simplicity would be highly desirable in such applications like sensor networks, since sensors (acting as relays) are individually weak. The drawback, however, is that the concatenated coding is prone to error propagation. To see this, recall that the framework discussed in Algorithm 2 applies the per-hop graph-matching LDPC-coding idea to each stage of the network trellis. Viewed from a sink code, the messages depart from the source nodes, and get encoded by an LDPC code each time they pass a stage in the trellis. With concatenated coding, the code-words from the previous LDPC code go direct to the next LDPC code. Since an in-edge symbol may participate in multiple check operations at different nodes, an error occurred during the transmission at time n will not only propagate to time $n+1$ and after, but is likely to cause multiple errors.

Transmission errors have been largely ignored in the theoretical study of network coding, but are non-negligible in practical systems. Our simulation experiments using practical wireless channel models show that the non-regenerative concatenated coding scheme is not suitable for transmission beyond two or three hops; otherwise the severe error propagation may completely defeat the purpose of network coding. On the other hand, the regenerative progressive coding scheme, by having intermediate relays decode and clean up the messages, can effectively stop error propagation, and is therefore fitful for large networks of many hops.

The biggest advantage for the hybrid coding scheme is its flexibility. Compared to the concatenated coding scheme, the progressive coding scheme has sacrificed complexity in exchange for robustness. The hybrid coding scheme allows individual nodes to choose a practice that best suits the global or local situation. For example, a node may by default perform mid-stage decoding, but can bypass it if it is short of resources.

Analysis

It is shown in (Mosk-Aoyama & Shah, 2006) that, for a network of $|V|$ nodes, the information dissemination time of a gossip algorithm using *random-mixing* linear codes (or, simply, random codes) on a finite Field **GF(q)** $(\mathbf{q} \geq |V|)$, depends on the evolution of the "dimension of the subspace" spanned by the messages at the various nodes during the course of the algorithm, but random codes in general outperforms plain-routing. For example, for an n-node fully-meshed network, the gossip algorithm using random codes disseminates $|M|$ messages to all the nodes in $O(|M|)$ time with probability $1-O(1/|M|)$ (Deb & Medard, 2004), whereas the gossip algorithm using the conventional replicate-and-randomly forward technique takes $\theta(|M| \log |M|)$ time (Karp, Schindelhauer, Shenker & Vocking, 2000). The improvement of the random codes over plain-routing comes from the fact that, in a finite Field **GF(q)**, if **q** is sufficiently large, then with a high probability, a randomly generated set of $|M|$ linear equations will be mutually independent, such that $|M|$ parity-symbols at the sink node suffice to extract $|M|$ source-symbols (Li, Yeung & Cai, 2003). In a plain routing case where a node randomly selects an in-edge source-symbol to put on the out-edge, analysis using the classical balls and bins process shows that a sink nodes needs $|M| \log(|M|/\delta)$ parity-symbols in order to deduce $|M|$ source-symbols with probability at least $1 - \delta$.

In the proposed LDPC network codes, each parity-symbol is constructed from "random-sparse-mixing" of source-symbols. This appears to be a case between the "random-once-mixing" of plain-routing and the "random-full-mixing" of random codes. From intuition, the number of parity-symbols required for a complete recovery of source-symbols is related to how sparse the mixing is. It may appear that a denser mixing implies a fewer number of parity-symbols, since each parity-symbols covers more source-symbols. Surprisingly, however, analysis shows that it does not necessarily take a lot of density for a random mixing to be efficient. Using results and analysis from the coding theory (Luby, Mitzenmacher, Shokrollahi & Spielman, 2001; Luby, 2002), we arrive at the following results.

Definition 2: (Degree distribution) Consider a linear code consisting of a set of parity-symbols, each of which is generated, independently of all other parity-symbols, from some random linear function of

$|M|$ source-symbols. For all $1 \leq d \leq |M|$, let $\rho(d)$ be the probability that a parity-symbol is incident to d randomly-selected source-symbols. The profile $\{\rho(d)\}$ is called the *degree distribution* of this code.

Theorem 2: (Lower bound on sum degree) For any degree distribution, the sum of the degrees of the parity-symbols must be at least $|M| \log(|M|/\delta)$ to cover all the source-symbols with probability at least $1 - \delta$.

Definition 3: (Soliton Distribution (Luby, 2002)) The *ideal soliton distribution* $\mu_0(\cdot)$ is defined as

$$\mu_0 = \begin{cases} 1 / |M|, & i = 1, \\ 1 / i(i-1), & i = 2, \cdots |M|. \end{cases} \qquad (6)$$

Let $R = c \cdot \log(|M|/\delta)\sqrt{|M|}$ for some suitable constant $c > 0$. Define

$$\tau(i) = \begin{cases} R / (i|M|), & i = 1, \cdots |M| / R - 1, \\ R \log(R / \delta) / |M|, & i = |M| / R, \\ 0, & i = |M| / R + 1, \cdots |M|. \end{cases} \qquad (7)$$

The *robust soliton distribution* $\mu_1(\cdot)$ is defined as:

$$\mu_1(i) = (\mu_0(i) + \tau(i)) / \kappa), \qquad i = 1, \cdots, |M|, \qquad (8)$$

where $\tau(\cdot)$ is the supplement, and κ is the normalization term satisfying $\kappa = \sum_{i=1}^{|M|} \mu_0(i) + \tau(i)$.

Theorem 3: (Efficiency of LDPC Network Codes) Consider a set of parity-symbols generated from $|M|$ source-symbols according to the robust soliton degree distribution. All the $|M|$ source-symbols can be successfully deduced from any $|M| + O(\sqrt{|M|} \log^2(|M| / \delta))$ parity-symbols with probability $1-\delta$.

Comment: The proofs are somewhat involved, so they are omitted here. We note that the name *Soliton distribution* is inspired by the fact that a Soliton wave is one where dispersion balances refraction perfectly (Luby, 2002). Theorem 3 states that the robust Soliton distribution ensures that just over $|M|$ parity-symbols with an average degrees being $O(\log(|M|/\delta))$ suffice to cover all the $|M|$ source-symbols with probability of at least $(1-\delta)$. The robust Soliton distribution can then be considered as the most efficient degree profile for distribute coding, since it achieves the lower bound specified in Theorem 2.

Theorem 3 bears an important implication on the proposed network coding framework: If the LDPC network code is constructed according to the robust Soliton distribution, then a sink node can recover all the $|M|$ source-symbols from any set of the parity-symbols whose aggregated size is only slightly more than $|M|$. Notice that the robust Soliton distribution can be completely determined by $|M|$, the source size, δ, the target performance parameter, and another non-negative parameter c. Hence, with a small control overhead, i.e. piggybacking $|M|\delta$ and c on each message, all the nodes in the network can generate and store the robust Soliton distribution, and then perform intelligent random-sparse-mixing:

1) Randomly choose a degree d from the robust Soliton distribution.
2) Choose uniformly at random d distinct source-symbols available locally.
3) The parity-symbol to spread on the out-edges is a random linear function of these d source-symbols. With this improved coding strategy, the proposed LDPC coding framework will be no less efficient, in terms of message dissemination time, than random codes. The results on dissemination time, developed in (Mosk-Aoyama & Shah, 2006) for random codes, will also be applicable to the proposed framework.

Here are additional comments about the difference between the proposed framework using LDPC network codes and the random-code-based network coding available in the literature (e.g. (Li, Yeung & Cai, 2003; Mosk-Aoyama & Shah, 2006; Deb & Medard, 2004)):

1) A check-sum symbol is incident to (almost) all the source-symbols in a random code, but only to a small number of source-symbols in an LDPC code. Hence, not only is the encoding complexity of an LDPC code smaller, but its Tanner graph is also much sparser, and hence contains far less short cycles, than that of a random code.
2) The sparseness makes it possible for an LDPC code to be decoded in linear time (in $|M|$) using a practical belief-propagation algorithm (MacKay, 1999), whereas a random code relies on matrix inversion, i.e. solving a set of linear equations, which has a complexity to the third power of $|M|$, where $|M|$ is number of source-symbols.
3) The random-mixing linear code in **GF(q)** requires $q > |M|$ in order for the decoder to successfully extract $|M|$ source-symbols from $|M|$ parity-symbols with a high probability (Li, Yeung & Cai, 2003). It should be noted that the encoding complexity and particularly the decoding complexity are prohibitively high for large **q**. An LDPC code in **GF(q)** does not need to impose any constraint on **q**, but an LDPC code generally requires more than $|M|$ parity-symbols to deduce $|M|$ source-symbols. For complexity consideration, we recommend binary LDPC codes (**q**=2), although high-order LDPC codes usually provide better performances than binary LDPC codes. If the degree distribution of the parity-symbols in a binary LDPC code follows the robust Soliton degree distribution, then the LDPC code requires only about $|M|$ parity-symbols to recover $|M|$ source-symbols for large $|M|$ (Luby, 2002).
4) Random codes whose decoding is based on matrix-inversion has no error-correction capability. Any one linear equation in error (e.g. a parity-symbol is flipped to a different symbol due to the channel noise) will cause serious conflicts in the set of linear equations, and since the decoder has no means of identifying (and discarding) the erroneous one, decoding will simple fail. An LDPC code typically has very strong error correction power. Its error-correction capability increases with the codeword length, and can asymptotically achieve the capacity for many noisy channels (MacKay, 1999).

Simulations

We now provide simulation results to verify the feasibility and efficiency of the proposed framework. The graph and the trellis model we used in the discussion is a convenient but simplified description of the wireless network, since the wireless channel is reduced to only two states: perfect and broken. In practice, a wireless channel has infinite states and can be anywhere between perfect and broken. To

Figure 5. Simulation comparison between the progressive coding scheme using LDPC codes and plain routing.

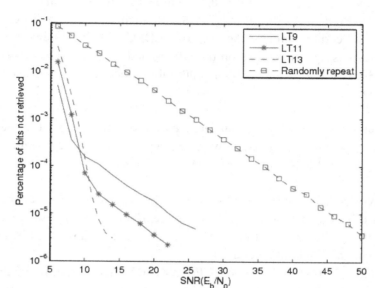

best reflect the reality, our simulations used the practical Rayleigh-faded wireless model rather than the erasure channel model. Since there is no "typical wireless network", to make our experiments easy to describe and informative, we arrange our network in the form of a multi-hop trellis. We consider a set of 100 source nodes, each equipped with an independent source-symbol, communication to the set of relays at the first hop, which then communicate to the relays at the second hop, and so on. For simplicity, we assume that there are sufficient numbers of relays in each hop, that the sets of relays at different hops do not overlap, and that the Rayleigh fading channels connecting them are homogeneous and spatially independent. We assume coarse synchronization, such that at each relay hop, the set of participating relays collaboratively form a random graph-matching LDPC code has discussed previously.

Figure 5 presents the performance of the progressive coding scheme using LDPC network codes. Three different classes of distributed LDPC codes, generated by random-sparse-mixing with an average degree of 9, 11 and 13, respectively, are tested, and compared with plain routing. In all the cases, an intermediate relay as well as a sink node waits until it accumulates 200 symbols before performing belief-propagation decoding. The plot shows the percentage of un-recovered source-symbols at a sink node as a function of the channel signal-to-noise ratio (SNR) measured in dB. We see that to ensure a sink node recovers all the source-symbols with a high probability, say, missing no more than $10-4$ source-symbols on average, the network coding based scheme saves at least 20 dB per node, compared to plain routing.

Figure 6 showed a similar case for the concatenated coding scheme. We tested the cases where the sink node collects 300 and 400 parity-symbols respectively before decoding. Disregarding the additional bandwidth expansion, their performances appear to be comparable to the progressive coding scheme. However, it should be noted that here we assumed that the sink node sits only three hops away from the sources. As soon as we increased the number of hops to four, the performance degrades drastically (due to severe error propagation). On the other hand, the performance of the progressive coding scheme

Figure 6. Simulation results of the concatenated coding scheme in a two-hop wireless network

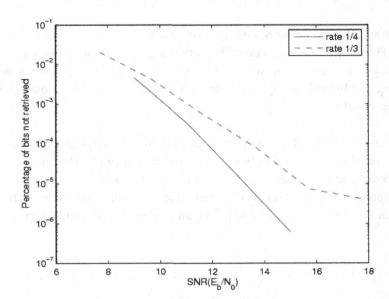

in Figure 5 is insensitive to the hop distance between the sources and the sink, thanks to the mid-stage decoding and cleaning-up process.

FUTURE TRENDS

Network codes are receiving increasing attention in systems with cooperating nodes, especially in data collection and data-dissemination applications, such as distributing legitimate content (security patches) in the Internet, downloading bulk or multimedia contents from multiple servers, collecting data over a wireless sensor network, teleconferencing, and exchanging information over an ad-hoc wireless network in a search-and-rescue scenario. Compared to the structured network codes, random network codes have the upside of flexibility, scalability, the ability to be constructed on-the-fly without the need to know the network topology beforehand. Random sparse-graph codes, such as the ones discussed in this chapter, are particularly attractive due to their low complexity, and the ability to combat errors. In additional to the specific network code that is being instantiated, recent research indicates that *scheduling*, namely, who relays first and who goes after whom, may significantly affect the system performance, especially in the case of 1-to-many receiver cooperation (Kaewprapha, Puttarak, Wang & Li, 2008). Additionally, the practice of integrating channel coding or source coding in networking coding, such as in the form of quasi-cyclic sparse-graph codes (Bao & Li, 2006), can greatly improve the overall coding gain. Several real-world applications, including Microsoft's *BitTorrent*, are also seriously exploring random network coding techniques to improve the efficiency of content distribution.

CONCLUSION

We have discussed an efficient network coding framework for multi-source multicasting in a general wireless network. The central idea is to describe network graphs as time-varying trellises, and to perform graph-matching LDPC coding at each trellis stage. Three different practices, progressive coding, concatenated coding and hybrid coding, are discussed and compared. The most notable feature of this framework is the practicality:

(1)　The network codes exploited are very practical and highly-efficient LDPC codes, which can be practically encoded and decoded in linear time, and which can support message dissemination as efficient as random codes in transmit time and bandwidth; and

(2)　The encoding process is completely distributed, requires no pre-knowledge of the link connectivity (although such knowledge would be helpful), and adapts to the instantaneous network topology on-the-fly.

REFERENCES

Ahlswede, R., Cai, N., Li, S.-Y. R., & Yeung, R. W. (2000, July). Network information flow. *IEEE Trans. Info. Theory*, 1204-1216.

Appuswamy, R., Franceschetti, M., & Zeger, K. (2006, July). Optimality of linear codes for broadcast-mode multicast networks. In *Proc. IEEE Intl. Symp. on Info. Theory*, Seattle, WA (pp. 50-53).

Bao, X., & Li, J. (2005, September). Matching code-on-graph with network-on-graph: Adaptive network coding for wireless relay networks. In *Proc. the 43rd Annual Allerton Conf. Communication, Control, Computing*, Champaign, IL.

Bao, X., & Li, J. (2006, July). A unified channel-network coding treatment for user cooperation in wireless ad-hoc networks. In *Proc. IEEE Int. Symp. Info. Theory*, Seattle, WA.

Bao, X., & Li, J. (2008, January). Adaptive network coded cooperation (ANCC) for wireless relay networks: Matching code-on-graph with network-on-graph. *IEEE Transaction on Wireless Communications*, 574-583.

Chung, S., Richardson, T., & Urbanke, R. (February 2001). Analysis of sum-product decoding of low-density parity-check codes using a Gaussian approximation. *IEEE Trans. Info. Theory*.

Deb, R., & Medard, M. (2004). Algebraic gossip: A network coding approach to optimal multiple rumor mongering. In *Proc. the 42nd Annual Allerton Conf. on Commun. Control and Computing*.

Hausl, C., & Dupraz, P. (2006). Joint network-channel coding for the multiple-access relay channel. In *Proc. of Int. Workshop Wireless Ad Hoc and Sensor Networks*.

Kaewgrapha, P., Puttarak, N., Wang, H., & Li, J. (2008, November). Receiver-cooperation: Network coding and distributed scheduling. In *Proc. IEEE Global Communications Conference*, New Orleans.

Karp, R., Schindelhauer, C., Shenker, S., & Vocking, B. (2000). Randomized rumor spreading. In *Proc. of the 41ˢᵗ Annual IEEE Symp. on Foundations of Comp. Sci.* (pp. 565-574).

Katti, S., Rahul, H., Hu, W., Katabi, D., Medard, M., & Crowcroft, J. (2006, September). XORs in the air: Practical wireless network coding. In *Proc. ACM SIGCOMM* (pp. 243–254).

Li, S.-Y. R., Yeung, R. W., & Cai, N. (2003, February). Linear network coding. *IEEE Trans. Info. Theory*, 371-381.

Luby, M. (2002). LT codes. In *Proc. IEEE Symp. on Foundations of Computer Sci.* (pp. 271-280).

Luby, M., Mitzenmacher, M., Shokrollahi, A., & Spielman, D. (2001, February). Efficient erasure correction codes. *IEEE Trans. Info. Theory*, 47.

MacKay, D. (1999). Good error-correcting codes based on very sparse matrices. *IEEE Trans. Info. Theory*.

Mosk-Aoyama, D., & Shah, D. (2006, July). Information dissemination via network coding. In *Proc. IEEE Int. Symp. Info. Theory*, Seattle, WA (pp. 1748-1752).

Peng, C., Zhang, Q., Zhao, M., & Yao, Y. (2007). SNCC: A selective network-coded cooperation scheme in wireless networks. In *Proc. of IEEE Int. Conf. Commun.*

Wang, T., & Giannakis, G. B. (2007). High-throughput cooperative communications with complex field network coding. In *Proc. Conf. Info. Sci. and Systems*.

Wu, Y. (2006, July). A trellis connectivity analysis of random linear network coding with buffering. In *Proc. Intl. Symp. Info. Theory*, Seattle, MA (pp. 768-772).

Zhang, S., Liew, S., & Lam, P. (2006, September). Physical layer network coding. In *Proc. 12ᵗʰ Annual International Conference on Mobile Computing and Networking*.

Section 3
Distributed Transmit and Receive Diversity Techniques for Cooperative Communications

Chapter 7
Cross–Layer Cooperative Beamforming for Wireless Networks

Lun Dong
Drexel University, USA

Athina P. Petropulu
Drexel University, USA

H. Vincent Poor
Princeton University, USA

ABSTRACT

Cooperative beamforming (CB) is a signal transmission technique that enables long-range communications in an energy efficient manner. CB relies on cooperation from a set of distributed network nodes, each carrying a single transmit antenna and acting as elements of a virtual antenna array. By appropriately weighting their transmissions, the cooperating nodes form one or more beams to cooperatively transmit one or more message signals to the desired destinations. In this chapter, a cross-layer framework is presented that brings the CB ideas closer to implementation in a wireless network setting. The process of sharing among the network nodes the information to be beamed is studied and evaluated in terms of its effect on the spectral efficiency of the overall system. Optimal or suboptimal beamforming weights are designed, and queuing analysis is provided to study delay characteristics of source messages.

DOI: 10.4018/978-1-60566-665-5.ch007

INTRODUCTION

Beamforming is a signal processing technique for directional signal transmission or reception. It applies to both radio and sound waves, and has been widely used in wireless communications, sonar, radar, medical imaging and other fields. In wireless systems, transmit beamforming involves a set of nodes that are typically elements of an antenna array. The signal to be beamed to a particular destination is made available to all array elements. Each antenna element multiplies the signal by a weight, and then transmits. The weighted signals from all antenna elements are then combined at the destination for further processing (e.g., equalization and symbol recovery). Suppose that N antennas transmit weighted versions of the same signal, each at power P, and the weights are such that all transmissions add up coherently at the destination (i.e., co-phasing). Then the power of the received signal is proportional to $N^2 P$. The received signal-to-noise ratio (SNR) increases proportionally to N^2, whereas the total transmit power increases only proportionally to N. Thus, beamforming can achieve *high energy efficiency*. By adjusting the weights among the various antennas, one is able to control the beampattern, i.e., the distribution of power in space. Thus, beamforming enables space-division multiple access (SDMA), controls interference and increases the reliability of the communication link. There is also receive beamforming, where a set of antenna elements weight the signal that they receive due to one of more sources/targets. The sum of the weighted signals is processed by a central station in order to obtain information about the target or the incoming signal.

Conventional beamformers, usually referred to as switched/fixed-beam antennas, use a fixed set of weights to combine the signals from the antennas and thus can only form a fixed set of beampatterns. A switch then selects the best beampattern, i.e., the one that yields the highest SNR, or signal-to-interference-plus noise ratio (SINR). The main advantage of the switched-beam approach is its simplicity, while its principle drawback is the lack of flexibility. Since the 1980s, there has been enormous interest in adaptive beamformers or smart antennas, in which signals from different antennas are combined by an adaptive algorithm (Compton, 1988). An adaptive beamformer is able to dynamically adapt its beampattern to satisfy various requirements in different scenarios.

Beamforming was initially studied in the context of an antenna array with elements at fixed locations (Godara, 1997; Krim & Viberg, 1996; Litva & Lo, 1996). Beamforming for linear arrays with randomly located antennas has also been studied in (Donvito & Kassam, 1979; Lo, 1964; Steinberg, 1972), where it was shown that with a large number of antennas a good beampattern can be formed with high probability. In all these cases, the signal to be beamed was provided to the antenna elements via a wireline connection. In a more recent work (Ochiai, Mitran, Poor & Tarokh, 2005), cooperative beamforming (also called collaborative or distributed beamforming) using a set of randomly distributed network nodes was proposed. Network nodes, each equipped with a single omni-directional antenna, were shown to be able to form a beam and collaboratively transmit a common signal to a far away destination. Due to the random placement of nodes, the beampattern could only be described in statistical terms. It was shown that, for uniformly distributed nodes over a disk, the directivity can approach. In (Ahmed & Vorobyov, 2008), it was shown that Gaussian distributed nodes can achieve even lower sidelobes than uniformly distributed nodes.

One of the major differences between cooperative beamforming and traditional beamforming with antenna arrays is that in the former the nodes locations are not known and need to be obtained dynamically. Also, since all nodes operate with independent antennas, there are always phase offsets due to mismatch between transmit and receive antenna oscillators. In (Ochiai, Mitran, Poor & Tarokh, 2005) two different scenarios were considered to deal with those issues, the open-loop scenario and the closed-loop scenario. In the closed-loop scenario, each cooperating node independently synchronizes itself to a beacon sent from the destination and adjusts its initial phase to it. Through this process it estimates its phase offset. In the open-loop scenario, each node estimates its position relative to some local reference point (e.g. the origin of the disk) within the cluster, which may be achieved via a beacon from the cluster head. Another possibility is for the destination to send a simple feedback only (a few bits) based on which each cooperating node iteratively adjust its initial phase. In (Mudumbai,Wild, Madhow & Ramchandran, 2006), each cooperating node independently makes a small random adjustment to its phase at each iteration, while the destination sends back one single bit of feedback, indicating whether the SNR improved or worsened with the independent adjustments of the current iteration. A distributed ascent algorithm is implemented by keeping the "good" phase adjustments and discarding the "bad" ones. It was shown that this algorithm converges with probability one for various practical scenarios and the convergence time is linear in the number of cooperating nodes.

Another important difference between cooperative beamforming and traditional beamforming is the means via which the signal is made available to all nodes. In a wireless network, the only means to relay information to the nodes is the wireless channel, which is susceptible to fading and interference. The work of (Ochiai, Mitran, Poor & Tarokh, 2005) did not consider how the common signals would be distributed to the network nodes, but did recognize that this step would involve some overhead. For example if in order to avoid multiuser interference the sources used a time-division multiple-access (TDMA) approach to distribute their signals to the cooperating nodes, the information-sharing time would increase proportionally to the number of source nodes. The additional time for information-sharing increases the transmission time, resulting in *information-sharing overhead*. As will become clear later, information-sharing significantly influences system design and network performance (e.g., spectral efficiency and delay), and needs to be taken into account before cooperative beamforming can be of practical interest. Information-sharing is usually considered to be a medium access control (MAC) layer problem, while beamforming is a physical (PHY) layer problem. In (Dong, Petropulu & Poor, 2008(a); Dong, Petropulu & Poor, 2008(b)), a random access (RA) as well as a TDMA information-sharing scheme was studied.

Cooperative beamforming can be particularly useful in wireless ad hoc sensor networks. For example consider nodes randomly distributed in a disaster area, which collect information locally, and then need to transmit it over long distances to some central station. Energy efficiency is of paramount importance as the sensors' lifetimes depend on it. Also, such nodes are likely to be equipped with a single antenna due to cost considerations and also due to size limitations.

This chapter discusses the ideas of (Dong, Petropulu & Poor, 2008(a); Dong, Petropulu & Poor, 2008(b)), referred to as PHY-MAC cross-layer cooperative beamforming. Two cross-layer CB approaches are presented, which differ on how the message signals are shared between sources and cooperating nodes, and on how they are weighted and forwarded. Both methods involve the MAC layer, during the information-sharing stage, and the physical layer, during the beamforming stage. The structure of the chapter is as follows. We first describe the system model. Then, the two cross-layer CB approaches are presented and their weight design and performance analysis are provided. The performance of the two approaches is compared in terms of spectral efficiency. Next, we extend the weight design into the scenario of multiple simultaneous beams. A queuing analysis is provided assuming that each potential source node has a queue and generates Poisson traffic. Finally, we present conclusions with some suggestions for future research.

Figure 1. System model

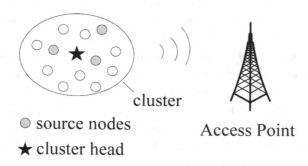

Notation

Bold uppercase letters denote matrices and bold lowercase letters denote column vectors. Conjugate, transpose and Hermitian transpose are represented by $(\cdot)^*$, $(\cdot)^T$ and $(\cdot)^H$ respectively; $\mathbb{E}\{\cdot\}$ denotes expectation; $\mathbf{I}_M$ is the identity matrix of size $M \times M$; denotes an all-zero matrix of size $M \times N$; $\text{diag}\{\mathbf{a}\}$ denotes a diagonal matrix with the elements of vector $\mathbf{a}$ along its diagonal; $CN\left(\mu, \sigma^2\right)$ denotes the circularly symmetric, complex Gaussian distribution with mean μ and variance σ^2.

SYSTEM MODEL

We consider a wireless network consisting of multiple network nodes and one common access point (AP) or central controller. Network nodes are divided into clusters according to their locations. Within each cluster one node is designated as a cluster head (CH). The geometry of a cluster can change dynamically. We assume that nodes within a cluster are uniformly located on a two-dimensional (2-D) disk with radius R. Each node is equipped with a single omni-directional antenna and operates in half-duplex mode.

Our focus is on a particular instantiation of uplink transmission in one cluster. Inter-cluster interference (i.e., interference from other clusters) at the AP is not considered here, as it can be avoided by assigning different time slots, frequency subcarriers or spreading codes to different clusters. The system model is shown in Figure 1. Suppose that, in this instantiation, K source nodes in the cluster of interest, denoted by $t_1, \ldots, t_k$, intend to transmit the symbols $s_1, \ldots, s_k$, to the AP. We assume that mutually uncorrelated symbols with unit power are used to transmit data, i.e., $\mathbb{E}\left\{\left|s_i\right|^2\right\} = 1$, and $\mathbb{E}\left\{s_i s_k\right\} = \mathbb{E}\left\{s_i\right\} \mathbb{E}\left\{s_k\right\}$ $(i \neq k)$. The total transmit power budget for transmitting s_i is P_i. The symbols are modulated into narrowband signals before transmission. In the mathematical formulation we will always deal with unmodulated baseband signals. For simplicity, and since the processing is done on a symbol by symbol basis, the symbol time index is not shown. All channels are frequency non-selective fading and remain unchanged during an instantiation of CB.

When implementing beamforming, a set of N distributed cooperating nodes in the cluster act as a virtual antenna array and form a beam to cooperatively transmit a signal to the AP. In this case, the N cooperating nodes are denoted by $c_1, \ldots, c_N$. We assume that the distances between nodes and the AP are much larger than the distances between nodes within the cluster. Thus, for intra-cluster channels, small-scale fading plays the dominant role. We model the baseband channels between nodes within the cluster as frequency-nonselective with Rayleigh distributed gains. Without loss of generality, the fading variance is normalized to one. Let a_{im} denote the channel gain between the source node t_i (subscript i for source nodes) and cooperating node c_m (subscript m for cooperating nodes) in the cluster, having zero mean and unit variance, i.e., $a_{im} \sim CN(0,1)$. The gains of different paths (i.e., different i's and/or m's) are assumed to be independent and identically distributed (i.i.d.). On the other hand, for the channels between cluster nodes and the AP, large-scale fading plays the dominant role. We model the effect of the baseband channels between nodes and the AP as multiplication by a complex scalar, whose magnitude represents path loss and its phase represents the phase offset due to distance (Tse & Viswanath, 2005). Furthermore, since the distances between cooperating nodes and destinations are much greater than the cluster size, we assume the path losses from all nodes in the cluster and the AP are the same. We represent the channel effect between node t_i and the AP as multiplication by $b_0 \cdot e^{j\theta_i}$, where b_0 represents the path loss between the origin of the disk containing cluster nodes and the AP, and θ_i represents the phase offset. The phase offset θ_i is determined by $\theta_i = 2\pi d_i / \lambda$, where λ is the signal wavelength and d_i is the distance between t_i and the AP. Similarly, we represent the channel effect between cooperating node c_m and the AP as multiplication by $b_0 \cdot e^{j\theta_m}$, where $\theta_m = 2\pi d_m / \lambda$ and d_m is the distance between c_m and the AP. Phase information can also be acquired by using local coordinates within the cluster. Let us denote the location of c_m in polar coordinates with respect to the origin of the disk by (r_m, ψ_m), where r_m is the distance between c_m and the origin and ψ_m is the azimuthal angle. Similarly, the location of the AP in polar coordinates is denoted by (d_0, ϕ_0), where d_0 represents the distance between the origin of the disk and the AP, and ϕ_0 is the azimuthal angle of the AP with respect to the origin of the disk. Under the far-field assumption due to $d \gg R$, it holds that

$$d_m \approx d_0 - r_m \cos(\phi_0 - \psi_m), \quad m = 1, \ldots, N .$$
(1)

Assuming the AP can compensate for the common phase e^{jd_0}, θ_m can also be written as

$$\theta_m = -\frac{2\pi}{\lambda} r_m \cos(\phi_0 - \psi_m), \quad m = 1, \ldots, N .$$
(2)

The illustration of notation is shown in Figure 2.

Direct Transmission (DT)

In DT, the K source nodes in the cluster transmit their signals directly to the AP one by one to avoid collisions. Obviously, K time units are needed due to this TDMA approach. Source node t_i transmits the signal $\sqrt{P_i}s_i$, where the scalar $\sqrt{P_i}$ is chosen to satisfy the power budget for the transmission of s_i to its destination. Then, the received signal, y_i, at the AP is given by

$$y_i = b_0 e^{j\theta_i} \sqrt{P_i}s_i + v_i, \quad i = 1, \ldots, K ,$$
(3)

Figure 2. Illustration of notation

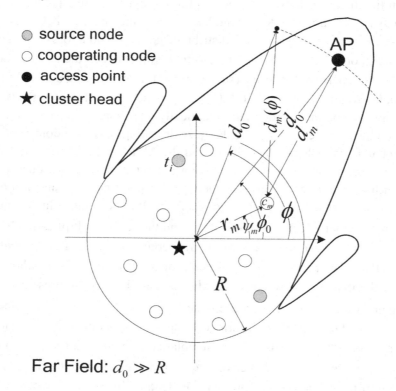

Far Field: $d_0 \gg R$

where v_i is the noise at the AP when s_i is transmitted, which is assumed to be zero-mean white complex Gaussian with variance σ^2, i.e., $v_i \sim CN\left(0, \sigma^2\right)$. The received SNR at the AP is

$$\gamma_i^{DT} = \frac{Pb_{i_0}^2}{\sigma^2}, \quad i = 1, \dots, K .$$

(4)

TDMA Cooperative Beamforming (TDMA-CB)

Under the above system model, we now consider an extension of the conventional CB approach (e.g., (Ochiai, Mitran, Poor & Tarokh, 2005)) into a two-stage scheme, where the second stage consists of beamforming, while the first stage consists of information-sharing that makes available the common signal to the cooperating nodes in a TDMA fashion. This scheme will be referred to as *TDMA cooperative beamforming* (TDMA-CB).

Stage 1-- Information-Sharing in a TDMA Fashion: The source nodes transmit their signals locally (within the cluster), one by one. Thus, K time units are required.

When t_i transmits, the cooperating nodereceives

$$x_m = a_{im}\sqrt{\tilde{P}_i}s_i + u_m \ , \tag{5}$$

where $u_m \sim CN\left(0, \sigma^2\right)$ is the noise at cooperating node c_m which is assumed to be white with respect to the index m. $\tilde{P}_i = \mu_i P_i$ is the transmit power, which is chosen so that the cooperating nodes can correctly decode the signal s_i with high probability, while the received signal power at the AP is negligible. Note that since the overall power budget for transmitting s_i is P_i, it holds that $\mu_i < 1$. In this chapter, for simplicity we assume that μ_i is known a priori.

Stage 2-- Beamforming: Cooperating nodes adjust their initial phases to form K beams in the direction of the AP one by one. K time units are required to form K beams, one for each source message signal. Specifically, to beam signal s_i, the cooperating node c_m transmits the weighted signal $\tilde{x}_m = w_m^i s_i$, where w_m^i is the weight of cooperating node c_m when beamforming s_i.

Weights of TDMA-CB

Since the power needed in Stage 1 is $\tilde{P}_i$, the total power of all cooperating nodes used in Stage 2 would be $P_i - \tilde{P}_i$. The weight w_m^i is given by

$$w_m^i = \sqrt{\frac{\left(1 - \mu_i\right)P_i}{N}}e^{-j\theta_m} \ , \tag{6}$$

where the square-root term on the right hand side (RHS) is needed to satisfy the power constraint, while the exponential term is the initial phase used to compensate for the phase offset of the channel between the cooperating node and the AP (i.e., co-phasing).

Received SNR

Since the path losses between all cooperating nodes and the AP are assumed to be the same, the weights of (6) correspond to the weights in maximal ratio transmission (Molisch, 2005), and thus are optimal in the sense that they maximize the received SNR.

Under the optimal weights of (6), the received signal at the AP is

$$y_i = \sqrt{\left(1 - \mu_i\right)NP_i}b_0 s_i + v_i \ . \tag{7}$$

and the maximal SNR is

$$\gamma_i^{\text{TDMA-CB}} = \left(1 - \mu\right)N\frac{P_i b_0^2}{\sigma^2}, \quad i = 1, \ldots, K \ . \tag{8}$$

From (4) and (8), one can see that the SNR gain in TDMA-CB is $(1 - \mu_i)N$ as compared to DT. For conventional models where information-sharing in Stage 1 is omitted, μ_i is equal to 0, and the SNR gain in TDMA-CB equals N.

Average Beampattern

Under the assumption that cooperating nodes are uniformly distributed on a disk with radius R, the average beampattern was analyzed in (Ochiai, Mitran, Poor & Tarokh, 2005). We here provide the average beampattern for our model based on the analysis of (Ochiai, Mitran, Poor & Tarokh, 2005).

Let us consider a receiving point with polar coordinate (d_0, ϕ), which is at distance $d_m(\phi)$ from each cooperating node c_m (see Figure 2). Under the weights of (6), the received signal is

$$y_i(\phi) = \left(\sum_{m=1}^{N} e^{-j\theta_m} e^{j\frac{2\pi}{\lambda} d_m(\phi)} \right) \sqrt{\frac{(1 - \mu_i) P_i}{N}} b_0 s_i + v_i . \tag{9}$$

where v_i represents the white Gaussian noise at the receiving point.

Under the far-field assumption and based on (2), the term within the bracket in (9) becomes

$$\sum_{m=1}^{N} e^{-j\theta_m} e^{j\frac{2\pi}{\lambda} d_m(\phi)} = \sum_{m=1}^{N} e^{j\alpha(\phi) z_m} \tag{10}$$

where $\alpha(\phi) = 4\pi (R/\lambda) \sin\left[\frac{1}{2}(\phi_0 - \phi)\right]$ and $z_m = (r_m/R) \sin\left[\psi_m - \frac{1}{2}(\phi_0 + \phi)\right]$. Under the assumption that cluster nodes are uniformly distributed in a disk, has the following probability density function (pdf):

$$f(z) = \frac{2}{\pi} \sqrt{1 - z^2}, \qquad -1 \leq z \leq 1 . \tag{11}$$

The average received power is

$$\begin{aligned}
\mathbb{E}\left\{ |y_i(\phi)|^2 \right\} &= \frac{(1 - \mu_i) P_i b_0^2}{N} \mathbb{E}\left\{ \left| \sum_{m=1}^{N} e^{j\alpha(\phi) z_m} \right|^2 \right\} + \sigma^2 \\
&= \frac{(1 - \mu_i) P_i b_0^2}{N} \mathbb{E}\left\{ \sum_{m=1}^{N} \sum_{l=1}^{N} e^{j\alpha(\phi)(z_m - z_l)} \right\} + \sigma^2 \\
&= (1 - \mu_i) N P_i b_0^2 \left(\frac{1}{N} + \frac{1}{N^2} \left| 2 \frac{J_1(\alpha(\phi))}{\alpha(\phi)} \right|^2 \right) + \sigma^2 .
\end{aligned} \tag{12}$$

where $J_1(\cdot)$ is the first-order Bessel function of the first kind.

Figure 3. Beampattern of TDMA-CB

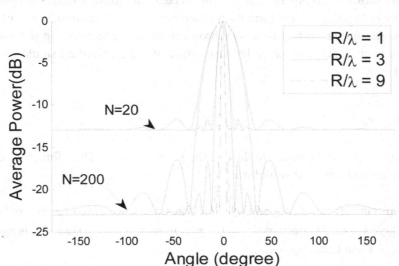

The average beampattern represents the average power distributed along different directions. Note that in (12), the term is independent of the azimuthal angle ϕ, so we can focus on the first term in the RHS of (12) only. From (12), the normalized average beampattern is expressed as

$$P_{av}(\phi) = \frac{1}{N} + \left(1 - \frac{1}{N}\right)\left|2\frac{J_1(\alpha(\phi))}{\alpha(\phi)}\right|^2 . \tag{13}$$

which equals one in the target direction ϕ_0, while the sidelobe level approaches $1/N$ as the angle moves away from the target direction.

Figure 3 shows analytical results of the (normalized) average beampattern under different values of N and R. As observed, increasing N reduces the sidelobe levels and increasing R narrows the mainlobe.

Remark: To apply the optimal weights of (6) in TDMA-CB, only local channel state information (CSI) is needed, i.e., a cooperating nodeneeds to have knowledge of its own phase offset, which in practice can be acquired by a beacon signal from the CH or the AP. Additional methods for adjusting phase for cooperative beamforming can be found in (Mudumbai, Barriac & Madhow, 2007; Mudumbai, Wild, Madhow & Ramchandran, 2006; Brown & Poor, 2008).

Random Access Cooperative Beamforming (RA-CB)

As compared to DT, TDMA-CB requires an additionaltime units for information-sharing in order that all cooperating nodes have access to the same message signals. To reduce the information-sharing time, a random access cooperative beamforming scheme, namely RA-CB, was proposed in (Dong, Petropulu & Poor, 2008(a); Dong, Petropulu & Poor, 2008(b)), which consists of the following stages:

Stage 1-- Information-Sharing in a Random Access Fashion: Thesource nodes transmit their signals simultaneously to share their information in a random access fashion, and thus Stage 1 requires only one time slot. Assuming perfect time synchronization between nodes in the same cluster, other nodes in the cluster hear a collision, i.e., a linear mixture of the transmitted signals. Cooperating node receives the signal

$$x_m = \sum_{i=1}^{K} a_{im} \sqrt{\tilde{P}_i} s_i + u_m \; . \tag{14}$$

There is no simple way for the nodes to decode the source signals. Therefore, the cooperating nodes will act only on the analog received waveform.

Stage 2-- Beamforming: Each cooperating node transmits a weighted version of the signal that the node received in Stage 1, i.e., it transmits $\tilde{x}_m = w_m^i x_m$, $m = 1, ..., N$, where w_m^i is again the weight of cooperating node c_m when beaming s_i.

Each cooperating node needs to determine how many sources and also which sources are present in the mixture that it received. One possible way to achieve this is via orthogonal pilot sequences that have been embedded in the beginning of each source's data packet. The cooperating node can cross-correlate the received mixture with each possible pilot sequence. Due to orthogonality of pilots corresponding to different sources, a non-zero cross-correlation will indicate the presence of the corresponding source. Once the sources are identified, the order in which each source will be beamed needs to be known by all cooperating nodes so that they all beam the same signal in each slot. This information can be provided by a higher network layer, or, it can be determined based on a list of source priorities that has been distributed to all nodes. Thus, once the sources in the mixture have been identified, based on the priority list, the nodes rank these sources and beam them in that order.

Weights Based on Local CSI

As was already discussed, in TDMA-CB the cooperating nodes select their weights so that their transmissions add coherently at the AP. The weight of a particular node requires knowledge of its phase offset with respect to the AP, i.e., local CSI. Similarly, we can design co-phasing weights for RA-CB based on local CSI. Here, local CSI for cooperating node c_m refers to as θ_m and a_{im} for $i = 1, ..., K$.

When beaming s_i, let us choose the weight of cooperating node c_m as

$$w_m^i = \rho_i \cdot a_{im}^* e^{-j\theta_m} \; . \tag{15}$$

where ρ_i is a scalar used to satisfy the power constraint in Stage 2, which is the same for all cooperating nodes, and a_{im}^* is the conjugate of the channel gain between the source t_i and the cooperating node c_m. From (15), the signal transmitted by node c_m in Stage 2 is

$$\tilde{x}_m = \rho_i e^{-j\theta_m} \left(|a_{im}|^2 \sqrt{\tilde{P}_i} s_i + a_{im}^* \sum_{\substack{k=1 \\ k \neq i}}^{K} a_{km} \sqrt{\tilde{P}_k} s_k + a_{im}^* u_m \right) . \tag{16}$$

Then, one can easily show that the average transmit power of c_m equals

$$\mathbb{E}\left\{ |\tilde{x}_m|^2 \right\} = \rho_i^2 \left(2\tilde{P}_i + \sum_{\substack{k=1 \\ k \neq i}}^{K} \tilde{P}_k + \sigma^2 \right) . \tag{17}$$

To satisfy the power constraint in Stage 2, i.e., $\left(1 - \mu_i\right) P_i$, ρ_i shall be

$$\rho_i = \sqrt{\frac{\left(1 - \mu_i\right) P_i}{N \left(2\tilde{P}_i + \sum_{\substack{k=1 \\ k \neq i}}^{K} \tilde{P}_k + \sigma^2 \right)}} . \tag{18}$$

Received SINR

From (16), the received signal at the AP is

$$y_i = \rho_i b_0 \sum_{m=1}^{N} \left(|a_{im}|^2 \sqrt{\tilde{P}_i} s_i + a_{im}^* \sum_{\substack{k=1 \\ k \neq i}}^{K} a_{km} \sqrt{\tilde{P}_k} s_k + a_{im}^* u_m \right) + v_i . \tag{19}$$

Within the square bracket of the right-hand side (RHS) of (19), the first term denotes the desired signal, the second one is the multi-user interference from other source nodes and the third one represents the contribution of noise at cooperating nodes.

Let us define η as

$$\eta \triangleq \sum_{\substack{k=1 \\ k \neq i}}^{K} a_{km} \sqrt{\tilde{P}_k} s_k + u_m . \tag{20}$$

It is easy to show that $\mathbb{E}\{\eta\} = 0$ and

$$\sigma_\eta^2 \triangleq \mathbb{E}\left\{ |\eta|^2 \right\} = \sum_{\substack{k=1 \\ k \neq i}}^{K} \tilde{P}_k + \sigma^2 . \tag{21}$$

Conditioned on a_{im}, the instantaneous SINR, $\gamma_i^{\text{RA-CB}}$, equals

$$\gamma_i^{\text{RA-CB}} = \frac{\rho_i^2 b_0^2 \tilde{P}_i \left(\sum_{m=1}^{N} |a_{im}|^2 \right)^2}{\rho_i^2 b_0^2 \sigma_\eta^2 \sum_{m=1}^{N} |a_{im}|^2 + \sigma^2} = \frac{\rho_i^2 b_0^2 \tilde{P}_i \xi_i^2}{\rho_i^2 b_0^2 \sigma_\eta^2 \xi_i + \sigma^2} .$$ (22)

where $\xi_i \triangleq \sum_{m=1}^{N} |a_{im}|^2$. Since $|a_{im}|$ is Rayleigh distributed, ξ_i is an Erlang distributed variable with the pdf

$$f_{\xi_i}(x) = \frac{x^{N-1} e^{-x}}{(N-1)!}, \qquad x \geq 0 .$$ (23)

From (22), an upper bound of the average SINR is given by

$$\mathbb{E}\left\{\gamma_i^{\text{RA-CB}}\right\} < \frac{\mathbb{E}\left\{\rho_i^2 b_0^2 \tilde{P}_i \xi_i^2\right\}}{\sigma^2}$$

$$= \frac{1 + \dfrac{1}{N}}{2 + \displaystyle\sum_{\substack{k=1 \\ k \neq i}}^{K} \dfrac{\tilde{P}_k}{\tilde{P}_i} + \dfrac{\sigma^2}{\tilde{P}_i}} \cdot \gamma_i^{\text{TDMA-CB}} < \gamma_i^{\text{TDMA-CB}} .$$ (24)

Therefore, the average SINR of RA-CB is always smaller than the SNR of TDMA-CB. Note that ρ_i^2 is of the order of $1/N$, is of the order of N and ξ_i^2 is of the order of N^2. One can see that the numerator of (22) is of the order of N, while the denominator of (22) is of the order of 1. Thus, the SINR of RA-CB is still of the order of N.

Outage Probability

Outage probability is of practical interest in some scenarios. It is easy to show that $\gamma_i^{\text{RA-CB}}$ is an increasing function of ξ_i. Then, the outage probability p_i^{out} is given by

$$p_i^{\text{out}} = \Pr\left(\gamma_i^{\text{RA-CB}} < \gamma_i^{\text{out}} = \frac{\rho_i^2 b_0^2 \tilde{P}_i \left(\xi_i^{\text{out}}\right)^2}{\rho_i^2 b_0^2 \sigma_\eta^2 \xi_i^{\text{out}} + \sigma^2} \right)$$

$$= \int_0^{\xi_i^{\text{out}}} \frac{x^{N-1} e^{-x}}{(N-1)!} dx$$

$$= 1 - e^{-\xi_i^{\text{out}}} \sum_{k=0}^{N-1} \frac{\left(\xi_i^{\text{out}}\right)^k}{k!} .$$ (25)

According to the central limit theorem, when N is large, we can also approximateas a Gaussian random variable, i.e., $\xi_i \sim \mathcal{N}(N, N)$, from which we obtain

$$p_i^{\text{out}} \approx \Phi\left(\frac{\xi_i^{\text{out}} - N}{\sqrt{N}}\right) . \tag{26}$$

where

$$\Phi(x) = \frac{1}{\sqrt{2\pi}} \int_{-\infty}^{x} e^{-\frac{x^2}{2}} dx . \tag{27}$$

Average Beampattern

Under the far-field assumption, the received signal at (d_0, ϕ) equals

$$y_i(\phi) = \rho_i b_0 e^{-j\alpha(\phi) z_m} \sum_{m=1}^{N} \left(|a_{im}|^2 \sqrt{\tilde{P}_i} s_i + a_{im}^* \sum_{\substack{k=1 \\ k \neq i}}^{K} a_{km} \sqrt{\tilde{P}_k} s_k + a_{im}^* u_m \right) + v_i \tag{28}$$

Where $\alpha(\phi)$ and z_m are the same as in (10). Taking into account the assumptions on the channel coefficients, it can be readily shown that the average received power at (d_0, ϕ) is given by

$$\mathbb{E}_{s,z,a,u,v}\{|y_i(\phi)|^2\} = N^2 \rho_i^2 b_0^2 \tilde{P}_i^2 \left(\frac{\beta}{N} + (1 - \frac{1}{N}) \left| 2 \frac{J_1(\alpha(\phi))}{\alpha(\phi)} \right|^2 \right) + \sigma^2 \tag{29}$$

Where

$$\beta = 2 + \sum_{\substack{k=1 \\ k \neq i}}^{K} \frac{\tilde{P}_k}{\tilde{P}_i} + \frac{\sigma^2}{\tilde{P}_i^2} . \tag{30}$$

The normalized average beampattern is then

$$P_{\text{av}}(\phi) = \frac{N}{N + \beta - 1} \left(\frac{\beta}{N} + (1 - \frac{1}{N}) \left| 2 \frac{J_1(\alpha(\phi))}{\alpha(\phi)} \right|^2 \right) \tag{31}$$

As $\beta > 1$, it holds that $\frac{N\beta}{N + \beta - 1} > 1$. Comparing (31) with (13), one can see that the average beampattern of RA-CB is similar to that of TDMA-CB, and the only difference is that the sidelobe power becomes higher. The peak/zero positions of the average beampattern, which depends only on $|J_1(\alpha(\phi)) / \alpha(\phi)|^2$, are exactly the same as in (13).

Weights Based on Global CSI

When global CSI, i.e., a_{im} and θ_m, for $i = 1, \ldots, K$, $m = 1, \ldots, N$, is available, optimal weights can be designed to maximize the received SINR at the AP.

Let us define the weight vector $\mathbf{w}_i = [w_1^i, \ldots, w_N^i]^H$, the vector $\mathbf{a}_i = [b_0\sqrt{\tilde{P}_i}a_{i1}e^{j\theta_1}, \ldots, b_0\sqrt{\tilde{P}_i}a_{iN}e^{j\theta_N}]^H$, and the noise vector $\mathbf{u} = [b_0 u_1 e^{j\theta_1}, \ldots, b_0 u_N e^{j\theta_N}]^H$.

When beaming s_i, the received signal at the AP equals

$$y_i = \sum_{m=1}^{N}\left(a_{im} b_0 e^{j\theta_m} w_m^i \sqrt{\tilde{P}_i} s_i + \sum_{\substack{k=1 \\ k \neq i}}^{K} a_{km} b_0 e^{j\theta_m} w_m^i \sqrt{\tilde{P}_k} s_k + b_0 e^{j\theta_m} w_m^i u_m \right) + v_i$$

$$= \mathbf{w}_i^H \mathbf{a}_i s_i + \mathbf{w}_i^H \sum_{\substack{k=1 \\ k \neq i}}^{K} \mathbf{a}_k s_k + \mathbf{w}_i^H \mathbf{u} + v_i . \tag{32}$$

We further define matrices $\mathbf{R}_S^i \triangleq \mathbf{a}_i \mathbf{a}_i^H$, $\mathbf{R}_I^i \triangleq \sum_{\substack{k=1 \\ k \neq i}}^{K} \mathbf{a}_k \mathbf{a}_k^H + \sigma^2 \mathbf{I}_N$ and

$$\mathbf{R} \triangleq \mathrm{diag}\left\{ \left[\sum_{k=1}^{K} \tilde{P}_k \mid a_{k1} \mid^2 + \sigma^2, \ldots, \sum_{k=1}^{K} \tilde{P}_k \mid a_{kN} \mid^2 + \sigma^2 \right] \right\} . \tag{33}$$

The power of the desired signal s_i is

$$\mathbf{E}\{\mathbf{w}_i^H \mathbf{a}_i s_i s_i^H \mathbf{a}_i^H \mathbf{w}_i\} = \mathbf{w}_i^H \mathbf{R}_S^i \mathbf{w}_i . \tag{34}$$

Similarly, the power of interference plus noise is $\mathbf{w}_i^H \mathbf{R}_I^i \mathbf{w}_i + \sigma^2$. For the given vectors $\mathbf{a}_i$ for $i = 1, \ldots, K$, the received SINR at the AP is given by

$$\gamma_i^{\mathrm{RA\text{-}CB}} = \frac{\mathbf{w}_i^H \mathbf{R}_S^i \mathbf{w}_i}{\mathbf{w}_i^H \mathbf{R}_I^i \mathbf{w}_i + \sigma^2} . \tag{35}$$

To maximize $\gamma_i^{\mathrm{RA\text{-}CB}}$ subject to the transmit power constraint, we can formulate the optimization problem

$$\arg\max_{\mathbf{w}_i} \gamma_i^{\mathrm{RA\text{-}CB}} = \frac{\mathbf{w}_i^H \mathbf{R}_S^i \mathbf{w}_i}{\mathbf{w}_i^H \mathbf{R}_I^i \mathbf{w}_i + \sigma^2}$$

$$\text{s.t. } \mathbf{w}^H \mathbf{R} \mathbf{w} = (1 - \mu_i) P_i . \tag{36}$$

Substituting the constraint into the objective function in (36), the optimization problem of (36) can be rewritten as

$$\arg\max_{\mathbf{w}_i} \gamma_i^{\mathrm{RA\text{-}CB}} = \frac{\mathbf{w}_i^H \mathbf{R}_S^i \mathbf{w}_i}{\mathbf{w}_i^H \widetilde{\mathbf{R}}_I^i \mathbf{w}_i} , \tag{37}$$

where

$$\widetilde{\mathbf{R}}_I^i \triangleq \mathbf{R}_I^i + \frac{\sigma^2}{(1-\mu_i)P_i}\mathbf{R} . \tag{38}$$

The problem in (37) is a Rayleigh quotient problem. The maximum of $\gamma_i^{\text{RA-CB}}$ corresponds to the maximal eigenvalue of the symmetric matrix $(\widetilde{\mathbf{R}}_I^i)^{-1}\mathbf{R}_S^i$, and the corresponding eigenvector is the optimal solution $\mathbf{w}_i^{\text{opt}}$ (Golub & Loan, 1996).

Noting that $\mathbf{R}_S^i = \mathbf{a}_i \mathbf{a}_i^H$, we further obtain

$$\mathbf{w}_i^{\text{opt}} = \beta \cdot (\widetilde{\mathbf{R}}_I^i)^{-1}\mathbf{a}_i , i = 1, \cdots, K \tag{39}$$

Where β is a scalar. To satisfy the transmit power constraint, β must satisfy

$$\beta = \sqrt{\frac{(1-\mu_i)P_i}{\mathbf{a}_i^H(\widetilde{\mathbf{R}}_I^i)^{-1}\mathbf{R}(\widetilde{\mathbf{R}}_I^i)^{-1}\mathbf{a}_i}} . \tag{40}$$

The maximal SINR under the optimal weights of (39) is

$$\gamma_i^{\text{RA-CB}} = \mathbf{a}_i^H(\widetilde{\mathbf{R}}_I^i)^{-1}\mathbf{a}_i . \tag{41}$$

Note that the maximal SINR is a random variable, as the entries of vectors $\mathbf{a}_k$ ($k = 1, \ldots, K$) are random.

Remarks:

- From (39), we see that global CSI, i.e., a_{im} and θ_m, for $i = 1, \ldots, K$, $m = 1, \ldots, N$, is needed for the computation of optimal weights. To acquire global CSI in practice, the CH can gather local CSI from all cooperating nodes, compute the optimal weights and then send the computed weights back to cooperating nodes for beamforming. Communication in this process would take place via a separate control channel, and certain higher layer protocols would also be involved. It is obvious that the cost for acquiring optimal weights of RA-CB is higher than that of TDMA-CB.

- When $K = 1$, the matrix $\widetilde{\mathbf{R}}_I^i$ becomes

$$\widetilde{\mathbf{R}}_I^i = \frac{\sigma^2}{(1-\mu_1)P_1}\begin{pmatrix} \tilde{P}_1 \mid a_{11} \mid^2 + \sigma^2 & & 0 \\ & \ddots & \\ 0 & & \tilde{P}_1 \mid a_{1N} \mid^2 + \sigma^2 \end{pmatrix} . \tag{42}$$

Substituting (42) into (41), the maximal SINR is

$$\gamma_1^{\text{RA-CB}} = \frac{b_0^2(1-\mu_1)P_1}{\sigma^2} \sum_{m=1}^{N} \frac{\tilde{P}_1 \mid a_{1m} \mid^2}{\tilde{P}_1 \mid a_{1m} \mid^2 + \sigma^2}$$
$$< \frac{b_0^2 N(1-\mu_1)P_1}{\sigma^2} = \gamma_1^{\text{TDMA-CB}} \ . \tag{43}$$

Therefore, in the case of one source node, the maximal SINR in RA-CB is always smaller than the maximal SNR in TDMA-CB. Intuitively, as the number of source nodes increases, the maximal SINR in RA-CB would decrease due to the additional interference. Therefore, as compared to TDMA-CB, RA-CB reduces the information-sharing time, at the expense of SINR reduction.

- Substituting (39) into the objective function in (36), one can see that the maximal SINR of RA-CB is an increasing function of the transmit power $(1-\mu_i)P_i$. Therefore, the equality power constraint in (36) is equivalent to the inequality power constraint $\mathbf{w}^H \mathbf{R} \mathbf{w} \leq (1-\mu_i)P_i$.

Performance Comparison

As compared to TDMA-CB, RA-CB reduces the information-sharing time, with the cost of reduced SINR. Both TDMA-CB and RA-CB exhibit a tradeoff between improved SNR/SINR gain and longer transmission time. As a result, to compare the performance of different schemes, *spectral efficiency* would be a fair figure of merit, as it takes into account both factors in the tradeoff. Spectral efficiency is defined using the sum rate:

$$C = \frac{1}{T} \sum_{i=1}^{K} C_i = \frac{1}{T} \sum_{i=1}^{K} \log_2(1+\gamma_i) \ (\text{bits/s/Hz}) \ , \tag{44}$$

where γ_i and C_i are the received SINR and transmission rate for transmitting the signal of source node t_i respectively, and T is the total transmission time for all the K signals.

Weights that Maximize Spectral Efficiency

To get an idea of how much data can at most be transmitted for TDMA-CB and RA-CB, we intend to compute the spectral efficiency based on the weights that achieve maximal spectral efficiency in each scheme.

Note that in Stage 2 of TDMA-CB and RA-CB, the signals are beamed to the AP in a TDMA fashion, so the design of $\mathbf{w}_i$ is independent of the design of $\mathbf{w}_k$ ($k \neq i$). Therefore, maximal spectral efficiency is achieved if and only if for every i the weights $\mathbf{w}_i$ are designed such that the received SNR/SINR is maximized. Therefore, the weights of (6) and (39) that maximize the received SNR/SINR also achieve maximal spectral efficiency.

Spectral Efficiency

In the following, we will compare the performance of DT, TDMA-CB and RA-CB based on spectral efficiency. Such analysis provides some guidelines for determining the best transmission scheme.

- **DT:**
  From (3), the spectral efficiency equals

$$C_{\mathrm{DT}} = \frac{1}{K} \sum_{i=1}^{K} \log_2 \left(1 + \frac{P_i b_0^2}{\sigma^2} \right) \tag{45}$$

where the scalar $1/K$ is due to the fact that K time units are required in DT.

- **TDMA-CB:**
  From (8), the spectral efficiency under optimal weights equals

$$C_{\mathrm{TDMA\text{-}CB}} = \frac{1}{2K} \sum_{i=1}^{K} \log_2 \left(1 + N(1-\mu_i) \frac{P_i b_0^2}{\sigma^2} \right) \tag{46}$$

where the scalar $1/(2K)$ reflects the fact that $2K$ time units are required in TDMA-CB.

- **RA-CB:**
  From (41), the spectral efficiency under optimal weights equals

$$C_{\mathrm{RA\text{-}CB}} = \frac{1}{K+1} \sum_{i=1}^{K} \log_2 \left(1 + \mathbf{a}_i^H (\widetilde{\mathbf{R}}_I^i)^{-1} \mathbf{a}_i \right) \tag{47}$$

where the scalar factor $1/(K+1)$ reflects that $K+1$ time units are required in RA-CB.

Determination of the Best Scheme

The expressions (45), (46) and (47) can be used to calculate the respective spectral efficiencies given the network parameters, such as power budget P_i, power needed in Stage 1 $\tilde{P}_i$, path loss b_0, noise level σ^2, number of sources K and number of cooperating nodes N. Figure 4 shows the simulation results of the average spectral efficiency versus the path loss b_0, where the number of source nodes is fixed at $K = 4$. One can determine the best transmission strategy among them in terms of spectral efficiency, by simply comparing the respective spectral efficiencies. In practice, before a transmission, the CH would compute these spectral efficiencies, and then notify cluster nodes of the best transmission strategy that they should use. Some remarks are as follows.
  Remarks:

- Simple criteria can be obtained for some special cases, e.g., when $K = 1$, RA-CB always achieves lower spectral efficiency than TDMA-CB due to the contribution of the noise at cooperating nodes [see (43)].

- To achieve maximal spectral efficiency, TDMA-CB requires that each cooperating node c_m knows its own phase information θ_m, while RA-CB requires that the CH has global CSI (i.e., a_{im} and θ_m, for $i = 1,...,K$, $m = 1,...,N$). In some cases, in addition to spectral efficiency, the overhead of obtaining optimal weights for TDMA-CB and RA-CB should also be taken into account. For example, to compare TDMA-CB and RA-CB, we can modify the comparison problem to: RA-CB outperforms TDMA-CB only if $(C_{\text{RA-CB}} - C_{\text{TDMA-CB}}) / C_{\text{TDMA-CB}}$ is higher than a certain threshold.

- In addition to spectral efficiency, in some scenarios other metrics may also be considered. For example, RA-CB may be preferable for security reasons in certain applications. In DT or TDMA-CB, sources or cooperating nodes transmit a single packet to the AP, and during this process eavesdroppers or hostile nodes may easily hear and decode it. On the other hand, in RA-CB, cooperating nodes transmit mixtures of multiple packets, which makes it harder for eavesdroppers to decode useful information.

Multiple Simultaneous Beams

In some cases, e.g., in wireless ad hoc networks, there may not exist a single AP or central controller. For such cases, we assume that K source nodes transmit signals to K *different* destinations, denoted by

Figure 4. Simulation results of average spectral efficiency versus path loss b_0. The number of source nodes is $K = 4$. A single beam is formed at a time towards the AP. The radius of the disk containing cluster nodes is $R = 10\lambda$. The number of cooperating nodes is $N = 10$ or 100.

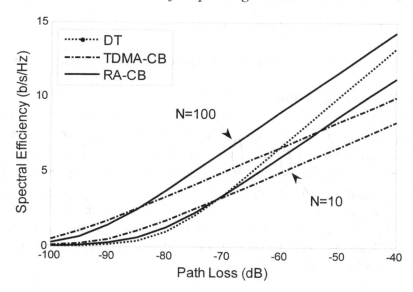

$q_1, \ldots, q_K$. If the destinations are located at distinct directions, we can form multiple beams simultaneously using TDMA-CB or RA-CB. Without loss of generality, in the following we will focus our analysis on the scenario in which $L \leq K$ simultaneous beams are to be formed towards distinct destinations at azimuthal angles $\phi_1, \ldots, \phi_L$. The phase offset between c_m and q_ℓ is denoted by

$$\theta_m^\ell = -\frac{2\pi}{\lambda} r_m \cos(\phi_\ell - \psi_m), \ell = 1, \ldots, L.$$ (48)

Also, the path loss between the origin of the disk and q_ℓ is denoted by b_ℓ, $\ell = 1, \ldots, L$.

In multiple simultaneous beams, a set of weights maximizing the SINR of one beam may not also maximize the SINR of another beam. Designing weights that maximize the spectral efficiency is in general a difficult problem. In the literature, different design objectives have been proposed to address similar problems, e.g., completely nulling all co-channel interference (Zhang & Dai, 2004), maximizing the signal-to-leakage ratio (Sadek, Tarighat & Sayed, 2005), minimizing the interference plus noise power at each destination (El-Keyi & Champagne, 2008) and maximizing the minimal SINR among all destinations (Li & Wang, 2008).

In the following, for TDMA-CB and RA-CB, we will consider two weight design objectives: (i) co-phasing weights that allow desired signals to add coherently at destinations; and (ii) nulling weights that completely null undesired interference at all the destinations.

Co-Phasing Weights

In this case, the co-phasing weights are straightforward extensions of the weights in (6) and (15) for single-beam beamforming.

- **TDMA-CB:** To beam $s_1, \ldots, s_L$ simultaneously, cooperating node c_m transmits $\tilde{x}_m = \sum_{\ell=1}^{L} w_m^\ell s_\ell$ where

$$w_m^\ell = \sqrt{\frac{(1 - \mu_\ell) P_\ell}{N}} e^{-j\theta_m^\ell}$$ (49)

- **RA-CB:** To beam $s_1, \ldots, s_L$ simultaneously, cooperating node c_m transmits $\tilde{x}_m = w_m x_m$, where the weights are chosen as

$$w_m = \sum_{\ell=1}^{L} \rho_\ell \cdot a_{\ell m}^* e^{-j\theta_m^\ell}$$ (50)

and x_m is given in (14). Note that we have dropped the superscript on w_m since the common weight w_m is used for all of the L beams. It should be pointed out that RA-CB with weights of (50) also corresponds to the "Transmit and Listen" scheme in (Dana & Hassibi, 2006).

It can be readily shown that the achievable SINR of TDMA-CB and RA-CB for multi-beam beamforming is still of the order of N, though smaller than that for single-beam beamforming (Dong, Petropulu & Poor, 2008(b)) due to the additional co-channel interference. Thus, any transmission rate can be achieved as long as the number of cooperating nodes is sufficiently large.

Note that the co-phasing weights in (49) or (50) require local CSI only. In particular, for TDMA-CB, the cooperating node c_m needs to know θ_m^ℓ for $\ell = 1, \ldots, L$. For RA-CB, the cooperating node c_m needs to know θ_m^ℓ for $\ell = 1, \ldots, L$ and a_{im} for $i = 1, \ldots, K$.

Nulling Weights

In this section, we consider the design objective of completely nulling undesired interference at all destinations, which yields simple closed-form solutions and requires global CSI.

Nulling Weights for TDMA-CB

Recall that cooperating node c_m transmits $\tilde{x}_m = \sum_{\ell}^{L} w_m^\ell s_\ell$. Let us define the vectors $\mathbf{a}^\ell = [b_\ell e^{j\theta_1^\ell}, \ldots, b_\ell e^{j\theta_N^\ell}]^H$ and $\mathbf{w}_\ell = [w_1^\ell, \ldots, w_N^\ell]^H$ for $\ell = 1, \ldots, L$. The received signal at the destination q_ℓ is

$$y_\ell = \mathbf{w}_\ell^H \mathbf{a}^\ell s_\ell + \sum_{\substack{i=1 \\ i \neq \ell}}^{L} \mathbf{w}_i^H \mathbf{a}^\ell s_i + v_\ell, \quad \ell = 1, \ldots, L, \tag{51}$$

where the first term of RHS is the desired signal, the second term represents undesired interference and the third term represents the noise at the destination.

To completely null the interference, the following linear constraints should be satisfied

$$\mathbf{w}_i^H \mathbf{a}^\ell = 0, \text{ for } \ell, i = 1, \ldots, L \text{ and } i \neq \ell. \tag{52}$$

To guarantee a non-zero solution for $\mathbf{w}_i$, we need $N \geq L$ which usually can be easily satisfied.

Under the constraints of (52), the weights maximizing spectral efficiency corresponds to the weights that maximize the SNR under a transmit power budget. The optimization problem can be formulated as

$$\arg\max_{\mathbf{w}_\ell} \gamma_\ell = \frac{\mathbf{w}_\ell^H \mathbf{a}^\ell (\mathbf{a}^\ell)^H \mathbf{w}_\ell}{\sigma^2}$$
$$\text{s.t.} \begin{cases} \mathbf{w}_\ell^H \mathbf{a}^i = 0 \\ \mathbf{w}_\ell^H \mathbf{w}_\ell = (1 - \mu_\ell) P_\ell \end{cases}$$
$$\text{for } \ell, i = 1, \ldots, L \text{ and } i \neq \ell. \tag{53}$$

Note that the constraints on $\mathbf{w}_\ell$ are independent of $\mathbf{w}_i$ ($i \neq \ell$), so that we can compute $\mathbf{w}_\ell, \ell = 1, \ldots, L$ separately. To solve (53), we first prove the following.

Proposition 1: $\mathbf{w}_\ell^{\text{opt}}$ is the optimal solution of (53), i.e., it achieves maximal SNR $\gamma_\ell^{\max}$ under the fixed transmit power $(1 - \mu_\ell) P_\ell$, if and only if $\mathbf{w}_\ell^{\text{opt}}$ minimizes transmit power under the fixed SNR $\gamma_\ell^{\max}$.

Proof: We prove Proposition 1 by contradiction. Suppose that weight vector $\mathbf{w}_\ell^{(1)}$ achieves maximal SNR $\gamma_\ell^{\max}$ under the fixed transmit power $(1-\mu_\ell)P_\ell$, and another weight vector $\mathbf{w}_\ell^{(2)} \neq \mathbf{w}_\ell^{(1)}$ achieves minimal transmit power under the fixed SNR $\gamma_\ell^{\max}$. Thus, it holds that $(\mathbf{w}_\ell^{(2)})^H \mathbf{w}_\ell^{(2)} < (1-\mu_\ell)P_\ell$. We can always find a scalar $\rho > 1$ such that under the weights $\rho\mathbf{w}_\ell^{(2)}$ the transmit power is $\rho^2(\mathbf{w}_\ell^{(2)})^H \mathbf{w}_\ell^{(2)} = (1-\mu_\ell)P_\ell$. However, under the weights $\rho\mathbf{w}_\ell^{(2)}$ the SNR is $\rho^2\gamma_\ell^{\max} > \gamma_\ell^{\max}$. In other words, $\mathbf{w}_\ell^{(1)}$ does not achieve the maximal SNR under the fixed transmit power $(1-\mu_\ell)P_\ell$, which contradicts our assumption. Therefore, $\mathbf{w}_\ell^{(1)}$ must be equal to $\mathbf{w}_\ell^{(2)}$. Q.E.D.

From Proposition 1, the optimization problem of (53) is equivalent to finding weights that minimize transmit power under the fixed SNR $\gamma_\ell^{\max}$, i.e.,

$$\arg\min_{\mathbf{w}_\ell} \mathbf{w}_\ell^H \mathbf{w}_\ell$$
$$\text{s.t.} \begin{cases} \mathbf{w}_\ell^H \mathbf{a}^i = 0 \\ \mathbf{w}_\ell^H \mathbf{a}^\ell = ge^{j\varphi} \end{cases}$$
$$\text{for } \ell, i = 1,...,L \text{ and } i \neq \ell \tag{54}$$

Where $g \triangleq \sqrt{\gamma_\ell^{\max}\sigma^2}$ and φ is an arbitrary angle within $[0, 2\pi)$.

Let us define the $L \times N$ matrix $\mathbf{A} = [\mathbf{a}^1,...,\mathbf{a}^L]^H$, and the $L \times 1$ column vector $\mathbf{e}_\ell = [\mathbf{0}_{1\times(\ell-1)}, 1, \mathbf{0}_{1\times(L-\ell)}]^T$. We can formulate the constraints in (54) as

$$\mathbf{A}\mathbf{w}_\ell = (ge^{j\varphi})\mathbf{e}_\ell, \ell = 1,...,L. \tag{55}$$

It is well known that the solution that minimizes the transmit power under (55) is the least-squares solution of (55) based on the pseudo-inverse of $\mathbf{A}$ (Boyd & Vandenberghe, 2004), i.e.,

$$\mathbf{w}_\ell^{\mathrm{opt}} = (ge^{j\varphi})\mathbf{A}^H(\mathbf{A}\mathbf{A}^H)^{-1}\mathbf{e}_\ell, \ell = 1,...,L. \tag{56}$$

From (56), the transmit power $(\mathbf{w}_\ell^{\mathrm{opt}})^H \mathbf{w}_\ell^{\mathrm{opt}}$ is proportional to $\gamma_\ell^{\max}$ and is independent of φ. Therefore, for convenience we can take $g = 1$ and $\varphi = 0$ in(56). The final nulling weights $\mathbf{w}_\ell^{\mathrm{opt}}$ are $\beta_\ell \mathbf{A}^H(\mathbf{A}\mathbf{A}^H)^{-1}\mathbf{e}_\ell$ where β_ℓ is given by

$$\beta_\ell = \sqrt{\frac{(1-\mu_\ell)P_\ell}{\mathbf{e}_\ell^H(\mathbf{A}\mathbf{A}^H)^{-1}\mathbf{e}_\ell}}. \tag{57}$$

Nulling Weights for RA-CB

Recall that cooperating node c_m transmits $\tilde{x}_m = w_m x_m$. Let us define the vectors $\mathbf{w} = [w_1,...,w_N]^H$, $\mathbf{a}_i^\ell = [b_\ell\sqrt{\tilde{P}_i}a_{i1}e^{j\theta_1^\ell},...,b_\ell\sqrt{\tilde{P}_i}a_{iN}e^{j\theta_N^\ell}]^H$ and $\mathbf{u}^\ell = [b_\ell u_1 e^{j\theta_1^\ell},...,b_\ell u_N e^{j\theta_N^\ell}]^H$ for $\ell = 1,...,L$ and $i = 1,...,K$. The received signal at the destination q_ℓ is

$$y_\ell = \mathbf{w}^H \mathbf{a}_\ell^\ell s_\ell + \sum_{\substack{i=1 \\ i \neq \ell}}^{K} \mathbf{w}^H \mathbf{a}_i^\ell s_i + \mathbf{w}^H \mathbf{u}^\ell + v_\ell . \tag{58}$$

To completely null out the interference (i.e., the second term on the RHS of (58)), the following linear constraints need to be satisfied:

$$\mathbf{w}^H \mathbf{a}_i^\ell = 0, i = 1, \dots, K, \ell = 1, \dots, L \text{ and } i \neq \ell . \tag{59}$$

To obtain the weights of RA-CB that maximize the spectral efficiency under the constraint of (59), we follow the strategy similar to that leads to the nulling weights of TDMA-CB. We here consider minimization of transmit power under fixed signal gains $g^2 b_1^2 (1 - \mu_1) P_1, \dots, g^2 b_L^2 (1 - \mu_L) P_L$ at the L destinations, where g is a scalar. It holds that

$$\mathbf{w}^H \mathbf{a}_\ell^\ell = g b_\ell \sqrt{(1 - \mu_\ell) P_\ell}, \text{ for } \ell = 1, \dots, L . \tag{60}$$

Let us define the $KL \times N$ matrix $\mathbf{A} = [\mathbf{a}_1^1, \dots, \mathbf{a}_L^1, \dots, \mathbf{a}_1^L, \dots, \mathbf{a}_L^L]^H$, the $K \times 1$ vector $\mathbf{e}_\ell = [\mathbf{0}_{1 \times \ell - 1}, 1, \mathbf{0}_{1 \times K - \ell}]^T$ and the $KL \times 1$ vector $\mathbf{e} = [b_1 \sqrt{(1 - \mu_1) P_1} \mathbf{e}_1^T, \dots, b_L \sqrt{(1 - \mu_L) P_L} \mathbf{e}_L^T]^T$. We can formulate the above linear constraints in the matrix form:

$$\mathbf{Aw} = g\mathbf{e}. \tag{61}$$

To guarantee a non-zero solution for $\mathbf{w}$, we need $N \geq KL$.

To find the solution to minimize the transmit power, the optimization problem can be formulated as

$$\arg\min_{\mathbf{w}} \mathbf{w}^H \mathbf{Rw}$$
$$\text{s.t. } \mathbf{Aw} = g\mathbf{e}, \tag{62}$$

Where $\mathbf{R}$ is defined in (33). Since $\mathbf{R}$ is a positive-definite matrix, the objective function of (62) is a convex function and has a unique global minimum (Boyd & Vandenberghe, 2004). The optimization problem of (62) can be solved by the method of Lagrange multipliers. By introducing Lagrange multipliers in (62), we obtain

$$2\mathbf{Rw} + \mathbf{A}^H \vartheta = 0 \text{ and } \mathbf{Aw} = g\mathbf{e} , \tag{63}$$

Where ϑ is a $2L \times 1$ column vector containing the Lagrange multipliers. Solving the above equations we obtain the optimal weights

$$\mathbf{w}^{\text{opt}} = g\mathbf{R}^{-1} \mathbf{A}^H (\mathbf{A} \mathbf{R}^{-1} \mathbf{A}^H)^{-1} \mathbf{e} . \tag{64}$$

The transmit power constraint is $\sum_{\ell=1}^{L} (1 - \mu_\ell) P_\ell$, so the scalar g in (64) equals

$$g = \sqrt{\frac{\sum_{\ell=1}^{L}(1-\mu_\ell)P_\ell}{\mathbf{R}^{-1}\mathbf{A}^H(\mathbf{A}\mathbf{R}^{-1}\mathbf{A}^H)^{-1}\mathbf{e}\mathbf{R}\mathbf{e}^H(\mathbf{A}\mathbf{R}^{-1}\mathbf{A}^H)^{-1}\mathbf{A}\mathbf{R}^{-1}}} . \tag{65}$$

DISCUSSION

The computation of nulling weights requires global CSI. Nulling weights completely null out interference at all the destinations; however, this comes at the expense of reducing the received power of the desired signals. The more nulls are formed, the smaller the received power is.

Note that the interference power at a destination is related to the distance or path loss, but the measurement noise power at destinations is a constant independent of the path loss. When the distance between the cluster and a destination is small, the received interference power is in general much larger than that of the measurement noise at the destination. Completely nulling out the interference can significantly increase the overall SINR, though the desired signal power is also reduced. On the other hand, when the distance is large, the interference power is small, and the noise at the destination plays the dominant role in the overall interference-plus-noise level. Nulling out interference does not significantly reduce the interference-plus-noise level, while it still yields reduced signal power. Thus, nulling weights may not be efficient under such scenarios, not to mention the requirement of knowledge of global CSI. Under such scenarios, the co-phasing weights in (49) and (50) may be more efficient due to the simplicity and small overhead (only local CSI is needed).

Figure 5 shows simulation results of the average spectral efficiency of TDMA-CB and RA-CB versus path loss in which both the nulling weights and the co-phasing weights are used. As expected, the nulling weights perform better in small path loss (small distance) regions, while the co-phasing weights achieve comparable/higher spectral efficiency in large path loss (large distance) regions.

Queuing Analysis

In this section, we provide queueing analysis for TDMA-CB and RA-CB based on the concept of *epochs* as defined in (Tsatsanis, Zhang & Banerjee, 2000; Zhang, Sidiropoulos & Tsatsanis, 2002). This analysis provides insight into how packet delays and the stability regions of traffic rate are influenced when each node has a queue.

In this section, source nodes are also referred to as *active* nodes. All data packets have the same length and each packet requires one time slot for transmission. Let J denote the total number of potential active nodes in a cluster. For mathematical tractability, and also to be able to use existing analytical tools, we assume that each of the J nodes generates symmetric Poisson traffic without packet loss. Since in practice packet loss occurs and packet retransmissions are needed, the obtained delay results represent lower bounds.

An epoch includes one or several consecutive slots that are dedicated to the transmission (including both information-sharing and beamforming) of the packets from the active nodes in the beginning of the epoch to their destinations. Idle slots, during which no packets are transmitted, also comprise epochs whose lengths are one slot. From the viewpoint of a particular node, say node i, two kinds of epochs can be distinguished, according to whether or not node i is active in this epoch. If node i sends a new packet during the first slot of an epoch, this epoch is a relevant epoch to node i. Otherwise, this epoch

Figure 5. Simulation results of average spectral efficiency versus path loss. The number of source nodes is $K = 4$. Destinations are at uniformly distributed azimuthal directions within $[0, 2\pi)$. Four multiple beams are formed simultaneously. The radius of the disk containing cluster nodes is $R = 10\lambda$. The number of cooperating nodes is $N = 100$.

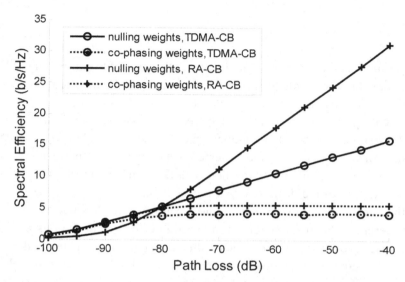

is an irrelevant epoch to node i.

Example 1: Suppose that the n th slot is not in an instantiation of RA-CB, and at slot n only nodes $\{1,2,3,4\}$ are active to transmit new packets. For RA-CB, the slots $n + 1, ..., n + 4$ are used for beamforming (assuming one beam per slot). Thus, the slots $n, ..., n + 4$ comprise an epoch (epoch length: 5 slots). Furthermore, for nodes $\{1,2,3,4\}$, the epoch is a relevant one, whereas for all other nodes it is an irrelevant epoch.

Example 2: Suppose that the n th slot is not in an instantiation of RA-CB, and at slot n no node has an available packet to transmit. Slot n still comprises an epoch (epoch length: 1 slot), which is an irrelevant epoch for all the J nodes.

We can model a node's buffer as an M/G/1 queue with vacation, in which the relevant epoch (length: h_R) and irrelevant epoch (length: h_I) play the roles of the service time and vacation time, respectively. Note that the M/G/1 queue model requires that the service time and vacation time are independent. However, the length of the relevant epoch and of the irrelevant epoch are both related to the traffic load. Thus, the M/G/1 model used here is only an approximation, which is valid under large user population and high transmit power (Tsatsanis, Zhang & Banerjee, 2000).

Let us assume that each node's buffer is fed with a Poisson source having rate λ_p. The average system delay (including the waiting time in the buffer and the transmission time in the channel) for a packet can be expressed as (Tsatsanis, Zhang & Banerjee, 2000)

$$D = \overline{h_R} + \frac{\lambda_p \overline{h_R^2}}{2(1 - \lambda_p \overline{h_R})} + \frac{\overline{h_I^2}}{2\overline{h_I}} \tag{66}$$

Where $\overline{h_R}$ and $\overline{h_R^2}$ are first and second moments of h_R, and $\overline{h_I}$ and $\overline{h_I^2}$ are first and second moments of h_I.

Let the probability that a node's buffer is empty be denoted by p_e. At the beginning of an epoch, the probability that K nodes are active is

$$p(K = k) = \binom{J}{K}(1 - p_e)^k p_e^{J-k}, 0 \le k \le J \tag{67}$$

Given λ_p, p_e can be determined by the following equation (Tsatsanis, Zhang & Banerjee, 2000):

$$1 = p_e \frac{1 + \lambda_p \overline{h_I} - \lambda_p \overline{h_R}}{1 - \lambda_p \overline{h_R}}. \tag{68}$$

In the following, two representative cases are considered for both TDMA-CB and RA-CB: (i) one beam is formed per slot (single destination); (ii) All beams are formed per slot (multiple destinations).

One Beam per Slot

TDMA-CB

For TDMA-CB, $2k$ slots are needed for transmissions of k simultaneous packets (k slots for information-sharing using TDMA, and another k slots for beamforming). For the relevant epoch of node i, the probability that $h_R = 2k$ equals

$$p(h_R = 2k) = p(k \text{ active nodes} \mid \text{node } i \text{ is active})$$
$$= \binom{J-1}{k-1}(1 - p_e)^{k-1} p_e^{J-k}, 1 \le k \le J \tag{69}$$

For the irrelevant epoch of node i, it holds that

$$p(h_I = 1) = p(\text{no active node} \mid \text{node } i \text{ is not active}) = p_e^{J-1}, \tag{70}$$

and

$$p(h_I = 2k) = p(k \text{ active nodes} \mid \text{node } i \text{ is not active})$$
$$= \binom{J-1}{k}(1 - p_e)^k p_e^{J-k-1}, 1 \le k \le J - 1. \tag{71}$$

Based on (69)-(71), we obtain

$$\overline{h_R} = \sum_{k=1}^{J} 2k \binom{J-1}{k-1} (1-p_e)^{k-1} p_e^{J-k}$$
$$= 2 \left[1 + \sum_{k=0}^{J-1} k \binom{J-1}{k} (1-p_e)^k p_e^{J-1-k} \right]$$
$$= 2 + 2(J-1)(1-p_e) \tag{72}$$

and

$$\overline{h_I} = p_e^{J-1} + \sum_{k=1}^{J-1} 2k \binom{J-1}{k} (1-p_e)^k p_e^{J-k-1}$$
$$= p_e^{J-1} + 2(J-1)(1-p_e) . \tag{73}$$

By substituting (72) and (73) into (68), we see that p_e is the unique solution in $[0,1]$ of the equation

$$\lambda_p p_e^J + (1 - 2\lambda_p J)p_e + 2\lambda_p J - 1 = 0 . \tag{74}$$

The maximal traffic flow is achieved by setting $p_e = 0$, i.e., $\lambda_p^{\max} = 1/(2J)$. $\lambda_p > 1/(2J)$ yields an unstable system in which the packet delay is unbounded. Finally, (66) can be used to obtain the average system delay.

RA-CB

$k+1$ slots are needed for transmissions of k simultaneous packets (one slot for random access and k slots for beamforming). For the relevant and irrelevant epochs it holds that

$$p(h_R = k) = \binom{J-1}{k-2} (1-p_e)^{k-2} p_e^{J-k+1}, 2 \le k \le J+1 \tag{75}$$

and

$$p(h_I = k) = \binom{J-1}{k-1} (1-p_e)^{k-1} p_e^{J-k}, 1 \le k \le J . \tag{76}$$

Then, we can obtain

$$\overline{h_R} = 2 + (J-1)(1-p_e) \tag{77}$$

and

$$\overline{h_I} = 1 + (J-1)(1-p_e).\tag{78}$$

Based on (68), it is easy to show that p_e is given simply by

$$p_e = 1 - \frac{\lambda_p}{1 - \lambda_p J}.\tag{79}$$

When $p_e = 0$, the maximal traffic flow is achieved, i.e. $\lambda_p^{\max} = 1/(J+1)$.

Figure 7 shows the analytical results for the two schemes when $J = 32$. One can see that the delay of the RA-CB is significantly smaller than the delay of TDMA-CB for high traffic load, while only slightly smaller than TDMA-CB for low traffic load. The reason is that for low traffic load the number of source nodes K is small, and the information-sharing time of TDMA-CB (i.e., K slots) is not significantly greater than that of RA-CB (i.e., 1 slot). For example, when the number of source nodes is $K = 1$, the packet delays of TDMA-CB and RA-CB are exactly the same. In addition, delay results are shown for only the stable traffic regions. It can be seen that the queue of TDMA-CB becomes unstable when the traffic load is greater than about 0.5. The stable region of traffic for RA-CB is much wider as compared to that of TDMA-CB.

All Beams in One Slot

TDMA-CB

$k + 1$ slots are needed for transmissions of k simultaneous packets. Thus, the system delay is the same as that of RA-CB when one beam is formed per slot.

RA-CB

2 slots are needed for transmissions of k simultaneous packets. We have the probabilities:

$$p(h_R = 2) = 1,\tag{80}$$

$$p(h_I = 1) = p_e^{J-1},\tag{81}$$

and

$$p(h_I = 2) = 1 - p_e^{J-1}.\tag{82}$$

It can be shown that

$$\lambda_p p_e^J - p_e - 2\lambda_p + 1 = 0.\tag{83}$$

Then, steps similar to those above can be followed to obtain the average system delay.

Figure 6. Average delay under symmetric Poisson traffic when one beam per slot is formed (analytical results, $J = 32$).

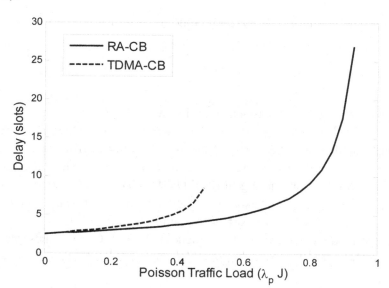

Figure 7 shows the analytical results for the two schemes ($J = 32$). One can see that the delay of RA-CB is much smaller than that of TDMA-CB.

CONCLUSIONS AND FUTURE RESEARCH DIRECTIONS

We have provided a cross-layer framework for cooperative beamforming, which brings the idea of cooperative beamforming closer to practice. Depending on how the message signal is shared, two cross-layer CB approaches, namely, TDMA-CB and RA-CB, have been presented. As compared to TDMA-CB, RA-CB reduces the information-sharing time at the cost of lower received SINR. We have presented weights that allow desired signals to add coherently, and weights that maximize the receive SNR/SINR. Performance comparison of different schemes has been conducted based on spectral efficiency. Further, we have extended weight design to multi-beam beamforming, for cases in which multiple network nodes transmit to distinct destinations. For such cases, we have presented two suboptimal weight design schemes. A queueing analysis for TDMA-CB and RA-CB for the case of Poisson traffic was also provided. As compared to TDMA-CB, RA-CB significantly reduces packet delays and extends the stability region of traffic rate.

There are several issues that need to be studied further in order to optimize the performance of the proposed approaches. For example, selection of cooperating nodes can be optimized based on various metrics, such as available power, location, and channel conditions with respect to the destination. Selection of which sources should transmit at a given time is also important in controlling queue lengths and end-to-end packet delays. Scheduling of sources for beamforming is another issue that can be optimized. Furthermore, the effects of incomplete CSI, lack of absolute time synchronization and the presence of carrier frequency offsets, need to be studied before the proposed approach can be of practical interest.

Figure 7. Average delay under symmetric Poisson traffic when all the K beams per slot are formed (analytical results, $J = 32$).

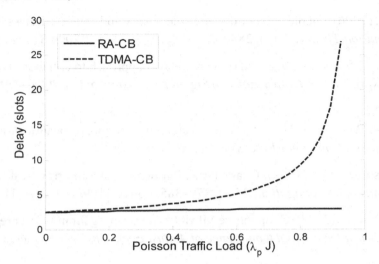

Further cross-layer CB algorithms that use different information sharing schemes in stage 1 can also be considered. TDMA-CB was used here as a representative fixed bandwidth allocation scheme via which the information is shared in an orthogonal multiple access fashion. Other orthogonal multiple access schemes, such as frequency division multiple access (FDMA), and code division multiple access (CDMA) could also be considered.

ACKNOWLEDGMENT

This chapter was prepared under the support of the National Science Foundation under Grants ANI-03-38807, CNS-06-25637 and CNS-04-35052, and under the support of the Office of Naval Research under Grant ONR-N-00014-07-1-0500.

REFERENCES

Ahmed, M., & Vorobyov, S. (2008). Performance characteristics of collaborative beamforming for wireless sensor networks with Gaussian distributed sensor nodes. In *Proceedings of 2008 IEEE International Conference of Acoustic Speech Signal Processing*, Las Vegas, NV (pp. 3249-3252).

Boyd, S., & Vandenberghe, L. (2004). *Convex optimization*. Cambridge, UK: Cambridge University Press.

Brown, D. R. III, & Poor, H. V. (2008). Time-slotted round-trip carrier synchronization for distributed beamforming. *IEEE Transactions on Signal Processing, 56*(11), 5630–5643. doi:10.1109/TSP.2008.927073

Compton, R. T. (1988). *Adaptive antennas-concepts and performance*. Englewood Cliffs, NJ: Prentice Hall.

Dana, A., & Hassibi, B. (2006). On the power efficiency of sensory and ad hoc wireless networks. *IEEE Transactions on Information Theory, 52*(7), 2890–2914. doi:10.1109/TIT.2006.876245

Dong, L., Petropulu, A. P., & Poor, H. V. (2008a). A cross-layer approach to collaborative beamforming for wireless ad hoc networks. *IEEE Transactions on Signal Processing, 56*(7), 2981–2993. doi:10.1109/TSP.2008.917352

Dong, L., Petropulu, A. P., & Poor, H. V. (2008b). Weighted cooperative beamforming for wireless networks. *IEEE Transactions on Signal Processing*, in revision.

Donvito, M. B., & Kassam, S. A. (1979). Characterization of the random array peak sidelobe. *IEEE Transactions on Antennas and Propagation, 27*(3), 379–385. doi:10.1109/TAP.1979.1142097

El-Keyi, A., & Champagne, B. (2008). Cooperative MIMO-beamforming for multiuser relay network. In *Proceedings of 2008 IEEE International Conference of Acoustic Speech Signal Processing*, Las Vegas, NV (pp. 2749-2752).

Godara, L. C. (1997). Application of antenna arrays to mobile communications, II: Beamforming and direction-of-arrival considerations. *Proceedings of the IEEE, 85*, 1195–1245. doi:10.1109/5.622504

Golub, G., & Loan, C. V. (1996). *Matrix computations, 3rd ed*. Baltimore, MD: The Johns Hopkins University Press.

Krim, H., & Viberg, M. (1996). Two decades of array signal processing research. *IEEE Signal Processing Magazine, 13*(4), 67–94. doi:10.1109/79.526899

Li, C., & Wang, X. (2008). Cooperative multibeamforming in ad hoc networks. *EURASIP Journal on Advances in Signal Processing*, Article ID 310247.

Litva, J., & Lo, T. K. Y. (1996). *Digital beamforming in wireless communications*. Boston, MA: Artech House.

Lo, Y. T. (1964). A mathematical theory of antenna arrays with randomly spaced elements. *IEEE Transactions on Antennas and Propagation, 12*(3), 257–268. doi:10.1109/TAP.1964.1138220

Molisch, A. F. (2005). *Wireless communications*. New York: Wiley-IEEE Press.

Mudumbai, R., Barriac, G., & Madhow, U. (2007). On the feasibility of distributed beamforming in wireless networks. *IEEE Transactions on Wireless Communications, 6*(5), 1754–1763. doi:10.1109/TWC.2007.360377

Mudumbai, R., Wild, B., Madhow, U., & Ramchandran, K. (2006). Distributed beamforming using 1 bit feedback: From concept to realization. In *Proceedings of 44th Allerton Conference on Communication, Control and Computing*, Monticello, IL.

Ochiai, H., Mitran, P., Poor, H. V., & Tarokh, V. (2005). Collaborative beamforming for distributed wireless ad hoc sensor networks. *IEEE Transactions on Signal Processing, 53*(11), 4110–4124. doi:10.1109/TSP.2005.857028

Sadek, M., Tarighat, A., & Sayed, A. H. (2007). A leakage-based precoding scheme for downlink multi-user MIMO channels. *IEEE Transactions on Wireless Communications, 6*(5), 1711–1721. doi:10.1109/TWC.2007.360373

Steinberg, B. D. (1972). The peak sidelobe of the phased array having randomly located elements. *IEEE Transactions on Antennas and Propagation, 20*(2), 129–136. doi:10.1109/TAP.1972.1140162

Tsatsanis, M. K., Zhang, R., & Banerjee, S. (2000). Network-assisted diversity for random access wireless networks. *IEEE Transactions on Signal Processing, 48*(3), 702–711. doi:10.1109/78.824666

Tse, D., & Viswanath, P. (2005). *Fundamentals of wireless communication*. Cambridge, UK: Cambridge University Press.

Zhang, H., & Dai, H. (2004). Cochannel interference mitigation and cooperative processing in downlink multicell multiuser MIMO networks. *EURASIP Journal on Wireless Communications and Networking, 2004*(2), 222-235.

Zhang, R., Sidiropoulos, N. D., & Tsatsanis, M. K. (2002). Collision resolution in packet radio networks using rotational invariance techniques. *IEEE Transactions on Communications, 59*(1), 146–155. doi:10.1109/26.975780

Chapter 8
Distributed Space–Time Block Coding for Amplify–and–Forward Cooperative Networks

Zhihang Yi
Queen's University, Canada

Il-Min Kim
Queen's University, Canada

ABSTRACT

This chapter focuses on distributed space-time codes (DSTCs) in cooperative networks. DSTCs can substantially improve the bandwidth efficiency of the cooperative network without requiring any feedback overheads from the destination to the relays. The first part of this chapter is dedicated to reviewing the existing works on DSTCs. In the second part of this chapter, distributed orthogonal space-time block codes (DOSTBCs) are presented in detail. It is shown that the DOSTBCs can achieve the single-symbol maximum likelihood decodability and full diversity order in an amplify-and-forward cooperative network. Then some special DOSTBCs, which generate uncorrelated noises at the destination, are introduced. Those codes are referred to as the row-monomial DOSTBCs. An upper bound of the data-rate of the row-monomial DOSTBC is derived and the codes achieving the upper bound are presented as well.

INTRODUCTION

A robust way to combat multipath fading phenomena in wireless communications is to implement diversity techniques. Diversity is achieved by providing the receivers multiple independent replicas of the same information-bearing symbols through space, time, or frequency. Recently, a new type of diversity, so-called cooperative diversity, has been proposed in order to provide diversity when the wireless terminals can not employ multiple antennas due to size or complexity constraints. The network that exploits the cooperative diversity is usually referred to as the cooperative network.

DOI: 10.4018/978-1-60566-665-5.ch008

Cooperative diversity is achieved by the cooperation of several single-antenna relays. Consequently, the coordination of the transmissions from the relays has a large impact on the performance and complexity of the network. The most well-known cooperative strategy is the repetition-based cooperative strategy, because it has very low complexity and achieves the full diversity order. However, researchers conclude that this strategy has very poor bandwidth efficiency. In order to improve the bandwidth efficiency of the cooperative networks, many other cooperative strategies have been proposed, including cooperative beamforming, relay selection scheme, and distributed space-time codes (DSTCs). Although the cooperative beamforming and relay selection scheme can improve the bandwidth efficiency, they are hard to implement because feedback overheads might be required from the destination to the relays.

This chapter will focus on the DSTCs, which can substantially improve the bandwidth efficiency without requiring any feedback overheads from the destination to the relays. The objective of this chapter is to provide the readers some insights into existing works of the DSTCs. Specifically, in the second Section of this chapter, we will review some of the most important works on DSTCs. It will be shown that, in the decode-and-forward (DF) cooperative networks, existing space-time codes (STCs), such as orthogonal STCs, quasi-orthogonal STCs and linear dispersion codes, can be used without any modification. This is because the use of cyclic redundancy check (CRC) at the relays makes the transmission from the relays to the destination essentially the same as that of a classic co-located multiple antenna system. On the other hand, the analysis and construction of the DSTCs in the amplify-and-forward (AF) cooperative networks is much harder. For example, the orthogonal space-time codes lose the single-symbol maximum likelihood (ML) decodability when they are used in the AF cooperative networks.

In the third Section of this chapter, we will discuss distributed orthogonal space-time block codes (DOSTBCs) in detail. The DOSTBCs are proposed by addressing the distributed nature of the cooperative networks. Due to this reason, those codes achieve the single-symbol ML decodability and the full diversity order in the AF cooperative networks. Then we will restrict our interests to some special DOSTBCs, which generate uncorrelated noises at the destination. Those codes are referred to as the row-monomial DOSTBCs. An upper bound of the data-rate of the row-monomial DOSTBC is derived and it shows that the codes achieve approximate twice higher bandwidth efficiency than the repetition-based cooperative strategy. Furthermore, we will also find the systematic construction methods that generate the row-monomial DOSTBCs achieving the upper bound. Numerical results demonstrate that the row-monomial DOSTBCs substantially outperform the repetition-based cooperative strategy when they have the same bandwidth efficiency.

BACKGROUND

In this section, we will first review some fundamental works on cooperative networks. After that, we will present the existing works on DSTCs.

Cooperative Networks

Multipath fading is one of the major impairments of wireless channels and it is seen as an obstacle to reliable data transmission. One effective method to mitigate fading is to implement *diversity* techniques in wireless communication systems. The fundamental idea of the diversity techniques is to transmit multiple independent replicas of the same information-bearing symbols to the receiver. By doing so, the probability that all the replicas suffer severe fading simultaneously is reduced considerably. There are several ways in which we can exploit the diversity. For example, we can employ *frequency diversity* by transmitting the same symbols on multiple carriers (Kaleh, 1996) or employ *time diversity* by transmitting the same symbols at multiple time slots (Dallal & Shamai, 1992). More attentions have been given to *spatial diversity* (Godara, 1997; Godara, 1997; Tarokh, Seshadri, & Calderbank, 1998; Winters, 1998). The spatial diversity is achieved by transmitting the same information-bearing symbols through multiple antennas, which can be deployed at the transmitter, or the receiver, or both ends. The antennas need to be spaced sufficiently far apart. Usually, a separation of a few wavelengths is required between two antenna elements in order to ensure that the symbols fade independently (Proakis, 2001). However, in some scenarios, such as in cellular, ad hoc, and sensor networks, there are strict size and complexity limitations on the wireless terminals, and hence, it may not be practical to implement multiple antennas and space them sufficiently far apart.

In order to cope with this difficulty, *cooperative diversity* has been proposed recently (Sendonaris, Erkip, & Aazhang, 2003; Sendonaris, Erkip, & Aazhang, 2003). The cooperative diversity exploits the broadcast nature and the inherent spatial diversity of wireless channels. The fundamental idea of the cooperative diversity is that several single-antenna terminals cooperate and achieve the spatial diversity by forming a distributed multi-antenna system. Specifically, a source, several relays, and a destination constitute a cooperative network, where the relays help the source transmit the information-bearing symbols to the destination. The same information-bearing symbols are transmitted over several independent channels, including the direct channel from the source to the destination and the indirect channels from the source through the relays to the destination. As a result, the system can exploit the spatial diversity and achieve better performance. The wireless networks exploiting the cooperative diversity are usually called the *cooperative networks*.

In order to coordinate the transmissions from different relays, Laneman *et al.* proposed *repetition-based cooperative strategy* (Laneman, Tse, & Wornell, 2004). In this strategy, the source broadcasted one information-bearing symbol at the first time slot. Then every relay was assigned one time slot to transmit this symbol to the destination. Furthermore, depending on how the relays processed the signals received from the source, two cooperative protocols, the AF protocol and the DF protocol, were proposed in (Laneman, Tse, & Wornell, 2004). The authors analyzed the outage probabilities of both protocols in a cooperative network using the repetition-based strategy (Laneman, Tse, & Wornell, 2004). The results showed that the repetition-based strategy could achieve the full diversity order in the number of cooperating terminals, i.e. $K + 1$ diversity order could be achieved if there were one source and K relays.

Due to its simplicity and full diversity order, the repetition-based cooperative strategy was used and studied in many publications (Anghel & Kaveh, 2004; Chen & Laneman, 2006; Hasna & Alouini, 2003; Ribeiro, Cai, & Giannakis, 2005; Wang & Giannakis, 2003). For the AF protocol, the ML detection scheme at the destination was proposed in (Laneman & Wornell, 2000). Later on, a maximum ratio combining (MRC) detection scheme was developed in (Anghel & Kaveh, 2004; Hasna & Alouini, 2003) and it achieved the same performance as the ML detection scheme. Furthermore, by using the moment generating function method, the authors derived the average symbol error rate (SER) of the repetition-based cooperative strategy in a Rayleigh fading environment (Anghel & Kaveh, 2004; Hasna & Alouini, 2003). For more general cooperative networks with an arbitrary number of cooperating branches and an arbitrary number of hops per branch, an asymptotic expression of the average SER was derived in (Ribeiro, Cai, & Giannakis, 2005) based on the technique developed in (Wang & Giannakis, 2003).

On the other hand, for the DF protocol, the ML detection scheme was proposed in (Sendonaris, Erkip, & Aazhang, 2003). Although it achieved the optimum performance, its complexity was very high, and hence, it was very hard to implement. In order to reduce the complexity, many sub-optimum detection schemes were developed, including λ-MRC (Sendonaris, Erkip, & Aazhang, 2003; Sendonaris, Erkip, & Aazhang, 2003), piecewise-linear detector (Chen & Laneman, 2006), cooperative MRC (C-MRC) (Wang, Cano, Giannakis, & Laneman, 2007), simple adaptive DF scheme (Herhold, Zimmermann, & Fettweis, 2004), and link-adaptive regeneration (Wang, Wang, & Giannakis, 2006). Every detection scheme has its own advantages and disadvantages. For example, the piecewise-linear detector can be used for both coherent and non-coherent cooperative networks; but it does not achieve the full diversity order. The λ-MRC achieves the full diversity order; but the parameter λ can not be specified analytically. The link-adaptive regeneration requires feedback overheads from the destination to the relays. Among all the sub-optimum detection schemes, the C-MRC and the simple adaptive DF scheme are the most preferable ones. It has been shown that the C-MRC could achieve almost the same performance as the ML detection scheme at a much lower complexity. However, the C-MRC requires a large amount of signaling overheads, which may be hard to fulfill in a fast fading environment. The simple adaptive DF scheme, on the other hand, requires much less signaling overheads; but its performance is slightly worse than the C-MRC.

All the works in (Anghel & Kaveh, 2004; Chen & Laneman, 2006; Hasna & Alouini, 2003; Herhold, Zimmermann, & Fettweis, 2004; Laneman & Wornell, 2000; Ribeiro, Cai, & Giannakis, 2005; Wang, Cano, Giannakis, & Laneman, 2007; Wang & Giannakis, 2003; Wang, Wang, & Giannakis, 2006) were based on the repetition-based cooperative strategy. Since only one relay is allowed to transmit the signals at each time slot, however, the repetition-based cooperative network has very poor bandwidth efficiency. Recently, many works are devoted to improving the bandwidth efficiency of the cooperative networks, including DSTCs (Laneman & Wornell, 2003; Nabar, Bölcskei, & Kneubühler, 2004; Yang & Belfiore, 2007; Jing & Hassibi, 2006; Yiu, Schober, & Lampe, 2006), cooperative beamforming (Hammerström, Kuhn, & Wittneben, 2004; Yi & Kim, 2007), and relay selection scheme (Zhao, Adve, & Lim, 2006; Bletsas, Khisti, Reed, & Lippman, 2006; Yi & Kim, 2008). All those schemes let more than one relay transmit the signals simultaneously to the destination, and hence, the bandwidth efficiency is improved.

The cooperative beamforming is to assign a proper weight coefficient to every relay in order to make the signals from different relays add up constructively at the destination. Some channel state information (CSI) of the channels from the relays to the destination is needed at the relays in order to compute the weight coefficients. In fact, the cooperative beamforming can achieve the full diversity order only when full CSI is at the relays (Hammerström, Kuhn, & Wittneben, 2004; Yi & Kim, 2007). This is hard to achieve even in a slow fading environment. Recently, many works have analyzed the cooperative beamforming when imperfect or partial CSI is known at the relays (Koyuncu, Jing, & Jafarkhani, 2008; Jing & Jafarkhani, 2007; Jing & Jafarkhani, 2008). On the other hand, the relay selection scheme always selects the relay with the best channel condition to transmit the signals (Zhao, Adve, & Lim, 2006; Bletsas, Khisti, Reed, & Lippman, 2006; Yi & Kim, 2008; Jing & Jafarkhani, 2008). The relay selection is accomplished at the destination, because the destination is usually assumed to have full CSI. Thus, the selection result must be fed back from the destination to the relays, which requires a small amount of feedback overheads.

Distributed Space-Time Codes

Unlike the cooperative beamforming and the relay selection scheme, the DSTCs do not need any feedback overheads, and hence, the DSTCs are more attractive especially in a fast fading environment. Furthermore, the DSTCs can considerably improve the bandwidth efficiency without losing any diversity order. In (Laneman & Wornell, 2003; Nabar, Bölcskei, & Kneubühler, 2004; Yang & Belfiore, 2007), for example, the authors proved that the DSTCs had higher bandwidth efficiency than the repetition-based cooperative strategy from the information theory aspect. After that, many works have been devoted to the design of practical DSTCs.

In the context of the DF protocol, it is usually assumed that CRC is performed at the relays (Yiu, Schober, & Lampe, 2006). The relays that correctly decode the information-bearing symbols form a decoding set and only the relays in this set can participate in the transmissions to the destination. Performing CRC can be readily accomplished by current wireless communication systems; but the extra hardware complexity of implementing CRC should not be neglected. Moreover, the relays need to inform the destination about which relays are in the decoding set and this certainly requires some extra signaling overhead. Since the relays in the decoding set have perfect copies of the information-bearing symbols, the transmission from the relays to the destination can be seen as that of a classic co-located multiple antenna system, and hence, existing STCs can be used without any modifications. When there are K relays, for example, one can choose an orthogonal space-time code $\mathbf{X}$ with K rows and assign each row to a specific relay. Once the relay can correctly decode the information-bearing symbols, it transmits those symbols to the destination using the pre-assigned row. This ensures that the decoding at the destination is still single-symbol ML decodable and full diversity order is maintained as well. Other STCs, such as quasi-orthogonal STCs and linear dispersion codes, can be used in a DF cooperative network as well in a similar manner.

Recently, many researchers have considered the construction of DSTCs by addressing the distributed nature of the cooperative networks. In (Sirkeci-Mergen & Scaglione, 2007; Savazzi & Spagnolini, 2007), for example, the authors developed randomized DSTCs, where each relay decided the code in a random and independent fashion. The randomized DSTCs had the advantage that they could be used in a cooperative network where no centralized control terminal existed and the number of relays in the network was random. Furthermore, in (Shang & Xia, 2006; Li & Xia, 2007; Damen & Hammons, 2007), the authors considered asynchronous cooperative networks. The proposed DSTCs could achieve the full diversity order even when the transmissions from the relays to the destination were not symbol aligned.

On the other hand, the AF protocol has a lower complexity than the DF protocol, since the relays working in the AF mode only needs to amplify the received signals from the source instead of decoding them. Due to the same reason, however, the relays can not perform CRC when they are working in the AF mode. This makes the analysis and construction of the DSTCs in the AF cooperative networks much harder than in the DF cooperative networks. In (Jing & Hassibi, 2006), Jing *et al.* applied the linear dispersion codes in the AF cooperative networks and studied the diversity order by deriving the average pairwise error probability (PEP). It was shown that the diversity order of the codes was $K(1 - \log\log E / \log E)$, where K was the number of relays and E was the transmission power. The authors noticed that the construction of the linear dispersion codes in the AF cooperative networks had different criteria from that of the STCs in the classic co-located multiple antenna systems. Motivated by this, many researchers have tried to design the optimum linear dispersion codes for the AF cooperative networks by minimizing the average PEP or outage probability (Ding, Zhang, & Wong, 2007; Liang, Wang, & Berenguer, 2007).

Recently, Mheidat *et al.* considered the non-coherent detection of the DSTCs in the AF cooperative networks in (Mheidat & Uysal, 2007). Some other researchers have investigated the construction of DSTCs in asynchronous AF cooperative networks (Guo and Xia, 2008; Li and Xia, 2007; Rajan and Rajan, 2008). They have provided some codes which could achieve full diversity order even in asynchronous AF cooperative networks. Moreover, the diversity-multiplexing tradeoff of the DSTCs in AF cooperative networks has been studied in (Elia, Oggier, & Kumar, 2007; Rajan & Rajan, 2006). In particular, Elia *et al.* have constructed a cyclic-division-algebra-based DSTC which was optimum in terms of the diversity-multiplexing tradeoff (Elia, Oggier, & Kumar, 2007).

Although extensive works have been accomplished to design DSTCs for the AF cooperative networks, very little of them address the problem of finding orthogonal DSTCs that are single-symbol ML decodable at the destination. In (Hua, Mei, & Cheng, 2003), Hua *et al.* applied the generalized orthogonal designs in the AF cooperative networks; but the codes were not single-symbol ML decodable at the destination anymore. The same results have been shown in (Jing & Jafarkhani, 2007). In (Kiran & Rajan, 2006; Rajan & Rajan, 2007; Rajan, Tandon, & Rajan, 2007; Maham & Hjørungnes, 2007), the authors have addressed the constructions of DSTCs with low decoding complexity at the destination. Moreover, in (Rajan & Rajan, 2007; Sreedhar, Chockalingam, & Rajan, 2008; Jagadeesh & Rajan, 2008), the authors constructed the DSTCs based on the generalized coordinate interleaved orthogonal designs. The proposed codes were single-symbol ML decodable and they could achieve the full diversity order in a pre-designed constellation. Very recently, Yi *et al.* proposed the DOSTBCs (Yi & Kim, 2007) for the AF cooperative networks. Those codes could achieve the single-symbol ML decodability and the full diversity order in an arbitrary constellation. In the next section, we will discuss the DOSTBCs in detail.

SINGLE-SYMBOL ML DECODABLE DISTRIBUTED SPACE-TIME BLOCK CODES

This section is dedicated to the DOSTBCs. We describe the cooperative network we study at first. Then we present the definition of the DOSTBCs and show that the codes achieve the single-symbol ML decodability and full diversity. After that, we restrict our interests to a subset of the DOSTBCs, which generate uncorrelated noises at the destination. We refer to the codes in this subset as the row-monomial DOSTBCs. An upper bound of the data-rate of the row-monomial DOSTBC is derived and the codes achieving this upper bound are also presented.

System Model

We consider a cooperative network with one source, K relays, and one destination. Every terminal is equipped with only one antenna and is constrained to be half-duplex. That is, a terminal can not receive and transmit signals at the same time. The channel from the source to the k-th relay and the channel from the k-th relay to the destination is denoted by h_k and f_k, respectively. The channel coefficients h_k and f_k are assumed to be spatially uncorrelated complex Gaussian random variables with zero mean and unit variance. We assume that the destination knows the instantaneous values of the channel coefficients h_k and f_k by using training sequences; but the source and relays do not have any knowledge of the instantaneous channel coefficients.

At the first N consecutive time slots, the source transmits N complex-valued symbols.[1] We let $\mathbf{s} = [s_1, \cdots, s_N]$ denote the symbol vector transmitted from the source and assume that the power of s_n is E_s. The coherence time of h_k is assumed to be larger than N. Thus, the received signal vector $\mathbf{y}_k$ at the k-th relay is given by

$$\mathbf{y}_k = h_k \mathbf{s} + \mathbf{n}_k, \tag{1}$$

where $\mathbf{n}_k = [n_{k,1}, \cdots, n_{k,N}]$ is the additive noise at the k-th relay and it is an uncorrelated complex Gaussian random vector with zero mean and identity covariance matrix. All the relays are working in the AF mode and the amplifying coefficient ρ is set to be

$$\rho = \sqrt{\frac{E_r}{1 + E_s}}, \tag{2}$$

where E_r denotes the transmission power per channel use at every relay. The amplifying coefficient ρ ensures that the average transmission power at every relays is E_r in a long term. In order to construct a distributed space-time code, the received signal vector $\mathbf{y}_k$ and its conjugate $\mathbf{y}_k^*$ are multiplied with $\mathbf{A}_k$ and $\mathbf{B}_k$, respectively.[2] The dimension of $\mathbf{A}_k$ and $\mathbf{B}_k$ is $N \times T$. As a result, the transmitted signal vector $\mathbf{x}_k$ from the k-th relay is given by

$$\begin{aligned}
\mathbf{x}_k &= \rho(\mathbf{y}_k \mathbf{A}_k + \mathbf{y}_k^* \mathbf{B}_k) \\
&= \rho h_k \mathbf{s} \mathbf{A}_k + \rho h_k^* \mathbf{s}^* \mathbf{B}_k + \rho \mathbf{n}_k \mathbf{A}_k + \rho \mathbf{n}_k^* \mathbf{B}_k.
\end{aligned} \tag{3}$$

The coherence time of f_k is assumed to be larger than T, and hence, the received signal vector $\mathbf{y}_D$ at the destination is given by

$$\begin{aligned}
\mathbf{y}_D &= \sum_{k=1}^{K} f_k \mathbf{x}_k + \mathbf{n}_D \\
&= \sum_{k=1}^{K} (\rho f_k h_k \mathbf{s} \mathbf{A}_k + \rho f_k h_k^* \mathbf{s}^* \mathbf{B}_k) + \sum_{k=1}^{K} (\rho f_k \mathbf{n}_k \mathbf{A}_k + \rho f_k \mathbf{n}_k^* \mathbf{B}_k) + \mathbf{n}_D,
\end{aligned} \tag{4}$$

where $\mathbf{n}_D = [n_{D,1}, \cdots, n_{D,T}]$ is the additive noise at the destination, and it is an uncorrelated complex Gaussian random vector with zero mean and identity covariance matrix. In order to write (4) in a compact form, we define $\mathbf{w}$, $\mathbf{X}$, and $\mathbf{n}$ as follows:

$$\mathbf{w} = [\rho f_1, \cdots, \rho f_K] \tag{5}$$

$$\mathbf{X} = [h_1 \mathbf{s} \mathbf{A}_1 + h_1^* \mathbf{s}^* \mathbf{B}_1; \cdots; h_K \mathbf{s} \mathbf{A}_K + h_K^* \mathbf{s}^* \mathbf{B}_K] \tag{6}$$

$$\mathbf{n} = \sum_{k=1}^{K} (\rho f_k \mathbf{n}_k \mathbf{A}_k + \rho f_k \mathbf{n}_k^* \mathbf{B}_k) + \mathbf{n}_D; \tag{7}$$

then (4) can be rewritten in the following way

$$\mathbf{y}_D = \mathbf{wX} + \mathbf{n}. \tag{8}$$

Note that the matrix $\mathbf{X}$ contains N information-bearing symbols, $s_1, \cdots, s_N$, and it lasts for T time slots. Thus, the data-rate of $\mathbf{X}$ equals to N / T.[3] Furthermore, it follows from (7) that mean of $\mathbf{n}$ is zero and the covariance matrix $\mathbf{R} = E\left\{\mathbf{n}^H \mathbf{n}\right\}$ of $\mathbf{n}$ is given by

$$\mathbf{R} = \sum_{k=1}^{K} \left[\mid \rho f_k \mid^2 \left(\mathbf{A}_k^H \mathbf{A}_k + \mathbf{B}_k^H \mathbf{B}_k \right) \right] + \mathbf{I}_{T \times T}. \tag{9}$$

Distributed Orthogonal Space-Time Block Codes

In this subsection, we will first define the DOSTBCs. Then, in order to evaluate the diversity order of the DOSTBCs, some fundamental properties of $\mathbf{A}_k$ and $\mathbf{B}_k$ are presented. Lastly, we show that the DOSTBCs can achieve the full diversity order.

It follows from (8) that the ML estimate $\hat{\mathbf{s}}$ of $\mathbf{s}$ is given by

$$\begin{aligned}
\hat{\mathbf{s}} &= \arg \min_{\mathbf{s} \in C} (\mathbf{y}_D - \mathbf{wX}) \mathbf{R}^{-1} (\mathbf{y}_D - \mathbf{wX})^H \\
&= \arg \min_{\mathbf{s} \in C} \left[\parallel \mathbf{y}_D \parallel^2 - 2\Re(\mathbf{wXy}_D^H) + \parallel \mathbf{wX} \parallel^2 \right]
\end{aligned} \tag{10}$$

$$= \arg \min_{\mathbf{s} \in C} \left[-2\Re\left(\mathbf{wXR}^{-1} \mathbf{y}_D^H \right) + \mathbf{wXR}^{-1} \mathbf{X}^H \mathbf{w}^H \right], \tag{11}$$

where C is the set containing all the possible symbol vector $\mathbf{s}$. Note that the ML metric (10) of a DSTC is different with that of a space-time code for co-located multiple antenna systems. In the latter case, the noise variance matrix is usually a scaled identity matrix, and hence, it can be neglected in the ML metric. For a DSTC, however, the noise variance matrix is not a scaled identity matrix for most cases as shown in (9). In fact, the noise variance matrix of a DSTC may be a non-diagonal matrix, i.e. the noise vector $\mathbf{n}$ at the destination may be correlated. This makes the analysis and construction of the DSTCs much harder than that of the space-time codes.

Inspired by the definition of the generalized orthogonal designs (Tarokh, Seshadri, & Calderbank, 1998; Wang & Xia, 2003), the DOSTBCs are defined in the following way.

Definition 1: *A* $K \times T$ matrix $\mathbf{X}$ is called a Distributed Orthogonal Space-Time Block Code (DOSTBC) in variables $s_1, \cdots, s_N$ if the following two conditions are satisfied:

D1.1) The entries of $\mathbf{X}$ *are* $0, \pm h_k s_n, \pm h_k^* s_n^*$, *or multiples of these indeterminates by* $\mathbf{j}$, *where* $\mathbf{j} = \sqrt{-1}$.
D1.2) The matrix $\mathbf{X}$ *satisfies the following equality*

$$\mathbf{X}\mathbf{R}^{-1}\mathbf{X}^{H} = \mid s_{1} \mid^{2} \mathbf{D}_{1} + \cdots + \mid s_{N} \mid^{2} \mathbf{D}_{N},$$ (12)

where $\mathbf{D}_{n} = diag[\mid h_{1} \mid^{2} D_{n,1}, \cdots, \mid h_{K} \mid^{2} D_{n,K}]$ and $D_{n,k}$ are non-zero for $k = 1, \cdots, K$. (Yi & Kim, 2007)

By substituting (12) into (11), it is not hard to see that the DOSTBCs are single-symbol ML decodable. Specifically, we can rewrite the first term in (11) as $-\Re\{\sum_{n=1}^{N}(\alpha_{n}s_{n} + \beta_{n}s_{n}^{*})\}$, where α_{n} and β_{n} are independent of s_{n}. It follows from (12) that the second term in (11) can be decomposed into a sum, where each term only depends on $\mid s_{n} \mid^{2}$. Therefore, (11) does not contain the terms $s_{n_{1}}s_{n_{2}}$, $s_{n_{1}}s_{n_{2}}^{*}$, $s_{n_{1}}^{*}s_{n_{2}}$, or $s_{n_{1}}^{*}s_{n_{2}}^{*}$, $n_{1} \neq n_{2}$. This means that the DOSTBCs defined in Definition I achieve the single-symbol ML decodability (Khan & Rajan, 2006; Tarokh, Seshadri, & Calderbank, 1998). Furthermore, the DOSTBCs can also achieve the full diversity order K as shown in the following lemma.

Lemma 1: *Assume* $T \geq K$, $E_{r} = c_{r}E$, and $E_{s} = c_{s}E$, where c_{r} and c_{s} are positive constants. The DOSTBCs can achieve the full diversity order K. (Yi & Kim, 2007)

Proof: See Appendix A in (Yi & Kim, 2007).

After obtaining the diversity order of the DOSTBCs, it is desirable to derive the upper bound of the data-rate of the DOSTBC and find the actual codes that achieve the upper bound. The upper bound of the data-rate of the DOSTBC has been found in (Yi & Kim, 2008) very recently. However, it is still an open problem to construct the actual code achieving the upper bound.

The major hindrance in the analysis and construction of the DOSTBCs comes from the fact that the noise covariance matrix $\mathbf{R}$ in (9) is not diagonal in general. Due to this reason, most of the existing methods used in the analysis and construction of space-time codes in multiple antenna systems can not be directly applied to the DOSTBCs. In order to facilitate the analysis and construction, we consider some special DOSTBCs, which have diagonal noise covariance matrices at the destination.

Row-Monomial Distributed Orthogonal Space-Time Block Codes

In this subsection, we first prove that, when $\mathbf{A}_{k}$ and $\mathbf{B}_{k}$ are row-monomial, the covariance matrix $\mathbf{R}$ becomes a diagonal matrix. Then we define the row-monomial DOSTBCs. Lastly, an upper bound of the data-rate of the row-monomial DOSTBC is derived.

As we stated in Section 2, the non-diagonality of $\mathbf{R}$ makes the analysis and construction of the DOSTBCs very hard. Thus, we restrict our interests to a special subset of the DOSTBCs, where $\mathbf{R}$ is diagonal. The diagonality of $\mathbf{R}$ is actually decided by the associated matrices $\mathbf{A}_{k}$ and $\mathbf{B}_{k}$ as shown in the following theorem.

Theorem 1: *The matrix* $\mathbf{R}$ *in* (9) *is a diagonal matrix if and only if* $\mathbf{A}_{k}$ *and* $\mathbf{B}_{k}$ *are row-monomial.* (Yi & Kim, 2007)

Proof: See Appendix B in (Yi & Kim, 2007).

Based on Theorem 3, the row-monomial DOSTBCs are defined in the following way.

Definition 2: *A* $K \times T$ *matrix* $\mathbf{X}$ *is called a row-monomial DOSTBC in variables* $s_{1}, \cdots, s_{N}$ *if it satisfies D1.1 and D1.2 in Definition 1 and its associated matrices* $\mathbf{A}_{k}$ *and* $\mathbf{B}_{k}$, $1 \leq k \leq K$, *are both row-monomial.* (Yi & Kim, 2007)

The row-monomial DOSTBCs certainly achieve the single-symbol ML decodability and the full diversity order, because they are in a subset of the DOSTBCs. Due to the same reason, all the results in Section 2 are still valid for the row-monomial DOSTBCs. Furthermore, the diagonality of $\mathbf{R}$ makes it possible to derive an upper bound of the data-rate of the row-monomial DOSTBC. By analyzing the properties of the associated matrices $\mathbf{A}_k$ and $\mathbf{B}_k$, an upper bound of the data-rate of the row-monomial DOSTBC is derived in the following theorem.

Theorem 2: *If a row-monomial DOSTBC* $\mathbf{X}$ *in variables* $s_1, \cdots, s_N$ *exists, its data-rate* $Rate_r$ *satis-fies the following inequality:*

$$
Rate_r = \frac{N}{T} \leq
\begin{cases}
\dfrac{1}{m}, & N = 2l, K = 2m \\[2ex]
\dfrac{2l+1}{2lm+2m}, & N = 2l+1, K = 2m \\[2ex]
\dfrac{1}{m+1}, & N = 2l, K = 2m+1 \\[2ex]
\min\left(\dfrac{2l+1}{2lm+2m+l+1}, \dfrac{2l+1}{2lm+2l+m+1}\right), & N = 2l+1, K = 2m+1
\end{cases}
$$

$$(13)$$

where l *and* m *are positive integers.* (Yi & Kim, 2007)

Proof: See Appendices C, D, and E in (Yi & Kim, 2007).

It is well-known that the data-rate of the repetition-based cooperative strategy is just $1/K$. Thus, the data-rate of the row-monomial DOSTBC is approximately twice higher than that of the repetition-based cooperative strategy according to (13). Note that the data-rate in (13) is the data-rate of the transmission from the relays to the destination. The comparison can be easily extended to the overall data-rate of the cooperative network, i.e. the data-rate of the entire transmission from the source to the destination. When N and K are both even, for example, the overall data-rate of the cooperative networks using the row-monomial DOSTBCs is $2/(2+K)$. On the other hand, the overall data-rate of the repetition-based cooperative strategy is $1/(1+K)$, which is smaller than $2/(2+K)$ for any non-negative K. Therefore, we can conclude that the row-monomial DOSTBCs have better bandwidth efficiency than the repetition-based cooperative strategy. Furthermore, the better bandwidth efficiency of the row-monomial DOSTBCs is achieved while the single-symbol ML decodability and the full diversity order are maintained.

It should be noted that the upper bound of the data-rate of the row-monomial DOSTBC decrease with K. For a cooperative network with many relays, the row-monomial DOSTBCs may not have good bandwidth efficiency. Therefore, although the row-monomial DOSTBCs are designed for a cooperative network with an arbitrary number of relays, it is preferable to implement them only for a cooperative network with a small K. When there are many relays in the cooperative network, the bandwidth efficiency can be improved by making the relays have some CSI. Very recently, we have shown in (Yi & Kim, 2008) that, when the k-th relay knows the phase of the channel coefficient h_k, the data-rate of the row-monomial DOSTBC becomes a constant and no longer decreases with K, which means much higher bandwidth efficiency. Similar results have been found in (Sreedhar, Chockalingam, & Rajan, 2008) as well.

Systematic Construction of the Row-Monomial DOSTBCs

In (Yi & Kim, 2007), a systematic construction method has been developed in order to construct the row-monomial DOSTBCs achieving the upper bound given in Theorem 2. In this subsection, we present some actual codes generated by this construction method. Given the values of N and K, we let $\mathbf{X}(N, K)$ denote such codes. We consider four different cases depending on the values of N and K.

$N = 2l$ **and** $K = 2m$

For example, when $N = 4$ and $K = 4$, the code constructed by Construction I in (Yi & Kim, 2007) is as follows:

$$\mathbf{X}(4,4) = \begin{bmatrix} h_1 s_1 & -h_1 s_2 & h_1 s_3 & -h_1 s_4 & 0 & 0 & 0 & 0 \\ h_2^* s_2^* & h_2^* s_1^* & h_2^* s_4^* & h_2 s_3^* & 0 & 0 & 0 & 0 \\ 0 & 0 & 0 & 0 & h_3 s_1 & -h_3 s_2 & h_3 s_3 & -h_3 s_4 \\ 0 & 0 & 0 & 0 & h_4^* s_2^* & h_4^* s_1^* & h_4^* s_4^* & h_4 s_3^* \end{bmatrix},$$

and it achieves the upper bound of the data-rate $1/2$.

$N = 2l + 1$ **and** $K = 2m$

For example, when $N = 5$ and $K = 4$, the code constructed by Construction II in (Yi & Kim, 2007) is given by $\mathbf{X}(5,4) = [\mathbf{X}_1, \mathbf{X}_2]$, where

$$\mathbf{X}_1 = \begin{bmatrix} h_1 s & -h_1 s_2 & h_1 s_3 & -h_1 s_4 & 0 & 0 & 0 & 0 \\ h_2^* s_2^* & h_2^* s_1^* & h_2^* s_4^* & h_2 s_3^* & 0 & 0 & 0 & 0 \\ 0 & 0 & 0 & 0 & h_3 s & -h_3 s_2 & h_3 s_3 & -h_3 s_4 \\ 0 & 0 & 0 & 0 & h_4^* s_2^* & h_4^* s_1^* & h_4^* s_4^* & h_4 s_3^* \end{bmatrix}$$

$$\mathbf{X}_2 = \begin{bmatrix} h_1 s_5 & 0 & 0 & 0 \\ 0 & h_2 s_5 & 0 & 0 \\ 0 & 0 & h_3 s_5 & 0 \\ 0 & 0 & 0 & h_4 s_5 \end{bmatrix}.$$

This code $\mathbf{X}(5,4)$ achieves the upper bound of the data-rate $5/12$.

$N = 2l$ **and** $K = 2m + 1$

For example, when $N = 4$ and $K = 5$, the code constructed by Construction III in (Yi & Kim, 2007) is given by $\mathbf{X}(4,5) = diag[\mathbf{X}_1, \mathbf{x}_2]$, where

$$\mathbf{X}_1 = \begin{bmatrix} h_1 s & -h_1 s_2 & h_1 s_3 & -h_1 s_4 & 0 & 0 & 0 & 0 \\ h_2^* s_2^* & h_2^* s_1^* & h_2^* s_4^* & h_2 s_3^* & 0 & 0 & 0 & 0 \\ 0 & 0 & 0 & 0 & h_3 s & -h_3 s_2 & h_3 s_3 & -h_3 s_4 \\ 0 & 0 & 0 & 0 & h_4^* s_2^* & h_4^* s_1^* & h_4^* s_4^* & h_4 s_3^* \end{bmatrix}$$

$$\mathbf{x}_2 = \begin{bmatrix} h_5 s_1 & h_5 s_2 & h_5 s_3 & h_5 s_4 \end{bmatrix}.$$

This code $\mathbf{X}(4,5)$ achieves the upper bound of the data-rate $1/3$.

$N = 2l+1$ **and** $K = 2m+1$

For example, when $N = 5$ and $K = 5$, the code constructed by Construction IV in (Yi & Kim, 2007) is given by $\mathbf{X}(5,5) = [\mathbf{X}_1, \mathbf{X}_2]$, where

$$\mathbf{X}_1 = \begin{bmatrix} h_1 s_2 & -h_1 s_3 & h_1 s_4 & -h_1 s_5 & 0 & 0 & 0 & 0 \\ h_2^* s_3^* & h_2^* s_2^* & h_2^* s_5^* & h_2 s_4^* & 0 & 0 & 0 & 0 \\ 0 & 0 & 0 & 0 & h_3 s_1 & -h_3 s_3 & h_3 s_4 & -h_3 s_5 \\ 0 & 0 & 0 & 0 & h_4^* s_3^* & h_4^* s_1^* & h_4^* s_5^* & h_4 s_4^* \\ 0 & 0 & 0 & 0 & 0 & 0 & 0 & 0 \end{bmatrix}$$

and

$$\mathbf{X}_2 = \begin{bmatrix} h_1^* s_1^* & h_1^* s_5^* & 0 & 0 & 0 & 0 & 0 \\ 0 & 0 & 0 & 0 & h_2 s_1 & -h_2 s_3 & 0 \\ 0 & 0 & h_3^* s_2^* & h_3^* s_4^* & 0 & 0 & 0 \\ 0 & 0 & 0 & 0 & 0 & 0 & h_4 s_2 \\ h_5 s_5 & -h_5 s_1 & h_5 s_4 & -h_5 s_2 & h_3^* s_3^* & h_5^* s_1^* & 0 \end{bmatrix}.$$

This code $\mathbf{X}(5,5)$ achieves the upper bound of the data-rate $1/3$.

Numerical Results

In this subsection, some numerical results are provided to compare the performance of the row-monomial DOSTBCs with that of the repetition-based cooperative strategy at first. Specifically, we compare the performance of the codes proposed in Subsection 4 with that of the repetition-based strategy.

In order to make the comparison fair, the modulation scheme and the transmission power per channel use of every relay need to be properly chosen for different circumstances. For example, when $N = 4$ and $K = 4$, the data-rate of $\mathbf{X}(4,4)$ is $1/2$. We choose quadrature phase shift keying (QPSK) as the modulation scheme, and hence, the bandwidth efficiency of $\mathbf{X}(4,4)$ is 1 bps/Hz. On the other hand, the data-rate of the repetition-based cooperative strategy is $1/4$. Therefore, 16-quadrature amplitude modulation (QAM) is chosen and it makes the bandwidth efficiency of the repetition-based cooperative strategy 1 bps/Hz as well. Similarly, in order to make the bandwidth efficiency equal to 2 bps/Hz, 16-QAM and 256-QAM are chosen for $\mathbf{X}(4,4)$ and the repetition-based cooperative strategy, respectively. Furthermore, we set the transmission power per channel use of every relay to be E_r for $\mathbf{X}(4,4)$. Since every relay transmits 4 times over 8 time slots, the average transmission power per time slot is $E_r / 2$. For the repetition-based cooperative strategy, the transmission power per channel use of every relay is set to be $2E_r$. Because every relay transmits once over 4 time slots, the average transmission power per time slot is $E_r / 2$ as well. When $N = 4$ and $K = 5$, proper modulation schemes and transmission power per channel use of every relay can be found by following the same way.[4]

When $N = 5$ and $K = 5$, 8-PSK and 64-QAM are chosen for $\mathbf{X}(5,5)$ to make it have the bandwidth efficiency 1 bps/Hz and 2 bps/Hz, respectively. On the other hand, 32-QAM and 1024-QAM are chosen for the repetition-based cooperative strategy to make it have the bandwidth efficiency 1 bps/Hz and 2 bps/Hz, respectively. For $\mathbf{X}(5,5)$, the transmission power per channel use of every relay is set to be E_r. Since the fourth relay transmits 5 times over 15 time slots, its average transmission power per time slot is $E_r / 3$. Every other relay transmits 6 times over 15 times slots, and hence, its average transmission power per time slot is $2E_r / 5$. For the repetition-based cooperative strategy, every relay transmits once over 5 time slots. Therefore, the transmission power per channel use of the fourth relay is set to be $5E_r / 3$, and the transmission power per channel use of the other relays is set to be $2E_r$.

The comparison results are given in Figs. 1–3. It can be seen that the performance of the DOSTBCs is much better than that of the repetition-based cooperative strategy in the whole signal-to-noise ratio (SNR) range. The performance gain of the DOSTBCs is more impressive when the bandwidth efficiency is 2 bps/Hz. For example, when $N = 4$ and $K = 4$, the performance gain of the DOSTBCs is ap-

Figure 1. Comparison of the row-monomial DOSTBCs with the repetition-based cooperative strategy, $N = 4$, $K = 4$. (Adapted from Fig. 2 in Yi & Kim (2008))

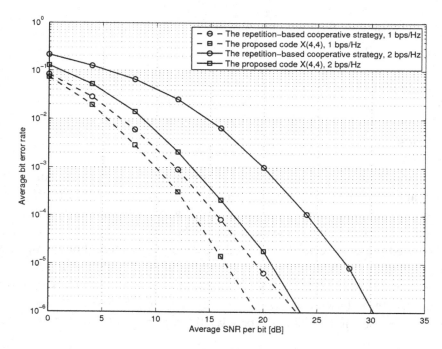

proximately 7 dB at 10^{-6} bit error rate (BER). This is because the DOSTBCs have higher data-rate than the repetition-based cooperative strategy, and hence, they can use the modulation schemes with smaller constellation size to achieve the same bandwidth efficiency. Furthermore, because the BER curves of the DOSTBCs are parallel with those of the repetition-based cooperative strategy, they should have the same diversity order. It is well-known that the repetition-based cooperative strategy can achieve the full diversity order K. Therefore, the DOSTBCs also achieve the full diversity order K as well.

Secondly, we compare the performance of the DOSTBCs with that of the linear dispersion codes in (Jing & Hassibi, 2006) and the quasi-orthogonal STCs in (Jing & Jafarkhani, 2007). The DOSTBCs have lower data-rates than the linear dispersion codes and the quasi-orthogonal STCs. The DOSTBCs need larger constellation sizes in order to achieve the same bandwidth efficiency as the linear dispersion codes and the quasi-orthogonal STCs. Therefore, it is not surprising that the DOSTBCs do not perform as well as the linear dispersion codes or the quasi-orthogonal space-time codes in Figs. 4 and 5. However, the decoding complexity of the DOSTBCs is much lower than that of the linear dispersion codes and the quasi-orthogonal space-time codes, which makes the DOSTBCs more attractive in practical implementations.

FUTURE TRENDS

An interesting topic might be the error performance analysis of the row-monomial DOSTBCs and the row-monomial DOSTBCs-CPI recently proposed in (Yi & Kim, 2008). One can start by considering the simplest case where there are only two relays in the cooperative network. In this case, the row-monomial DOSTBCs and the row-monomial DOSTBCs-CPI are usually referred to as the distributed Alamouti's code. There have been some works devoted to the error performance analysis of this code. For example,

Figure 2. Comparison of the row-monomial DOSTBCs with the repetition-based cooperative strategy, $N = 4$, $K = 5$.

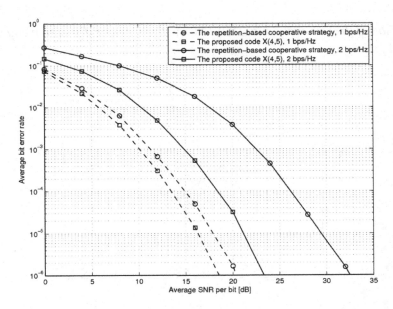

Figure 3. Comparison of the row-monomial DOSTBCs with the repetition-based cooperative strategy,
$N = 5$, $K = 5$.

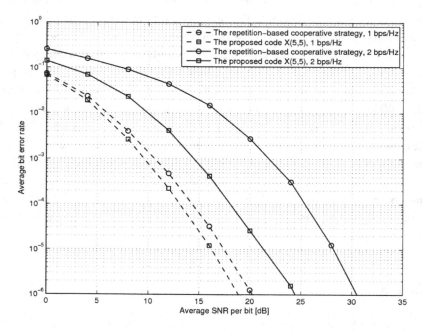

Anghel et al. (Anghel & Kaveh, 2006) found an asymptotic BER expression of the distributed Alamouti's code. Furthermore, in (Chang & Hua, 2004), the authors also considered an Alamouti's-coded cooperative network and they presented a BER expression of the code at high SNR range. At low and moderate SNR range, however, the BER expressions in (Anghel & Kaveh, 2006; Chang & Hua, 2004) were not accurate. Very recently, Ju et al. found the exact BER expression of the distributed Alamouti's code (Ju, Song, & Kim, 2008) in an AF cooperative network. However, the authors of (Ju, Song, & Kim,

Figure 4. Comparison of the row-monomial DOSTBCs with the linear dispersion codes in (Jing & Hassibi, 2006), $N = 4$, $K = 4$.

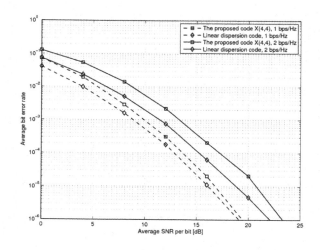

Figure 5. Comparison of the row-monomial DOSTBCs with the quasi-orthogonal STCs in (Jing & Jafarkhani, 2007), $N = 4$, $K = 4$.

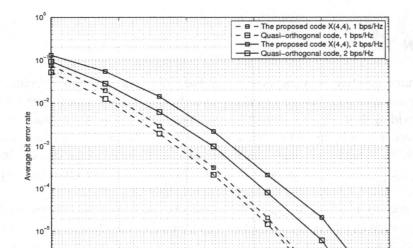

2008) assumed that the first-hop channels, i.e. the channels from the source to the relays, had the same variance and the second-hop channels, i.e. the channels from the relays to the destination, also had the same variance. In practice, it is more important to consider a dissimilar cooperative network, where all the channels possibly have different variances. Unfortunately, it is very hard to extend the results in (Ju, Song, & Kim, 2008) to a dissimilar cooperative network.

Therefore, it will be interesting to analyze the error performance of the distributed Alamouti's code in a dissimilar cooperative network. It might be very hard to obtain the exact BER or symbol error rate (SER) expression when the channels have different variances. However, the instantaneous SNR expression of the distributed Alamouti's code can be written as a summation of two terms. Although those two terms are dependent on each other, one may ignore such dependency and find approximate BER and SER expressions. Certainly, it should be analyzed how the dependency of the two terms in the instantaneous SNR expression affects the accuracy of the approximate BER and SER expressions.

Most of the previous publications on DSTBCs always assume that the *k*-th relay does not know the amplitude of the channel coefficient h_k. *This is because, under this assumption, the amplifying coefficient at every relay is set as a constant, which greatly simplifies the analysis of the DSTCs. Since the amplifying coefficient is a constant, however, the amplifiers at the relays may go into saturation when the first-hop channels are just in slight fading (Anghel & Kaveh, 2006). Thus, it is necessary to investiga*te the DSTCs when the k-th relay knows the phase and amplitude of h_k, i.e. the k-th relay knows the instantaneous value of . For this case, the authors of (Anghel & Kaveh, 2004) proposed an amplifying coefficient for the relays, which was a function of , and it could avoid the saturation of the amplifiers. However, the amplifying coefficient proposed in (Anghel & Kaveh, 2004) is for the repetition-based cooperative strategy, and hence, it might not be suitable for DSTCs. It has been noticed that, when the amplifying coefficient proposed in (Anghel & Kaveh, 2004) is used by DSTCs, the codes can not even achieve the full diversity order. Therefore, one may consider the design of new amplifying coefficients for the DSTCs in order to avoid the saturation of the amplifiers and maintain the full diversity order.

CONCLUSION

Diversity techniques are effective means to combat the fading phenomenon inherent in wireless communications. In particular, spatial diversity is achieved by deploying multiple antennas at the wireless terminals and providing the receiver with multiple independent replicas of the same information-bearing symbols. The antennas must be spaced sufficiently far apart in order to eliminate the correlations between the signals. When the wireless terminals can not employ multiple antennas due to size or complexity constraints, cooperative diversity has been proposed by exploiting the broadcast nature of wireless channels.

Cooperative diversity is achieved by several single-antenna terminals cooperate and form a distributed multiple antenna system. Specifically, the relays help the source transmit the information-bearing symbols to the destination. The coordination of the transmissions from the relays is critical to a cooperative network and it has a large impact on the performance and complexity of the network. The most well-known cooperative strategy is the repetition-based cooperative strategy, where only one relay is allowed to transmit the signal to the destination at each time slot. At a price of very poor bandwidth efficiency, the repetition-based strategy makes the decoding at the destination very simple and it achieves the full diversity order. In order to improve the bandwidth efficiency, many schemes, such as DSTCs, cooperative beamforming, and relay selection scheme, have been proposed. Among them, the DSTCs might be the most attractive because no feedback overheads are required.

When the relays works in the DF mode, they can perform CRC, and hence, existing space-time codes designed for the co-located multiple antenna system can be used in the cooperative networks without any modifications. As a result, existing works focus on the construction of DSTCs by considering the distributive nature of the cooperative network. For example, DSTCs have been proposed for asynchronous cooperative networks and cooperative networks without any centralized control terminal. On the other hand, the analysis and construction of DSTCs in the AF cooperative networks are much harder. Most of the works focus on constructing the optimum DSTCs by minimizing the average error probability or outage probability.

Among all the DSTCs proposed for the AF cooperative networks, the codes achieving the single-symbol ML decodability are of our particular interests. We first notice that orthogonal space-time codes, which are single-symbol ML decodable in co-located multiple antenna systems, lose the single-symbol ML decodability when they are directly used in the AF cooperative networks. In fact, due to the distributive nature, the cooperative networks have many unique properties that the co-located multiple antenna systems do not have. By addressing those unique properties, the DOSTBCs have been proposed for the AF cooperative networks. Those codes are single-symbol ML decodable and they achieve the full diversity order in any constellations. Then we do some further investigation on the row-monomial DOSTBCs, which are some special DOSTBCs generating uncorrelated noises at the destination. An upper bound of the data-rate of the row-monomial DOSTBC has been obtained and it shows that the row-monomial DOSTBCs can achieve approximately twice higher bandwidth efficiency than the repetition-based cooperative strategy. The row-monomial DOSTBCs achieving the upper bound have been found as well.

REFERENCES

Anghel, P. A., & Kaveh, M. (2004). Exact symbol error probability of a cooperative network in a Rayleigh-fading environment. *IEEE Transactions on Wireless Communications, 3*, 1416–1421. doi:10.1109/TWC.2004.833431

Anghel, P. A., & Kaveh, M. (2006). On the performance of distributed space-time coding systems with one and two nongenerative relays. *IEEE Transactions on Wireless Communications, 5*, 682–692.

Bletsas, A., Khisti, A., Reed, D. P., & Lippman, A. (2006). A simple cooperative diversity method based on network path selection. *IEEE Journal on Selected Areas in Communications, 24*, 659–672. doi:10.1109/JSAC.2005.862417

Chang, Y., & Hua, Y. (2004). Diversity analysis of orthogonal space-time modulation for distributed wireless relays. *Proceedings of the IEEE, ICASSP2004*, 561–564.

Chen, D., & Laneman, J. N. (2006). Modulation and demodulation for cooperative diversity in wireless systems. *IEEE Transactions on Wireless Communications, 5*, 1785–1794. doi:10.1109/TWC.2006.1673090

Dallal, Y. E., & Shamai, S. (1992). Time diversity in DPSK noisy phase channels. *IEEE Transactions on Communications, 40*, 1703–1715. doi:10.1109/26.179934

Damen, M. O., & Hammons, A. R. (2007). Delay-tolerant distributed TAST codes for cooperative diversity. *IEEE Transactions on Information Theory, 53*, 3755–3773. doi:10.1109/TIT.2007.904773

Ding, Y., Zhang, J.-K., & Wong, K. M. (2007). The amplify-and-forward half-duplex cooperative system: Pairwise error probability and precoder design. *IEEE Transactions on Signal Processing, 55*, 605–617. doi:10.1109/TSP.2006.885761

Elia, P., Oggier, F., & Kumar, P. V. (2007). Asymptotically optimal cooperative wireless networks with reduced signaling complexity. *IEEE Journal on Selected Areas in Communications, 25*, 258–267. doi:10.1109/JSAC.2007.070203

Godara, L. C. (1997). Applications of antenna arrays to mobile communications, part I: Performance improvement, feasibility, and system considerations. *Proceedings of the IEEE, 85*, 1031–1060. doi:10.1109/5.611108

Godara, L. C. (1997). Applications of antenna arrays to mobile communications, part II: Beamforming and direction-of-arrival considerations. *Proceedings of the IEEE, 85*, 1195–1245. doi:10.1109/5.622504

Guo, X., & Xia, X.-G. (2008). A distributed space-time coding in asynchronous wireless relay networks. *IEEE Transactions on Wireless Communications, 7*, 1812–1816. doi:10.1109/TWC.2008.070042

Hammerström, I., Kuhn, M., & Wittneben, A. (2004). Impact of relay gain allocation on the performance of cooperative diversity networks. In *Proc. IEEE VTC'04* (pp. 1815–1819).

Hasna, M. O., & Alouini, M. S. (2003). End-to-end performance of transmission systems with relays over rayleigh–fading channels. *IEEE Transactions on Wireless Communications, 2*, 1126–1131. doi:10.1109/TWC.2003.819030

Herhold, P., Zimmermann, E., & Fettweis, G. (2004). A simple cooperative extension to wireless relaying. In *Proc. International Zurich Seminar on Communications* (pp. 36–39).

Hua, Y., Mei, Y., & Cheng, Y. (2003). Wireless antennas-making wireless communications perform like wireline communications. In *Proc. IEEE AP-S Topical Conference on Wireless Communication Technology* (pp. 47–73).

Jagadeesh, H., & Rajan, B. S. (2008). Single-symbol ML decodable precoded DSTBCs for cooperative networks. In *Proc. IEEE ICC'08* (pp. 991—995).

Jing, Y., & Hassibi, B. (2006). Distributed space-time coding in wireless relay networks. *IEEE Transactions on Wireless Communications, 5,* 3524–3536. doi:10.1109/TWC.2006.256975

Jing, Y., & Jafarkhani, H. (2007). Using orthogonal and quasi-orthogonal designs in wireless relay networks. *IEEE Transactions on Information Theory, 53,* 4106–4118. doi:10.1109/TIT.2007.907516

Jing, Y., & Jafarkhani, H. (2007). Network beamforming using relays with perfect channel information. In *Proc. IEEE ICASSP'07* (pp. 15—20).

Jing, Y., & Jafarkhani, H. (2008). Single and multiple relay selection schemes and their diversity orders. In *Proc. IEEE ICC'08* (pp. 349—353).

Jing, Y., & Jafarkhani, H. (2008). Network beamforming with channel mean and covariance at relays. In *Proc. IEEE ICC'08* (pp. 19—23).

Ju, M.-C., Song, H.-K., & Kim, I.-M. (2008). Exact BER analysis of distributed Alamouti's code for cooperative diversity networks. *IEEE Trans. Commun.,* in second round of revision.

Kaleh, G. K. (1996). Frequency-diversity spread-spectrum communication system to counter bandlimited gaussian interference. *IEEE Transactions on Communications, 44,* 886–893. doi:10.1109/26.508308

Khan, M. Z., & Rajan, B. S. (2006). Single-symbol maximum likelihood decodable linear STBCs. *IEEE Transactions on Information Theory, 52,* 2062–2091. doi:10.1109/TIT.2006.872970

Kiran, T., & Rajan, B. S. (2006). Distributed space-time codes with reduced decoding complexity. In *Proc. IEEE ISIT'06* (pp. 542—546).

Koyuncu, E., Jing, Y., & Jafarkhani, H. (2008). Distributed beamforming in wireless relay networks with quantized feedback. *IEEE Journal on Selected Areas in Communications, 26,* 1429–1439. doi:10.1109/JSAC.2008.081009

Laneman, J. N., Tse, D. N., & Wornell, G. W. (2004). Cooperative diversity in wireless networks: Efficient protocols and outage behavior. *IEEE Transactions on Information Theory, 50,* 3062–3080. doi:10.1109/TIT.2004.838089

Laneman, J. N., & Wornell, G. W. (2000). Energy-efficient antenna sharing and relaying for wireless networks. *Proc. of Wireless Communications and Networking Conference, 1,* pp. 7–12.

Laneman, J. N., & Wornell, G. W. (2003). Distributed space-time-coded protocols for exploiting cooperative diversity in wireless networks. *IEEE Transactions on Information Theory, 49,* 2415–2425. doi:10.1109/TIT.2003.817829

Li, Y., & Xia, X.-G. (2007). A family of distributed space-time trellis codes with asynchronous cooperative diversity. *IEEE Transactions on Communications, 55*, 790–800. doi:10.1109/TCOMM.2007.894117

Li, Z., & Xia, X.-G. (2007). A simple Alamouti space-time transmission scheme for asynchronous cooperative systems. *IEEE Signal Processing Letters, 14*, 804–807. doi:10.1109/LSP.2007.900224

Liang, K.-C., Wang, X., & Berenguer, I. (2007). Minimum error rate linear dispersion codes for cooperative relays. *IEEE Transactions on Vehicular Technology, 6*, 2143–2157. doi:10.1109/TVT.2007.897636

Maham, B., & Hjørungnes, A. (2007). Distributed GABBA space-time codes in amplify-and-forward cooperation. In *Proc. IEEE ITW '07* (pp. 1-5).

Mheidat, H., & Uysal, M. (2007). Non-coherent and mismatched-coherent receivers for distributed ST-BCs with amplify-and-forward relaying. *IEEE Transactions on Wireless Communications, 6*, 4060–4070. doi:10.1109/TWC.2007.060180

Nabar, R. U., Bölcskei, H., & Kneubühler, F. W. (2004). Fading relay channels: Performance limits and space-time signal designs. *IEEE Journal on Selected Areas in Communications, 22*, 1099–1109. doi:10.1109/JSAC.2004.830922

Proakis, J. G. (2001). *Digital Communications*. New York: McGraw-Hill.

Rajan, G. S., & Rajan, B. S. (2006). A non-orthogonal distributed space-time coded protocol, part-II: Code construction and DM-G tradeoff. In *Proc. IEEE ITW '06* (pp. 488—492).

Rajan, G. S., & Rajan, B. S. (2007). Distributed space-time codes for cooperative networks with partial CSI. In *Proc. IEEE WCNC '07* (pp. 902–906).

Rajan, G. S., & Rajan, B. S. (2007). Algebraic distributed space-time codes with low ML decoding complexity. In *Proc. IEEE ISIT '07* (pp. 1516—1520).

Rajan, G. S., & Rajan, B. S. (2008). OFDM based distributed space time coding for asynchronous relay networks. In *Proc. IEEE ICC '08* (pp. 1118—1122).

Rajan, G. S., Tandon, A., & Rajan, B. S. (2007). On four-group ML decodable distributed space time codes for cooperative communication. In *Proc. IEEE WCNC '07* (pp. 11—15).

Ribeiro, A., Cai, X., & Giannakis, G. B. (2005). Symbol error probabilities for general cooperative links. *IEEE Transactions on Wireless Communications, 4*, 1264–1273. doi:10.1109/TWC.2005.846989

Savazzi, S., & Spagnolini, U. (2007). Distributed orthogonal space time coding: Design and outage analysis for randomized cooperation. *IEEE Transactions on Signal Processing, 6*, 4546–4557.

Sendonaris, A., Erkip, E., & Aazhang, B. (2003). User cooperation diversity–part I: System description. *IEEE Transactions on Communications, 51*, 1927–1938. doi:10.1109/TCOMM.2003.818096

Sendonaris, A., Erkip, E., & Aazhang, B. (2003). User cooperation diversity–part II: Implementation aspects and performance analysis. *IEEE Transactions on Communications, 51*, 1939–1948. doi:10.1109/TCOMM.2003.819238

Shang, Y., & Xia, X.-G. (2006). Shift-full-rank matrices and applications in space-time trellis codes for relay networks with asynchronous cooperative diversity. *IEEE Transactions on Information Theory, 52*, 3153–3167. doi:10.1109/TIT.2006.876222

Sirkeci-Mergen, B., & Scaglione, A. (2007). Randomized space-time coding for distributed cooperative communication. *IEEE Transactions on Signal Processing, 55*, 5003–5017. doi:10.1109/TSP.2007.896061

Sreedhar, D., Chockalingam, A., & Rajan, B. S. (2008). Single-symbol ML decodable distributed STBCs for partially-coherent cooperative networks. *IEEE ICC, 08*, 1029–1033.

Tarokh, V., Seshadri, N., & Calderbank, A. R. (1998). Space-time codes for high data rate wireless communication: Performance criterion and code construction. *IEEE Transactions on Information Theory, 44*, 744–765. doi:10.1109/18.661517

Wang, H., & Xia, X.-G. (2003). Upper bounds of rates of complex orthogonal space-time block codes. *IEEE Transactions on Information Theory, 49*, 2788–2796. doi:10.1109/TIT.2003.817830

Wang, T., Cano, A., Giannakis, G. B., & Laneman, J. N. (2007). High-performance cooperative demodulation with decode-and-forward relays. *IEEE Transactions on Communications, 55*, 1427–1438. doi:10.1109/TCOMM.2007.900631

Wang, T., Wang, R., & Giannakis, G. B. (2006). Smart regenerative relays for link-adaptive cooperative communications. In *Proc. 40th Conference on Information Sciences and Systems 2006 (CISS'06)* (pp. 1038-1043).

Wang, Z., & Giannakis, G. B. (2003). A simple and general parameterization quantifying performance in fading channels. *IEEE Transactions on Communications, 51*, 1389–1398. doi:10.1109/TCOMM.2003.815053

Winters, J. H. (1998). Smart antennas for wireless systems. *IEEE Personal Commun., 5*, 23–27. doi:10.1109/98.656155

Yang, S., & Belfiore, J.-C. (2007). Optimal space-time codes for the MIMO amplify-and-forward cooperative channel. *IEEE Transactions on Information Theory, 53*, 647–663. doi:10.1109/TIT.2006.888998

Yi, Z., & Kim, I.-M. (2007). Joint optimization of relay-precoders and decoders with partial channel side information in cooperative networks. *IEEE Journal on Selected Areas in Communications, 25*, 447–458. doi:10.1109/JSAC.2007.070219

Yi, Z., & Kim, I.-M. (2007). Single-symbol ML decodable distributed STBCs for cooperative networks. *IEEE Transactions on Information Theory, 53*, 2977–2985. doi:10.1109/TIT.2007.901177

Yi, Z., & Kim, I.-M. (2008). Diversity order analysis of the decode-and-forward cooperative networks with relay selection. *IEEE Transactions on Wireless Communications, 7*, 1792–1799. doi:10.1109/TWC.2008.061041

Yi, Z., & Kim, I.-M. (2008). Row-monomial distributed orthogonal space-time block codes with channel phase information. In *IEEE ICC'08*.

Yi, Z., & Kim, I.-M. (2008). The impact of noise correlation on the single-symbol ML decodable distributed STBCs. In *IEEE ICC '08*.

Yiu, S., Schober, R., & Lampe, L. (2006). Distributed space-time block coding. *IEEE Transactions on Communications*, *54*, 1195–1206. doi:10.1109/TCOMM.2006.877947

Zhao, Y., Adve, R., & Lim, T. J. (2006). Symbol error rate of selection amplify-and-forward relay systems. *IEEE Communications Letters*, *10*, 757–759. doi:10.1109/LCOMM.2006.060774

ENDNOTES

[1]   If the transmitted symbols are real-valued, it has been shown in (Hua, Mei, & Cheng, 2003) that the rate-one generalized real orthogonal design proposed in (Tarokh, Seshadri, & Calderbank, 1998) can be used in AF cooperative networks without any changes. The codes maintain the single-symbol ML decodability and achieve the full diversity order. Therefore, we focus on the complex-valued symbols in this chapter.

[2]   This construction method originates from the construction of a linear space-time code for co-located multiple-antenna systems, where the transmitted signal vector from the k-th antenna is $\mathbf{s}\mathbf{A}_k + \mathbf{s}^*\mathbf{B}_k$ (Wang & Xia, 2003). Since we consider the amplify-and-forward cooperative networks, the relays do not have the estimate of $\mathbf{s}$. Therefore, they use $\mathbf{y}_k$ and $\mathbf{y}_k^*$, which contain the information of $\mathbf{s}$, to construct the transmitted signal vector.

[3]   Note that N time slots are needed by the source to transmit the symbol vector $\mathbf{s}$ to the relays. Thus, the data-rate of the entire transmission is actually $N/(N+T)$. In this chapter, we focus on the design of $\mathbf{X}$, and hence, we will use the data-rate N/T of $\mathbf{X}$ as the metric to evaluate the bandwidth efficiency.

[4]   When $N=5$ and $K=4$, the date-rate of $\mathbf{X}(5,4)$ is $5/12$. In order to make the bandwidth efficiency of $\mathbf{X}(5,4)$ equal to 1 bps/Hz or 2 bps/Hz, the size of the modulation scheme should be $2^{12/5}$ or $2^{24/5}$, which can not be implemented in practice. Therefore, we do not evaluate the performance of $\mathbf{X}(5,4)$ in this chapter.

Chapter 9
Relay Selection in Cooperative Networks

Elzbieta Beres
University of Toronto, Canada

Raviraj Adve
University of Toronto, Canada

ABSTRACT

Cooperative diversity has the potential of implementing spatial diversity and mitigating the adverse effects of channel fading without requiring multiple antennas at transmitters and receivers. Traditionally, cooperative diversity is implemented using maximal ratio combining (MRC), where all available nodes relay signals between the source and destination. It has recently been proposed, however, that for each source-destination transmission, only a single best node should be selected to act as a relay. The resulting scheme, referred to as selection cooperation or opportunistic relaying, outperforms MRC schemes and can be implemented in a distributed fashion with limited feedback. This result is not unexpected, as selection requires some (though very limited) information regarding instantaneous channel conditions, while MRC does not. When implemented in a distributed network, however, MRC does require feedback for the synchronization of nodes, rendering a comparison of the two schemes worthwhile and fair. In this chapter, we provide a detailed overview of selection. We begin with a single source-destination pair, and discuss its implementation and performance under various constraints and scenarios. We then discuss a less-common scenario, a multisource network where nodes act both as sources and as relays.

INTRODUCTION

In distributed wireless systems, cooperative diversity and relaying can harness the advantages of multiple-input multiple-output (MIMO) systems without requiring multiple antennas at each receiver and/or transmitter. For practical networks this is motivated by the need for simple, inexpensive nodes with limited processing power and a single receive antenna. Examples of such networks, of particular interest

DOI: 10.4018/978-1-60566-665-5.ch009

in this chapter, include mesh networks comprising a mesh of access points and sensor networks comprising cheap, potentially battery operated, wireless nodes. In recent years there has been an enormous effort invested in the development of cooperative diversity schemes and protocols. However, most of this effort has been for the 3-node case: a source node communicating to a destination with the help of a single relay node. This chapter deals with the more practical case wherein a source communicates to a destination with many relays available to help in this communication. We then generalize this scenario to the case of multiple sources and multiple destinations communicating simultaneously. Given that a source may avail of multiple relays, a natural question arises: how many relays should help the source? The answer to this question is the central theme of this chapter – only a single best one should help.

In distributed networks, scaling cooperation to multiple nodes presents important challenges. The most straightforward approach is to ask all relays to cooperate. The maximal ratio combining (MRC) scheme requires the coherent addition of multiple, independently-faded, copies of the transmitted signal and is analogous to MRC in traditional MIMO transmission with co-located antennas. In general, this combination can be achieved through pre-coding (Sendonaris & Erkip & Aazhang, 2003), orthogonal transmissions and a maximal-ratio-combiner (Boyer & Falconer & Yanikomeroglu, 2004; Laneman & Tse & Wornell, 2004; Herhold & Zimmermann & Fettweis, 2005), or through distributed space-time codes (DSTC) (Barbarossa & Scutari, 2004; Laneman & Wornell, 2003; Uysal & Canpolat, 2005). Orthogonal transmissions are inefficient and suffer from a bandwidth penalty (Laneman & Wornell, 2003), and precoding and DSTC require accurate channel state information (CSI) and strict symbol-level synchronization, respectively. Neither solution, therefore, seems practical from an implementation point of view.

To overcome these difficulties, selection has been proposed as a practical and efficient method of implementing diversity in distributed systems. Selection restricts cooperation to one "best" relay. In this regard, selection in cooperative networks could be considered the analog of selection combining in a traditional receive or transmit diversity scheme using co-located antennas. In a traditional MIMO system, selection suffers a significant performance loss compared to MRC. However, due to the distributed nature of the networks of interest here, as we will see, selection makes effective use of limited power and bandwidth resources and outperforms MRC-based schemes, including DSTC.

Selection omits the problem of multiple relay transmissions by requesting only a single relay to forward the information from the source. Because a single best relay is selected each time the channel conditions change, the diversity order is clearly equivalent to that of MRC. The process of relay selection does involve some feedback. If the decision is made by the destination, the system requires $log_2(m-1)$ bits of feedback, where $(m-1)$ is the number of relays. If the decision is made by the relays, the channel magnitude must be quantized and fed back to the relays. This feedback, however, is significantly lower than that required for precoding (which requires the channel phase), and arguably lower than that required for symbol-level node synchronization necessary to implement DSTC. Recently selection was shown to be the optimal linear beamforming scheme under the constraint of $log_2(m-1)$ bits of feedback (Zhao & Adve & Lim, 2007).

Selection schemes have been analyzed theoretically in terms of outage probability and multiplexing diversity trade-off for both amplify-and-forward and decode-and-forward transmissions and under various power constraints and signal-to-noise ratio (SNR) regimes (Beres & Adve, 2007; Beres & Adve, 2008; Bletsas & Khisti & Win, 2007; Michalopoulos & Karagiannidis, 2008; Zhao & Adve & Lim, 2007). In addition, one implementation of selection using repeat-accumulate codes and demodulate-and-forward is available in the literature (Chu & Adve & Eckford, 2007). In this chapter, we present a detailed review of these works and their conclusions under various scenarios. We begin with an exposition of selection protocols in a network with a single source communicating with its destination with the help of a group of relays; a scenario that has been well explored the literature. We then discuss a much less explored area, the multi-source scenario where multiple sources communicate with their respective destination with the help of a group of relays, who must occasionally relay for more than one source.

Our goal with this chapter is not to present all details of the various schemes proposed in connection with selection in a distributed network. Rather our goal is to survey the available literature, discuss and compare implementation details and point the reader to the relevant sources for detailed reading.

SELECTION IN A SINGLE SOURCE NETWORK

System Model

In this section, we consider a single source node communicating with a destination with the help of m-1 potential relays. The instantaneous channels a_{ij} between all nodes are modeled as independent, slow and flat - fading Rayleigh with mean power of $1 / \lambda_{ij}$, i.e., $| a_{ij} |^2 \sim \lambda_{ij} \exp[-\lambda_{ij} | a_{ij} |^2]$. We assume the relays cannot transmit and receive simultaneously; this half-duplex constraint requires that communication between source and destination be divided into two time-slots. The source and relay transmit with power P_s and P_r, respectively and we assume that time-slots are of unit duration, such that transmitted power and energy are equivalent.

In a decode-and-forward (DF) system, the relays attempt to decode the sources' transmission, re-encode the data for relaying. In this chapter we assume the relay uses repetition coding, i.e., the same code structure as the source. Note that the relay could use an independent codebook (Laneman & Wornell, 2003), however, that scheme does not change our discussion significantly. In an amplify-and-forward (AF) system, the relay simply amplifies its received signal before relaying. In this case, both the signal and noise terms are amplified. The amplification factor can be chosen to meet an instantaneous power constraint or a long-term power constraint. Each scheme results in its own achievable data rate or outage scheme.

Selection Protocols for DF and AF

Selection protocols differ in their figure of merit used to select the relay from the available m-1 relays. Generally these figures of merit are tied to the cooperation protocol in use. Selection has been proposed independently for DF (Beres & Adve, 2008; Bletsas & Khisti & Reed & Lippman, 2006) and for AF (Bletsas & Khisti & Reed & Lippman, 2006; Zhao & Adve & Lim, 2007). In the context of cooperative diversity, selection implies that a **single** relay out of the available m-1 nodes is selected to aid the source in its communication with the destination. This relay is selected based on *instantaneous* channel conditions, and a new relay is selected each time the channel conditions change. In the above cited works, different terms have been used to denote similar schemes. In what follows, we discuss the various proposed selection schemes and clarify the terminology used to denote them.

In the context of decode-and-forward, we have proposed selection and have named our scheme *Selection Cooperation* (Beres & Adve, 2008). In independent works, Bletsas et al. have proposed selection as *Opportunistic Relaying,* with a further distinction between two similar schemes which differ in their relay selection criterion: Reactive Opportunistic Relaying and Proactive Opportunistic Relaying (Bletsas, Khisti, Reed, Lippman, 2006; Bletsas, Khisti, Win, 2007). In decode-and-forward, Selection Cooperation uses the same relay-selection metric as Reactive Opportunistic Relaying; the two schemes are thus identical. In this chapter we use the term DF Selection Cooperation to refer to *Proactive* Opportunistic Relaying and DF Opportunistic Relaying to refer exclusively to Proactive Opportunistic Relaying.

For amplify-and-forward systems, selection has been presented by Zhao et al (Zhao & Adve & Lim, 2007) and Bletsas et al. (Bletsas, Khisti, Reed, Lippman, 2006; Bletsas, Khisti, Win, 2007). The schemes, though identical, vary slightly in their presentation, as they start with different assumptions and constraints. They have been assigned different names by their authors: AF Selection Cooperation (Zhao & Adve & Lim, 2007) and AF Opportunistic Relaying (Bletsas & Khisti & Reed & Lippman, 2006; Bletsas & Khisti & Win, 2007). In (Zhao & Adve & Lim, 2007b), the authors investigate selection in amplify-and-forward systems as a special instance of a distributed system with limited feedback. In this framework, they show that selection is the optimal scheme, in the sense of received SNR when the number of feedback bits is restricted to $log_2(m\text{-}1)$; essentially, selection minimizes the noise amplification in a amplify-and-forward system.

In all cases, the authors present protocols in the context of the half-duplex constraint, which precludes the relay from transmitting and receiving simultaneously on the same channel. The transmission is thus divided into two times slots: the source transmits in the first with power P_s Joules/symbol, and the relay transmits in the second time slot with power P_r Joules/symbol. The destination then uses maximum ratio combining to decode the symbols from the source and the relay. The information transmitted by the relay depends on the nature of the cooperative diversity scheme. DF Selection Cooperation and DF Opportunistic Relaying differ in how the relay is selected. AF Selection Cooperation and AF Opportunistic Relaying, on the other hand, are identical: both algorithms select the relay that maximizes the mutual information at the destination. The differences in their presentation in literature arise from different assumptions and constraints.

We discuss these protocols below. Our discussion largely focuses on mutual information as a measure of the quality of a communication link. This choice reflects the fact that there have been few implementations of selection in cooperative distributed networks. We will briefly review the work of (Chu & Adve & Eckford, 2007) to highlight some of the issues that arise.

DF Selection Cooperation: In decode-and-forward, the relays attempt to decode and re-encode the data from the source before retransmission. Clearly, the relay only contributes useful information if the decoding process is successful. In DF Selection, therefore, the relay selection is made from the nodes that have decoded successfully, a decoding set $D(s)$. The relaying node r_i is selected from this decoding set such that it maximizes the relay-destination mutual information (equivalently channel power):

$$r_i = \arg \max_{r_k \in D(s)} \left| a_{r_k d} \right|^2 . \tag{1}$$

Essentially, the relay that can decode correctly and provide the greatest information to the destination is chosen. The decoding set $D(s)$ is formed from all relays which correctly decode the source information, i.e., a relay $r_k \in D(s)$ if its source-relay channel mutual information is higher than the required rate:

$$r_k \in D(s) \quad \text{if} \quad I_{sr_k} = \frac{1}{2}\log\left(1 + \frac{P_s}{N_0}\left|a_{sr_k}\right|^2\right) > R, \tag{2}$$

where P_s Joules/symbol is the power transmitted by the source, N_0 is the noise power spectral density, and P_s/N_0 is the SNR at a relaying node without accounting for fading. The factor of ½ models the two time slots required for relaying given the half duplex constraint.

Note that, in order to select the best relay, the fact that an individual relay has decoded correctly must be communicated to the node making the selection, i.e., that node must have knowledge of the decoding set.

DF Opportunistic Relaying: The selected relaying node r_i maximizes the minimum mutual information between the source-relay and relay destination channels, i.e.,

$$r_i = \arg\max_{r_k=1\ldots m-1}\min\left(\left|a_{s,r_k}\right|^2, \left|a_{r_k d}\right|^2\right) \tag{3}$$

Note that in DF opportunistic relaying, the relay is picked from the entire set of available relays, not just from a decoding set. The constraint of successful decoding is reintroduced *after* the relay selection, i.e., the relay transmits only if the both the source-relay and relay-destination mutual information is above the required rate:

$$\frac{1}{2}\log\left(1 + \frac{P}{N_0}\min\left(\left|a_{s,r_i}\right|^2, \left|a_{r_i d}\right|^2\right)\right) > R, \tag{4}$$

where $P_s = P_r = P$. The max-min selection criterion described above accounts for the two factors that impact relay selection. The maximum information a relay can contribute is limited by the relay-destination channel. The source, therefore, can maximize its data rate (or minimize its outage probability) by choosing the relay with the best relay-destination channel. However, to be useful, the relay needs to decode the source's transmission. Its ability to decode is limited by the source-relay channel. A relay's ability to help is, therefore, limited by the minimum of the source-relay and relay-destination channels. The opportunistic relaying selection criterion picks the best of the available minima. Note, however, that the constraint in (4) requires both the source-relay channel and the relay-destination channel mutual information to be above the required rate, while in (2) this constraint is necessary only for the source-relay channel. As we will see, this overly-constrained relay selection leads to some performance degradation as compared to DF Selection Cooperation.

AF Selection Cooperation and Opportunistic Relaying

In the first time slot, the source transmits symbol x_s to the relay and destination. The signals received at the relay and destination are

$$\begin{aligned}
y_{s,r_i} &= \sqrt{P_s}\,a_{sr_i}x_s + n_{s,ri}, \\
y_{s,d} &= \sqrt{P_s}\,a_{sd}x_s + n_{s,d},
\end{aligned} \tag{5}$$

where as before, the source transmits with power P_s Joules/symbol and $n_{s,ri}$ and $n_{s,ri}$ are the instantaneous noise terms with power spectral density N_0. Before retransmitting, the relay normalizes the signal using a short-term or long-term power constraint. With a short-term power constraint, the normalization is $\sqrt{E^n \left\{ |y_{s,r}|^2 \right\}} = \sqrt{P_s |a_{sr}|^2 + N_0}$; with a long-term constraint, the normalization is $\sqrt{E^{n,c} \left\{ |y_{s,r}|^2 \right\}} = \sqrt{P_s / \lambda_{s,r} + N_0}$. The operands $E^n \left\{ \ \right\}$ and $E^{n,c} \left\{ \ \right\}$ denote the expectation over the noise statistics and noise and channel statistics, respectively. Note that with the short term power constant, the amplification factor is a function of the instantaneous channel condition while, with the long-term power constraint, the amplification factor is a function of the average channel power. The noise amplification factors are therefore different in the two cases.

Assuming no source-destination link and a long-term power constraint, the mutual information at the destination with relay r_k selected is (Bletsas, Khisti, Reed, Lippman, 2006)

$$I_{AF-sel-long-term} = \frac{1}{2} \log_2 \left[1 + \frac{\frac{P_s}{N_0} \frac{P_r}{N_0} |a_{s,r_k}|^2 |a_{r_k d}|^2}{\frac{P_s}{N_0} \frac{1}{\lambda_{sr_k}} + \frac{P_r}{N_0} |a_{r_k d}|^2 + 1} \right]. \tag{6}$$

Including the source-destination channel and using a short-term power constraint results in the following expression for the mutual information at the destination ((Zhao & Adve & Lim, 2007)

$$I_{AF-sel-short-term} = \frac{1}{2} \log_2 \left[1 + \frac{P_s}{N_0} |a_{sd}|^2 + \frac{\frac{P_s}{N_0} \frac{P_r}{N_0} |a_{s,r_k}|^2 |a_{r_k d}|^2}{\frac{P_s}{N_0} |a_{s,r_k}|^2 + \frac{P_r}{N_0} |a_{r_k d}|^2 + 1} \right] \tag{7}$$

A relay r_i is selected to maximize the mutual information in (6) and (7). It therefore maximizes the metric w_k, i.e., $r_i = \arg \max_{r_k = 1 \ldots m-1} w_k$, where

$$w_k = \frac{\frac{P_s}{N_0} \frac{P_r}{N_0} |a_{s,r_k}|^2 |a_{r_k d}|^2}{\frac{P_s}{N_0} \frac{1}{\lambda_{sr_k}} + \frac{P_r}{N_0} |a_{r_k d}|^2 + 1} , \text{ under the long-term power constraints, and}$$

$$w_k = \frac{\frac{P_s}{N_0} \frac{P_r}{N_0} |a_{s,r_k}|^2 |a_{r_k d}|^2}{\frac{P_s}{N_0} |a_{s,r_k}|^2 + \frac{P_r}{N_0} |a_{r_k d}|^2 + 1} , \text{ under the short-term power constraint.}$$

Without a detailed study, not currently available in the literature, it is not clear which approach leads to better performance.

AF Selection as a Special Case of Beamforming

In the context of amplify-and-forward, Zhao et al. analyzed selection as a special case of beamforming with limited feedback of $\log_2(m\text{-}1)$ bits of feedback and beamforming vectors as columns of an $(m\text{-}1)$ by $(m\text{-}1)$ identity matrix (Zhao & Adve & Lim, 2007b). In a traditional MIMO system with co-located antennas, the performance achievable with such an identity matrix is identical to any unitary matrix, i.e., any rotation of the codebook of beamforming vectors results in the same performance. In AF schemes, on the other hand, this does not hold true.

Unsurprisingly, selection suffers significant performance loss in terms of outage probability and bit error rate as compared to optimal beamforming with unlimited feedback. However, optimal beamforming requires knowledge of the coherent channel values at the transmitter. This information is usually fed-back and the relays use one beamforming vector from a codebook. As shown in (Zhao & Adve & Lim, 2007b; Zhao, 2008) this codebook design is extremely complicated and optimal codebooks must be created *given the channel conditions* and random codebooks are the only practical solution. With a similar number of feedback bits, selection outperforms beamforming with random codebooks in terms of outage probability and diversity order. Although it is possible to improve the performance of the random codebooks beamforming by adding more feedback, selection is optimal under the constraint of $\log_2(m\text{-}1)$ bits of feedback, i.e., amongst the class of unitary $(m\text{-}1)$ by $(m\text{-}1)$ matrices, the identity matrix is optimal. Because selection exhibits robust performance under few bits of feedback and, unlike all other beamforming schemes, has the significant benefit of not requiring synchronization, the authors argue that selection is the most attractive currently available beamforming scheme for distributed networks.

Evaluation of Protocols

Outage Probability

This communication between the source and destination targets an end-to-end data rate of R. The outage probability P_{out} is defined as the probability that the mutual information I between source and destination, including the relaying obtained through selection, falls below the required rate R, i.e., $P_{out} = \Pr[I < R]$. Outage probability is an important metric, as it is analytically tractable and provides a bound on the symbol error rate. The outage probability depends on the specific relay selection scheme used and the power constraints imposed on the source and the relays. It has been analyzed under several scenarios (Beres & Adve, 2008; Bletsas & Khisti & Reed & Lippman, 2006; Bletsas & Khisti & Win, 2007; Michalopoulos & Karagiannidis, 2008).

Decode-and-Forward

In (Beres & Adve, 2008) we give the approximate outage probability of *DF Selection Cooperation* in the high SNR regime, assuming a single source-destination pair with $m\text{-}1$ potential relay nodes and an individual energy constraint of P on each node, i.e. $P_s = P_r = P$ Joules/symbol:.

$$P_{out-DF} = \Pr[I_{sel} < R] \approx \left(\frac{2^{2R}-1}{P/N_0}\right)^m \lambda_{sd} \left(\frac{1}{|D(s)|+1}\right) \sum_{|D(s)|} \prod_{r_i \in |D(s)|} \lambda_{r_i,d} \prod_{r_i \notin |D(s)|} \lambda_{s,r_i}, \tag{8}$$

Note that, as expected, selection cooperation provides the full diversity order of *m*. These results are verified through simulations in Figure 1 below.

Bletsas et al. neglect the source-destination channel and give the exact outage probability of *DF Selection Cooperation* under an aggregate power constraint (Bletsas & Khisti & Win, 2007).

$$P_{out-DF} = \prod_{k=1}^{m-1} \left[1 - \exp\left\{ -\frac{2^{2R}-1}{P_{tot}/N_o} \left(\frac{\lambda_{sr_k}}{\zeta} + \frac{\lambda_{r_k,d}}{(1-\zeta)} \right) \right\} \right], \tag{9}$$

where ζ determines the amount of power allocated to the source and relay: the source and the relay transmit with power $P_s = \zeta P_{tot}$ and $P_r = (1-\zeta)P_{tot}$, respectively. The diversity order here is *(m-1)* because the source-destination channel is neglected. The authors show that under an aggregate power constraint, selection is optimal in terms of outage probability, i.e., minimizing the outage probability as a function of power when using all *m-1* relays leads to the selection scheme. This result is intuitive: because the SNR gains are added linearly, all resources should be allocated to the strongest link. The authors also show that when the source-destination channel is neglected, DF Selection Cooperation and DF Opportunistic Cooperation have identical performance in terms out outage probability.

We further analyze Selection by removing the high-SNR assumption and providing an approximation to outage probability in the low-to-medium SNR regime (Bletsas & Khisti & Win, 2007). Michalopoulos and Karagiannidis (2008) advance this analysis by providing the exact outage probability expressions at arbitrary SNR, of the DF Selection Cooperation and DF Opportunistic Relaying schemes. The authors show that DF Selection Cooperation slightly outperforms *DF Opportunistic Relaying* in terms of outage probability. The reason for this is best illustrated with an example.

Figure 1. Outage probability for selection cooperation with a single source-destination pair. R = 1b/s/ Hz, $\lambda_{i,j}$ =1, m = 3, 4, ..., 6. (©2008, IEEE. Used with permission).

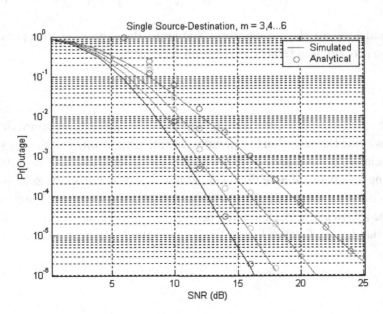

Consider a two relay network with an outage threshold $T = \dfrac{2^{2R} - 1}{P/N_0} = 3$ and the following channel powers: $\left|a_{sr_1}\right|^2 = 3.5, \left|a_{r_1 d}\right|^2 = 1.5, \left|a_{sr_2}\right|^2 = 2.5, \left|a_{r_2 d}\right|^2 = 2, \left|a_{sd}\right|^2 = 2.$ In this case, DF Opportunistic Relaying selects the second relay, r_2, with the corresponding channel $\left|a_k\right|^2 = \min\left\{ \left|a_{sr_2}\right|^2, \left|a_{r_2 d}\right|^2 \right\} = 2.$ Because $\left|a_k\right|^2 < T$, the selected relay remains silent resulting in an outage, since the source-destination channel is in outage. DF Selection Cooperation, on the other hand, selects r_1, the only member of the decoding set. The resulting channel $\left|a_k\right|^2 = \left|a_{sr_1}\right|^2 = 1.5$ adds coherently to the source-destination and prevents an outage. From this example, it is clear that DF Selection Cooperation outperforms DF Opportunistic Relaying because it allows for more frequent relaying by requiring that only the source-relay channel be above the threshold.

Amplify-and-Forward

The outage probability of amplify-and-forward assuming a source-destination link, and *m-1* potential relay nodes and the selection criteria given by (3) is given by Zhao et al. (Zhao & Adve & Lim, 2007).

$$P_{out-AF} = \frac{\dfrac{1}{\lambda_{sd} P_s} \displaystyle\prod_{k=1}^{m-1}\left(\dfrac{1}{P_s \lambda_{sr_k}} + \dfrac{1}{P_r \lambda_{r_k d}} \right)}{m} \left(\frac{2^{2R} - 1}{1/N_o} \right)^m \tag{10}$$

Bletsas et al. assume an aggregate power constraint and give the exact outage probability of *AF Opportunistic Relaying* by neglecting the source-destination channel (Bletsas & Khisti & Win, 2007).

$$P_{out\_AF} = \prod_{k=1}^{m-1}\left[1 - \lambda_{r_k d} \int_0^\infty \exp\left\{ -\lambda_{sr_k} \frac{2^{2R} - 1}{P_s/N_o}\left(1 + \left(\frac{P_s/N_0}{P_r/N_0}\frac{1}{\lambda_{sr_k}} + \frac{1}{P_r/N_0}\right)\frac{1}{z} \right) - \lambda_{r_k,d} z \right\} dz \right]. \tag{11}$$

As with decode-and-forward, the authors show that under the aggregate power constraint, amplify-and-forward selection is optimal in terms of outage probability.

Bit-Error Rate

In (Michalopoulos & Karagiannidis, 2008) the authors provide closed-form approximations to the bit error rate probabilities of the DF Selection Cooperation and DF Opportunistic Relaying schemes. In terms of bit error rate, the relative performance of the two schemes depends on the threshold used to accept a relay into the decoding set. We note here that evaluating cooperative diversity schemes in terms of bit error rates, where we allow for the possibility of error propagation, is a broad research area. We point the interested reader to the following works and references therein (Atay Onat, & Adinoyi & Yanikomeroglu & Thompson & Marsland; Atay Onat & Fan & Yanikomeroglu & Thompson).

Diversity Multiplexing Trade-Off

In our discussion so far, we have focused on reliability, measured here through the outage probability. On the other hand, the use of multiple antennas suggests that the transmit data rate, the so called multiplexing gain, can be increased as the SNR increases. The diversity order and normalized multiplexing gain are defined as (Zheng & Tse, 2003).

$$\Delta_{sel} = - \lim_{SNR \to \infty} \frac{P_{out}(SNR)}{\log(SNR)} \text{ and } \Delta_{sel} = - \lim_{SNR \to \infty} \frac{P_{out}(SNR)}{\log(SNR)} \quad R_{norm} = \lim_{SNR \to \infty} \frac{R(SNR)}{\log(SNR)} \text{ and}$$

$$R_{norm} = \lim_{SNR \to \infty} \frac{R(SNR)}{\log(SNR)}, \tag{12}$$

where $P_{out}(SNR)$ and $R(SNR)$ are outage probability and the achievable data rate at a signal-to-noise ratio of SNR. Note that if a transmission scheme achieves the maximum possible multiplexing gain, it achieves zero diversity.

For the single source-destination pair, the diversity-multiplexing trade-off was derived by Bletsas et al. (2006) for the Opportunistic Relaying scheme for both DF and AF:

$$\Delta_{sel} R_{norm} = m(1 - 2R_{norm}) \cdot \tag{13}$$

We leave a comparison of this trade-off with DSTC for the section on multiple-source transmissions.

PRACTICAL IMPLEMENTATIONS OF THE PROTOCOLS

The implementation of selection in distributed networks has been scarcely researched. In this section, we discuss two implementation issues: the implementation of relay selection in general networks, and the design of relay selection metrics in coded, complexity constrained networks.

Implementation of Relay Selection in General Networks

In the above section, we discussed the metrics used to select a relay. Once the metrics are established and evaluated, from a theoretical perspective, the problem is solved. Practically speaking, however, we must address the issue of decision making: which entity evaluates the metrics and selects the relay? Clearly, Selection Cooperation and Opportunistic Relaying in both decode-and-forward and amplify-and-forward require channel information to make the relay selection. The implementation of the selection protocol thus depends on the network and the ability for nodes to communicate with each other. For the single source-destination pair, two decision protocols have been presented in the literature. Both protocols are distributed, in the sense that they do not necessitate a central unit to organize the transmissions. Here, we refer to them as *Relay-Driven* and *Destination-Driven* protocols.

Destination-Driven (Beres & Adve, 2008)

This protocol is more suited to DF Selection Cooperation. The source distributes its data in the first time slot, while the destination and each of the *m-1* relays decode this information. If a relay correctly decodes the source information, it sends a bit of feedback to the destination to indicate its participation in the decoding set *D(s)*. Using this bit of feedback for channel estimation, the destination selects the relay which maximizes the relay-destination channel. The *destination-driven* implementation is simple and practical for Selection Cooperation; it is, however, unsuited to Opportunistic Relaying where the relay selection metric includes the source-relay channel. Because this channel would need to be communicated to the destination, Opportunistic Relaying has a more efficient implementation with the *Relay-Driven* decision scheme.

Relay-Driven (Bletsas & Khisti & Reed & Lippman, 2006).

This protocol can implement both Selection Cooperation and Opportunistic Relaying. Each relay evaluates the selected metric and starts a timer with initial value T_i, inversely proportional to the metric. The timer of the relay with the largest metric reduces to zero first. This relay, thus having selected itself, sets off a flag and transmits the source information to the destination. The remaining relays overhear the flag and do not initiate their own transmissions. This implementation necessitates channel feedback to the relays. Furthermore, there is a non-zero probability, analyzed by the authors, that the timers of multiple relays will expire essentially simultaneously, causing a collision at the destination.

We note here that if cooperative diversity is implemented in a centralized network where the source (or some central unit) has access to channel information between all nodes, the relay-selection decision can be made before the source transmission, in which case only the selected relay decodes the source information while the other relays may conserving their circuit energy and go into sleep-mode.

Relay Selection in Coded, Complexity-Constrained Systems

In (Chu & Adve & Eckford, 2007), Chu et al. investigate the topic of relay selection for coded cooperation in wireless sensor networks, where complexity constraints on the nodes prevent the relay from fully decoding the source data. Instead, the relay uses repeat-accumulate (RA) codes to form codewords from demodulated bits. In this scenario with multiple potential relays which do not decode the source message, the authors show that optimal relay selection, based on the source-relay, relay-destination, and source-destination channel, is too impractical for low-complexity sensor nodes. The mutual information of the equivalent relay channel, however, is a good heuristic to approximate the optimal selection method. This choice, based on the notion that the mutual information is a good measure of the quality of the channel, performs well in terms of BER, and significantly outperforms the max-min criterion of DF Opportunistic Selection. More recent work has improved on this approach by accounting for the code structure using the Bhattacharya parameter (Chu & Adve & Eckford, 2008). Note that in demodulate-and-forward, the relay only demodulates, but does not decode the source's codeword.

Figure 2 shows the frame error rate (FER) for the low-complexity relay selection scheme using the Bhattacharya parameter. The FER is compared to that obtained by an exhausted search method, which serves as a lower bound on the FER performance. The low-complexity relay selection schemes exhibits performance very similar to the exhaustive search, demonstrating its effectiveness in minimizing FER.

SELECTION IN A MULTI-SOURCE NETWORK

System Model

In this section, we extend the concept of selection cooperation to network settings. The results presented in this section can be found in (Beres & Adve, 2008). The network comprises a set M of m nodes. Each node $s \in M$ has information to transmit to its own destination, $d \notin M$, and acts as a potential relay for other nodes in M. The transmission of each source occurs in orthogonal channels, and each channel is further subdivided into two time-slots to satisfy the half-duplex constraint. While we use the notation s for a source node and r for a relay node, we emphasize that every node in M is a source node and is potentially also a relay node. The channel between any two nodes is assumed independent of all other channels. This model is appropriate for networks where each node may have its own destination. As in the section above, the instantaneous channels a_{ij} between all nodes are modeled as independent, slow and flat - fading and Rayleigh. In this section we compare selection to DSTC as the closest alternative scheme that can address the multi-source problem.

Because the network comprises multiple sources, each node can potentially relay for several other nodes. This raises the question of relay selection and power allocation and motivates the various selection schemes discussed below. Note that the performance of the relay selection schemes from the previous section with only one source-destination node is not clear a-priori; under a peak power constraint, the relay selected by one source-destination pair can potentially influence the performance of another source-destination pair if it selects the same relay. The multi-source scenario therefore adds significant complexity to the relay selection problem.

Each node has a peak power constraint of P Joules/symbol. In DSTC, every node expects to relay

Figure 2. FER for the relay selection scheme (solid lines) and exhaustive search (dash-dot lines) for 1,2,3 relays. (©2008, IEEE. Used with permission).

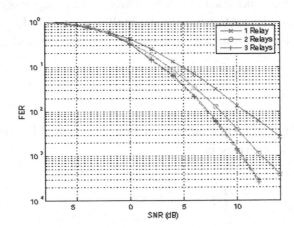

for all other *m-1* nodes and thus uses *2P/m* Joules/symbol per source in both phases. In our case, in the first phase, each source sends its data using full power *P* Joules/symbol. In the second phase, each relay divides its power evenly between the source nodes it is supporting. A relay node supporting *N* source nodes will thus use *P/N* Joules/symbol for each source. Note that a relay node does not know *a priori* how many nodes it will relay for and does not know the channel to the destination for these sources; it thus cannot pre-compute a better power distribution. Clearly, with additional feedback to the relays, a better power distribution may be possible.

We consider both centralized and decentralized versions of the network. A centralized network is governed by a central unit (CU) with knowledge of all network parameters. A CU makes all assignment decisions. In the absence of a CU, the network is decentralized and decisions are made locally by the nodes, with limited information regarding the rest of the network.

Selection Cooperation Schemes

The concept of selection in a network is identical to that presented in the preceding sections. In the first time-slot, all nodes use orthogonal channels to transmit information to their respective destinations, and each node decodes the information from the other *m-1* sources. Each node determines if it has decoded the information correctly. If node s_j has decoded the information from source s_i correctly, it declares itself as a member of the decoding set $D(s_i)$ of nodes eligible to relay for node s_i. Such a decoding set, $D(s_i)$, is formed for each source $s_i \in M$.

In the second time-slot, for each source s_i, a relay is chosen from its decoding set $D(s_i)$, and each relay forwards the information for the source. The activity of a node s_i can thus be summarized as follows: in the first time-slot, it transmits its information and decodes the information of the other *m-1* nodes; in the second time-slot, it forwards the information of those nodes for which it was chosen as a relay.

We present three relay selection schemes based on varying degrees of centralization and tolerance for numerical complexity. While in the above section the relay was selected as the one with the best instantaneous channel to the destination, in a network setting the per-node power constraint motivates search for a more sophisticated schemes. For example, suppose two source nodes, s_1 and s_2 are assigned to the same relay *r* with the best instantaneous channel to both $d(s_1)$ and $d(s_2)$. The power available at node *r* for each source is *P/2*. However, performance could potentially be improved by assigning one of the source nodes to a different "free" relay node with available power *P*. The problem is thus to assign relays to source nodes to minimize some figure of merit which depends on channel conditions as well as available power at the relays. However, as we shall see, the simple assignment scheme of the above section, extended to multiple sources, remains an effective tradeoff between complexity and performance.

Optimal Relay Assignment

We state here the optimal relay choice in a network setting. The mutual information between source s_i and destination d_{s_i} if using r_i as the relay is:

$$I_{s_i d_{s_i}; r_j} = \frac{1}{2} \log \left(1 + SNR \left| a_{s_i, d(s_i)} \right|^2 + \frac{SNR}{N_j} \left| a_{r_i, d(s_i 0)} \right|^2 \right), \tag{14}$$

where N_j is the number of sources that choose node r_i as their relay, and $SNR = P/N_0$ is the signal-to-noise ratio, without accounting for fading, at the relay and destination. One optimal approach, in max-min sense, calculates the mutual information between source and destination of all m transmissions for all possible relay assignments, and picks the relay assignment which maximizes the minimum mutual information of these m transmissions:

$$\{r(s_1),\ldots,r(s_m)\} = \arg \max_{\forall i_1 \in D(s_1),\ldots,i_m \in D(s_m)} \min \left\{ I_{s_1 d_{s_1} ; r_{i_1}}, \ldots, I_{s_m d_{s_m} ; r_{i_m}} \right\} \tag{15}$$

where, for example, $I_{s_1 d_{s_1} ; r_{i_1}}$ denotes the mutual information between source s_1 and its destination $d(s_1)$, with node r_{i_1} taken from $D(s_1)$ used as a relay. This optimal scheme requires a CU with global knowledge of all channels to make m^m comparisons and choose the best one in max-min sense. Hence, though optimal, this scheme is clearly impractical.

Sequential Relay Selection

The optimal algorithm described can be simplified considerably by performing this search sequentially for each source node in M. The sub-optimal algorithm thus works as follows. The relay of the first node, $r(s_1)$, is chosen from $D(s_1)$, the decoding set of s_1 independently of the other sources. This relay, $r(s_1)$, is the node with the highest channel power to the destination of s_1, $d(s_1)$. For the second source node s_2, two potential relaying nodes from $D(s_2)$ are picked for consideration: nodes r_j and r_k, with the best and second-best channels to the destination, respectively. If r_j is not already used as a relay for s_1, i.e., $r_j \neq r_{s_1}$, it is automatically chosen as the relay for s_2. If r_j is already relaying, however, the CU decides between r_j and r_k by considering that r_j would need to halve its power to accommodate both source nodes s_1 and s_2. The CU thus compares $I_{s_2 d_{s_2} ; r_k}$ to $\min \left\{ I_{s_1 d_{s_1} ; r_j}, I_{s_2 d_{s_2} ; r_j} \right\}$, where

$$I_{s_2 d_{s_2} ; r_k} = \frac{1}{2} \log \left(1 + SNR \left| a_{s_2, d(s_2)} \right|^2 + \frac{SNR}{1} \left| a_{r_k, d(s_2)} \right|^2 \right),$$

$$I_{s_1 d_{s_1} ; r_j} = \frac{1}{2} \log \left(1 + SNR \left| a_{s1, d(s_1)} \right|^2 + \frac{SNR}{2} \left| a_{r_j, d(s_1)} \right|^2 \right),$$

$$I_{s_2 d_{s_2} ; r_j} = \frac{1}{2} \log \left(1 + SNR \left| a_{s_2, d(s_2)} \right|^2 + \frac{SNR}{2} \left| a_{r_j, d(s_2)} \right|^2 \right). \tag{16}$$

If $I_{s_2 d_{s_2} ; r_k}$ is the larger value, r_k is chosen as the relay for s_2; otherwise, r_j is chosen. This process repeats until each source has been assigned a relay, with potentially one more relay added to the comparison for each source node considered. For each source node s_i the CU begins with the relay r_j with the best channel to $d(s_i)$. If that relay is free, it is automatically assigned to s_i. Otherwise, the CU calculates the mutual information for s_i and for all N_j source nodes already supported by r_j using $P/(N_j+1)$ as the available relay power. The minimum of all the N_j+1 values of mutual information is associated with relaying node r_j. The CU repeats this process for the relay with the next best channel, until it finds a relay that is not claimed by any source node. It then assigns the source s_i to the relay node with the highest relay-destination mutual information.

The iterative scheme is significantly simpler than the exhaustive search of the optimum scheme, but is yet of $O(m!)$ complexity. Also, like the optimum scheme, it requires a centralized node with knowledge of all inter-node channels. Therefore, while more efficient than the optimum scheme, this scheme is also likely to be impractical.

Distributed Relay Assignment

The simple selection scheme extends the selection approach in first section to the network setting considered here. The destination of s_i, $d(s_i)$, picks as its relay the node with the highest instantaneous relay-destination channel power, $r(s_i)$:

$$r(s_i) = \arg \max_{r_k \in D(s_i)} \left\{ \left| a_{rk} \right|^2 \right\}; k = 1 \ldots \left| D(s_i) \right|, \tag{17}$$

i.e., each relay is picked independently of the other source-destination pairs. The only penalty is that a node relaying for N_j sources uses power P/N_j for each forwarding link. As the numerical results given below show, this scheme is extremely effective, achieving near optimum performance for small network sizes. The rest of this section is largely focused on this scheme.

This simple scheme is of $O(m^2)$ complexity in the worst case, where each of the m destinations must make $m-1$ comparisons to find a maximum out of (at most) $m-1$ channels. The scheme can also be implemented in fully decentralized manner, as each destination picks its relay independently of all other nodes.

Evaluation of Relay Selection Schemes

Outage Probability

The outage probability of Simple Selection scheme is given by

$$P_{out} = \left(\frac{2^{2R}-1}{SNR} \right)^m \lambda_{sd} \sum_{|D(s)|} \frac{1}{\left| D(s) \right| + 1} \prod_{r_i \in |D(s)|} \lambda_{r_i,d} \prod_{r_i \notin |D(s)|} \lambda_{s,r_i} \sum_{n=1}^{m-2} K_m (n+1)^{\left| D(s) \right|}, \tag{18}$$

where

$$K_m = \binom{m}{n} \left(\frac{1}{m-2} \right)^n \left(\frac{m-3}{m-2} \right)^{m-2-n} \tag{19}$$

Figure 3 compares the outage probability of all three selection schemes via simulation. Not surprisingly, performance improves with increasing system intelligence. Finally, the difference in performance between the three schemes also increases with increasing network sizes. This is to be expected, since with more nodes there is a higher potential for power splitting, and the optimal and sub-optimal approaches reduce this problem. Note, however, that even for a network size of $m = 5$, the performance loss of the simplest scheme is not very large. Our conclusion is that the simple selection scheme is a very good and practical choice

The outage probability of DSTC and Simple Selection Cooperation is compared in Figure 4 for channels with $\lambda = 1$ and $R = 1$ b/s/Hz. In this scenario, the Simple Selection Cooperation always outperforms DSTC for increasing network sizes. This improvement comes from efficient use of power in Selection, which allocates all available power to the best channel. The higher the network size, the larger the probability that the selected channel is "good". DSTC, on the other hand, allocates the power evenly across all channels, and thus cannot take advantage of the strongest channel.

Figure 5 shows the ratio (independent of SNR) of the outage probability of Selection Cooperation to the outage probability of DSTC for various rates R and network sizes m. Clearly,

Selection Cooperation outperforms DSTC when this ratio is less than one, as is the case for all shown values of R when $m > 3$. DSTC, on the other hand, perform better for the small network size of $m = 3$ when $R > 2$. The entire function falls sharply with increasing m, while for larger values of m the dependence on R is negligible.

Diversity Multiplexing Trade-Off

Using the same definitions as in the preceding section, we have derived the Diversity Multiplexing Trade-off for the Simple Selection Scheme.

$$\Delta_{sel} R_{norm} = m(1 - 2R_{norm}) \tag{19}$$

Figure 3. Outage probabilities of simple, sub-optimal and optimal selection. R = 1b/s/Hz, $\lambda_{i,j}$ =1, m = 3, 5. (©2008, IEEE. Used with permission).

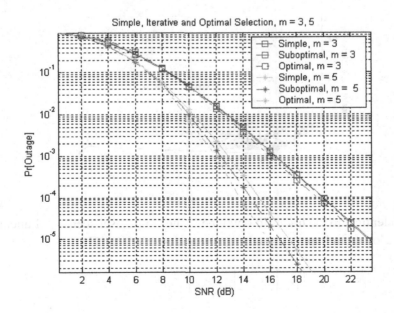

Figure 4. Outage probabilities of DSTC and Simple Selection Cooperation. R = 1b/s/Hz, $\lambda_{i,j}$ =1, m = 3, 5, 8. (©2008, IEEE. Used with permission).

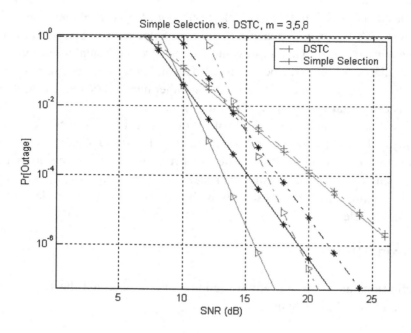

Figure 5. Ratio of Selection outage probability to DSTC outage probability for R = 1 − 5 and m = 3,4,...,10. . (©2008, IEEE. Used with permission).

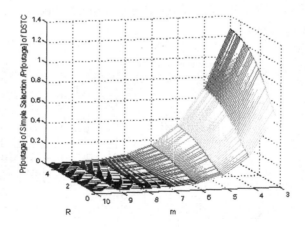

This trade-off is identical to the trade-off obtained in the single-user scenario. Laneman and Wornell (2003) determine the lower and upper bounds for the trade-off of DSTC as

$$m(1 - 2R_{norm}) \leq \Delta_{DSTC}R_{norm} \leq m\left(1 - \left\lceil \frac{m-1}{m} \right\rceil 2R_{norm}\right) \tag{20}$$

The selection diversity-multiplexing trade-off is thus exactly the lower bound of the DSTC trade-off. For increasing network sizes *m*, however, the upper and lower bound converge, as does the performance of selection cooperation and DSTC. Furthermore, the flat curves as a function of *R* for higher *m* in Figure 5 suggests that DSTC reaches only the lower bound of the trade-off curve for higher *m*, and the performance difference between the two schemes is dominated by the SNR-gain of their outage probabilities.

Overhead and System Requirements

Due to the geographical distribution of nodes, any cooperative transmission scheme suffers some overhead. In this section, we discuss the overhead, centralization and complexity requirements of selection cooperation.

After the first time slot, each of the *m* relays attempts to decode *m-1* messages. For each message, the relay must indicate its success or failure with one bit, resulting in a total of *m(m-1)* feed-forward bits. For each source, the destination (or the CU) must select one of the *m-1* relays using *log(m-1)* feedback bits, resulting in a total of *mlog(m-1)* feedback bits. The overhead is thus *m[log(m-1) +(m-1)]* bits per network, and *[log(m-1) + (m-1)]* bits per source-destination pair.

The simple selection scheme is fully decentralized (i.e. the destination only needs to know the channels from its relays), and thus the above calculated overhead is sufficient to implement this scheme. The transmission of the overhead bits must occur on orthogonal channels, decreasing the available bandwidth for the transmission and affecting system performance. We assume, however, that cooperation will be implemented in low-mobility environments, such as mesh networks, where channels change slowly in time. In such a case, the channel coherence time spans many code blocks, and the *m[log(m-1) + (m-1)]* overhead bits are assumed negligible in comparison to the transmitted information bits.

It is difficult to compare the implementation overhead of DSTC and selection cooperation. In both schemes, only the relay nodes that have correctly decoded the message proceed to encode and transmit the data. Because the relays must inform each other whether they have correctly decoded the message of each of the sources, a feed-forward overhead also exists in DSTC systems, and is identical to that of selection cooperation: *(m-1)* bits per source-destination pair. The incremental overhead of the simple selection scheme is thus only *log(m-1)* bits.

In a traditional MIMO system, space-time coding may be preferable to selection diversity because its implementation does not require feedback. In a distributed system, however, DSTC incurs a significant overhead, generally not accounted for in any analysis. DSTC requires overhead for synchronization and it is likely that it will be larger than the *log(m-1)* bits required to implement simple selection cooperation. The performance comparison of selection cooperation and DSTC thus becomes fair in terms of overhead requirements. Additionally, there may be feedback required to organize the participating nodes into indexes of the DSTC matrix. Given that the overhead for synchronization and this organization is difficult to quantize, for this purposes of this paper, we do not assume any further centralization or overhead requirements of DSTC. In summary, the network requirements of each scheme are shown in Table 1.

Table 1. Complexity and centralization requirements for the simple, sub-optimal and optimal selection schemes and DSTC

Scheme	Complexity	Centralization	Overhead
Simple	m^2	No	$m[log(m-1) + (m-1)]$
Sub-Optimal	$m!$	Yes	$m[log(m-1) + (m-1)] + Lm^m$
Optimal	m^m	Yes	$m[log(m-1) + (m-1)] + Lm^m$
DSTC	--	No	$m(m-1)$

CONCLUSION

In this chapter, we have explored Selection in cooperative diversity systems in a variety of scenarios. In a single source-destination network under an aggregate power constraint, Selection is optimal in terms of outage probability for both DF and AF systems. In this context, Selection can be interpreted as a special case of beamforming with columns of the identity matrix, where it is optimal for AF under the constraint of $log_2(m-1)$ bits of feedback. In a multi-source network, Selection outperforms distributed space-time codes in most scenarios even under a per-node constraint. This strong performance of Selection is due to its efficient use of power and bandwidth as compared to other available schemes, and has motivated the search for suitable relay selection schemes in practical, low-complexity coded networks.

REFERENCES

Atay Onat, F., Adinoyi, A., Yanikomeroglu, H., Thompson, J. S., & Marsland, I. D. (Accepted for publication). Threshold selection for SNR-based selective digital relaying in cooperative wireless networks. [retrieved from http://www.sce.carleton.ca/faculty/yanikomeroglu/cv/publications.shtml]. *IEEE Transactions on Wireless Communications*.

Atay Onat, F., Fan, Y., Yanikomeroglu, H., & Thompson, J. S. (Accepted for publication). Asymptotic BER analysis of threshold digital relaying schemes in cooperative wireless systems. [retrieved from http://www.sce.carleton.ca/faculty/yanikomeroglu/cv/publications.shtml]. *IEEE Transactions on Wireless Communications*.

Barbarossa, S., & Scutari, G. (2004). Distributed space-time coding for multi-hop networks. In . *Proceedings of IEEE International Conference on Communications, ICC*, 2004.

Beres, E., & Adve, R. S. (2007). Outage probability of selection cooperation in the low to medium SNR regime. *IEEE Communications Letters, 11*(7), 589–591. doi:10.1109/LCOMM.2007.070097

Beres, E., & Adve, R. S. (2008). Selection cooperation in multi-source cooperative networks. *IEEE Transactions on Wireless Communications, 7*(1), 118–127. doi:10.1109/TWC.2008.060184

Bletsas, A., Khisti, A., Reed, D. P., & Lippman, A. (2006). A simple cooperative diversity method based on network path selection. *IEEE Journal on Selected Areas of Communications, Special Issue on 4G Wireless Systems, 24*(3), 659-672.

Bletsas, A., Shin, H., & Win, M. Z. (2007). Cooperative communications with outage-optimal opportunistic relaying. *IEEE Transactions on Wireless Communications*, *6*(9), 3450–3460. doi:10.1109/TWC.2007.06020050

Boyer, J., Falconer, D., & Yanikomeroglu, H. (2004). Multihop diversity in wireless relaying channels. *IEEE Transactions on Communications*, *52*(10), 1820–1830. doi:10.1109/TCOMM.2004.836447

Chu, J., Adve, R. S., & Eckford, A. (2007). Relay selection for low-complexity coded cooperation. In *IEEE Globecom 2007*.

Chu, J., Adve, R. S., & Eckford, A. (2008). Relay selection for low-complexity coded cooperation using the Bhattacharyya parameter. In *IEEE Int. Conf. on Comm. 2008*.

Herhold, P., Zimmermann, E., & Fettweis, G. (2005). Cooperative multi-hop transmission in wireless networks. *Computer Networks Journal*, *49*(3), 229–324.

Laneman, J. N., Tse, D. N. C., & Wornell, G. W. (2004). Cooperative diversity in wireless networks: Efficient protocols and outage behavior. *IEEE Transactions on Information Theory*, *50*(12), 3062–3080. doi:10.1109/TIT.2004.838089

Laneman, J. N., & Wornell, G. W. (2003). Distributed space-time-coded protocols for exploiting cooperative diversity in wireless networks. *IEEE Transactions on Information Theory*, *49*(10), 2415–2425. doi:10.1109/TIT.2003.817829

Michalopoulos, D. S., & Karagiannidis, G. K. (2008). (Accepted for publication). Performance analysis of single relay selection in Rayleigh fading. [retrieved from http://users.auth.gr/~dmixalo/]. *IEEE Transactions on Wireless Communications*.

Sendonaris, A., Erkip, E., & Aazhang, B. (2003). User cooperation diversity-part I, II. *IEEE Transactions on Communications*, *51*(11), 1927–1948. doi:10.1109/TCOMM.2003.818096

Uysal, M., & Canpolat, O. (2005). On the distributed space-time signal design for a large number of relay terminals. In *Proceedings of IEEE Wireless Communications and Networking Conference*, 2005.

Zhao, Y. (2008). Cooperative networks with multiple relays: Selection, repetition, and beamforming. Unpublished doctoral dissertation, University of Toronto, Toronto, ON.

Zhao, Y., Adve, R. S., & Lim, T. J. (2007). Improving amplify-and-forward relay networks: Optimal power allocation versus selection. *IEEE Transactions on Wireless Communications*, *6*(8), 3114–3123.

Zhao, Y., Adve, R. S., & Lim, T. J. (2007b). Beamforming with limited feedback in amplify-and-forward cooperative networks. In *IEEE Globecom 2007*.

Zheng, L., & Tse, D. N. C. (2003). Diversity and multiplexing: A fundamental tradeoff in multiple-antenna channels. *IEEE Transactions on Information Theory*, *49*(5), 1073–1096. doi:10.1109/TIT.2003.810646

Chapter 10
Energy Efficient Communication with Random Node Cooperation

Zhong Zhou
University of Connecticut, USA

Jun-Hong Cui
University of Connecticut, USA

Shengli Zhou
University of Connecticut, USA

Shuguang Cui
Texas A&M University, USA

ABSTRACT

In this chapter, we focus on the energy efficient cooperative communication with random node cooperation for wireless networks. By "random," we mean that the cooperative nodes for each communication event are randomly selected based on the network and channel conditions. Different from the conventional deterministic cooperative communication where cooperative nodes are determined prior to the communication, here the number of cooperative nodes and the cooperation pattern may be random, which is more practical given the random nature of the channels among the source nodes, relay nodes, and destination nodes. In addition, it is more robust to the dynamic wireless network environment. Starting with a thorough literature survey, we then discuss the challenges for random cooperative communication systems. Afterwards, two examples are presented to illustrate the design methodologies. In the first example, we analyze a simple scheme for clustered wireless networks, where cooperative communication is deployed in the long-haul inter-cluster transmissions to improve the energy efficiency. We quantify the energy performance and emphasize its difference from the conventional deterministic ones. In the second example, we consider the cross-layer design between the physical layer and the medium access control (MAC) layer for the one-hop random single-relay networks. We unify the power control and the relay selection at the physical layer into the MAC signaling in a distributed fashion. This example clearly shows the strength of cross-layer design for energy-efficient cooperative systems with random node collaboration. Finally, we conclude with discussions over possible future research directions.

DOI: 10.4018/978-1-60566-665-5.ch010

INTRODUCTION

In energy-constrained wireless networks, such as wireless sensor networks, energy is normally non-renewable and is thus of paramount importance in the network design and operation. It is well known that multi-input multi-output (MIMO) techniques based on antenna arrays can dramatically reduce the required transmission energy in wireless fading environments under a certain throughput constraint due to spatial diversity. However, many wireless devices, for example, wireless sensor nodes, are usually limited in size such that it is impossible to mount multiple antennas at one node. Consequently, the traditional MIMO techniques can not be used in such networks to save energy. Nevertheless, it has been found that cooperative communication schemes, where multiple nodes collaborate on forming virtual antenna arrays, can be deployed to improve energy efficiency.

In this chapter, we focus on energy efficient cooperative communications with random node cooperation for wireless networks. By "random", we mean that the cooperative nodes for each communication event are randomly selected based on the network and channel conditions. Different from the conventional deterministic cooperative communication where cooperative nodes are determined prior to the communication, here the number of cooperative nodes and the cooperation pattern may be random, which is more practical given the random nature of the channels among the source nodes, relay nodes, and destination nodes. The random nature of cooperators incurs many new design challenges over those in the deterministic cooperative communication. In this chapter, we elaborate on these issues. Two typical examples are given to illustrate the general methodology and analysis techniques for random cooperative systems. We emphasize on the system energy efficiency to show the benefits of random cooperation for energy constrained wireless networks.

The rest of the chapter is organized as follows. We first introduce the related background with a thorough literature survey over cooperative communication. Then, we discuss in depth the design challenges for random cooperative schemes. After that, we give two design examples. The first example is for clustered wireless networks, where nodes in one cluster randomly participate in the long-haul inter-cluster communication to improve the overall energy efficiency. We will highlight the analysis of the system energy efficiency and clearly show its difference from the conventional deterministic ones. In the second example, we discuss the cross-layer design between the physical layer and the medium access control (MAC) layer for the one-hop random single-relay networks. We unify the power control and the relay selection at the physical layer into the MAC signaling in a distributed fashion. This example clearly illustrates the strength of cross-layer design for random cooperative systems. We conclude with some observations and a discussion of future research directions.

BACKGROUND

A large amount of research work has been done recently investigating various cooperative relay schemes. Generally speaking, these schemes can be classified into the following two categories: amplify-and-forward schemes and decode-and-forward schemes. For the first category, cooperative nodes do not decode their received packets, they just amplify and relay the received signal to the destination. Schemes such as those in [Sendonaris et al., 2003, Ochiai et al., 2005] belong to this category. In [Sendonaris et al., 2003, Ochiai et al., 2005], the source and cooperative nodes adjust the phase of their transmissions such that their signal can add coherently at the destination (i.e., distributed beamforming). Amplify-and-forward schemes usually impose strict requirements over the RF front-ends, which increases system complexity and costs.

For the decode-and-forward schemes, cooperative nodes decode their received packets, then re-encode these packets and relay them to the destination. No beamforming capabilities are assumed for these schemes. In [Laneman et al., 2004, Stefanov and Erkip, 2004], the authors analyzed cooperative protocols under the framework of diversity-multiplexing tradeoffs. The basic setup includes one sender, one receiver, and one intermediate relay node. In [Laneman and Wornell, 2003], the authors studied general cooperative systems with multiple relay nodes. In this work, the authors investigated the diversity-multiplexing tradeoff and proposed distributed space-time coding based on orthogonal space-time codes [Tarokh et al., 1999]. In order to apply such distributed codes in practice, certain code distribution protocols need to be deployed to assign code matrix columns to individual cooperators. A closely-related study was presented in [Sirkeci-Mergen and Scaglione, 2005], where the number of relaying nodes is modeled as a random variable and each relay node randomly selects a column from an orthogonal space-time code matrix. This random code selection scheme simplifies the code distribution process. A thorough performance analysis along with asymptotic properties (over SNR and number of nodes) is also presented in [Sirkeci-Mergen and Scaglione, 2005]. It is shown that full diversity (equal to the total number of columns of the chosen space-time code matrix) can be achieved even with this simple random coding scheme, if the total number of cooperative nodes is large enough.

System performance of cooperative schemes can be further improved if energy is optimally allocated across the collaborative system. In [Hasna and Alouini, 2004], the authors optimized the energy allocation among relay nodes with the objective of minimizing the link outage probability. This energy allocation problem is converted into a convex optimization problem, which can be efficiently solved. It should be noted that the collaborative pattern in [Hasna and Alouini, 2004] is deterministic. In [Su et al., 2005], the authors presented an optimal power allocation scheme for a decode-and-forward cooperation protocol based on the symbol error rate (SER) analysis. The results therein are for a two-node cooperation system, where the source and the relay need to know the link state information. With such channel state information both at the source and the destination, the outage performance of different cooperative protocols has been analyzed in [Ahmed et al., 2004a]. It is shown that power control could provide significant performance improvements. For two-node amplify-and-forward cooperation protocols, significant energy savings can be obtained through power control even when a few bits of quantized channel information are available at the source [Ahmed et al., 2004b, Ahmed et al., 2006]. Their analysis is based on the average power constraint where the average transmission power over all channel realizations is limited by a predefined value. The energy efficiency of deterministic cooperative communication has particularly been investigated in [Cui et al., 2004, Cui and Goldsmith, 2006]. The authors of [Cui et al., 2004] investigated the energy issues in a clustered sensor network, where sensors collaborate on signal transmission and/or reception in a deterministic way. It is shown that if the long-haul transmission distance (between clusters) is large enough, cooperative communication can dramatically reduce the total energy consumption even when all the collaboration overhead is considered.

In practical networks, with multiple potential cooperative nodes at hand, it is possible to select only one or several of them to reduce the cost and improve the system performance. Cooperative node selection schemes have been investigated recently [Madan et al., 2006, Zorzi and Rao, 2003, Bletsas et al., 2006]. The authors of [Madan et al., 2006] took the cooperative overhead into consideration and proposed an optimal relay-selection rule. Training sequences are needed in their scheme in order for the destination to perform channel state information (CSI) estimation. Node location and distance information is used for cooperative node selection in [Zorzi and Rao, 2003]. In [Bletsas et al., 2006], the authors proposed a new selective cooperation scheme based on the instantaneous channel strength, where potential relays compete with each other based on certain policies and only one winner is selected to aid the communication process.

Cooperative communication has also been investigated together with upper-layer network protocols [Cui and Goldsmith, 2006, Zheng et al., 2005, Scaglione et al., 2006, Liu et al., 2006, Jakllari et al., 2006, Zhang and Zhang, 2008]. Based on [Cui et al., 2004], the authors in [Cui and Goldsmith, 2006] combined the cooperative communication scheme with a cross-layer design framework for multi-hop clustered sensor networks. The system is optimized to improve the overall energy efficiency and reduce the network delay. In [Zheng et al., 2005], the authors examined the performance of cooperative strategies in interference-limited ad hoc networks, where the cooperative scheme at the physical layer is jointly analyzed with the MAC layer. It is shown that cooperative schemes can improve the overall network performance when jointly designed with fair and efficient resource allocation schemes. In [Scaglione et al., 2006], the authors clarified the gap between the link abstraction used in the traditional wireless networking and its broader definition in the cooperative communication, and showed the complexity tradeoff among the physical and higher layers for cooperative communication. Liu et al. [Liu et al., 2006] pointed out the necessity of cross-layer design for cooperative communication and propose a new MAC protocol for IEEE 802.11 networks. Jakllari et al. [Jakllari et al., 2006] proposed a new framework for cooperative communication in multi-hop wireless networks. Their approach spans the physical, MAC, and routing layers. The key physical layer property that they exploited is the increased transmission range due to the diversity gain of cooperative communications. They showed that significant improvements in the end-to-end performance such as throughput and delay can be achieved in multi-hop wireless environments. The proposed cross-layer design approach is robust to mobility and interference-induced link failures. In [Zhang and Zhang, 2008], the authors investigated cooperation-aware routing in multi-source multi-destination multi-hop wireless networks. In this work, virtual-link-based contention graphs are used in the routing process to optimize the network performance.

Much research work has been done for cooperative communication in special network scenarios such as clustered wireless networks and homogenous networks. The authors of [Dohler et al., 2006] analyzed the distributed space time block coding (**DSTBC**) based cooperative communication for multi-tier clustered wireless sensor networks. Through their analysis on the SER and throughput performance, the authors showed that cooperative communication is more energy efficient than direct communication. However, in this work, the number of cooperative nodes in each cluster is fixed and the inherent circuit energy consumption of wireless transceivers is ignored, which is reported recently to be important for low-power wireless sensor networks [Cui et al., 2004]. Yong et al. [Yong et al., 2006] extended the LEACH protocol [Wendi et al., 2000] to enable the multi-hop transmissions among clusters by incorporating a cooperative MIMO scheme into hop-by-hop transmissions. Effective performance improvements in terms of energy efficiency and reliability can be achieved through the adaptive selection of cooperative nodes and the joint consideration over multi-hop routing and cooperative communication.

Although previous research shed much light on the analysis and the design process of cooperative communication, much of them ignore the distributed random nature of cooperative communication systems in practice and assume deterministic cooperation among nodes, which may somewhat mislead the system design. In random cooperative communication schemes, the cooperative nodes for each communication event are randomly chosen based on the network and channel conditions. They are more robust and adaptive to the dynamic network and channel conditions, and thus are expected to be more

Figure 1. Comparison between random and deterministic cooperation

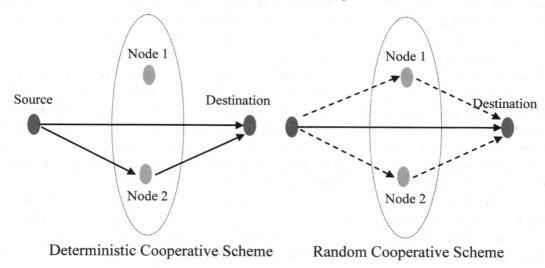

Deterministic Cooperative Scheme   Random Cooperative Scheme

efficient. Up to now, limited research has been done in this area. In the next section, we will discuss the challenges for the analysis and design of a random cooperative communication system. We hope that our analysis could motivate more research interests over this interesting subject.

CHALLENGES

Compared with the deterministic counterparts, random cooperative communication incurs dynamic cooperative patterns and thus imposes many new design challenges. As such, the analysis and design process for deterministic schemes cannot be directly used here.

First, cooperative nodes now randomly participate in each communication event according to the network and channel conditions. Performance analysis for such random distributed systems is quite complex. For deterministic cooperative communication, the outage as well as SER performance has been well explored [Laneman and Wornell, 2003, Ahmed et al., 2004a]. Because of the pre-defined cooperative pairs, performance analysis for such deterministic cooperative systems is simplified. However, the story for random cooperative communication is different. For random schemes, cooperative pairs are randomly formed based on the current network and channel conditions. The same source may utilize different cooperative partners at different times even for the same destination. Let us use one simple example to make things clear. As shown in Figure 1, for the deterministic scheme, with a pre-defined pattern, node 2 will be used as the cooperative node and will not change over time. In this case, node 1 has no impact on the system, even sometimes node 1 may be a better cooperator than node 2. While for the random scheme, cooperating nodes are randomly determined based on the current network and channel conditions. Both node 1 and node 2 have certain probabilities to serve as the cooperator in the system. Thus, the system performance now is jointly related to node 1 and node 2.

The corresponding analysis for such a random system is thus much more complicated than the deterministic one. Note that in practical networks, both the number of cooperative nodes and the wireless channels are random, which further complicates the system analysis. Some pioneering work such as those in [Sirkeci-Mergen and Scaglione, 2005] has been done to analyze random cooperative communications. But more theoretical analysis is needed for the further understanding of random cooperative schemes.

Second, in order to optimize system performance for cooperative systems, delicate distributed power control strategies are needed at each cooperative node. For deterministic cooperative schemes, power control has been well exploited [Hasna and Alouini, 2004, Su et al., 2005, Ahmed et al., 2004a], either to minimize the outage probability or to maximize the energy efficiency. However, as for the random case, although some work such as [Madan et al., 2006] has been done, many issues remain un-addressed. The randomness of cooperation imports one more degree of freedom into the design space. Here, not only the transmitting power, but also the actual cooperative nodes need to be adjusted on-the-fly. System performance can be optimized with a joint design of cooperative node selection and power control. For example, in [Madan et al., 2006], the authors showed the great energy advantage of the joint node selection and power control for random beamforming cooperative systems in homogeneous networks. However, purely distributed power control is quite difficult for general random distributed networks. First of all, in order to optimize the performance of the whole network, the global optimum and convergence should be guaranteed. In addition, for distributed networks, communication and energy overhead at all layers need to be minimized. In this sense, power control for general random cooperative systems is no longer an isolated issue at the physical layer, but an involved problem across the physical layer, the MAC layer, and even the routing layer.

Third, new cooperative protocols are needed for the random interaction among cooperative nodes. For deterministic schemes, cooperative nodes are pre-defined such that the cooperative protocol could be straightforward. For example, the training sequences in the cooperative beamforming schemes or the space-time code columns in the **DSTBC** schemes are assumed to be fixed and known *a prior* to each cooperative node. While in random cases, these assumptions do not hold any more and new dynamic protocols need to be devised. For example, in random STBC schemes, the number of space-time code columns is usually much smaller than the number of nodes in the network, thus it is impossible to assign a unique code to each node beforehand. New cooperative protocols are thus indispensable for the real-time distribution of the space-time code columns to active cooperative nodes. Compared with deterministic ones, these dynamic cooperative protocols will inevitably incur extra communication overhead and degrade the system performance. How to minimize these negative effects is a challenging research problem.

At last, we re-emphasize that random cooperative communication schemes involve not only the physical layer, but also the MAC layer, and even the routing layer. Cross-layer design is necessary for the performance optimization. Different from deterministic cases, random cooperative communication demands dynamic cooperation and interactions among multiple random nodes, which cannot be solved solely by the physical layer. MAC layer and even routing layer are inevitably involved in the cooperative process. The performance of these upper layer protocols will affect the performance of the cooperative scheme, while the underlying cooperative scheme also has great impact on the behavior of the upper-layer protocols [Jakllari et al., 2006, Cui and Goldsmith, 2006]. Joint optimization across multiple protocol layers could exploit the synergies between different protocol elements and avoid negative interactions to improve the network performance, if designed properly.

In conclusion, the design of a random efficient cooperative communication system is definitely not a trivial extension of the conventional deterministic one. The randomness of the cooperation pattern makes the performance analysis quite complex. Protocols for deterministic schemes cannot be applied directly here and new protocols are needed. Besides, the design for a random cooperative system should jointly consider multiple layers for high efficiency. In the next section, we will give two design examples to illustrate some general principles for the design and analysis of energy-efficient random cooperative systems.

DESIGN EXAMPLES

In the first example, we consider an energy efficient random system for clustered wireless networks, where multiple cooperative nodes randomly participate in the long-haul inter-cluster communication. In this example, we will highlight the analysis for the system energy efficiency and clearly show its difference from the conventional deterministic ones. In the second example, we explore the cross-layer design between the physical layer and the MAC layer for one-hop random single-relay networks. Here, one and only one optimal relay is selected from a random pool of cooperative nodes to aid the communication process. We unify the power control and the relay selection of the physical layer into the MAC signaling in a distributed fashion. Through this example, we demonstrate the strength of cross-layer design. More details on these design examples can be found in [Zhou et al., 2006, Zhou et al., 2007].

Design Example 1: Energy Efficient Random Cooperative Communication for Clustered Networks

System Model. We consider a wireless network composed of multiple clusters of nodes, as shown in Figure 2, with each node having only one antenna. However, the nodes within the same cluster are closely spaced and can cooperate on signal transmission and/or reception. We assume that the average distance between the adjacent clusters is much larger than the average distance among the intra-cluster nodes. This is a typical wireless network scenario, and has been widely used in the literature [Cui et al., 2004, Laneman and Wornell,2003, Cui and Goldsmith, 2006, Zheng et al., 2005, Li, 2004, Li et al., 2005].

Suppose that the cluster head in a cluster wants to send a message to the cluster head of a nearby cluster. In other words, the cluster head of the first cluster is the source, and the cluster head of the second cluster is the destination. Since the transmission distance is relatively large between clusters, the source can first broadcast the message to other nodes in its cluster and then uses cooperative transmission to reduce the transmission energy. This approach has been shown to be energy-efficient relative to direct

Figure 2. Network model

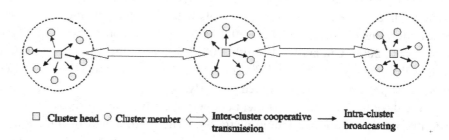

286

(non-cooperative) transmission under certain conditions [Cui et al., 2004, Cui and Goldsmith, 2006]. Specifically, the system works in two phases as follows:

- **Phase 1 (Intra-cluster broadcasting):** The source node broadcasts the packet with certain energy (E_{t1} per symbol) to the nodes within the same cluster. All the nodes in the cluster decode the received packet simultaneously. With CRC parity check bits, we assume that each node knows exactly whether the reception is successful or not.
- **Phase 2 (Inter-cluster cooperative transmission):** The source node and all the nodes that decode the packet correctly will "cooperatively" transmit the packet simultaneously with the same energy (E_{t2} per symbol) to the destination node (the cluster head of the receiving cluster). In the inter-cluster transmission, we use the cooperative schemes based on distributed space-time coding [Laneman and Wornell, 2003].

This model can maximize the achievable spatial diversity by probing all available nodes in the same cluster for joint relaying and is simple to implement. Compared with the deterministic cooperative schemes that assume perfect intra-cluster communication and cooperative relaying by all the nodes in the cluster, this scheme considers a more realistic scenario where packet errors may occur during the intra-cluster communication. Thus, the number of cooperative nodes in the next inter-cluster transmission stage is random. Compared with the optimal schemes in [Ahmed et al., 2004a, Ahmed et al., 2004b], the performance of our scheme will be degraded to some extent. However, since no complex signaling or feedback processes are assumed here, our scheme greatly simplifies the communication process and reduces the corresponding signaling overhead.

In the analysis, we consider a practical reception model, where the PER is a continuous function of the channel SNR [Liu et al., 2004]. Compared with the schemes based on SER analysis [Sirkeci-Mergen and Scaglione, 2005, Li, 2004, Li et al., 2005,], our PER-based scheme is more practical in wireless networks since error detection is usually done at the packet level.

For the intra-cluster transmission phase, let r denote the distance between the source node and a cooperative node in the same cluster. We assume that the source node is in the center of the cluster, and there are N nodes uniformly distributed in a circular area with radius R_1 around the source node. The distance r is then a random variable with p.d.f.

$$f(r) = \frac{2r}{R_1^2}, \ 0 < r \le R_1. \tag{1}$$

We assume an average path loss that is proportional to r^α, where α is the propagation constant. The average received SNR $\overline{\gamma}_1$ at a particular node can be written as

$$\overline{\gamma}_1 = \frac{GE_{t1}}{N_0 r^\alpha}, \tag{2}$$

where E_{t1} denotes the average transmission energy per symbol, N_0 is the one-sided AWGN spectral density at the receiver, and G is a constant that is defined by the signal frequency, antenna gains, and other parameters [Cui et al., 2004]. We assume Rayleigh fading wireless channels and thus the instantaneous received SNR γ_1 has an exponential distribution $f(\gamma_1)$. Conditional on each instantaneous value of γ_1, we have an AWGN channel. The exact closed-form PER in AWGN channels is difficult to obtain. An approximate expression for PER is provided as follows [Liu et al., 2004]:

$$PER(\gamma_1) = \begin{cases} 1 & 0 < \gamma_1 < \gamma_{pn} \\ a_n e^{-g_n \gamma_1} & \gamma_1 \geq \gamma_{pn} \end{cases},$$
(3)

where a_n, g_n and γ_{pn} are parameters that are dependent on the packet length, modulation, coding, and other factors. Tables are provided in [Liu et al., 2004] for the values of these parameters under different modulation schemes. Numerical results in [Liu et al., 2004] show that this approximation is quite accurate in general setups. Thus, we will use (3) for PER calculation in the following analysis. Given an average SNR $\overline{\gamma}_1$, the PER averaged over Rayleigh fading is given as

$$\overline{PER_1}(\overline{\gamma}_1) = \int_0^\infty PER(\gamma_1) f(\gamma_1) d\gamma_1$$
$$= \frac{a_n}{1 + g_n \overline{\gamma}_1} e^{-\gamma_{pn}(g_n + \frac{1}{\overline{\gamma}_1})} + (1 - e^{-\frac{\gamma_{pn}}{\overline{\gamma}_1}}).$$
(4)

Substituting $\overline{\gamma}_1$ in (2) into (4) and averaging it over the distance (c.f. (1)), we have

$$\overline{PER_1}(E_{t1}) = \int_0^{R_1} \overline{PER_1}(E_{t1}, r) \frac{2r}{R_1^2} dr$$
$$= \int_0^{R_1} \frac{a_n r^\alpha N_0}{r^\alpha N_0 + g_n E_{t1} G} e^{-\gamma_{pn}(g_n + \frac{N_0 r^\alpha}{E_{t1} G})}$$
$$+ (1 - e^{-\frac{\gamma_{pn} N_0 r^\alpha}{E_{t1} G}}) \frac{2r}{R_1^2} dr .$$
(5)

Let us denote the number of nodes that correctly decode the packet as M. Since the channels from the source to different relays are independent, the event that one node receives a packet successfully is independent of others. Thus, the number of successful nodes actually is subject to a binomial distribution. The probability that M nodes can decode the packet correctly is

$$P(M) = \binom{N}{M} [1 - \overline{PER_1}(E_{t1})]^M [\overline{PER_1}(E_{t1})]^{N-M} .$$
(6)

The mean value of M is thus

$$M_E = E(M) = N[1 - \overline{PER_1}(E_{t1})].$$
(7)

For the inter-cluster transmission phase, given the transmission energy per symbol per node E_{t2}, the average received SNR $\overline{\gamma}_2$ corresponding to each relay node is

$$\overline{\gamma}_2 = \frac{GE_{t2}}{N_0 D^\alpha},$$

(8)

where D is the inter-cluster distance between the transmitting and the receiving clusters (from center to center). Since we assume that D is much larger than the intra-cluster distance R, we can approximate the transmission distances between all the transmitting nodes and the receiving node (the cluster head of the receiving cluster) as D. The distributed space-time coding is built upon the orthogonal space-time block codes [Maure and Tarokh, 2001, Tarokh et al., 1999], where each cooperative transmitting node is assigned a column from the space-time code matrix. We assume symbol-level synchronization and perfect channel knowledge at the receiver, as in [Laneman and Wornell, 2003, Sirkeci-Mergen and Scaglione, 2005]. Due to the orthogonal property, the effective received SNR γ_2 conditional on channel realizations is

$$\gamma_2 = \left(\sum_{i=1}^{M_0} |h_i|^2\right) \overline{\gamma}_2$$

(9)

where h_i denotes the channel between the ith transmitting node and the receiving node. We assume that h_i's are independent and identically distributed with a Rayleigh fading distribution. As such, γ_2 is subject to a central chi-square distribution with $2M_0$ degrees of freedom [Proakis, 2000]

The average PER with M_0 transmitting nodes then can be derived as

$$\overline{PER_2}(\overline{\gamma}_2, M_0) = \frac{a_n \Gamma(M_0, (g_n + \frac{D^\alpha N_0}{GE_{t2}})\gamma_{pn})}{\Gamma(M_0)(1 + \frac{GE_{t2}}{D^\alpha N_0})^{M_0}}$$

$$+ \frac{\Gamma(M_0) - \Gamma(M_0, \frac{\gamma_{pn} N_0 D^\alpha}{GE_{t2}})}{\Gamma(M_0)}$$

(10)

where $\Gamma(x, y)$ is the incomplete Gamma function.

The overall end-to-end average PER can then be obtained as

$$\overline{PER}(E_{t1}, E_{t2}) = \sum_{M=0}^{N} P(M)\overline{PER_2}(E_{t2}, M + 1)$$

$$= \sum_{M=0}^{N} \binom{N}{M} [1 - \overline{PER_1}(E_{t1})]^M [\overline{PER_1}(E_{t1})]^{N-M} \overline{PER_2}(E_{t1}, M + 1).$$

(11)

We now analyze the overall energy consumption of this random cooperative scheme. We consider both the transmission energy and the associated circuit energy consumption. In the intra-cluster broadcasting phase, the source node transmits with average energy E_{t1} per symbol and N nodes try to receive the signal. During the second phase, on average, $M_E + 1$ nodes (including the source node) transmit with average energy E_{t2} per symbol per node. Then, the average total energy consumption $\bar{E}_{symbol}$ in the network per symbol is

$$\bar{E}_{symbol} = E_{t1} + E_{ct} + (M_E + 1)(E_{t2} + E_{ct}) + (N+1)E_{cr}$$
$$= E_{t1} + E_{ct} + (N(1 - \overline{PER_1(E_{t1})}) + 1)(E_{t2} + E_{ct}) + (N+1)E_{cr}. \qquad (12)$$

where E_{ct} is the transmitter circuit energy consumption per symbol, and E_{cr} is the receiver circuit energy consumption per symbol.

To minimize the overall energy consumption for a packet under a PER requirement, the optimization problem over the transmission energy (E_{t1} and E_{t2}) can be cast as

$$\min_{E_{t1}, E_{t2}} \quad \bar{E}_{packet} = \frac{1}{b} L_b \{ E_{t1} + E_{ct} + (N+1)E_{cr}$$
$$+ [N(1 - \overline{PER_1(E_{t1})}) + 1](E_{t2} + E_{ct}) \}$$
$$subject \; to \quad 0 \le E_{t1} \le E_{max},$$
$$0 \le E_{t2} \le E_{max},$$
$$\overline{PER(E_{t1}, E_{t2})} \le \overline{PER_0} \qquad (13)$$

Where L_b is the number of bits in a packet, and b denotes how many bits each symbol conveys, which is decided by the choices on modulation and coding [Liu et al., 2004]. And E_{max} is the maximum transmitting energy per symbol allowed at each node, $\overline{PER_0}$ is the PER target that is specified by the system. Eq. (13) is a two-dimensional optimization problem. However, for each given E_{t1}, the value of E_{t2} can be uniquely determined by the PER constraint using (11). Hence, we can solve (13) by a one-dimensional numerical search on E_{t1}.

Numerical Results We set the power loss constant $\alpha = 2$, $N_0 = 10^{-10}$ W/Hz, $R_1 = 20$ m, $D = 200$ m, $G = 1$, and $N = 10$. The parameters a_n, g_n, and γ_{pn} for the PER calculation are obtained from [Liu et al., 2004]. The information bit rate is set to be 10 kb/s. The transmitter circuit power is $P_{ct} = 150$ mW and the receiver circuit power is $P_{cr} = 100$ mW. The corresponding circuit energy consumption is equal to the power multiplied by the corresponding transmission time.

Figure 3 clearly shows us that an optimal E_{t1} does exist to minimize the total energy consumption. In this particular example, if $E_{t1} = 2.5 \times 10^{-7}$ J, $\bar{E}_{packet}$ reaches the minimum value of 3.16×10^{-1} J. If $E_{t1} < 2.5 \times 10^{-7}$ J, no enough nodes can correctly decode the packets, thus the energy benefit from the cooperation is limited. If $E_{t1} > 2.5 \times 10^{-7}$ J, too many nodes can correctly decode the packets and relay them. The energy benefit from the cooperation is canceled by the increase of E_{t1} and the additional circuit energy consumed in the cooperating nodes. Therefore, the source can adjust its broadcasting energy to control the average number of cooperating nodes for the overall energy minimization.

In Figure 4, we change N from 1 to 10. It shows that $\overline{E}_{packet}$ does not decrease monotonically with N. Instead, there exists an optimal value for N. For example, when $\overline{PER}_0 = 0.01$, the optimal N is 3. This can be explained as follows. With the increase of N, there will be more potential cooperative transmitting nodes. Although a larger N can provide more potential nodes for cooperative communication, it also introduces additional circuit energy cost in the listening mode during source broadcasting. When N increases, the additional circuit energy consumption may surpass the savings on the inter-cluster transmitting energy.

Figure 4 also shows us that the optimal N is different under different $\overline{PER}_0$ requirements. In this example, when $\overline{PER}_0 = 0.01$, the optimal N is 3, and when $\overline{PER}_0$ is reduced to 0.001, the optimal N equals 5. The increase of the optimal N with the decrease of $\overline{PER}_0$ is due to the following facts. The smaller the required $\overline{PER}_0$ is, the larger is the portion of the overall energy that will be consumed in the inter-cluster transmission. Since the inter-cluster transmission energy will decrease when the number of cooperative nodes increases, the optimal number of nodes N will increase as $\overline{PER}_0$ becomes smaller.

From this design example, we see that the analysis for random cooperative systems is quite complicated. Many factors such as network node distribution, PER requirements and the distances among nodes affect the system performance. To evaluate the performance of this random cooperative system, we first derive the distribution of active cooperative nodes and then, obtain the final result by averaging the performance over the derived distribution. For deterministic cooperative schemes, the evaluation process is much simpler since the active cooperative nodes are fixed.

Design Example 2: Energy Efficient Random Single-relay Cooperative Communication

In this example, we consider energy efficient random single-relay cooperative communication schemes for one-hop wireless sensor networks. Compared with multi-node cooperative schemes, single-relay

Figure 3. Energy consumption over intra-cluster transmission energy

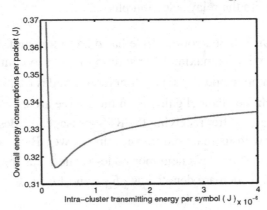

Figure 4. Energy consumption over the number of network nodes

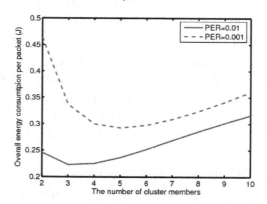

cooperation requires neither cooperative beamforming nor distributed space-time coding, where only the "best" relay out of a set of candidate nodes participates in the data transmission. Single-relay co-operative schemes are also easy to implement and incur less cooperation overhead. Besides, they can potentially achieve the same diversity-multiplexing tradeoff as that of multi-node cooperative schemes [Bletsas et al. 2006].

In this example, we *jointly* design the MAC protocol and the physical-layer power control strategy, where the power control and the node selection are unified into the MAC signaling in a *distributed* fashion. We aim to maximize the network lifetime, which is defined as the network operation time until the first node drains out its energy, by solving several distributed linear programming problems. The optimization elements are nicely embedded in the MAC protocol.

System model. As shown in Figure 5(a), the single-relay-selection cooperative communication scheme selects the "best" relay from a set of potential relay nodes, and then uses this "best" relay to aid the source-to-destination communication. For example, in Figure 5(a), there are n potential cooperative nodes. The single-relay-selection cooperative scheme only selects the best relay node i for cooperation and all other nodes keep silent during the communication process.

For such a random selective relay system, the main problem is to design the best strategy for relay selection. Multiple potential relays need to cooperate with each other to find the optimal relay, which is definitely not an easy task in a distributed environment. As shown in Figure 5 (b), we divide the overall communication process for one data burst into the following three phases, where the distributed power control is done concurrently with the relay selection phase.

- **Phase 1: Relay selection** When a source node has data to transmit, it first sends out a RTS (request-to-send) message with its maximal transmitting power to contend for the shared wireless channel. The destination node and the source's neighbor nodes (denoted as the set N_s) hear this message, based on which the channel gains from the source are estimated (denoted by $| h_{sd} |$ and $| h_{sk} | \ k \in N_s$ respectively). After receiving the RTS message, the destination node replies a CTS (clear-to-send) message with its maximal transmitting power. Based on the information in CTS, the source node and the destination's neighbor nodes (denoted as the set N_d) estimate the channels from the destination, which are denoted as $| h_{sd} |$ and $| h_{kd} | \ k \in N_d$, respectively.

Figure 5. Single-relay cooperative transmission system

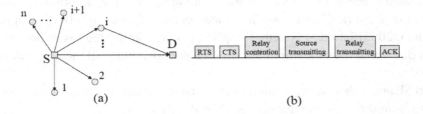

After a successful RTS/CTS exchange, all neighboring nodes of both the source and the destination become aware of this transmission event and refrain from transmitting data to avoid collisions. At this time, all overhearing nodes will calculate their priorities according to some predefined policies based on the information collected from RTS/CTS packets (Here, our policy is to maximize the network lifetime). Then they compete with each other within a time window, which is called the "relay contention period". The competition process is executed as follows. Node i will listen to the channel in the relay contention period. If it has not heard any beacon messages from other nodes for time t_i (where the higher its priority is, the smaller t_i is), it will broadcast one beacon message to grab the channel. In this way, the node with the highest priority will transmit first and "win" the competition to serve as the relay for cooperative data transmission. The competition is done within a fixed competition window of length T_{max}. In some cases, it is possible that no relays can support the data transmission or multiple beacon messages collide. Under such circumstances, the source cannot decode beacon messages from others and no cooperative communication can be formed. If the source node cannot support the data transmission by itself, it will back off and wait for some random time before initiating another RTS. If the source node can support the data transmission by itself, it will transmit its packet directly to the destination.

In this phase, each relay node determines the required transmission energy per symbol E_{t1} for the source and the transmission energy per symbol E_{t2} for itself based on its perceived channel conditions. Since the computation is done locally at the relay node, it needs to inform the source node the calculated E_{t1} in its beacon message.

- **Phase 2: Source transmission** The source node sends out data with transmission energy E_{t1} per symbol. The selected "best" relay decodes the received data. The destination stores the received signal from the source and defers the decoding to the next phase.
- **Phase 3: Relay transmission** The "best" relay forwards the decoded data to the destination with transmission energy E_{t2} per symbol. The destination combines the received data and the stored signal (received in Phase 2) for joint decoding. The transmitted signals in Phase 2 and Phase 3 will have the same length and format, but with different power. If the destination can decode the message correctly, it will send back an ACK message with its maximal transmitting power to the source; note that relay nodes do not need to be acknowledged.

This scheme is fully distributed and easy to implement. Since the RTS/CTS exchange is implemented in the 802.11-like MAC protocols anyway, the additional overhead is mainly due to the relay competition. Compared with data transmission, the relay competition period is short and may be negligible.

In addition, since there is only one node selected to relay the transmission, it is much simpler than the traditional distributed space-time coding or beam-forming that requires multi-node cooperation.

To maximize the network lifetime, we here incorporate a simple but effective approach proposed in [Chen and Zhao, 2005] into this system. Simply speaking, the strategy is to find an optimal power control solution that maximizes the minimum residual energy among all the nodes during each packet transmission.

According to Shannon theory, the minimum required transmission energy to support a data rate R (bits per symbol) from the source to destination should satisfy

$$R \leq \frac{1}{2} \log_2(1 + E_{t1}G_{sd} + E_{t2}G_{id}), \tag{14}$$

if node i is selected as the best relay, where $G_{sd} = |h_{sd}|^2$ and $G_{id} = |h_{id}|^2$. The factor $\frac{1}{2}$ in (14) is due to time sharing between the source and relay transmissions. On the other hand, in order for node i to successfully decode the source signal, it should satisfy

$$R \leq \frac{1}{2} \log_2(1 + E_{t1}G_{si}). \tag{15}$$

Thus, for each potential relay node, to maximize the network lifetime, the power control optimization is formulated as follows:

$$\max_{E_{t1}, E_{t2}} g_i(E_{t1}, E_{t2}) := \min\{E_{src} - N_s E_{t1}, E_i - N_s E_{t2}\}$$

$$subject \ to \quad \frac{(2^{2R} - 1)}{G_{si}} \leq E_{t1} \leq E_{max},$$

$$E_{t1}G_{sd} + E_{t2}G_{id} \geq (2^{2R} - 1),$$

$$0 \leq E_{t2} \leq E_{max} \quad , \tag{16}$$

where E_{src} is the current energy supply in the source node, E_i is the current energy supply in the ith relay node, E_{max} is the maximal transmission energy per symbol, and $N_s = N_b / R$ is the packet length in symbols.

From (16), it is clear that to find the power control solution that maximizes the network lifetime, node i needs to know parameters $E_{src}, E_i, G_{sd}, G_{si}$, and G_{id}. In this scheme, we can take a full advantage of the existing RTS/CTS hand-shaking signals. In the RTS message, the source node includes its energy level E_{src} as a parameter. After receiving the RTS message, node i retrieves E_{src} and measures G_{si}. The destination node estimates G_{sd} based on RTS, and then includes G_{sd} in its CTS message. When the relay node receives the CTS message from the destination, it can estimate G_{id} and decode the contained G_{sd}. Each relay then calculates its $g_i(E_{t1}, E_{t2})$ according to (10). If a valid solution is found, node i will participate in the competition process. It will attempt a beacon message if it does not receive any beacon message from others within a time delay

$$t_i = T_{\max} \times \frac{E_{src} - g_i(E_{t1}, E_{t2})}{E_{src}} . \tag{17}$$

As such, the larger $g_i(E_{t1}, E_{t2})$ is, the less waiting delay for node i, and the more likely it will be selected as the best relay.

Simulation Results We simulate the proposed cooperative scheme using the network simulator ns-2. We modify its 802.11 MAC protocol implementation to incorporate the proposed scheme. In the simulation, multiple network nodes are uniformly distributed in a circular area with radius 100 meters. Each node randomly chooses another node as its destination and generates the traffic with an average inter-packet arrival time randomly chosen within the interval (2, 6). The bit rate is set to be 10 kbps, and the packet length is 1000 bits. The length of the RTS/CTS/Beacon message is set to be 80 bits. The competition window of beacon message is set to be 10 ms, which is one tenth of the packet transmission time. The initial energy of every node is set to be 10 Joules. All energy consumption including the energy spent at RTS/CTS and the circuit is considered.

We simulate two other schemes for comparison. One is the direct transmission scheme with power control at the source node; and the other is the two-hop routing scheme with power control at both the source and the relay. For the two-hop routing scheme, the best two-hop route (from the source to the relay and then from the relay to the destination) that minimizes the transmission energy is selected based on G_{sd}, G_{si}, and G_{id}.

We observe from Figure 6 that with the increase of the number of nodes, the network lifetime will increase correspondingly. This is reasonable since more nodes means larger cooperation diversity to explore. When the number of nodes is relatively small, e.g., from 5 to 12, the network lifetime increases rapidly. But with further increase of the number of nodes, the network lifetime will not change much. This is mainly due to the collisions and the energy consumption of all control packets such as RTS/CTS. Figure 6 also shows us that this random scheme outperforms other schemes considerably, even when collisions and other sources of energy consumption are considered.

Figure 7 depicts the total number of packets transmitted in the network versus the number of nodes. Compared with other schemes, the total number of packets transmitted in the random scheme is much larger. For example, when the number of nodes in the network equals 10, this random scheme can transmit at least 3 times more packets than the direct scheme before the first node dies.

Figure 6. Network lifetime over network size

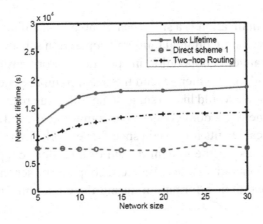

Figure 7. Total number of packets transmitted over network size

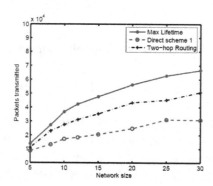

From this design example, we can clearly see the benefits that lie in the cross-layer design for random cooperation communication. In this example, by combining the relay selection and power control into the MAC layer signaling, little communication and energy overhead is added. The proposed random scheme can significantly prolong the network lifetime. Other cross-layer design strategies such as joint design between the physical layer, the routing layer, and even the transport layer can also be explored to further improve the system performance.

CONCLUSION

In this chapter, we provided an overview of the random cooperative communication for wireless networks. Unlike its deterministic counterpart, the random cooperative communication does not specify cooperative nodes in advance. Instead, cooperative nodes are chosen during the communication process based on the given channel and the network conditions. Such a random cooperative scheme is expected to be more robust to dynamic wireless network environments. We discuss the challenges that lie in designing general random cooperative schemes. The random nature makes the analysis and design process much more complicated than the deterministic case. Two examples are given to show the peculiarities of random cooperative systems. The first is for clustered wireless networks, where nodes in one cluster randomly participate in the long-haul inter-cluster communication to improve the overall energy efficiency. The second is to explore cross-layer design between the physical layer and the MAC layer for the one-hop random single-relay network.

Random cooperative communication is a still research area in its infancy. Thus, we still have limited insight or experience over developing optimal random cooperation schemes. At this stage, much can be learned by investigating the random cooperation in specific network environments (e.g., homogenous networks and clustered networks), and then extend the result to more general settings.

For future work in this area, we would like to suggest the following research directions; 1) Theoretical modeling and analysis for general random cooperative communication. To the best of our knowledge, there are only limited theoretical results for certain specific network environments. More general results are needed to provide guidelines for the system design of a random cooperation system; 2) Efficient interaction schemes among cooperative nodes. Random cooperative schemes are distributed and cooperative nodes need to exchange their information during the communication process for cooperation.

Efficient node interaction schemes are prerequisites for an efficient random cooperative system; 3) Cross-layer optimization for random cooperative schemes. Random cooperative communication demands dynamic cooperation and interactions among multiple random nodes, which cannot be solved solely at the physical layer. MAC layer and even routing layer are inherently involved in this dynamic process. A good cross-layer design framework is imperative for the further development of efficient random cooperative systems.

REFERENCES

Ahmed, N., Khojastepour, M., & Aazhang, B. (2004a). Outage minimization and optimal power control for the fading relay channel. In *IEEE Information Theory Workshop*, Houston, TX (pp. 458–462).

Ahmed, N., Khojastepour, M., Sabharwal, A., & Aazhang, B. (2004b). On power control with finite rate feedback for cooperative relay networks. In *International Symposium on Information Theory and Its Applications*.

Ahmed, N., Khojastepour, M. A., Sabharwal, A., & Aazhang, B. (2006). Outage minimization with limited feedback for the fading relay channel. *IEEE Transactions on Communications, 54*(4), 659–669. doi:10.1109/TCOMM.2006.873074

Akyildiz, I. F., Su, W., Sankarasubramaniam, Y., & Cayirci, E. (2002). A survey on sensor networks. *IEEE Communications Magazine, 40*(8), 102–114. doi:10.1109/MCOM.2002.1024422

Al-Karaki, J. N., & Kamal, A. E. (2004). Routing techniques in wireless sensor networks: A survey. *IEEE Personal Communications, 11*, 6–28.

Barbarossa, S., & Scutari, G. (2004a). Distributed space-time coding for multihop networks. In . *Proceedings of IEEE International Conference on Communications, 2*, 916–920.

Barbarossa, S., & Scutari, G. (2004b). Distributed space-time coding strategies for wideband multihop networks. In *Proceedings of ICASSP '04* (Vol. 4, pp. 501–504).

Bletsas, A., Khitsi, A., Reed, D. P., & Lippman, A. (2006). A simple cooperative diversity method based on network path selection. *IEEE Journal on Selected Areas in Communications, 24*(3), 659–672. doi:10.1109/JSAC.2005.862417

Chang, Y.-C., Lin, Z.-S., & Chen, J.-L. (2006). Cluster based self-organization management protocols for wireless sensor networks. *IEEE Transactions on Consumer Electronics, 52*(1), 75–80. doi:10.1109/TCE.2006.1605028

Chen, Y., & Zhao, Q. (2005). On the lifetime of wireless sensor networks. *IEEE Communications Letters, 9*(11), 976–978. doi:10.1109/LCOMM.2005.11010

Cui, S., & Goldsmith, A. (2006). Cross-layer design in energy constrained networks using cooperative MIMO techniques. *EURASIP Signal Processing Journal, 86*(8), 1804–1814.

Cui, S., Goldsmith, A. J., & Bahai, A. (2004). Energy efficiency of MIMO and cooperative MIMO in sensor networks. *IEEE Journal on Selected Areas in Communications, 22*(6), 1089–1098. doi:10.1109/JSAC.2004.830916

Cui, S., Goldsmith, A. J., & Bahai, A. (2005). Energy constrained modulation optimization. *IEEE Transactions on Wireless Communications, 4*(5), 2349–2360. doi:10.1109/TWC.2005.853882

Dohler, M., Li, Y., Vucetic, B., Aghvami, A. H., Arndt, M., & Barthel, D. (2006). Performance analysis of distributed space-time block encoded sensor networks. *IEEE Transactions on Vehicular Technology, 55*(7), 1776–1789. doi:10.1109/TVT.2006.878729

Hasna, M. O., & Alouini, M.-S. (2004). Optimal power allocation for relayed transmissions over Rayleigh-fading channels. *IEEE Transactions on Wireless Communications, 3*(6), 1999–2004. doi:10.1109/TWC.2004.833447

Jakllari, G., Krishnamurthy, S. V., Faloutsos, M., Krishnamurthy, P. V., & Ercetin, O. (2006). A framework for distributed spatio-temporal communications in mobile ad hoc networks. In *Proceedings of the 25th IEEE International Conference on Computer Communications*, Barcelona, Spain (pp. 1–13).

Laneman, J. N., Tse, D. N. C., & Wornell, G. W. (2004). Cooperative diversity in wireless networks: Efficient protocols and outage behavior. *IEEE Transactions on Information Theory, 22*(12), 3062–3080. doi:10.1109/TIT.2004.838089

Laneman, J. N., & Wornell, G. W. (2003). Distributed space time-coded protocols for exploiting cooperative diversity in wireless networks. *IEEE Transactions on Information Theory, 49*(10), 2415–2425. doi:10.1109/TIT.2003.817829

Li, X. (2004). Space-time coded multitransmission among distributed transmitters without perfect synchronization. *IEEE Signal Processing Letters, 11*(12), 948–952. doi:10.1109/LSP.2004.838213

Li, X., Chen, M., & Liu, W. (2005). Application of STBC-encoded cooperative transmissions in wireless sensor networks. *IEEE Signal Processing Letters, 12*(2), 134–138. doi:10.1109/LSP.2004.840870

Liu, P., Tao, Z., Lin, Z., Erkip, E., & Panwar, S. (2006). Cooperative wireless communications: A cross-layer approach. *IEEE Wireless Communications, 13*(4), 84–92. doi:10.1109/MWC.2006.1678169

Liu, Q., Zhou, S., & Giannakis, G. B. (2004). Cross-layer combining of adaptive modulation and coding with truncated ARQ over wireless links. *IEEE Transactions on Wireless Communications, 22*(5), 1746–1755. doi:10.1109/TWC.2004.833474

Madan, R., Metha, N., Molisch, A., & Zhang, J. (2006). *Energy-efficient cooperative relaying over fading channels with simple relay selection* (Tech. Rep. No. 2006.075). Mitsubishi Electrical Research Lab.

Maure, P., & Tarokh, V. (2001). Transmit diversity when receiver does not know the number of transmit antennas. In *Proceedings of International symposium on Wireless Personal Multimedia Communications*.

Nabar, R. U., Bolcskei, H., & Kneubuhler, F. W. (2004). Fading relay channels: Performance limits and space-time signal design. *IEEE Journal on Selected Areas in Communications, 22*(6), 1099–1109. doi:10.1109/JSAC.2004.830922

Ochiai, H., Mitran, P., Poor, H. V., & Tarokh, V. (2005). Collaborative beamforming for distributed wireless ad hoc sensor networks. *IEEE Transactions on Signal Processing, 53*(11), 4110–4124. doi:10.1109/TSP.2005.857028

Proakis, J. G. (2000). *Digital communications*. New York: McGrwaw-Hill.

Qien, G. E., Holm, H., & Hole, K. J. (2004). Impact of channel prediction on adaptive coded modulation performance in Rayleigh fading. *IEEE Transactions on Vehicular Technology, 53*(3), 758–769. doi:10.1109/TVT.2004.827156

Rabiner, W., & Heizelman, A. Chandrakasan, & Hari, B. (2000). Energy-efficient communication protocol for wireless microsensor networks. In *Proceedings of the Hawaii International Conference on System Science*.

Scaglione, A., Geockel, D. L., & Laneman, J. N. (2006). Cooperative communications in mobile ad hoc networks: Rethinking the link abstraction. *IEEE Signal Processing Magazine, Special Issues on Signal Processing for Wireless Ad hoc Communication Networks, 23*(5), 18–29.

Sendonaris, A., Erkip, E., & Aazhang, B. (2003). User cooperation diversity-part I: System description. *IEEE Transactions on Communications, 51*(11), 1927–1938. doi:10.1109/TCOMM.2003.818096

Simon, M. K., & Alouini, M.-S. (2000). *Digital communication over fading channels: A unified approach to performance analysis*. New York: Wiley-Interscience.

Sirkeci-Mergen, B., & Scaglione, A. (2005). Randomized distributed space-time coding for cooperative communication in self organized networks. In *IEEE Proceedings of SPAWC* (pp. 500–504).

Stefanov, A., & Erkip, E. (2004). Cooperative coding for wireless networks. *IEEE Transactions on Communications, 52*(9), 2415–2425. doi:10.1109/TCOMM.2004.833070

Su, W., & Sadek, K. A., & Liu, K. J. R. (2005). SER performance analysis and optimum power allocation for decode-and-forward cooperation protocol in wireless networks. In *Proceedings of IEEE Wireless Communication and Networking Conference* (Vol. 2, pp. 984–989).

Tarokh, V., Jafarkhani, H., & Calderbank, A. R. (1999). Space-time block codes from orthogonal designs. *IEEE Transactions on Information Theory, 45*(5), 1456–1467. doi:10.1109/18.771146

Yan, Y., Chen, M., & Kwon, T. (2006). A novel cluster-based cooperative MIMO scheme for multi-hop wireless sensor networks. *EURASIP Journal on Wireless Communications and Networking, 2*(4), 1–9. doi:10.1155/WCN/2006/72493

Ye, W., Heidemann, J., & Estrin, D. (2004). Medium access control with coordinated adaptive sleeping for wireless sensor networks. *IEEE/ACM Transactions on Networking, 12*(3), 493–506.

Zhang, J., & Zhang, Q. (2008). Cooperative routing in multi-source multi-destination multi-hop wireless networks. In *Proceeding of 27th IEEE International Conference on Computer Communications* (mini symposium).

Zheng, H., Zhu, Y., Shen, C., & Wang, X. (2005). On the effectiveness of cooperative diversity in ad hoc networks: A MAC layer study. In *IEEE Proceedings of ICASSP* (pp. 509–512).

Zhou, Z., Zhou, S., Cui, J.-H., & Cui, S. (2007), Energy-efficient cooperative communication based on power control and selective relay in wireless sensor networks. In *IEEE Proceedings of Military Communications*.

Zhou, Z., Zhou, S., Cui, S., & Cui, J.-H. (2006). Energy efficient cooperative communication in clustered wireless sensor networks. In *IEEE Proceedings of Military Communications*.

Zorzi, M., & Rao, R. R. (2003). Geographic random forwarding for ad hoc and sensor networks: Energy and latency performance. *IEEE Transactions on Mobile Computing, 2*(4), 349–365. doi:10.1109/TMC.2003.1255650

Chapter 11
Diversity Combining for Cooperative Communications

Diomidis S. Michalopoulos
Aristotle University of Thessaloniki, Greece

George K. Karagiannidis
Aristotle University of Thessaloniki, Greece

ABSTRACT

A major advantage of cooperative communications is the potential for forming distributed antenna arrays, that is arrays whose elements are not collocated, but carried by independent relaying terminals. This allows for a study and design of cooperative communications under a novel perspective, where the inherent end-to-end paths between the source and destination terminal constitute the multiple branches of a virtual, distributed diversity receiver. As a result, the well-known combining methods used in conventional diversity receivers can be implemented in a distributed fashion, resulting in novel relaying protocols and, generally, in new ways for exploiting the resources available in cooperative relaying setups. This chapter provides an overview of this distributed diversity concept, as well as a performance analysis of the corresponding distributed diversity schemes, with particular emphasis on the analysis of distributed switch-and-stay combining. Further insights regarding the potential of implementing the distributed diversity concept in practical applications are obtained.

INTRODUCTION

The benefits of cooperative communications in terms of link reliability and coverage extension are becoming more and more known to the telecommunication community, and have attracted the interest of both academia and industry in the last few years. Through the utilization of spatially distributed single-antenna terminals, cooperative communications are designed to achieve the beneficial effects of diversity in future networks without the need for employing multiple antennas at either the transmitter or the receiver end. As has been demonstrated in many research works in this area, cooperating terminals

DOI: 10.4018/978-1-60566-665-5.ch011

can form virtual antenna arrays that ensure multiple transmissions through independent channels, avoiding thus the practical limitations that multiple antenna employment entails, especially when referring to small, hand-held devices.

Taking the above into account, it naturally follows that the well-known spatial combining techniques that have been already proposed in the literature and concern the "classical" diversity concept can now be implemented in a distributed fashion through the aforementioned cooperation among the participating terminals. To be more precise, the end-to-end paths inherent in any cooperative formulation can represent the diversity branches of a virtual, distributed diversity scheme, following the concept of virtual antenna arrays. Hence, cooperative communications can be studied as well as designed under this point of view, by utilizing well-known combining methods which are already used in multi-branch diversity receivers.

Intuitively, this usually leads to more or less the same advantages and disadvantages of these techniques over each other, in the sense that they are expected to follow the same performance complexity tradeoff, i.e., offering simpler practical implementations with some cost in performance and vice versa. Nevertheless, the relative performance of the distributed diversity techniques over each other in some cases may differ from that of traditional diversity implementation, since the overall consumed power in cooperative applications is distributed among the participating terminals. Hence, the performance of distributed diversity highly depends on the assumptions made regarding the overall transmitted power, since, given a total power constraint, techniques that activate fewer relays may be preferable in the sense that significant amounts of energy are saved. Apparently, in order to gain insight into the performance of the distributed diversity concept, the above intuitions need to be verified mathematically, allowing also for further future extension. Therefore, this chapter aims at

a) illustrating relaying protocols that constitute distributed implementations of well-known diversity combining techniques over cooperative networks and

b) providing easily-tractable mathematical expressions for certain performance metrics related to the above distributed diversity schemes, such as the outage or the symbol error probability.

BACKGROUND

The main idea behind the concept of cooperative communications is the utilization of a number of relaying terminals, represented here by R_i, for wirelessly forwarding the information sent by a source terminal S to a destination one, D. This cooperative setup is illustrated in Figure 1. In general, there are two main relaying modes which the relays may operate in: The decode-and-forward (DF) and the amplify-and-forward (AF). The former relaying mode (a.k.a. regenerative relaying) consists of fully decoding the message received by the source, then remodulating the detected symbols into the same or different alphabet and forwarding the resulting data to the destination. The latter mode (a.k.a. non-regenerative relaying) consists of simply amplifying and forwarding to the destination the received signal without any further process (Laneman & Tse & Wornell, 2004).

Depending on the type of the amplification gain employed, AF relaying can be further categorized into variable-gain and fixed-gain relaying. Specifically, the relays may utilize channel state information (CSI) knowledge of the channel in its receiver end, so as to be able to amplify the received signal

Figure 1. The typical cooperative setup, where a set of relaying terminals assists the communication between a source and a destination terminal.

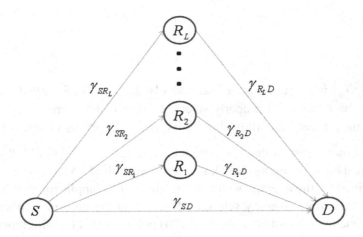

accordingly and thus maintain a limited transmission power. In such case, the relay gain, G_i, employed by the R_i relay has the form of (Laneman & Tse & Wornell, 2004)

$$G_i^2 = \frac{P_{R_i}}{P_S a_{SR_i}^2 + N_0} \tag{1}$$

where the general notation a_{AB} corresponds to the channel amplitude between terminals A and B (with $A, B \in \{S, R_i, D\}$), P_A denotes the power transmitted by terminal A, and N_0 represents the noise power, which is assumed identical in each link throughout this chapter. Notice that in order for the above gain to be employed, the relay requires CSI of the source-relay link, thereby the variable gain relays are also known as CSI-assisted AF relays. The equivalent signal-to-noise-ratio (SNR) of the dual-hop source-relay-destination channel associated with R_i is then derived as (Hasna & Alouini, 2004)

$$\gamma_i = \frac{\gamma_{SR_i} \gamma_{R_i D}}{\gamma_{SR_i} + \gamma_{R_i D} + 1} \tag{2}$$

where $\gamma_{AB} = \frac{P_A a_{AB}^2}{N_0}$ denotes the instantaneous SNR of the corresponding A-B link. For the case of multihop relaying transmissions (i.e. the case where the signal reaches the destination through multiple - greater than two - cascaded links) with CSI-assisted AF relays, closed-form bounds for the end-to-end SNR and the resulting performance when operating over generalized fading channels can be found in (Karagiannidis & Tsiftsis & Mallik, 2006).

On the other hand, if such CSI knowledge is not available or if simply the complexity of the relay needs to be kept low, the relay gain employed may take a predefined fixed value, independent of the instantaneous source-relay channel conditions. This relaying mode is also known as "blind" AF relaying. The resulting end-to-end SNR is then expressed as (Hasna & Alouini, 2004)

$$\gamma_i = \frac{\gamma_{SR_i} \gamma_{R_i D}}{C_i + \gamma_{R_i D}} \tag{3}$$

where $C_i = P_S / \left(G_i^2 N_0 \right)$ is a constant for a fixed value of the gain G. Should this be the case, however, the fixed relay gain value has to be properly selected, since if the source-relay channel is strong the transmission power may exceed the allowable limits. Hence, a proper choice for the fixed relay gain value is the average of the corresponding CSI-assisted one, i.e., $G_i^2 = E \left\langle P_R \middle/ \left(P_S a_{SR_i}^2 + N_0 \right) \right\rangle$ with $E \langle \cdot \rangle$ denoting expectation. This relaying mode is known as "semi-blind" AF relaying, since in this case the relay gain is determined by the average source-relay channel. In applications where the relays form a somewhat distributed transmit diversity setup, by forwarding the multiple replicas of the transmitted signal in orthogonal channels, Hasna & Alouini (2004) point out that the difference in the performance of CSI-assisted and semi-blind AF relaying in repetition-based relaying is not considerable. In fact, this performance difference is much smaller than one might expect considering the complexity differences of these types of AF relaying. Regarding the case of multihop transmissions with fixed-gain AF relays, the end-to-end SNR as well as bounds on the resulting overall performance are available in (Karagiannidis, 2006).

A major constraint in both DF and AF relaying is the fact that the relays cannot receive and transmit simultaneously and on the same frequency band, because of the inevitable coupling effects between the transmitter and receiver circuitries. This is referred to as the half-duplex constraint, and is actually the main reason why the transmissions from the source to the destination terminal in cooperative communications are divided into two phases: In the former, the source broadcasts the message to the relays and the destination, where it is stored for future usage. In the latter phase, the relays forward the previously received message to the destination, performing either simple signal amplification (case of AF relaying) or decoding and re-encoding in cascade (case of DF relaying). As a result, cooperative communications suffer from an inherent reduction in the overall spectral efficiency by a factor of $1 / 2$, as compared to conventional point-to-point communications (Laneman & Tse & Wornell, 2004). Some recent works devoted to alleviating the problem of spectral inefficiency in cooperative communications include (Rankov & Wittneben, 2007; Fan & Wang & Thompson & Poor, 2007; Michalopoulos & Karagiannidis, 2008-{4}).

Generally speaking, the advantage of employing DF relaying lies in the fact that freshly remodulated signals are transmitted from the relays, so that the noise at the source-relay link is completely eliminated. On the other hand, DF relaying may lead to error propagations, particularly in cases where the source-relay link is weak and no error identification coding is employed. In DF relaying, however, the fact that the signal is completely decoded and reencoded at the relays allows for advanced relaying protocols which may determine, for instance, whether or not a relay is activated (Lee & Kim, 2007). Moreover, multiple DF relays can be utilized so as to implement channel coding in a somewhat distributed fashion, known also as coded cooperation (Hunter & Nosratinia, 2006). Overall, in cases where a single relay is employed DF relaying results in slightly better error performance as compared to AF relaying (Hasna & Alouini, 2003). Nevertheless, in cases where multiple relaying terminals are activated the resulting multiple transmissions need to be appropriately combined, accounting for potential error propagations by the relays. The concept of mitigating error propagations is further analyzed in the ensuing section.

DISTRIBUTED DIVERSITY TECHNIQUES

Distributed Maximal Ratio Combining (Benchmark)

Maximal Ratio Combining (MRC) has been shown to offer the maximum attainable performance in receiver diversity systems, since it takes advantage of all the available branches, combining them according to the corresponding instantaneous receive SNRs (Simon & Alouini, 2005). Consequently, it follows that the best, in terms of performance, way of combining the signals arriving from multiple relays is to attribute them weights analogous to the corresponding end-to-end SNRs. This technique is dubbed here as *distributed MRC*, and serves as a performance benchmark of all the possible distributed diversity techniques.

AF Relaying

Depending on the forwarding mode employed by the relays (e.g., AF or DF), the MRC weights at the destination are appropriately adjusted so as to meet the corresponding end-to-end SNR expressions. In particular, for the CSI-assisted AF relaying scenario, the signal arriving to the destination from a particular relay, say the ith one, is weighted according to $w_i = a_i / N_i$, where $a_i = a_{SR_i} G_i a_{R_i D}$ is the equivalent channel gain of the ith end-to-end branch and $N_i = N_{SR_i} G_i^2 a_{R_i D}^2 + N_{R_i D}$ is the total noise power at the destination input. Similarly, the signal arriving to the destination directly from the source terminal is attributed a weight that equals $w_0 = a_0 / N_0$. This results in an overall output SNR having the form of

$$\gamma_i = \gamma_0 + \sum_i \frac{\gamma_{SR_i} \gamma_{R_i D}}{\gamma_{SR_i} + \gamma_{R_i D} + 1} \tag{4}$$

That is, the SNR at the output of the destination's receiver equals the sum of the individual end-to-end SNRs corresponding to the signals incident from the different relays, plus the signal received directly from the source.

Anghel & Kaveh (2004) conducted an approximate Symbol error rate (SER) analysis of the distributed MRC scheme, deriving analytical expressions for the overall SER for arbitrary number of participating relays, as well as closed-form bounds on the SER performance, for Rayleigh fading environments. The approximate analysis is based upon approximating the instantaneous end-to-end SNR γ_i of the ith branch as

$$\gamma_i = \frac{\gamma_{SR_i} \gamma_{R_i D}}{\gamma_{SR_i} + \gamma_{R_i D}} \tag{5}$$

which actually represents an upper bound on the exact end-to-end SNR that is very tight in the medium and high SNR regime. The resulting SER expressions are then derived using the well-known Moment Generating Function (MGF)-based approach (Simon & Alouini, 2005). Additionally, a lower bound on the SER of distributed MRC with multiple CSI-assisted AF relays has been derived in (Tsiftsis & Karagiannidis & Kotsopoulos & Pavlidou, 2004), using an inequality between the harmonic and geometric mean of positive random variables.

Asymptotic expressions for the outage probability of distributed MRC have been also derived in (Li & Kishore, 2007), for both the cases of CSI-assisted gain and semi-blind fixed gain relays, when assuming the versatile Nakagami-*m* fading distribution. In the latter work it was also shown that when fixed-gain relays are utilized higher average SNRs are attained, compared to those achieved through CSI-assisted gain employment. Additionally, the case of multiple relay cooperative diversity with MRC combining at the destination has been considered and analyzed in (Ikki & Ahmed, 2007), when operating over Nakagami-*m* fading channels. In the latter work, the MGF-based approach was used to derive an approximate analytical SER, as well as an outage probability expression. More recently, a closed-form error analysis of AF fixed-gain relaying when operating over dissimilar Nakagami-*m* fading channels was conducted in (Suraweera & Karagiannidis, 2008), where the effects of unbalanced fading conditions were also pointed out. The reader may also find in (Ribeiro & Cai & Giannakis, 2005) asymptotic expressions for the SER of the distributed MRC case; those expressions are derived through the McLaurin series approximation and hold for various channel fading models. This reveals that, as expected, the diversity order achieved via cooperative diversity does not depend on the fading distribution.

DF Relaying

A major difference on the nature of AF and DF relaying lies in the fact that, in the former, each of the end-to-end channels associated with a single relay is actually a cascaded link where both the source-relay and the relay-destination channels affect symmetrically the overall performance, whereas in the latter the presence of decoding at the relay divides the end-to-end link into two separate channels. To put it another way, the derivation of the MRC combining weights in the AF case is straightforward since these weights are directly related to the overall cascaded channel and the overall noise at the destination input, however this may not be the case in DF cooperative diversity systems.

The question that follows the distributed implementation of MRC in DF relaying systems is how to combine the signals received at the destination, given that these signals convey freshly modulated symbols at the relays hence the SNR seen by the destination does not encompass the source-relay channels. Needless to say, combining the available branches in a multi-relay uncoded DF cooperative diversity according to the relay-destination SNRs would result in considerable performance degradation, particularly when the source-relay channels are relatively weak, since the destination would not be able to identify if the SNRs seen by the destination correspond to correctly re-modulated symbols. To this end, several methods have been proposed which aim at limiting (or if possible eliminating) the deteriorating effects of error propagations, so as the error performance of the end-to-end links involved be reflected by the corresponding relay-destination links. These methods are usually referred to as *adaptive DF relaying*, since their implementation involves activation of the available relays only if some conditions are satisfied.

In particular, one of these methods is the use of the so-called threshold DF relaying (Herhold & Zimmermann & Fettweis, 2004; Hu & Beaulieu, 2006; Suraweera & Smith & Armstrong, 2006), which is actually based upon the primal selection relaying protocol proposed in (Laneman & Tse & Wornell, 2004). According to this DF relaying method, each of the relays involved is activated only if the source-relay SNR is higher than a predetermined threshold value, which usually depends on the target transmission data rate. The subset of relays with input SNR higher than this threshold is often referred to as the *decoding set*. Another method of reducing the risk of error propagation is the employment of Cyclic Redundancy Check (CRC), which enables the relays to identify any possible detection errors and hence refrain from forwarding (Lee & Kim, 2007). This, however, comes at the cost of expanded bandwidth required in order to embody the redundancy packet into the transmission frame.

One major disadvantage of adaptive DF relaying is the reduced diversity benefit followed by the conditional activation of the available relaying terminals, particularly when the threshold value in threshold DF relaying is too high. The choice of the appropriate threshold is therefore of high importance in these systems; works devoted to optimizing this threshold include (Siriwingpairat & Himsoon & Su & Liu, 2006; Onat & Adinoyi & Fan & Yanikomeroglu & Thompson, 2007). Apart from adaptive DF relaying, however, the destructive effects of error propagation can be mitigated through the use of the so-called cooperative MRC (C-MRC) (Wang & Cano & Giannakis & Laneman, 2007). According to this method, all the available relays are activated in each transmission period while the destination attributes appropriate weights to the incoming symbols, which depend on both the source-relay and the relay-destination links. This overcomes the problems of reduced diversity gain and increased bandwidth consumption that the employment of threshold DF relaying and CRC entail, achieving thus full diversity order with low complexity equipment. Similar to C-MRC, the relays may employ a family of strategies where the decoded symbols at the relays are scaled in power before forwarded to the destination, resulting in increased diversity benefits regardless of the underlying modulation (Wang & Wang & Giannakis, 2006).

Distributed Selection Combining (SC): The Concept of PHY-Layer Fairness

The *distributed SC* concept is equivalent to selecting a single relay among the set of available ones according to some appropriately defined policies. As one may expect, the benefits of employing distributed SC instead of distributed MRC is the attainment of the same diversity order as that of MRC (i.e., on the order of the number of available relays) while activating only a single relay; that is, considerably reducing the extra resources that multiple relay activation entails while achieving the same diversity gain. Moreover, one major advantage of activating only a single relay stems from the fact that, as is pointed out above, the multiple relay activation is followed by a reduction in the overall spectral efficiency, due to the fact that the multiple relay transmissions must occur on orthogonal channels. Additionally, the hardware complexity at the destination is considerably reduced through distributed SC since the number of receiver chains employed is reduced as well.

A particular method for implementing the distributed SC concept based on local channel measurements has been proposed and analyzed in (Bletsas & Khisti & Reed & Lippman, 2006). Specifically, the authors proposed a method for selecting the participating relay in a fashion which does not necessarily require the presence of a central unit. This is achieved by having the relays perform local channel estimations (that is, each relay estimates the corresponding source-relay and relay-destination links) and then calculate a certain metric which actually reflects the instantaneous end-to-end channel conditions. These metrics may be either the minimum of the source-relay and relay-destination SNRs, i.e.

$$m_i = \min\left(\gamma_{SR_i}, \gamma_{R_iD}\right) \tag{6}$$

or the harmonic mean of the above quantities, i.e.

$$m_i = \frac{\gamma_{SR_i}\gamma_{R_iD}}{\gamma_{SR_i} + \gamma_{R_iD}}. \tag{7}$$

Note that the latter metric represents an approximation to the end-to-end SNR of the CSI-assisted relaying channel in the medium to high SNR regime (Hasna & Alouini, 2003). Each of the potential relays then sets a time-counter which is inversely proportional to the above metrics; as soon as the shortest counter expires, this relay (that is actually the one that is going to be selected) broadcasts a flag packet informing the rest of the relays about its presence. The same method can be also extended to the case where no direct communication among the relays is possible. In that case, the destination terminal collects all the metrics formed by the potential relays and makes a decision on the selected terminal, informing the cooperating nodes about their potential activation through a flag packet. The relay selection scheme described above is referred to as *opportunistic relaying*, owing to the somewhat "opportunistic" usage of the available resources, and can be applied to both the AF and DF relaying mode.

A similar yet distinct implementation of the distributed SC concept is the so-called *selection cooperation* (Beres & Adve, 2008), which however applies exclusively to the DF relaying mode. According to this scheme, the potential relays are first categorized into those which can successfully decode the incoming message and those which cannot, depending on the value of SNR at their input. Those relays with input SNR higher than a predetermined threshold form the so-called *decoding set*, similar to the threshold DF relaying mode described above. Then, the relay selection decision is made on a central unit (which could be, e.g., the destination terminal), depending on the SNR of the relay-destination links associated with the decoding set. The reader may find closed-form expressions for the outage and the bit error probability of threshold-based DF opportunistic relaying (i.e., opportunistic relaying when threshold DF-relaying is used) and selection cooperation in (Michalopoulos & Karagiannidis, 2008-{2}). A general comment derived in the latter work is that both schemes may outperform one another, depending on the threshold value that determines the number of activated relays.

The outage probability of the distributed SC scheme is defined as the probability that the end-to-end SNR associated with the selected relay is below a given threshold, γ_{th}, which is related to the target rate, r, through (Bletsas & Khisti & Reed & Lippman, 2006)

$$\gamma_{th} = 2^{2r} - 1. \tag{8}$$

Since the "best" relay corresponds to the highest end-to-end SNR, the overall outage probability will be given as

$$P_{out} = \prod_{i=1}^{L} F_{\gamma_i}\left(\gamma_{th}\right) \tag{9}$$

where $F_{\gamma_i}\left(\bullet\right)$ denotes the cumulative density function (CDF) of the i_{th} end-to-end SNR, γ_i.

The distributed SC concept has been also shown to offer improved outage performance over the conventional scheme where all the available relays participate in the forwarding process, assuming AF relaying and equal amount of total transmitted power in both cases (Zhao & Adve & Lim, 2007). This is mainly due to the aforementioned spectral inefficiency caused by the multiple orthogonal transmissions. That is, denoting with L the number of participating relays, the channel capacity per unit bandwidth (a.k.a. spectral efficiency) for the distributed MRC case is written as

$$C_{MRC} = \frac{1}{L+1} \log_2(1 + \sum_{i=1}^{L} \gamma_i)$$ (10)

whereas that of distributed SC has the form of

$$C_{SC} = \frac{1}{2} \log_2 \left(1 + \max_{i=1,...,L} \left(\gamma_i\right)\right)$$ (11)

i.e., the pre-log factor on the capacity formula is lower on distributed MRC, hence the overall capacity is also lower for a wide SNR region, particularly for large number of participating relays. Consequently, assuming the same total transmitted power in both the distributed MRC and distributed SC cases, it follows that the outage probability (i.e., the probability that a certain data rate cannot be supported by the system) is lower for distributed SC. It was also showed in (Zhao & Adve & Lim, 2007) that even if optimal power allocation is employed in the distributed MRC case, selecting a single relay still leads to lower outage probability. Other works on the distributed SC concept include (Tsiftsis & Karagiannidis & Mathiopoulos & Kotsopoulos, 2006), where expressions for the outage probability of distributed SC under fixed-gain AF relaying were derived; (Krikidis & Thompson & McLaughlin & Goertz, 2008), where the relay selection concept under partial CSI knowledge was studied.

The concept of PHY-layer fairness: In cooperative diversity applications where the potential relays are mobile terminals, the relay selection methods described above can be modified so as to encompass the concept of fairly dealing with the relays in terms of energy consumption. To be more precise, since any relay activation involves usage of that relay's energy in order to improve the quality of the source-destination communication, it naturally follows that cooperative diversity should studied and thereafter designed under a fairness (in terms of energy consumption) perspective. This type of fairness is dubbed as physical layer (PHY-layer) fairness (Michalopoulos & Karagiannidis, 2008-{1}), so as to diversify from other type of fairness associated with the higher levels of the protocol stack. This PHY-layer fairness-based selection of the participating relay is implemented by attributing a weight coefficient to each potential relay that depends only on the average end-to-end channel conditions. The relay selection metrics are then modified as

$$b_i = \frac{m_i}{\mu_i}$$ (12)

Figure 2. Outage probability, normalized on the outage threshold SNR, γ_{th}, of (a) the equal energy consumption (PHY-fairness) scheme, (b) the best path selection and (c) the scheme where the relays are selected for an identical fraction of time. The cases where the number of available relays, L, equals 3 and 5, respectively, are considered.

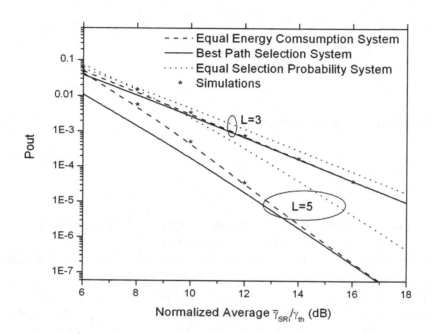

where μ_i is the aforementioned weight coefficient and m_i is the metric for selecting the "best" relay, i.e., when the PHY-layer fairness concept in not taken into account. The selected relay then is that with the highest b_i. Note that the μ_i coefficients are derived so as equal energy consumption among the available relays is ensured; they are obtained by numerically equating the expressions for the average power consumption provided in (Michalopoulos & Karagiannidis, 2008-{1}). In the general fading scenario where the average channel conditions of the potential relays are not necessarily identical with one another, it has been shown that, for small number of available relays or for high SNR values, similar outage and error performances as those of opportunistic relaying are achieved. The major benefit of the above scheme, however, is the fact that the average power consumptions of all relays are (approximately) equal with each other.

Figure 2 depicts the outage probability of the scheme that selects a single relay based on the afore-mentioned PHY-fairness rule, as compared to the best relay selection scheme and the scheme where the relays are selected for identical fraction of time. Two cases regarding the number of available relays, L, are considered, namely $L = 3$ and $L = 5$. One may notice that the cost in outage performance of taking the average power consumption under account is not high enough, particularly for high SNRs and small number of available relays. As expected, both the best path selection and the equal energy consumption scheme outperform the system that selects the relays for an identical fraction of time, since in the latter scheme the performance optimization is not considered at all.

Distributed Switch and Stay Combining

In conventional diversity systems with two available branches, the simplest, as well as easiest to implement diversity method is the so-called switch and stay combining (SSC). According to SSC, the receiver switches and stays connected to any of the available branches for as long as the SNR of that branch is above a predetermined threshold. If the SNR drops below that threshold, the receiver switches to the other branch and the above procedure is repeated. SSC generally serves as a suboptimal yet simpler substitute of SC, since the SNR of only a single branch needs to be estimated each time, with some cost on performance (Simon & Alouini, 2005).

Depending on the number of relays involved, the distributed implementation of SSC can be categorized into single-relay distributed SSC (DSSC) and two-relay DSSC. Generally speaking, the main idea behind DSSC in relaying applications lies in switching from an end-to-end branch to another if the SNR of the active branch drops below a given threshold. This leads to suboptimal performance compared to distributed SC, however the overall complexity is reduced since the SNR of only a single end-to-end branch needs to be estimated each time.

i) **Two-Relay DSSC:** In cases where two relaying terminals participate in the cooperative process, the two input branches of the virtual SSC combiner are the corresponding end-to-end channels. Hence, two-relay DSSC can be thought of as a method of activating a single relay out of two in a SSC fashion; that is, the same relay remains active for as long as the corresponding end-to-end SNR is sufficiently high. This can be implemented in practice by having the destination performing continuous comparison of the end-to-end SNR of the active branch with some predefined threshold value, known as the switching threshold T. If the overall received SNR is lower than T, the destination indicates that a switching must occur at the ensuing transmission period via appropriate feedback sent to the relays. Should this be the case, the relays then switch from the active to idle mode and vice versa; then, the above procedure is repeated. Of course, the direct source-destination channel may be combined together with that associated with the active relay, into a "local" combiner such as MRC. Moreover, the transmission periods are considered small enough so that constant fading conditions during two consecutive periods are assumed.

As one may expect, the performance of DSSC is strongly affected by the value of the switching threshold T. Let the fading in each link be Rayleigh distributed, the two end-to-end branches be represented by $\mathcal{R}_1$ and $\mathcal{R}_2$, respectively. The outage probability of two-relay DSSC can be shown to be given by (Michalopoulos & Karagiannidis, 2008-{3})

$$P_{out}\left(\gamma_{th}\right) = \begin{cases} \dfrac{F_{\gamma_1}\left(T\right)F_{\gamma_2}\left(T\right)\left(F_{\gamma_1}\left(\gamma_{th}\right)+F_{\gamma_2}\left(\gamma_{th}\right)\right)}{F_{\gamma_1}\left(T\right)+F_{\gamma_2}\left(T\right)}, \gamma_{th} < T \\[4mm] \dfrac{F_{\gamma_1}\left(T\right)F_{\gamma_2}\left(T\right)\left(F_{\gamma_1}\left(\gamma_{th}\right)+F_{\gamma_2}\left(\gamma_{th}\right)-2\right)}{F_{\gamma_1}\left(T\right)+F_{\gamma_2}\left(T\right)} \\[4mm] +\dfrac{F_{\gamma_1}\left(\gamma_{th}\right)F_{\gamma_2}\left(T\right)+F_{\gamma_2}\left(\gamma_{th}\right)F_{\gamma_1}\left(T\right)}{F_{\gamma_1}\left(T\right)+F_{\gamma_2}\left(T\right)}, \gamma_{th} \geq T \end{cases}$$

(13)

where $F_{\gamma_i}\left(x\right)$ denotes the cumulative distribution function of γ_i and γ_{th} is the outage threshold SNR. If the relays operate in the DF mode, the end-to-end SNR of the ith branch, γ_i, can be defined as

$$\gamma_i := \min\left(\gamma_{SR_i}, \gamma_{R_iD}\right)$$

(14)

which corresponds actually to an outage-based definition of γ_i, so that the CDF of γ_i evaluated at γ_{th} coincides with the outage probability of the $\mathcal{R}_i$ branch. Consequently, $F_{\gamma_i}\left(x\right)$ is given by $F_{\gamma_i}\left(x\right) = 1 - \exp\left(-x / \overline{\gamma}_{SR_i}\right)\exp\left(-x / \overline{\gamma}_{R_iD}\right)$. On the other hand, if the relays operate in the AF mode, $F_{\gamma_i}\left(x\right)$ is obtained as (Anghel & Kaveh, 2004)

$$F_{\gamma_i}\left(x\right) = 1 - \frac{2x}{\sqrt{\rho_i}}e^{-\frac{\sigma_i x}{\rho_i}}K_1\left(\frac{2x}{\sqrt{\rho_i}}\right)$$

(15)

where $K_v\left(\cdot\right)$ stands for the modified Bessel function of the second kind and order v. Setting $T = \gamma_{th}$ yields the optimum outage performance of DSSC, i.e.,

$$P_{out}\left(\gamma_{th}\right) = F_{\gamma_1}\left(\gamma_{th}\right)F_{\gamma_2}\left(\gamma_{th}\right).$$

(16)

The above equation demonstrates that, although DSSC is much simpler to implement than distributed SC since only one end-to-end channel needs to be estimated for each transmission period, *the outage performances of two-relay DSSC and distributed SC with two available relays are identical with one another, regardless of the relaying mode* (i.e., DF or AF). This represents the main advantage of two-relay DSSC, i.e., the fact that the channel estimation of one of the two end-to-end branches can be eliminated without any cost on outage performance.

In case the direct source-destination channel is taken into account, equation (16) again gives an expression for the overall outage probability, however $F_{\gamma_1}\left(\gamma_{th}\right)$ and $F_{\gamma_2}\left(\gamma_{th}\right)$ are now modified so as to correspond to the probability that the combined channel from the source and one of the relays is in outage. Figure 3 depicts the normalized outage probability, normalized on the outage threshold SNR, γ_{th}, for the two-relay DSSC scheme. Both the cases of AF and DF relaying are considered, as well as the cases where the destination receives from a single relaying terminal or combines the relayed branch with the source-destination one.

Let $\mathcal{A}$ denote the active branch (i.e., the event of $\mathcal{R}_1$ being the active branch is expressed as $\mathcal{A} = \mathcal{R}_1$). Considering the mode of operation of two-relay DSSC, the overall error probability can be expressed as

$$\Pr\{\mathcal{E}\} = \sum_{i=1}^{2} p_{\mathcal{R}_i} \left[F_{\gamma_i}(T) \Pr\{\mathcal{E}|\mathcal{A} = \mathcal{R}_j\} + \left(1 - F_{\gamma_i}(T)\right) \Pr\{\mathcal{E}|(\mathcal{A} = \mathcal{R}_i \text{ and } \gamma_i \geq T)\} \right] \qquad (17)$$

where $p_{\mathcal{R}_i}$ denotes the steady-state probability of selecting the ith relay given in (Tellambura & Annamalai & Bharghava, 2001). The conditional BER for uncoded BPSK modulation is derived as

$$\Pr\{\mathcal{E}|(\mathcal{A} = \mathcal{R}_i, \gamma_i \geq T)\} = \frac{1}{2\overline{\gamma}_{SR_i}} I\left(\frac{1}{\overline{\gamma}_{SR_i}}, 1, T\right) + \frac{1}{2\overline{\gamma}_{R_iD}} I\left(\frac{1}{\overline{\gamma}_{R_iD}}, 1, T\right)$$

$$- \frac{1}{2\overline{\gamma}_{SR_i}\overline{\gamma}_{R_iD}} I\left(\frac{1}{\overline{\gamma}_{SR_i}}, 1, T\right) I\left(\frac{1}{\overline{\gamma}_{R_iD}}, 1, T\right) \qquad (18)$$

where the auxiliary function $I\left(\alpha, \beta, \omega\right)$ is defined as (Michalopoulos & Karagiannidis, 2007)

Figure 3. Outage probability, normalized on the outage threshold SNR, γ_{th}, of two-relay DSSC, for the cases where AF or DF relays are employed, with or without MRC employment at the destination terminal.

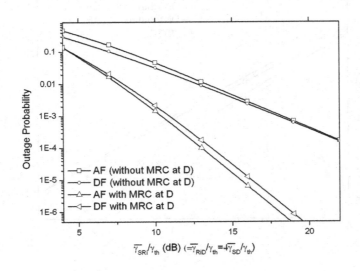

$$I\left(\alpha,\beta,\omega\right) \;=\frac{1}{\alpha}\,e^{-\alpha\omega}\mathrm{erfc}\left(\sqrt{\beta}\omega\right)-\frac{\sqrt{\beta}}{\alpha\sqrt{\alpha+\beta}}\,\mathrm{erfc}\left(\sqrt{(\alpha+\beta)}\omega\right). \tag{19}$$

Note that the probabilities $\Pr\left\{\mathcal{E}\middle|\mathcal{A}=\mathcal{R}_i\right\}$ can be also expressed as shown in (18), by setting $T=0$.

Figure 4 depicts the BER of two-relay DSSC and distributed SC with two available relays, when uncoded BPSK modulation and either DF or AF relaying is employed, versus the average values of γ_{SR_i} and γ_{R_iD}. Each of the DSSC curves in this Figure 4 was generated by using the optimal switching threshold, T, which is derived numerically by minimizing the corresponding BEP expressions with respect to T. One may observe from Figure 4 the cost of simplifying the relay selection procedure by employing DSSC instead of distributed SC, in terms of BER.

ii) **Single-Relay DSSC:** In cases where a single relay is available, the two branches of the "virtual" SSC can be represented by the source-relay-destination (i.e., the relayed) and the direct source-destination one, which are denoted here by $\mathcal{D}$ and $\mathcal{R}$, respectively. The single-relay DSSC can be thus thought of as a method of deciding in a switch and stay fashion whether the destination would receive from the relay or just from the source in a cooperative relaying system with a single available relay. In particular, in each transmission period the destination compares the received SNR (which equals to either $\gamma_{\mathcal{D}}$ or $\gamma_{\mathcal{R}}$, depending on which was the active branch in the previous transmission period) with a preselected fixed switching threshold T. If the received SNR is lower than T, then a branch-switching occurs. This is implemented by appropriate feedback sent to both the source and the relay, indicating a switching on the transmission path (from the direct to the relayed one and vice versa) during the ensuing transmission period.

Figure 4. BEP performance of two-relay DSSC and distributed SC, for uncoded BPSK modulation.

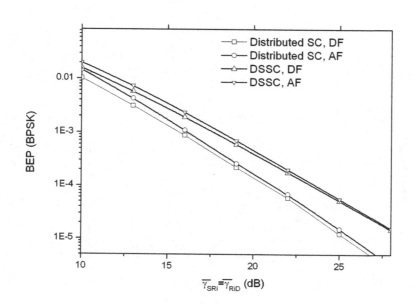

Let us focus on the case where the relay operates in the DF mode, adopting the outage-based definition for the end-to-end SNR of the relayed branch, i.e. $\gamma_{\mathcal{R}} := \min\left(\gamma_{SR}, \gamma_{RD}\right)$. The outage probability of single-relay DSSC can be derived as (Michalopoulos & Karagiannidis, 2007)

$$
\begin{aligned}
\Pr\left\{\mathcal{O}\right\} = {}& p_{\mathcal{D}}\left[\left(1 - F_{\gamma_{\mathcal{D}}}\left(T\right)\right)\Pr\left\{\mathcal{O}\middle|\left(\mathcal{A} = \mathcal{D}, \gamma_{\mathcal{D}} > T\right)\right\} + F_{\gamma_{\mathcal{D}}}\left(T\right)\Pr\left\{\mathcal{O}\middle|\mathcal{A} = \mathcal{R}\right\}\right] \\
& + p_{\mathcal{R}}\left[\left(1 - F_{\gamma_{\mathcal{R}}}\left(T\right)\right)\Pr\left\{\mathcal{O}\middle|\left(\mathcal{A} = \mathcal{R}, \gamma_{\mathcal{R}} > T\right)\right\} + F_{\gamma_{\mathcal{R}}}\left(T\right)\Pr\left\{\mathcal{O}\middle|\mathcal{A} = \mathcal{D}\right\}\right]
\end{aligned}
\tag{20}
$$

where for Rayleigh fading $F_{\gamma_{\mathcal{R}}}\left(x\right) = 1 - \exp\left(-x / \overline{\gamma}_{SR}\right)\exp\left(-x / \overline{\gamma}_{RD}\right)$; $p_{\mathcal{D}}$ and $p_{\mathcal{R}}$ represent the steady state probabilities of receiving from the direct (i.e., source-destination) and the relayed branch, respectively, which are given in (Tellambura & Annamalai & Bharghava, 2001).

Let $\gamma_{th}^{\mathcal{D}}$ represent the outage threshold SNR of the direct branch (i.e., $\gamma_{th}^{\mathcal{D}} = 2^r - 1$ with r representing the target data rate per spectrum usage) which can be expressed in terms of $\gamma_{th}^{\mathcal{R}}$ through $\gamma_{th}^{\mathcal{R}} = \gamma_{th}^{\mathcal{D}}\left(\gamma_{th}^{\mathcal{D}} + 2\right)$. The conditional outage probabilities in (20) can then be taken from the general form

$$
\Pr\left\{\mathcal{O}\middle|\left(\mathcal{A} = \mathcal{Z}, \gamma_{\mathcal{Z}} > w\right)\right\} = \begin{cases} \dfrac{F_{\gamma_{\mathcal{Z}}}\left(\gamma_{th}^{\mathcal{Z}}\right) - F_{\gamma_{\mathcal{Z}}}\left(w\right)}{1 - F_{\gamma_{\mathcal{Z}}}\left(w\right)}, & w \leq \gamma_{th}^{\mathcal{Z}} \\ 0, & w > \gamma_{th}^{\mathcal{Z}} \end{cases}
$$

by substituting $\mathcal{Z}$ and w with the appropriate values, i.e. $\mathcal{Z} \in \left\{\mathcal{D}, \mathcal{R}\right\}, w \in \left\{T, 0\right\}$.

The BER of single-relay DSSC is obtained as

$$
\begin{aligned}
\Pr\left\{\mathcal{E}\right\} = {}& p_{\mathcal{D}}\left[\left(1 - F_{\gamma_{\mathcal{D}}}\left(T\right)\right)\Pr\left\{\mathcal{E}\middle|\left(\mathcal{A} = \mathcal{D}, \gamma_{\mathcal{D}} > T\right)\right\} + F_{\gamma_{\mathcal{D}}}\left(T\right)\Pr\left\{\mathcal{E}\middle|\mathcal{A} = \mathcal{R}\right\}\right] \\
& + p_{\mathcal{R}}\left[\left(1 - F_{\gamma_{\mathcal{R}}}\left(T\right)\right)\Pr\left\{\mathcal{E}\middle|\left(\mathcal{A} = \mathcal{R}, \gamma_{\mathcal{R}} > T\right)\right\} + F_{\gamma_{\mathcal{R}}}\left(T\right)\Pr\left\{\mathcal{E}\middle|\mathcal{A} = \mathcal{D}\right\}\right].
\end{aligned}
\tag{21}
$$

Note that the conditional BEP corresponding to the relayed branch, conditioned on the events that $\mathcal{A} = \mathcal{R}$ and $\gamma_{\mathcal{R}} > T$ is given in (18); the conditional error probability $\Pr\left\{\mathcal{E}\middle|\left(\mathcal{A} = \mathcal{D}, \gamma_{\mathcal{D}} > T\right)\right\}$ is expressed as

$$
\Pr\left\{\mathcal{E}\middle|\left(\mathcal{A} = \mathcal{D}, \gamma_{\mathcal{D}} > T\right)\right\} = \frac{1}{2\overline{\gamma}_{SD}}I\left(\frac{1}{\overline{\gamma}_{SD}}, 1, T\right).
\tag{22}
$$

Figures 5 and 6 illustrate the outage probability and BPSK-modulated BER of single-relay DSSC, for several channel realizations. Specifically, the source-relay and relay-destination channels are assumed to experience independent, identically distributed Rayleigh fading, with average SNR equal to λ-times the average SNR of the source-destination link, i.e., $\overline{\gamma}_{SR} = \overline{\gamma}_{RD} = \lambda\overline{\gamma}_{SD}$, where $\lambda \in \left\{8, 15, 40\right\}$. Along with the performance of single-relay DSSC, the performance of a similar yet different scheme is also depicted, that is dubbed here as simplified incremental relaying (S-IR). This scheme resembles the incremental relaying transmission protocol proposed in (Laneman & Tse & Wornell, 2004), that is, the destination sends request-to-forward to the relay only if the source-destination channel is not sufficiently high, i.e., if $\gamma_D < T$. However, the destination terminal does not utilize a MRC in this model, hence in cases where the relayed branch is active the destination receives only via the relayed path, and not directly from the source.

In general, the BER performances of these two systems are close with one another; however, the DSSC scheme offers a lower BEP, which becomes more evident as the source-destination channel becomes relatively bad. Loosely speaking, this can be explained by considering the fact that in the DSSC scheme the relayed branch is used for a greater fraction of time, hence the higher the $\overline{\gamma}_{SR}$ and $\overline{\gamma}_{RD}$ (as compared to $\overline{\gamma}_{SD}$), the greater the difference in the corresponding error probabilities. Regarding the outage performance, the S-IR scheme slightly outperforms the DSSC one, since the relayed branch is less frequently used and thus the expected spectral efficiency is higher. Furthermore, observing the slope of the curves in both figures, it is evident that, as expected, both the DSSC and the S-IR scheme attain diversity order of two, since the number of independent paths from which the destination can receive is two. Finally, we note that the curves labeled as "DF relaying" in both figures correspond to the case where the destination receives only through the $\mathcal{R}$ (and never through the $\mathcal{D}$) branch, where, apparently, diversity order one is attained.

Finally, it should be noted that the two-relay and single-relay DSSC schemes represent special cases of the versatile "distributed switch diversity" concept. In the latter scheme, only a single relay out of the set of available ones is activated in each transmission period in a fashion similar to switch diversity (Simon & Alouini, 2005). That is, a relay may be active for as long as the corresponding end-to-end SNR is sufficiently high; then the system may switch to another relaying terminal which is selected based on

Figure 5. Average BEP of the DSSC and the S-IR scheme for several channel realizations, assuming BPSK modulation.

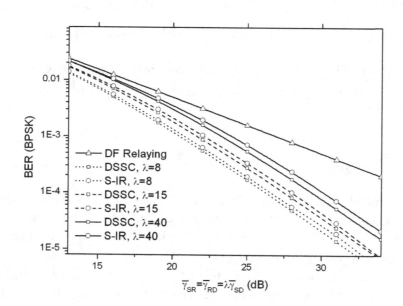

Figure 6. Outage probability of the DSSC and the S-IR scheme for several channel realizations.

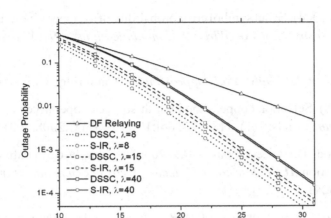

the average channel conditions, or simply through a round-robin procedure. This actually represents a generalization of the DSSC concept to multiple relay scenarios, aiming at reducing the overall complexity that selection based on instantaneous channel conditions entails. The main drawback of such a scheme, however, lies in the fact that in order to achieve full diversity order the destination needs to examine all the available paths. Apparently, such a distributed implementation of switch diversity is difficult to implement in practice in a manner that does not incur latency to the system, which sometimes may be destructive in terms of data rate. In other words, the translation of the switching diversity concept with multiple branches into cooperative diversity setups includes several practical constraints, which need to be taken into account before implementing in practical applications. Such constraints are absent in DSSC setups, where the switching between only two end-to-end links takes place.

CONCLUSION

Through the formulation of virtual, distributed antenna arrays, cooperative communications bring about novel ways of designing future wireless networks in which the inherent end-to-end paths between the source and the destination terminal are treated as the multiple branches of a virtual diversity receiver. As a result, the well-known combining methods applied to conventional receivers can be implemented in a distributed fashion, motivating for novel relaying protocols and generally allowing for a novel design under this distributed diversity perspective. This unexplored aspect of diversity combining includes the distributed implementation of Switch and Stay Combining (SSC), as well as those of Maximal Ratio Combining (MRC) and Selection Combining (SC) which have constituted the first attempts of studying cooperative communications from this point of view. This chapter has provided some relaying protocols relevant to the distributed diversity framework, as well as a performance analysis for the aforementioned schemes. Further insights regarding the relative performance of the above schemes, as well as the potential for implementing the distributed diversity concept in practical relaying applications have been derived.

REFERENCES

Anghel, P. A., & Kaveh, M. (2004). Exact symbol error probability of a cooperative network in a Rayleigh-fading environment. *IEEE Transactions on Wireless Communications*, *3*(9), 1416–1421. doi:10.1109/TWC.2004.833431

Baker, G., & Graves-Morris, P. A. (1996). *Padé approximants*. Cambridge University Press.

Beres, E., & Adve, R. (2008). Selection cooperation in multisource cooperative networks. *IEEE Transactions on Wireless Communications*, *7*(1), 118–127. doi:10.1109/TWC.2008.060184

Bletsas, A., Khisti, A., Reed, D. P., & Lippman, A. (2006). A simple cooperative diversity method based on network path selection. *IEEE Journal on Selected Areas in Communications*, *24*(3), 659–672. doi:10.1109/JSAC.2005.862417

Fan, Y., Wang, C., Thompson, J., & Poor, V. (2007). Recovering multiplexing loss through successive relaying using repetition coding. *IEEE Transactions on Wireless Communications*, *6*(12), 4484–4493. doi:10.1109/TWC.2007.060339

Hasna, M. O., & Alouini, M.-S. (2003). End-to-end performance of transmission systems with relays over Rayleigh-fading channels. *IEEE Transactions on Wireless Communications*, *2*(6), 1126–1131. doi:10.1109/TWC.2003.819030

Hasna, M. O., & Alouini, M.-S. (2004). A performance study of dual-hop transmissions with fixed gain relays. *IEEE Transactions on Wireless Communications*, *3*(6), 1963–1968. doi:10.1109/TWC.2004.837470

Herhold, P., Zimmermann, E., & Fettweis, G. (2004). A simple cooperative extension to wireless relaying. *International Zurich Seminar on Communications* (pp. 36-39).

Hu, J., & Beaulieu, N. C. (2006). A closed-form expression for the outage probability of decode-and-forward relaying in dissimilar Rayleigh fading channels. *IEEE Communications Letters*, *10*(12), 813–815. doi:10.1109/LCOMM.2006.061048

Hunter, T. E., & Nosratinia, A. (2006). Diversity through coded cooperation. *IEEE Transactions on Wireless Communications*, *5*(2), 283–289. doi:10.1109/TWC.2006.1611050

Ikki, S., & Ahmed, M. H. (2007). Performance analysis of cooperative diversity wireless networks over Nakagami-m fading channel. *IEEE Communications Letters*, *11*(4), 334–336. doi:10.1109/LCOM.2007.348292

Karagiannidis, G. K. (2006). Performance bounds of multihop wireless communications with blind relays over generalized fading channels. *IEEE Transactions on Wireless Communications*, *5*(3), 498–503. doi:10.1109/TWC.2006.1611077

Karagiannidis, G. K., Tsiftsis, T. A., & Mallik, R. K. (2006). Bounds for multihop relayed communications in Nakagami-*m* fading. *IEEE Transactions on Communications*, *54*(1), 18–22. doi:10.1109/TCOMM.2005.861679

Krikidis, I., Thompson, J., McLaughlin, S., & Goertz, N. (2008). Amplify-and-forward with partial relay selection. *IEEE Communications Letters*, *12*(4), 235–237. doi:10.1109/LCOMM.2008.071987

Laneman, J. N., Tse, D. N. C., & Wornell, G. W. (2004). Cooperative diversity in wireless networks: Efficient protocols and outage behavior. *IEEE Transactions on Information Theory, 50*(12), 3062–3080. doi:10.1109/TIT.2004.838089

Lee, I.-H., & Kim, D. (2007). BER analysis for decode-and-forward relaying in dissimilar Rayleigh fading channels. *IEEE Communications Letters, 11*(1), 52–54. doi:10.1109/LCOMM.2007.061375

Li, Y., & Kishore, S. (2007). Asymptotic analysis of amplify-and-forward relaying in Nakagami-fading environments. *IEEE Transactions on Wireless Communications, 6*(12), 4256–4262. doi:10.1109/TWC.2007.060368

Michalopoulos, D. S., & Karagiannidis, G. K. (2007). Distributed switch and stay combining (DSSC) with a single decode and forward relay. *IEEE Communications Letters, 11*(5), 408–410. doi:10.1109/LCOMM.2007.070018

Michalopoulos, D. S., & Karagiannidis, G. K. (2008). -{1}). PHY-layer fairness in amplify and forward cooperative diversity systems. *IEEE Transactions on Wireless Communications, 7*(3), 1073–1082. doi:10.1109/TWC.2008.060825

Michalopoulos, D. S., & Karagiannidis, G. K. (2008). -{2}). Performance analysis of single relay selection in Rayleigh fading. *IEEE Transactions on Wireless Communications, 7*(10), 3718–3724. doi:10.1109/TWC.2008.070492

Michalopoulos, D. S., & Karagiannidis, G. K. (2008). -{3}). Two-relay distributed switch and stay combining. *IEEE Transactions on Communications, 56*(11), 1790–1794. doi:10.1109/TCOMM.2008.070017

Michalopoulos, D. S., & Karagiannidis, G. K. (2008-{4}). Spectral efficient cooperative communications via spatial signal separation. *International Conference on Communications (ICC'08), IEEE* (pp. 1039-1043).

Onat, F. A., Adinoyi, A., Fan, Y., Yanikomeroglu, H., & Thompson, J. S. (2007). Optimum threshold for SNR-based selective digital relaying schemes in cooperative wireless networks. *Wireless Communications and Networking Conference, 2007, WCNC 2007, IEEE* (pp. 969-974).

Rankov, B., & Wittneben, A. (2007). Spectral efficient protocols for half-duplex fading relay channels. *IEEE Journal on Selected Areas in Communications, 25*(2), 379–389. doi:10.1109/JSAC.2007.070213

Ribeiro, A., Cai, X., & Giannakis, G. B. (2005). Symbol error probabilities for general cooperative links. *IEEE Transactions on Wireless Communications, 4*(3), 1264–1273. doi:10.1109/TWC.2005.846989

Simon, M. K., & Alouini, M.-S. (2005). *Digital communications over fading channels, 2nd edition.* New York: Wiley

Siriwongpairat, W. P., Himsoon, T., Su, W., & Liu, K. J. R. (2006). Optimum threshold-selection relaying for decode-and-forward cooperation protocol. *Wireless Communications and Networking Conference, 2006, WCNC 2006, IEEE* (pp. 1015-1020).

Suraweera, H. A., & Karagiannidis, G. K. (2008). Closed-form error analysis of the non-identical Nakagami-m relay fading channel. *IEEE Communications Letters, 12*(4), 259–261. doi:10.1109/LCOMM.2008.071922

Suraweera, H. A., Smith, P. J., & Armstrong, J. (2006). Outage probability of cooperative relay networks in Nakagami-m fading channels. *IEEE Communications Letters, 10*(12), 834–836. doi:10.1109/LCOMM.2006.060834

Tellambura, C., Annamalai, A., & Bhargava, V. K. (2001). Unified analysis of switched diversity systems in independent and correlated fading channels. *IEEE Transactions on Communications, 49*(11), 1955–1965. doi:10.1109/26.966072

Tsiftsis, T. A., Karagiannidis, G. K., Kotsopoulos, S. A., & Pavlidou, F.-N. (2004). BER analysis of collaborative dual-hop wireless transmissions. *IET Electronics Letters, 40*(11), 679–681. doi:10.1049/el:20040393

Tsiftsis, T. A., Karagiannidis, G. K., Mathiopoulos, T., & Kotsopoulos, S. A. (2006). Nonregenerative dual-hop cooperative links with selection diversity. *EURASIP Journal on Wireless Communications and Networking, 2006 . Article ID, 17862*, 1–8.

Wang, T., Cano, A., Giannakis, G. B., & Laneman, J. N. (2007). High-performance cooperative demodulation with decode-and-forward relays. *IEEE Transactions on Communications, 55*(7), 1427–1438. doi:10.1109/TCOMM.2007.900631

Wang, T., Wang, A., & Giannakis, G. B. (2006). Smart regenerative relays for link-adaptive cooperative communications. *40th Conference on Information Sciences and Systems 2006, CISS 2006, IEEE* (pp. 1038-1043).

Zhao, Y., Adve, R., & Lim, T. J. (2007). Improving amplify-and-forward relay networks: Optimal power allocation versus selection. *IEEE Transactions on Wireless Communications, 6*(8), 3114–3123.

Chapter 12
Single and Double–Differential Coding in Cooperative Communications

Manav R. Bhatnagar
University of Oslo, Norway

Are Hjørungne
University of Oslo, Norway

ABSTRACT

In this chapter, we discuss single and double-differential coding for a two-user cooperative communication system. The single-differential coding is important for the cooperative systems as the data at the destination/relaying node can be decoded without knowing the channel gains. The double-differential modulation is useful as it avoids the need of estimating the channel and carrier offsets for the decoding of the data. We explain single-differential coding for a cooperative system with one relay utilizing orthogonal transmissions with respect to the source. Next, we explain two single-differential relaying strategies: active user strategy (AUS) and passive users relaying strategy (PURS), which could be used by the base-station to transmit data of two users over downlink channels in the two-user cooperative communication network with decode-and-forward protocol. The AUS and PURS follow an improved time schedule in order to increase the data rate. A probability of error based approach is also discussed, which can be used to reduce the erroneous relaying of data by the regenerative relay. In addition, we also discuss how to implement double-differential (DD) modulation for decode-and-forward and amplify-and-forward based cooperative communication system with single source-destination pair and a single relay. The DD based systems work very well in the presence of random carrier offsets without any channel and carrier offset knowledge at the receivers, where the single differential cooperative scheme breaks down. It is further shown that optimized power distributions can be used to improve the performance of the DD system.

DOI: 10.4018/978-1-60566-665-5.ch012

INTRODUCTION

Cooperative communication has several promising features to become a main technology in future wireless communication systems. It has been shown in literature (Sendonaris et al., 2003a, Nosratinia et al., 2004) that the cooperative communication can avoid the difficulties of implementing actual antenna array and convert the single-input single-output (SISO) systems into a virtual multiple-input multiple-output (MIMO) system. In this way, cooperation between the users allows them to exploit the diversity gain and other advantages of MIMO system at a SISO wireless network. Many cooperative protocols have been proposed in last seven years. There exist three key protocols (1) Decode-and-Forward (DAF) (2) Amplify-and-Forward (AAF) (3) Coded Cooperation. We concentrated over the cooperative systems with DAF and AAF protocols in this chapter.

In the DAF protocol with a single relay, a user (source) needs to select another user which agrees to relay its data to the destination (Sendonaris et al., 2003a, Sendonaris et al., 2003b). The source sends information to the destination and the relay as well. The relay decodes the data sent by the source and retransmits the decoded data to the destination. Hence, the destination has two received replicas (in the case of perfect relaying) of the same data and the quality of reception is expected to improve. To avoid the wrong relaying of data, some intelligence could be included into the relay terminal to make a decision about the quality of reception and for this the source may apply *ideal* cyclic redundancy code (CRC) (Merkey & Posner, 1984) over the transmitted data. Using the CRC, the relay can judge whether it has received the signal correct or not, hence, it can stop relaying wrong data and the performance of the cooperative system is improved. In (Su et al., 2005), the symbol error rate (SER) performance analysis for the coherent DAF relaying protocol for M-PSK and M-QAM modulation is performed.

To avoid the problem of error propagation in DAF scheme, a selection relaying protocol is proposed in (Laneman et al., 2001, Laneman et al., 2004), where depending upon the channel between the source and the relay, the relay decides whether it should relay the data or not. Another method to avoid the error propagation called incremental relaying is proposed in (Laneman et al., 2004), where the destination decides whether it needs relaying or not based on the channel between the source and the destination.

In AAF protocol, the relaying terminal acts *non-regeneratively* over the received data. It amplifies the received data such that an instantaneous or average power constraint is satisfied and forwards toward the destination. This protocol was proposed and analyzed in (Laneman et al., 2004). It has been shown in (Laneman et al., 2004) that for the two-user case, this method achieves diversity order of two, which is the best possible outcome at high SNR.

A cooperative network can be implemented coherently, but it requires that the destination node must possess the information about the channel coefficients of all links (involved in cooperation) in the network in case of AAF protocol. For DAF protocol based cooperative system the destination needs the information about the channel between the source and the destination and the channels between the relays and the destination. In order to satisfy this requirement, lots of training data and feed forward transmissions from the relays are required. One attractive solution to avoid this requirement is differential modulation. In a differential cooperative system it is not required for a node to possess information about the channels of the other links. Hence, differential modulation can save a significant amount of the training data and avoid the difficulty of practical implementation of the cooperative systems with a small loss in the performance. It is worth mentioning here that DAF protocol in particular is more suitable for realization of a differential modulation based cooperative network as it completely avoids the need of any information about the channel (even of the channel statistics which is required by AAF based relays in differential cooperative system) at all nodes in the cooperative network. Differential modulation for cooperative system with *single source-destination* pair with a relay using DAF protocol with BPSK constellation is proposed for Rayleigh channels in (Zhao & Li, 2005b) and Nakagami-*m* channels in (Zhao & Li, 2006). A maximum-likelihood (ML) decoder for differentially modulated BFSK signal transmitted through multiple regenerative relays is found in (Chen & Laneman, 2004a, Chen & Laneman, 2004b). Differential modulation with AAF protocol over Rayleigh fading channels is proposed in (Zhao & Li, 2005b, Himsoon et al., 2006) and over Nakagami-*m* channels is proposed in (Zhao & Li, 2005a). In (Zhao & Li, 2005a, Zhao & Li, 2005b, Zhao & Li, 2007) a limited case of BPSK modulation is treated, whereas, in (Himsoon et al., 2006), an expression of approximate bit error rate (BER) of AAF based differential cooperative communications system with *M*-PSK constellation is derived. All these differential schemes utilize orthogonal transmissions from the source and the relay, which reduces the effective data rate. Therefore, a better time scheduling (in the case of TDMA) can be implemented to improve the data rate of differential cooperative system. A differential scheme based on Alamouti space-time block code (STBC) (Alamouti, 1998) utilizing better than orthogonal time scheduling, is proposed in (Tarasak et al., 2004, Tarasak et al., 2005). The method of (Tarasak et al., 2004, Tarasak et al., 2005) is applicable for *uplink* channels in the cooperative network. The BS appends the CRC bits to a frame before differential encoding so that the relay can check whether the whole frame is received correctly or not. However, this method reduces the spectral efficiency. Moreover, if the CRC bits are received erroneously, the error floor in the detection increases. Most of the above discussed differential schemes can be extended to multiple relay scenarios with some added signal processing complexity at the receiver/relays/source.

Multiple relays in a cooperative network can act as additional spatial dimensions for transmission of the data to the destination. Therefore, STBC is a natural candidate for implementing differential coding for the multiple relays in a distributed manner. STBC is important in a sense that it provides much more flexibility in terms of signal processing as it has already been explored rigorously in the literature for multiple-input multiple-output (MIMO) systems. For example, Alamouti STBC can be used to obtain low complexity differential decoding at the destination (Jing & Jafarkhani, 2008). STBC ensures that multiple relays in the cooperative system are used in an optimized manner. In order to improve the performance of the cooperative system, distributed algebraic STBC can be used in a non-coherent (Rajan & Rajan, 2007) or partially coherent (Kiran & Rajan, 2007) manner. However, it is difficult to implement differential STBC in a distributed manner if there is also a direct link between the source and the destination. In (Wang et al., 2006), STBC belonging to a unitary group is implemented using differential modulation for cooperative system with multiple relays by including the direct link between the source and the destination. The cooperative network though virtually resembles MIMO system, but it is physically much more different than MIMO system. The biggest problem is the loss of the effective data rate due to the sequential broadcasting transmissions from the source to the relays such that the relays can obtain the copies of the data sequence to be relayed to the destination in next phase, which is different from MIMO system that does not need such transmissions. In general, half of the useful bandwidth is wasted in broadcasting phase in cooperative networks.

The single-differential schemes assume that the channel is constant over at least *two* consecutive time intervals. However, in the presence of carrier offsets, the flat fading wireless channel does *not* remain constant over two consecutive time intervals and these differential schemes experience substantial performance loss. Double-differential (DD) modulation (Simon & Divsalar, 1992, Stoica et al., 2004, Simon et al., 1994, Bhatnagar & Hjørungnes, 2007b) is a key differential technique to remove the effect of carrier offset from the communication system. It differs from single-differential modulation in a sense that the decoder uses *three* consecutively received data samples for decoding the current symbol. Two levels of single-differential modulation are employed at the transmitter and a simple heuristic decoder (Stoica et al., 2004, Eq. (15)) is used at the receiver to find the estimate of the transmitted data. It has been shown in (Stoica et al., 2004, Section III) that the heuristic decoder coincides with the maximum likelihood decoder (MLD) under the assumption that the product of two zero-mean white circularly symmetric Gaussian noise terms in the decision variable is also zero-mean white circularly symmetric Gaussian. Symbol error rate (SER) expressions for the double-differentially modulated data over Rayleigh and Ricean fading SISO channels with carrier offsets are provided in (Bhatnagar & Hjørungnes, 2007b). In (Liu et al., 2001), a double-differential orthogonal space-time block code for time-selective MIMO channels is proposed. A distributed double-differential modulation based on (Liu et al., 2001) with *regenerative* relays over cooperative Rayleigh channels is proposed and an upper bound of the pairwise error performance (PEP) is obtained in (Cano et al., 2007). A distributed DD coding based on full and square OSTBC is proposed in (Bhatnagar et al., 2008f) for *amplify-and-forward* based cooperative system. This scheme extends the double-differential coding for full and square OSTBC of (Bhatnagar et al., 2007d, Bhatnagar et al., 2008g, Bhatnagar et al., 2008h) into cooperative networks. The DD coding of (Bhatnagar et al., 2008f) provides better data rate than (Cano et al., 2007).

In this book chapter, we concentrate on conveying a thorough idea of the basic implementation issues about the single-and double-differential modulation in cooperative networks. In order to make the presentation simple and easy to understand, we consider a basic cooperative system with a single source destination pair and a single relay. The protocols used for differential cooperative systems are DAF and AAF which are the same as used in their coherent counterparts. Therefore, the results presented in this book chapter can be extended to a general scenario with multiple relaying nodes with the help of results available for the coherent cooperative networks in the literature. The key idea of this book chapter is how to improve the spectral efficiency of cooperative networks by using single-and double-differential modulation and a better time scheduling.

This chapter is organized as follows: In Section 2, the single and double-differential modulations are explained. Section 3 discusses the differential modulation for two user cooperative system in detail and summarizes the distributed differential coding based on STBC. The double-differential coding for the cooperative system with single source destination pair and one relay is explained in Section 4. Some conclusions and future directions are provided in Section 5.

Notation: Lower-and upper-case letters x, X are used for scalar quantity, boldface lower-and upper-case letters $\mathbf{x}$, $\mathbf{X}$ are used for vector and matrix representation, respectively, $(\bullet)^*$ represents conjugation of matrix or scalar, $E\{\bullet\}$ stands for expectation, $(\bullet)^T$ provides transpose of a matrix or vector, arg stands for argument of a complex quantity, and max represents the maximum value.

BACKGROUNDS

In this section, we will review single-differential and double-differential modulation for SISO system.

Single-Differential Modulation

Let $b[n]$ denotes the symbols belonging to the unit-norm M-PSK constellation Ξ to be transmitted at the time $n = 1, 2, \ldots\ldots$. A differentially modulated signal $s[n]$ is obtained from $b[n]$ as

$$s[n] = s[n-1]b[n],\tag{1}$$

with $s[1] \in \{\pm 1, \pm j\}$, $j = \sqrt{-1}$, as initialization symbol. As $\left|b[n]\right| = 1$ for the unit-norm M-PSK symbols, it follows from (1) that $\left|s[n]\right| = 1$, " n . We consider a flat fading SISO channel described by

$$x[n] = \sqrt{P}hs[n] + e[n],\tag{2}$$

where $x[n]$ is the received signal, P is the transmitted signal power, h is the channel gain which is assumed to be constant over at least two consecutive time intervals (n-1 and n), and $e[n]$ is complex-valued AWGN noise. The decoding of $b[n]$ is performed as follows [Proakis, 2001]:

$$\hat{b}[n] = \arg \max_{b \in \Xi} \mathrm{Re}\left\{x[n]x^*[n-1]b^*\right\}.\tag{3}$$

It can be seen from (3) that the decoding of $b[n]$ can be performed without any channel knowledge in a differential modulation based system. This property is very useful for cooperative systems as estimation of channel coefficients of different links requires transmission of training data, which reduces effective data rate. In addition, these channel estimates are also needed to be feed forwarded from the relays to the destination which also further reduces the effective data rate. In Section 3, the cooperative communication systems employing differential modulation are discussed.

Double-Differential Modulation

In a DD modulation based system (Simon & Divsalar, 1992, Stoica et al., 2004, Simon et al., 1994, Bhatnagar & Hjørungnes, 2007b), the transmitted signal $v[n]$, $n = 1, 2, \ldots\ldots$, is obtained from $b[n]$, $n = 1, 2, \ldots\ldots$ as

$$\begin{aligned}
s[n] &= s[n-1]b[n], \\
v[n] &= v[n-1]s[n],
\end{aligned}\tag{4}$$

with $v[1], s[2] \in \{\pm 1, \pm j\}$ as initialization symbols. As $|b[n]| = 1$ for the unit-norm M-PSK symbols, it follows from (4) that $|v[n]| = |s[n]| = 1$, " n. We consider a flat fading SISO channel with carrier offset described by

$$x[n] = \sqrt{P} h e^{jwn} v[n] + e[n], \tag{5}$$

where $w \in [-\pi, \pi)$ is the unknown frequency offset. The channel and frequency offset are assumed to be constant over at least three consecutive time intervals (n-2, n-1, and n). The receiver makes a decision variable, $d[n] = X[n] X^*[n-1]$, where $X[n] = x[n] x^*[n-1]$. The decoding of $b[n]$ is performed in the following way (Stoica et al., 2004, Eq. (15)):

$$\hat{b}[n] = \arg \max_{b \in \Xi} \mathrm{Re} \left\{ X[n] X^*[n-1] b^* \right\}. \tag{6}$$

The decoding of (6) corresponds to maximum-likelihood decoding (MLD) in (Stoica et al., 2004) under the assumption that the product of two zero-mean white circularly symmetric Gaussian noise terms in the decision variable is also zero-mean white circularly symmetric Gaussian. It can be seen from (6) that the double-differential modulation avoids the need of the knowledge of carrier offsets and channel gains for decoding of the data. The estimation of the carrier offset in general is a computationally complex problem and a training based estimator requires fast Fourier transformation (FFT) and maxima searching procedure. For implementation of a coherent cooperative network the destination needs information about all carrier offsets present in cooperative links in addition to their channel gains. In Section 4, it is shown how to avoid this impractical requirement by using double-differential modulation such that the cooperative system can enjoy full diversity in the presence of random carrier offsets.

SINGLE-DIFFERENTIAL SCHEMES FOR COOPERATIVE COMMUNICATION SYSTEMS

In this section, we will review the single-differential modulation based schemes for cooperative communication systems. In Subsection 3.1, it is shown how to implement differential modulation for the cooperative systems where the source and the single relay use orthogonal transmissions. In Subsection 3.2, it is shown that better data rates can be achieved over downlink channels by choosing a better time scheduling, which partially utilizes the orthogonal transmissions. In Subsection 3.3, single-differential schemes for cooperative networks based on STBC are summarized.

Differential Modulation for Cooperative System with a Single Orthogonal Relay

Let us consider a basic cooperative communication system, which consists of one source (s), one relay (r), and one destination (d) terminal as shown in Fig. 1, where $h_{p,q}$, $p \neq q$, is channel coefficient between node $p \in \{s, r\}$ and $q \in \{r, d\}$ with variance $\sigma_{p,q}^2 = E\left\{|h_{p,q}|^2\right\}$. Each of them can either transmit or receive a signal at a time. The transmission of the data from the source to the destination terminal is

Figure 1. Cooperative communication system with a single orthogonal relay

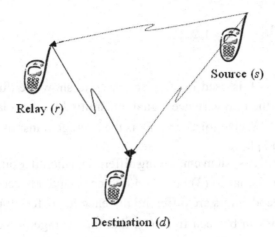

furnished in two phases. In the first phase, the source broadcasts data to the destination and the relay. The relay decodes/amplifies the received data and retransmits it to the destination, in the second phase. To avoid the interference, source and relay use orthogonal channels for transmission. For ease of presentation we assume that in both phases, the source and relay transmit stream of data through time division multiple access (TDMA). In the TDMA scheme, the source has to remain silent in the second phase in order to maintain the orthogonality between the transmissions. However, in the frequency division multiple access (FDMA) or the code division multiple access (CDMA) schemes, the source and the relay can transmit at the same time.

Differential Transmission for Cooperative System with a Single Orthogonal Relay

If we use differential modulation in the cooperative communications system, the data received during the *first phase* at the destination is

$$x_{s,d}\left[n\right] = \sqrt{P_1}h_{s,d}s\left[n\right] + e_{s,d}\left[n\right], \quad n = 1,2,\dots, \tag{7}$$

and at the relay is

$$x_{s,r}\left[n\right] = \sqrt{P_1}h_{s,r}s\left[n\right] + e_{s,r}\left[n\right], \quad n = 1,2,\dots, \tag{8}$$

where P_1 is the power transmitted in first phase by the source per channel use, and $e_{s,d}\left[n\right]$ and $e_{s,r}\left[n\right]$ are complex-valued AWGN noise on the two links, and $s\left[n\right]$ is differentially modulated signal obtained from $b\left[n\right]$ belonging to the unit-norm M-PSK constellation Ξ and to be transmitted at the time n, similar to (1).

During the *second phase* the source remains silent for TDMA and the relay demodulates (DAF protocol) or amplifies/scales (AAF protocol) the received data of (8) and retransmits such that the received signal by the destination in the second phase is:

$$x_{r,d}[l] = \sqrt{P_2} h_{r,d} s_{r,d}[l] + e_{r,d}[l], \qquad l = 1, 2, \ldots, \tag{9}$$

where l is the time index which is used in the place of n to show the difference in time of the first and second phases, $s_{r,d}[l]$ is the unit variance signal transmitted by the relay, $h_{r,d}$ is the channel gain, $e_{r,d}[l]$ is the complex-valued AWGN noise, and P_2 is the average transmission power by the realy per channel use during the second phase.

In the case of DAF cooperative system employing differential modulation, the relay first differentially decodes the transmitted data similar to (3) based on two consecutively received data samples $x_{s,r}[n]$ and $x_{s,r}[n-1]$. Next, the decoded bits are differentially encoded before transmission.

As $|s[n]| = 1$ for all n, it can be seen from (8), that the average power of $x_{s,r}[l]$ is $P_1 \sigma_{s,r}^2 + \sigma^2$, where $\sigma_{s,r}^2$ is the variance of $h_{s,r}$, and σ^2 is the variance of the AWGN noise $e_{s,r}[l]$, hence, in the case of AAF cooperative system $s_{r,d}[l]$ is given as

$$s_{r,d}[l] = \frac{x_{s,r}[l]}{\sqrt{P_1 \sigma_{s,r}^2 + \sigma^2}}. \tag{10}$$

It can be seen from (10) that $s_{r,d}[l]$ is a unit-variance signal.

Differential Decoding for Cooperative System with a Single Orthogonal Relay

A maximum likelihood (ML) decoder for differential DAF cooperative system with BPSK is derived in (Zhao & Li, 2007, Chen & Laneman, 2004a, Chen & Laneman, 2004b) as

$$\hat{b}[n] = \text{sign}\left(g(t_1) + t_0\right) \tag{11}$$

where $g(t_1) = \ln \dfrac{(1-\varepsilon)e^{t_1} + \varepsilon}{\varepsilon e^{t_1} + 1 - \varepsilon}$, $\varepsilon = \dfrac{1}{2 + 2\bar{\gamma}}$, $t_1 = \dfrac{q_1}{\sigma^2}$, $q_1 = x_{r,d}^*[l-1] x_{r,d}[l] + x_{r,d}[l-1] x_{r,d}^*[l]$, $t_0 = \dfrac{q_0}{\sigma^2}$, $q_0 = x_{s,d}^*[l-1] x_{s,d}[l] + x_{s,d}[l-1] x_{s,d}^*[l]$, where $\bar{\gamma} = P_1 \sigma_{s,r}^2 / \sigma^2$ is the average SNR of the link between the source and the relay. The non-linear function $g(t_1)$ effectively clips large inputs to $\pm \ln[1-\varepsilon]/\varepsilon$ and it can be approximated by a piecewise-linear (PL) function as

$$g(t_1) \approx g_{\text{PL}}(t_1) \triangleq \begin{cases} -T_1, & t_1 < -T_1, \\ t_1, & -T_1 \leq t_1 \leq T_1, \\ T_1, & t_1 > T_1, \end{cases} \tag{12}$$

where $T_1 = \ln\left[(1-\varepsilon)/\varepsilon\right]$ assuming $\varepsilon \leq 0.5$. It is shown in (Zhao & Li, 2007) that the PL approximation has negligible performance degradation as compared to the ML decoder. It can be seen from (11) and

(12) that the destination can decode the data without any channel knowledge in a differential cooperative system utilizing DAF scheme.

The following linear combiner can be used for an AAF based differential cooperative system (Zhao & Li, 2007, Eq. (14)):

$$d[k] = x_{s,d}^*[n-1]x_{s,d}[n] + \frac{1+\overline{\gamma}_{s,r}}{1+\overline{\gamma}_{s,r}+\overline{\gamma}_{r,d}}x_{r,d}^*[l-1]x_{r,d}[l],$$ (13)

where $k = n = 1$, $\overline{\gamma}_{s,r} = P_1\sigma_{s,r}^2/\sigma^2$ is the average SNR of the link between the source and the relay, and $\overline{\gamma}_{r,d} = P_2\sigma_{r,d}^2/\sigma^2$ is the average SNR of the link between the relay and the destination. The decision is made as follows:

$$\hat{b}[n] = \arg\max_{b\in\Xi} \mathrm{Re}\left\{d[n]b^*\right\}.$$ (14)

It can be seen from (13) and (14) that decoding of $b[n]$ can be done without exact channel knowledge at the receiver. However, the receiver needs the knowledge of the channel statistics for decoding. The performance of differential cooperative system with DAF and AAF protocols using BPSK is shown in Fig. 2. The DAF based scheme with the PL approximation (12) works better than AAF based scheme with linear combiner (12) for SNR in the range [0 – 25] dB.

Figure 2. Performance of differential cooperative systems

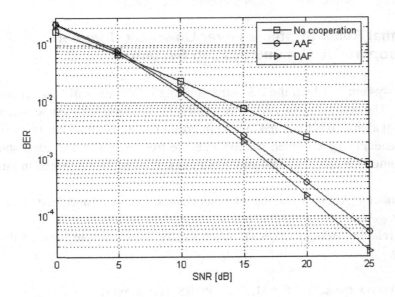

Power Allocation for Differential Cooperative System with Single Orthogonal Relay

The performance of the differential cooperative system can be improved by optimizing the power allocation in terms of P_1 (power transmitted by the source) and P_2 (power transmitted by the relay) with the total power constraint $P_1 + P_2 = P$. The optimized power allocation can be obtained by several methods. A few of them are maximizing the received SNR at the destination, minimizing the pairwise error probability (PEP), minimizing the mean square error (MSE), minmizing the outage probability, and minimizing the probability of error of the decoder at the destination. In (Himsoon et al., 2006), optimized values of P_1 and P_2 are found by numerically minimizing an upper bound of the bit error rate (BER) of the AAF based differential cooperative system. An optimized power distribution is found in (Zhao & Li, 2007) by numerically minimizing the average probability of error and outage probability of DAF and AAF based differential cooperative system with BPSK constellation.

In Fig. 3, we have plotted the analytical SER of differential AAF cooperative system with a single orthogonal relay as a function of P_1 under different link conditions and QPSK constellation. The constraint $P_1 + P_2 = P = 2$ is used. The minimum SER values of P_1 are also shown for different link conditions in Fig. 2. The expression of SER of the differential AAF cooperative system can be obtained by the results given in (Hasna and Alouini, 2004, Simon & Alouini, 2005). For more details see (Bhatnagar et al., 2008e). Fig. 2 suggests that the uniform power allocation, i.e., $P_1 = P_2 = 1/2$ is not an optimized solution when all links are similar in the cooperative system. Better performance can be obtained by choosing $P_1 = 1.4$ and $P_2 = 0.6$. However, an approximately uniform power distribution is an optimized solution when the link between the source and relay is good, i.e., $\sigma_{sr}^2 = 100$.

The most of the above mentioned single-differential schemes can be implemented for multiple relays with some modifications. However, the transmission of one symbol is performed in two time intervals in the above mentioned schemes, which reduces the effective data rate. In the following subsection, we will show that it is possible to obtain differential cooperative schemes for downlink channels for two user cooperative systems with the DAF protocol to achieve better data rate by utilizing a modified time scheduling with partially orthogonal relaying. Interestingly, these differential schemes outperform the AAF based differential scheme in a two user's cooperative system.

Single-Differential Relay Strategies over Downlink Channel in Two-User Cooperative Communication System

Consider a cellular system in which the BS is transmitting information of two cooperative users over downlink channels. The source, relay, and destinations are equipped with single antenna and can only receive or transmit at a time. It is assumed that all interuser channels, BS-relay, relay-user, and BS-user channels are independent of each other. All channels are frequency flat and quasi static, i.e., they are fixed for multiple time intervals. Two differential strategies can be implemented in this system:

1) Active user strategy (AUS) - when the interuser channels are good and the users may work as relay for each other.
2) Passive users relaying strategy (PURS) - when a relay is also present between the users and base-station.

Fig. 4(a) and (b) show the setups for AUS and PURS, respectively.

Figure 3. Numerical minimization of the SER of the amplify-and-forward based differential cooperative system with single orthogonal relay. $\sigma_{s,r}^2 \in \{10, 100\}$ *and* $\sigma_{r,d}^2 \in \{10, 100\}$ *correspond to the cooperative systems with* $\sigma_{s,d}^2 = \sigma_{r,d}^2 = 1$, $\sigma_{s,r}^2 \in \{10, 100\}$ *and* $\sigma_{s,d}^2 = \sigma_{s,r}^2 = 1$, $\sigma_{r,d}^2 = 10, 100$, *respectively.*

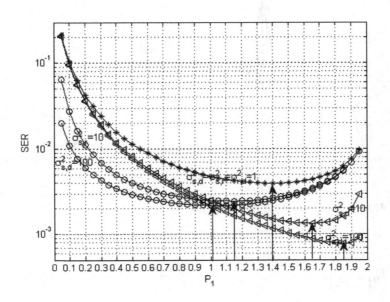

Active User Strategy

In this scheme, we consider the cooperative communication system shown in Fig. 4(a). The BS has to transmit $b_1[n]$ and $b_2[n] \in$ BPSK, $1 \le n \le N / 3$ to User 1 and User 2, respectively, where N is the length of one frame and assume that 3 is a factor of N. Consider a transmission scheme, where $N / 3$ time slots are devoted to each transmission and the channel is assumed to be fixed over these $N / 3$ time intervals. The BS first encodes $b_1[n]$ and $b_2[n]$ into differential symbols $s_1[n]$ and $s_2[n]$, $1 \le n \le N / 3$ respectively, before transmission as follows:

$$
\begin{aligned}
s_1[n] &= b_1[n] s_1[n-1], \\
s_2[n] &= j b_2[n] s_2[n-1],
\end{aligned}
\tag{15}
$$

where $s_1[0], s_2[0] \in \{\pm 1, \pm j\}$ are initialization symbols. In the first $N / 3$ time intervals, the BS transmits a super positioned/combined version of $s_1[n]$ and $s_2[n]$ as $s_1[n] + j s_2[n]$. In the next $N / 3$ time intervals, the BS *transmits* $s_1[n]$ and User 2 *relays* the estimated $-j \hat{s}_1^*[n]$, to User 1 and in the further $N / 3$ time intervals BS and User 1 transmit $s_2[n]$ and $-j \hat{s}_2^*[n]$, respectively to User 2. The additional multiplication of j and conjugation of $\hat{s}_1[n]$ and $\hat{s}_2[n]$ is done to maintain the normal differential decoding at the receiver. The transmission schedule of this scheme is shown in Table 1. The received signal at User 1 in the first $N / 3$ time intervals corresponding to the direct transmission is

Figure4. Two-user cooperative communication system for (a) active user strategy (AUS) (b) passive users relaying strategy (PURS).

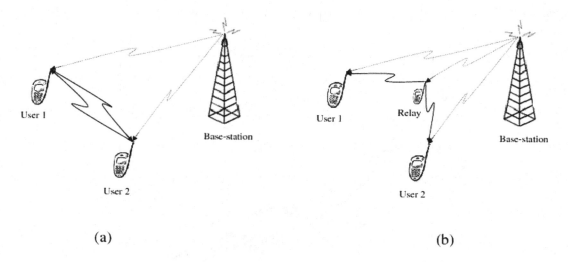

(a)   (b)

$$x_{s,1}[n] = \sqrt{P_1} h_{s,1} \left(s_1[n] + js_s[n] \right) + e_{s,1}[n], \ 1 \le n \le N/3, \tag{16}$$

where $h_{s,1}$ is the channel between BS and User 1 during $1 \le n \le N/3$, P_1 is the average power transmitted by BS per channel use, and $e_{s,1}[n]$ is additive white Gaussian noise (AWGN) received by User 1. The data received at the User 2 in the first $N/3$ time intervals at any $1 \le n \le N/3$ will be

$$x_{s,2}[n] = \sqrt{P_1} h_{s,2} \left(s_1[n] + js_2[n] \right) + e_{s,2}[n], \tag{17}$$

Table 1. Transmission schedule of AUS

	First Time Period	**Second Time Period**	**Third Time Period**
Transmission	$1 \le n \le N/3$	$N/3 + 1 \le n \le 2N/3$	$2N/3 + 1 \le n \le N$
BS-U1	$s_1[n] + js_2[n]$	$s_1[n]$	—
BS-U2	$s_1[n] + js_2[n]$	—	$s_2[n]$
U1-U2	—	—	$-j\hat{s}_2^*[n]$
U2-U1	—	$-j\hat{s}_2^*[n]$	—

where $h_{s,2}$ is the channel between BS and User 2 and $e_{s,2}[n]$ is AWGN noise received by User 2. In the next $N/3$ time intervals, the received data at User 1 at $N/3+1 \leq n \leq 2N/3$ is

$$x_{s,2,1}[n] = \sqrt{\frac{P_2}{2}} h'_{2,1} s_1[n-N/3] - j\sqrt{\frac{P_2}{2}} h_{2,1} \hat{s}_1^*[n-N/3] + e_{2,1}[n],\tag{18}$$

where $\hat{s}_1[n-N/3]$ is the differential symbol transmitted by User 2 corresponding to the estimate of the User 1's data obtained by it at $1 \leq n \leq N/3$, $h'_{2,1}$ is the channel between BS and User 1 during $N/3+1 \leq n \leq 2N/3$, $h_{2,1}$ is the channel between User 2 and User 1 during $N/3+1 \leq n \leq 2N/3$, and P_2 is the total average power transmitted per channel use during $N/3+1 \leq n \leq 2N/3$. The additional multiplication of j and conjugation in (18) is done to maintain the normal differential decoding at the receiver. The receiver at User 1 computes the following for decision variable:

$$d_c[n] = \left(x_{x,2,1}[n] x_{s,2,1}^*[n-1]\right) + \left(x_{s,1}[n-N/3] x_{s,1}^*[n-N/3-1]\right),$$
$$N/3+1 \leq n \leq 2N/3.\tag{19}$$

It can be shown that the real part of $d_c[n]$ can be used to decode the data of User 1 (Bhatnagar & Hjørungnes, 2007a, Bhatnagar & Hjørungnes, 2008d, Appendix I). Therefore, for deciding about the BPSK symbol, the receiver only checks the sign of the real part of $d_c[n]$. We call this differential scheme *active user strategy* (AUS) as the users are active, i.e., they act as relays for each other.

Passive Users Relaying Strategy

In this scheme, the users act passively, i.e., they do not act as relay for each other and data is relayed by a relay. The cooperative communication system of PURS is shown in Fig. 4(b). In this case, the BS selects a relay for relaying the data to the users. Assume $s_1[n]$ and $s_2[n]$ are differential symbols of User 1 and User 2, respectively obtained from (14). We also assume that the data of both users belongs to BPSK constellation $\{\pm 1\}$. In this scheme, the BS *transmits* the symbols of User 1 in the first $N/3$ time intervals and the symbols of User 2 in the next $N/3$ time intervals. In the final $N/3$ time intervals of the frame, only the relay transmits a *combined* version of the data of both the users. The transmission schedule of this scheme is shown in Table 2. The data received by User 1 in the first $N/3$ time intervals is

$$x_{s,1}[n] = \sqrt{P_1} h_{s,1} s_1[n] + e_{s,1}[n], \ 1 \leq n \leq N/3.\tag{20}$$

In the same time intervals, the data received by the relay is

$$x_{s,r}[n] = \sqrt{P_1} h_{s,r} s_1[n] + e_{s,r}[n], \ 1 \leq n \leq N/3.\tag{21}$$

Table 2. Transmission schedule for PURS

Transmission	First Time Period $1 \leq n \leq N / 3$	Second Time Period $N / 3 + 1 \leq n \leq 2N / 3$	Third Time Period $2N / 3 + 1 \leq n \leq N$
BS-U1	$s_1[n]$	—	—
BS-U2	—	$s_2[n]$	—
BS-R	$s_1[n]$	$s_2[n]$	—
R-U1	—	—	$\hat{s}_1[n] + j\hat{s}_2[n]$
R-U2	—	—	$\hat{s}_1[n] + j\hat{s}_2[n]$

where $h_{s,r}$ is the channel between BS and relay during $1 \leq n \leq N / 3$ and P_1 is the average power transmitted by BS per channel use during $1 \leq n \leq N / 3$. User 2 and the relay receive the following data the next $N / 3$ time intervals:

$$x_{s,2}[n] = \sqrt{P_2}h_{s,2}s_2[n] + e_{s,2}[n], \ N / 3 + 1 \leq n \leq 2N / 3, \tag{22}$$

$$x_{s,r}[n] = \sqrt{P_2}h'_{s,r}s_2[n] + e_{s,r}[n], \ N / 3 + 1 \leq n \leq 2N / 3. \tag{23}$$

where $h'_{s,r}$ is the channel between BS and relay and P_2 is the average power transmitted by BS per channel use during $N / 3 + 1 \leq n \leq 2N / 3$. The relay first computes the following for the decision about the data of User 1:

$$d_{r,1}[n] = x_{s,r}[n]x^*_{s,r}[n - 1], \ 1 \leq n \leq N / 3, \tag{24}$$

and decides about the BPSK data by comparing the real part of $d_{r,1}[n]$ with zero. Similarly, it decodes the data of the second user. The relay encodes the estimated symbols of User 1 and User 2 into differential form through (1) and makes a combined version of these symbols and retransmits this new symbol to the users in the final $N / 3$ time intervals. The received data at User 1 is

$$x_{r,1}[n + 2N / 3] = \sqrt{P_3}h_{r,1}\left(\hat{s}_1[n] + j\hat{s}_2[n + N / 3]\right) + e_{r,1}[n + 2N / 3], \ 1 \leq n \leq N / 3, \tag{25}$$

where $\hat{s}_1[n]$ and $\hat{s}_2[n + N / 3]$ are the differential symbols transmitted by the relay corresponding to the estimates of data of User 1 and User 2 and P_3 is the average power transmitted by the relay per channel use. The receiver at User 1 computes the following before decision:

$$d_c\left[n + 2N / 3\right] = x_{s,1}\left[n\right]x_{s,1}^*\left[n - 1\right] + x_{r,1}\left[n + 2N / 3\right]x_{r,1}^*\left[n + 2N / 3 - 1\right], \ 1 \leq n \leq N / 3.$$

$$(26)$$

It is shown in (Bhatnagar & Hjørungnes, 2007a, Bhatnagar & Hjørungnes, 2008d, Appendix II) that the real part of $d_c\left[n + 2N / 3\right]$ can be used to decode the data of User 1.

It is seen from Table 1 and Table 2 that the AUS and PURS schemes transmit the data of both users in three time intervals. Whereas, the differential schemes with single orthogonal relay (Himsoon et al., 2006, Zhao & Li, 2005a, Zhao & Li, 2005b, Zhao & Li, 2007) assume completely orthogonal transmissions and there the data of two users is transmitted in four time intervals to the destination. Therefore, the AUS and PURS schemes achieve better data rate than these differential schemes.

Relay Selection Criterion

In the selection relaying protocol (Laneman et al., 2004), transmission or non-transmission of signal from relay depends upon the condition of the channel. In AUS and PURS, the source or destination does not have knowledge about the channel gains. Therefore, we will discuss a criterion which can be used by the relay to determine the quality of reception of the data. This criterion is based on a simple method of channel power estimation and calculation of the conditional probability of error in making the decision about the differentially encoded data. Let the channel remain stationary over $N / 3$ time intervals, i.e., it is a flat fading channel. It can be seen from Tables 1 and 2 that the relay/relaying user will receive the data from the BS over at least $N / 3$ time intervals in both strategies. Let $x = hs + e$ be the row vector of data received over the $N / 3$ time intervals by the relay, where $s = \left[s_0, s_1, \ldots, s_{N/3-1}\right]$, $E\left\{|s_i|^2\right\} = 1$, denotes the data vector, h is the channel gain between the BS and relay, and $e = \left[e_0, e_1, \ldots, e_{N/3-1}\right]$ denotes the AWGN noise vector. Then $E\left\{xx^H\right\} = \left(N / 3\right)E\left\{|h|^2\right\} + \left(N / 3\right)\sigma^2$, where σ^2 is the variance of AWGN noise. Hence, an estimator of SNR $\gamma = E\left\{|h|^2\right\} / \sigma^2$ is as follows (Hwang et al., 2003):

$$\gamma = \frac{E\left\{xx^H\right\} - N / 3 \cdot \sigma^2}{\left(N / 3\right)\sigma^2}$$

$$(27)$$

Assume that $\sigma^2 = 1$ and known at the receiver, we can obtain an approximate estimate of the SNR between the BS and relay from (27) as

$$\hat{\gamma} \approx \frac{xx^H}{\left(N / 3\right)}.$$

$$(28)$$

The probability of error in the received data is always a function of SNR of the link. The expression of conditional probability of error for binary differential PSK is given in (Proakis, 2001, Eq. (5.2.69)) as

$$P_b\left(\gamma\right) = \frac{1}{2}\exp\left(-\gamma\right),$$

$$(29)$$

where γ is the SNR of the link. The closed form expression of the conditional probability of error in four phase differentially encoded PSK is given in (Proakis, 2001, Eq. (5.2.70)) as

$$P_b\left(\gamma\right) = Q_1\left(a,b\right) - \frac{1}{2}I_0\left(ab\right)\exp\left[-\frac{1}{2}\left(a^2+b^2\right)\right]$$

(30)

where $Q_1\left(a,b\right)$ is the Marcum Q function defined as (Proakis, 2001, Eq. (2.1.123))

$$Q_1\left(a,b\right) = \exp\left\{-\frac{1}{2}\left(a^2+b^2\right)\right\}\sum_{k=0}^{\infty}\left(\frac{a}{b}\right)^k I_k\left(ab\right)$$

(31)

and I_k is the modified Bessel function, defined as (Proakis, 2001, Eq. (2.1.120))

$$I_k\left(x\right) = \sum_{m=0}^{\infty}\frac{\left(x/2\right)^{k+2m}}{m!\,\Gamma\left(k+m+1\right)},\quad x \geq 0,$$

(32)

and $a = \sqrt{\gamma\left(2-\sqrt{2}\right)}$ and $b = \sqrt{\gamma\left(2+\sqrt{2}\right)}$, where $\Gamma\left(\cdot\right)$ is the gamma function (Grandshteyn & Ryzhik, 2000, Eq. (8.310.1)).

In order to decide about the possible relaying of the received data the relay must know whether it can decode the data with permissible average error or not. One rough method for this is to compare the instantaneous SNR with the average SNR of the link. However, it can be seen from (29) and (30) that the probability of error in the received data is not directly proportional to instantaneous SNR but it is monotonic nonlinear function of SNR. Therefore, comparison of the conditional probability of error with the average probability of error leads to more accurate criterion than simply comparing the SNRs.

In the selection relaying, the relay/relaying user calculates the probability of error in the received data by (29) and (30), and if it is greater than the average probability of error it stops the transmission, and if it is less than or equal to the average probability of error it relays the data to the users. Similarly, the destination can also decides about whether it needs relaying of data or not after checking the quality of the received data using the proposed criterion, in incremental relaying protocol (Laneman et al., 2004).

Figs. 5 and 6 show the performance of AUS and PURS with selection and incremental relaying, respectively. It is seen that the performance of both schemes improves a lot when the SNR of link between the BS and the relay is high (25-35 dB).

Fig. 7 shows the comparison of AUS and PURS with decode-and-forward, selection, and incremental relaying protocols. We have plotted the performance results for both schemes assuming equal average SNR at all links. It is seen that the AUS and PURS always perform better than the direct DPSK (no cooperation). The PURS performs better than the AUS with all protocols in general. However, AUS scheme seems to have more diversity gain than the PURS, with selection relaying, hence, the AUS performs better than PURS strategy at higher SNRs (>25 dB). Both the proposed relaying schemes also perform better than the conventional same rate (approx.) amplify-and-forward differential relaying scheme (Himsoon et al., 2006) at all SNRs in the case of selection and incremental relaying. However, with the decode-

Figure 5. Performance of AUS and PURS with selection relaying

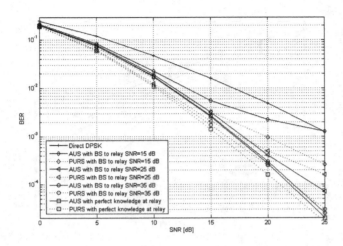

Figure 6. Performance of AUS and PURS with incremental relaying

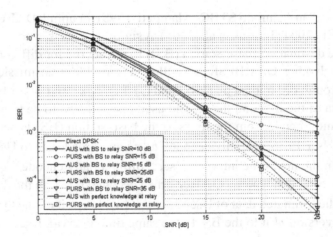

and-forward protocol, both relaying schemes perform better than the amplify-and-forward relaying at lower SNRs. We have also plotted the performance of the bench-mark scheme and AUS and PURS with perfect knowledge at the relay in Fig. 7. It can be seen from Fig. 7 that the PURS performs approximately 2 dB poor than the bench-mark scheme (Bhatnagar et al., 2007a, Bhatnagar et al., 2008d).

The performance of AUS and PURS schemes can be further improved by an optimized power distribution with a constraint on the total transmitted power.

Figure 7. Comparison of performances of AUS and PURS

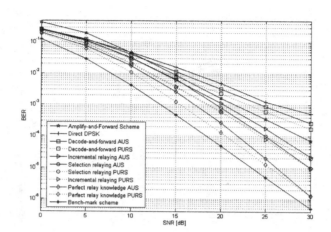

Differential Modulation for Cooperative System Based on STBC

A differential scheme based on Alamouti STBC (Alamouti, 1998) for uplink DAF cooperative channels is proposed in (Tarasak et al., 2004, Tarasak et al., 2005). A simple DAF based cooperative system consisting of two users and one destination (base-station (BS)) is considered for differential modulation. Both users act as relays for each other. Only BPSK constellation is used for transmission. The signals of both users are transmitted through orthogonal phases (I and Q). Let $b_1[n]$ and $b_2[n]$ belonging to the BPSK constellation, be the data of the first and the second user, respectively, to be transmitted at time n. The transmission of the Alamouti code from the users to the base-station (BS) is performed in *three* time-slots. In first time-slot, User 1 differentially encodes $b_1[n]$ and broadcasts to User 2 and the BS. Then User 2 decodes $b_1[n]$. In the next time interval, User 2 differentially encodes $b_2[n]$ and broadcasts to User 1 and the BS on the orthogonal phase Q. User 1 then decodes $b_2[n]$. In the third time interval, User 1 differentially encode the estimated data of User 2, i.e., $\hat{b}_2[n]$, keeps it on the Q phase, and transmits a negative and conjugate version of it to the BS. In the same time interval, User 2 differentially encodes $\hat{b}_1[n]$ and transmits a conjugate version of it to the BS. A differential decoder based on the received data in six consecutive time intervals is derived in (Tarasak et al., 2004, Tarasak et al., 2005). In the case of using the DAF protocol, the BER versus SNR performance of this scheme suffers from a error floor when the direct transmission SNR is greater than the interuser SNR. To avoid the problem with the error floor, it can be assumed that each transmitted message uses CRC bits such that the relay can perfectly know whether it has decoded the transmitted signal correctly or not. The performance of the differential scheme improves with the CRC bits and also with selection and incremental relaying.

In (Zhang, 2005), a generalized differential scheme based on unitary space-time block codes is proposed assuming single source destination pair with more than two ($U > 2$) relays in between. The whole transmission is divided into two phases consisting of several time intervals. In the first phase, symbols are transmitted using differential encoding to the relays. The relays decode the symbols differentially. Then N relays are selected which have perfectly decoded the transmitted symbols, where $N \leq U$. The symbol vector is mapped into a $N \times N$ unitary space-time matrix $\boldsymbol{G}_k$ which belongs to a finite unitary matrix group G, where k denotes the k-th block. Each of the N relays separately encodes $\boldsymbol{G}_k$ into the corresponding row of the differential code matrix $\boldsymbol{D}_k$. Let $\mathbf{D}_{k-1} = \left[\mathbf{d}_{k-1,1}^T, \mathbf{d}_{k-1,2}^T, \dots, \mathbf{d}_{k-1,N}^T \right]^T$, then the i-th relay obtains $\mathbf{d}_{k,i}$ as $\mathbf{d}_{k,i} = \mathbf{d}_{k-1,i}\mathbf{G}_k$. A PEP upper bound is also obtained for the differential cooperative system in (Zhang, 2005). The idea of (Zhang, 2005), is generalized in (Wang et al., 2006) by including a direct link between the source and the destination. In (Jing & Jafarkhani, 2008), a distributed single-differential coding for multiple relays based on OSTBC is proposed, which is an extension of the distributed coherent scheme of (Jing & Hassibi, 2006). A partially coherent distributed single-differential STBC code for multiple relays is proposed in (Kiran & Rajan, 2007). In (Kiran & Rajan, 2007), a code design criterion is explained by assuming that the destination terminal has perfect knowledge of the channels between the relays and the destination. A low decoding complexity distributed single-differential STBC based on scaled unitary matrices is proposed for multiple relays in (Rajan & Rajan, 2007).

It is obvious from the discussion above that the distributed differential schemes based on STBC require N sequential transmissions from the source to the relays such that all N relays can have the estimate of the N symbols to be transmitted to the destination in next $T \geq N$ time intervals through a distributed STBC matrix. Therefore, $T + N$ time intervals are needed for the transmission of N symbols. For example, by using the Alamouti STBC in a distributed manner in a differential cooperative system with two relays and no direct link between the source and the destination, two symbols can be transmitted in four time intervals. Whereas, by using better time scheduling as done in AUS and PURS two symbols can be transmitted in three time intervals. It is not possible to compare the performance of AUS and PURS with the distributed coding schemes based on STBC since AUS and PURS are limited to BPSK constellation only. However, it is a topic of future research that the time scheduling of AUS and PURS can be combined with STBC to achieve higher data rates with optimized performance in a cooperative network with multiple relays.

DOUBLE-DIFFERENTIAL CODING FOR COOPERATIVE COMMUNICATION SYSTEM

In this section, we will explain how to implement double-differential modulation in the cooperative system shown in Fig. 1 consisting one source-destination pair and one relay. It is assumed that each link is now also perturbed by random carrier offsets, which are generated because of the mismatch between the receive and the transmit oscillators or relative movement of the transmitters and the receivers. The source and the relay follow orthogonal transmissions as explained in Subsection 3.1. For the comparison we will also consider the system models of AUS and PURS shown in Fig. 4, now perturbed with random carrier offsets. Let us first explain the channel model thoroughly.

Channel Model

All three links are assumed to be Nakagami-m distributed with the following probability density function (p.d.f.) (Simon & Alouini, 2005, Eq. (2.21))

$$f_{\gamma_{p,q}}\left(\gamma_{p,q}\right) = \frac{m_{p,q}^{m_{p,q}} \gamma_{p,q}^{m_{p,q}-1}}{\overline{\gamma}_{p,q}^{m_{p,q}} \Gamma\left(m_{p,q}\right)} \exp\left(-\frac{m_{p,q}\gamma_{p,q}}{\overline{\gamma}_{p,q}}\right), \qquad \gamma_{p,q} \geq 0, \tag{33}$$

where $m_{p,q} \geq 1/2$ is the Nakagami-m fading parameter, $\gamma_{p,q} = P_i\left|h_{p,q}\right|^2 \big/ \sigma^2$ is the instantaneous SNR, $P_i \in \left\{P_1, P_2\right\}$ is the power transmitted in first or second time interval, $h_{p,q}$ is a *zero-mean* Nakagami-m channel coefficient, and $\overline{\gamma}_{p,q} = P_p \sigma_{p,q}^2 \big/ \sigma^2$ is the average SNR over the link between the p and the q terminals in the cooperative system. The channel of each link is assumed to be a block fading channel, which remains constant for *at least three* consecutive time intervals and all the channel coefficients are assumed to be independent of each other. It is assumed that all three links are perturbed by different carrier offsets $w_{p,q} = 2\pi f_{p,q} T_s$ (Larsson & Stoica, 2003), where $f_{p,q}$ is the physical carrier frequency offset (CFO) in Hertz, $f_{p,q} \in \left[-1 \big/ \left(2T_s\right)\right], 1 \big/ \left(2T_s\right)$, and T_s is the sampling period in seconds. Apparently, $w_{p,q} \in \left[-\pi, \pi\right)$ and the maximum value of $w_{p,q}$ corresponds to the offset of 50% of the carrier. We assume that the carrier offsets $w_{p,q}$ are random and uniformly distributed over $\left[-\pi, \pi\right)$, however, in general, there is no restriction over the probability distribution of the carrier offsets and they could have any probability distribution. We have assumed that these carrier offsets remain fixed for *at least three* consecutive sampling instances. The presence of carrier offsets makes all three block-fading channels behave as *time-varying channels*, which do *not* remain stationary over two consecutive time intervals. Since the phase term $e^{jw_{p,q}n}$ is multiplied with the channel coefficient $h_{p,q}$, and the result is time-varying even though $w_{p,q}$ and $h_{p,q}$ stays constant for the same three consecutive time instants.

DD Modulation for Cooperative System

If we use DD modulation in the cooperative communication system, the data received during the *first phase* at the destination is

$$x_{s,d}\left[n\right] = \sqrt{P_1} h_{s,d} e^{jw_{s,d}n} v\left[n\right] + e_{s,d}\left[n\right], \qquad n = 1, 2, \ldots, \tag{34}$$

and at the relay is

$$x_{s,r}\left[n\right] = \sqrt{P_1} h_{s,r} e^{jw_{s,r}n} v\left[n\right] + e_{s,r}\left[n\right], \qquad n = 1, 2, \ldots, \tag{35}$$

where $v\left[n\right]$ is double-differentially modulated data obtained from (4) and $w_{s,d}$ and $w_{s,r}$ are the carrier offsets between source and destination, and source and relay, respectively. The relay can now amplify/scale (AAF protocol) or decode (DAF protocol) the received data before retransmitting it to the destination. We will consider both AAF and DAF protocols for DD modulation in rest of the section.

If AAF protocol is used, then during the *second phase*, the relay amplifies/scales the received data of (35) and retransmits such that the received signal by the destination in the *second phase* is:

$$x_{r,d}[l] = \sqrt{P_2} h_{r,d} e^{jw_{r,d}l} s_{r,d}[l] + e_{r,d}[l], \qquad l = 1, 2, \dots, \tag{36}$$

where $w_{r,d}$ is the carrier offset between relay and destination and $s_{r,d}[l]$ is given in (10). It is also assumed that $P_1 + P_2 = P$, where P is the total transmitted power. We can call this system as double-differential amplify-and-forward (DDAAF) cooperative system.

In the DAF based cooperative system the relay decodes the data through (35), encodes it double-differentially through (4) and retransmits during *second phase* as follows:

$$x_{r,d}[l] = \sqrt{P_2} h_{r,d} e^{jw_{r,d}m} \hat{v}[l] + e_{r,d}[l], \qquad l = 1, 2, \dots, \tag{37}$$

where $\hat{v}[l] \in \{0, v[n]\}$, $l = n$, is the DD encoded data corresponding to the estimate of the data transmitted by the source during the first phase. It is assumed here that the source adds ideal CRC (Merkey & Posner, 1984) bits while transmitting the information symbols such that the relay could know, whether, it has received the information correctly or not (for selection relaying). The following maximal ratio combining (MRC) (Brennan, 2003) based receiver is used at the destination for amplify based system (Bhatnagar et al., 2008b):

$$d[k] = \alpha_1 \left(x_{s,d}[n] x_{s,d}^*[n-1] \right) \left(x_{s,d}[n-1] x_{s,d}^*[n-2] \right)^* + \\ \alpha_2 \left(x_{r,d}[l] x_{r,d}^*[l-1] \right) \left(x_{r,d}[l-1] x_{r,d}^*[l-2] \right)^*, \tag{38}$$

where $k = n = l$, i.e., the data received by the destination during the same time interval with respect to the beginning of the each phase is combined, and α_1 and α_2 are given by

$$\alpha_1 = \frac{1}{\left(2P_1 \left| h_{s,d} \right|^2 + \sigma^2 \right) \sigma^2}, \tag{39}$$

and

$$\alpha_2 = \frac{\left(P_1 \sigma_{s,r}^2 + \sigma^2 \right)^2}{k}, \tag{40}$$

where

$$k = 2P_1 P_2^2 \left| h_{r,d} \right|^4 \left| h_{s,r} \right|^2 \sigma^2 + 2P_1 P_2 \left(P_1 \sigma_{s,r}^2 + \sigma^2 \right) \left| h_{r,d} \right|^2 \left| h_{s,r} \right|^2 \sigma^2 + P_2^2 \left| h_{r,d} \right|^4 \sigma^4 + 2P_2 \times \\ \left(P_1 \sigma_{s,r}^2 + \sigma^2 \right) \left| h_{r,d} \right|^2 \sigma^4 + \left(P_1 \sigma_{s,r}^2 + \sigma^2 \right)^2 \sigma^4. \tag{41}$$

The normalization factors can be found as $\alpha_1 = 1 / E_1$ and $\alpha_2 = 1 / E_2$, where E_1 and E_2 are the average noise powers of $X_{s,d}[n] = x_{s,d}[n]x_{s,d}^*[n-1]$ and $X_{r,d}[n] = x_{r,d}[n]x_{r,d}^*[n-1]$, respectively. However, as we intend to use DD modulation, the destination and relay are not expected to have knowledge of the exact channel coefficients, therefore, we can use emulated maximum ratio combining (EMRC) by replacing the channel coefficients $\left|h_{s,r}\right|^2$, $\left|h_{r,d}\right|^2$, and $\left|h_{s,d}\right|^2$ by their variances $\sigma_{s,r}^2$, $\sigma_{r,d}^2$, and $\sigma_{s,d}^2$, respectively, in (39) and (40).

For decode based system a decision variable can be made as follows (Bhatnagar & Hjørungnes, 2007c):

$$d[k] = \left(x_{s,d}[n]x_{s,d}^*[n-1]\right)\left(x_{s,d}[n-1]x_{s,d}^*[n-2]\right)^* + \left(x_{r,d}[l]x_{r,d}^*[l-1]\right)\left(x_{r,d}[l-1]x_{r,d}^*[l-2]\right)^*, \qquad (42)$$

where $k = n = l$. Then, the data is decoded as

$$\hat{b}[n] = \arg\max_{b\in\Xi}\mathrm{Re}\left\{d[k]b^*\right\}, \qquad k = n. \qquad (43)$$

Contrary to the DDAAF system, the DDDAF system avoids the need of knowing channel statistics at the destination.

Comparison of DD Cooperative System with Single-Differential Schemes and Trained Cooperative System

Fig. 8 shows the performance of DDAAF and DDDAF schemes over Rayleigh fading channels. It can be seen that DDDAF (with perfect relaying) performs better than DDAAF. The amplify-and-forward based single-differential scheme of (Himsoon et al., 2006) and decode-and-forward based single-differential PURS scheme of Subsection 3.2.2 break down because of the random carrier offsets. The reason for the failure of the single-differential schemes is that the channel does not remain constant over two consecutive time intervals because of the random carrier offsets, which is the basic assumption for the single-differential modulation. Moreover, both DD cooperative schemes perform better than the direct transmission DD scheme which utilizes only the direct link between the source and the destination for transmission of DD data.

A training based AAF cooperative system which utilizes the two initialization symbols $v[1]$ and $v[2]$ as training data to find the estimates of the channel gains and carrier offsets before decoding $b[n]$ is implemented in (Bhatnagar et al., 2008b). A comparison using equal rate trained and DDAAF cooperative system with a single source-destination pair and a single relay is shown in Fig. 9 over flat fading Rayleigh fading channels with random carrier offsets. It can be seen that the DD based cooperative system outperforms the equal rate trained cooperative system at all SNRs shown in the figure. The performance of a coherent decoder is also shown in Fig. 9 which has perfect channel state information (CSI) and carrier offsets knowledge (COK). It can be seen that the DDAAF system works approximately 7 dB poorer than the coherent system.

Figure 8. Comparison of DDAAF, DDDAF, direct transmission DD scheme, PURS (Subsection 3.2.2), and amplify-and-forward based differential cooperative scheme of (Himsoon et al., 2006) using BPSK constellation.

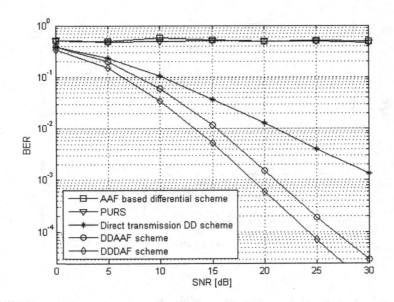

Apparently, DD cooperative systems are more suitable for the practical implementation as it is almost impossible to have a carrier offsets free network. By using DD modulation, a lot of training data and feed forward transmissions can be saved and as a result the effective data rate can be improved in a practical system.

Performance Analysis of DD Cooperative System

It can be seen by comparing (3) to (6) that the decoding of *double-differentially* modulated signal depends upon $X[n]$ in the similar manner as the decoding of *single-differentially* modulated signal depends upon $x[n]$. Therefore, the performance of double-differential M-PSK (DDMPSK) can be approximated by the BER expressions of differential M-PSK (DMPSK) with the SNR of $X[n]$ under the assumption that the product of two zero-mean white circularly symmetric Gaussian noise terms in $X[n]$ is also zero-mean white circularly symmetric Gaussian. This connection is shown in more detail in (Simon et al., 1994, Simon & Divsalar, 1992, Bhatnagar & Hjørungnes, 2007b).

Average BER of DDAAF System

Using the analogy between the single and double-differential modulation approximate BER of DDAAF system over general (in terms of m) Nakagami-m channels with carrier offsets is obtained in (Bhatnagar et al., 2008b).

Figure 9. Comparison of DDAAF with AAF based trained and coherent cooperative systems

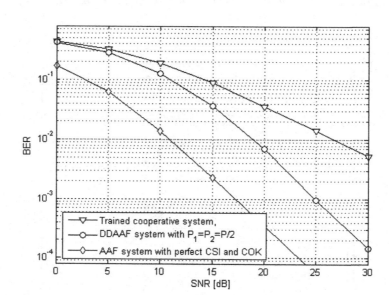

Lemma 1.*The approximate BER of binary amplify-and-forward based cooperative system with double-differential modulation over Rayleigh channels is given by [Bhatnagar et al., 2008b]*

$$
P_b = \frac{\exp\left(\beta + 0.5\right)}{16}\left\{\begin{array}{l}
\dfrac{56\sigma^6 + 36P_1\sigma_{s,d}^2\sigma^4}{\left(2\sigma^2 + P_1\sigma_{s,d}^2\right)^2\left(2\sigma^2 + P_1\sigma_{s,r}^2\right)}W_{-1,0.5}\left(2\beta\right) + \\[2ex]
\dfrac{\sqrt{2}\left(\sigma^2 + P_1\sigma_{s,r}^2\right)^{1/2}\left(28\sigma^5 + 18P_1\sigma_{s,d}^2\sigma^3\right)}{\sqrt{P_2}\sigma_{r,d}\left(2\sigma^2 + P_1\sigma_{s,d}^2\right)\left(2\sigma^2 + P_1\sigma_{s,r}^2\right)^{1/2}}W_{-0.5,0}\left(2\beta\right) + \\[2ex]
\dfrac{16P_1\sigma_{s,r}^2\sigma^4}{\left(2\sigma^2 + P_1\sigma_{s,r}^2\right)^2\left(2\sigma^2 + P_1\sigma_{s,d}^2\right)}W_{-2,0.5}\left(2\beta\right) + \\[2ex]
\dfrac{4\sqrt{2}P_1\sigma_{s,r}^4\left(\sigma^2 + P_1\sigma_{s,r}^2\right)^{1/2}}{\sqrt{P_2}\sigma_{r,d}\left(2\sigma^2 + P_1\sigma_{s,r}^2\right)^{3/2}\left(2\sigma^3 + P_1\sigma_{s,d}^2\sigma\right)}W_{-1.5,0}\left(2\beta\right)
\end{array}\right\}, \tag{44}
$$

where $\beta = \dfrac{\left(\sigma^2 + P_1\sigma_{s,r}^2\right)\sigma^2}{P_2\sigma_{r,d}^2\left(2\sigma^2 + P_1\sigma_{s,r}^2\right)}$ and $W_{\lambda,\mu}\left(\bullet\right)$ is Whittaker function (Grandshteyn & Ryzhik, 2000, Eq. (9.220.4)).

Proof: Refer (*Bhatnagar et al., 2008b*) for a proof.

Let us consider a symmetric case when $P_1 = P_2 = P$, $\sigma_{s,r} = \sigma_{r,d} = \sigma_{s,d} = \sigma_s$, and the SNR of each link is $\gamma_s = P_s^2/\sigma^2$. From (Abramowitz & Stegun, 1972, Eq. (13.1.33)), the Whittaker function can be expressed as

$$W_{\lambda,\mu}(z) = e^{-z} z^{\mu+0.5} U(0.5 + \mu - \lambda, 1 + 2\mu, z),\tag{45}$$

where $U(\cdot,\cdot,\cdot)$ is the confluent hypergeometric function of second kind. If $\gamma_s \to \infty$, (44) can be written using (45) as

$$P_b \approx \frac{\exp(1/2)}{16} \left[\begin{array}{c} 36\gamma_s^{-3} U\left(2,2,2\gamma_s^{-1}\right) + 18\sqrt{2}\gamma_s^{-2} U\left(1,1,2\gamma_s^{-1}\right) + \\ 16\gamma_s^{-3} U\left(3,2,2\gamma_s^{-1}\right) + 4\sqrt{2}\gamma_s^{-2} U\left(2,1,2\gamma_s^{-1}\right) \end{array} \right].\tag{46}$$

At high SNR, the probability of error of the DDAAF system can be further approximated by using (Abramowitz & Stegun, 1972, Eqs. (13.5.7) and (13.5.9)) as

$$P_b \approx \varepsilon \gamma_s^{-2},\tag{47}$$

where is a positive constant which is independent of γ_s. It can be seen from (47) that $\lim_{\gamma_s \to \infty} \varepsilon \gamma_s^{-2} = 0$, therefore, the DDAAF system achieves diversity of the order of *two* over the Rayleigh channels.

Analytical Performance of DDDAF System

The analytical expression for symbol error rate (SER) of DDDAF system with flat fading Nakagami-channels with carrier offsets is obtained in (Bhatnagar et al., 2008c).

Lemma 2. *The SER of DAF cooperative communication system with DDMPSK modulation over Nakagami- channels with carrier offsets can be upper bounded as (Bhatnagar et al., 2008c)*

$$P_s \leq \frac{(M-1)}{M^2} \sigma^{2m_{s,d}} \exp\left(\frac{1}{4}\sin^2\frac{\pi}{M}\right) \frac{1}{\Pi}\left[\mu + M\sigma^{2m_{r,d}}\left(\sigma^2 + \frac{P_1}{4m_{s,r}}\sigma_{s,r}^2 \sin^2\frac{\pi}{M}\right)^{m_{s,r}}\right],\tag{48}$$

where

$$\Pi = \left(\sigma^2 + \frac{P_1}{4m_{s,d}}\sigma_{s,d}^2 \sin^2\frac{\pi}{M}\right)^{m_{s,d}} \left(\sigma^2 + \frac{P_1}{4m_{s,r}}\sigma_{s,r}^2 \sin^2\frac{\pi}{M}\right)^{m_{s,r}} \left(\sigma^2 + \frac{P_2}{4m_{r,d}}\sigma_{r,d}^2 \sin^2\frac{\pi}{M}\right)^{m_{r,d}},$$

and

$$\mu = \left(M - 1\right)\sigma^{2m_{s,r}} \left[\left(\sigma^2 + \frac{P_2}{4m_{r,d}}\sigma_{r,d}^2 \sin^2 \frac{\pi}{M}\right)^{m_{r,d}} - \sigma^{2m_{r,d}} \exp\left(\frac{1}{8}\sin^2 \frac{\pi}{M}\right)\right].$$

Proof: Refer (*Bhatnagar et al., 2008c*) for a proof.

Power Allocation for DDDAF System

It can be seen from (48) that the upper bound of SER of the DD cooperative system with the DAF protocol depends non-linearly upon P_1 and P_2. Hence, if the total transmit power $P = P_1 + P_2$ is fixed, it is possible to find a power allocation for the source and the relay terminals which minimizes the SER. However, for the case of general Nakagami-m fading it is difficult to find a closed-form solution of the power distribution. Therefore, a power allocation over each link which minimizes the SER over general Nakagami-m channels with carrier offsets can be found numerically. Nevertheless, we can find a closed-form optimized power distribution for Rayleigh fading channels with carrier offsets.

Lemma 3. *If all the channel gains are available, the optimized power allocation for high SNR over Rayleigh channels with carrier offsets is*

$$P_1 = \frac{\sigma_{s,r} + \sqrt{\sigma_{s,r}^2 + 8\left(\frac{M-1}{M}\right)\sigma_{r,d}^2}}{3\sigma_{s,r} + \sqrt{\sigma_{s,r}^2 + 8\left(\frac{M-1}{M}\right)\sigma_{r,d}^2}} P,$$

$$P_2 = \frac{2\sigma_{s,r}}{3_{s,r} + \sqrt{\sigma_{s,r}^2 + 8\left(\frac{M-1}{M}\right)\sigma_{r,d}^2}} P.$$

$$(49)$$

Proof: Refer (*Bhatnagar & Hjørungnes, 2007c*) for proof.

Fig. 10 shows the performance of DDDAF scheme using QPSK constellation with uniform ($P_1 = P_2$) and optimized power distribution over Rayleigh fading channels with random carrier offsets. The optimized powers are calculated from (49).

Distributed Double-Differential Coding for Cooperative System

STBC can be used for distributed double-differential coding in a cooperative network with more than one relay. In (Cano et al., 2007), a distributed DD coding for DAF cooperative system based on the DD codes for MIMO system of (Liu et al., 2001) is explained for arbitrary number of relays. This scheme utilizes the double-differential encoding of (Liu et al., 2001), where each symbol is first mapped to a diagonal unitary matrix before DD encoded. It is assumed that no transmission error occurs and all relays can decode the data successfully, therefore, all relays has correct copy of the data sequence required by the destination. Two different relaying strategies are explained in (Cano et al., 2007): (1) time division multiplexing (TDM) based transmission and (2) simultaneous transmission. In the TDM based scheme,

Figure 10. Power allocations for DDDAF scheme

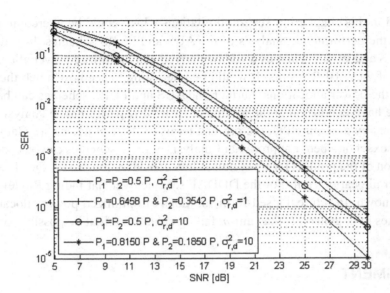

the relays are allowed to transmit the data in a particular order such that a simple heuristic decoder can be realized. Whereas, in the simultaneous transmission based scheme, all relays can transmit simultaneously, however, in this case, the decoder only works well for extremely large lengths ($\geq 10^3$) of interleaver. In (Bhatnagar et al., 2008f), the double-differential coding for OSTBC of (Bhatnagar et al., 2007d, Bhatnagar et al., 2008g, Bhatnagar et al., 2008h) is utilized to obtain a distributed DD coding for amplify-and-forward based cooperative systems. The scheme of (Bhatnagar et al., 2008f) is able to achieve better data rate than that of (Cano et al., 2007) for the same SNR.

CONCLUSIONS AND FUTURE RESEARCH DIRECTIONS

In this chapter, we discussed practical differential schemes for two user cooperative networks. Firstly, assuming a carrier offset-free cooperative network we discussed how to implement differential modulation for a simple cooperative network with orthogonal relay. It is further shown that the effective data rate can be improved in downlink channels by utilizing better time schedules with differential modulation. A selection method based on the conditional probability of error was also discussed, which can be implemented practically for decode-and-forward based cooperative networks. The performance of the differential cooperative schemes can be improved with this criterion to a large extent.

Secondly, we discussed a practical case, when all links in a cooperative network perturbed by different carrier offsets. In order to avoid the practically infeasible assumption of having the exact estimates of all carrier offsets and channel gains available at all nodes in the network, we explained how to implement double-differential modulation in the cooperative system with amplify-and-forward and decode-and-forward protocols. The cooperative system achieves full diversity at high SNR along with double-differential modulation.

FUTURE DIRECTIONS

The work presented in this chapter has several limitations which could be overcome in the future. We have not presented the performance analysis of the AUS and PURS schemes. In future, the BER expressions of the AUS and PURS schemes could be derived. An optimized power allocation can also be found on the basis of these BER results. It would be interesting to see how much these strategies can gain by optimizing the power allocations. Further, the AUS and PURS schemes can be extended to the multi-user case. We have conducted an approximate BER/SER performance analysis of the DDAAF and DDDAF schemes by assuming that the product terms of AWGN noises are Gaussian distributed. However, in future, exact or a better approximate analysis of these schemes can be conducted by taking the actual distribution of the product terms of noises into consideration. So far the closed-form solution for optimized power allocations only for the DDDAF scheme over flat fading Rayleigh channels with carrier offsets are known. In future, closed-form solutions of optimized power allocation for DDAAF and DDDAF schemes over general Nakagami-*m* fading channels might be possibly obtained.

ACKNOWLEDGMENT

This work was supported by the Research Council of Norway project 176773/S10 called OptiMO, which belongs to the VERDIKT program.

REFERENCES

Abramowitz, M., & Stegun, I. A. (1972). *Handbook of mathematical runctions*. New York: Dover Publications, Inc.

Alamouti, S. M. (1998). A simple transmit diversity technique for wireless communications. *IEEE Journal on Selected Areas in Communications, 16*(8), 1451–1458. doi:10.1109/49.730453

Bhatnagar, M. R., & Hjørungnes, A. (2007a). Downlink cooperative communication using differential relaying. In *Proc. 4th International Symposium on Wireless Communication Systems (ISWCS)*, Trondheim, Norway (pp. 1–5).

Bhatnagar, M. R., & Hjørungnes, A. (2007b). SER expressions for double-differential modulation. In *Proc. IEEE Inform. Theory Workshop*, Bergen, Norway (pp. 203–207).

Bhatnagar, M. R., & Hjørungnes, A. (2007c). Double-differential modulation for decode-and-forward cooperative communications. In *Proc. IEEE International Conference on Signal Processing and Communication (ICSPC)*, Dubai, United Arab Emirates (pp. 1–4).

Bhatnagar, M. R., & Hjørungnes, A. (2008b). Differential relay strategies over downlink channel in two-user cooperative communication system. *Wireless Personal Communications* (accepted for publication 18.09.2008).

Bhatnagar, M. R., & Hjørungnes, A. (2008f). Distributed double-differential orthogonal space-time code for cooperative networks. In *Proc. IEEE GLOBECOM,* New Orleans, LA.

Bhatnagar, M. R., Hjørungnes, A., & Bose, R. (2008h). Precoded DDOSTBC with non-unitary constellations over correlated Rayleigh channels with carrier offsets. In *Proc. IEEE International Symposium on Information Theory, (ISIT'08)*, Toronto, Canada.

Bhatnagar, M. R., Hjørungnes, A., & Song, L. (2007d). Double-differential orthogonal space-time block codes for arbitrarily correlated Rayleigh channels with carrier offsets. *IEEE Trans. Wireless Comm.* (conditionally accepted for publication subject to the reviewers' and editor's final decision after third round of review of the revised manuscript submitted 14.09.2008).

Bhatnagar, M. R., Hjørungnes, A., & Song, L. (2008a). Amplify-and-forward cooperative communications with double-differential modulation over Nakagami-*m* channels. In *Proc. IEEE Wireless Communications and Networking Conference (WCNC)*, Las Vegas, NV (pp. 1–5).

Bhatnagar, M. R., Hjørungnes, A., & Song, L. (2008b). Cooperative communications over flat fading channels with carrier offsets: A double-differential modulation approach. *EURASIP Journal on Advances in Signal, 2008*(531786), 1-11.

Bhatnagar, M. R., Hjørungnes, A., & Song, L. (2008e). Amplify based double-differential modulation for cooperative communications. In *Proc. for IEEE International Conference on Communication Systems and Networks (COMSNETS'09)*, Bangalore, India.

Bhatnagar, M. R., Hjørungnes, A., & Song, L. (2008g). Double differential orthogonal space-time modulation. In *Proc. IEEE Conference on Acoustics, Speech, and Signal Processing (ICASSP'08)*, Las Vegas, Nevada.

Bhatnagar, M. R., Hjørungnes, A., Song, L., & Bose, R. (2008c). Double-differential decode-and-forward cooperative communications over Nakagami-*m* channels with carrier offsets. In *Proc. IEEE Sarnoff Symposium*, Princeton, NJ.

Brennan, D. G. (2003). Linear diversity combining techniques. *Proceedings of the IEEE, 91*(2), 331–356. doi:10.1109/JPROC.2002.808163

Cano, A., Morgado, E., Caamano, A., & Ramos, J. (2007). Distributed double-differential modulation for cooperative communications under CFO. In *Proc. IEEE GLOBECOM*, Washington, D.C. (pp. 3447–3441).

Chen, D., & Laneman, J. N. (2004a). Cooperative diversity for wireless fading channels without channel state information. In *Proc. 48th Asilomar Conf. Signals, Systems, and Computers*, Monterey, CA (pp. 1307–1312).

Chen, D., & Laneman, J. N. (2004b). Noncoherent demodulation for cooperative diversity in wireless systems. In *Proc. of Globecom*, Dallas, TX (pp. 32–35).

Grandshteyn, I. S., & Ryzhik, I. M. (2000). *Table of integrals, series, and products, 6th edition*. San Diego, CA: Academic Press.

Hasna, M. O., & Alouini, M.-S. (2004). A performance study of dual-hop transmissions with fixed gain relays. *IEEE Transactions on Wireless Communications, 3*(6), 1963–1968. doi:10.1109/TWC.2004.837470

Himsoon, T., Su, W., & Liu, K. J. R. (2006). Differential transmission for amplify-and-forward cooperative communications. *IEEE Signal Processing Letters*, *12*(9), 597–600. doi:10.1109/LSP.2005.853067

Hwang, C. S., Nam, S. H., Chung, J., & Tarokh, V. (2003). Differential space time block codes using nonconstant modulus constellations. *IEEE Transactions on Signal Processing*, *51*(11), 2955–2964. doi:10.1109/TSP.2003.818157

Jing, Y., & Hassibi, B. (2006). Distributed space-time coding in wireless relay networks. *IEEE Transactions on Wireless Communications*, *5*(12), 3524–3536. doi:10.1109/TWC.2006.256975

Jing, Y., & Jafarkhani, H. (2008). Distributed differential space-time coding for wireless relay networks. *IEEE Transactions on Wireless Communications*, *56*(7), 1092–1100.

Kiran, T., & Rajan, B. S. (2006). Partially-coherent distributed space-time codes with differential encoder and decoder. *IEEE Journal on Selected Areas in Communications*, *25*(2), 426–433.

Laneman, J. N., Tse, D. N. C., & Wornell, G. W. (2004). Cooperative diversity in wireless networks: Efficient protocols and outage behaviour. *IEEE Transactions on Information Theory*, *50*(12), 389–400. doi:10.1109/TIT.2004.838089

Laneman, J. N., Wornell, G. W., & Tse, D. N. C. (2001). An efficient protocol for realizing cooperative diversity in wireless networks. In *Proc. Int. Symp. On Information Theory*, Washington, D.C. (pp. 294).

Larsson, E. G., & Stoica, P. (2003). *Space-time block coding for wireless communications*. Cambridge, UK: Cambridge University Press.

Liu, Z., Giannakis, G. B., & Hughes, B. L. (2001). Double-differential space-time block coding for time-selective fading channels. *IEEE Transactions on Communications*, *49*(9), 1529–1539. doi:10.1109/26.950340

Merkey, P., & Posner, E. C. (1984). Optimum cyclic redundancy codes for noisy channels. *IEEE Transactions on Information Theory*, *30*(6), 865–867. doi:10.1109/TIT.1984.1056971

Nosratinia, A., Hunter, T. E., & Hedayat, A. (2004). Cooperative communication in wireless networks. *IEEE Communications Magazine*, *42*, 74–80. doi:10.1109/MCOM.2004.1341264

Papoulis, A. (1991). *Probability, random variables, and stochastic processes, 3rd edition*. Singapore: McGraw-Hill Book Company.

Proakis, J. G. (2001). *Digital communications, 4th edition*. Singapore: McGraw-Hill.

Rajan, G. S., & Rajan, B. S. (2007). Noncoherent low-decoding-complexity space-time codes for wireless relay networks. In *Proc. IEEE International Symposium on Information Theory (ISIT'07)*, Nice, France (pp.1521–1525).

Sendonaris, A., Erkip, E., & Aazhang, B. (2003a). User cooperation diversity — part I: System description. *IEEE Transactions on Communications*, *51*(11), 1927–1938. doi:10.1109/TCOMM.2003.818096

Sendonaris, A., Erkip, E., & Aazhang, B. (2003b). User cooperation diversity — part II: Implementation aspects and performance analysis. *IEEE Transactions on Communications*, *51*(11), 1939–1948. doi:10.1109/TCOMM.2003.819238

Simon, M. K., & Alouini, M.-S. (2005). *Digital communication over fading channels: A unified approach to performance analysis*. New York: John Wiley & Son Inc.

Simon, M. K., & Divsalar, D. (1992). On the implementation and performance of single and double-differential detection schemes. *IEEE Transactions on Communications, 40*(2), 278–291. doi:10.1109/26.129190

Simon, M. K., Hinedi, S. M., & Lindsey, W. C. (1994). *Digital communication techniques: Signal design and detection*. NJ: Prentice-Hall.

Stoica, P., Liu, J., & Li, J. (2004). Maximum-likelihood double-differential detection clarified. *IEEE Transactions on Information Theory, 50*(3), 572–576. doi:10.1109/TIT.2004.825501

Su, W., Sadek, A. K., & Liu, K. J. R. (2005). SER performance analysis and optimal power allocation for decode-and-forward cooperation protocol in wireless networks. In *Proc. IEEE Wireless Communications and Networking Conf. (WCNC)*, New Orleans, LA (pp. 984–989).

Tarasak, P., Minn, H., & Bhargava, V. K. (2004). Differential modulation for two-user cooperative diversity systems. In *Proc. of Globecom*, Dallas, TX (pp. 3389–3393).

Tarasak, P., Minn, H., & Bhargava, V. K. (2005). Differential modulation for two-user cooperative diversity systems. *IEEE Journal on Selected Areas in Communications, 23*(9), 1891–1900. doi:10.1109/JSAC.2005.853792

Tsiftsis, T. A., Karagiannidis, G. K., Mathiopoulos, P. T., & Kotsopoulos, S. A. (2006). Nonregenerative dual-hop cooperative links with selection diversity. *EURASIP Journal on Wireless Comm. and Net., 2006*(17862), 1–8.

Wang, G., Zhang, Y., & Amin, M. (2006). Differential distributed space-time modulation for cooperative networks. *IEEE Transactions on Wireless Communications, 5*(11), 3097–3108. doi:10.1109/TWC.2006.04714.

Zhang, Y. (2005). Differential modulation schemes for decode-and-forward cooperative diversity. In *Proc. IEEE Int. Conf. Acoust., Speech, SignalProc. (ICASSP'05)*, Philadelphia, PA (pp. 918–920).

Zhao, Q., & Li, H. (2005a). Performance analysis of an amplify-based differential modulation for wireless relay networks under Nakagami-m fading channels. In *Proc. IEEE Int. Workshop on Signal Process. Advances in Wireless Communications*, New York (pp. 211–215).

Zhao, Q., & Li, H. (2005b). Performance of differential modulation with wireless relays in Rayleigh fading channels. *IEEE Communications Letters, 9*(4), 343–345. doi:10.1109/LCOMM.2005.1413628

Zhao, Q., & Li, H. (2006). Performance of decode-based differential modulation for wireless relay networks in Nakagami-m channels. In *Proc. 2006 IEEE Int. Conf. Acoustics, Speech, Signal Processing (ICASSP)*, Toulouse, France (pp. 673–675).

Zhao, Q., & Li, H. (2007). Differential modulation for cooperative wireless systems. *IEEE Transactions on Signal Processing, 55*(5), 2273–2283. doi:10.1109/TSP.2006.890922

Chapter 13
Space–Time Coding For Non–Coherent Cooperative Communications

J. Harshan
Indian Institute of Science, India

G. Susinder Rajan
Indian Institute of Science, India

B. Sundar Rajan
Indian Institute of Science, India

ABSTRACT

Cooperative communication in a wireless network can be based on the relay channel model where a set of users act as relays to assist a source terminal in transmitting information to a destination terminal. Recently, the idea of space-time coding (STC) has been applied to wireless networks wherein the relay nodes cooperate to process the received signal from the source and forward them to the destination such that the signal received at the destination appears like a space-time block code (STBC). Such STBCs (referred as distributed space time block codes [DSTBCs]) when appropriately designed are known to offer spatial diversity. It is known that separate classes of DSTBCs can be designed based on the destination's knowledge of various fading channels in the network. DSTBCs designed for the scenario when the destination has either the knowledge of only a proper subset of the channels or no knowledge of any of the channels are called non-coherent DSTBCs. This chapter addresses the problems and results associated with the design, code construction, and performance analysis (in terms of pairwise error probability [PEP]) of various noncoherent DSTBCs.

DOI: 10.4018/978-1-60566-665-5.ch013

Figure 1. MIMO Channel model

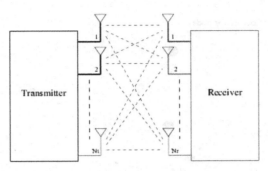

INTRODUCTION

For a point to point communication in a wireless fading channel, deploying multiple antennas at the source and the destination terminals (referred as a collocated Multiple Input and Multiple Output (MIMO) channel: (See Figure 1)) has been proved to be effective in combating the degrading effects of multi-path fading in addition to providing higher capacities than the Single Input Single Output (SISO) fading channel (single antenna at both the source and the destination). In particular, it is well known that the capacity of a collocated MIMO channel scales linearly with the minimum of the number of transmit and receive antennas for high receive Signal to Noise Ratio (SNR) (The ratio of the received signal power to the additive noise variance at every antenna of the receiver is referred as the receive SNR. Receive SNR has been defined in detail in Section on *MIMO channel model*) when perfect estimates of the channel are available at the receiver [Teletar, 1999]. Also, a spatial diversity order equal to the product of the number of the transmit antennas and receive antennas can potentially be obtained in slow-fading scenarios. When perfect estimates of the channel are available at the destination, the channel is referred as a coherent collocated MIMO channel [Foschini, 1996; Tarokh, Seshadri & Calderbank, 1998]. However, the overhead and/or the difficulty involved in obtaining these perfect estimates at the receiver leads to the need for designing signaling schemes assuming that the receiver doesn't have the knowledge of the channel. Such a channel is referred as non-coherent collocated MIMO channel and the corresponding signaling schemes are called non-coherent signaling schemes. Signal design for coherent and non-coherent collocated MIMO channels has been fairly well developed. Throughout the chapter, unless specified, MIMO channels refer to collocated MIMO channels.

In a wireless network, if the source terminal and the destination terminal are engaged in a point to point communication and are precluded from using multiple antennas, then spatial diversity is forbidden. Recently, a promising technique called '**cooperative communication**' has attracted a lot of attention in the research community wherein several users in the network which are geographically separated support the source in transmitting information to the destination (see Figure 2). Since, the destination receives the source signal through several independent paths, a potential spatial diversity order of at most the number of relays (including the source terminal) is promised. Such a method of obtaining spatial diversity is termed as '**cooperative diversity**' [Jing & Hassibi, 2006; Laneman, Tse & Wornell, 2004; Laneman & Wornell, 2003; Nabar, Bolcskei & Kneubehler, 2004; Sendonaris, Erkip & Aazhang, 2003]. Moving one step further, the idea of space-time coding which was originally devised for a MIMO channel has been applied to wireless networks under the frame-work of cooperative communication. In this scenario, the

Figure 2. General cooperative relay network model

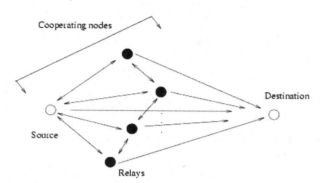

source broadcasts its signal to several terminals in the network referred as relays excluding/including the destination terminal. The half duplex constrained relay nodes cooperate to process the received source signal and forward them to the destination such that the signal received at the destination appears like a Space-Time Block Code (STBC) (In general, a finite set of complex matrices is referred as a STBC. In particular, a design in complex variables can be used to construct a STBC by making the variables take values form an underlying complex signal set). Such a cooperative scenario is referred as a two-hop cooperative communication, the signaling scheme is referred as '**Distributed Space-Time Coding**' (DSTC) and the corresponding codes are called Distributed Space Time Block Codes (DSTBC). Apart from using DSTBCs, there are several other techniques of obtaining spatial diversity in a wireless network. One such method is to select a single best relay among the several users in the network based on some criterion to support the communication between the source and the destination [Bletsas, Khisti, Reed & Lippman, 2006]. Such schemes are also shown to provide spatial diversity at the cost of using additional resources (such as power and bandwidth) on the training sequences in selecting the best relay.

Several protocols such as '**Amplify and Forward protocol**' (AF), Decode and Forward (DF), Compress and Forward (CF) and Demodulate and Forward (DF) [Chen & Laneman, 2004; Chen & Laneman, 2006] protocols are available for user cooperation. Among them, the prominent ones are AF and DF protocols [Laneman, Tse & Wornell, 2004]. With the DF protocol, it is difficult to achieve full diversity because of the possible erroneous decoding at some of the relays. To mitigate such scenarios, use of relay selection strategies or Cyclic Redundancy Check (CRC) codes or Fountain codes [Luby, 2002] becomes essential which in turn increases the overhead on the system and/or requires more resources such as power and bandwidth. The former protocol is attractive considering the simplicity of the underlying technique wherein, every relay node normalizes its received signal and forwards an appropriately scaled version to the destination. A generalized protocol to AF is Linear-Process and Forward (LPF) protocol wherein every relay linearly processes the received signal before transmitting to the destination [Jing & Hassibi, 2006]. The choice of the linear processing at every relay will determine the error performance of the overall protocol. Henceforth, throughout the chapter, we consider only LPF protocol. Further, we assume an orthogonal LPF protocol where the source and the relays do not transmit simultaneously in the same frequency band.

In this chapter, cooperative communication in two-hop wireless networks is considered wherein the source signal reaches the destination terminal through two sets of channels. The first set consists of the channels from the source to the relays/destination and the second set consists of the channels from the relays to the destination (see Figure 2 for a two-hop model). Similar to coherent and non-coherent communication in MIMO systems, in DSTC, the destination may or may not have the knowledge of the above set/ subset of channels when decoding for the source signal. DSTC when the destination node has the knowledge of all the channels in the network is referred as coherent DSTC [Jing & Hassibi, 2006]. To avoid the overhead or the difficulty in conveying all the channel information in the network to the destination, we consider designing DSTBCs when one of the two sets of channels or both sets of channels are not available at the destination terminal. Such a cooperative channel is called non-coherent cooperative channel. Further, DSTC when the destination has the knowledge of only a proper subset of the channels is referred as partially-coherent DSTC. The scenario when the destination doesn't have the knowledge of at least one of the channels in the network will be called fully non-coherent DSTC. Henceforth, throughout the chapter; we consider designing DSTBCs for non-coherent co-operative channels. We address the problems and solutions involving the design, construction and performance analysis (in terms of PEP) of DSTBCs for non-coherent cooperative communications.

ORGANIZATION OF THE CHAPTER

In the Section, *Background on Non Coherent MIMO Communications*, a short review of non-coherent MIMO channels is provided. For a thorough review, we refer the reader to [Belfiore, Cipriano, 2006]. In the Section, *DSTC for relay networks with AF protocol*, we introduce DSTC for amplify and forward wireless relay networks. Classification of DSTC based on the knowledge of various sets of channels at the destination is considered and a code design criterion for non-coherent DSTC is provided along with differential coding/decoding techniques. In the Section, *Differential code constructions for fully non-coherent relay channels*, several code constructions based on the differential techniques are presented. In the Section, *Codes based on training for non-coherent relay channel*, we show how coherent DSTBCs can be used for non-coherent scenario along with training and channel estimation. In the Section *Numerical results*, simulation results comparing the performance of various differential DSTBCs are provided. Possible directions of future work and concluding remarks constitute the Section, *Future trends* and the Section, *Conclusion* respectively.

Notations

For a complex matrix $\mathbf{X}$, the matrices $\mathbf{X}^*, \mathbf{X}^T, \mathbf{X}^H, |\mathbf{X}|$, Re $\mathbf{X}$, Im $\mathbf{X}$ and $\|\mathbf{X}\|$ denote, respectively, the conjugate, transpose, conjugate transpose, determinant, real part, imaginary part and the Frobenius norm of $\mathbf{X}$. The $T \times T$ identity matrix is denoted by $\mathbf{I}_T$. We use $|x|$ to denote the absolute value of a complex number x and $E[x]$ to denote the expectation of the random variable x. We write $x \sim \mathcal{CN}(\mu, \Gamma)$ when x is a circularly symmetric complex Gaussian random vector with mean μ and covariance matrix Γ. The set of integers, real numbers and complex numbers are, respectively, denoted by $\mathbb{Z}$, $\mathbb{R}$ and $\mathbb{C}$. The symbol i is used to denote square root of -1.

BACKGROUND ON NON COHERENT MIMO COMMUNICATIONS

In this section, a brief background on the signal model, code design criterion and code constructions for non-coherent MIMO channels are presented.

MIMO Channel Model

A MIMO channel with N_T antennas at the source (transmitter) and N_R antennas at the destination (receiver) as shown in Figure 1 is considered. The channel between every pair of the source and the destination antennas is assumed to be i.i.d quasi static and flat fading. The channels are assumed to be constant over a block of T complex channel uses and take independent realizations every block. The MIMO channel equation at the t-th block (one block corresponds to T complex channel uses) is given by,

$$\mathbf{Y}(t) = \sqrt{\rho}\mathbf{X}(t)\mathbf{H}(t) + \mathbf{N}(t),$$

where $\mathbf{Y}(t) \in \mathbb{C}^{T \times N_R}$ is the received matrix, $\mathbf{X}(t) \in \mathbb{C}^{T \times N_T}$ is the transmitted codeword matrix, $[\mathbf{H}]_{i,j}(t)$, is the fade coefficient between the $i-$th transmit antenna and the $j-$th receive antenna which is distributed as $\mathcal{CN}(0,1)$ for all $i = 1, 2, \cdots N_T$ and $j = 1, 2, \cdots, N_R$. The additive Gaussian noise matrix at the destination is $\mathbf{N}(t) \in \mathbb{C}^{T \times N_R}$ whose components are also distributed as $\mathcal{CN}(0,1)$ and ρ is the receive SNR at each of the destination antenna.

It is assumed that the codeword matrix $\mathbf{X}(t)$ belongs to a finite set of $T \times N_T$ matrices called a STBC denoted by $\mathcal{C}$. The source and the destination are assumed not to have the knowledge of the channel matrix $\mathbf{H}(t)$. However, both the terminals are assumed to know the statistics of the channel. The channel matrix is referred as the Channel State Information (CSI) throughout this chapter. The class of STBCs designed for the MIMO channel when the destination doesn't have the knowledge of CSI is called non-coherent STBCs. Throughout this section, a special class of STBCs called unitary STBCs is considered. A STBC $\mathcal{C}$ is called unitary if every $\mathbf{X}(t) \in \mathcal{C}$ satisfies the condition that $\mathbf{X}^H(t)\mathbf{X}(t) = m\mathbf{I}_{N_T}$ for some positive real number m which is a constant independent of the codeword $\mathbf{X}(t)$. Unless mentioned, throughout the chapter, we consider codes wherein $m = 1$. An information theoretic reasoning on why unitary STBCs are important for the non-coherent MIMO channels is provided by [Hochwald & Marzetta, 1999]. They made the initial investigations on the ergodic capacity of the non-coherent MIMO channels and showed that for all values of SNR at the receiver, the capacity achieving signal matrix is of the form $\mathbf{X}(t) = \mathbf{U}\boldsymbol{\Delta}$ where $\mathbf{U}$ is an isotropically distributed unitary matrix of order $T \times N_T$ and $\boldsymbol{\Delta}$ is an independent non-negative diagonal matrix. Henceforth, throughout this section, only unitary matrices are considered. A background on the design and construction of non-coherent STBCs is provided in the next subsection.

Non-Coherent ML Decoder

When the transmitted codeword is $\mathbf{X}(t)$ and the destination doesn't have the CSI, $\mathbf{Y}(t)$ is a Gaussian matrix with mean $\mathbf{0}$ and variance $\boldsymbol{\Sigma}_y = \rho\mathbf{X}(t)\mathbf{X}^H(t) + \mathbf{I}_T$. The conditional PDF (Probability Density

Function) $P(\mathbf{Y}(t) \mid \mathbf{X}(t))$ is given by

$$P(\mathbf{Y}(t) \mid \mathbf{X}(t)) = (1/\mid \Sigma_y \mid^{N_R} \pi^{TN_R}) \exp(-\text{Trace}(\mathbf{Y}(t)^H \Sigma_y^{-1} \mathbf{Y}(t))).$$

The ML decoder decodes to a codeword $\widehat{\mathbf{X}}(t)$ where $\widehat{\mathbf{X}}(t) = arg\max_{\mathbf{X} \in \mathcal{C}} P(\mathbf{Y}(t) \mid \mathbf{X}(t))$.

Using the matrix identity, $\mid \mathbf{I} + \mathbf{AB} \mid = \mid \mathbf{I} + \mathbf{BA} \mid$, the matrix inversion lemma, $(\mathbf{A} + \mathbf{BCD})^{-1} = \mathbf{A}^{-1} - \mathbf{A}^{-1}\mathbf{B}(\mathbf{C}^{-1} + \mathbf{DA}^{-1}\mathbf{B})^{-1}\mathbf{DA}^{-1}$ and the definition of unitary STBCs, the following theorem provides an ML decoder for the non-coherent MIMO channel.

Theorem 1: [Hughes, 2000] For a unitary STBC, a non-coherent ML decoder is given by,

$$\widehat{\mathbf{X}}(t) = arg\max_{\mathbf{X}(t) \in \mathcal{C}} \text{Trace}\left(\mathbf{Y}^H(t)\mathbf{X}(t)\mathbf{X}^H(t)\mathbf{Y}(t)\right). \tag{1.2.1}$$

Code Design Criteria for the Non-Coherent MIMO Channel

In this subsection, we briefly state the design criterion on the STBC, $\mathcal{C}$ by analyzing the Chernoff bound on the Pairwise Error Probability (PEP) of the decoder in (1.2.1) for unitary STBCs. In particular, unitary STBCs considered for the PEP analysis satisfy $m = T$. However, unitary STBCs with $m = 1$ also continue to satisfy the full diversity criterion which is an outcome of the PEP analysis.

Theorem 2: [Section II.C, Section II.C, [Hughes, 2000]]. If $\mathbf{X}_0(t), \mathbf{X}_1(t) \in \mathcal{C}$ (such that $\mathbf{X}_0(t)$ and $\mathbf{X}_1(t)$ are distinct codewords), the probability that the decoder chooses $\mathbf{X}_1(t)$ as the most likely transmitted codeword when $\mathbf{X}_0(t)$ is transmitted from the source is upper-bounded by

$$\Pr\{\mathbf{X}_0(t) \to \mathbf{X}_1(t)\} \leq \frac{1}{\mid \mathbf{I}_{N_T} + \frac{\rho^2 T^2}{4(1+\rho T)}[\mathbf{I}_{N_T} - \frac{1}{T^2}\mathbf{X}_1^H(t)\mathbf{X}_0(t)\mathbf{X}_0^H(t)\mathbf{X}_1(t)]\mid^{N_R}}. \tag{1.2.2}$$

For large values of ρ, the above bound on the PEP depends essentially on the cross-product matrix $(1/T)\mathbf{X}_1^H(t)\mathbf{X}_0(t)$. Note that, the transmit diversity order will be equal to the rank of the matrix $\mathbf{I}_{N_T} - (1/T^2)\mathbf{X}_1^H(t)\mathbf{X}_0(t)\mathbf{X}_0^H(t)\mathbf{X}_1(t)$ which is upper-bounded by N_T. The maximum diversity order is achieved when 1 is not a singular value of $(1/T)\mathbf{X}_1^H(t)\mathbf{X}_0(t)$. In other words, the maximum transmit diversity order is achieved when $\mid T^2\mathbf{I}_{N_T} - \mathbf{X}_1^H(t)\mathbf{X}_0(t)\mathbf{X}_0^H(t)\mathbf{X}_1(t) \mid = \mid \mathbf{X}^H(t)\mathbf{X}(t) \mid \neq 0$ where $\mathbf{X}(t) = [\mathbf{X}_0(t) \ \mathbf{X}_1(t)]$. Therefore, the transmit diversity order is N_T if and only if the columns of the matrices $\mathbf{X}_0(t)$ and $\mathbf{X}_1(t)$ are linearly independent. Hence, a necessary condition for full diversity is $T \geq 2N_T$. Therefore the criterion for full diversity in a non-coherent MIMO channel is to design a finite set of $T \times N_T$ unitary matrices such that the columns of any pair of matrices should be linearly independent, whereas, for a coherent MIMO channel, the criterion for full diversity is to design a finite set of $T \times N_T$ matrices such that the difference matrix of any pair of matrices should have full rank [Tarokh, Seshadri & Calderbank, 1998].

In the literature, two approaches have been followed for designing codes in a non-coherent MIMO channel. The first one is to explicitly generate $T \times N_T$ unitary matrices satisfying the full diversity criterion. Such codes can be viewed as Grassmann manifolds wherein the columns of every codeword are viewed as the generators of a N_T - dimensional subspace in a T - dimensional vector space (since $N_T \leq T$). Hence, the design criterion is to generate non-intersecting N_T - dimensional subspaces keeping the distance between the subspaces in terms of principle angles as large as possible. The second approach is to use '**differential encoding and decoding techniques**' which is considered in detail in the next subsection [Hassibi & Hochwald, 2002; Hochwald & Sweldens, 2000; Hughes, 2000; Jing & Hassibi, 2003; Oggier & Hassibi, 2007; Yuen, Guan & Tjhung, 2006]. For a cooperative relay channel, as far as non-coherent communication is concerned, there is only one work that addresses non-differential code construction (See the last paragraph of the Subsection 0) whereas there are several works which consider differential code construction. Hence, this chapter predominantly deals with codes based on differential techniques for non-coherent cooperative channels. As a result, only differential methods are considered for MIMO channels also. However, codes based on Grassmann manifolds are referred in the chapter whenever necessary. Details on such codes can be found in [Belfiore & Cipriano, 2006; Kammoun & Belfiore, 2003; Tarokh & Kim, 2002; Oggier, Sloane, Diggavi & Calderbank, 2005].

Differential Coding Techniques for Non-Coherent MIMO Channel

In this subsection, we provide a flavor of the differential coding/decoding techniques for the non-coherent MIMO channel and show how such codes meet the full diversity criterion [Hughes, 2000]. We continue to refer the MIMO channel model in Figure 1.

Within a quasi-static interval, the transmission of matrices is considered in several blocks. The source is equipped with a finite set of $N_T \times N_T$ unitary matrices $\mathcal{C}$ (in this case $T = N_T$). The differential encoding is performed at the source as follows. A $N_T \times N_T$ matrix $\mathbf{X}(0)$ is transmitted at the beginning of the quasi-static interval i.e. $\mathbf{X}(0)$ is transmitted at $t = 0$. The matrix $\mathbf{X}(0)$ is chosen such that $\mathbf{X}(0)^H \mathbf{X}(0) = \mathbf{X}(0)\mathbf{X}(0)^H = \mathbf{I}_{N_T}$. Subsequently, at the $t - $ th block, the matrix transmitted to the destination is given by

$$\mathbf{X}(t) = \mathbf{U}(t)\mathbf{X}(t-1) \text{ for } t = 1, 2, 3... \tag{1.2.3}$$

where $\mathbf{U}(t) \in \mathcal{C}$ and $\mathbf{X}(t-1)$ refers to the matrix transmitted in the $(t-1)-$ th block. It is to be noticed that the unitary matrices (with $m = 1$) are essential in meeting the power constraint at the source. The differential decoding is performed as follows. The received matrix at the $t-$ th and $(t-1)-$ th block are given by

$$\mathbf{Y}(t-1) = \mathbf{X}(t-1)\mathbf{H}(t-1) + \mathbf{N}(t-1) , \tag{1.2.4}$$

$$\mathbf{Y}(t) = \mathbf{X}(t)\mathbf{H}(t) + \mathbf{N}(t) = \mathbf{U}(t) \ \mathbf{X}(t-1) \ \mathbf{H}(t) + \mathbf{N}(t) . \tag{1.2.5}$$

Since the channel remains same for any two consecutive blocks with in the coherence interval, we have $\mathbf{H}(t) = \mathbf{H}(t-1) = \mathbf{H}$. Using (1.2.4) in (1.2.5), we write,

$$\mathbf{Y}(t) = \mathbf{U}(t)\mathbf{Y}(t-1) + \mathbf{N}'(t), \tag{1.2.6}$$

where $\mathbf{N}'(t) = \mathbf{N}(t) - \mathbf{U}(t)\mathbf{N}(t-1)$. For the above equivalent channel equation, the destination can use the following decoder

$$\widehat{\mathbf{U}}(t) = arg\min_{\mathbf{U}(t)\in\mathcal{C}} \| \mathbf{Y}(t) - \mathbf{U}(t)\mathbf{Y}(t-1) \|^2. \tag{1.2.7}$$

Note that the above decoder doesn't require the knowledge of the channel. Henceforth, throughout the chapter, the above decoder is referred as a ML 1-lag detector. In general, a true ML decoder is a sequential detector having exponential complexity in the block length wherein all the received matrices $\mathbf{Y}(0), \mathbf{Y}(1)\cdots\mathbf{Y}(t)$ are used for detecting the codewords $\mathbf{U}(1), \mathbf{U}(2)....\mathbf{U}(t)$. However, recently in [Dimitris & Karystinos, 2008], it has been proved that for orthogonal STBCs, non-coherent sequence detection can be performed in polynomial time with respect to the block length. Note that for a true ML detector, the duration of the quasi-static block has to be $t+1$ codeword uses. In the next subsection, we show that if the unitary codebook $\mathcal{C}$ is designed with the full diversity criteria for the coherent MIMO channel, then the decoder in (1.2.7) gives full diversity.

Connection Between the Design Criterion for Coherent and Non-Coherent MIMO Channel with Differential Decoder

The decoder introduced in (1.2.7) can be further simplified as

$$\widehat{\mathbf{U}}(t) = arg\min_{\mathbf{U}(t)\in\mathcal{C}} \text{Trace } (\mathbf{Y}(t) - \mathbf{U}(t)\mathbf{Y}(t-1))^H(\mathbf{Y}(t) - \mathbf{U}(t)\mathbf{Y}(t-1))$$

$$= arg\max_{\mathbf{U}(t)\in\mathcal{C}} \text{Trace } (\mathbf{Y}(t)^H \mathbf{U}(t)\mathbf{Y}(t-1) + \mathbf{Y}(t-1)^H \mathbf{U}(t)^H \mathbf{Y}(t)). \tag{1.2.8}$$

Considering the received matrices $\mathbf{Y}(t-1)$ and $\mathbf{Y}(t)$, a new equivalent MIMO channel using two successive codeword uses can be written as,

$$\overrightarrow{\mathbf{Y}} = \overrightarrow{\mathbf{X}}\overrightarrow{\mathbf{H}} + \overrightarrow{\mathbf{N}} = \begin{bmatrix} \mathbf{Y}(t-1) \\ \mathbf{Y}(t) \end{bmatrix} = \begin{bmatrix} \mathbf{I}_{N_T} \\ \mathbf{U}(t) \end{bmatrix}\mathbf{X}(t-1)\mathbf{H} + \begin{bmatrix} \mathbf{N}(t-1) \\ \mathbf{N}(t) \end{bmatrix}, \tag{1.2.9}$$

where $\overrightarrow{\mathbf{Y}} = \begin{bmatrix} \mathbf{Y}(t-1) \\ \mathbf{Y}(t) \end{bmatrix}, \overrightarrow{\mathbf{H}} = \mathbf{X}(t-1)\mathbf{H}, \overrightarrow{\mathbf{X}} = \begin{bmatrix} \mathbf{I}_{N_T} \\ \mathbf{U}(t) \end{bmatrix}$ and $\overrightarrow{\mathbf{N}} = \begin{bmatrix} \mathbf{N}(t-1) \\ \mathbf{N}(t) \end{bmatrix}$.

Since $\mathbf{X}(t-1)$ is unitary, the new channel $\overrightarrow{\mathbf{H}}$ has the same distribution as that of $\mathbf{H}$. When $\mathbf{H}$ is not known at the destination, the non-coherent ML decoder and the code design criteria studied in the Section *Non-coherent ML decoder* and the Section *Code design criteria for the non-coherent MIMO channel* can be applied to the channel in (1.2.9). The ML 1-lag detector provided in (1.2.8) can be re-written in the equivalent form of trace maximization as,

$$\widehat{\overset{\leftrightarrow}{\mathbf{X}}} = arg\max_{\overset{\leftrightarrow}{\mathbf{X}} \in \mathcal{C}} \ \text{Trace}\left(\overset{\leftrightarrow}{\mathbf{Y}}^H \overset{\leftrightarrow}{\mathbf{X}} \overset{\leftrightarrow}{\mathbf{X}}^H \overset{\leftrightarrow}{\mathbf{Y}}\right),$$

where $\overset{\leftrightarrow}{\mathcal{C}} = \left\{[\mathbf{I} \ \mathbf{U}^T]^T\right\}$ for all $\mathbf{U} \in \mathcal{C}$. From the error analysis of the detector given in (1.2.1) in the Section *Code design criteria for the non-coherent MIMO channel*, for full diversity, the unitary codebook $\mathcal{C}$ must be such that the columns of the two matrices $[\mathbf{I}_{N_T} \ \mathbf{U}_i^T]^T$ and $[\mathbf{I}_{N_T} \ \mathbf{U}_j^T]^T$ are linearly independent for any $\mathbf{U}_i, \mathbf{U}_j \in \mathcal{C}$ such that $i \neq j$. It is important to note that if the codebook $\mathcal{C}$ is designed satisfying the full diversity criterion for the coherent MIMO channel, i.e. if $\mathbf{U}_i$ and $\mathbf{U}_j$ are such that $\mathbf{U}_i - \mathbf{U}_j$ has full rank for any $\mathbf{U}_i, \mathbf{U}_j \in \mathcal{C}$, then the columns of two matrices $[\mathbf{I}_{N_T} \ \mathbf{U}_i^T]^T$ and $[\mathbf{I}_{N_T} \ \mathbf{U}_j^T]^T$ are linearly independent. Therefore, unitary codebooks designed for full diversity in the coherent MIMO channel provides full diversity when used for the non-coherent differential MIMO channel using the decoder in (1.2.8). Various techniques for construction of unitary codebooks can be found in [Hassibi & Hochwald, 2002; Hochwald & Sweldens, 2000; Hughes, 2000; Jing & Hassibi, 2003; Oggier & Hassibi, 2007; Yuen, Guan & Tjhung, 2006].

Distributed Space Time Coding (DSTC) for Relay Networks with AF Protocol

In this section, we provide details on the design of STBCs for cooperative networks. The readers can refer [Jing & Hassibi, 2006; Laneman, Tse & Wornell, 2004; Laneman & Wornell, 2003; Nabar, Bolcskei & Kneubehler, 2004; Sendonaris, Erkip & Aazhang, 2003] for more details on cooperative communication strategies in relay networks. Throughout this section, we consider the design of STBCs based on AF protocol and its generalizations (LPF protocol) although STBCs based on DF protocol have been well studied [Yiu, Schober & Lampe, 2005; Yiu, Schober & Lampe, 2006]. We introduce the network model and provide various ingredients required by the users to construct STBCs in a distributed way.

Signal Model

The wireless network considered consists of $R+2$ nodes each having a single antenna as shown in Figure 3[1]. There is one source node and one destination node. All the other R nodes work as '**relays**'. The channel from the source node to the $j-$th relay and the $j-$th relay to the destination node are denoted by f_j and g_j respectively for $j = 1, 2, \cdots, R$. The fading coefficients f_j, g_j are i.i.d $\mathcal{CN}(0,1)$ with coherence time interval of at least $2T$ (which includes the two phases of transmission from the source to the destination). Symbol level synchronization is assumed at the destination [Jing & Hassibi, 2006]. We assume that there is no direct link between the source and the destination. The relays do not have the knowledge of the channels from (i) the source to the relays and (ii) relays to the destination.

Every transmission from the source to the destination comprises of two phases. In the first phase the source transmits a T-length complex vector from the codebook $\mathcal{S} = \left\{\mathbf{s}_1, \mathbf{s}_2, \mathbf{s}_3, \cdots, \mathbf{s}_L\right\}$ consisting of information vectors $\mathbf{s}_l \in \mathbb{C}^T$ such that $E\left[\mathbf{s}_l^H \mathbf{s}_l\right] = 1$. Let P_1 be the average power used at the source node every complex channel use. When the information vector $\mathbf{s}$ is transmitted, the received vector at the $j-$th relay is given by,

Figure 3. Wireless relay network model

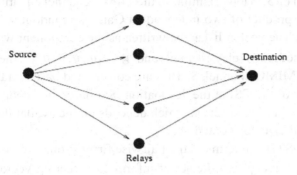

$$r_j = \sqrt{P_1 T} f_j s + n_j, \quad j = 1, 2, \cdots, R.$$

where $\mathbf{n}_j$ is the additive noise vector at the $j-$ th relay distributed as $\mathcal{CN}\left(0, \mathbf{I}_T\right)$. In the secondphase, all the relay nodes are scheduled to transmit T - length vectors to the destination simultaneously.[2] Each relay is equipped with a fixed $T \times T$ unitary matrix $\mathbf{A}_j$ and is allowed to linearly process the received vector.[3] We have assumed the number of channel uses in the first and the second phase to be same; in general, they can be different. The $j-$ th relay is scheduled to transmit

$$t_j = \sqrt{\frac{P_2}{(1+P_1)}} \mathbf{A}_j \tilde{\mathbf{r}}_j ,$$

where for any vector $\mathbf{s}$, $\tilde{\mathbf{s}}$ denotes either $\mathbf{s}$ or its conjugate $\mathbf{s}^*$. The term P_2 is the average power transmitted by each relay per channel use. The total power used from all the nodes per channel use is $P = P_1 + RP_2$. The vector received at the destination is given by,

$$\mathbf{y} = \sum_{j=1}^{R} (g_j \mathbf{t}_j) + \mathbf{w} = \sqrt{\frac{P_1 P_2 T}{(1+P_1)}} \mathbf{Sh} + \mathbf{n}, \tag{1.3.1}$$

where $\mathbf{w}$ is the additive noise vector at the destination distributed as $\mathcal{CN}\left(0, \mathbf{I}_T\right)$, $\mathbf{S} = \begin{bmatrix} \mathbf{A}_1 \tilde{\mathbf{s}} & \mathbf{A}_2 \tilde{\mathbf{s}} & \cdots & \mathbf{A}_R \tilde{\mathbf{s}} \end{bmatrix} \in \mathbb{C}^{T \times R}$, $\mathbf{n} = \sqrt{\frac{P_2}{(1+P_1)}} \left[\sum_{j=1}^{R} (g_j \mathbf{A}_j \mathbf{n}_j) \right] + \mathbf{w}$ and $\mathbf{h} = [\tilde{f}_1 g_1 \cdots \tilde{f}_R g_R]^T$.

Definition 1: The collection $\mathcal{C}$ of matrices shown below, when $\mathbf{s}$ runs over $\mathcal{S}$,

$$\mathcal{C} = \left\{ \begin{bmatrix} \mathbf{A}_1 \tilde{\mathbf{s}} & \mathbf{A}_2 \tilde{\mathbf{s}} & \cdots & \mathbf{A}_R \tilde{\mathbf{s}} \end{bmatrix} \right\} \tag{1.3.2}$$

is called the Distributed Space-Time Block code (DSTBC).

The channel equation in (1.3.1) looks similar to the MIMO channel equation for $N_R = 1$ except that each component of $\mathbf{h}$ is a product of two independent Gaussian random variables instead of a single Gaussian random variable. The vector $\mathbf{h}$ can be written as the component wise product of two vectors $\mathbf{f} = [\tilde{f}_1, \cdots \tilde{f}_R]^T$, the set of source to relay channels and $\mathbf{g} = [g_1, g_2 \cdots g_R]^T$, the set of relay to destination channels. Recall that for a MIMO channel, STBCs are constructed based on the design criterion which depends on the availability of the CSI at the destination. Similarly, in a cooperative channel, DSTBCs are constructed based on the design criterion which depends on the availability of $\mathbf{h}$ at the destination through the knowledge of the vectors $\mathbf{f}$ and $\mathbf{g}$.

The performance of a DSTBC in terms of the Pairwise Error Probability (PEP) is determined by the relay matrix set, $\mathcal{A} = \{\mathbf{A}_1, \cdots \mathbf{A}_R\}$ and the set of information bearing vectors $\mathcal{S}$. In particular, for a given extent of the channel knowledge at the destination, the sets $\mathcal{A}$ and $\mathcal{S}$ must be chosen such that the DSTBC provides a diversity order equal to the number of relays, R. Towards that direction, DSTC has to be first classified in to different groups depending on the knowledge of $\mathbf{h}$ at the destination. The following definition partitions DSTC into two classes based on the knowledge of $\mathbf{f}$ and $\mathbf{g}$ at the destination.

Definition 2: DSTC is referred as coherent DSTC, when the destination has the knowledge of both $\mathbf{f}$ and $\mathbf{g}$, otherwise, it is called '**non-coherent DSTC**'. In non-coherent DSTC, if the destination has the knowledge of only $\mathbf{g}$ but not $\mathbf{f}$, then it is called partially coherent DSTC. When the destination has no knowledge of both $\mathbf{f}$ and $\mathbf{g}$, then it is called fully non-coherent DSTC.

Non Coherent Distributed Space Time Coding

In this subsection, we study partially coherent DSTC followed by fully non-coherent DSTC. First, we introduce a partially coherent ML decoder and discuss the code design criterion to achieve full diversity.

Partially-Coherent ML Decoder and Code Design Criterion

In a partially coherent set-up, the destination has the knowledge of only $\mathbf{g}$ for every codeword use. Since $\mathcal{A}$ is a unitary matrix set, $\mathbf{n}$ is a Gaussian random vector with $E[\mathbf{n}] = \mathbf{0}$ and $E[\mathbf{nn}^H] = (1 + (P_2 / (P_1 + 1))\sum_{j=1}^{R}(|g_j|^2))\mathbf{I}_T$. The received vector, $\mathbf{y}$ is a Gaussian random vector with $E[\mathbf{y} | \mathbf{S}, g_j] = \mathbf{0}$ and $E[\mathbf{yy}^H | \mathbf{S}, g_j] = \Sigma_y = \rho \mathbf{S} \Sigma_h \mathbf{S}^H + \gamma \mathbf{I}_T$ where $\rho = P_2 P_1 T / (P_1 + 1)$, $\gamma = (1 + (P_2 / (P_1 + 1))\sum_{j=1}^{R}(|g_j|^2))$ and $\Sigma_h = \text{diag}(|g_1|^2, \cdots |g_R|^2)$. The partially-coherent ML decoder decodes to a codeword $\hat{\mathbf{S}} = arg \max_{\mathbf{s} \in \mathcal{C}} P(\mathbf{y} | \mathbf{S}, g_j) = (1 / |\Sigma_y| \pi^T) \exp(-(\mathbf{y}^H \Sigma_y^{-1} \mathbf{y}))$. On the similar lines of the definition of unitary STBCs, we define unitary DSTBCs as follows: A DSTBC $\mathcal{C}$ is called a unitary DSTBC if every $\mathbf{S} \in \mathcal{C}$ satisfies the condition that $\mathbf{S}^H \mathbf{S} = m\mathbf{I}_R$, where m is a constant real number independent of the codeword $\mathbf{S}$ [Kiran & Rajan, 2007].

Theorem 3: [Kiran & Rajan, 2007; Harshan & Rajan, 2008]. For a unitary DSTBC, the partially-coherent ML decoding is given by

$$arg \max_{\mathbf{s} \in \mathcal{C}} \left(\mathbf{y}^H \mathbf{SGS}^H \mathbf{y} \right), \tag{1.3.3}$$

where $\mathbf{G} = \mathrm{diag}\left(\beta_1, \cdots \beta_R \right)$ and $\beta_j = | g_j |^2 \left(| g_j |^2 + \gamma \rho^{-1} \right)^{-1}$ for $j = 1, \cdots, R$.

For the ML decoder in (1.3.3), in [Kiran & Rajan, 2007], a Chernoff bound on the PEP is derived and a code design criterion has been proposed for minimizing the PEP. It is shown that the code design criterion is essentially the same as that for non-coherent STBCs in a collocated MIMO channel. We briefly revisit the design criterion for the DSTC framework: Let $\mathbf{S}_n, \mathbf{S}_m \in \mathcal{C}$ and let $\mathbf{S}_{nm} = \begin{bmatrix} \mathbf{S}_n & \mathbf{S}_m \end{bmatrix} \in \mathbb{C}^{T \times 2R}$ be the juxtaposed matrix of $\mathbf{S}_n$ and $\mathbf{S}_m$. For achieving a diversity order of R, the determinant of $\mathbf{S}_{nm}^H \mathbf{S}_{nm}$ has to be non zero for all m, n such that $n \neq m$. Since $\mathbf{S}_{nm} \in \mathbb{C}^{T \times 2R}$, it can be verified that $\mathbf{S}_{nm}^H \mathbf{S}_{nm} \in \mathbb{C}^{2R \times 2R}$. Therefore, for full diversity, a necessary condition is $T \geq 2R$. Since the code design criterion for a partially coherent cooperative channel is same as that of a non-coherent MIMO channel, codes existing for the latter channel can potentially be extended to work in the former set-up. In that direction, we provide details on how to extend the differential scheme studied in the Section, *Differential coding techniques for non-coherent MIMO channel* to the partially coherent scenario. It is also shown how the above differential scheme can be used in fully non-coherent DSTC.

Differential Coding for Non-Coherent Relay Channels

In the differential encoding scheme proposed for the MIMO channel, the source conveys information to the destination by transmitting one of the several unitary matrices (from a finite set of unitary matrices) using two successive codeword uses. However, in the cooperative set-up considered in this chapter, the source node is restricted to transmit only vectors (along the time axis) instead of matrices (this is because the source node is assumed to have only one antenna) and hence extending the techniques presented in the Section, *Differential coding techniques for non-coherent MIMO channel* to the cooperative set-up is not straightforward. In this section, a differential coding scheme is presented for the co-operative channel which is a generalized version of the scheme presented for MIMO channels. We assume that the coherence time interval of all the fading channels in the network is at least $4T$.

Towards explaining the differential coding scheme for cooperative networks, various ingredients required at the source and the relay matrices are presented [Kiran & Rajan, 2007]. The source node is equipped with a code $\{\mathbf{s}, \mathcal{G}\}$ where $\mathbf{s}$ is a fixed $T \times 1$ complex vector and $\mathcal{G} = \{\mathbf{G}_1, \mathbf{G}_2, \cdots \mathbf{G}_L\}$ is a finite set of $T \times T$ unitary matrices such that $\mathbf{G}_i \mathbf{s} \neq \mathbf{G}_j \mathbf{s}$ for all $\mathbf{G}_i \neq \mathbf{G}_j$. The sets $\mathcal{G}$ and $\mathcal{A}$ should satisfy the following conditions [Kiran & Rajan, 2007; Jing & Jafarkhani, 2008]; (i) $\mathbf{G}_j \mathbf{A}_k = \mathbf{A}_k \widetilde{\mathbf{G}}_j$ for all j, k and (ii) $\mathbf{S}^H \mathbf{S} = \mathbf{I}_R$ where $\mathbf{S} = \begin{bmatrix} \mathbf{A}_1 \tilde{\mathbf{s}} & \mathbf{A}_2 \tilde{\mathbf{s}} & \cdots & \mathbf{A}_R \tilde{\mathbf{s}} \end{bmatrix}$ is called the initial matrix which is the matrix corresponding to the vector $\mathbf{s}$ used in the code. The transmission follows the two stages per block model with differential encoding performed at the source node. If $\mathbf{s}(t-1)$ denotes the transmitted vector by the source in the $(t-1)$-th block, the vector transmitted in the t-th block is of the form $\mathbf{s}(t) = \mathbf{G}(t)\mathbf{s}(t-1)$ where $\mathbf{G}(t) \in \mathcal{G}$ is chosen based on the information to be transmitted. The received vector at the destination in the t-th block is $\mathbf{y}(t) = \mathbf{S}(t)\mathbf{h} + \mathbf{n}(t)$ where $\mathbf{S}(t) = \begin{bmatrix} \mathbf{A}_1 \tilde{\mathbf{s}}(t) & \mathbf{A}_2 \tilde{\mathbf{s}}(t) & \cdots & \mathbf{A}_R \tilde{\mathbf{s}}(t) \end{bmatrix}$. If the code $\mathcal{G}$ and the set of relay matrices satisfy the condition, $\mathbf{G}_j \mathbf{A}_k = \mathbf{A}_k \widetilde{\mathbf{G}}_j$ for all j, k, then

$$\mathbf{S}(t) = \mathbf{G}(t) \begin{bmatrix} \mathbf{A}_1 \tilde{\mathbf{s}}(t-1) & \mathbf{A}_2 \tilde{\mathbf{s}}(t-1) & \cdots & \mathbf{A}_R \tilde{\mathbf{s}}(t-1) \end{bmatrix} \text{ for some } \mathbf{G}(t) \in \mathcal{G} \text{ [Kiran \& Rajan, 2007; Jing \& }$$

Jafarkhani, 2008]. Therefore,

$\mathbf{y}(t) = \mathbf{G}(t)\mathbf{S}(t-1)\mathbf{h} + \mathbf{n}(t) = \mathbf{G}(t)\mathbf{y}(t-1) + \mathbf{n'}(t)$, where $\mathbf{n'}(t) = \mathbf{n}(t) - \mathbf{G}(t)\mathbf{n}(t-1)$. Thus, the DSTBC is given by,

$$\mathcal{D} = \{\mathbf{GS} : \mathbf{G} = \mathbf{I}_R \text{ or } \prod_k \mathbf{G}_k, \mathbf{G}_k \in \mathcal{G}\} \qquad (1.3.4)$$

where $\prod_k \mathbf{G}_k$ denotes arbitrary product of matrices in $\mathcal{G}$. Since any $\mathbf{G}_j \in \mathcal{G}$ is unitary and the initial matrix is also unitary, any matrix $\mathbf{X} \in \mathcal{D}$ is also unitary. Hence, the distributed code is a unitary code.

With differential techniques in a partially coherent set-up, the optimal partially coherent ML detector would use the entire sequence $\{\mathbf{y}(0), \mathbf{y}(1) \cdots \mathbf{y}(t-1)\}$ for detecting the sequence of codewords $\mathbf{G}(1)$, $\mathbf{G}(2) \dots \mathbf{G}(t-1)$. In general, this decoder has exponential complexity. On the similar lines of the equivalent MIMO channel in (1.2.9), we have an equivalent cooperative channel,

$$\begin{bmatrix} \mathbf{y}(t-1) \\ \mathbf{y}(t) \end{bmatrix} = \begin{bmatrix} \mathbf{S}(t-1) \\ \mathbf{G}(t)\mathbf{S}(t-1) \end{bmatrix} \mathbf{h} + \begin{bmatrix} \mathbf{n}(t-1) \\ \mathbf{n}(t) \end{bmatrix}. \qquad (1.3.5)$$

With the knowledge of only the $\mathbf{g}$ vector, the decoder for the truncated partially coherent receiver is given by Theorem 3 where the unitary code consists of codewords of the form $[\mathbf{S}(t-1)^T \ (\mathbf{G}(t)\mathbf{S}(t-1))^T]^T$. On the similar lines of the argument made in the Section, *Connection between the design criterion for coherent and non-coherent MIMO channel with differential encoder,* it can be shown that full diversity can be achieved if $\mathbf{G} - \mathbf{G'}$ is invertible for any $\mathbf{G}, \mathbf{G'} \in \mathcal{G}$ such that $\mathbf{G} \neq \mathbf{G'}$.

Extending the Differential STBC Construction to Fully Non-Coherent Relay Channels

In partially coherent DSTC, the destination must have the knowledge of the vector $\mathbf{g}$ for every quasi-static interval. This requires the transmission of training sequences from all the relays to the destination every quasi-static interval which in turn results in the reduction of the overall transmission rate. Motivated by the use of the decoder in (1.2.8) for non-coherent MIMO channels, an alternate decoder to the truncated partially coherent ML decoder is proposed (referred as ML 1-lag detector) which is given by,

$$\widehat{\mathbf{G}}(t) = arg \min_{\mathbf{G}(t) \in \mathcal{G}} \| \mathbf{y}(t) - \mathbf{G}(t)\mathbf{y}(t-1) \|^2 . \qquad (1.3.6)$$

Notice that this decoder is completely non-coherent since it does not use the knowledge of the channel vector $\mathbf{h}$. In [Oggier & Hassibi, 2006] a Chernoff bound on the PEP is derived for the above decoder and it is shown that, for full diversity, the design criterion on the codebook $\mathcal{G}$ is the same as that on the codebook for coherent DSTC [Jing & Hassibi, 2006]. Also, it is shown that the CER (Codeword Error rate) performance is the same for the decoders in (1.3.3) (when used as truncated ML decoder) and (1.3.6). Thus, distributed differential codes can be used both in partially coherent channel as well as in fully non-coherent channel. The only difference being the use of an appropriate decoder. Henceforth, without loss of generality, we discuss differential codes assuming the use of the decoder in (1.3.6) which in turn avoids the need for any channel knowledge at the destination. The decoder in (1.3.6) has also been used in [Jing & Jafarkhani, 2008] where real orthogonal designs and *SP* (2) have been used for differential encoding/decoding in a fully non-coherent distributed space time coding up.

One of the key requirements in the distributed construction of differential codes is to generate two commuting sets of unitary matrices. Even though several full diversity achieving unitary codes are available for MIMO channels, such codes cannot be directly applied in the distributed set-up since the codewords need to commute with the relay matrices. An example for a distributed differential code is presented before getting to the details of the construction of such codes in the Section, *Differential Code construction for fully non-coherent relay channels*.

Example 1: [Kiran & Rajan, 2007; Tarokh & Kim, 2002] For a relay network with two relay nodes, the generalized PSK codes for non-coherent point to point communication can be adapted to the relay network in a differential encoding set-up. For $R = T = 2$ and L being the cardinality of the code, the generalized PSK code is,

$$\mathcal{G} = \left\{ \mathbf{G}_k = \begin{bmatrix} \cos(\theta_k) & -\sin(\theta_k) \\ \sin(\theta_k) & \cos(\theta_k) \end{bmatrix} : \theta_k = \frac{\pi k}{L}, 0 \leq k \leq L-1 \right\}.$$

The initial vector is $\mathbf{s} = [1 \ 0]^T$ and the relay matrices are $\mathbf{A}_1 = \mathbf{I}_2$ and $\mathbf{A}_2 = \begin{bmatrix} 0 & -1 \\ 1 & 0 \end{bmatrix}$. It is to be noticed that $\mathbf{G}_j \mathbf{A}_k = \mathbf{A}_k \mathbf{G}_j$ for all j and k.

DIFFERENTIAL CODE CONSTRUCTION FOR FULLY NON-COHERENT RELAY CHANNELS

Different tools and methods to generate the necessary ingredients, $\mathcal{A}$ and $\{\mathcal{G}, \mathbf{s}\}$ are provided in this section. In particular, this section considers the construction of two classes of differential codes; the first one deals with the construction of codes having exponential encoding and decoding complexity (see the Section, *Codes with exponential encoding and decoding complexity*) in the rate of the code whereas the second one deals with codes having low encoding and decoding complexity (see the Section, *Codes with low encoding and decoding complexity*).

Codes with Exponential Encoding and Decoding Complexity

In this subsection, we present several methods of constructing the sets $\mathcal{A}$ and $\{\mathcal{G}, \mathbf{s}\}$ such that the differential codes using the above ingredients have encoding and decoding complexity which are exponential in the rate of the codes. The methods presented include algebraic tools such as division algebras, Cayley algebras and cyclic groups.

Constructing Commuting Sets of Matrices from Division Algebras

The authors in [Kiran & Rajan, 2007] have introduced commuting sets of unitary matrices from division algebras for the sets $\mathcal{A}$ and $\mathcal{G}$. We give only a flavor on how to choose such matrices and refer the readers to [Elia, Kumar, Pawar, Kumar & Lu, 2006; Kiran & Rajan, 2005; Oggier, 2005] for a detailed treatment on space time code constructions based on division algebras. A division algebra D is a vector space over a field, $\mathbb{F}$ which is also an associative ring, not necessarily commutative and every element

$d \in D$ has a multiplicative inverse. Let n denote the dimension of the vector space D over $\mathbb{F}$. For an arbitrary element, $d \in D$ the map $L_d(x) = dx$ is a $\mathbb{F}$ linear isomorphism from D to D. This defines a one-one map from d to the set of all $\mathbb{F}$ linear isomorphism of D which associates d to L_d. With a fixed basis of D, every d can be associated with an $n \times n$ invertible matrix over $\mathbb{F}$ denoted as L_d. This is referred as the left regular representation of D. Similarly, right regular representation is denoted by R_d. The most important fact about these two representations is that these two commute. Thus, two commuting sets of $n \times n$ matrices each of them isomorphic to the division algebra have been identified. The set $\mathcal{G}$ is chosen as a subset of unitary matrices from $\mathcal{L}$ (set of matrices from left regular representation of D) and $\mathcal{A}$ as a subset from $\mathcal{R}$ (set of matrices from right regular representations of D). This ensures that any matrix from $\mathcal{G}$ commutes with any matrix from $\mathcal{A}$. Furthermore, since $\mathcal{L}$ is a division algebra, full diversity requirement is also guaranteed.

Codes from Cyclic Groups

A distributed differential coding scheme has been proposed in [Oggier & Hassibi, 2006] wherein the sets $\mathcal{G}$ and $\mathcal{A}$ consists of diagonal unitary matrices. Since the matrices in each set are diagonal, the two sets will trivially satisfy the commutative property. Additionally, $\mathcal{G}$ is given the structure of a cyclic group. It is shown that full diversity can be obtained by choosing a generator of the group with an appropriate structure. The relay matrix set, $\mathcal{A}$ is constructed from a Generalized Butson Hadamard (GBH) matrix $\mathbf{M} \in \mathbb{C}^{R \times R}$ (See Definition 3).

Definition 3: [Horadam, 2005]. A Generalized Butson-Hadamard (GBH) matrix is a $T \times T$ matrix $\mathbf{M}$ with entries such that $\mathbf{M}\mathbf{M}^H = \mathbf{M}^H\mathbf{M} = T\mathbf{I}_T$ and conjugate of every entry $[\mathbf{M}]_{i,j}$ is its inverse i.e. $[\mathbf{M}]_{i,j}^* = [\mathbf{M}]_{i,j}^{-1}$.

The matrix $\mathbf{M}$ is used to construct $\mathcal{A}$ where the elements of the $i-$ th column of $\mathbf{M}$ are used as the diagonal elements of $\mathbf{A}_i$. An example for such a code is provided below for a network with $R = 3$ and $|\mathcal{G}| = L = 63$.

Example 2:

$$\mathcal{G} = \left\{ \mathbf{D}^i, \mathbf{D} = \begin{bmatrix} w_{63} & 0 & 0 \\ 0 & w_{63}^{17} & 0 \\ 0 & 0 & w_{63}^{26} \end{bmatrix} \right\} \text{ for } i = 1, \cdots, 63 \text{ and } \mathbf{s} = \frac{1}{\sqrt{R}}[1 \ 1 \cdots 1]^T \in \mathbb{C}^R, \text{ where}$$

$$w_{63} = \exp\left(\frac{2\pi i}{63}\right).$$

The corresponding GBH matrix is

$$\mathbf{M} = \begin{bmatrix} 1 & 1 & 1 \\ 1 & w_3 & w_3^2 \\ 1 & w_3^2 & w_3 \end{bmatrix} \text{ where } w_3 = \exp\left(\frac{2\pi i}{3}\right).$$

The relay matrices are constructed from the matrix $\mathbf{M}$ as follows:

$$\mathbf{A}_1 = \begin{bmatrix} 1 & 0 & 0 \\ 0 & 1 & 0 \\ 0 & 0 & 1 \end{bmatrix}, \mathbf{A}_2 = \begin{bmatrix} 1 & 0 & 0 \\ 0 & w_3 & 0 \\ 0 & 0 & w_3^2 \end{bmatrix} \text{ and } \mathbf{A}_3 = \begin{bmatrix} 1 & 0 & 0 \\ 0 & w_3^2 & 0 \\ 0 & 0 & w_3 \end{bmatrix}$$

It is to be noted that the encoding and the decoding complexity of this code is exponential in the rate of the code.

Recently, in [Harshan & Rajan, 2008], DSTBCs are constructed based on non-intersecting subspaces for a partially coherent cooperative channel. The proposed codes also make use of a unitary matrix group at the source and diagonal matrices at the relays. When the group is cyclic, a necessary and sufficient condition on the generator of the cyclic group to achieve full diversity and to minimize the PEP is proved. Certain conditions on the choice of a generator of the cyclic group are provided to reduce the decoding complexity at the destination.

Codes from Cayley Algebra

Recently, differential DSTBCs have been proposed using unitary Cayley transforms [Oggier & Lequeu, 2008] using which the set $\mathcal{G}$ can be constructed. For a Hermitian matrix $\mathbf{X}$, the unitary Cayley transform $\mathbf{U}$ is defined by $\mathbf{U} = (\mathbf{I} + i\mathbf{X})^{-1}(\mathbf{I} - i\mathbf{X})$. A Cayley code $\mathcal{G}$ is a family of unitary matrices given by,

$$\mathcal{G} = \left\{ \mathbf{U}_j = (\mathbf{I} + i\mathbf{X}_j)^{-1}(\mathbf{I} - i\mathbf{X}_j), j = 1, 2, \cdots, L \right\} \tag{1.4.1}$$

for a family $\mathcal{X} = \left\{ \mathbf{X}_j \ j = 1, 2, \cdots L \right\}$ of Hermitian matrices. Cayley codes were originally introduced for MIMO channels where it has been proved that the set $\mathcal{G}$ is fully diverse if the set $\mathcal{X}$ is fully diverse [Hassibi & Hochwald, 2002]. For full diversity, the set $\mathcal{X}$ is generally designed using division algebras [Oggier & Hassibi, 2007]. However, in [Oggier & Lequeu, 2008], due to the additional commuting property, $\mathcal{X}$ is designed using the multiplication matrices from number fields. Furthermore, it is shown that multiplication matrices are jointly diagonalizable using which, one can obtain commuting families of matrices. Note that the above construction is available for any number of relays.

Codes from Circulant Matrices

Circulant matrices are used in [Jing & Jafarkhani, 2008] to construct differential DSTBCs. The authors use a $R \times R$ matrix $\mathbf{A}$ which is a primitive $R - \text{th}$ root of $\mathbf{I}_R$ to construct the j-th relay matrix $\mathbf{A}_j$ as $\mathbf{A}_j = \mathbf{A}^{j-1}$ for all $j = 1$ to R. The relay matrix set $\mathcal{A}$ is given by,

$$\mathcal{A} = \left\{ \mathbf{A}_j = \mathbf{A}^{j-1} \ | \ \mathbf{A} = \begin{bmatrix} 0 & 1 & 0 & \cdots & 0 \\ 0 & 0 & 1 & \cdots & 0 \\ \vdots & \vdots & \cdots & \ddots & \vdots \\ 1 & 0 & 0 & \cdots & 0 \end{bmatrix} \right\}.$$

A well known result that a matrix commutes with the matrices in set $\mathcal{A}$ if and only if it is a circulant matrix has been used to construct $\mathcal{G}$. Since, in general, circulant matrices are not unitary, special unitary circulant matrices are constructed as follows:

$$\mathcal{G} = \left\{ \mathrm{u}_1\mathbf{A}_1, \mathrm{u}_2\mathbf{A}_2 \; .... \; \mathrm{u}_R\mathbf{A}_R \mid \mathrm{u}_j \in \mathcal{F}_j \text{ for } j = 1, 2, ...R \right\}$$

where $\mathcal{F}_j$ is a finite set of unit norm elements. An example for $\mathcal{F}_j$ is M-PSK signal set. The circulant codes can be designed with $\mathcal{F}_j$ being chosen as M-PSK signal sets rotated by an angle θ_j. In general, such circulant codes are not fully diverse for arbitrary values of θ_j. A necessary and sufficient condition on the θ_j's is presented in [Jing & Jafarkhani, 2008] to achieve full diversity. It is to be noted that the above circulant codes also have exponential encoding and decoding complexity since $\mathcal{F}_j$'s are entangled.

Example 3: For a relay network with $R = 4$, the relay matrices are given by,

$$\mathcal{A} = \left\{ \mathbf{A}_j = \mathbf{A}^{j-1} \mid \mathbf{A} = \begin{bmatrix} 0 & 1 & 0 & 0 \\ 0 & 0 & 1 & 0 \\ 0 & 0 & 0 & 1 \\ 1 & 0 & 0 & 0 \end{bmatrix} \right\}.$$

The circulant code $\mathcal{G}$ is given by $\left\{ \mathrm{u}_1\mathbf{A}_1, \mathrm{u}_2\mathbf{A}_2, \mathrm{u}_3\mathbf{A}_3 \text{ and } \mathrm{u}_4\mathbf{A}_4 \mid \mathrm{u}_j \in \mathcal{F}_j \text{ for } j = 1, 2, ...4 \right\}$. The signal set $\mathcal{F}_j$ is chosen to be 64 PSK with an angle of rotation θ_j for a transmission rate of 1 bpcu (bits per channel use). The rotations angles of $\theta_1 = 0$, $\theta_2 = 1.5$, $\theta_3 = 3$ and $\theta_4 = 4.5$ guarantees full diversity for the above code.

Codes with Low Encoding and Decoding Complexity

In [Rajan & Rajan, 2008], unitary STBCs from linear designs having low encoding and decoding complexity is proposed for distributed differential coding. The readers can refer [Hassibi & Hochwald, 2002; Tarokh, Jafarkhani & Calderbank, 1999] for more details on linear designs.

Definition 4: A linear design $\mathbf{S}(x_1, x_2, \cdots x_K)$ of size $n \times n$ in K real variables is a $n \times n$ matrix with entries being a complex linear combination of the variables $x_1, \cdots, x_K$. The above design can be written as

$$\mathbf{S}(x_1, x_2, \cdots x_K) = \sum_{i=1}^{K} x_i\mathbf{B}_i,$$

where $\mathbf{B}_i \in \mathbb{C}^{n \times n}$ are called the weight matrices.

Towards constructing unitary DSTBCs from linear designs, codebooks $\mathcal{G}$ consisting of unitary matrices are generalized to allow codebooks consisting of scaled unitary matrices [Rajan & Rajan, 2008]. In this section, the changes required to the existing protocol (in the Section, *Differential coding for non-coherent relay channels*) in order to accommodate the scaled unitary property of the set $\mathcal{G}$ are pointed. The matrices in $\mathcal{G}$ satisfy the condition $\mathbf{G}(t)^H \mathbf{G}(t) = a^2(t)\mathbf{I}_R$ for some $a^2(t) \in \mathbb{R}^+$. In order to maintain the average power constraint at the source, the scaled unitary matrices must satisfy the condition $E\left[a^2(t)\right] = 1$. With the above generalization, the vector transmitted from the source at the *t*-th block is

$$\mathrm{s}(t) = \frac{1}{a(t-1)}\mathbf{G}(t)\mathrm{s}(t-1). \tag{1.4.2}$$

With this, the average power transmitted by the source will continue to be unity and hence the average power constraint at the source is met. The new decoding metric is given by,

$$\widehat{\mathbf{G}}(t) = arg\min_{\mathrm{G}(t)\,\in\,\mathcal{G}} \parallel \mathbf{y}(t) - \frac{1}{a(t-1)}\mathbf{G}(t)\mathbf{y}(t-1)\parallel^2, \tag{1.4.3}$$

where $a(t-1)$ is estimated from the decision at the $(t-1)-$th block. Apart from the above changes, the differential scheme remains as in the Section, *Differential coding for non-coherent relay channels*. The use of scaled unitary matrices in $\mathcal{G}$ provides an opportunity to use linear designs in the differential set-up. It is to be noted that linear designs have been extensively used for coherent MIMO channels [Jafarkhani, 2001; Karmakar & Rajan, 2006; Khan & Rajan, 2006; Liang, 2003; Tarokh, Jafarkhani & Calderbank, 1999; Yuen, Guan & Tjhung, 2005] and coherent co-operative channels to construct STBCs with low decoding complexity. Also, STBCs from Quasi-Orthogonal Designs (QODs) have been applied using the differential technique for MIMO channels [Yuen, Guan & Tjhung, 2006].

A linear STBC is a finite set of complex matrices obtained using a linear design $\mathbf{S}(x_1, x_2, \cdots x_K)$ when the variables $x_1, x_2, \cdots x_K$ take values from a finite signal set $\mathcal{M} \subseteq \mathbb{C}$. Complex Orthogonal Designs (CODs) [Liang, 2003; Tarokh, Jafarkhani & Calderbank, 1999] are the most celebrated examples for linear designs. A linear STBC obtained using the Alamouti design with a M-PSK signal set (with unit average energy) and a M-QAM signal set (with unit average energy) correspondingly results in a unitary codebook and a scaled unitary codebook having the full diversity criterion. The authors in [Jing & Jafarkhani, 2008] have used Alamouti and $SP\,(2)$ codes in the differential set-up for networks with two and four relays. They have also employed square real orthogonal designs for networks with 2, 4 and 8 relays where it is shown that the Alamouti code and real orthogonal designs continue to maintain the linear decoding complexity while $SP\,(2)$ codes can be decoded using a sphere decoder. Motivated by the low decoding complexity of the Alamouti design, four group decodable (See Definition 5) codes were proposed recently in [Rajan & Rajan, 2008] for networks with number of relays of the form 2^m for any positive integer m.

Definition 5: A unitary STBC, $\mathcal{G}$ is said to be 4-group decodable if the decoding metric in (1.3.6) splits as follows,

$$\hat{x}_1, \hat{x}_2, \cdots \hat{x}_K = arg \min_{x_1, x_2, \cdots x_K \in \mathcal{M}} \| \mathbf{y}(t) - \left[\sum_{i=1}^{K} x_i \mathbf{B}_i \right] \mathbf{y}(t-1) \|^2$$

$$= arg \min_{x_1, x_2, \cdots x_K \in \mathcal{M}} \sum_{j=1}^{4} \| \mathbf{y}(t) - \left[\sum_{i=1}^{|G_j|} x_{i,j} \mathbf{B}_{i,j} \right] \mathbf{y}(t-1) \|^2,$$

where the variables $x_1, x_2, \cdots x_K$ are partitioned in to four groups $G_1, G_2, \cdots G_4$. Further, $x_{i,j}$ is the $i-$th variable belonging to the $j-$th group and the corresponding weight matrix is $\mathbf{B}_{i,j}$ for all $j = 1, \cdots 4$.

Notice that the 4-group decodability is only the property of the linear design. Linear designs with four-group decodable property are constructed using '**extended Clifford algebras**' in [Rajan & Rajan, 2008]. However, the STBCs generated from the proposed designs are not fully diverse for an arbitrary signal set and moreover, all the columns of the design are not orthogonal. Hence, such designs cannot be directly used to construct scaled unitary matrices using arbitrary signal set $\mathcal{M}$. Therefore, the authors explicitly construct signal sets for the proposed linear designs such that the resulting linear STBCs (i) contain scaled unitary matrices meeting the power constraint and (ii) provide full diversity as well.

Example 4: For $R = 4$, the unitary STBC, $\mathcal{G}$ is obtained using the following design,

$$\mathbf{S}(x_1, x_2, \cdots x_4) = \frac{1}{\sqrt{4}} \begin{bmatrix} x_1 & x_2 & -x_3^* & -x_4^* \\ x_2 & x_1 & -x_4^* & -x_3^* \\ x_3 & x_4 & x_1^* & x_2^* \\ x_4 & x_3 & x_2^* & x_1^* \end{bmatrix}.$$

Note that the first column of the above design is not orthogonal to the second column. The appropriate signal set for the above design can be obtained using Construction 4.4 in [Rajan & Rajan, 2008]. The relay matrices are given by,

$$\mathbf{A}_1 = \mathbf{I}_4, \ \mathbf{A}_2 = \begin{bmatrix} 0 & 1 & 0 & 0 \\ 1 & 0 & 0 & 0 \\ 0 & 0 & 0 & 1 \\ 0 & 0 & 1 & 0 \end{bmatrix}, \ \mathbf{A}_3 = \begin{bmatrix} 0 & 0 & -1 & 0 \\ 0 & 0 & 0 & -1 \\ 1 & 0 & 0 & 0 \\ 0 & 1 & 0 & 0 \end{bmatrix} \text{ and } \mathbf{A}_4 = \begin{bmatrix} 0 & 0 & 0 & -1 \\ 0 & 0 & -1 & 0 \\ 0 & 1 & 0 & 0 \\ 1 & 0 & 0 & 0 \end{bmatrix}.$$

The initial vector is $\mathbf{s} = \begin{bmatrix} 1 & 0 & \ldots & 0 \end{bmatrix}^T$ and the initial matrix is $\mathbf{G}(0) = \mathbf{I}_4$. This code is 4- group decodable or single complex symbol decodable (since $K = 8$).

Alternate Differential Coding for Fully Non-Coherent Channels

The differential technique introduced in the Section, *Differential coding for non-coherent relay channels* is one of the ways of differential coding for non-coherent wireless networks. In this section, we briefly introduce another differential method presented in [Wang, Zhang & Amin, 2006]. In this set-up, the primary modification in the network model is the assumption of the existence of a direct link between the source and the destination. Such a link is assumed to be quasi-static and flat fading. It is also assumed that the source transmits in the second phase along with the relays. In this set-up, the source node is equipped with $\mathcal{G}$, a finite set of fully diverse $(R+1) \times (R+1)$ unitary matrices and differential encoding is performed on $\mathcal{G}$. (The source contains $(R+1) \times (R+1)$ matrices instead of $R \times R$ matrices due to an additional channel from the source to the destination). Unlike the method in the Section, *Differential coding for non-coherent relay channels,* in the $t-$th block, the source transmits a different row of the codeword $\mathbf{G}(t)$ to each of the R relay terminals which involves $(R+1)R$ complex channel uses. In the second phase, all the relays and the source will transmit $R+1$ rows of $\mathbf{G}(t)$ simultaneously to the destination. Note that, in the second phase, the source transmits the $(R+1)-$th row of $\mathbf{G}(t)$ whereas the relays together forward the corrupted version of the first R columns of $\mathbf{G}(t)$. It is to be observed that the relay matrix set $\mathcal{A}$ is a set of identity matrices. The vector received at the destination is $\mathbf{y}(t) = \mathbf{S}(t)\mathbf{h} + \mathbf{n}(t) \in \mathbb{C}^{(R+1)^2 \times 1}$ where

$$\mathbf{S}(t) = \mathbf{G}'(t)\mathbf{S}(t-1) = \Big(\mathbf{G}'(t)\mathbf{G}'(t-1)...\mathbf{G}'(1)\Big)\mathbf{S}(0) \in \mathbb{C}^{(R+1)^2 \times (R+1)}$$

with $\mathbf{G}'(t) = \text{diag}[\mathbf{G}(t)\ \mathbf{G}(t) \cdots \mathbf{G}(t)] = \mathbf{I}_{R+1} \otimes \mathbf{G}(t)$ and $\otimes$ denotes the Kronecker product. Since $\mathbf{G}(t)$ is unitary, $\mathbf{G}'(t)$ also turns out to be unitary. Also note that $\mathbf{S}(t)$ is not unitary, but since $\mathbf{G}'(t)$ is unitary we have,

$$\mathbf{S}(t)^H \mathbf{S}(t) = \mathbf{S}(t-1)^H \mathbf{G}'(t)^H \mathbf{G}'(t)\mathbf{S}(t-1) \tag{1.4.4}$$

$$= \mathbf{S}(t-1)^H \mathbf{S}(t-1)$$

$$= \mathbf{S}(0)^H \mathbf{S}(0).$$

where $\mathbf{S}(0)$ is the codeword corresponding to the initial matrix $\mathbf{G}(0)$ at the source. Therefore, the transmit power at all the nodes is unchanged for different time slots and subsequently the power constraint on all the nodes is met. Based on the assumption that $\mathbf{h}$ remains same for two consecutive codeword uses, we can write $\mathbf{y}(t) = \mathbf{G}(t)\mathbf{S}(t-1)\mathbf{h} + \mathbf{n}(t) = \mathbf{G}(t)\mathbf{y}(t-1) + \tilde{\mathbf{n}}(t)$. Thus the following differential detector can be used

$$\widehat{\mathbf{G}}(t) = arg\min_{\mathbf{G}(t) \in \mathcal{G}} \| \mathbf{y}(t) - \mathbf{G}'(t)\mathbf{y}(t-1) \|^2. \tag{1.4.5}$$

Notice that unlike the differential scheme in the Section, *Differential coding for non-coherent relay channels*, the relay matrices and the code $\mathcal{G}$ need not satisfy the commutative property and hence the technique holds good for any full diversity achieving unitary code existing for coherent MIMO channels. But, this technique loses out in rate as more number of channel uses is required in the first phase, wherein the source transmits a different row of the codeword to every relay.

Codes Based on Training for Non-Coherent Relay Channel

Recently, a combination of channel training, channel estimation and coherent DSTC is proposed for non-coherent relay communications [Rajan & Rajan, 2008]. The proposed technique which is based on the work in [Dayal, Brehler & Varanasi, 2004] is divided in to two cycles, (i) the training cycle and (ii) the data transmission cycle. The training cycle comprises of $R + 1$ channel uses for channel estimation at the destination before the start of the data transmission cycle (one channel use from the source to the relays and R channel uses from the relays to the destination). It is assumed that the quasi-static duration of the channel is at least $3R + 1$ channel uses where the additional $2R$ channel uses are used for data transmission. Unless specified, the assumptions on the network model remain as in the Section, *Signal model*.

In the first phase of the training cycle, the source transmits the complex number, 1 to all the relays. The received symbol at the j-th relay is $r_j = \sqrt{P_1^t} f_j + n_j, j = 1, 2, \cdots, R$. The second phase of the training cycle comprises of R channel uses, out of which one channel use is assigned for every relay to forward a scaled version of r_j. Without loss of generality, we may assume that the j-th slot is assigned for the j-th relay. At the end of the training cycle, the received vector at the destination is given by,

$$\mathbf{y} = \sqrt{\frac{P_1^t P_2^t}{(1 + P_1^t)}} \mathbf{I}_R \mathbf{h} + \mathbf{n} \tag{1.4.6}$$

Treating the entries of $\mathbf{h}$ as independent i.i.d complex Gaussian random variables and $\mathbf{n}$ as complex Gaussian with mean $\mathbf{0}$ and the covariance matrix $((P_2^t / (P_1^t + 1)) + 1)\mathbf{I}_R$ the authors propose to estimate the equivalent channel matrix $\mathbf{h}$ as follows:

$$\hat{\mathbf{h}} = \sqrt{\frac{P_1^t P_2^t}{(1 + P_1^t)}} \left(1 + \frac{P_1^t P_2^t + P_2^t}{1 + P_1^t} \right)^{-1} \mathbf{y} . \tag{1.4.7}$$

The total power used from all the nodes in the training cycle is $P^t = P_1^t + R P_2^t$. In the data transmission cycle, codewords of a coherent DSTBC are distributively constructed and transmitted by all the relays. An additional total power of P^d is used in the data transmission cycle per codeword use from all the nodes. Hence, the total power used by all the nodes in this strategy per codeword use per channel realization is $P_t = P^t + P^d$. Using the estimate $\hat{\mathbf{h}}$, coherent decoding can be done as $\hat{\mathbf{X}} = arg \min_{\mathbf{x} \in \mathcal{C}} || \mathbf{y} - \sqrt{\frac{P_1^d P_2^d}{(1 + P_1^d)}} \mathbf{X} \hat{\mathbf{h}} ||^2$. Thus coherent DSTBCs can be employed in non-coherent relay networks via this training scheme. Simulations results show that the above estimator is good enough to outperform the best known differential codes using the decoder in (1.3.6).

Numerical Results

In this section, performance comparison of various differential codes such as the cyclic codes [Oggier, Hassibi, 2006], the circulant codes [Jing & Jafarkhani, 2008] and the Extended Clifford Algebras (ECA) codes [Rajan & Rajan, 2008] are provided in terms of the Codeword Error Rate (CER) (CER which corresponds to errors in decoding every codeword is considered as error events of our interest). The simulation results are presented for a relay network with $R = 4$ using codes having rates 1 bits per channel use (bpcu) and 1.5 bpcu. The ECA code presented in Example 4 is considered for simulations. The circulant code of Example 3 is chosen for the rate of 1 bpcu. However, for a rate of 1.5 bpcu, 1024-PSK is considered for the signal set $\mathcal{F}_i$ in the circulant code. The following cyclic codes are used in the simulations for rate 1 bpcu and rate 1.5 bpcu respectively:

$$\left\{ \begin{bmatrix} w_{256} & 0 & 0 & 0 \\ 0 & w_{256}^{11} & 0 & 0 \\ 0 & 0 & w_{256}^{67} & 0 \\ 0 & 0 & 0 & w_{256}^{101} \end{bmatrix}^i, i = 0, \ldots, 255 \right\} \text{ and } \left\{ \begin{bmatrix} w_{4096} & 0 & 0 & 0 \\ 0 & w_{4096}^{43} & 0 & 0 \\ 0 & 0 & w_{4096}^{877} & 0 \\ 0 & 0 & 0 & w_{4096}^{2039} \end{bmatrix}^i, i = 0, \ldots, 4095 \right\}$$

where $w_{256} = e^{\frac{2\pi i}{256}}$ and $w_{4096} = e^{\frac{2\pi i}{4096}}$. Figure 4 shows the error performance curves of the above three codes as a function of the total power used. It can be observed that for a transmission rate of 1 bpcu, the ECA code outperforms the cyclic code by about 5 dB and the circulant code by more than 10 dB. Similar results can be observed for a transmission rate of 1.5 bpcu also. It is worthwhile to note that

Figure 4. Error performance comparison of several differential codes. (Adapted from [Rajan & Rajan, 2008] (46))

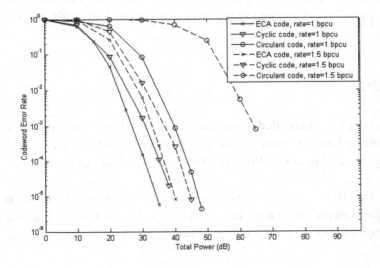

for rate 1 bpcu, the decoding search space for the ECA code is only 4 whereas it is 256 for the other two codes. Similarly for rate 1.5 bpcu, the decoding search space for the ECA code is only 8 whereas it is 4096 for the other two codes.

We also provide simulation results comparing the error performance of the training based codes (studied in the Section, *Codes based on training for non-coherent relay channel*) against the best known differential code for 4 relays (which is the ECA code). (See Figure 5) For simulations, the coherent DSTBC to be employed in the data transmission cycle is given by

$$
\mathbf{X} = \begin{bmatrix} x_1 & x_2 & -x_3^* & -x_4^* \\ x_2 & x_1 & -x_4^* & -x_3^* \\ x_3 & x_4 & x_1^* & x_2^* \\ x_4 & x_3 & x_2^* & x_1^* \end{bmatrix}
$$

where the 2-tuples $\left\{\mathrm{Re}(x_1), \mathrm{Re}(x_2)\right\}, \left\{\mathrm{Re}(x_3), \mathrm{Re}(x_4)\right\}, \left\{\mathrm{Im}(x_1), \mathrm{Im}(x_2)\right\}$ and $\left\{\mathrm{Im}(x_3), \mathrm{Im}(x_4)\right\}$ take values from Quadrature Amplitude Modulation (QAM) signal set rotated by $166.7078°$ (QAM constellation size is chosen depending on the transmission rate). Throughout this section, the DSTBC from the above design when used in the training based strategy is referred as the Training Based code (TB code). We choose $P_t = (1 + \alpha)P^d$, where α denotes what additional percentage of the total power used in the data transmission cycle, P^d is used for the training cycle. In order to quantify the loss in error performance due to channel estimation errors in the training based strategy, the performance of the corresponding coherent STBC is considered as the reference. Fig. 5 shows the error performance for $\alpha = 0$ and $\alpha = 0.4$. The transmission rates (when calculating transmission rate, the rate loss due to initial few channel uses for training is ignored- $R + 1$ for TBC and $2R$ for ECA codes) of 1 bpcu and 2 bpcu are considered. It can be observed that for a rate of 1 bpcu and codeword error rate (CER) of 10^{-5}, the TB codes outperform the ECA codes by approximately 2 dB for $\alpha = 0$. For a transmission rate of 2 bpcu, the performance gap between the TB code and the ECA code increases to 8 dB.

Thus, we infer that the performance advantage of the TB codes over ECA codes increases as the transmission rate increases. In spite of the simple channel estimation method employed, note that the performance loss due to channel estimation errors is only about 3 dB for transmission rates of 1 and 2 bpcu respectively.

FUTURE TRENDS

Among the several classes of distributed differential codes till date, the 4-group decodable codes from ECA are shown to outperform all other codes both in error performance as well as in encoding and decoding complexity. Several possible directions for future work are as follows:

- An interesting direction for future work is to design single-symbol decodable differential codes (if they exist) for networks with any number of relays. This will further reduce both the encoding and decoding complexity. Deriving an upper bound on the maximum rate of such codes is also challenging.

Figure 5. Error performance comparison of training based codes with ECA codes. (Adapted from [Rajan & Rajan, 2008 (47)])

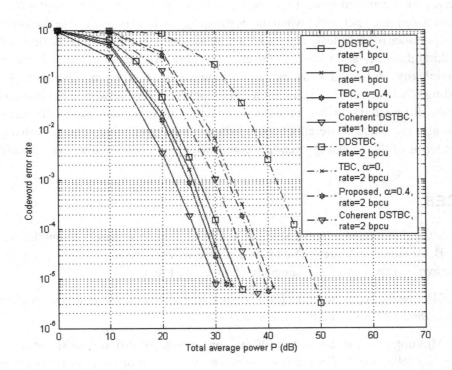

- It is reported in the Section, *Extending the differential STBC construction to fully non-coherent relay channels*, that the error performance of the differential codes is the same when either the receiver uses only the knowledge of channels from relays to destination or when it doesn't use any channel knowledge. Investigations on why the knowledge of a subset of channels doesn't give benefit in the error performance compared to fully non-coherent scenario can be made.
- Diversity Multiplexing Tradeoff (DMT) [Zheng & Tse, 2003] analysis of non-coherent co-operative channels is another interesting direction of future work.
- It is also interesting to study the design of DSTBCs when the relays are allowed to co-operate among themselves before transmitting simultaneously to the destination in the second phase i.e. when interaction among the relays is allowed.
- Non-coherent DSTBCs can be studied when a relay selection strategy is employed in the network. In such a scenario, the design of DSTBCs must be such that the use of any non-empty subset of the relays must provide full diversity.
- Construction of DSTBCs can be studied for a non-coherent relay channel when symbol level synchronization is not assumed among all the relays in the network.

CONCLUSION

A thorough survey of distributed space-time coding has been presented in this chapter for amplify and forward protocol based relay networks when the destination terminal doesn't have the knowledge of all the channels. Importance of code designs for such a scenario is attributed to the reduction in the overhead involved in making the destination learn all the channels both in terms of the overall transmission rate and also the complexity. Motivated by the existing coding techniques for non-coherent MIMO channels, differential coding schemes have been studied for non-coherent relay networks. Several code constructions based on the differential schemes have also been pointed. Error performance comparison of some of the codes has also presented. Differential codes from Clifford algebras have been shown to be the best till date both in terms of the encoding-decoding complexity and also in terms of error performance.

REFERENCES

Belfiore, J. C., & Cipriano, A. M. (2006). Space-time coding for non-coherent channels. In H. Bolcskei, G. Gesbert, C. B. Papadias & V. Veen (Eds.), *Space time wireless systems: From array processing to MIMO communication* (pp. 198-217). Cambridge University Press.

Bhatnagar, M., Hjorunges, A., & Song, L. (2007). Precoded DOSTBCs over Rayleigh channels. *EUR-ASIP Research Letters in Communications, 2007*, 16438.

Bhatnagar, M., Hjorunges, A., & Song, L. (2007). Non orthogonal differential space time block code with nonunitary constellations. In *Proc of IEEE international Workshop on Signal Processing Advances in Wireless Communications*.

Bhatnagar, M., Hjorunges, A., & Song, L. (2008). Precoded differential orthogonal space-time modulation over correlated Ricean MIMO channels. *IEEE Journal of Selected Topics in Signal Processing, 2*(2), 124–134.

Bletsas, A., Khisti, A., Reed, D., & Lippman, A. (2006). A simple cooperative diversity method based on network path selection. *IEEE JSAC, 24*(3), 659–672.

Chen, D., & Laneman, N. (2004). Noncoherent demodulation for cooperative diversity in wireless systems. In *Proceedings of IEEE GLOBECOM* (pp. 31-35).

Chen, D., & Laneman, N. (2006). Modulation and demodulation for cooperative diversity in wireless systems. *IEEE Transactions on Wireless Communications, 5*(07), 1785–1794.

Dayal, P., Brehler, M., & Varanasi, M. K. (2004). Leveraging coherent space time codes for noncoherent communication via training. *IEEE Transactions on Information Theory, 50*(09), 2058–2080.

Dimitris, P., & Karystinos, G. (2008). Efficient ML noncoherent orthogonal STBC detection. In *Proceedings of Allerton Conference*.

Elia, P., Kumar, K. R., Pawar, S. A., Kumar, P. V., & Lu, H. F. (2006). Explicit, minimum delay space time codes achieving the diversity multiplexing gain trade off. *IEEE Transactions on Information Theory, 52*(09), 896–900.

Foschini, G. J. (1996). Layered space time architecture for wireless communication in a fading environment when using multielement antennas. *Bell Labs Technical Journal, 01*(2).

Harshan, J., & Rajan, B. S. (2007). Coordinate interleaved distributed space time coding for two antenna relay networks. In *Proceedings of IEEE GLOBECOM* 2007, Washington D.C. (pp. 1719-1723).

Harshan, J., & Rajan, B. S. (2008). A nondifferential distributed space-time coding for partially-coherent cooperative communication. *IEEE Trans. Wireless Communication Letters, 07*(11), 4076–4081.

Harshan, J., & Rajan, B. S. (2008, May 19-23). Nondifferential DSTBCs for partially-coherent cooperative communication. In *Proceedings of ICC 2008*, Beijing, China (pp.996-1000).

Hassibi, B., & Hochwald, B. (2002). Cayley differential unitary space time codes. *IEEE Transactions on Information Theory, 48*(06), 1485–1503.

Hassibi, B., & Hochwald, B. M. (2002). High-rate codes that are linear in space and time. *IEEE Transactions on Information Theory, 48*(7), 1804–1824.

Hassibi, B., & Marzetta, T. L. (2002). Multiple antennas and isotropically random inputs: The received signal density in closed form. *IEEE Transactions on Information Theory, 48*(06), 1473–1484.

Hochwald, B., & Marzetta, T. L. (1999). Capacity of a mobile multiple antenna communication links in Rayleigh flat fading. *IEEE Transactions on Information Theory, 45*(01), 139–157.

Hochwald, B., & Marzetta, T. L. (2000). Unitary space time modulation for multiple antenna communication in Rayleigh flat fading. *IEEE Transactions on Information Theory, 46*(02), 543–564.

Hochwald, B., & Sweldens, W. (2000). Differential unitary space-time modulation. *IEEE Transactions on Communications, 48*(12), 2041–2052.

Horadam, K. J. (2005). A generalized Hadamard transform. *ISIT, Adelaide* (pp. 1006-1008).

Hughes, B. L. (2000). Differential space-time modulation. *IEEE Transactions on Information Theory, 46*(07), 2567–2578.

Jafarkhani, H. (2001). A quasi-orthogonal space-time block code. *IEEE Transactions on Communications, 49*(1), 1–4.

Jing, Y., & Hassibi, B. (2003). Unitary space-time modulation via Cayley transforms. *IEEE Transactions on Signal Processing, 51*(11), 2891–2904.

Jing, Y., & Hassibi, B. (2006). Distributed space-time coding in wireless relay networks. *IEEE Transactions on Wireless Communications, 5*(12), 3524–3536.

Jing, Y., & Hassibi, B. (2008). Diversity analyses of distributed space time codes in relay networks with multiple transmit/receive antennas. *Eurosip Journal on Advances in Signal Processing, 2008*.

Jing, Y., & Jafarkhani, H. (2008). Distributed differential space-time coding for wireless relay networks. *IEEE Transactions on Communications, 56*(7), 1092–1100.

Kammoun, I., & Belfiore, J. C. (2003). A new family of Grassmann space time codes for non-coherent MIMO systems. *IEEE Trans. Communication Letters, 07*(11), 528–530.

Karmakar, S., & Rajan, B. S. (2006, July 9-14). Minimum-decoding complexity, maximum-rate space-time block codes from Clifford algebras. In *Proc. IEEE International Symposium on Information Theory*, Seattle, WA (pp. 788-792).

Khan, Z., & Rajan, B. S. (2006). Single-symbol maximum-likelihood decodable linear STBCs. *IEEE Transactions on Information Theory*, 52(5), 2062–2091.

Kiran, T., & Rajan, B. S. (2005). STBC schemes with nonvanishing determinant for certain number of transmit antennas. *IEEE Transactions on Information Theory*, 51(08), 2984–2992.

Kiran, T., & Rajan, B. S. (2007). Partially-coherent distributed space-time codes with differential encoder and decoder. *IEEE Journal on Selected Areas in Communications*, 25(2), 426–433.

Laneman, L., Tse, D., & Wornell, G. W. (2004). Cooperative diversity in wireless networks: Efficient protocols and outage behavior. *IEEE Transactions on Information Theory*, 50(12), 3062–3080.

Laneman, N., & Wornell, G. W. (2003). Distributed space-time-coded protocols for exploiting cooperative diversity in wireless networks. *IEEE Transactions on Information Theory*, 49(10), 2415–2425.

Liang, X. (2003). Orthogonal designs with maximal rates. *IEEE Transactions on Information Theory*, 49(10), 2468–2503.

Luby, M. (2002). LT codes. In *Proc. 43rd Annual IEEE Symposium on Foundations of Computer Science (FOCS)* (pp. 271-282).

Nabar, R. U., Bolcskei, H., & Kneubehler, F. W. (2004). Fading relay channels: Performance limits and space time signal design. *IEEE Journal on Selected Areas in Communications*, 22(06), 1099–1109.

Oggier, F. (2005). First application of cyclic algebras to non-coherent MIMO channels. In *Allerton Conference*.

Oggier, F., & Hassibi, B. (2006). *A coding strategy for wireless networks with no channel information*. Presented at the Forty-Fourth Annual Allerton Conference on Communication, Control, and Computing.

Oggier, F., & Hassibi, B. (2007). Algebraic Cayley differential space time codes. *IEEE Transactions on Information Theory*, 53(05).

Oggier, F., & Lequeu, E. (2008, July 6-11). Differential distributed space-time coding based on Cayley codes. In *Proceedings of IEEE International Symposium on Information Theory (ISIT) 2008*, Toronto, Canada (pp. 2548-2552).

Oggier, F., Sloane, A., Diggavi, S., & Calderbank, A. R. (2005). Nonintersecting subspaces based on finite alphabets. *IEEE Transactions on Information Theory*, 51(12), 4320–4325.

Rajan, G. S., & Rajan, B. S. (2006, October 22-26). A nonorthogonal distributed space time coded protocol, part-1: Signal model and design criteria. In *Proceedings of ITW 2006*, Chengdu, China (pp. 385-389).

Rajan, G. S., & Rajan, B. S. (2006, October 22-26). A nonorthogonal distributed space time coded protocol, part-2: Code construction and DM-G trade-off. In *Proceedings of IEEE Information Theory Workshop*, Chengdu, China (pp. 488-492).

Rajan, G. S., & Rajan, B. S. (2007). Signal set design for full-diversity low-decoding-complexity differential scaled-unitary STBCs. In *Proceedings of IEEE International Symposium on Information Theory, (ISIT 2007)*, Nice, France (pp. 1616-1620).

Rajan, G. S., & Rajan, B. S. (2008). Algebraic distributed differential space time codes with low decoding complexity. *IEEE Transactions on Wireless Communications, 7*(10), 3962–3971.

Rajan, G. S., & Rajan, B. S. (2008). Leveraging coherent distributed space-time codes for noncoherent communication in relay networks via training. To appear in *IEEE Trans. Wireless Communication*. Also available at arXiv: 0804.3155v2 [cs.IT] 3 Sep 2008.

Sendonaris, A., Erkip, E., & Aazhang, B. (2003). User cooperation diversity—part I: System description. *IEEE Transactions on Communications, 51*(11), 1927–1938.

Sendonaris, A., Erkip, E., & Aazhang, B. (2003). User cooperation diversity—part II: Implementation aspects and performance analysis. *IEEE Transactions on Communications, 51*(11), 1939–1948.

Tarokh, V., Jafarkhani, H., & Calderbank, A. R. (1999). Space-time block codes from orthogonal designs. *IEEE Transactions on Information Theory, 45*(5), 1456–1467.

Tarokh, V., & Kim, I. M. (2002). Existence and construction of noncoherent unitary space-time codes. *IEEE Transactions on Information Theory, 48*(12), 3112–3117.

Tarokh, V., Seshadri, N., & Calderbank, A. R. (1998). Space time codes for high data rate wireless communication: Performance criterion and code construction. *IEEE Transactions on Information Theory, 44*(02), 744–765.

Teletar, I. E. (1999). Capacity of multiantenna Gaussian channels. *European Transactions on Telecommunications, 10*(06), 1999.

Wang, G., Zhang, Y., & Amin, M. (2006). Differential distributed space-time modulation for cooperative networks. *IEEE Transactions on Communications, 05*(11).

Yiu, S., Schober, R., & Lampe, L. (2005). Noncoherent distributed space-time block coding. *Vehicular Technology Conference* (pp. 947-951).

Yiu, S., Schober, R., & Lampe, L. (2006). Distributed space-time block coding. *IEEE Transactions on Communications, 54*(07), 1195–1206.

Yuen, C., Guan, Y. L., & Tjhung, T. T. (2005). Quasi-orthogonal STBC with minimum decoding complexity. *IEEE Transactions on Wireless Communications, 4*(5), 2089–2094.

Yuen, C., Guan, Y. L., & Tjhung, T. T. (2006). Single-symbol decodable differential space-time modulation based on QO-STBCs. *IEEE Transactions on Wireless Communications, 05*, 3329–3335.

Zheng, L., & Tse, D. N. C. (2002). Communication on the Grassmann manifold: A geometric approach to the noncoherent multiple antenna channels. *IEEE Transactions on Information Theory, 48*(2), 359–383.

Zheng, L., & Tse, D. N. C. (2003). Diversity and multiplexing: A fundamental tradeoff in multiple-antenna channels. *IEEE Transactions on Information Theory*, *49*(05), 1073–1096.

ENDNOTES

[1]   For distributed space time coding with multiple antennas at the source, the relays and the destination, we refer the reader to [Harshan & Rajan, 2007; Jing & Hassibi, 2008].

[2]   If a direct link is assumed between the source and the destination, the source can also transmit along with the relays simultaneously in the second phase. DSTC with such an assumption has also been well studied [Rajan & Rajan, 2006].

[3]   Note that the relays are assumed to have unitary matrices to keep the co-variance matrix of the additive noise vector, a scaled identity matrix which in turn keeps the performance analysis of the protocol simpler. In general, non-unitary matrices can also be used at the relays [Rajan & Rajan, 2006].

Section 4
Broadband Cooperative Communications

Chapter 14
Resource Allocation for a Cooperative Broadband MIMO–OFDM System

Ibrahim Y. Abualhaol
Broadcom Corporation, USA

Mustafa M. Matalgah
University of Mississippi, USA

ABSTRACT

In this chapter, a cooperative broadband relay-based resource allocation technique is proposed for adaptive bit and power loading multiple-input-multiple-output/orthogonal frequency division multiplexing (MIMO-OFDM) system. In this technique, sub-channels allocation, M-QAM modulation order, and power distribution among different sub-channels in the relay-based MIMO-OFDM system are jointly optimized according to the channel state information (CSI) of the relay and the direct link. The transmitted stream of bits is divided into two parts according to a suggested cooperative protocol that is based on sub-channel-division. In this protocol, the first part is sent directly from the source to the destination, and the second part is relayed to the destination through an indirect link. Such a cooperative relay-based system enables us to exploit the inherent system diversities in frequency, space and time to maximize the system power efficiency. The BER performance using this cooperative sub-channel-division protocol with adaptive sub-channel assignment and adaptive bit/power loading are presented and compared with a noncooperative ones. The use of cooperation in a broadband relay-based MIMO-OFDM system showed high performance improvement in terms of BER.

DOI: 10.4018/978-1-60566-665-5.ch014

Figure 1. Relay-based wireless communication system

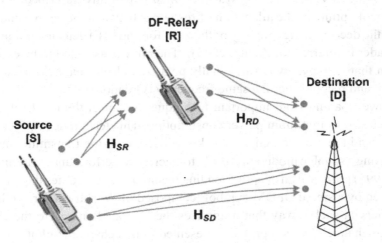

INTRODUCTION

It is well-known that, due to space separation, multiple-input-multiple-output (MIMO) systems have the advantages of improving the received signal-to-noise ratio (SNR) and suppressing the co-channel interference (CCI) (Winters, 1994). The use of orthogonal frequency division multiplexing (OFDM) gives the system the advantage of inter-symbol-interference (ISI) resistance, due to the use of a cyclic prefix, in addition to the advantage of its spectrum efficiency as compared to the conventional frequency division multiplexing (FDM) (Keller, 2000). When MIMO technique is combined with OFDM, the advantages of both techniques, such as CCI rejection and ISI resistance, can be jointly utilized. Furthermore, the MIMO-OFDM adaptive resource allocation benefits from combined frequency- and space-domain freedom as well as multiuser diversity (Zhang, 2003). Converting a frequency-selective MIMO channel into a number of parallel flat-fading MIMO sub-channels using linear precoding/decoding - based on single value decomposition (SVD) - enables the MIMO processing on a per subcarrier basis. Linear precoding/decoding requires full channel state information (CSI) at both the transmitter and receiver (Sampath, 2001; Scaglione, 2002).

The first formulation of general relaying problems appeared in the information theory community in (Cover, 1979; Van der Meulen, 1971) and served as the inciting cause of the concurrent development of the ALOHA system at the University of Hawaii. The traditional relay channel model is comprised of three nodes: a source that transmits information, a destination that receives information, and a relay that both receives and transmits information to enhance the communication between the source and the destination, as shown in Fig. 1. Recently, many new models with multiple relays have been examined (Kramer, 2005; Schein, 2000).

Adaptive resource allocation for cooperative MIMO-OFDM benefits from the combined frequency- and space-domain freedom as well as cooperative multiuser diversity due to the spatial parallelism and frequency selectivity of the channel. Cooperative relay-based MIMO-OFDM systems will allow the reuse of OFDM subcarriers by means of optimal precoder and decoder matrix transformations that partitions the MIMO channel into parallel non-interfering single-input-single-output (SISO) channels. In this chapter, a cooperative sub-channel division protocol is presented to improve the bit error rate

(BER) performance of the MIMO-OFDM system, which is partially described in (Abualhaol, 2008). The proposed protocol optimizes the allocation of resources between the source and the relay by using SVD coupled with the decorrelating property of the zero-forcing (ZF) detector (for more information on ZF detector, the reader is referred to (Zhang, 2005)). The relay is assumed to be equipped with a dual-polarization system that employs two orthogonally polarized electromagnetic waves, hence, enabling the relay to transmit and receive over the same frequency (Ordano, 1997).

Adaptive bit/power loading is an important technique to control the resultant data rate to meet a predefined QoS metric with minimum power consumption. This technique requires CSI knowledge at both the transmitter and the receiver. Given this knowledge, both the transmitter and the receiver can coordinate for choosing a suitable modulation order for increased performance. The authors in (De Souza, 1999; Chow et al, 1995) proposed an adaptive loading technique that assumes that the desired number of bits, to be transmitted by a single OFDM-symbol (composed of N sub-carrier symbols), is distributed among the subcarriers in such a way that minimizes the allocated energy to the entire transmission. The cooperative sub-channel division protocol presented in this chapter exploit this loading technique to boost the performance of the proposed cooperative resource allocation technique. In summary, the main focus of the proposed technique is to optimize the allocation of resources (i.e., MIMO-OFDM subchannels) between the source and the relay by using SVD technique coupled with the decorrelation property of ZF detector.

The remainder of this chapter is organized as follows. Section 8.2 presents the adaptive bit/power loading technique for MIMO-OFDM system. Section 8.3 introduces the system and channel model under consideration. Section 8.4 presents the MIMO-OFDM cooperative sub-channel division protocol. Section 8.5 provides a brief description for feasible signaling model. After that, simulation results that show the system performance are given in section 8.6. Finally, the chapter is concluded in section 8.7.

Adaptive Bit/Power Loading for MIMO-OFDM System

Adaptive modulation is an important technique that yields increased data rates over non-adaptive ones. An inherent assumption in channel adaptation is some form of channel knowledge at both the transmitter and the receiver. Given this knowledge of the CSI, both the transmitter and receiver can agree upon modulation scheme for increased performance. In this chapter, we consider adaptive bit and power allocation schemes (De Souza, 1999; Chow et al, 1995). The adaptive loading technique employed in this chapter is an efficient technique to achieve power and rate optimization based on knowledge of the subchannel gains.

In the discrete bit-loading algorithm of (De Souza, 1999), we are given a set of $K_{min} \times N$ increasing convex functions $e_n(b)$ that represent the amount of energy necessary to transmit b bits on subchannel n, where $n = 1, 2, ..., K_{min} \times N$, at the desired BER. We assumed $e_n(0) = 0$. The allocation problem of B bits that minimizes the total assigned transmission power can be formulated as

$$\min \sum_{n=1}^{K_{min} \times N} e_n(b_n) \tag{1}$$

$$\text{Subject to} \sum_{n=1}^{K_{min} \times N} b_n = B \tag{2}$$

Figure 2. Initialization bit/power loading algorithm

```
Input: SNR(i), i = 1, 2, ..., K_min × N, Tuning parameter (TUN)
Output: Energy increments table △e_i(b),  i = 1, 2, ..., K_min × N
i ← 1;
while i ≤ K_min × N do
    b̂(i) = log₂(1+SNR(i)/GAP);
    b(i) = {⌊b̂(i)⌋ | b(i) = 0, 1, 2, 4, 6, 8} (available modulation order);
    if b(i) = 0 then
        e_i(b(i)) = 0;
    else
        e_i(b(i)) = (2^b(i) − 1)× TUN/SNR(i);
    end
    i ← i + 1;
end
ind = 1;
while ind ≤ 9 do
    if ind ≤ 8 then
        △e_i(ind) = e_i(ind) − e_i(ind − 1);
    else
        △e_i(ind) = ∞;
    end
    ind ← ind + 1;
end
```

$$b_n \in \mathbf{Z}, b_n \geq 0, n = 1, 2, ..., K_{min} \times N. \tag{3}$$

The procedural algorithm that initializes the bit allocation is given in Fig 2.

In this chapter, we considered $K_{min} \times N$ MIMO-OFDM sub-channels. If the channel gain and noise power spectral density (PSD) for all sub-channels are known, then we can generate the energy increment table ($\Delta e_i(b(i))$) that provides the incremental energies required for the sub-channel to transition from supporting 0 bits to 1 bit, from 1 bit to 2 bits, from 2 bits to 3 bits and so on. Since we require our system to have a maximum of 8 bits, the energy increment required to go from 8 bits to 9 bits is set to a very high value. Also, we require the subchannel to have only 0, 1, 2, 4, 6 or 8 bits. Thus, odd numbers of bits are not supported. In order to take care of this, the energy increment table has to be changed using an averaging technique. Given the initial bit allocation, and the energy increment table, the algorithm in Fig. 3 optimizes the bit allocation (De Souza, 1999).

Finally, in order to deal with a single violated bit constraint, the algorithm in Fig. 4 is used.

Given these three algorithms, we have a complete characterization of the bit/power loading procedure for MIMO-OFDM frequency-selective fading channels.

System and Channel Model

Consider a cooperative relay-based MIMO-OFDM wireless communication system as in Fig. 1, where S is the source communicating directly with a destination D and indirectly through terminal R that acts as detect-and-forward relay (DF-relay) (here, the relay perform channel decoding and precoding and not the full detection of the transmitted symbols). Let M_t and M_r denote the number of transmitted and received antennas, respectively, for all S, R, and D. The frequency band is divided into N sub-channels (equivalently N subcarriers). Assume that the source S has a data rate requirement of R_s bits to be transmitted per OFDM symbol. In each symbol duration, a data stream composed of R_s bits is fed into the S/P block, as shown in Fig. 6, which in turn segments the data stream into $K_{min} \times N$ parallel streams,

Figure 3. Optimized bit allocation with minimum transmission power algorithm

```
Input: Initial bit allocation (b_ini), the total number of bits to be
        allocated (B)
Output: Optimum bit allocation (b_opt)
Ḃ ← 1 ;
b ← b_ini;
n ← 1;
while n ≤ K_min × N do
  |  Ḃ ← Ḃ + b(n);
  |  while Ḃ ≠ B do
  |    |  if Ḃ > B then
  |    |    |  m = arg max_{1≤i≤K_min×N} △e_i(b(i));
  |    |    |  B ← B − 1;
  |    |    |  b(n) ← b(m) − 1;
  |    |  else
  |    |    |  m = arg min_{1≤i≤K_min×N} △e_i(b(i));
  |    |    |  B ← B + 1;
  |    |    |  b(n) ← b(m) + 1;
  |    |  end
  |  end
  |  n ← n + 1;
end
b_opt ← b;
```

Figure 4. Resolving one bit violation algorithm

```
Input: Optimum bit allocation (b), sub-channel with one bit violation
        (v) , set of sub-channels with 0 or 1 bit allocation (A)
Output: Optimum bit allocation (b) without violation is sub-channel (v)
z ← 1;
while z ≤ |A| do
  |  T1(z) = △e_v(b(v)) − △e_z(b(z));
  |  T2(z) = △e_z(b(z) + 1) − △e_v(b(v));
end
D1 = arg min_{1≤z≤|A|} T1(z);
D2 = arg min_{1≤z≤|A|} T2(z);
if D1 ≤ D2 then
  |  b(v) ← b(v) + 1;
  |  b(D1) ← b(D1) − 1;
else
  |  b(v) ← b(v) − 1;
  |  b(D2) ← b(D1) + 1;
end
```

where the number of bits in the sub-channels $1, \cdots, K_{min} \times N$, are $b_{s,1}, b_{s,2}, ..., b_{s,K_{min} \times N}$, respectively. Here, $K_{min} = min\{M_t, M_r\}$. An adaptive loading technique (i.e., bit and power loading) is employed in this model as given in (De Souza, 1999; Chow et al, 1995). The adaptive bit loading is implemented in the S/P block at the transmitter and in the P/S block at the receiver. The adaptive power loading is implemented after the modulator bank at the transmitter and at the equalization block at the receiver.

The MIMO-OFDM system as shown in Fig. 5 is used identically in all the three links S-D, S-R and R-D. The relay R is assumed to have R_r bits to be sent at the same time when an OFDM-symbol from S is sent. The relay R forwards its information by segmenting the data stream into $K_{min} \times N$ parallel streams, where, again, the number of bits in the sub-channels $1, \cdots, K_{min} \times N$, are $b_{r,1}, b_{r,2}, ..., b_{r,K_{min} \times N}$. These data streams in both S and R are modulated, using six different M-QAM signal constellations

Figure 5. Transmitter/receiver structure of MIMO-OFDM resource allocation

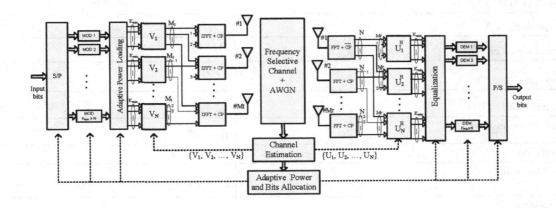

(as given in Table 1) with gray coding, into symbols $\{X_{s,1}, X_{s,2},..., X_{s,K_{min} \times N}\}$ and $\{X_{r,1}, X_{r,2},..., X_{r,K_{min} \times N}\}$ for S and R, respectively.

Referring to Fig. 1, $\mathbf{H}_{SR}$, $\mathbf{H}_{RD}$, and $\mathbf{H}_{SD}$ are the channel fading envelope matrices of the S-R, R-D, and S-D links, respectively. Under the assumption of perfect channel CSI estimation of $\mathbf{H}_{SR}$, $\mathbf{H}_{RD}$, and $\mathbf{H}_{SD}$ at S, R, and D, it was shown in (Sampath, 2001) that using optimal precoder and decoder matrix transformation can partition the MIMO channel of the k_{th} subcarrier into K_{min} non-interfering parallel SISO sub-channels. The SVD of the channel matrix of the k_{th} subcarrier can be written as

$$\mathbf{H}_k = \begin{bmatrix} h_{1,1} & h_{1,2} & \cdots & h_{1,M_t} \\ h_{2,1} & h_{2,2} & \cdots & h_{2,M_t} \\ \vdots & \vdots & \ddots & \vdots \\ h_{M_r,1} & h_{M_r,2} & \cdots & h_{M_r,M_t} \end{bmatrix} = \mathbf{U}_k \mathbf{\Sigma}_k \mathbf{V}_k^H, \tag{4}$$

Figure 6. Relay-based MIMO-OFDM cooperative division protocol

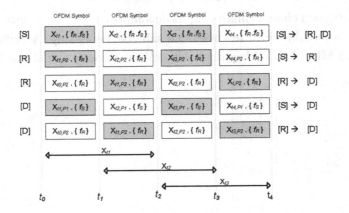

Table 1. System parameters

Bandwidth (BW)	20 MHz
No. of subcarriers (N)	64
Subcarrier separation (Δf)	312.5 KHz
Symbol time (T_s)	3.2 μs
Cyclic prefix time (T_{cyc})	0.8 μs
Symbol rate (symbol/sec)	16×10^6
M-QAM order	0, 1, 2, 4, 6, 8
Power delay profile	$[1, 1/e, 1/e^2]$
Noise variance	1×10^{-3}
Coherence time	200 μs

where $\mathbf{U}_k \in \mathbf{C}^{M_r \times K_{min}}$, $\mathbf{V}_k^H \in \mathbf{C}^{K_{min} \times M_t}$ are unitary matrices, and $\boldsymbol{\Lambda}_k \in \mathbf{C}^{K_{min} \times K_{min}}$ is a diagonal matrix consisting of singular values of $\mathbf{H}_k$. Then, the concept of the SVD channel coding can be given by the following matrix transformations:

$$\mathbf{U}_k^H \mathbf{H}_k \mathbf{V}_k = \mathbf{U}_k^H \mathbf{U}_k \boldsymbol{\Lambda}_k \mathbf{V}_k^H \mathbf{V}_k = \boldsymbol{\Lambda}_k. \tag{5}$$

If the k_{th} subcarrier, where $k = 1, 2, ..., N$, is used to send K_{min} symbols ($\mathbf{X}_k$), then the received symbols $\mathbf{Y}_k$ after precoding and decoding may be given as

$$\mathbf{Y}_k = (\mathbf{U}_k^H)(\mathbf{H}_k)(\mathbf{V}_k \mathbf{X}_k) + (\mathbf{U}_k^H) \mathbf{n}_k = \boldsymbol{\Lambda}_k \mathbf{X}_k + \mathbf{n}_k', \tag{6}$$

where $\mathbf{X}_k = [X_{k,1}, X_{k,2}, ..., X_{k,K_{min}}]^T$, $\mathbf{Y}_k = [Y_{k,1}, Y_{k,2}, ..., Y_{k,K_{min}}]^T$, and $\mathbf{n}_k' = \mathbf{U}_k^H \mathbf{n}_k$ is a white Gaussian noise vector with elements of equal variances. As a consequence of the transmit precoding and receive-shaping shown in Fig. 6, each channel matrix $\mathbf{H}_k$ is decomposed into K_{min} uncorrelated parallel SISO sub-channels, and the total number of the available sub-channels using N subcarriers is $K_{min} \times N$. The overall transmitted MIMO-OFDM symbol using all subcarriers over all the uncorrelated channels can be written as

$$\begin{bmatrix} \mathbf{Y}_1 \\ \mathbf{Y}_2 \\ \vdots \\ \mathbf{Y}_N \end{bmatrix} = \begin{bmatrix} \boldsymbol{\Lambda}_1 & \mathbf{0} & \cdots & \mathbf{0} \\ 0 & \boldsymbol{\Lambda}_1 & \cdots & \mathbf{0} \\ \vdots & \vdots & \vdots & \vdots \\ \boldsymbol{\Lambda} & 0 & \cdots & \end{bmatrix}_N \begin{bmatrix} \mathbf{X}_1 \\ \mathbf{X}_2 \\ \vdots \\ \mathbf{X}_N \end{bmatrix} + \begin{bmatrix} \mathbf{n}_1' \\ \mathbf{n}_2' \\ \vdots \\ \mathbf{n}_N' \end{bmatrix}, \tag{7}$$

and in more compact form as

$$\mathbf{Y} = \rangle \mathbf{X} + \mathbf{n}. \tag{8}$$

The links S-R, R-D and S-D, in the cooperative relay-based system shown in Fig. 1, are associated with singular value matrices $\mathbf{\Lambda}_{SR}$, $\mathbf{\Lambda}_{RD}$, and $\mathbf{\Lambda}_{SD}$, respectively. In this chapter, the $K_{min} \times N$ uncorrelated sub-channels are optimally divided between the direct link (i.e., S-D) and the relay link (i.e., S-R-D). We define $\mathbf{I}_{SD}$ and $\mathbf{I}_{SRD}$ to be $(N \times K_{min}) \times (N \times K_{min})$ assignment diagonal matrices for the direct and the relay links, respectively. The diagonal elements of these matrices take the value of one if the corresponding sub-channel is assigned to be used over that link and zero otherwise.

The channel matrix $\mathbf{H}$ is part of a spatial multiplexing system where the m_{th} data symbol d_m is directly transmitted on the m_{th} transmit antenna. At a given instant, this leads to the well-known base-band model

$$\mathbf{r} = \mathbf{H}\mathbf{d} + \mathbf{w}, \tag{9}$$

with the $M_t \times 1$ transmit vector $\mathbf{d} = (d_1, d_2, , d_{M_t})^T$; the $M_r \times M_t$ channel matrix $\mathbf{H}$; the $M_r \times 1$ received vector $\mathbf{r} = (r_1, r_2, , r_{M_r})^T$; and the $M_r \times 1$ noise vector $\mathbf{w} = (w_1, w_2, , w_{M_r})^T$.

In linear equalization-based detection, an estimate of the transmitted data vector $\mathbf{d}$ is formed as $\mathbf{y} = \mathbf{G}\mathbf{r}$ with an equalization matrix $\mathbf{G}$. The detected data vector is then obtained as $\hat{\mathbf{d}} = Q\{\mathbf{y}\}$, where $Q\{.\}$ denotes componentwise quantization.

For the zero-forcing (ZF) equalizer, $\mathbf{G}$ is given by the pseudo-inverse (Golub, 1996) of $\mathbf{H}$, i.e., $\mathbf{G} = \mathbf{H}^{\#} = (\mathbf{H}\mathbf{H}^H)^{-1}\mathbf{H}^H$. Thus, the result of ZF equalization is

$$\mathbf{y}_{ZF} = \mathbf{H}^{\#}\mathbf{r} = (\mathbf{H}^H\mathbf{H})^{-1}\mathbf{H}^H\mathbf{r} = \mathbf{d} + \tilde{\mathbf{w}}, \tag{10}$$

which is the transmitted data vector $\mathbf{d}$ corrupted by the transformed noise $\tilde{\mathbf{w}} = \mathbf{H}^{\#}\mathbf{w}$. This means that the interference caused by the channel $\mathbf{H}$ is completely removed (i.e., forced to zero). However, in general, the transformed noise $\tilde{\mathbf{w}} = \mathbf{H}^{\#}\mathbf{w}$ is larger than $\mathbf{w}$ (i.e., noise enhancement).

The noise enhancement effect in the ZF equalizer can be reduced by using the minimum mean-square error (MMSE) equalizer. The equalization matrix in MMSE can be given as $\mathbf{G} = (\mathbf{H}^H\mathbf{H} + \sigma_w^2\mathbf{I})^{-1}\mathbf{H}^H$ which minimize the mean-square error $E\{\mathbf{H}\mathbf{G}\mathbf{r} - \mathbf{d}\mathbf{H}^2\}$ (Kay, 1993). Thus, the result of MMSE equalization is

$$\mathbf{y}_{MMSE} = (\mathbf{H}^H\mathbf{H} + \sigma_w^2\mathbf{I})^{-1}\mathbf{H}^H\mathbf{r} + \mathbf{G}\mathbf{n}. \tag{11}$$

This means that MMSE will prevent noise enhancement, which is the case in ZF, but on the other hand, will prevent $\mathbf{y}$ from growing too large. While ZF or MMSE equalization alone does not, in general, imply a loss of information (i.e., an optimal detector could still be based on $\mathbf{y}_{ZF}$ or y_{MMSE}), the subsequent componentwise quantization of $\mathbf{y}_{ZF}$ or $\mathbf{y}_{MMSE}$ is suboptimal since it does not take into account the correlation of the components of the transformed noise $\tilde{\mathbf{w}}$ (Zheng, 2003).

MIMO-OFDM Cooperative Division Protocol

The idea of MIMO-OFDM cooperative division protocol in a relay-based system can be descried as shown in Fig. 6. In this figure, four OFDM-symbols are transmitted using $K_{min} \times N$ sub-channels directly from S to D and indirectly through R. According to the CSI of $\mathbf{H}_{SD}$, $\mathbf{H}_{SR}$, and $\mathbf{H}_{RD}$, the available $K_{min} \times N$ sub-channels are divided into two sets: the direct transmission set ($\{f_S\}$) and the relay transmission set ($\{f_R\}$). The direct transmission set is used to send the first part ($P1$) of the transmitted OFDM-symbol (i.e., $X_{t_u,P1}$, where u is the time slot index). The second part ($P2$) of the transmitted OFDM-symbol (i.e., $X_{t_u,P2}$) is sent using the relay transmission subset. The procedural algorithm of the protocol starts when R and D receive X_{t_u} where the first set of symbols $\mathbf{X}_{t_u,P1} = \mathbf{I}_{SD}\mathbf{X}_{t_u}$ is sent using the $\{f_S\}$ subset. At the same time, the relay makes use of the other set of sub-channels $\{f_R\}$ to receive and relay (in the next time slot) the second part of symbols $\mathbf{X}_{t_u,P2} = \mathbf{I}_{SRD}\mathbf{X}_{t_u}$. The source needs two time slots period to send one OFDM-symbol, one part directly and the second part by relaying. Meanwhile, D receives the first part using $\{f_S\}$ subset and gets the relayed part of the second part of the previous time slot(i.e, $\mathbf{X}_{t_{u-1},P2}$) using $\{f_R\}$ subset. It is assumed that the relay can send and receive symbols over the same sub-channels set $\{f_R\}$, which can be achieved by using separable polarization system. The number of received OFDM-symbols using T time slots is $T-1$, as is clear from Fig. 6.

Problem Formulation

The problem formulation starts by analyzing the MIMO-OFDM system according to the OFDM system prospective. Here we denote $\mathbf{Y}_j^{(o)}$ to be the j_{th} OFDM-symbol that consists of N modulated symbols, $j = 1, 2, ..., K_{min}$, and (o) is used for OFDM formulation. If we decompose Y_i and X_i, where $i = 1, 2, ..., N$, then we can write

$$\mathbf{Y}_i = \left[Y_{i,1}, Y_{i,2}, ..., Y_{i,K_{min}} \right]^T, \tag{12}$$

$$\mathbf{X}_i = \left[X_{i,1}, X_{i,2}, ..., X_{i,K_{min}} \right]^T; \tag{13}$$

then $\mathbf{Y}_j^{(o)}$ and $\mathbf{X}_j^{(o)}$ are given as

$$\mathbf{Y}_j^{(o)} = \left[Y_{1,j}, Y_{2,j}, ..., Y_{N,j} \right]^T = \left[\mathbf{Y}_1(j), \mathbf{Y}_2(j), ..., \mathbf{Y}_N(j) \right]^T, \tag{14}$$

$$\mathbf{X}_j^{(o)} = \left[X_{1,j}, X_{2,j}, ..., X_{N,j} \right]^T = \left[\mathbf{X}_1(j)\mathbf{X}_2(j), ..., \mathbf{X}_N(j) \right]^T. \tag{15}$$

Define the transformation matrix $\mathbf{e}_j$, that transforms $\mathbf{Y}$ and $\mathbf{X}$ (as given in (7) and (8)) into $\mathbf{Y}_j^{(o)}$ and $\mathbf{X}_j^{(o)}$, as follows:

$$\mathbf{Y}_j^{(o)} = \mathbf{e}_j \mathbf{Y}, \tag{16}$$

$$\mathbf{X}_j^{(o)} = \mathbf{e}_j \mathbf{X}. \tag{17}$$

After careful observation of the above equations, we can express $\mathbf{Y}$ and $\mathbf{X}$ in terms of $\mathbf{Y}_j^{(o)}$ and $\mathbf{X}_j^{(o)}$ as

$$\mathbf{Y} = \sum_{j=1}^{K_{min}} \mathbf{e}_j^T \mathbf{Y}_j^{(o)}, \tag{18}$$

$$\mathbf{X} = \sum_{j=1}^{K_{min}} \mathbf{e}_j^T \mathbf{X}_j^{(o)}. \tag{19}$$

The discrete fourier transform (DFT) operation on $\mathbf{Y}_j^{(o)}$ and $\mathbf{X}_j^{(o)}$ can be represented by matrix multiplication as follows:

$$\mathbf{Y}_j^{(o)} = \mathbf{Q}\mathbf{y}_j^{(o)}, \tag{20}$$

$$\mathbf{X}_j^{(o)} = \mathbf{Q}\mathbf{x}_j^{(o)}, \tag{21}$$

where $\mathbf{Q}$ is $N \times N$ matrix given by

$$\mathbf{Q} = \frac{1}{\sqrt{N}} \begin{bmatrix} 1 & 1 & 1 & \dots & 1 \\ 1 & W_N & W_N^2 & \dots & W_N^{N-1} \\ \vdots & \vdots & \vdots & \ddots & \vdots \\ 1 & W_N^{N-1} & W_N^{2(N-1)} & \dots & W_N^{(N-1)^2} \end{bmatrix}, \tag{22}$$

where $W_N = e^{-j\frac{2\pi}{N}}$ and $\mathbf{Q}^{-1} = \mathbf{Q}^H$. The time domain overall received vector ($\mathbf{y}$) can be derived as follows:

$$\mathbf{Y}_j^{(o)} = \mathbf{\Lambda}_j^{(o)} \mathbf{X}_j^{(o)} \Rightarrow \mathbf{y}_j^{(o)} = \mathbf{Q}^H \mathbf{\Lambda}_j^{(o)} \mathbf{Q}\mathbf{x}_j^{(o)}, \tag{23}$$

$$\mathbf{y} = \sum_{j=1}^{K_{min}} \mathbf{e}_j^T \mathbf{y}_j^{(o)} = \sum_{j=1}^{K_{min}} \mathbf{e}_j^T \mathbf{Q}^H \rangle_j^{(o)} \mathbf{Q}\mathbf{x}_j^{(o)} = \left[\sum_{j=1}^{K_{min}} \mathbf{e}_j^T \mathbf{Q}^H \rangle_j^{(o)} \mathbf{Q}\mathbf{e}_j \right] x. \tag{24}$$

By referring to Fig. 6 that $\rangle_j^{(o)} = e_j \rangle e_j^T$, then the overall received time domain MIMO-OFDM symbol ($\mathbf{y}$) can be given as

$$\mathbf{y} = \left[\sum_{j=1}^{K_{min}} \mathbf{e}_j^T \mathbf{Q}^H e_j \rangle e_j^T \mathbf{Q}\mathbf{e}_j \right] x = \left[\sum_{j=1}^{K_{min}} \mathbf{B}_j^H \rangle \mathbf{B}_j \right] x = \tilde{\mathbf{H}}(\rangle) x, \tag{25}$$

where $\mathbf{B}_j = \mathbf{e}_j^T \mathbf{Q} \mathbf{e}_j$, and $\tilde{\mathbf{H}}(\boldsymbol{\Lambda})$ is the effective time domain channel matrix. The noise effect is omitted because unitary transformation does not affect the statistical properties of Gaussian noise.

Optimum Sub-Channel Assignment

Mathematically, the MIMO-OFDM symbol ($\mathbf{X}$) is divided into two parts, $\mathbf{X}_{P1}$ and $\mathbf{X}_{P2}$, that are to be sent using direct link (S-D) relay link (S-R-D), respectively. The the received symbols $\mathbf{Y}_{P1}$ and $\mathbf{Y}_{P2}$ can be given as

$$\mathbf{Y}_{P1} = \boldsymbol{\Lambda}_{SD}\mathbf{I}_{SD}\,\mathbf{X} \tag{26}$$

$$\mathbf{Y}_{P2} = \boldsymbol{\Lambda}_{SR}\boldsymbol{\Lambda}_{RD}\mathbf{I}_{SRD}\mathbf{X} \tag{27}$$

where these equations are valid if $\mathbf{H}_{SR} = \mathbf{H}_{SD}$, which can be achieved by applying a signal processing technique at the relay (e.g., for square matrices we multiply the received signal at the relay by $\mathbf{H}_{SD}\mathbf{H}_{SR}^{-1}$) . Under the assumption that each sub-channel is assigned either to direct link (S-D) or the relay link (S-R-D), then we can describe it mathematically as $\mathbf{I}_{SD} + \mathbf{I}_{SRD} = \mathbf{I}$, where $\mathbf{I}$ is $(N \times K_{min}) \times (N \times K_{min})$ identity matrix. The overall received symbol $\mathbf{Y}$ may be given as

$$\mathbf{Y} = \mathbf{Y}_{P1} + \mathbf{Y}_{P1} = (\boldsymbol{\Lambda}_{SD}\mathbf{I}_{SD} + \boldsymbol{\Lambda}_{SR}\boldsymbol{\Lambda}_{RD}\mathbf{I}_{SRD})\,\mathbf{X} \tag{28}$$

Equivalently, using time domain representation in (25), the overall time domain received signal $\mathbf{y}$ can be written as

$$\mathbf{y} = \left[\sum_{j=1}^{K_{min}}\mathbf{B}_j^H \left(\boldsymbol{\gt}_{SR}\boldsymbol{\gt}_{RD}\mathbf{I}_{SRD} + \boldsymbol{\gt}_{SD}\mathbf{I}_{SD}\right)\mathbf{B}_j\right]x = \left[\sum_{j=1}^{K_{min}}\mathbf{B}_j^H \boldsymbol{\gt}_{eq}\mathbf{B}_j\right]x = \tilde{\mathbf{H}}(\boldsymbol{\gt}_{eq})x, \tag{29}$$

where $\boldsymbol{\gt}_{eq} = \boldsymbol{\gt}_{SR}\boldsymbol{\gt}_{RD}\mathbf{I}_{SRD} + \boldsymbol{\gt}_{SD}\mathbf{I}_{SD}$ is the overall effective eigenvalue matrix. As a consequence, the optimization problem can be formulated as follows:

$$\arg\max_{\{\mathbf{I}_{SRD},\mathbf{I}_{SD}\}} \left[\sum_{j=1}^{K_{min}}\mathbf{B}_j^H \left(\boldsymbol{\gt}_{SR}\boldsymbol{\gt}_{RD}\mathbf{I}_{SRD} + \boldsymbol{\gt}_{SD}\mathbf{I}_{SD}\right)\mathbf{B}_j\right]$$

Subject to:

It is clear from the above formulation that such optimization problem can be solved to maximize the received vector $\mathbf{y}$. The following simple procedure can be used to get the optimal channel assignment:

- Let $\mathbf{I}_{SD} = \mathbf{I}_{SRD}$ = zero matrix

- $" = \rangle_{SD} - \rangle_{SR}, \rangle_{RD}$
- If $"(i,i) > 0 \Rightarrow \mathbf{I}_{SD}(i,i) = 1$, $i = 1, 2, .., K_{min} \times N$
- $\mathbf{I}_{SRD} = \mathbf{I} - \mathbf{I}_{SD}$

where I is an $(K_{min} \times N) \times (K_{min} \times N)$ identity matrix. The resultant assignment matrices (i.e., $\mathbf{I}_{SD}, \mathbf{I}_{SRD}$) (by applying the previous procedure) maximize the components of the received vector $\mathbf{y}$. Particularly, in (30), this kind of optimization requires a reliable channel estimation. The next section introduces a feasible signaling model to utilize such channel estimation.

SIGNALING MODEL

The signaling model can be considered modification on the signaling model given in (Zhang, 2005). The relay (R) and the source (S) obtain the transmission parameters which are the assignment matrices (i.e., $\mathbf{I}_{SD}, \mathbf{I}_{SRD}$) and the bit/power allocations for S and R, respectively. Knowledge of CSI at both R and S, as well as the destination (D), is essential for the implementation of the sub-channel division MIMO-OFDM relay-based cooperative system. Fig. 7 illustrates the signaling model in a TDD (time division duplex) system, assuming the transmission time slots are of equal length and are denoted by t_o. The relay predicts the channel between the source and the relay $\mathbf{H}_{SR}(t - t_o)$, based on the time slot received at time $t - 3t_o$. At the same time, the destination predicts the channel between the source and itself (i.e., $\mathbf{H}_{SD}(t - t_o)$), based on the same time slot, which is received at time $t - 3t_o$. At the next time slot, which is $t - 2t_o$, the destination predicts $\mathbf{H}_{RD}$. Then, the source sends the transmission parameters which are shown in Fig. 7 as $\Psi(t - t_o)$, that contain information about sub-channels, power and bits allocation for both R and S. Those transmission parameters can be sent via control channel. If the channel can be considered as quasi-static, and the variation is relatively slow for at least $3t_o$, then the quality of channel prediction can be accepted.

SIMULATION RESULTS

The cooperative relay-based MIMO-OFDM wireless system is assumed as shown in Fig. 1. The entire system is considered as a discrete-time system where pulse-shaping and matched-filtering are not considered. The OFDM system parameters are given in Table 1. Six different square M-QAM signal constellations are used with orders $0, 2, 4, 6, 8$ (i.e., number of bits per symbol), where 0 means no transmission. In this simulation, identically independent distributed (iid) Rayleigh block-fading channels with exponential power delay profile $[1, e^{-1}, e^{-2}]$ for three modulated symbols are assumed. The channel coherence time is assumed to be 50 OFDM-symbols. It is assumed that the MIMO-OFDM system parameters are identical for the three links (i.e., S-D, S-R, and R-D) and are given in Table 1. Adaptive bit and power loading are performed as given in (De Souza, 1999) and are compared with the case of fixed bit loading where the total number of bits per sub-channel are fixed but the power is adaptively changing based on the same adaptive power loading technique. Throughout the simulation and for fair comparison, the system is simulated to send 100 bits/OFDM-symbol/antenna. Furthermore, 5×10^4 simulation runs were conducted for each point (i.e., average BER, average Eb/No [dB]).

The average BER comparisons of fixed (i.e. fixed bit loading) and an adaptive MIMO-OFDM system

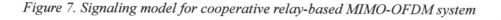

Figure 7. Signaling model for cooperative relay-based MIMO-OFDM system

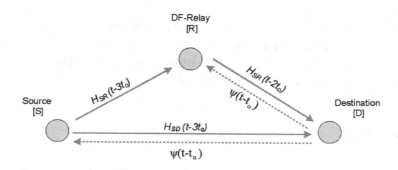

of 2x2 and 4x4 MIMO are given in Fig. 8. It is clear from Fig. 8 that the advantage of adaptive loading in MIMO-OFDM increases with MIMO system order for the case of independent Rayleigh fading channels. For example, at 10^{-1} average BER, the average Eb/No improvement using adaptive loading was 4.9 dB and 6 dB for 2x2 and 4x4 MIMO, respectively. Furthermore, the improvement is expected to be more for a higher Eb/No range.

Fig. 9 compares the average BER of cooperative relay-based using optimal sub-channel division and non-cooperative MIMO-OFDM for 1x1 (i.e., SISO), 2x2 and 4x4 MIMO systems. The MIMO-OFDM cooperative division protocol is applied to generate the cooperative curves in Fig. 9. the advantage of cooperation is well-noticed by improving the system performance where the improvement increases with the increase of average Eb/No. For example, to have 10^{-2} average BER in a 4x4 MIMO, we need 7 dB more than that required without cooperation. Moreover, the advantage of cooperation increases with the increase of the MIMO system order.

The average BER comparisons of the noncooperative and cooperative MIMO-OFDM systems of 2x2 and 4x4 MIMO using ZF and MMSE detection techniques are given in Fig. 10 and Fig. 11, respectively. The advantage of using ZF detection is obvious for high Eb/No values in both cooperative and noncooperative schemes. But due to the enhancement of the received signal in the cooperative scheme, both MMSE and ZF have almost the same performance in the range for Eb/No< 20 dB. As a final demonstration of the behavior of the proposed cooperative system, the available sub-channels are split permanently between the direct link and the relay link. The BER comparisons of split and cooperative MIMO-OFDM systems of 2x2 and 4x4 MIMO using ZF detection technique are given in Fig. 12. Moreover, random splitting of sub-channels between the direct link and the relay link using a uniform random variable of each channel (here, the uniform random variable is defined from -1/2 to 1/2, and whenever the generated sample > 0 the sub-channel is assigned to the direct link and whenever the sample $<= 0$, the sub-channel is assigned to the relay link) is also considered in Fig. 13. Both 12 and 13 show the superiority of the proposed optimum cooperative scheme in terms of average BER.

Figure 8. Average BER comparison between fixed and adaptive loading MIMO-OFDM systems for 2x2 and 4x4 MIMO.

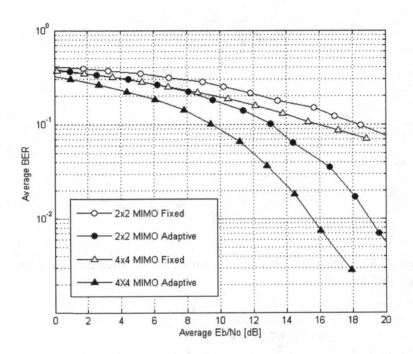

Figure 9. Average BER comparison between cooperative relay-based and non-cooperative MIMO-OFDM systems for 1x1, 2x2, 4x4 MIMO.

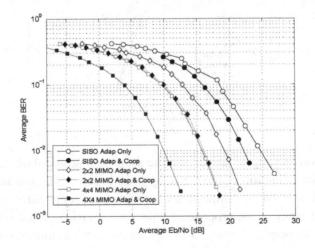

CONCLUSION

In this chapter, cooperative sub-channel division protocol for the relay-based MIMO-OFDM system

Figure 10. Average BER comparison between ZF and MMSE detection techniques for 1x1, 2x2, 4x4 MIMO.

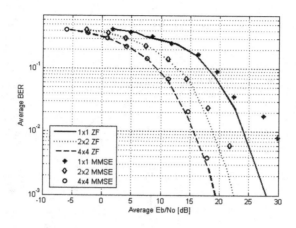

Figure 11. Average BER comparison between cooperative relay-based ZF and MMSE detection techniques for 1x1, 2x2, 4x4 MIMO.

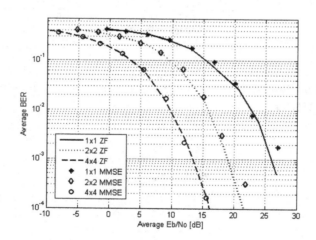

with adaptive bit and power loading is suggested. The suggested protocol divides the available MIMO-OFDM sub-channels between the direct and the relay links to send the information from source to destination. The sub-channel division is optimized to maximize the overall MIMO-OFDM received signal. Our results show the superiority of the proposed cooperative division protocol in improving the system performance in terms of decreased average BER. In addition, the simulation shows that the involved complexity in precoding and matrix shaping boosts up the performance, and increasing the order of the MIMO system results in increased system performance. These preliminary results can contribute to the development of MIMO-OFDM standards to improve the the system throughput and extend the range by utilizing the suggested cooperative division protocol.

Figure 12. Average BER comparison between optimum sub-channel division and split protocol for 1x1, 2x2, 4x4 MIMO.

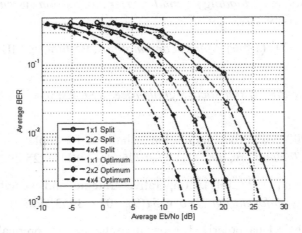

Figure 13. Average BER comparison between optimum sub-channel division and random protocol for 1x1, 2x2, 4x4 MIMO.

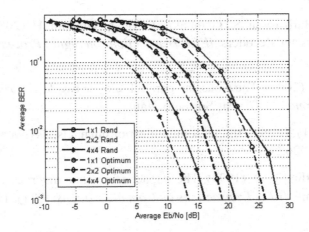

REFERENCES

Abualhaol, I., & Matalgah, M. (2008). Subchannel-division adaptive resource allocation technique for cooperative relay-based MIMO-OFDM wireless communication systems. In *Proceedings IEEE Wireless Communications and Networking Conference,* Las Vegas, NV (pp. 1002-1007).

Chow, P. (1995). Practical discrete multitone transceiver loading algorithm for data transmission over spectrally shaped channels. *IEEE Transactions on Communications, 43*(2), 773–775. doi:10.1109/26.380108

Cover, T., & Gamal, A. (1979). Capacity theorems for the relay channel. *IEEE Transaction on Informormation Theory, 25*(5), 572–584. doi:10.1109/TIT.1979.1056084

De Souza, J. C. (1999). *Discrete bit loading for multicarrier modulation systems*. Unpublished doctoral dissertation, Stanford University, CA.

Golub, G. H., & Van Loan, C. F. (1996). *Matrix computations*. Baltimore, MD: Johns Hopkins University Press.

Kay, S. M. (1993). *Fundamentals of statistical signal processing: Estimation theory*. Englewood Cliffs, NJ: Prentice Hall.

Keller, T., & Hanzo, L. (2000). Adaptive modulation techniques for duplex OFDM transmission. *IEEE Transactions on Vehicular Technology, 49*(5), 1893–1906. doi:10.1109/25.892592

Kramer, G., Gastpar, M., & Gupta, P. (2005). Cooperative strategies and capacity theorems for relay networks. *IEEE Transactions on Information Theory, 51*(9), 3037–3063. doi:10.1109/TIT.2005.853304

Ordano, L., & Tallone, F. (1997). Dual polarized propagation channel: Theoretical model and experimental results. In *Proceeding 10th Int. Conf. Antennas and Propagation* (Vol. 2, pp. 363–366).

Sampath, H., Stoica, P., & Paulraj, A. (2001). Generalized linear precoder and decoder design for MIMO channels using the weighted MMSE criterion. *IEEE Transactions on Communications, 49*(12), 2198–2206. doi:10.1109/26.974266

Scaglione, A., Stoica, P., Barbarossa, S., Giannakis, G. B., & Sampath, H. (2002). Optimal designs for space-time linear precoders and decoders. *IEEE Transactions on Signal Processing, 50*(5), 1051–1064. doi:10.1109/78.995062

Schein, B., & Gallager, R. (2000). The Gaussian parallel relay network. In *Proceedings IEEE International Symposium on Information Theory,* Sorrento, Italy (p. 22).

Van der Meulen, E. C. (1971). Three-terminal communication channels. *Advances in Applied Probability, 3*, 120–154. doi:10.2307/1426331

Winters, J., Salz, J., & Gitlin, R. D. (1994). The impact of antenna diversity on the capacity of wireless communication systems. *IEEE Transactions on Communications, 42*(1), 1740–1751. doi:10.1109/TCOMM.1994.582882

Zhang, Y. J., & Ben Lataief, K. B. (2005). An efficient resource-allocation scheme for spatial multiuser access in MIMO-OFDM systems. *IEEE Transactions on Communications, 53*(1), 107–116. doi:10.1109/TCOMM.2004.840666

Zhang, Y. J., & Letaief, K. B. (2003). Optimizing power and resource management for multiuser MIMO-OFDM systems. *IEEE Global Telecommunications Conference,* San Francisco, CA (Vol. 1, pp. 179-183).

Zheng, L., & Tse, D. (2003). Diversity and multiplexing: A fundamental tradeoff in multiple antenna channels. *IEEE Transactions on Information Theory, 49*(5), 1073–1096. doi:10.1109/TIT.2003.810646

Chapter 15
Single–Carrier Frequency Domain Equalization for Broadband Cooperative Communications

Tae-Won Yune
POSTECH, Republic of Korea

Dae-Young Seol
POSTECH, Republic of Korea

Dongsik Kim
POSTECH, Republic of Korea

Gi-Hong Im
POSTECH, Republic of Korea

ABSTRACT

Cooperative diversity is an effective technique to combat the fading phenomena in wireless communications without additional complexity of multiple antennas. Multiple terminals in the network form a virtual antenna array in a distributed fashion. Even though each of them is equipped with only one antenna, spatial diversity gain can be achieved through cooperation. In this chapter, we discuss relay-assisted single carrier transmissions extending conventional transmit diversity schemes. We focus on distributed space-frequency block coded single carrier transmission, in order to operate over fast fading channels. A pilot design technique is also discussed for channel estimation of this single carrier cooperative system, which shows better channel tracking performance than conventional block-type channel estimations. In addition, spectral efficient cooperative diversity protocols are presented, where the users access a relay simultaneously or transmit superposed data blocks. Interference from the other user is effectively removed by using an iterative detection technique.

DOI: 10.4018/978-1-60566-665-5.ch015

INTRODUCTION

Single carrier frequency-domain equalization (SC-FDE) has similar performance and essentially the same overall complexity as orthogonal frequency division multiplexing (OFDM) (Falconer, 2002; Prasad, 2004). This SC-FDE is preferred for the uplink transmission, because its transmit structure at mobile equipment is very simple. Moreover, the peak-to-average power ratio (PAPR) of transmitted sequences of SC-FDE is much lower than that of OFDM, and thus the required dynamic range of power amplifier is smaller and the battery life time is longer than that of OFDM (Kwon, in press). SC frequency division multiple access (SC-FDMA), which is a combination of FDMA and SC-FDE, is currently a strong candidate for the uplink multiple access scheme in 3rd Generation Partnership Project Long Term Evolution (3GPP-LTE) (Myung, 2006).

Space-time coding is a communication technique for wireless systems, which realizes spatial diversity by introducing temporal and spatial correlations into the signals transmitted from different antennas (Alamouti, 1998; Agrawal, 1998; Al-Dhahir, 2001; Choi, 2007). Most significantly, Alamouti (1998) proposed a simple space-time block code (STBC) for two transmit antennas, guaranteeing full spatial diversity and full rate over frequency-flat channels. Since Alamouti coded system does not require any channel state feedback information, it has been widely adopted in wireless standards such as IEEE 802.11n WLAN and 802.16e mobile WiMAX. STBC combined with SC-FDE was presented by Al-Dhahir (2001), which is based on the Alamouti scheme for frequency selective channels. The major drawback of the STBC SC-FDE, however, is that the channel is assumed to be constant for two consecutive symbol intervals (Jang, 2006; Bauch, 2003). Therefore, the system breaks down when used in a time-varying mobile environment. In order to mitigate the fast fading distortion caused by high-speed mobility, Jang *et al.* (2006) recently proposed space-frequency block code (SFBC) combined with SC-FDE. Although the SFBC SC-FDE achieves spatial diversity gain over fast fading channels, it introduces 3dB PAPR increase over two transmit antennas and additional computational complexity at the transmitter.

Employing multiple antennas in uplink communications is restricted, due to the limitation of size and complexity of the mobile equipment. Cooperative diversity overcomes these problems without additional complexity of multiple antennas and provides an effective means of improving spectral and power efficiencies (Sendonaris, 2003; Pabst, 2004). Recently, relay and cooperative networks have been proposed for applications of several emerging systems. IEEE 802.16j (LMSC, 2008) is a developing standard for 802.16-based multihop networks. The multihop relay is a promising solution to expand coverage and to enhance throughput and system capacity. It is expected that the complexity of relay stations will be considerably less than the complexity of legacy base stations. The gains in coverage and throughput can be leveraged to reduce total deployment cost. Relaying is also considered in IEEE 802.11s, a developing mesh networking standard. There have been extensive works on cooperative diversity, but most of them assume frequency-flat fading environment. For broadband wireless applications, frequency-selective channels must be considered. Mheidat *et al.* (2007) extended conventional STBC SC-FDE in a distributed fashion, so called D-STBC SC-FDE, for practical implementation of cooperative networks. They have considered all underlying links experience frequency selective fading. For fast fading channels, the conventional SFBC SC-FDE can also be extended in a distributed fashion, but still holds the disadvantages of increased PAPR and computational complexity. In practice, the relay works in the half-duplex mode, i.e., the relay cannot simultaneously transmit and receive in the same frequency channel. It means that the transmission of one information symbol from source to destination requires two channel uses (Rankov, 2007).

Figure 1. TDMA-based cooperative protocols for (a) D-STBC and (b) D-SFBC.

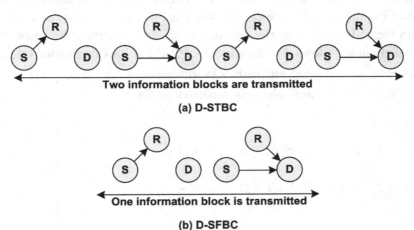

In this chapter, we discuss relay-assisted single carrier transmissions extending conventional transmit diversity schemes over frequency-selective fading channels. In particular, we present a new D-SFBC single carrier system over fast fading channels. The D-SFBC SC-FDE achieves the diversity gain without any increase of PAPR and computational complexity at the mobile equipment, while only the transmitted sequences of relay have 3 dB PAPR increase. It is desirable especially for the infrastructure-based relay scenario. A pilot design technique is also discussed for channel estimation of D-SFBC SC-FDE. In addition, spectral efficient cooperative diversity protocols are followed, which compensate the loss caused by the half-duplex constraint.

Relay-Assisted Single-Carrier Transmission over fast Fading Channels

Distributed STBC and SFBC SC-FDEs

Mheidat *et al.* (2007) extended an STBC SC-FDE system in a distributed fashion with single relay. They have considered all underlying links experience frequency selective fading, and analyzed maximum achievable diversity order through the derivation of the pariwise error probability expression. TDMA-based cooperative protocol is used as shown in Fig. 1(a). In the system, the source communicates with the relay during the first signaling interval, while the destination does not receive the direct signals from the source. Then, both the relay and source communicate with the destination in the second signaling interval, using a protocol proposed by Nabar *et al.* (2004). This protocol is motivated by the practical consideration that the destination terminal may be engaged in data transmission to another terminal during this period. Since STBC system codes across the consecutive data blocks, the D-STBC SC-FDE requires four time slots for the transmission of two information blocks.

The performance of the D-STBC SC-FDE deteriorates in a fast fading environment, since the channels are assumed to be constant for four consecutive blocks (Mheidat, 2007). In order to overcome this problem, the conventional SFBC SC-FDE (Jang, 2006) can be extended in a distributed fashion. The SFBC system codes across two transmit antennas and over two adjacent subcarriers instead of two consecutive block intervals. Thereby, the SFBC system gives robust performance over fast fading channels. Unlike an OFDM, the SFBC, however, cannot be directly applicable to a single carrier system (Jang, 2006). Note that the transmit sequences of single carrier system are processed in time domain, not in frequency domain. To obtain the SFBC single carrier signal, the transmit sequences for respective antennas are processed in time domain as follows. Firstly, a block of QAM or PSK modulated information symbols, $\mathbf{x} = [x(0),...,x(N-1)]^T$, is separated into two blocks with the length $N/2$, i.e., $[x(0),...,x(N/2-1)]^T$ and $[x(N/2),..., x(N-1)]^T$. Each block is repeated to form $\mathbf{x}^{h1} = [x^{h1}(0),...,x^{h1}(N-1)]^T$ and $\mathbf{x}^{h2} = [x^{h2}(0),..., x^{h2}(N-1)]^T$ as

$$\mathbf{x}^{h1} = cat([x(0),...,x(N/2-1)]^T, [x(0),...,x(N/2-1)]^T),$$

$$\mathbf{x}^{h2} = cat([x(N/2),...,x(N-1)]^T, [x(N/2),...,x(N-1)]^T) \tag{1}$$

where $cat(\mathbf{a}, \mathbf{b})$ denotes the vertical concatenation of two column vectors, $\mathbf{a}$ and $\mathbf{b}$. The frequency domain representations of $\mathbf{x}^{h1}$ and $\mathbf{x}^{h2}$ can be written as $\mathbf{X}^{h1} = \mathbf{W}\mathbf{x}^{h1}$ and $\mathbf{X}^{h2} = \mathbf{W}\mathbf{x}^{h2}$, where $\mathbf{W}$ is an $N \times N$ orthonormal discrete Fourier transform (DFT) matrix. Since $\mathbf{x}^{h1}$ and $\mathbf{x}^{h2}$ are formed by a concatenation of identical column vectors, odd components of $\mathbf{X}^{h1}$ and $\mathbf{X}^{h2}$ are zero, i.e.,

$$X^{h1}(2k+1) = X^{h2}(2k+1) = 0, \qquad 0 \le k \le \frac{N}{2} - 1. \tag{2}$$

The transmit sequence of each antenna in the frequency domain, $\mathbf{X}_i$, i=1,2, is formed as

$$X_1(k) = \frac{1}{\sqrt{2}}(X^{h1}(k) + X^{h2}(k-1)_N),$$

$$X_2(k) = \frac{1}{\sqrt{2}}(-X^{h2*}(k) + X^{h1*}(k-1)_N) \tag{3}$$

where $(\cdot)^*$ denotes the complex-conjugate operation. $(n)_N$ represents $n \pmod N$. Note from (2) and (3) that the adjacent subcarriers of each transmit antenna are SFBC coded as follows

$$\begin{bmatrix} X_1(2l) & X_1(2l+1) \\ X_2(2l) & X_2(2l+1) \end{bmatrix} = \frac{1}{\sqrt{2}} \begin{bmatrix} X^{h1}(2l) & X^{h2}(2l) \\ -X^{h2*}(2l) & X^{h1*}(2l) \end{bmatrix}. \tag{4}$$

Using the DFT symmetry and frequency shift properties (Oppenheim, 1989), i.e, $x^*(-n)_N \Leftrightarrow X^*(k)$, and $W_N^n x(n) \Leftrightarrow X(k+1)$, $n, k = 0, 1, ..., N-1$, the single carrier SFBC sequences of two transmit antennas, $\mathbf{x}_1 = [x_1(0),...,x_1(N-1)]^T$ and $\mathbf{x}_2 = [x_2(0),..., x_2(N-1)]^T$, are obtained in time domain as follows

$$x_1(n) = \frac{1}{\sqrt{2}}(x^{h1}(n) + W_N^{-n}x^{h2}(n)),$$

$$x_2(n) = \frac{1}{\sqrt{2}}(-x^{h2*}(-n)_N + W_N^{-n}x^{h1*}(-n)_N) \tag{5}$$

where $W_N^n = e^{-j(2\pi n/N)}$. In order to mitigate the fast fading distortion caused by high-speed mobility, the conventional SFBC SC-FDE can be extended in a distributed fashion. D-SFBC SC-FDE can complete the transmission of one information block within two time slots, while the D-STBC SC-FDE requires four time slots for the transmission of two information blocks. Although the D-SFBC SC-FDE is robust against fast fading channels, it has disadvantages relative to the D-STBC SC-FDE, in view of PAPR and computational complexity. Note from (5) that each transmit sequence of SFBC SC-FDE shows 3dB PAPR increase by adding two time domain signals, and requires N complex multiplications and additions. High PAPR requires large dynamic range of the transmit power amplifier and reduces the power efficiency, and thus the cost of mobile equipment is increased and the battery life time is decreased. To cope with these problems, we discuss a new relay-assisted D-SFBC SC-FDE for uplink transmission in the next subsection.

Efficient Implementation of D-SFBC SC-FDE

Conceptual block diagram for efficient implemention of the D-SFBC SC-FDE is shown in Fig. 2(a). The channel impulse responses (CIRs) from the transmitting node (α) to receiving node (β) is given by $\mathbf{h}_{\alpha\beta} = [h_{\alpha\beta}(0),..., h_{\alpha\beta}(L_{\alpha\beta})]^T$, where $L_{\alpha\beta}$ denotes the channel memory length and $(\cdot)^T$ is the transpose. In this chapter, subscripts S, R, and D stand for the source, relay, and destination nodes, respectively. The transmit sequence of the source (mobile equipment), $\mathbf{x}_S^1$, is equal to the information block $\mathbf{x}$, i.e., $\mathbf{x}_S^1 = \mathbf{x}$. In the first time slot, the source transmits $\mathbf{x}_S^1$ after appending a cyclic prefix (CP) with length L_{SR}, making the channel matrix circulant. At the relay, removing the CP, the received signal is given by

$$\mathbf{r}_R = \sqrt{E_{SR}}\mathbf{H}_{SR}\mathbf{x}_S^1 + \mathbf{n}_R \tag{6}$$

where $\mathbf{n}_R$ is a complex additive white Gaussian noise (AWGN) vector with each entry having a zero-mean and variance of $\sigma_n^2 / 2$ per dimension. $E_{\alpha\beta}$ represents the average energy available at the receiving node (β), and $\mathbf{H}_{\alpha\beta}$ is an $N \times N$ circulant channel matrix with entries $[\mathbf{H}_{\alpha\beta}]_{k,l} = h_{\alpha\beta}((k-l)_N)$. Path loss and shadowing effects in $\alpha \to \beta$ link are included into $E_{\alpha\beta}$ for simplicity. The received signal, $\mathbf{r}_R$, is normalized as $\tilde{\mathbf{r}}_R \triangleq \mathbf{r}_R / \sqrt{E_{SR} + \sigma_n^2}$ to ensure unit average energy, and the DFT of $\tilde{\mathbf{r}}_R$ is processed like below

$$\begin{bmatrix} X_R(2l) \\ X_R(2l+1) \end{bmatrix} = \begin{bmatrix} -\tilde{R}_R^*(2l+1) \\ \tilde{R}_R^*(2l) \end{bmatrix}, l = 0, 1, ..., \frac{N}{2} - 1 \tag{7}$$

Figure 2. Transceiver block diagram of D-SFBC SC-FDE, (a) Conceptual block diagram (b) Receiver structure.

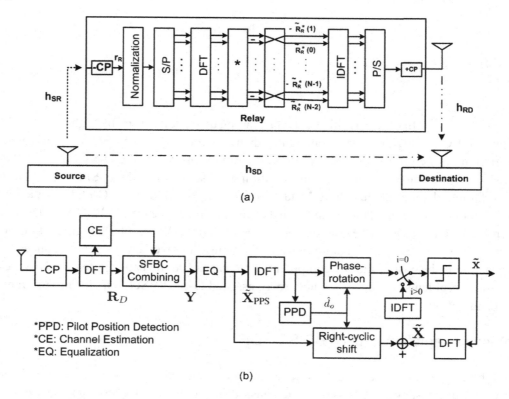

(a)

(b)

*PPD: Pilot Position Detection
*CE: Channel Estimation
*EQ: Equalization

where $\mathbf{X}_R$ is the transmit signal of the relay in frequency domain, and $\tilde{\mathbf{R}}_R = \mathbf{W}\tilde{\mathbf{r}}_R$. Changing the signs of the odd components and permutating the even and odd components can be conducted by multiplying the following matrices, $\mathbf{S}$ and $\mathbf{P}$, respectively.

$$\mathbf{S} = \mathbf{I}_{\frac{N}{2} \times \frac{N}{2}} \otimes \begin{bmatrix} 1 & 0 \\ 0 & -1 \end{bmatrix}, \quad \mathbf{P} = \mathbf{I}_{\frac{N}{2} \times \frac{N}{2}} \otimes \begin{bmatrix} 0 & 1 \\ 1 & 0 \end{bmatrix} \tag{8}$$

where $\mathbf{I}_{k \times k}$ is the $k \times k$ identity matrix, and $\otimes$ denotes the Kronecker product. Then, the transmit signal of the relay can be represented as

$$\mathbf{x}_R = \mathbf{W}^H \mathbf{P} \mathbf{S} \{\mathbf{W}\tilde{\mathbf{r}}_R\}^* = \sqrt{\frac{E_{SR}}{E_{SR} + \sigma_n^2}} \mathbf{W}^H \mathbf{P} \mathbf{S} \{\mathbf{W}\mathbf{H}_{SR}\mathbf{x}_S^1\}^* + \mathbf{n}_R^{'} \tag{9}$$

where $(\cdot)^H$ denotes the complex-conjugate transpose, and $\mathbf{W}^H$ is an inverse DFT (IDFT) matrix. In Fig. 2(a), the relay requires DFT/IDFT operations to code over adjacent subcarriers. We now give an efficient implementation method of the relay, generating the same transmit sequence without DFT/IDFT operations, by processing the received signals in time domain. The procedure of time domain processing is explained as follows.

Step 1,Conjugation: Using the DFT symmetry property, the time domain representation of $\tilde{\mathbf{R}}_R^*$ in (7) is given by

$$r_c(n) = \tilde{r}_R^*(-n)_N. \tag{10}$$

Step 2. Separation of even and odd subcarriers: To code over adjacent subcarriers, we separate the even and odd components of the $\mathbf{R}_c \ (= \tilde{\mathbf{R}}_R^*)$. Even (odd) components of the $\mathbf{R}_c$ can be easily separated by multiplying a sequence $\mathbf{D}_e \ (\mathbf{D}_o)$ in frequency domain as follows

$$R_e(k) = R_c(k)D_e(k), \qquad R_o(k) = R_c(k)D_o(k) \tag{11}$$

where $\mathbf{D}_e = [1,0,1,0,...,1,0]_{1 \times N}^T$, $\mathbf{D}_o = [0,1,0,1,...,0,1]_{1 \times N}^T$. Using the fact that the multiplication of sequences in frequency domain corresponds to the circular convolution in time domain, the time domain representation of $\mathbf{R}_e$ and $\mathbf{R}_o$ can be obtained as follows

$$r_e(n) = r_c(n) \circledast_N \delta_e(n), \qquad r_o(n) = r_c(n) \circledast_N \delta_o(n) \tag{12}$$

where $\circledast_N$ represents the N-point circular convolution, $\mathbf{d}_e = \mathbf{W}^H \mathbf{D}_e$, and $\mathbf{d}_o = \mathbf{W}^H \mathbf{D}_o$. Note that $\delta_e(n) = 0$ and $\delta_o(n) = 0$ except for $n = 0, \dfrac{N}{2}$ as follows

$$\mathbf{d}_e(n) = \begin{cases} \dfrac{1}{2} & , \quad n = 0 \\ \dfrac{1}{2} & , \quad n = \dfrac{N}{2} \\ 0 & , \quad otherwise \end{cases}, \qquad \mathbf{d}_o(n) = \begin{cases} \dfrac{1}{2} & , \quad n = 0 \\ -\dfrac{1}{2} & , \quad n = \dfrac{N}{2} \\ 0 & , \quad otherwise \end{cases}. \tag{13}$$

Therefore, instead of circular convolution, the $\mathbf{r}_e$ and $\mathbf{r}_o$ in (12) can be obtained by adding two time domain signals as follows

$$r_e(n) = \frac{1}{2}r_c(n) + \frac{1}{2}r_c(n - N/2)_N, \qquad r_o(n) = \frac{1}{2}r_c(n) - \frac{1}{2}r_c(n - N/2)_N. \tag{14}$$

Step 3. Shift and Summation of sequences: The transmit sequence of relay can be obtained from $\mathbf{r}_e$ and $\mathbf{r}_o$ by using the frequency shift property as follows

$$x_R(n) = r_e(n) \times W_N^{-n} - r_o(n) \times W_N^n. \tag{15}$$

Further, substituting (14) into (15), $\mathbf{x}_R$ can be rewritten as

$$x_R(n) = \frac{1}{2} r_c(n)(W_N^{-n} - W_N^n) + \frac{1}{2} r_c(n - N/2)_N(W_N^{-n} + W_N^n)$$

$$= r_c(n) \times j \sin(2\pi n / N) + r_c(n - N/2)_N \times \cos(2\pi n / N). \tag{16}$$

Note from (16) that we have directly obtained the transmit sequence of relay, $\mathbf{x}_R$, in the time domain, using the received signals from the source. It requires $2N$ multiplications of real by complex values, instead of DFT and IDFT operations. The overall complexity for (16) is half of the conventional SFBC (see (5)). Note that there is no additional complexity at the source node. Due to the addition of two time domain signals, $\mathbf{x}_R$ may have peak power at the sample time, $n = N(2k+1)/8, k = 0, 1, 2, 3$, where the $|\cos(2\pi n / N)|$ and $|\sin(2\pi n / N)|$ are $1/\sqrt{2}$, and the PAPR is increased by 3dB.

In the second time slot, the source and the relay transmit $\mathbf{x}_S^2$, remaining the same as $\mathbf{x}_S^1$, and $\mathbf{x}_R$ after appending a CP with length $L = \max(L_{SD}, L_{RD})$. At the destination, removing the CP, the received signal is given by

$$\mathbf{r}_D = \sqrt{E_{RD}} \mathbf{H}_{RD} \mathbf{x}_R + \sqrt{E_{SD}} \mathbf{H}_{SD} \mathbf{x}_S^2 + \mathbf{n}_D$$

$$= \sqrt{\frac{E_{RD} E_{SR}}{E_{SR} + \sigma_n^2}} \mathbf{H}_{RD} \mathbf{W}^H \mathbf{PS}\{\mathbf{W} \mathbf{H}_{SR} \mathbf{x}\}^* + \sqrt{E_{SD}} \mathbf{H}_{SD} \mathbf{x} + \sqrt{E_{RD}} \mathbf{H}_{RD} \mathbf{n}_R' + \mathbf{n}_D \tag{17}$$

where $\mathbf{n}_D$ is a complex AWGN vector with each entry having a zero-mean and variance of $\sigma_n^2 / 2$ per dimension.

For the equalization in the frequency domain, the first term of (17) can be written, with omitting $\sqrt{E_{RD} E_{SR} / (E_{SR} + \sigma_n^2)}$, as

$$\mathbf{H}_{RD} \mathbf{W}^H \mathbf{PS}\{\mathbf{W} \mathbf{H}_{SR} \mathbf{x}\}^* = \mathbf{W}^H \mathbf{\Lambda}_{RD} \mathbf{PS} \mathbf{\Lambda}_{SR}^*\{\mathbf{W} \mathbf{x}\}^* \tag{18}$$

where $\mathbf{\Lambda}_{SR}(\triangleq \mathbf{W} \mathbf{H}_{SR} \mathbf{W}^H)$ and $\mathbf{\Lambda}_{RD}(\triangleq \mathbf{W} \mathbf{H}_{RD} \mathbf{W}^H)$ are $N \times N$ diagonal matrices. Both $\mathbf{\Lambda}_{SR}^*$ and $\mathbf{S}$ are diagonal matrices, and thus (18) can be rewritten as

$$\mathbf{W}^H \mathbf{\Lambda}_{RD} \mathbf{PS} \mathbf{\Lambda}_{SR}^*\{\mathbf{W} \mathbf{x}\}^* = \mathbf{W}^H \mathbf{\Lambda}_{RD} \mathbf{P} \mathbf{\Lambda}_{SR}^* \mathbf{S}\{\mathbf{W} \mathbf{x}\}^*. \tag{19}$$

Since the channel frequency responses (CFRs) between adjacent subcarriers are approximately constant, i.e., $\Lambda_{SR}(2k) \cong \Lambda_{SR}(2k+1)$ for large N (Jang, 2006), (19) can be changed as

$$\mathbf{W}^H \mathbf{\Lambda}_{RD} \mathbf{P} \mathbf{\Lambda}_{SR}^* \mathbf{S}\{\mathbf{W} \mathbf{x}\}^* \cong \mathbf{W}^H \mathbf{\Lambda}_{RD} \mathbf{\Lambda}_{SR}^* \mathbf{PS}\{\mathbf{W} \mathbf{x}\}^* = \mathbf{W}^H \mathbf{\Lambda}_{EQ} \mathbf{PS}\{\mathbf{W} \mathbf{x}\}^* \tag{20}$$

where $\Lambda_{EQ}(\triangleq \Lambda_{RD}\Lambda_{SR}^*)$ is the equivalent CFR.

With the normalization as in (Mheidat, 2007), the noise can be handled as complex AWGN having a zero-mean and variance of $\sigma_n^2 / 2$ without affecting the signal-to-noise ratio (SNR). After normalization, we have

$$\mathbf{r}_D' = \sqrt{\gamma_1}\mathbf{W}^H \Lambda_{EQ}\mathbf{PS}\{\mathbf{Wx}\}^* + \sqrt{\gamma_2}\mathbf{H}_{SD}\mathbf{x} + \mathbf{n} \tag{21}$$

where γ_1 and γ_2 are defined as

$$\gamma_1 = \frac{(E_{SR} / \sigma_n^2)E_{RD}}{1 + E_{SR} / \sigma_n^2 + \sum_{m=0}^{L_{RD}} \left|h_{RD}(m)\right|^2 E_{RD} / \sigma_n^2}, \quad \gamma_2 = \frac{(1 + E_{SR} / \sigma_n^2)E_{SD}}{1 + E_{SR} / \sigma_n^2 + \sum_{m=0}^{L_{RD}} \left|h_{RD}(m)\right|^2 E_{RD} / \sigma_n^2}. \tag{22}$$

Then, the frequency domain representation of $\mathbf{r}_D'$, $\mathbf{R}_D'$, can be divided into even and odd components as

$$R_D'(2k) = -\sqrt{\gamma_1}\Lambda_{EQ}(2k)X^*(2k+1) + \sqrt{\gamma_2}\Lambda_{SD}(2k)X(2k) + N(2k),$$

$$R_D'^*(2k+1) = \sqrt{\gamma_1}\Lambda_{EQ}^*(2k+1)X(2k) + \sqrt{\gamma_2}\Lambda_{SD}^*(2k+1)X^*(2k+1) + N^*(2k+1), \tag{23}$$

where $\Lambda_{SD} = \mathbf{W}\mathbf{H}_{SD}\mathbf{W}^H$, $\mathbf{X} = \mathbf{Wx}$, and $\mathbf{N} = \mathbf{Wn}$. Rewriting (23) into the matrix form, we get

$$\mathbf{R}_k' \triangleq \begin{bmatrix} R_D'(2k) \\ R_D'^*(2k+1) \end{bmatrix} \cong \begin{bmatrix} \Lambda_{SD}'(2k) & -\Lambda_{EQ}'(2k) \\ \Lambda_{EQ}'^*(2k) & \Lambda_{SD}'^*(2k) \end{bmatrix} \begin{bmatrix} X(2k) \\ X^*(2k+1) \end{bmatrix} + \begin{bmatrix} N(2k) \\ N^*(2k+1) \end{bmatrix}$$

$$\triangleq \Lambda_k'\mathbf{X}_k' + \mathbf{N}_k' \tag{24}$$

where $\Lambda_{EQ}' \triangleq \sqrt{\gamma_1}\Lambda_{EQ}$ and $\Lambda_{SD}' \triangleq \sqrt{\gamma_2}\Lambda_{SD}$. From (24), we can now derive a linear combination receiver under the minimum mean square error (MMSE) criterion (Proakis, 2001; Falconer, 2002). The resultant equation is similar to that of the maximal ratio combining (MRC) receiver with the two branch diversity system, which is

$$\mathbf{Y}_k' \triangleq \Lambda_k'^H\mathbf{R}_k' = \begin{bmatrix} \tilde{\Lambda}(k) & 0 \\ 0 & \tilde{\Lambda}(k) \end{bmatrix} \mathbf{X}_k' + \Lambda_k'^H\mathbf{N}_k'$$

$$\tilde{\Lambda}(k) \triangleq \gamma_1 \left|\Lambda_{EQ}(2k)\right|^2 + \gamma_2 \left|\Lambda_{SD}(2k)\right|^2$$

$$\tilde{\mathbf{X}}_k' \triangleq (\textgreater{}_k^{'H} \textgreater{}_k' + \frac{1}{SNR} \mathbf{I}_{2\times2})^{-1} \mathbf{Y}_k = \begin{bmatrix} \tilde{X}(2k) \\ \tilde{X}^*(2k+1) \end{bmatrix}. \tag{25}$$

Note that decisions of the SC-FDE are made in the time domain, although channel equalization is performed in the frequency domain. Therefore, the estimates of the information symbols can be obtained as $\tilde{\mathbf{x}} = \mathbf{W}^H \tilde{\mathbf{X}}$, where $\tilde{\mathbf{X}} = [\tilde{X}(0), \tilde{X}(1), ..., \tilde{X}(N-1)]^T$.

Throughout this chapter, the performances are evaluated by Monte Carlo simulation. For each point on the performance curves, at least 1000 errors are collected. Fig. 3 compares the symbol-error rate (SER) performance of D-STBC SC-FDE and D-SFBC SC-FDE over slow fading (normalized Doppler frequency ($f_D N T_s$) = 0.001) and fast fading ($f_D N T_s$ = 0.04) channels, which correspond to the mobile equipment speeds of 4 and 160 km/h, respectively. We consider uncoded systems with the block size N =256, QPSK constellation, 3MHz bandwidth, and 3.1GHz carrier frequency. All underlying links experience frequency-selective channels, where $S \to R$ and $S \to D$ links are modeled as 4-path, and $R \to D$ link is 2-path with a uniform delay power profile. Jakes' model is used for generation of time-varying channel coefficients (Stüber, 2000). We assume a fixed relay scenario for the uplink transmission. The perfect channel state information (CSI) is assumed at the destination, and the $S \to D$ and $R \to D$ links are balanced, i.e., perfect power control. Note that the D-SFBC SC-FDE significantly outperforms the D-STBC SC-FDE, when there exists severe Doppler spread. An analytic comparison of two systems can also be found in (Seol, in press).

Channel Estimation for Fast Fading Channels

Both OFDM and SC-FDE require precise CSI for the equalization. To estimate the CSI, frequency domain multiplexed (FDM) pilots have been typically used for OFDM systems, while the pilots of SC-FDE are time domain multiplexed (TDM) on dedicated blocks. However, with the conventional block-type channel estimation using TDM pilots, the SC-FDE deteriorates as the mobile equipment speed increases. In (Lam, 2006), instead of TDM pilots for SC-FDE systems, two techniques of FDM pilot insertion have been proposed to estimate fading channels. One is superimposing the pilot symbols on the scaled data-carrying frequencies, called frequency domain superimposed pilot technique (FDSPT). The other is shifting groups of data frequencies for insertion of the pilot symbols, named frequency expanding technique (FET). Although these techniques facilitate flexible and efficient assignment of data and pilots on available spectrum, both of them have limitations. FDSPT degrades the bit error performance, and FET lowers the spectral efficiency depending on pilot overhead ratio, compared to the SC-FDE without pilots. In order to overcome these problems of FET and FDSPT, a pilot position selection technique (PPST) for the SC-FDE was proposed (Kim, 2008).

We discuss a pilot position selection/detection technique for channel estimation of D-SFBC SC-FDE. This scheme superimposes pilots on data-carrying tones whose positions are selected to minimize the distortion of time domain signals caused by the loss of useful data tones in frequency domain. In addition, it effectively tracks the channel variations caused by high-speed mobility, thus preserving the diversity benefit of D-SFBC SC-FDE for fast fading channels. Although we focus on the D-SFBC SC-FDE, this scheme can also be applied to the D-STBC SC-FDE in a similar manner. The corresponding destination structure and frequency domain equalization are also presented, where the pilot positions are blindly

Figure 3. SER performance of D-STBC and D-SFBC SC-FDEs (N=256, QPSK, E_{SR}/N_0=20dB).

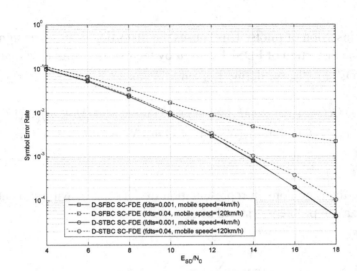

detected and the distorted data symbols are iteratively reconstructed.

From (25) and (26), the destination of D-SFBC SC-FDE requires CFRs, $\sqrt{\gamma_1}$, $_{EQ}$ and $\sqrt{\gamma_2}$, $_{SD}$, for the SFBC combining and channel equalization. The FDSPT for SC-FDE can be extended to the D-SFBC SC-FDE to estimate the CFRs over fast fading environments. In order to estimate $\sqrt{\gamma_1}$, $_{EQ}$ and $\sqrt{\gamma_2}$, $_{SD}$ at the same time, we design pilots which are superimposed on data tones. The data tones $\mathbf{X}$ are given by taking N-point DFT on the length N information sequence $\mathbf{x}$. The assignment of data and pilots can be given by

$$\mathbf{X}_{\text{FDSPT}} = [C(0), 0, X(2), \ldots, X(M-1), C(1), 0, X(M+2), \ldots, X(N-1)]^T \tag{26}$$

where $M = N / N_p$ is the spacing of adjacent pilots, N_p is the number of pilots per block, and the data tone scaling factor is set to zero (Lam, 2006). The nulled elements preserve the orthogonality of pilots between $\mathbf{x}_S^2$ and $\mathbf{x}_R$ in the second time slots. Thus, the CFRs, $\sqrt{\gamma_1}$, $_{EQ}$ and $\sqrt{\gamma_2}$, $_{SD}$, can be completely separated and estimated in frequency domain at the destination. A length N_p Chu sequence $\{C(k)\}_{k=0}^{N_p-1}$ used as the pilot is represented as (Chu, 1972)

$$C(k) = \begin{cases} e^{j\pi rk^2/N_p}, & \text{for even } N_p \\ e^{j\pi rk(k+1)/N_p}, & \text{for odd } N_p \end{cases} \tag{27}$$

where r is relatively prime to N_p. In (27), we can see that

$$\{X(0), X(1), X(M), X(M+1), \ldots, X((N_p-1)M), X((N_p-1)M+1)\}$$

are eliminated for the insertion of pilots. This causes the distortion of original signals in time domain. In order to minimize the distortion of FDSPT caused by the loss of useful data tones, we present PPST for the D-SFBC SC-FDE. Firstly, we define $\digamma_m$ and $\ddot{}_m$ as

$$\digamma_m = [X(m), X(m+1), X(M+m), X(M+m+1), \ldots, X((N_p-1)M+m), X((N_p-1)M+m+1)]^T$$
$$\ddot{}_m = \{m, m+1, M+m, M+m+1, \ldots, (N_p-1)M+m, (N_p-1)M+m+1\}, \quad 0 \le m < M-1.$$

$$(28)$$

Let $\{\overline{x}^m(n)\}_{n=0}^{N-1}$ be the time domain representation of $\{X(k)\}_{k=0}^{N-1}$ with nulls at the positions Ψ_m. The distortion minimization problem is stated as

$$d_o = \arg\min_m \sum_{n=0}^{N-1} \left| x(n) - \overline{x}^m(n) \right|^2 = \arg\min_m \sum_{n=0}^{N-1} \left| \frac{1}{\sqrt{N}} \sum_{k \in \Psi_m} X(k) e^{j\frac{2\pi nk}{N}} \right|^2$$

$$= \arg\min_m \left(\sum_{k \in \ddot{}_m} |X(k)|^2 + \frac{1}{N} \sum_{n=0}^{N-1} \sum_{k \in \ddot{}_m} \sum_{\substack{l \in \ddot{}_m \\ l \ne k}} X(k)X^*(l) e^{j\frac{2\pi n(k-l)}{N}} \right).$$

$$(29)$$

In (29), since the second term on the righthand side is zero, d_o can be simply found in frequency domain as follows

$$d_o = \arg\min_m \sum_{k \in \ddot{}_m} |X(k)|^2 = \arg\min_m \digamma_m^H \digamma_m.$$

$$(30)$$

From (30), we can select pilot position offset d_o with N multiplication operations, and null the elements $\{X(d_o), X(d_o+1), X(M+d_o), X(M+d_o+1), \ldots, X((N_p-1)M+d_o), X((N_p-1)M+d_o+1)\}$ on the $\{X(k)\}_{k=0}^{N-1}$ to insert pilots on those positions. Then, by left cyclic-shifting d_o samples, the frequency domain transmit sequence $\mathbf{X}_{PPS}$ is given by

$$\mathbf{X}_{PPS} = [C(0), 0, X(d_o+2)) \ldots, X(M+d_o-1), C(1), 0, X(M+d_o+2), \ldots, X(d_o-1)]^T.$$

$$(31)$$

After N-point IDFT of $\mathbf{X}_{PPS}$, the time domain signals $\mathbf{x}_{PPS}$ are generated. Then, the $\mathbf{x}_{PPS}$ is set to the transmit sequence of source $\mathbf{x}_s^1$ in (6).

Fig. 2(b) shows the destination structure of PPST. At the destination, the received signal at the pilot position in (23) is expressed as

$$R_D'(k) = \begin{cases} \sqrt{\gamma_2}\Lambda_{SD}(k)C(k) + N(k), & k \pmod{M} = 0 \\ \sqrt{\gamma_1}\Lambda_{EQ}(k)C^*(k) + N(k), & k \pmod{M} = 1. \end{cases} \tag{33}$$

With the knowledge of $\{C(k)\}_{k=0}^{N_p-1}$, the CFRs $\{\sqrt{\gamma_2}\Lambda_{SD}(k)\}$ and $\{\sqrt{\gamma_1}\Lambda_{EQ}(k)\}$ at the pilot positions are estimated as $\sqrt{\gamma_2}\hat{\Lambda}_{SD}(k) = R_D'(k) / C(k)$ and $\sqrt{\gamma_1}\hat{\Lambda}_{EQ}(k) = R_D'(k) / C^*(k)$. The whole CFRs, $\sqrt{\gamma_2}\hat{\ }_{SD}$ and $\gamma_{S_D} = \eta\sqrt{E_{S_D}(E_{S_R} + E_{S_R} + \sigma_w^2)}$, are obtained by using DFT/IDFT-based interpolation (Lam, 2006; Kim, 2008). With the estimated CFRs, SFBC combining and MMSE equalization are performed. By nulling the elements on Ψ_0 of $\tilde{\mathbf{X}}_{PPS}$, the equalized output is given as

$$\tilde{\mathbf{X}}_{PPS} = [0, 0, \tilde{X}(d_o + 2))..., \tilde{X}(M + d_o - 1), 0, 0, \tilde{X}(M + d_o + 2), ..., \tilde{X}(d_o - 1)]^T. \tag{34}$$

Note that the cyclic-shift of samples in the frequency domain corresponds to the phase rotation in the time domain. Using this fact, we blindly detect the pilot position d_o as

$$\hat{d}_o = \arg\min_m \frac{1}{N} \sum_{n=0}^{N-1} \left[\min_{a_i \in \chi} \left| \tilde{x}_{PPS}(n)e^{j\frac{2\pi mn}{N}} - a_i \right|^2 \right] \tag{35}$$

where χ is a finite alphabet set. Pilot position detection technique (PPDT) in (35) requires $2(M-1)N$ multiplications. Then, the initial estimates of information sequence $\{\hat{x}^{(0)}(n)\}_{n=0}^{N-1}$ are obtained as

$$\hat{x}^{(0)}(n) = \arg\min_{a_i \in \chi} \left| \tilde{x}_{PPS}(n)e^{j\frac{2\pi\hat{d}_o n}{N}} - a_i \right|, 0 \le n < N. \tag{36}$$

We iteratively reconstruct the signals with the aid of tentative data decisions as follows

$$\tilde{X}^{(i)}(k) = \begin{cases} \hat{X}^{(i-1)}(k), & k \pmod{M} = \hat{d}_o \text{ or } \hat{d}_o + 1 \\ \tilde{X}_{PPS}(k - \hat{d}_o), & \text{otherwise} \end{cases} \tag{37}$$

where i represents the iteration number starting with $i = 1$, and $\{\hat{X}^{(i-1)}(k)\}_{k=0}^{N-1}$ are DFT of $\{\hat{x}^{(i-1)}(n)\}_{n=0}^{N-1}$. From Fig. 2(b) and the discussion above, each iteration requires N-point DFT/IDFT operations.

Figure 4 shows the bit error rate (BER) performance of the PPST/PPDT-based channel estimation for D-SFBC SC-FDE for different mobile equipment speeds at $E_{SR} / N_0 = E_{SD} / N_0 = 20$ dB. We consider a frame-based transmission with 6 information blocks where each data block consists of 1024 QPSK modulated symbols. $N_p = 32$, 5 MHz bandwidth and 3 GHz carrier frequency are used. All underlying links experience frequency-selective channels, where $S \to R$ and $S \to D$ links are modeled as six-tap typical urban channel (COST 207, 1998), and $R \to D$ link is 2-path with a uniform delay power profile. For the comparison, we show the performance of block-type channel estimation, where the Chu sequence with length N_p is inserted as a training block before the first and after the sixth blocks. The channel estimates for the block without pilots are linearly interpolated using those blocks. It is shown that the PPST/PPDT approaches the block-type channel estimation (with pilot overhead: $2N_p \times 2$) of D-SFBC SC-FDE system over slow fading channel with mobile equipment speed of 2 km/h. It is also observed that the PPST/PPDT significantly outperforms the conventional block-type channel estimation, when there exists a severe Doppler spread. Compared with FDSPT, at the initial iteration (i=0), the BER performance of PPST is better than that of FDSPT. The performance gap between PPST and FDSPT is reduced with iterations.

Spectral Efficient Cooperative Diversity Protocols

Relay usually works in half-duplex mode, since the large difference in the power of transmitted and received signals drives the relay's analog amplifier of the receive chain into saturation in the full-duplex mode. Because of the half-duplex constraint, the cooperation for data transmission of one user is taken over two time slots, which leads to the loss of data rate. We present spectral efficient protocols with iterative multiuser detection scheme for relay-assisted and user cooperative transmission systems. In both systems, the data block of each user is interleaved by user-specific interleaver which can be used to distinguish signals from different users (Moher, 1998). Interference caused by the other user is removed by using the iterative multiuser detection with frequency-domain equalization (IMD-FDE) (Lim, 2007).

Iterative Multiuser Detection with Frequency-Domain Equalization

In IMD-FDE, simultaneous users share the same time-frequency resources. Since the signals of multiple users are superimposed in a common subcarrier, multiple access interference (MAI) is occurred. IMD-FDE

Figure 4. BER performance of the PPST/PPDT and block-type channel estimation over different mobile equipment speeds.

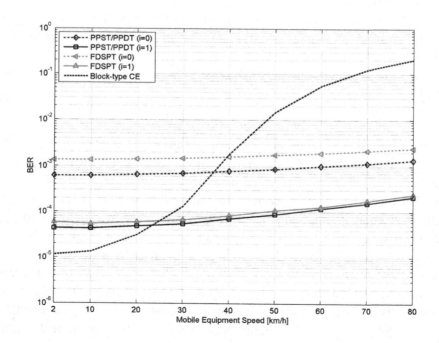

employs distinct chip-level interleavers combined with low-rate channel coding and frequency-domain multiuser detection to separate the signals transmitted from different users. The transmitter of IMD-FDE inserts the CP, as in SC-FDMA. With help of the CP, the receiver simply performs both multiuser detection and channel equalization in the frequency domain.

In the transmitter of IMD-FDE, the input data sequence of the uth user is encoded, and then the coded bits are spread. The spread bits are independently interleaved using a user-specific interleaver. The interleaved bits are mapped to a multilevel sequence such as 2^Q-QAM symbols with the size of N. After inserting the CP, each block is transmitted over a channel. The receiver of IMD-FDE uses the multiuser detection and the channel equalization in the frequency domain, to remove MAI and ISI. The received block in the frequency domain is given by $\mathbf{R} = \sum_{u=1}^{U} \mathbf{\Lambda}_u \mathbf{W} \mathbf{x}_u + \mathbf{N}$, where $\mathbf{\Lambda}_u$ is the $N \times N$ diagonal channel matrix, $\mathbf{x}_u$ is the $N \times 1$ transmit vector of the uth user. The IMD-FDE can be described briefly as the following iterative procedures:

- To perform MMSE equalization, the mean $\overline{\mathbf{x}}_u$ and the variance ν_u of the transmitted symbols are required. These mean and variance are functions of the *a priori* log likelihood ratios (LLRs) obtained from the maximum *a posteriori* (MAP) decoder. For the initial iteration, the *a priori* information provided by the decoder is not available, and thus the mean and the variance of the transmitted symbols are zero and σ_s^2, respectively.

- Multiuser interference cancellation for the uth user is performed as $\mathbf{Z}_u = \mathbf{R} - \sum_{u'=1, u' \neq u}^{U} \mathbf{\Lambda}_{u'} \mathbf{W} \overline{\mathbf{x}}_{u'}$.

- The equalized symbols are obtained by carrying out the equalization block as $\hat{\mathbf{X}}_u = \mathbf{G}_u^H \mathbf{Z}_u - (\mathbf{G}_u^H \mathbf{\Lambda}_u - \dfrac{1}{N} trace(\mathbf{G}_u^H \mathbf{\Lambda}_u)) \mathbf{W} \overline{\mathbf{x}}_u$. The coefficients of the MMSE equalizer for IMD-FDE are given by

$$G_{u,k} = \dfrac{\sigma_s^2 \Lambda_{u,k}}{\sum_{u'=1}^{U} \nu_{u'} \left| \Lambda_{u',k} \right|^2 + \sigma_N^2}, \; k = 0, 1, \cdots, N-1. \tag{38}$$

- After performing the N-point IDFT, the extrinsic LLRs are obtained, deinterleaved, despread, and fed to the MAP decoder. The MAP decoder computes the extrinsic information for the coded bits. The extrinsic information of the coded bit is used as the *a priori* information of the multiuser interference canceller, and the one-tap MMSE equalizer.

Spectral Efficient Protocol for Relay-Assisted SC-FDE

Protocol and System Model

Figure 5(a) presents a spectral efficient protocol for relay-assisted SFBC system. In the first slot, two sources communicate with the relay simultaneously, while the destination does not receive the direct signal from the sources. In the second slot, the relay and sources communicate with the destination. With this protocol and the aid of iterative multiuser detection, the spectral loss can be compensated.

Figure 6(a) depicts a transmitter block diagram for the uth user. The coded bits are independently interleaved using user-specific interleaver, and then they are mapped to a 2^Q-PSK or 2^Q-QAM symbol

Figure 5. Spectral efficient protocols for (a) relay-assisted and (b) superposed cooperative transmissions.

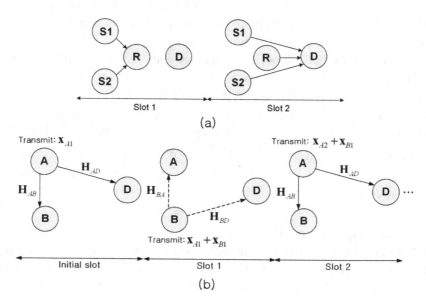

(a)

(b)

sequence with size of N, where N is the size of the DFT. We assume that QPSK $(Q = 2)$ modulation is used, i.e., $x_u(n) \in \{\xi_1 = \frac{1}{\sqrt{2}}(1 + j),\ \xi_2 = \frac{1}{\sqrt{2}}(1 - j),\ \xi_3 = \frac{1}{\sqrt{2}}(-1 + j),\ \xi_4 = \frac{1}{\sqrt{2}}(-1 - j)\}$. In the first time slot, a CP of length $L = max(L_{S_1R}, L_{S_2R})$ is appended in the head of $\{x_u(n)\}_{n=0}^{N-1}, u = 1, 2$, and then the symbol block is transmitted over a fading channel. At the relay, removing the CP, the received signal is given by

$$\mathbf{r}_R = \sqrt{E_{S_1R}}\mathbf{H}_{S_1R}\mathbf{x}_1 + \sqrt{E_{S_2R}}\mathbf{H}_{S_2R}\mathbf{x}_2 + \mathbf{n}_R \tag{39}$$

where $\mathbf{x}_u = [x_u(0)x_u(1)\cdots x_u(N-1)]^T, u = 1, 2$, denotes the transmit vector of the uth user. $\mathbf{n}_R$ is a complex AWGN vector with covariance matirx of $\sigma_n^2\mathbf{I}_N$. The received signal, $\mathbf{r}_R$, is scaled as $\tilde{\mathbf{r}}_R \triangleq \sqrt{2 / (E_{S_1R} + E_{S_2R} + \sigma_n^2)}\mathbf{r}_R \triangleq \gamma_R\mathbf{r}_R$ to ensure balance of power at the destination. To have an Alamouti's orthogonal structure in the frequency domain, the transmit signal of the relay can be represented as

$$\mathbf{x}_R = \mathbf{W}^{-1}\mathbf{PS}\{\mathbf{W}\tilde{\mathbf{r}}_R\}^* = \gamma_R\{\sqrt{E_{S_1R}}\mathbf{W}^{-1}\mathbf{PS}\{\mathbf{WH}_{S_1R}\mathbf{x}_1\}^* + \sqrt{E_{S_2R}}\mathbf{W}^{-1}\mathbf{PS}\{\mathbf{WH}_{S_2R}\mathbf{x}_2\}^*\} + \mathbf{n}'_R \tag{40}$$

where $\mathbf{n}'_R = \dfrac{\sqrt{2}\mathbf{W}^{-1}\mathbf{PS}\{\mathbf{Wn}_R\}^*}{\sqrt{E_{S_1R} + E_{S_2R} + \sigma_n^2}}$.

The transmit sequence of relay, $\mathbf{x}_R$, can be efficiently obtained in the time domain as (see Section 2)

$$x_R(n) = r_c(n) \times j\sin(2\pi n / N) + r_c(n - N / 2)_N \times \cos(2\pi n / N). \tag{41}$$

where $r_c(n) = \tilde{r}_R^*(-n)_N$.

In the second time slot, the sources and relay transmit $\mathbf{x}_u, u = 1, 2$, and $\mathbf{x}_R$ after appending CPs with length $L = \max(L_{S_1 D}, L_{S_2 D}, L_{RD})$, respectively. At the destination, removing the CP, the received signal is given by

$$
\begin{aligned}
\mathbf{r}_D &= \sqrt{E_{S_1 D}}\mathbf{H}_{S_1 D}\mathbf{x}_1 + \sqrt{E_{S_2 D}}\mathbf{H}_{S_2 D}\mathbf{x}_2 + \sqrt{E_{RD}}\mathbf{H}_{RD}\mathbf{x}_R + \mathbf{n}_D \\
&= \sqrt{E_{S_1 D}}\mathbf{H}_{S_1 D}\mathbf{x}_1 + \gamma_R\sqrt{E_{RD}E_{S_1 R}}\mathbf{H}_{RD}\mathbf{W}^{-1}\mathbf{PS}\{\mathbf{WH}_{S_1 R}\mathbf{x}_1\}^* \\
&\quad + \sqrt{E_{S_2 D}}\mathbf{H}_{S_2 D}\mathbf{x}_2 + \gamma_R\sqrt{E_{RD}E_{S_2 R}}\mathbf{H}_{RD}\mathbf{W}^{-1}\mathbf{PS}\{\mathbf{WH}_{S_2 R}\mathbf{x}_2\}^* + \sqrt{E_{RD}}\mathbf{H}_{RD}\mathbf{w}_R' + \mathbf{n}_D
\end{aligned}
\tag{42}
$$

By omitting $\gamma_R\sqrt{E_{RD}E_{S_u R}}, u = 1, 2$, the second term and fourth term of (42) can be written as

$$
\begin{aligned}
\mathbf{H}_{RD}\mathbf{W}^{-1}\mathbf{PS}\{\mathbf{WH}_{S_u R}\mathbf{x}_u\}^* &= \mathbf{W}^{-1}\Lambda_{RD}\mathbf{PS}\Lambda_{S_u R}^*\{\mathbf{Wx}_u\}^* \\
&= \mathbf{W}^{-1}\Lambda_{RD}\mathbf{P}\Lambda_{S_u R}^*\mathbf{S}\{\mathbf{Wx}_u\}^* \cong \mathbf{W}^{-1}\Lambda_{RD}\Lambda_{S_u R}^*\mathbf{PS}\{\mathbf{Wx}_u\}^*, u = 1, 2
\end{aligned}
\tag{43}
$$

where $\Lambda_{RD}(= \mathbf{WH}_{RD}\mathbf{W}^{-1})$ and $\Lambda_{S_u R}(= \mathbf{WH}_{S_u R}\mathbf{W}^{-1})$ are $N \times N$ diagonal matrices. We assume that the CFRs between adjacent subcarriers are approximately constant, i.e., $\Lambda_{S_u R}(2k) \cong \Lambda_{S_u R}(2k + 1)$, where $\Lambda_{S_u R}(k)$ is the kth diagonal element of $\Lambda_{S_u R}$. With the normalization, $\mathbf{r}_D$ can be rewritten as

$$\mathbf{r}_D' = \gamma_{S_1 D}\mathbf{H}_{S_1 D}\mathbf{x}_1 + \gamma_{S_2 D}\mathbf{H}_{S_2 D}\mathbf{x}_2 + \gamma_{S_1 RD}\mathbf{W}^{-1}\Lambda_{S_1 RD}\mathbf{PS}\{\mathbf{Wx}_1\}^* + \gamma_{S_2 RD}\mathbf{W}^{-1}\Lambda_{S_2 RD}\mathbf{PS}\{\mathbf{Wx}_2\}^* + \mathbf{n} \tag{44}$$

where

$$\gamma_{S_u D} = \eta\sqrt{E_{S_u D}(E_{S_1 R} + E_{S_2 R} + \sigma_w^2)}, \quad \gamma_{S_u RD} = \eta\sqrt{2E_{S_u R}E_{RD}}, u = 1, 2,$$

$$\mathbf{n} = \eta\sqrt{E_{S_1 R} + E_{S_2 R} + \sigma_n^2}(\mathbf{n}_D + \sqrt{E_{RD}}\mathbf{H}_{RD}\mathbf{n}_R'), \quad \eta = \sqrt{1 / (E_{S_1 R} + E_{S_2 R} + \sigma_n^2 + 2E_{RD}\sum_{l=0}^{L_{RD}}|h_{RD}(l)|^2)}, \quad \text{and}$$

$\Lambda_{S_u RD} = \Lambda_{S_u R}^*\Lambda_{RD}$. Here, $\Lambda_{S_u RD}$ is the equivalent CFR for the $S_u \to R \to D$ link. Then, by performing DFT on $\mathbf{r}_D'$, $\mathbf{R}_D'(= \mathbf{Wr}_D')$ can be obtained as follows:

$$\mathbf{R}_D' = \gamma_{S_1 D}\Lambda_{S_1 D}\mathbf{X}_1 + \gamma_{S_2 D}\Lambda_{S_2 D}\mathbf{X}_2 + \gamma_{S_1 RD}\Lambda_{S_1 RD}\mathbf{PSX}_1^* + \gamma_{S_2 RD}\Lambda_{S_2 RD}\mathbf{PSX}_2^* + \mathbf{N} \tag{45}$$

where $\mathbf{X}_u = \mathbf{Wx}_u, \Lambda_{S_u D} = \mathbf{WH}_{S_u D}\mathbf{W}^{-1}, u = 1, 2, \triangleq \Lambda_{1,k}\mathbf{X}_{1,k}' + \Lambda_{2,k}(\mathbf{X}_{2,k}' - \bar{\mathbf{X}}_{2,k}') + \mathbf{N}'$ and $\mathbf{N} = \mathbf{Wn}$.

Iterative Detection Scheme

Figure 6(b) shows a receiver structure employing the iterative multiuser detection scheme. After successive multiuser interference cancellation (SMIC), the received vector $\mathbf{Z}_1$ of the first user can be expressed in the frequency domain as

$$\mathbf{Z}_1 = \gamma_{S_1 D} \, {}_{S_1 D}\mathbf{X}_1 + \gamma_{S_1 RD} \, {}_{S_2 RD}\mathbf{PSX}_1^* + \gamma_{S_2 D} \, {}_{S_2 D}(\mathbf{X}_2 - \bar{\mathbf{X}}_2) + \gamma_{S_2 RD} \, {}_{S_2 RD}\mathbf{PS}(\mathbf{X}_2^* - \bar{\mathbf{X}}_2^*) + \mathbf{N} \tag{46}$$

where $\bar{\mathbf{X}}_2 = \mathbf{W}\bar{\mathbf{x}}_2$ and $\bar{\mathbf{x}}_2$ is the mean of the transmit vector $\mathbf{x}_2$. $\bar{\mathbf{x}}_2$ is a function of the *a priori* LLRs $\{L_{EA}(c_2(m))\}_{m=0}^{2N-1} = \pi_A(\{L_D(c_2(l))\}_{l=0}^{2N-1})$, which can be obtained from soft mapping as $\bar{x}_2(n) = \frac{1}{\sqrt{2}}(\tanh(L_{EA}(c_2(2m))/2) + j\tanh(L_{EA}(c_2(2m+1))/2))$. By taking complex-conjugate operation on odd elements, we can divide (47) into even and odd components, and (47) can be changed into the matrix form as

$$
\begin{aligned}
\mathbf{Z}'_{1,k} &\triangleq [Z_1(2k) \quad Z_1^*(2k+1)]^T \\
&= \begin{bmatrix} \Lambda'_{S_1 D}(2k) & -\Lambda'_{S_1 RD}(2k) \\ \Lambda'^*_{S_1 RD}(2k) & \Lambda'^*_{S_1 D}(2k) \end{bmatrix} \begin{bmatrix} X_1(2k) \\ X_1^*(2k+1) \end{bmatrix} \\
&+ \begin{bmatrix} \Lambda'_{S_2 D}(2k) & -\Lambda'_{S_2 RD}(2k) \\ \Lambda'^*_{S_2 RD}(2k) & \Lambda'^*_{S_2 D}(2k) \end{bmatrix} \begin{bmatrix} X_2(2k) - \bar{X}_2(2k) \\ X_2^*(2k+1) - \bar{X}_2^*(2k+1) \end{bmatrix} + \begin{bmatrix} N(2k) \\ N^*(2k+1) \end{bmatrix} \\
&\triangleq \, {}_{1,k}\mathbf{X}'_{1,k} + \, {}_{2,k}(\mathbf{X}'_{2,k} - \bar{\mathbf{X}}'_{2,k}) + \mathbf{N}'
\end{aligned} \tag{47}
$$

where $\Lambda'_{S_u D}(2k) = \gamma_{S_u D}\Lambda_{S_u D}(2k)$, and $\Lambda'_{S_u RD}(2k) = \gamma_{S_u RD}\Lambda_{S_u RD}(2k), u = 1, 2$. From (47), we can now derive a linear combining receiver under the MMSE criterion, which is

$$
\begin{aligned}
\tilde{\mathbf{Z}}_{1,k} &\triangleq [\tilde{Z}_1(2k) \quad \tilde{Z}_1(2k+1)]^T = \, {}_{1,k}^H \mathbf{Z}'_{1,k} \\
&= \begin{bmatrix} \tilde{\Lambda}_1(2k) & 0 \\ 0 & \tilde{\Lambda}_1(2k) \end{bmatrix} \begin{bmatrix} X_1(2k) \\ X_1^*(2k+1) \end{bmatrix} \\
&+ \begin{bmatrix} \breve{A}_2(2k) & \breve{B}_2(2k) \\ \breve{C}_2(2k) & \breve{D}_2(2k) \end{bmatrix} \begin{bmatrix} X_2(2k) - \bar{X}_2(2k) \\ X_2^*(2k+1) - \bar{X}_2^*(2k+1) \end{bmatrix} + \, {}_{1,k}^H \begin{bmatrix} N(2k) \\ N^*(2k+1) \end{bmatrix}
\end{aligned} \tag{48}
$$

where

$\tilde{\Lambda}_1(2k) = |\Lambda'_{S_1 D}(2k)|^2 + |\Lambda'_{S_1 RD}(2k)|^2$. In (48), $\breve{A}_2(2k) = \Lambda'^*_{S_1 D}(2k)\Lambda'_{S_2 D}(2k) + \Lambda'_{S_1 RD}(2k)\Lambda'^*_{S_2 RD}(2k) = \breve{D}_2^*(2k)$, and $\breve{B}_2(2k) = -\Lambda'^*_{S_1 D}(2k)\Lambda'_{S_2 RD}(2k) + \Lambda'_{S_1 RD}(2k)\Lambda'^*_{S_2 D}(2k) = -\breve{C}_2^*(2k)$. We rewrite $\tilde{Z}_1(2k)$ and $\tilde{Z}_1(2k+1)$

Figure 6. Transceiver block diagram for spectral efficient relay-assisted SC system, (a) Transmitter structure. (b) Receiver structure.

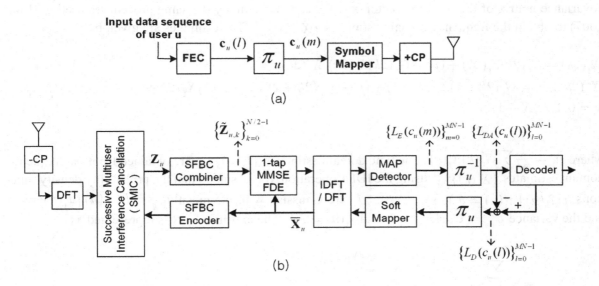

(a)

(b)

as follows:

$$\tilde{Z}_1(2k) = \tilde{\Lambda}_1(2k)X_1(2k) + \breve{A}_2(2k)(X_2(2k) - \bar{X}_2(2k))$$

$$+ \breve{B}_2(2k)(X_2^*(2k+1) - \bar{X}_2^*(2k+1)) + \Lambda_{S_1D}'^*(2k)N(2k)$$

$$+ \Lambda_{S_1RD}'(2k)N^*(2k+1),$$

$$\tilde{Z}_1(2k+1) = \tilde{\Lambda}_1(2k)X_1^*(2k+1) - \breve{B}_2^*(2k)(X_2(2k) - \bar{X}_2(2k))$$

$$+ \breve{A}_2^*(2k)(X_2^*(2k+1) - \bar{X}_2^*(2k+1)) - \Lambda_{S_1RD}'^*(2k)N(2k)$$

$$+ \Lambda_{S_1D}'(2k)N^*(2k+1). \tag{49}$$

Then, we can apply the same method used in IMD-FDE to derive the coefficients of the frequency domain equalizer for the SFBC system. The coefficients of the one-tap MMSE equalizer are obtained as follows:

$$G_1(2k) = G_1(2k+1) = \frac{\tilde{\Lambda}_1(2k)}{\nu_1 \mid \tilde{\Lambda}_1(2k) \mid^2 + \nu_2 \mid \breve{A}_2(2k) \mid^2 + \nu_2 \mid \breve{B}_2(2k) \mid^2 + \sigma_n^2 \tilde{\Lambda}_1(2k)},$$

$$k = 0, 1, ..., N - 1 \tag{50}$$

where $\nu_u = \dfrac{1}{N} trace(\mathbf{V}_u), u = 1, 2$, and $\mathbf{V}_u = Cov(\mathbf{x}_u, \mathbf{x}_u) = Diag(\nu_{u,0}, \nu_{u,1}, \cdots, \nu_{u,N-1})$ represents the covariance matrix of the transmit vector $\mathbf{x}_u$. We can also employ the same procedures used in (Lim, 2007) to obtain the frequency domain estimates $\{\hat{X}_1(k)\}_{k=0}^{N-1}$. The estimates are given by

$$\hat{X}_1(2k) = G_1^*(2k)\tilde{Z}_1(2k) + (\mu_1 - G_1^*(2k)\tilde{\Lambda}_1(2k))\bar{X}_1(2k),$$
$$\hat{X}_1^*(2k+1) = G_1^*(2k+1)\tilde{Z}_1(2k+1) + (\mu_1 - G_1^*(2k+1)\tilde{\Lambda}_1(2k+1))\bar{X}_1^*(2k+1),$$
$$k = 0, 1, ..., N-1 \tag{51}$$

where $\mu_1 = \dfrac{1}{N} \sum_{k=0}^{N-1} G_1^*(k)\Lambda_1(k)$. The time domain estimates $\{\hat{x}_1(n)\}_{n=0}^{N-1}$ are obtained from the frequency domain estimates $\{\hat{X}_1(k)\}_{k=0}^{N-1}$ using the IDFT operation. It is assumed that the probability density functions $p(\hat{x}_1(n) \mid x_1(n) = \xi_q), q = 1, 2, ..., 2^Q$, are Gaussian with the mean $\mu_{1,q} = E(\hat{x}_1(n) \mid x_1(n) = \xi_q)$ and the variance $\sigma_{1,q}^2 = Cov(\hat{x}_1(n), \hat{x}_1(n) \mid x_1(n) = \xi_q)$. The extrinsic LLR is represented as

$$L_E(c_1(m)) = \ln \frac{\displaystyle\sum_{\forall \xi_q : c_1(m)=0} \exp(-\dfrac{\| \hat{x}_1(n) - \mu_{1,q}) \|^2}{\sigma_{1,q}^2})}{\displaystyle\sum_{\forall \xi_q : c_1(m)=1} \exp(-\dfrac{\| \hat{x}_1(n) - \mu_{1,q}) \|^2}{\sigma_{1,q}^2})}, \tag{52}$$

where $\forall \xi_q : c_1(m) = 0$ and $\forall \xi_q : c_1(m) = 1$ are the mutually exclusive subsets of all possible transmitted symbols, which comprise the symbols with '0' and '1' in the m th bit position, respectively. The statistics $\mu_{1,q}$ and $\sigma_{1,q}^2$ of $\hat{x}_1(n)$ are given by

$$\mu_{1,q} = \frac{\xi_q}{N} \sum_{k=0}^{N-1} G_1^*(k)\Lambda_1(k) = \xi_q \mu_1, \text{ and } \mu_{1,q} = \frac{\xi_q}{N} \sum_{k=0}^{N-1} G_1^*(k)\Lambda_1(k) = \xi_q \mu_1 \; \sigma_{1,q}^2 = \mu_1 - \nu_1 \mu_1^2, \text{ and}$$
$$\sigma_{1,q}^2 = \mu_1 - \nu_1 \mu_1^2. \tag{53}$$

Assuming that the QPSK modulation is used, we can obtain the extrinsic LLRs as follows:

$$L_E(c_u(2n)) = \frac{2\sqrt{2}\text{Re}\{\hat{x}_u(n)\}\mu_u}{\sigma_u^2}, \text{ and } L_E(c_u(2n)) = \frac{2\sqrt{2}\text{Re}\{\hat{x}_u(n)\}\mu_u}{\sigma_u^2},$$
$$L_E(c_u(2n+1)) = \frac{2\sqrt{2}\text{Im}\{\hat{x}_u(n)\}\mu_u}{\sigma_u^2}. \text{ and } L_E(c_u(2n+1)) = \frac{2\sqrt{2}\text{Im}\{\hat{x}_u(n)\}\mu_u}{\sigma_u^2}. \tag{54}$$

The *a priori* LLRs $\{L_{DA}(c_u(l))\}_{l=0}^{2N-1} = \pi_u^{-1}(\{L_E(c_u(m))\}_{m=0}^{2N-1})$ are fed to the MAP decoder. The MAP decoder computes the extrinsic information for both the coded and decoded bits. The extrinsic information $L_D(c_u(l))$ of the coded bit is used as the *a priori* information of SMIC and the one-tap MMSE equalizer. For computer simulations, we consider a block transmission with each data block consisting of 256 QPSK symbols, and a 1/2-rate convolutional code with constraint length of 3. Random interleavers are used. All underlying links experience frequency-selective channels, where $S_u \rightarrow R$, $S_u \rightarrow D$, $u = 1, 2$, and $R \rightarrow D$ links are modeled as EDGE TU (Furuskar, 1999) channels with $L = 2$. It is assumed that perfect channel estimation is available at the receiver, and the $S_u \rightarrow R, u = 1, 2$, and $R \rightarrow D$ links are balanced, i.e., perfect power control. Figure 7 shows the block error rate (BLER) performance at SNR($S_u \rightarrow R, u = 1, 2$) = 10 dB with $f_D N T_s$ of 0.001/0.04. The single-user performance is included as a lower bound. We observe from Fig. 7 that this system achieves diversity gain and approaches the lower bound for $I = 7$, where I is the number of iterations. This result means that the scheme can increase spectral efficiency of relay-assisted system by accessing a relay simultaneously. It is also shown that this system has the advantage of SFBC even for fast fading environments. Note that this scheme can be extended to the general case of multiple sources, with low-rate channel coding. Spreading, which can be regarded as a repetition coding, is a simple method to support more than two users (Lim, 2007).

Spectral Efficient Protocol for Superposed Cooperative Transmission

Protocol and System Model

Figure 5(b) presents a spectral efficient protocol for superposed cooperative transmission. The protocol is similar to that of (Ishii, 2007), except that the data block of each user is interleaved by user-specific interleaver. To compensate the spectral loss of conventional cooperative decode and forward schemes, we transmit the superposed data blocks consisting of two cooperative users. Detailed protocols are described as follows:

Initial slot: In the initial slot, a CP of length $L = max(L_{AB}, L_{AD})$ is appended in the head of $\{x_{A1}(n)\}_{n=0}^{N-1}$, and then the symbol block is transmitted with full transmit power over a fading channel. After removing the CP, the received signals at nodes D and B are, respectively,

$$\mathbf{r}_{D0} = \mathbf{H}_{AD}\mathbf{x}_{A1} + \mathbf{n}_{D0}, \tag{55}$$

$$\mathbf{r}_{B0} = \mathbf{H}_{AB}\mathbf{x}_{A1} + \mathbf{n}_{B0}, \tag{56}$$

where $\mathbf{x}_{A1} = [x_{A1}(0)x_{A1}(1)\cdots x_{A1}(N-1)]^T$ denotes the transmit vector of the user A. $\mathbf{n}_{B0}$ and $\mathbf{n}_{D0}$ are complex AWGN vectors with covariance matrix of $\sigma_n^2 \mathbf{I}_N$. Node B decodes $\mathbf{x}_{A1}$ from the received signal $\mathbf{r}_{B0}$.

Slot 1: In the first slot, node B encodes its own data and the data of node A which is decoded in the previous time slot. It is assumed that the decoding of $\mathbf{x}_{A1}$ at node B was successful. After user-specific interleaving, the interleaved data of two users are superposed and transmitted. Node B uses power $1 - \gamma^2$ for its own data, and power γ^2 for $\mathbf{x}_{A1}$. The total transmit power is equal to one. The received signals at D and A are

$$\mathbf{r}_{D1} = \sqrt{1-\gamma^2}\mathbf{H}_{BD}\mathbf{x}_{B1} + \gamma\mathbf{H}_{BD}\mathbf{x}_{A1} + \mathbf{n}_{D1}, \tag{57}$$

$$\mathbf{r}_{A1} = \sqrt{1-\gamma^2}\,\mathbf{H}_{BA}\mathbf{x}_{B1} + \gamma\mathbf{H}_{BA}\mathbf{x}_{A1} + \mathbf{n}_{A1}. \tag{58}$$

Node A decodes $\mathbf{x}_{B1}$ from the received signal $\mathbf{r}_{A1}$. $\mathbf{x}_{A1}$ is known at node A (because the $\mathbf{x}_{A1}$ originated at node A two time slots ago), so it can be directly subtracted from the received signal $\mathbf{r}_{A1}$.

Slot 2: Assuming that the decoding of $\mathbf{x}_{B1}$ at node A was successful, node A transmits its own data $\mathbf{x}_{A2}$ and the node B's data $\mathbf{x}_{B1}$. Similar to slot 1, the power level used for $\mathbf{x}_{A2}$ is $1-\gamma^2$ and that used for $\mathbf{x}_{B1}$ is γ^2. Then, the received signal at D is

$$\mathbf{r}_{D2} = \sqrt{1-\gamma^2}\,\mathbf{H}_{AD}\mathbf{x}_{A2} + \gamma\mathbf{H}_{AD}\mathbf{x}_{B1} + \mathbf{n}_{D2}, \tag{59}$$

Iterative Detection Scheme

The receiver structure is similar to Fig. 6(b), except that pre-cancellation is performed after DFT and linear combiner precedes SMIC. After DFT on $\mathbf{r}_{D1}$ and $\mathbf{r}_{D2}$, $\mathbf{R}_{D1}(=\mathbf{W}\mathbf{r}_{D1})$ and $\mathbf{R}_{D2}(=\mathbf{W}\mathbf{r}_{D2})$ can be obtained as follows:

$$\mathbf{R}_{D1} = \sqrt{1-\gamma^2}\,\rangle_{BD}\mathbf{X}_{B1} + \gamma\rangle_{BD}\mathbf{X}_{A1} + \mathbf{N}_{D1}, \tag{60}$$

Figure 7. The block error rate performance of the relay-assisted SC-FDE system on EDGE TU channels (N=256, QPSK, f_DNT_s=0.001).

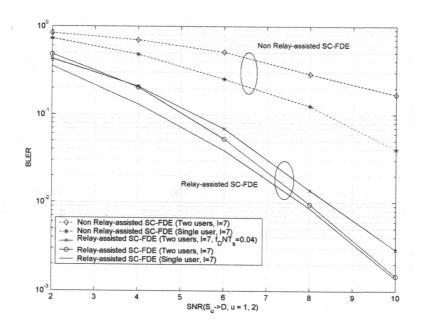

$$\mathbf{R}_{D2} = \sqrt{1-\gamma^2}\, \Lambda_{AD}\mathbf{X}_{A2} + \gamma\, \Lambda_{AD}\mathbf{X}_{B1} + \mathbf{N}_{D2}, \tag{61}$$

where $\mathbf{X}_{ui} = \mathbf{W}\mathbf{x}_{ui}$, $u = A, B$, $i = 1, 2$ and $\mathbf{N}_{Di} = \mathbf{W}\mathbf{n}_{Di}$. $\Lambda_{TR}(= \mathbf{W}\mathbf{H}_{TR}\mathbf{W}^{-1})$ is an $N \times N$ diagonal matrix. In (60), $\mathbf{X}_{A1}$ can be cancelled using the previously detected $\bar{\mathbf{X}}_{A1}$ from (55) and (57). To distinguish this cancellation process from the soft interference cancellation during iteration, we call this process *pre-cancellation*. After pre-cancellation, $\mathbf{R}'_{D1}$ can be expressed as

$$\mathbf{R}'_{D1} = \sqrt{1-\gamma^2}\, \Lambda_{BD}\mathbf{X}_{B1} + \gamma\, \Lambda_{BD}(\mathbf{X}_{A1}-\bar{\mathbf{X}}_{A1}) + \mathbf{N}_{D1} \approx \sqrt{1-\gamma^2}\, \Lambda_{BD}\mathbf{X}_{B1} + \mathbf{N}_{D1}. \tag{62}$$

It is also assumed that the decoding of $\mathbf{X}_{A1}$ is perfect. Since the power of $\mathbf{X}_{A2}$ is less than that of $\mathbf{X}_{B1}$ in (61) and (62), it motivates us to apply the SMIC to reconstruct $\mathbf{X}_{A2}$. Thus, we decode $\mathbf{X}_{B1}$ first. Then we proceed to subtract the soft decoded version of $\mathbf{X}_{B1}$ from the received signal $\mathbf{R}_{D2}$, and decode $\mathbf{X}_{A2}$. To decode $\mathbf{X}_{B1}$ from the linearly combined form of (61) and (62), we change (61) and (62) into the matrix form as

$$\mathbf{R}'_{B1} \triangleq [\mathbf{R}'_{D1} \quad \mathbf{R}_{D2}]^T = \begin{bmatrix} \sqrt{1-\gamma^2}\, \Lambda_{BD} \\ \gamma\, \Lambda_{AD} \end{bmatrix} \mathbf{X}_{B1} + \begin{bmatrix} 0 \\ \sqrt{1-\gamma^2}\, \Lambda_{AD} \end{bmatrix} \mathbf{X}_{A2} + \begin{bmatrix} \mathbf{N}_{D1} \\ \mathbf{N}_{D2} \end{bmatrix}. \tag{63}$$

From (63), we derive a linear combining receiver, which is

$$\begin{aligned}
\tilde{\mathbf{R}}_{B1} &= [\sqrt{1-\gamma^2}\, \Lambda_{BD}^H \quad \gamma\, \Lambda_{AD}^H]\mathbf{R}'_{B1} \\
&= ((1-\gamma^2)\Lambda_{BD}^H\Lambda_{BD} + \gamma^2\, \Lambda_{AD}^H\Lambda_{AD})\mathbf{X}_{B1} + \gamma\sqrt{1-\gamma^2}\, \Lambda_{AD}^H\Lambda_{AD}\mathbf{X}_{A2} + \sqrt{1-\gamma^2}\, \Lambda_{BD}^H\mathbf{N}_{D1} + \gamma\, \Lambda_{AD}^H\mathbf{N}_{D2} \\
&\triangleq \Lambda_{B1}\mathbf{X}_{B1} + \Lambda_{A2}\mathbf{X}_{A2} + \mathbf{N}_{B1},
\end{aligned} \tag{64}$$

where $\Lambda_{B1}(k) = (1-\gamma^2)\,|\,\Lambda_{BD}(k)\,|^2 +\gamma^2\,|\,\Lambda_{AD}(k)\,|^2$, and $\Lambda_{A2}(k) = \gamma\sqrt{1-\gamma^2}\,|\,\Lambda_{AD}(k)\,|^2$. After SMIC, the linearly combined vector for $\mathbf{X}_{B1}$ can be expressed as

$$\mathbf{Z}_{B1} = \Lambda_{B1}\mathbf{X}_{B1} + \Lambda_{A2}(\mathbf{X}_{A2} - \bar{\mathbf{X}}_{A2}) + \mathbf{N}_{B1}, \tag{65}$$

where $\bar{\mathbf{X}}_{A2} = \mathbf{W}\bar{\mathbf{x}}_{A2}$ and $\bar{\mathbf{x}}_{A2}$ is the mean of the transmit vector $\mathbf{x}_{A2}$. The coefficients of the one-tap MMSE equalizer are obtained as

$$G_{B1}(k) = \frac{\Lambda_{B1}(k)}{\nu_{B1}\,|\,\Lambda_{B1}(k)\,|^2 +\nu_{A2}\,|\,\Lambda_{A2}(k)\,|^2 +\sigma_n^2\Lambda_{B1}(k)}, \quad k = 0,1,...,N-1, \tag{66}$$

where $\nu_{uj} = \frac{1}{N} trace(\mathbf{V}_{uj}), u = A, B, j = 1, 2$, and $\mathbf{V}_{uj} = Cov(\mathbf{x}_{uj}, \mathbf{x}_{uj}) = Diag(\nu_{uj}(0), \nu_{uj}(1), \cdots, \nu_{uj}(N-1))$ represents the covariance matrix of the transmitted vector $\mathbf{x}_{uj}$. In the initial iteration, we set $\mathbf{V}_{B1}$ to the values used in the previous detection and $\overline{\mathbf{x}}_{A2} = \mathbf{0}$, $\mathbf{V}_{A2} = \mathbf{I}_N$. Because there is no *a priori* information of $\mathbf{x}_{A2}$ available from the decoder in the initial iteration, the performance of MMSE equalizer with linear combining is even worse than that of MMSE equalizer without linear combining. Thus, to improve the initial performance of the system, we use the pre-cancelled $\mathbf{R}'_{D1}$ for equalization in the initial iteration. The coefficients of MMSE equalizer for the initial iteration are obtained as

$$G_{B1}(k) = \frac{\sqrt{1-\gamma^2}\Lambda_{BD}(k)}{(1-\gamma^2)\nu_{B1} \mid \Lambda_{BD}(k) \mid^2 + \sigma_n^2}, \quad k = 0, 1, ..., N-1. \tag{67}$$

We also employ the procedures used in (Lim, 2007) to obtain the frequency domain estimates $\{\hat{X}_{B1}(k)\}_{k=0}^{N-1}$. The estimates are given by

$$\hat{X}_{B1}(k) = G_{B1}^*(k)Z_{B1}(k) + (\mu_{B1} - G_{B1}^*(k)\Lambda_{B1}(k))\overline{X}_{B1}(k), \quad k = 0, 1, ..., N-1, \tag{68}$$

where $\mu_{B1} = \frac{1}{N}\sum_{k=0}^{N-1}G_{B1}^*(k)\Lambda_{B1}(k)$. The time domain estimates $\{\hat{x}_{B1}(n)\}_{n=0}^{N-1}$ are obtained from the frequency domain estimates $\{\hat{X}_{B1}(k)\}_{k=0}^{N-1}$ using the IDFT operation. Then, we can obtain the extrinsic LLRs and the *a priori* LLRs $\{L_{DA}(c_{B1}(l))\}_{l=0}^{2N-1} = \pi_B^{-1}(\{L_E(c_{B1}(m))\}_{m=0}^{2N-1})$ are fed to the MAP decoder.

Because there is no $\mathbf{x}_{A2}$ component in $\mathbf{R}_{D1}$, $\mathbf{x}_{A2}$ is detected from $\mathbf{R}_{D2}$ without linear combining. After SMIC, $\mathbf{R}_{D2}$ can be expressed as

$$\mathbf{Z}_{A2} = \sqrt{1-\gamma^2}, \,_{AD}\mathbf{X}_{A2} + \gamma, \,_{AD}(\mathbf{X}_{B1} - \overline{\mathbf{X}}_{B1}) + \mathbf{N}_{D2}. \tag{69}$$

Then, the coefficients of the one-tap MMSE equalizer are obtained as

$$G_{A2}(k) = \frac{\sqrt{1-\gamma^2}\Lambda_{AD}(k)}{(1-\gamma^2)\nu_{A2} \mid \Lambda_{AD}(k) \mid^2 + \gamma^2\nu_{B1} \mid \Lambda_{AD}(k) \mid^2 + \sigma_n^2}, \quad k = 0, 1, ..., N-1. \tag{70}$$

For computer simulations, we used a 1/2-rate convolutional code with a generator polynomial $G = (7,5)_8$ and $f_D N T_s = 0.001$. Random interleavers with different interleaving patterns are employed for user separation. Each data block consists of 256 BPSK or QPSK symbols. $\mathbf{H}_{AB}$, $\mathbf{H}_{BA}$, $\mathbf{H}_{AD}$, and $\mathbf{H}_{BD}$ are modelled as EDGE TU (Furuskar, 1999) channels with $L = 2$. We assume that a node, which acts as a relay, can perfectly decode the data block from the source node.

Figure 8 shows the BER performance with the optimum weighting factor obtained from simulations ($\gamma^2 = 0.15$). It is seen that the superposed cooperative transmission system can obtain diversity gain over frequency selective channels. The performance of perfect pre-cancellation is also included for comparison, where the previously detected data block is assumed to be correct. It is shown that the pre-cancellation process effectively removes the interference by using the previously detected data block, which leads to a further reduction in the computational complexity (see (62)).

FUTURE TRENDS

The half-duplex constraint of cooperative transmission leads to a substantial loss in spectral efficiency. There have been some approaches to compensate the half-duplex loss. One way to avoid the pre-log factor 1/2 is to use a full-duplex relay that may receive and transmit at the same time and frequency. Oscillation caused by RF output feedback between transmit and receive antennas extremely limits the transmit and receive outputs. RF repeater, which directly relays the received signal, has been used for restricted applications such as in-building or subway. An interference cancellation system (ICS) repeater has an additional feature solving this isolation/interference problem, which uses digital signal processor to cancel the feedback interference. However, the large difference in the power of transmitted and received signals makes it difficult to cancel the self-interference.

Network coding over the physical-layer is another approach to cope with the half-duplex problem. Network coding is an arbitrary operation on messages arriving at nodes to generate the outputs which can be decoded at the final destinations. It was originally developed for routing in wired networks (Chou, 2007). For example, we assume that nodes S1 and S2 intend to exchange information using two-way relaying, which is first introduced by Shannon (1961). In the first (second) phase, node S1

Figure 8. BER performance of superposed cooperative transmission with pre-cancellation (QPSK, EDGE TU channels, $\gamma^2 = 0.15$).

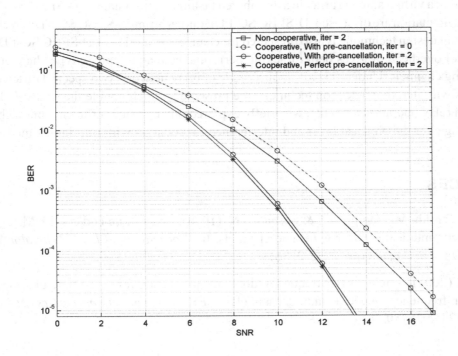

(S2) transmits its own signal s_1 (s_2) to relay, and then the relay decodes the signal s_1 (s_2). In the third phase, the relay broadcasts the signal generated by bit-wise XOR operation, $s_3 = s_1 \oplus s_2$, to nodes S1 and S2. By decoding the received signal s_3, nodes S1 and S2 can then recover the desired signal s_2 by $s_1 \oplus s_3$ and s_1 by $s_2 \oplus s_3$, respectively. Note that network coding requires only three transmission time slots to complete transmissions (Rankov, 2007; Liang, 2008), while four time slots are required in conventinal approaches.

Using the property of electromagnetic waveforms, a physical-layer network coding (PNC) was introduced for two-way relaying (Zhang, 2006). In PNC, we need only two time slots for nodes S1 and S2 to exchange two messages. In the first phase, both S1 and S2 send their own signals, and the relay receives the mixed signal from the two nodes. In the second phase, the relay broadcasts the mixed signal to both S1 and S2. Similar to the network coding method discussed above, S1 (S2) can remove its own transmitted signal in the received signal, and then recover the desired message from S2 (S1). Recently, an adaptive network-coded cooperation (ANCC) over Galois field was proposed to match the instantaneous network graph with the channel code graph (Bao, 2008). To further improve the network throughput, complex field netwok coding (CFNC) approach has been developed, which entails symbol-level operations at the physical-layer (Wang, 2008). CFNC achieves higher throughput than ANCC and PNC schemes while attaining the maximum possible diversity gain with multiuser detection. With the advance of channel and source coding schemes, the network coding can give more benefits to future wirless relay networks.

CONCLUSION

Cooperative diversity is an effective technique to combat the fading phenomena in wireless communications, by forming a virtual antenna array in a distributed fashion. In this chapter, D-SFBC SC-FDE and its efficient implementation are discussed. D-SFBC SC-FDE outperforms D-STBC SC-FDE over fast fading channels. A pilot design technique is also presented for channel estimation of D-SFBC SC-FDE. Without additional pilot overhead, it gives better tracking performance than the block-type channel estimation over fast fading channels. Under the half-duplex mode of relay network, two spectral efficient cooperative protocols with iterative detection technique are presented, which double the spectral efficiency of TDMA-based relay operations. Finally, we briefly explored other spectral efficient approaches, such as network coding over the physical layer and interference cancellation with full-duplex mode.

REFERENCES

Agrawal, D., Tarokh, V., Naguib, A., & Seshadri, N. (1998). Space-time coded OFDM for high data-rate wireless communication over wideband channels. In *Proceedings of IEEE Vehicular Technology Conference,* Ottawa, Canada (pp. 2232-2236).

Al-Dhahir, N. (2001). Single-carrier frequency-domain equalization for space-time block-coded transmissions over frequency-selective fading channels. *IEEE Communications Letters, 5*(7), 304–306. doi:10.1109/4234.935750

Alamouti, S. M. (1998). A simple transmit diversity technique for wireless communications. *IEEE Journal on Selected Areas in Communications, 16*(10), 1451–1458. doi:10.1109/49.730453

Bao, X., & Li, J. (2008). Adaptive network coded cooperation (ANCC) for wireless relay networks: Matching code-on-graph with network-on-graph. *IEEE Transactions on Wireless Communications, 7*(2), 574–583. doi:10.1109/TWC.2008.060439

Bauch, G. (2003). Space-time block codes versus space-frequency block codes. In *Proceedings of IEEE Vehicular Technology Conference,* Jeju, South Korea (pp. 567-571).

Choi, C. H., Lim, J. B., & Im, G. H. (2007). Unique-word-based single carrier system with decision feedback equalization for space-time block coded transmissions. *IEEE Communications Letters, 11*(1), 28–30. doi:10.1109/LCOMM.2007.060875

Chou, P. A., & Wu, Y. (2007). Network coding for the Internet and wireless networks. *IEEE Signal Processing Magazine, 24*(5), 77–85. doi:10.1109/MSP.2007.904818

Chu, D. (1972). Polyphase codes with good periodic correlation properties. *IEEE Transactions on Information Theory, 18*(7), 531–532. doi:10.1109/TIT.1972.1054840

COST 207 TD(86)51-REV 3(WG1). (1998). Proposal on channel transfer functions to be used in GSM tests late 1986.

Falconer, D., Ariyavisitakul, S. L., Benyamin-Seeyar, A., & Eidson, B. (2002). White paper: Frequency domain equalization for single-carrier broadband wireless systems. Retrieved from http://www.sce.carleton.ca/bbw/papers/whitepaper2.pdf

Furuskar, A., Mazur, S., Muller, F., & Olofsson, H. (1999). EDGE: Enhanced data rates of GSM and TDMA/136 evolution. *IEEE Personal Communications Magazine, 6*(3), 56–66. doi:10.1109/98.772978

Ishii, K. (2007). Coded cooperation protocol utilizing superposition modulation for half-duplex scenario. In *Proceedings of IEEE Personal, Indoor, and Mobile Radio Communications*, Athens, Greece.

Jang, J. H., Won, H. C., & Im, G. H. (2006). Cyclic prefixed single carrier transmission with SFBC over mobile wireless channels. *IEEE Signal Processing Letters, 13*(5), 261–264. doi:10.1109/LSP.2006.870374

Kim, D., Kwon, U. K., & Im, G. H. (2008). Pilot position selection and detection for channel estimation of SC-FDE. *IEEE Communications Letters, 12*(5), 350–352. doi:10.1109/LCOMM.2008.080071

Kwon, U. K., Kim, D., & Im, G. H. (in press). Amplitude clipping and iterative reconstruction of MIMO-OFDM signals with optimum equalization. *IEEE Transactions on Wireless Communications*.

Lam, C.-T., Falconer, D., Danilo-Lemoine, F., & Dinis, R. (2006). Channel estimation for SC-FDE systems using frequency domain multiplexed pilots. In *Proceedings of IEEE Vehicular Technology Conference*, Montreal, Canada.

Liang, Y.-C., & Zhang, R. (2008). Optimal analogue relaying with multiple-antennas for physical layer network coding. In *Proceedings of IEEE International Conference on Communications*, Beijing, China.

Lim, J. B., Choi, C. H., Yune, T. W., & Im, G. H. (2007). Iterative multiuser detection for single-carrier frequency-domain equalization. *IEEE Communications Letters*, *11*(6), 471–473. doi:10.1109/LCOMM.2007.070063

LMSC. (2008). *Draft amendment to IEEE standard for local and metropolitan area networks-part 16: Air interface for fixed and mobile broadband wireless access systems-multihop relay specification.* Retrieved from http://www.ieee802.org/16/

Mheidat, H., Uysal, M., & Al-Dhahir, N. (2007). Equalization techniques for distributed space-time block codes with amplify-and-forward relaying. *IEEE Transactions on Signal Processing*, *55*(5), 1839–1852. doi:10.1109/TSP.2006.889974

Moher, M., & Guinand, P. (1998). An iterative algorithm for asynchronous coded multi-user detection. *IEEE Communications Letters*, *2*(8), 229–231. doi:10.1109/4234.709440

Myung, H. G., Lim, J., & Goodman, D. J. (2006). Single carrier FDMA for uplink wireless transmission. *IEEE Vehicular Technology Magazine*, *1*(3), 30–38. doi:10.1109/MVT.2006.307304

Nabar, R. U., Bölcskei, H. M., & Kneübhler, F. W. (2004). Fading relay channels: Performance limits and space-time signal design. *IEEE Journal on Selected Areas in Communications*, *22*(6), 1099–1109. doi:10.1109/JSAC.2004.830922

Oppenheim, A., & Schafer, R. (1989). *Discrete-time signal processing.* Englewood Cliffs, NJ: Prentice-Hall.

Pabst, R., Walke, B. H., Schultz, D. C., Herhold, P., Yanikomeroglu, H., & Mukherjeee, S. (2004). Relay-based deployment concepts for wireless and mobile broadband radio. *IEEE Communications Magazine*, *42*(9), 80–89. doi:10.1109/MCOM.2004.1336724

Prasad, R. (2004). *OFDM for wireless communications systems.* London: Artech House.

Rankov, B., & Wittneben, A. (2007). Spectral efficient protocols for half-duplex fading relay channels. *IEEE Journal on Selected Areas in Communications*, *25*(2), 379–389. doi:10.1109/JSAC.2007.070213

Sendonaris, A., Erkip, E., & Aazhang, B. (2003). User cooperation diversity. Part I. System description. *IEEE Transactions on Communications*, *51*(11), 1927–1938. doi:10.1109/TCOMM.2003.818096

Seol, D. Y., Kwon, U. K., & Im, G. H. (in press). Performance of single carrier transmission with cooperative diversity over fast fading channels. *IEEE Transactions on Communications*.

Shannon, C. E. (1961). Two-way communication channels. In *Proceedings of the 4th Berkeley Symposium on Mathematical Statistics and Probability* (pp. 611-614).

Stüber, G. L. (2000). *Principle of mobile communication*, 2nd ed. Kluwer Academic Publishers.

Wang, T., & Giannakis, G. B. (2008). Complex field network coding for multiuser cooperative communications. *IEEE Journal on Selected Areas in Communications*, *26*(3), 561–571. doi:10.1109/JSAC.2008.4481380

Zhang, S., Liew, S. C., & Lam, P. P. (2006). Hot topic: Physical-layer network coding. In *Proceedings of 12ᵗʰ Annual International Conference on Mobile Computing and Networking,* Los Angeles, CA (pp. 23-26).

Section 5
Methematical Tools for Analysis and Design of Cooperative Networks

Chapter 16
Applications of Majorization Theory in Space–Time Cooperative Communications

Aydin Sezgin
Stanford University, USA

Eduard A. Jorswieck
TU Dresden, Germany

ABSTRACT

This chapter discusses important aspects in cooperative communications such as power allocation and node distributions using majorization theory, spanning both theoretical foundations and practical issues. Majorization theory provides a large amount of tools and techniques which can be used in order to accelerate the pace of developments in this fascinating research area of cooperative communications. The aim of the chapter is to build good intuition and insight into this important field of cooperative communications and how majorization theory can be used in order to solve quite complex problems in a very efficient and elegant way. Although we focus on some specific applications, the tools can be also applied to other setups and processing techniques.

INTRODUCTION

In recent years, the goal of providing high speed wireless data services has generated a huge amount of interest among the research community. Recent information theoretic results have demonstrated that the ability of a system to support a high link quality and higher data rates in the presence of Rayleigh fading improves significantly with the use of multiple transmit and receive antennas (Telatar 1999; Foschini & Gans 1998; Tarokh, Seshadri, & Calderbank 1998), so called MIMO (multiple input multiple output) systems. However, due to the size and power requirements the number of antennas at the mobile units is limited. Furthermore, shadowing and the impact of scatterers distribution limits the available degrees of freedom. With this restriction in addition to a disturbed link between mobile and destination, reliable communication can not be guaranteed by using a conventional point-to-point connection.

DOI: 10.4018/978-1-60566-665-5.ch016

Therefore, (Sendoris, Erkip, & Aazhang 1998; Sendoris, Erkip, & Aazhang 2003) proposed to allow cooperation between users and relays. For example, they can create a virtual MIMO system in order to achieve transmit diversity and to reduce the detrimental effect of pass-loss and shadowing in wireless environments. From an information theoretic point of view the cooperative scheme is founded in the relay channel. Recent studies on relay channels can be found in, e.g., (Kramer, Gastpar, & Gupta 2003) and references therein.

Among others, the capacity was investigated by T. M. Cover and A. El Gamal in their fundamental work (Cover and El Gamal 1979) based on work of E.C. van der Meulen (van der Meulen 1971), who introduced this channel model. Note that cooperative communication is a generalization of relay channels, since cooperative users act both as sources as well as relays. From a different perspective, cooperative transmission schemes can be seen as a network code for a special scenario (Fragouli and Soljanin 2007; Yeung, Li, Cai, & Zhang 2005).

The use of cooperation diversity has been shown to result in a reduction of transmit power and increase of spectral efficiency. Several methods are available in order to realize cooperative cooperation, including decode-and-forwarding and amplify-and-forwarding. Depending on the method used and on the quality of the link between the cooperating nodes significant gains can be achieved. This was shown e.g. in (Oechtering & Sezgin 2004). In (Nosratinia and Hunter 2007), it was shown that network-wide diversity gains can be achieved even with distributed grouping and partner selection procedures in such networks.

Relaying, so far only for the extension of coverage, has already been introduced in the 802.16j (WiMAX) standard. However, this is promising that cooperative methods will be adopted in more commercial products in the future, perhaps in a much more sophisticated way.

As discussed in the introduction of (Kramer, Berry, El Gamal, El Gamal, Franceschetti, Gastpar, & Laneman 2007), analyzing the performance of such networks appears be much more difficult than single point-to-point links. Thus, although several important results in cooperative communications have been obtained, many open questions remain in both the theory as well as the practice. Majorization theory (Marshall & Olkin 1979, Palomar & Jiang 2006; Jorswieck & Boche 2006) is a useful tool in order to derive the optimal transmit strategy and to characterize the impact of different systems parameters on the performance of a wireless network. It provides some interesting insights into the behavior of the system, even when the performance function is rather difficult to analyze, as e.g. for non-convex functions. Majorization itself has been used in a vast field of applications. It was already applied successfully in point-to-point MIMO communication links (Palomar & Jiang 2006; Jorswieck & Boche 2006). For example, Majorization theory was used in (Palomar & Jiang 2006) to optimize classes of objective functions in multiple antenna and cellular systems and in (Jorswieck & Boche 2007) in order to investigate the impact of spatial correlation and user distribution on the capacity of such systems.

This chapter discusses important aspects in cooperative communications using Majorization theory, spanning both theoretical foundations and practical issues. Majorization theory provides a large amount of tools and techniques which can be used in order to accelerate the pace of developments in this fascinating research area of cooperative communications. We first discuss different ways how cooperation among users and relays might take place, both for bidirectional communication as well as for unidirectional communication. We focus on Decode-and Forward and Amplify-and-Forward Relay Processing, however, the tools can be also applied to other relay processing techniques. We then introduce the theory of Majorization with some definitions, properties and examples. We also show how functions can be characterized with respect to Schur-convexity and discuss some proof techniques. Furthermore, the

Figure 1. Cooperative system with decode-and-forward at the relay

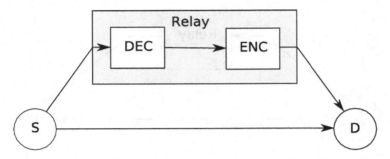

optimization of Schur-convex functions is discussed. Next, we describe cooperative wireless system structures and show, how Majorization theory can be used in order to derive optimal transmit strategies. Further, the impact of different system parameters (as e.g. the location of the relays) on the system performance is investigated. The chapter has copious illustrations with a reasonable mathematical content. The aim is to build good intuition and insight into this important field of cooperative communications and how Majorization theory can be used in order to solve quite complex problems in a very efficient and elegant way.

Cooperative Communication Systems

This section serves as an introduction on how cooperation among users and relays might take place, both for bidirectional communication as well as for unidirectional communication. Here, we focus on basic elements of cooperative communication systems. In general, every cooperative communication system can be represented as a connection of those basic elements. We start with the discussion on relay processing strategies. Afterwards, the aspect of information flow is briefly enlightened.

Relay Processing

A fundamental issue in cooperative communication systems is how the relay is processing the received data from the source. There are many techniques available. The most common techniques are Decode-and-forward (DF) and amplify-and-forward (AF) as well as compress-and-forward (CF). Depending on the scenario, one or another technique might outperform the other strategies. In this chapter, we focus on the decode-and-forward and amplify-and-forward techniques, although the analysis and techniques used in the chapter are not restricted to those.

Decode and Forward

A cooperative system using the Decode-and-forward protocol is presented in Figure. 1.

The source transmits data, which is then received at the relay and, if there is direct link available, also at the destination. At the relay, the received message is first decoded. The link from the source to the relay is in average of higher quality as the source-destination link, the decoding at the relay is much more reliable. For simplicity, it is often assumed that the decoding at the relay is successful with probability one. In practice, also the decoding at the relay is imperfect and error propagation has to be

Figure. 2. Unidirectional cooperative system

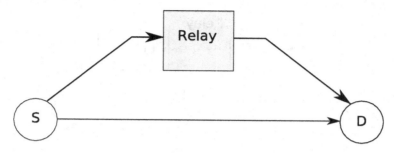

taken into account. Afterwards encoding is performed using the same encoder as the source encoder. Alternatively, different coding strategies may be applied. This newly encoded data is then transmitted to the destination in a consecutive time slot (or a different frequency). Assuming that the decoding process was successful (to a certain degree) at the relay, the additional information from the relay can be used in order to improve the performance.

Amplify-and-Forward

Differently from the case above, the relay is amplifying the received data and forwarding it to destination in a consecutive time slot. If the relay is equipped with multiple antennas, than linear precoding strategies (i.e. beamforming in combination with power allocation) can be applied.

The amount of amplification depends on the availability of channel state information at the relay and the performance criterion, which has to be optimized. As already mentioned, there are also other relay processing strategies possible, however the chapter discusses only the two above mentioned approaches.

Information Flow

In addition to partitioning cooperative systems by aspects of relay processing, it is also possible to divide them into unidirectional and bidirectional systems. In unidirectional systems, the information flow is solely from the source to the destination, while for the bidirectional case the flow is in both directions.

Unidirectional Systems

An example for a unidirectional system is depicted in Figure 2. In general, it is assumed that the relays are working in a half-duplex mode, i.e. the signal is received in the first time slot and transmitted in the next time slot. The communication can also take place on two different non-interfering subcarriers.

Bidirectional Systems

In the bidirectional case, both the source and the destination have data to transmit to each other, i.e. there is an exchange of information. Assuming full-duplex nodes might be unrealistic and thus, in the first time slot the source and the destination are transmitting data to the relay. Although the source signal might be available at the destination trough a direct link and vice versa, the transmitting nodes are not able to detect the signal due to the half-duplex constraints.

From the relay point of view, the channel looks like a multiple access channel (MAC) with two transmitters. The relay may be equipped with single or multiple antennas and might use linear or non-linear detection schemes. Another possibility is the application of network coding. Once the data is processed at the relay, it is transmitted back to the source and the destination, either simultaneously or consecutively. In the relay transmit mode, the channel looks like a broadcast channel.

In the following sections of this book chapter, we analyze the different cooperative communication systems.

Majorization Theory

In this chapter, we provide the basic definitions for majorization and order preserving functions, namely Schur-convex and Schur-concave functions. For later application, we list a number of important properties and show how one can check whether a function is Schur-convex and Schur-concave. Finally, we provide some examples to illustrate the abstract notion. Note that parts of this section are discussed in more detail in [Jorswieck&Boche2007]. However, the section is self-contained such that no further reference is needed for the basic understanding. Rigorous proofs can be found in [Marshall&Olkin1979] and [Jorswieck&Boche2007]. Furthermore, the section contains some novel material needed for the specific applications in later chapters.

Definitions, Properties and Examples

There are many ways to define a partial order on vectors, e.g., majorization. In this section, we take the intuitive way and motivate the order by its operational meaning which we use in the applications later. In order to assure that the concept of majorization is well understood, we present three equivalent definitions and discuss their relationship.

Everybody is familiar with the natural order of non-negative real numbers, i.e. for $a, b \in \Re_+$ we say that a is greater than or equal to b if $a - b$ is non-negative and write $a \geq b$. This is a total order since it is anti-symmetric, transitive, and complete.

Let us consider the set of non-negative real vectors $\mathbf{a}, \mathbf{b} \in \Re_+^n$ of dimension n. There are many possible orders for these two vectors. One very natural order is similar to the order for the real numbers from above. First, sort the entries of the two vectors in non-increasing order, i.e. $a_1 \geq a_2 \geq ... \geq a_n \geq 0$. $\mathbf{a}$ is greater than or equal to $\mathbf{b}$ if the difference $\mathbf{a} - \mathbf{b}$ has non-negative entries. It is obvious that this is only a partial order and not all vectors can be compared, e.g. $\mathbf{a} = \begin{bmatrix} 5 & 1 \end{bmatrix}$ and $\mathbf{b} = \begin{bmatrix} 4 & 2 \end{bmatrix}$ because the difference is $\begin{bmatrix} 5 & 1 \end{bmatrix} - \begin{bmatrix} 4 & 2 \end{bmatrix} = \begin{bmatrix} 1 & -1 \end{bmatrix} \notin \Re_+^2$.

If we stick to the example and study the two (unordered) vectors of length two in more detail we can observe that the vector $\mathbf{b}$ has *more blurred* components than the vector $\mathbf{a}$ because both components of $\mathbf{b}$ can be expressed as a linear combination of the components of $\mathbf{a}$ as

$$b_1 = \frac{3}{4} a_1 + \frac{1}{4} a_2 \tag{1}$$

and

$$b_2 = \frac{1}{4}a_1 + \frac{3}{4}a_2 \tag{2}$$

Note that the vector $\mathbf{a}$ cannot be expressed by a non-negative linear combination of $\mathbf{b}$, where the coefficients sum to one.

This example leads directly to the first definition of majorization:

Definition 1: Let $\mathbf{a}, \mathbf{b} \in \Re_+^n$. We say that the vector $\mathbf{a}$ majorizes the vector $\mathbf{b}$ and write $\mathbf{a} \succ \mathbf{b}$ if there exists a doubly-stochastic matrix $\mathbf{P}$ such that $\mathbf{b} = \mathbf{Pa}$.

A doubly-stochastic matrix is a matrix with non-negative entries and row and column sum equal to one, i.e.

$$\sum_{i=1}^{n} P_{ij} = 1, \ 1 \le j \le n, \ \text{ and } \sum_{j=1}^{n} P_{ij} = 1, \ 1 \le i \le n \tag{3}$$

Note that by the definition, it follows automatically that the sum of all components of both vectors is equal.

The doubly-stochastic matrix for the example of the two vectors above is given by

$$\mathbf{b} = \begin{bmatrix} 4 \\ 2 \end{bmatrix} = \begin{bmatrix} \frac{3}{4} & \frac{1}{4} \\ \frac{1}{4} & \frac{3}{4} \end{bmatrix} \begin{bmatrix} 5 \\ 1 \end{bmatrix} = \mathbf{Pa} \tag{4}$$

We could also rewrite the equation using the two permutations $\pi_1 = \begin{bmatrix} 1 & 2 \end{bmatrix}$ and $\pi_2 = \begin{bmatrix} 2 & 1 \end{bmatrix}$ and probabilities $t(\pi_1) = \frac{3}{4}$ and $t(\pi_2) = \frac{1}{4}$ as

$$\mathbf{b} = \begin{bmatrix} 4 \\ 2 \end{bmatrix} = t(\pi_1)\pi_1 \circ \mathbf{a} + t(\pi_2)\pi_2 \circ \mathbf{a} = \frac{3}{4}\begin{bmatrix} 5 \\ 1 \end{bmatrix} + \frac{1}{4}\begin{bmatrix} 1 \\ 5 \end{bmatrix}, \tag{5}$$

in which the notion $\pi_1 \circ \mathbf{a}$ means that the permutation operator π_1 is applied to the vector $\mathbf{a}$. This observation can be formalized in the next Definition of majorization:

Definition 2: Two vectors $\mathbf{a}, \mathbf{b} \in \Re_+^n$ fulfill the majorization inequality $\mathbf{a} \succ \mathbf{b}$ if there exists a probability distribution t on the symmetric group S_n such that

$$\mathbf{b} = \sum_{\pi \in S_n} t(\pi) \pi \circ \mathbf{a} \tag{6}$$

Note that S_n is the symmetric group on $\{1, 2, ..., n\}$. It has the order $n!$.

For these two basic definitions of majorization we do not need a certain order of the components of the two vectors. The third equivalent definition needs the elements of the two vectors ordered in non-increasing order.

Definition 3: Two vectors $\mathbf{a}, \mathbf{b} \in \Re_+^n$ fulfill the majorization inequality $\mathbf{a} \succ \mathbf{b}$ if after sorting the entries of each vector into non-increasing order, the sum of all entries is equal and the sum of the entries in every prefix of $\mathbf{a}$ is no less than the corresponding sum in $\mathbf{b}$, i.e.,

$$\sum_{k=1}^{m} a_k \geq \sum_{k=1}^{m} b_k, \ 1 \leq m < n, \ \text{ and } \ \sum_{k=1}^{n} a_k = \sum_{k=1}^{n} b_k. \tag{7}$$

For the two-dimensional example from above the definition leads to the inequality $a_1 = 5 \geq 4 = b_1$ and equality $a_1 + a_2 = 6 = b_1 + b_2$.

Note that majorization is anti-symmetric, transitive but *not complete* (for vectors with more than two components). One counterexample of vectors which cannot be compared by majorization is given by $\mathbf{a} = \begin{bmatrix} 7 & 3 & 1 \end{bmatrix}$ and $\mathbf{b} = \begin{bmatrix} 6 & 5 & 0 \end{bmatrix}$. Therefore majorization is a partial order.

However, we can always compare the extreme vectors $\check{s} = \left[\frac{1}{n}, \frac{1}{n}, ..., \frac{1}{n} \right]$ and $\hat{A} = \begin{bmatrix} 1, 0, ..., 0 \end{bmatrix}$ with an arbitrary vector $\mathbf{b}$ as follows $\check{s} \prec \mathbf{b} \prec \hat{A}$.

Schur-Convexity: Characterizations and Proof Techniques

In this section, we study a function that takes as an argument a vector and maps to the non-negative real numbers, i.e. $f : \Re_+^n \to \Re_+$. Our interest is in the class of functions which are monotonic with respect to the order of majorization – the class of order-preserving functions. In the scalar case $n = 1$, it is very easy to check whether a function is monotonic increasing or decreasing by studying the first derivative. The general case with arbitrary $n > 1$ is more complicated. First, we need the definition of monotone functions with respect to majorization. These functions are called Schur-convex (monotonic increasing) or Schur-concave (monotonic decreasing).

Definition 4: A function $f : \Re_+^n \to \Re_+$ is called Schur-convex of order n if for $\mathbf{a}, \mathbf{b} \in \Re_+^n$ $\mathbf{a} \succ \mathbf{b}$ implies $f(\mathbf{a}) \geq f(\mathbf{b})$. If the last inequality is reversed the function is called Schur-concave.

From the definition and the last remark in Section 1.5.1 follows for any Schur-convex function f and the extreme vectors $\check{s}$ and $\hat{A}$ with arbitrary $\mathbf{b}$ and $\check{s} \prec \mathbf{b} \prec \hat{A}$ that $f(\check{s}) \leq f(\mathbf{b}) \leq f(\hat{A})$.

The class of Schur-convex functions depends on the dimension n. If a function is Schur-convex for all orders $n \geq 1$ it is simply called Schur-convex.

From an application point of view we need now simple criteria to check whether a given function is Schur-convex or Schur-concave. In other words, we need to give necessary (and sufficient) conditions for Schur-convexity and Schur-concavity.

Theorem 1: Consider a convex[1] and symmetric[2] function $f : \Re_+^n \to \Re_+$. Then the function is Schur-convex.

Proof: Take some $\mathbf{a} \succ \mathbf{b}$ with $\mathbf{b} = \sum_{\pi \in S_n} t(\pi) \pi \circ \mathbf{a}$.

$$f(\mathbf{b}) = f\left(\sum_{\pi \in S_n} t(\pi) \pi \circ \mathbf{a} \right) \leq \sum_{\pi \in S_n} t(\pi) f(\pi \circ \mathbf{a}).$$
$$= \sum_{\pi \in S_n} t(\pi) f(\mathbf{a}) = f(\mathbf{a}) \tag{8}$$

where the inequality follows from Jensen's inequality and the convexity of the function and the second equality follows from the symmetry.

Interestingly, it is sufficient to check only one step in the transition from **a** to **b**, i.e. to consider only two components which are *blurred* when going from **a** to **b**. *Blur* means here that part from the larger component is transferred to the weaker component. Since the function is symmetric, we can consider the first and second component. The monotony behavior is checked for this transfer by considering *Schur's condition*

$$\frac{\partial f(\mathbf{x})}{x_1} - \frac{\partial f(\mathbf{x})}{x_2} \geq 0. \tag{9}$$

Theorem 2: Consider a symmetric function $f : \Re_+^n \to \Re_+$. If Schur's condition in equation (9) is satisfied then the function is Schur-convex. If the function fulfils $\frac{\partial f(\mathbf{x})}{x_1} - \frac{\partial f(\mathbf{x})}{x_2} \leq 0$ it is Schur-concave.

A third way to check whether a composed function f is Schur-convex is to deconstruct the function in its building blocks and carefully check for monotonic behavior, e.g. a monotonic increasing function of a Schur-convex function will be Schur-convex.

As an example, the function $f(\mathbf{x}) = \log\left(1 + \sum_{k=1}^{n} x_k\right)$ is Schur-concave because the function is symmetric and concave.

Note that the methods to check for Schur-convexity can be also applied to functions in which an expectation with respect to some random variables occurs. In this context, the notion of exchangeable random variables is useful. A set of random variables is exchangeable if the probability distribution is symmetric.

Theorem 3: If $X_1, \ldots, X_n$ are exchangeable random variables, $\mathbf{a}, \mathbf{b} \in \Re_+^n$ with $\mathbf{a} \succ \mathbf{b}$, and $f : \Re_+^n \to \Re_+$ is a continuous convex and symmetric function then,

$$\mathrm{E}\left[f\left(a_1 X_1, a_2 X_2, \ldots, a_n X_n\right)\right] \geq \mathrm{E}\left[f\left(b_1 X_1, b_2 X_2, \ldots, b_n X_n\right)\right] \lim_{x \to \infty}. \tag{10}$$

If f is strictly convex, equality occurs only when $\mathbf{a} = \mathbf{b}$, possible after reordering components, or X_i is zero with probability one.

As an example, the function $f(\mathbf{a}) = \mathrm{E}\left[\log\left(1 + \sum_{k=1}^{n} a_k X_k\right)\right]$ with exchangeable random variables $X_1, \ldots, X_n$ is Schur-concave.

As another example, the function $g(\mathbf{a}) = \mathrm{E}\left[\log\left(1 + \max_{1 < k \leq n}\left(a_k X_k\right)\right)\right]$ is Schur-convex. This follows from carefully checking Schur's condition (see Section 2.2.5 in [Jorswieck&Boche2007]).

Optimization of Schur-Convex Functions

In this section, we consider programming problems which involve a Schur-convex or Schur-concave function and use the properties of those functions to solve the problems.

The simplest case is a programming problem in which a Schur-concave or Schur-convex function f

is to be maximized and $\mathbf{x}$ is constrained to be non-negative and its sum is constrained to some constant $\sum_{k=1}^{n} x_k = c$, i.e.

$$\max f(\mathbf{x}) \quad \text{s.t.} \quad \mathbf{x} \geq 0, \quad \sum_{k=1}^{n} x_k = c. \tag{11}$$

1. If the function f is Schur-convex, the optimum is achieved with $\mathbf{x}_1^* = \begin{bmatrix} c & 0 & \cdots & 0 \end{bmatrix}$.
2. If the function f is Schur-concave, the optimum is achieved with $\mathbf{x}_2^* = \frac{c}{n}\begin{bmatrix} 1 & 1 & \cdots & 1 \end{bmatrix}$.

It is important to note that the optimality of $\mathbf{x}_1^*$ is necessary for Schur-convexity of f but not sufficient. Equally, the optimality of $\mathbf{x}_2^*$ is necessary for Schur-concavity of f but not sufficient. The statement about Schur-convexity and Schur-concavity says more about the function than where the maximum or minimum is located. It characterizes the behavior of the function for any (comparable) pair of vectors. To stress the difference, consider the following example.

In the example it is sufficient to take a scalar parameter $\frac{1}{2} < \lambda < 1$ and a variable $0 \leq \varsigma$. Consider the objective function

$$\phi(\lambda, \varsigma) = 1 - \frac{\lambda \exp(-\frac{\varsigma}{\lambda}) - (1 - \lambda)\exp(-\frac{\varsigma}{1-\lambda})}{2\lambda - 1}. \tag{12}$$

Whether the function is monotonic increasing (Schur-convex) with respect to λ or decreasing (Schur-concave) depends on the parameter ς. Consider also the minimum of the function.

1. The function is monotonic increasing if $\varsigma_1 \leq 1$. (This can be easily checked by taking the first derivative with respect to λ.
2. The function is monotonic decreasing if $\varsigma_2 \geq \frac{3}{2}$.
3. But the minimum of the function is obtained with $\lambda = 1$ for all
    $$\varsigma_* \leq -\frac{1}{2}L_W\left(-1, -\frac{1}{2}\exp\left(-\frac{1}{2}\right)\right) \approx 1.2564312.$$

It can be observed that the region where the function is Schur-convex is smaller than the region where the minimum is obtained for $\lambda = 1$. This example will get an operational meaning in the application section when the outage probability of some cooperative scheme is analyzed.

We refer the interested reader to the standard textbook about majorization (Marshall & Olkin 1979) and to the recent comprehensive survey with applications (Jorswieck & Boche 2007).

Interferer Geometry

Introduction to the Problem

The phenomenon of interference is of central importance of future wireless systems. Interference is an impairment which occurs in cellular networks when e.g. the system is in the downlink mode and the mobile station receives data from the desired base station and also interfering signals form the adjacent basestations. Interference plays also a role in decentralized networks, where a node tries to communicate to a different node and neighboring nodes are also active. Interference is also an issue in cognitive radio environments with respect to the process of spectrum sharing.

When investigating the impact of interference, it is often assumed that the interferers are located on the circle around the receiving element of the desired communication link. This assumption simplifies the analysis of such systems, since several issues as e.g. near-far effects can be avoided. In general, however, the interferers can be located arbitrary within the proximity of the receiving end. It is thus unclear, what kind of discrepancy the theoretical results obtained based on this assumption will have in comparison to measurement campaigns or even practical implementations. Another question is whether such an assumption is rather too optimistic.

In this section, we investigate the impact of the interferers' location by using the outage probability of the desired link as the performance criterion. The results could also be used by system designers for the optimal placement (whenever applicable) of base stations or sensors etc. in a network. Surprisingly, it turns out that the assumption of having all interferers' located at the same distance from the receiving unit is rather pessimistic. That means, if the location of the interferers' is highly spread out, then the outage probability is minimized.

System Model

Suppose that there is a source node which wants to communicate with its respective destination node by using a relay node as depicted in Figure 3, since the direct link to the destination is too weak. As shown in the figure, there are several relay nodes the source can communicate with by assuming that the quality of the channels to the relays is comparable. The communication is disturbed by other nodes, which are also communicating at the same time using the same frequency. Each relay receives on average the same interference level as a result of the transmission activity of the interfering nodes. However, the interferer geometry of each relay is different. One relay might get strong interference from one node, while another get gets interference from several nodes, which are individually not so strong. The question here is which relay should be selected for cooperation, i.e. which geometry is more preferable.

The setup is shown in Figure 3 from a different perspective in Figure 4. We assume that the channel between the pair can be described by the zero-mean complex Gaussian random variable h with $h{\sim}CN(0,1)$.

Figure 3. First phase communication in a wireless network

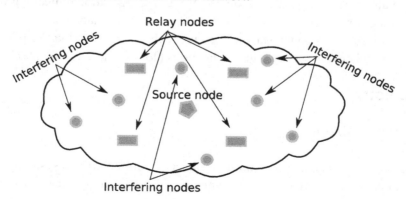

The power of the transmitter is constrained to P. Within the proximity of the receiving unit are several interferers, in total K. We model the channel between the interferer k, $1 \leq k \leq K$ and the receiving unit again as the zero-mean complex Gaussian random variable g_k with $g_k \sim CN(0,1)$. The average received power from the kth interferer at the receiving unit is given by c_k. The vector $c=[c_1, ..., c_K]$ contains the average received power from all interferers and each c_K captures the path loss and shadowing.

The system equation is given by

$$y = \sqrt{P}h + \sum_{k=1}^{K} \sqrt{c_k} g_k + n$$

where n is the additive white Gaussian noise with variance N_0.

At the receiver, a simple detection scheme is employed, where the interference is regarded as an additional noise term. One advantage of such a strategy is the low complexity processing effort at the receiver.

The resulting signal-to-interference-and-noise ratio (SINR) at the receiving unit is thus given by

$$\mathsf{SINR} = \frac{P \, | \, h \, |^2}{N_0 + I},$$

where the interference term I caused by the K interferers is given by

$$I = \sum_{k=1}^{K} c_k \, | \, g_k \, |^2 . \tag{13}$$

The instantaneous rate of this system is thus given by

$$C = \log_2 \left(1 + \mathsf{SINR} \right). \tag{14}$$

The performance measure we are interested in is the outage probability, i.e. the probability that the instantaneous rate C is below a certain threshold rate R_{th}

Figure 4. Interference network

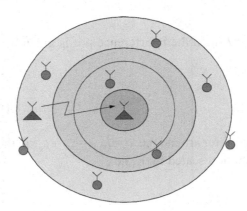

$$Pr\left[C \leq R_{th}\right] = Pr\left[C \leq \log_2\left(1 + \gamma_{th}\right)\right] = Pr\left[\mathsf{SINR} \leq \gamma_{th}\right],$$

where γ_{th} is the threshold SINR in order to achieve R_{th}.

An important issue in such a multi-user environment is the impact of the location of the interferers, which is expressed by the power levels $c_1, ..., c_k, ..., c_K$ received at the receiving unit, on the outage probability. In order to make a fair comparison between different scenarios, we assume that the received interference power level at the receiver is equal for all scenarios, i.e.

$$\sum_{k=1}^{K} c_K = K. \tag{15}$$

Although the transmit power of the interferers might be equal, depending on the distance to the receiving unit, the received power level from each interferer is in general different.

Note that the power cannot be traded between the interferers, thus changes in the c_k translate in varying the location of the interferers by keeping the sum constant according to equation (15).

Impact of Interference Location

In this section, we derive the impact of the interferers' location on the outage probability. To this end, we need the cumulative distribution function (cdf) of the interference term in equation (13), which is given by (Mallik, Win, Shao, Alouini, & Goldsmith 2004)

$$F_I(\nu) = 1 - \frac{1}{\prod_{k=1}^{K} c_k} \sum_{k=1}^{K} \frac{c_k \exp(-\nu / c_k)}{\prod_{l=1,l\neq k}^{K}\left(\frac{1}{c_l} - \frac{1}{c_k}\right)} \tag{16}$$

where the $c_k > 0$ are all distinct. Analogously the probability density function (pdf) is given by

$$f_I(\nu) = \frac{1}{\prod_{k=1}^{K} c_k} \sum_{k=1}^{K} \frac{\exp(-\nu / c_k)}{\prod_{l=1,l\neq k}^{K}\left(\frac{1}{c_l} - \frac{1}{c_k}\right)}$$

If the c_k are all equal, i.e. $c_k = 1$, then the interference pdf is given by a chi-square distribution with K degrees of freedom, i.e.

$$f_I(\nu) = \frac{1}{(K-1)!} \nu^{K-1} \exp(-\nu). \tag{17}$$

If there is only one strong interferer around, i.e. $c_1 = K$, $c_2 = ... = c_k = 0$, then the interference pdf reduces to a standard exponential distribution scaled by K, i.e.,

$$f_I(\nu) = \frac{1}{K} \exp(-\nu / K). \tag{18}$$

As mentioned above, the probability of an outage is given by

$$F_{\mathsf{SINR}}(\gamma_{th}) = Pr\left[\mathsf{SINR} \leq \gamma_{th}\right] \tag{19}$$

Equivalently, the probability for successful transmission is given by (Sezgin, Jorswieck, & Paulraj 2008)

$$Pr\left[\mathsf{SINR} \geq \gamma_{th}\right] = \int_0^\infty F_I(I = \nu) \frac{\gamma_{th}}{P} \exp(-\frac{\gamma_{th}}{P}(N_0 + \nu)) \, d\nu \tag{20}$$

In the following, we investigate whether the function in equation (20) is Schur-convex or Schur-concave.

Due to the lack of space, we present the analysis by first considering the case, in which $K=2$ interferers are present. The analysis to cases with more than two interferers can be found in (Sezgin, Jorswieck, & Paulraj 2008) and (Pereira, Sezgin, Paulraj, & Papanicolaou 2008).

For $K=2$ interferers, the cdf equation (16) of the interference reduces to

$$F_I(I = \nu) = 1 - \frac{c_1 \exp(-\nu / c_1) - c_2 exp(-\nu / c_2)}{c_1 - c_2}$$

In the following, we analyze in equation (20) with respect to Schur-concavity using the conditions in the previous section. For this, we have to compute

$$\Delta(\mathbf{c}) = \frac{\partial Pr\left[\mathsf{SINR} \geq \gamma_{th}\right]}{\partial c_1} - \frac{\partial Pr\left[\mathsf{SINR} \geq \gamma_{th}\right]}{\partial c_2}$$

resulting in

$$\Delta(\mathbf{c}) = \frac{\gamma_{th}}{P} \exp(-\frac{\gamma_{th}}{P} N_0) \int_0^\infty f(\mathbf{c}, \nu) \underbrace{\exp(-\frac{\gamma_{th}}{P})}_{g(\nu)} \tag{21}$$

with

$$f(\mathbf{c}, \nu) = \frac{\partial F_I(I = \nu)}{\partial c_1} - \frac{\partial F_I(I = \nu)}{\partial c_2}.$$

Obviously $g(\nu)$ in equation (21) is a strictly decreasing function with

$$g(0) = 1 \qquad \lim_{\nu \to \infty} g(\nu) = 0$$

The function $f(c, v)$ has for all c the following properties:

$$f(\mathbf{c}, 0) = 0, \qquad \lim_{\nu \to \infty} f(\mathbf{c}, \nu) = 0 \;,$$

and there is one zero point $v^*(c)$ with $f(c, v \le v^*) \ge 0$ and $f(c, v \ge v^*) \le 0$. An example for $f(c, v)$ for $K = 2$ users with $c = [1.2, 0.8]$ is plotted in Figure 5.

Thus, we have the following inequality for equation (21)

$$\int_0^\infty f(\mathbf{c}, \nu) g(\nu) \, d\nu \ge g(\nu^*) \int_0^\infty f(\mathbf{c}, \nu) \, d\nu$$

Interestingly, it can be shown that

$$\int_0^\infty f(\mathbf{c}, \nu) \, d\nu = 0$$

Thus,

$$\Delta(\mathbf{c}) \ge 0.$$

From this it follows that equation (20) is a Schur-convex function, which in turn means, that the outage probability in equation (19) is a Schur-concave function.

Figure 5. $f(c, v)$ for $K = 2$ users with $c_1 = 1.2$ and $c_2 = 0.8$

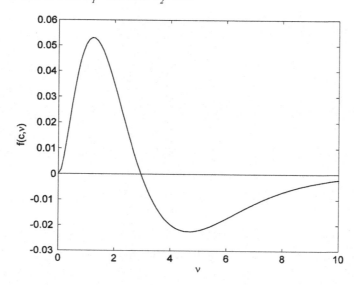

Illustration

In this section, the theoretical results are illustrated by means of numerical simulations. In Figure 6, the outage probability versus the average received power level from the strongest interferer of the one-to-one communication link corrupted by in total $K=2$ interferers is presented. The average receive power level corresponds directly to the location of the interferer. The threshold SINR γ_{th} is set to 7dB. On the abscissa, the point at the extreme left corresponds to having both interferers equidistantly away from the receiving unit, while the point at the extreme right corresponds to having the strong interferer very close to the receiving unit and the second interferer very far away. The figure basically presents the impact of the movement of the strong interferer towards the receiving unit on the outage probability. In order to have a fair comparison, at the same time the second interferer is moving away from the receiving unit. From the figure, we observe that the outage probability is a monotonically decreasing function, i.e. having the interferers more spread out within the cell provides lower outages than having them equidistantly away from the receiving unit. This result agrees with the theoretical findings in the previous sections.

Two-Hop Relaying Network

New network architectures are needed in order to meet the demand for ubiquitous access to wireless high date rate communications. For example, the usage of relays is a way to establish the communication between source and destination pairs, which do not have a direct connection. In the following, we analyze the performance of a two-hop relay network. In more details, we are interested in the impact of different node location on the ergodic sum rate of the two-hop system. Relay networks have been analyzed recently also in (Dana & Hassibi 2006, Morgenshtern & Bölcskei 2007) from a large system limit point of view with multiple relays. Here, we focus on the single relay scenario.

Introduction to the Problem

We consider a two-hop relaying network where multiple sources are communicating with multiple destinations using a relay. It is assumed that there is no direct link between sources and destinations. The communication takes place in two steps. In the first step, the sources transmit data to the relay. The received data is buffered at the relay, which is using a Decode-and-forward protocol. Thus, the relay has always packets for the destination node, which is currently served. In the second step, the relay is forwarding the received messages to one or more destination nodes.

System Model

The two-hop network with K source-destination pairs and one relay is shown in Figure7. We assume that each node as well as the relay has an individual power constraint. It is further assumed that the node at the receiving end of a communication link knows the channel perfectly.

The source nodes have no information about the channel at all in the first hop. In the second hop, we are considering two cases. In the first case, we assume that the relay is uninformed about the channels to the destination nodes, while for the second case, we assume that the relay knows the channel statistics to each destination node.

Figure 6. Outage probability vs. user location, K=2. Threshold SINR γ_{th} is set to 5 (or 7dB in dB scale). The received power level from the second interferer is given by $c_2=K-c_1$.

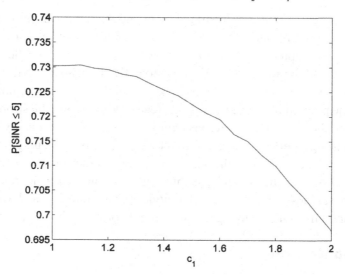

Figure 7. Two-hop network with K source-destination pairs and one relay

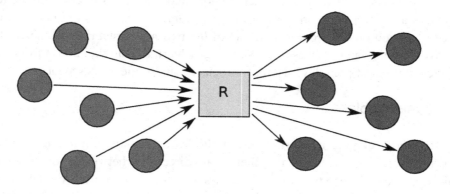

First-Hop

The system equation for the first hop is given by

$$y(t) = \sum_{k=1}^{K} c_k h_k x_k(t) + n(t) \tag{22}$$

where $x_k(t)$ is the transmit signal from source k, $1 \leq k \leq K$ at time instant t, h_k is the quasi-static block flat-fading channel from the k-th source to the relay station. The h_k, $1 \leq k \leq K$, are complex Gaussian distributed with zero mean and with unit variance. The variables c_k are the variance profiles and capture the antenna gain, path loss and shadowing. Without loss of generality we order the source nodes in a decreasing way according to their variance profiles c_k, i.e. $c_1 \geq c_2 \geq ... \geq c_K$. The average transmit power from each source is normalized to P. The additive Gaussian white noise is given by $n(t)$ with variance σ^2, while $y(t)$ is the received signal at the relay at time instant t. In the following, we omit the time index t for convenience. We assume that the relay is performing successive interference cancellation. The ergodic sum rate is then given by

$$R = \mathrm{E}\left[\log\left(1 + \frac{P}{\sigma^2}\sum_{k=1}^{K} c_k \left|h_k\right|^2\right)\right] \tag{23}$$

A good way in order to compare different node location scenarios is to study the impact of the channel variances on the ergodic sum rate in equation (23) using majorization theory. The channel variances hereby capture the path loss and shadowing. In order to guarantee a fair comparison, we constrain the sum variance to be equal to the number of users, i.e.

$$\sum_{k=1}^{K} c_k = K .$$

We start by verifying Schur's condition for the ergodic sum rate by computing the first derivative of R with respect to c_1 and c_2, which are given by

$$\frac{\partial R}{\partial c_1} = \int_0^\infty e^{-t}\left(\mathrm{E}\left[\left|h_1\right|^2 e^{-\frac{P}{\sigma^2}c_1\left|h_1\right|^2}\right]\mathrm{E}\left(e^{-\frac{P}{\sigma^2}c_2\left|h_2\right|^2}\right)\mathrm{E}\left(g(t)\right)\right)dt \tag{24}$$

and

$$\frac{\partial R}{\partial c_2} = \int_0^\infty e^{-t}\left(\mathrm{E}\left(e^{-\frac{P}{\sigma^2}c_1\left|h_1\right|^2}\right)\mathrm{E}\left[\left|h_2\right|^2 e^{-\frac{P}{\sigma^2}c_2\left|h_2\right|^2}\right]\mathrm{E}\left(g(t)\right)\right)dt \tag{25}$$

where $g(t) \geq 0$ is given by

$$g(t) = e^{-t\frac{P}{\sigma^2}\sum_{l=3}^{K}c_l\left|h_l\right|^2} .$$

Thus, we have

$$\frac{\partial R}{\partial c_1} - \frac{\partial R}{\partial c_2} = \int_0^\infty e^{-t}D(t)\mathrm{E}\left(g(t)\right)dt \tag{26}$$

where $exp(-t) \geq 0$ and $E(g(t)) \geq 0$. What remains to show is for $D(t)$ it holds that

$$D(t) = \mathrm{E}\left[\left|h_1\right|^2 e^{-\frac{P}{\sigma^2}c_1\left|h_1\right|^2}\right]\mathrm{E}\left(e^{-\frac{P}{\sigma^2}c_2\left|h_2\right|^2}\right) - \mathrm{E}\left(e^{-\frac{P}{\sigma^2}c_1\left|h_1\right|^2}\right)\mathrm{E}\left[\left|h_2\right|^2 e^{-\frac{P}{\sigma^2}c_2\left|h_2\right|^2}\right] \leq 0 \tag{27}$$

Since $|h_i|^2$, for $i=1,2$, are standard exponential distributed random variables, we have

$$D(t) = \frac{1}{\left(t\frac{P}{\sigma^2}c_1+1\right)^2\left(t\frac{P}{\sigma^2}c_2+1\right)} - \frac{1}{\left(t\frac{P}{\sigma^2}c_2+1\right)^2\left(t\frac{P}{\sigma^2}c_1+1\right)} \tag{28}$$

Thus

$$D(t) = \frac{t\frac{P}{\sigma^2}\left(c_2-c_1\right)}{\left(t\frac{P}{\sigma^2}c_1+1\right)^2\left(t\frac{P}{\sigma^2}c_2+1\right)^2} \leq 0 \tag{29}$$

since $c_1 \geq c_2$. It follows that the ergodic sum rate is a Schur-concave function. That means the ergodic sum rate is maximized when the channels from the source nodes to the relay have equal variances, i.e. the source nodes are equidistantly away from the relay.

Second-Hop

Similar to the first hop, here it is also interesting to understand the impact of the channel variances and thus location of the destination nodes on the ergodic sum rate of the second-hop.

The received signal during the second hop at destination node k is in general given by

$$y_k\left(t+1\right) = g_k\sum_{l=1}^{K}x_l\left(t+1\right) + w_k\left(t+1\right) \tag{30}$$

where the channels g_k, $1 \leq k \leq K$, are complex Gaussian distributed with zero mean and variance c_k. The noise at the destination node k is given by w_k which is modeled as zero-mean complex Gaussian distributed with variance σ^2. Thus, communication in the first hop takes place during even time instances t, while communication in the second hop takes place in the odd time instances $t+1$. As mentioned above, we distinguish between two cases of channel state information knowledge at the relay station. First, we assume that the statistics of the channels are known, i.e. the channel variances. Afterwards, we assume that the transmitter has no knowledge at all.

Statistical Channel Knowledge at the Relay Station

It turns out, that for the case, in which both the relay as well as the destination nodes are equipped with a single-antenna each and the variances of the channels are known at the relay, the best strategy is to transmit to the destination node with the highest channel variance (Jorswieck & Boche 2006). The ergodic sum rate is then given as

$$R = \mathrm{E}\left[\log\left(1 + \frac{P}{\sigma^2}|h|^2 \max_{k=1...K}c_k\right)\right] \tag{31}$$

Since the max-operator is a Schur-convex function, it turns out that the ergodic sum rate for the case, where statistical knowledge at the relay station is available, increases with more spread out destination nodes.

No Channel Knowledge at the Relay Station

By assuming no channel knowledge at the relay station, it turns out, that for the case, in which both the relay as well as the destination nodes are equipped with a single-antenna each, the best strategy is time-sharing (Sezgin, Jorswieck & Charafeddine 2008; Knopp & Humblet 1995), i.e. the relay divides the overall transmission time T into several slots and transmits data only to user k within time slot $t+2k-1$. Since the channel is unknown to the relay station, it turns out that equal power allocation to the time slots is optimal. Thus, the ergodic sum rate is given by

$$R = \frac{1}{K} \sum_{k=1}^{K} \mathrm{E}\left[\log\left(1 + \frac{P}{\sigma^2} c_k \left|h_k\right|^2\right)\right] \qquad (32)$$

It turns out that the expression in equation (32) is Schur-concave with respect to the channel variances $c=[c_1, c_2, ..., c_K]$. In order to show that, we compute the first derivative of the ergodic sum rate with respect to c_k

$$\frac{1}{K} \frac{\frac{P}{\sigma^2}\left|h_k\right|^2}{1 + \frac{P}{\sigma^2} c_k \left|h_k\right|^2} \qquad (33)$$

Since the second derivation with respect to c_k is given by

$$-\frac{1}{K} \frac{\left(\frac{P}{\sigma^2}\left|h_k\right|^2\right)^2}{\left(1 + \frac{P}{\sigma^2} c_k \left|h_k\right|^2\right)^2} \qquad (34)$$

and negative, the function in equation (32) is concave. Since the random variables $h_1, h_2, ..., h_K$ are independent and thus exchangeable, the ergodic sum rate is symmetric with respect to $c_1, c_2, ..., c_K$. Thus, concavity and symmetry, which are necessary conditions for Schur-concavity are fulfilled and thus the ergodic sum rate in equation (32) is a Schur-concave function. Thus, if no channel state information is available at the relay station then the best scenario is when all destination nodes are equidistantly away from the relay station.

Illustration

The ergodic sum rate for the first hop with $K=2$ nodes is depicted in Figure 8 as a function of the channel with the largest variance, i.e. c_1. The transmit SNR at the nodes was set to $SNR=10dB$.

From the figure, we observe that the ergodic sum rate is decreasing for increasing c_1, which confirms our theoretical result.

In Figure 9, the ergodic sum rate for the second hop as a function of the variance c_1 is shown. From the simulations, we observe that the ergodic sum rate is increasing with increasing c_1 when the relay knows the variances of the channels. However, if the relay has no information about the channels, then the ergodic rate is decreasing, i.e. here a scenario is preferred in which all destination nodes are equidistantly away from the relay node. Note that the achievable ergodic rates for no CSI and statistical CSI are equal if the users are symmetrically distributed. This corresponds to $c_1=1$.

WIRELESS SENSOR NETWORK CAPACITY WITH THE MMSE AMPLIFY AND FORWARD RELAY STRATEGY

Introduction to the Problem

In this section, relays are used in order to improve the performance of a wireless sensor network. Without the relays, the propagation loss from the source to the destination node can reduce the signal strength significantly, such that detection is not possible anymore. Beside other strategies, there are three main categories of relaying: Amplify-and-forward, compress-and-forward, and decode-and-forward. In this section, the amplify-and-forward strategy is adopted. For conventional amplify-and-forward strategies, the relay nodes merely compensate for the phase of the signals in order to achieve a constructive addition of the signals at the receiver by using their full power. Here, by adopting the modified strategy used in (Khajehnouri & Sayed 2007), the relays are able to adjust their power in order to achieve a certain SNR target, increase the transmit range and power efficiency of the communication link between the source and the destination. In (Khajehnouri & Sayed 2007), it was assumed that all relays are equidistantly placed between the source and destination, which we refer to as the symmetric scenario. It was not analyzed how the distance of the individual relays to the source and to the destination influences the capacity of

Figure 8. First hop: The ergodic sum rate as a function of the variance c_1 of the channel to the first node. c_2 is given as $c_2=2-c_1$

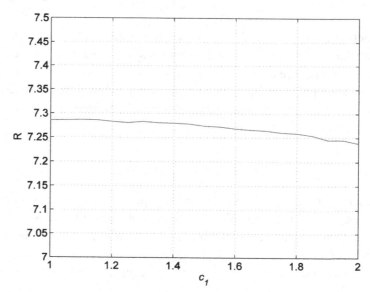

the system under consideration.

System Model

The system under consideration is shown in Figure10. The source intents to submit the data to the destination, however, the communication link between the source and destination is corrupted due the propagation loss and perhaps obstacles and scatterers, such that it might not be possible for the destination to decode the data reliably. The relays, however, are assumed to have a more reliable communication link to both the source and destination. In the first time slot, the data is received from the source, which is then amplified and forwarded in the next time slot to the destination. The relays, in total N, are operating in a half-duplex mode, i.e. they are not able to receive and transmit data simultaneously. Thus, the source is only transmitting half of the time. Of course, it is also possible to optimize the amount of time dedicated to the first hop and the second hop, but this fact is neglected in the following analysis. It is assumed that each node is equipped with a single antenna.

The received signal vector at the N relays is given by

$$\mathbf{r} = \mathbf{h}s + \mathbf{v} \tag{35}$$

where s is the transmit signal with variance σ_s^2 and $\mathbf{h}=[h_1, h_2, \dots, h_N]^{\mathrm{T}}$ is the channel vector . The entries h_i are independent random Gaussian distributed variables with zero mean and variance c_i. Thus, the individual received SNRs at the relays depend directly on the variance of the channel. For our analysis, it is assumed that the overall average received SNR is kept constant, which is reflected by the fact that

Figure 9. Second Hop: Ergodic sum rate as a function of the variance c_1 of the channel to the first node. c_2 is given as $c_2=2-c_1$. Either statistical CSI at the relay or no CSI at the relay is assumed.

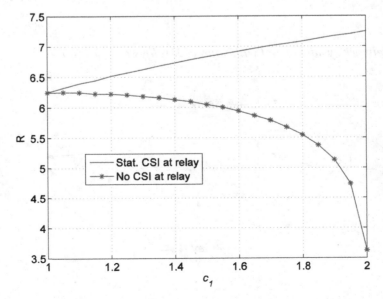

$$\sum_{i=1}^{N} c_i = N$$

is kept constant. A lower c_i means that the relay i is located further away from the source and thus closer to the destination. If one of the c_i equals N, then the relay i is very close to the receiver, while the other ones are far away. If all c_i are equal to $1/N$, then they are all equidistantly away from the source.

The noise vector v in equation (35) is a zero-mean random complex Gaussian distributed variable with variance $\sigma_v^2 I$.

After receiving the signal, the relays apply a transformation given by

$$\mathbf{x} = \mathbf{Fr}, \qquad (36)$$

and transmit the modified signal to the destination. The transformation matrix F has to be determined in order to optimize the performance. Since no cooperation among the relays is assumed, a diagonal structure is imposed on F. The received signal at the destination is given by

$$t = \mathbf{g}^{\mathrm{T}}\mathbf{x} + n, \qquad (37)$$

where the vector $g=[g_1, g_2, \dots, g_N]^{\mathrm{T}}$ describes the random channel between relays and the destination and n is the Gaussian noise with variance σ_n^2. Using equations (35) and (36) in equation (37) results in

$$t = \underbrace{\mathbf{g}^{\mathrm{T}}\mathbf{Fh}}_{h} s + \underbrace{\mathbf{g}^{\mathrm{T}}\mathbf{Fv}}_{w} + n. \qquad (38)$$

By assuming that w is Gaussian distributed, the capacity of the resulting source-destination links is given by

Figure10. Relay Network system model with source, N relays and destination

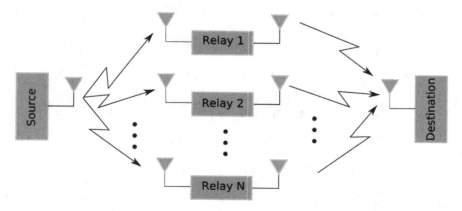

$$C = \frac{1}{2} \mathrm{E} \left[\log_2 \left(1 + \frac{\sigma_s^2 |h|^2}{\sigma_w^2} \right) \right],$$

(39)

where σ_w^2 is the variance of w and the factor ½ comes from the fact that the relays are working in a half-duplex mode, i.e. they are not able to receive and transmit simultaneously.

Capacity Analysis

We are interested in how the distance of the individual relays to the source and to the destination is influencing the capacity of the system. Since the relays are in general only of limited capability, it is assumed that only linear processing is possible at the relays. Thus, following the analysis of (Khajehnouri & Sayed 2007) F is selected such that the mean square error (MSE) between the uncorrupted receive signal at the destination, i.e. $g^T x$, and the transmit signal s is minimized. Three power constraints for the relays have been discussed in (Khajehnouri & Sayed 2007). Those were an individual power constraint for each relay, a global power constraint over all relays and the case with QoS requirements. In this section, we focus on the last case, which has the following advantages. It guarantees a certain quality of service (QoS) based on a certain target SNR_t, which has to be fulfilled. Furthermore, only the necessary amount of power is spent in order to achieve a certain QoS. The optimization is thus expressed by

$$\hat{\mathbf{F}} = \arg \min_{\mathbf{F}} J\left(\mathbf{F}\right),$$

where

$$J\left(\mathbf{F}\right) = \mathrm{E}\left[\left| \eta s - g^T \mathrm{x} \right|^2\right]$$

is the MSE. As shown in (Khajehnouri & Sayed 2007), it turns out that in order to ensure a certain target SNR_t at the destination node, η has be chosen as

$$\eta = \sqrt{SNR_t \frac{\sigma_v^2}{\sigma_s^2}} \; .$$

After some manipulations, it has been shown in (Khajehnouri & Sayed 2007) that the optimal diagonal $\hat{F}$ has to satisfy the following condition

$$\left(\mathbf{g}^{\mathrm{T}}\hat{\mathbf{F}}\right)^H = \eta \left(\frac{\sigma_s^2}{\sigma_v^2 + \sigma_s^2 \left\| \mathbf{h} \right\|^2} \right) \mathbf{h} \, ,$$

where $||.||^2$ denotes the l2-norm. By using this in order to compute σ_w^2 in equation (39), we have

$$\sigma_w^2 = \mathrm{E}\left[\left|\mathbf{g}^T\hat{\mathbf{F}}\mathbf{v} + n\right|^2\right] = \sigma_n^2 + \frac{\eta^2\sigma_s^4\left\|\mathbf{h}\right\|^2}{\left(\sigma_v^2 + \sigma_s^2\left\|\mathbf{h}\right\|^2\right)^2}\sigma_v^2. \tag{40}$$

Similarly, $|h|^2$ is given by

$$\left|h\right|^2 = \frac{\eta^2\sigma_s^4\left\|\mathbf{h}\right\|^4}{\left(\sigma_v^2 + \sigma_s^2\left\|\mathbf{h}\right\|^2\right)^2} \tag{41}$$

Substituting equation (40) and (41) into equation (39) gives

$$C = \frac{1}{2}\mathrm{E}\left[\log_2\left(1 + \frac{\sigma_s^6\eta^2\left\|\mathbf{h}\right\|^4}{\sigma_n^2\left(\sigma_v^2 + \sigma_s^2\left\|\mathbf{h}\right\|^2\right)^2 + \eta^2\sigma_s^4\sigma_v^2\left\|\mathbf{h}\right\|^2}\right)\right]. \tag{42}$$

The pdf of $\left\|h\right\|^2 = \sum_{i=1}^{N}c_i\left|h_i\right|^2$ is given by

$$f_{\left\|\mathbf{h}\right\|^2}\left(x\right) = \frac{1}{\prod_{i=1}^{N}c_i}\sum_{i=1}^{N}\frac{\exp\left(-x\,/\,c_i\right)}{\prod_{l=1,l\neq i}^{N}\left(\frac{1}{c_l} - \frac{1}{c_i}\right)}.$$

In the following, we show that equation (42) is maximized by having all relays equidistantly apart from the source, i.e. the expression in equation (42) is Schur-concave. The Schur-concavity of equation (42) is shown directly by using the Schur-condition. To this end, we have to determine the derivate of

$$C = \frac{1}{2}\int_0^\infty \log_2\left(1 + \frac{\sigma_s^6\eta^2 x^2}{\sigma_n^2\left(\sigma_v^2 + \sigma_s^2 x\right)^2 + \eta^2\sigma_s^4\sigma_v^2 x}\right)\frac{1}{\prod_{i=1}^{N}c_i}\sum_{i=1}^{N}\frac{\exp\left(-x\,/\,c_i\right)}{\prod_{l=1,l\neq i}^{N}\left(\frac{1}{c_l} - \frac{1}{c_i}\right)}dx$$

with respect to c_1 and c_2 and analyze the difference of those derivatives, i.e.,

$$\frac{\partial C}{\partial c_1} - \frac{\partial C}{\partial c_2} = \frac{1}{2}\int_0^\infty \log_2\left(1 + \frac{\sigma_s^6\eta^2 x^2}{\sigma_n^2\left(\sigma_v^2 + \sigma_s^2 x\right)^2 + \eta^2\sigma_s^4\sigma_v^2 x}\right)\left(\frac{\partial f_{\left\|\mathbf{h}\right\|^2}\left(x\right)}{\partial c_1} - \frac{\partial f_{\left\|\mathbf{h}\right\|^2}\left(x\right)}{\partial c_2}\right)dx. \tag{43}$$

Interestingly, the derivative of the log-expression in equation (43) with respect to x given by

$$\frac{\sigma_s^6\eta^2 x^2\sigma_v^2\left(2\sigma_n^2\sigma_v^2 + 2\sigma_s^2\sigma_n^2 x + \eta^2\sigma_s^4 x\right)}{\left(\sigma_n^2\left(\sigma_v^2 + \sigma_s^2 x\right)^2 + \eta^2\sigma_s^4\sigma_v^2 x + \eta^2\sigma_s^6\sigma_v^2 x^2\right)\left(\sigma_n^2\left(\sigma_v^2 + \sigma_s^2 x\right)^2 + \eta^2\sigma_s^4\sigma_v^2 x\right)}$$

shows that the log-expression is monotonically increasing with *x*, i.e. $||h||^2$. Furthermore, it can be shown that

$$C = \lim_{x \to \infty} \frac{1}{2} \log_2 \left(1 + \frac{\sigma_s^6 \eta^2 x^2}{\sigma_n^2 \left(\sigma_v^2 + \sigma_s^2 x \right)^2 + \eta^2 \sigma_s^4 \sigma_v^2 x} \right) = \frac{1}{2} \log_2 \left(1 + \frac{\eta^2 \sigma_s^2}{\sigma_n^2} \right).$$

Thus, equation (43) can be upper bounded by

$$\frac{\partial C}{\partial c_1} - \frac{\partial C}{\partial c_2} \leq \frac{1}{2} \log_2 \left(1 + \frac{\eta^2 \sigma_s^2}{\sigma_n^2} \right) \int_0^\infty \left(\frac{\partial f_{\|\mathbf{h}\|^2} (x)}{\partial c_1} - \frac{\partial f_{\|\mathbf{h}\|^2} (x)}{\partial c_2} \right) dx$$

$$= \frac{1}{2} \log_2 \left(1 + \frac{\eta^2 \sigma_s^2}{\sigma_n^2} \right) \left(\frac{\partial \int_0^\infty f_{\|\mathbf{h}\|^2} (x) dx}{\partial c_1} - \frac{\partial \int_0^\infty f_{\|\mathbf{h}\|^2} (x) dx}{\partial c_2} \right) = 0$$

since $\int_0^\infty f_{\|\mathbf{h}\|^2} (x) dx = 1$. Thus the capacity in equation (42) is a Schur-concave function. From that it follows that having all relays equidistantly away from the source is optimal.

Simulation

Figure 11 shows the ergodic capacity of the relay network as a function of c_1. In the example below, two relays were used, i.e. *N=2*, the noise variances at the relays and the destination were set to $\sigma_v^2 = \sigma_n^2 = 1$ and the source power was set to *20dB*. The parameter η was set to $\eta=1$.

The left-most point $c_1=1$ and $c_2=1$, since $c_2=2-c_1$, corresponds to the case, where both relays are equidistantly away from the source node, while the right-most point $c_1=2$ and $c_2=0$ corresponds to the case, where one relay is very close to the source and the other one is very far away, such that effectively only one relay is present. From the figure we observe that the ergodic capacity is monotonically decreasing by increasing c_1, which verifies our analytical result.

Bidirectional Transmission Systems

In this section, we consider the bidirectional relay channel. This type of channel is a typical example for the advantages of network coding (Yeung, Li, Cai, & Zhang, 2005; Fragouli & Soljanin, 2007). Therefore, we will first introduce the system model and review the network coding result. Next, we study the cooperation on the physical layer first from an information theoretic point of view (Oechtering, Schnurr, Bjelakovic & Boche, 2008) and then from a signal processing perspective. It will turn out that the network coding corresponds to the XOR coding on the physical layer (Hammerström, Kuh, Esli, Zaho, Wittneben, & Bauch, 2007). As we will discuss later on, we have two phases in the bidirectional relay channel, namely the MAC phase and the BC phase. We derive the capacity achieving signal processing at the relay node for the BC phase and illustrate the rate regions of the MAC and BC phase by numerical experiments. These show the significant performance gain by cooperation at the relay and source nodes.

We restrict the communication to take part in two phases. Therefore we first discuss the achievable rate regions for the first phase which corresponds to the uplink or multiple access channel (MAC) transmission. Then the communication for the second phase is studied which corresponds to the downlink or broadcast channel (BC) with common information[3]. Next, we discuss the transceiver optimization if CSI is available at the transmitters and the relay channels Finally, numerical simulations illustrate the results and the developed transceiver optimization algorithm. The optimal scheme is compared to baseline approaches as time-sharing, superposition coding, and XOR coding.

System Model

We study the flat-fading time-division bidirectional two-phase relay channel with two nodes and a single relay node. Each node as well as the relay is equipped with multiple antennas, node one has n_1 antennas, node two has n_2 antennas, and the relay has n_r antennas, respectively. We explicitly exclude the possibility for feedback between relay and nodes to keep the protocol simple. Furthermore, we allocate the same amount of time for the two phases. Finally, we restrict our attention to the scenario in which the relay applies decode-and-forward[4]. The system is shown in Figure 12.

Suppose that node one transmits a message to node two and vice versa. The two messages from node one to node two and from node two to node one, respectively, are independent. The channel between node one and the relay is denoted by $\mathbf{H}_1$ and the channel between node two and the relay by $\mathbf{H}_2$. The nodes have individual transmit power constraints P_1 and P_2. The relay has a power constraint P_r. The noise power at the three receivers is equal to σ^2.

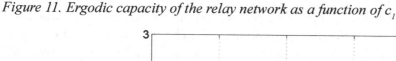

Figure 11. Ergodic capacity of the relay network as a function of c_1

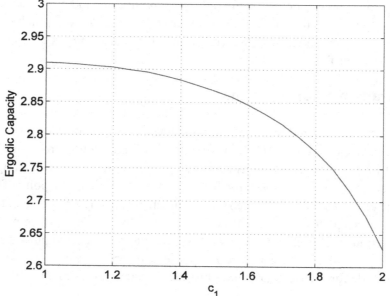

In the following, we consider the two phases: Phase I, in which the nodes are allowed to transmit and phase II, in which the relay transmits.

Rate Regions for the First and Second Phase

In this section, we review the recent information theoretic results for the MAC phase (Tse & Viswanath, 2005) and for the BC phase (Oechtering, Schnurr, Bjelakovic & Boche, 2008) and discuss them in terms of cooperative signal processing at the relay.

Rate Region for the First Phase

In the first phase, both nodes transmit simultaneously to the relay. This is called the Gaussian MAC for which the optimal coding strategy and its associated achievable rate region, the capacity region, is well known (see e.g. (Tse & Viswanath, 2005, chapter 6)). The optimal processing at the relay is successive interference cancellation (SIC). The rates achievable for communications between node one and relay R_1 and between node two and relay R_2 are characterized by the following three inequalities:

$$R_1 \leq \log \det \left(\mathbf{I} + \tfrac{1}{\sigma^2} \mathbf{H}_1 \mathbf{Q}_1 \mathbf{H}_1^H \right) \tag{44}$$

$$R_2 \leq \log \det \left(\mathbf{I} + \tfrac{1}{\sigma^2} \mathbf{H}_2 \mathbf{Q}_2 \mathbf{H}_2^H \right) \tag{45}$$

$$R_1 + R_2 \leq \log \det \left(\mathbf{I} + \tfrac{1}{\sigma^2} \mathbf{H}_1 \mathbf{Q}_1 \mathbf{H}_1^H + \tfrac{1}{\sigma^2} \mathbf{H}_2 \mathbf{Q}_2 \mathbf{H}_2^H \right) \tag{46}$$

where the normalized transmit covariance matrices $\mathbf{Q}_1$ and $\mathbf{Q}_2$ corresponds to the linear precoding at the two nodes. The strategy which achieves capacity consists of a Gaussian codebook, followed by serial/parallel conversion, power allocation of the spatial streams, and finally beamforming. The rank of the transmit covariance matrix corresponds to the number of parallel spatial data streams, see Figure 13.

The achievable rate region is given as the union of all feasible transmit strategies

$$R_{MAC} = \bigcup_{\substack{tr(\mathbf{Q}_1) \leq P_1 \\ tr(\mathbf{Q}_2) \leq P_2}} dpch \left\{ R_1, R_2 \right\} \tag{47}$$

where $dpch \left\{ \bullet \right\}$ denotes the downward positive comprehensive hull which is defined for the vector $\mathbf{x} \in R_+^2$ by the set $dpch \left\{ \mathbf{x} \right\} = \left\{ \mathbf{y} \in R_+^2 : y_i \leq x_i, i = 1, 2 \right\}$.

Certain operating points on the region can be found by optimization of the weighted sum rate or power minimization under rate constraints.

Rate Region for the Second Phase

The basic idea for the broadcast phase is well known in telephony as echo cancellation. Since each node knows some part of the message which is sent by the relay (the echo) it can compensate for that before decoding the new message. Hence, each receiver node performs also SIC. The own message is subtracted and then the message from the other node detected.

Figure 12. Flat-fading bidirectional MIMO relay-channel

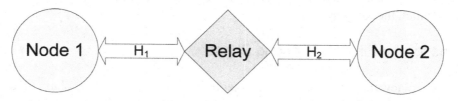

Define the rates

$$C_k\left(\mathbf{Q}_r\right) = \log\det\left(I + \frac{P_r}{\sigma^2}\mathbf{H}_k^H\mathbf{Q}_k\mathbf{H}_k\right) \tag{48}$$

for $k = 1, 2$. Then, the maximum achievable rate region of the broadcast (or multicast) phase of the bidirectional relay channel is characterized by

$$R_{BC}\left(\mathbf{Q}_r\right) = \bigcup_{tr(\mathbf{Q}_r)\leq 1} dpch\left\{C_1\left(\mathbf{Q}_r\right), C_2\left(\mathbf{Q}_r\right)\right\}. \tag{49}$$

The proof of this result can be found in (Oechtering, Schnurr, Bjelakovic & Boche, 2008).

The overall achievable rate region of the two-phase bidirectional relay channel is the intersection of the two rate regions weighted by one-half due to the two time slots which are needed for communication

$$R_{BRC}\left(\mathbf{Q}_1, \mathbf{Q}_2, \mathbf{Q}_r\right) = \frac{1}{2}R_{MAC} \cap R_{BC}. \tag{50}$$

The achievable rate region depends on the three linear precoding matrices at the two nodes and at the relay node.

Before we are going to optimize these covariance matrices, note that different extensions of this bidirectional relay channel are possible. If unequal time sharing between MAC and BC phase is allowed, the time sharing parameter τ could be optimized. Furthermore, more than two phases could be considered, leading to a larger rate region and more complicated protocols and expressions.

Optimization

In this subsection, we optimize the transmit covariance matrices $\mathbf{Q}_1, \mathbf{Q}_2, \mathbf{Q}_r$ under the individual power constraints with respect to different objective functions. The properties of the optimization problems as well as their algorithmic solution is presented.

Optimal Transmit Strategies

In [Yu et al. 2004] an iterative algorithm is proposed which solves the sum rate optimization problem on the MIMO MAC rate region

$$\max_{\substack{tr(\mathbf{Q}_1)\leq P_1 \\ tr(\mathbf{Q}_2)\leq P_2}} R_1 + R_2 = \max_{\substack{tr(\mathbf{Q}_1)\leq P_1 \\ tr(\mathbf{Q}_2)\leq P_2}} \log\det\left(\mathbf{I} + \frac{1}{\sigma^2}\mathbf{H}_1\mathbf{Q}_1\mathbf{H}_1^H + \frac{1}{\sigma^2}\mathbf{H}_2\mathbf{Q}_2\mathbf{H}_2^H\right). \tag{51}$$

This optimization problem is a convex optimization problem in the transmit covariance matrices and can therefore be efficiently solved. In addition, due to the nice problem structure an efficient and simple iterative algorithm can be applied.

The basic idea is to fix the transmit strategy of one node and optimize the covariance matrix of the other node. It can be shown that in each such step the sum rate is increased and that after convergence to the attraction point the individual Karush-Kuhn-Tucker (KKT) as well as the joint KKT conditions are fulfilled.

The so called iterative waterfilling algorithm is described below.

Initialize $\mathbf{Q}_1 = \dfrac{P_1}{n_1}\mathbf{I}$ and $\mathbf{Q}_2 = \dfrac{P_2}{n_2}\mathbf{I}$.

While (accuracy not reached) do

1. *solve* $\mathbf{Q}_1 = \arg\max\limits_{tr(\mathbf{Q}_1)\leq P_1} \log\det\left(\mathbf{I} + \frac{1}{\sigma^2}\mathbf{H}_1\mathbf{Q}_1\mathbf{H}_1^H + \frac{1}{\sigma^2}\mathbf{H}_2\mathbf{Q}_2\mathbf{H}_2^H\right)$ *for fixed* $\boldsymbol{Q}_2$ *by simple single-user waterfilling;*

2. *solve* $\mathbf{Q}_2 = \arg\max\limits_{tr(\mathbf{Q}_2)\leq P_2} \log\det\left(\mathbf{I} + \frac{1}{\sigma^2}\mathbf{H}_1\mathbf{Q}_1\mathbf{H}_1^H + \frac{1}{\sigma^2}\mathbf{H}_2\mathbf{Q}_2\mathbf{H}_2^H\right)$ *for fixed* $\boldsymbol{Q}_1$ *by simple single-user waterfilling;*

end while.

Each step in the while-loop of the iterative waterfilling algorithm is simple single-user waterfilling, i.e. the eigenvectors of the transmit covariance matrix is chosen to diagonalize the effective channel and the eigenvalues are chosen according to waterfilling (Cover & Thomas, 1991). Usually, a small number of iterations is sufficient to achieve high accuracy.

The generalization of the sum rate optimization to the weighted sum rate optimization problem

$$\max_{\substack{tr(\mathbf{Q}_1)\leq P_1 \\ tr(\mathbf{Q}_2)\leq P_2}} \lambda R_1 + (1-\lambda)R_2 \tag{52}$$

and its algorithmic solution is discussed in (Boche & Wiczanowski, 2007). The optimization problem can be rewritten in convex form and the iterative waterfilling skeleton applied to solve the problem. Therefore, the complete capacity region of the MAC phase can be computed efficiently.

MIMO BC Phase Algorithm Development

The optimization problem of the weighted sum rate for the BC phase is given by (Wyrembelski, Oechtering, Bjelakovic, Schnurr, & Boche, 2008)

$$\max_{\substack{\mathbf{Q}_r\geq 0 \\ tr(\mathbf{Q}_r)\leq 1}} \lambda \log\det\left(\mathbf{I} + \frac{P_r}{\sigma^2}\mathbf{H}_1^H\mathbf{Q}_r\mathbf{H}_1\right) + (1-\lambda)\log\det\left(\mathbf{I} + \frac{P_r}{\sigma^2}\mathbf{H}_2^H\mathbf{Q}_r\mathbf{H}_2\right) \tag{53}$$

which is also a convex optimization problem since the objective is concave and the constraint set is a convex set. However, the difference to the single-user rate optimization is that the optimal eigenvectors

Figure 13. Capacity approaching structure at the transmitters.

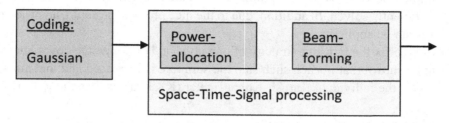

of the transmit covariance matrix cannot be characterized in closed form. A characterization for the 2x1 MISO channel is given in (Oechtering, Wyrembelski, & Boche, 2008). In this subsection, we will solve algorithmically the MIMO optimization problem by a fixed point algorithm originally proposed and developed in (Jorswieck, Sezgin, Ottersten, & Paulraj, 2008) in the context of chunk-processing for MIMO MAC.

The idea is to study the necessary and sufficient Karush-Kuhn-Tucker conditions to develop an iterative fixed-point algorithm (Jorswieck, Sezgin, Ottersten, & Paulraj, 2008) The resulting optimization algorithm is described as follows.

Initialize $\mathbf{Q}_r^0 = \dfrac{1}{n_r}\mathbf{I}$, $\ell = 0$

While (accuracy not reached) do

1. $$\mathbf{Q}_r^{\ell+1} = \lambda \mathbf{Q}_r^{\ell,\frac{1}{2}} \mathbf{H}_1 \left[\mathbf{I} + \frac{P_r}{\sigma^2}\mathbf{H}_1^H \mathbf{Q}_r^\ell \mathbf{H}_1\right]^{-1} \mathbf{H}_1^H \mathbf{Q}_r^{\ell,\frac{1}{2}}$$
    $$+ \left(1-\lambda\right)\mathbf{Q}_r^{\ell,\frac{1}{2}} \mathbf{H}_2 \left[\mathbf{I} + \frac{P_r}{\sigma^2}\mathbf{H}_2^H \mathbf{Q}_r^\ell \mathbf{H}_2\right]^{-1} \mathbf{H}_2^H \mathbf{Q}_r^{\ell,\frac{1}{2}}$$

2. $$\mathbf{Q}_r^{\ell+1} = \frac{\mathbf{Q}_r^{\ell+1}}{tr\left(\mathbf{Q}_r^{\ell+1}\right)}$$

3. $\ell = \ell + 1$

end while.

It can be shown that the update rule always converges to the fixed point which solves the weighted sum rate optimization problem. Note that the fixed point iteration has only linear convergence and any Newton style algorithm has local super-linear convergence. However, the update rule could be further refined to include spectral power allocation. If a Newton style algorithm is used, this extension is not directly possible and an alternating optimization approach would lead to slower convergence. For further discussions, the interested reader is referred to (Jorswieck, Sezgin, Ottersten, & Paulraj, 2008).

The idea to show that the fixed point algorithm described above converges is to show that it improves the objective function in each step and is bounded. The problem boils down to show that for any given positive semi-definite matrix $\mathbf{Q}$ the function g is for all positive semi-definite matrices $\mathbf{B}$ greater than or equal to zero, i.e.,

$$g\left(\mathbf{Q},\mathbf{B}\right) = tr\left(\mathbf{Q}^{1/2}\mathbf{B}\mathbf{Q}^{1/2}\mathbf{B}\right) - tr\left(\mathbf{Q}\mathbf{B}\right)^2 \geq 0. \tag{54}$$

It can be shown that the minimum and maximum of the function g is obtained for *commuting* matrices $\mathbf{Q}$ and $\mathbf{B}$. Therefore, the function is reduced to a vector function

$$g\left(\alpha,\beta\right) = \sum_{k=1}^{n}\alpha_k\beta_k^2 - \left(\sum_{k=1}^{n}\alpha_k\beta_k\right)^2 = \sum_{k=1}^{n}\alpha_k\left(\beta_k - \sum_{l=1}^{n}\alpha_l\beta_l\right)^2 \tag{55}$$

where $\alpha_1,...,\alpha_n$ are the eigenvalues of $\mathbf{Q}$ and $\beta_1,...,\beta_n$ are the eigenvalues of $\mathbf{B}$.

Let us order the eigenvalues of $\mathbf{B}$ in non-increasing order, i.e., $\beta_1 \geq \beta_2 \geq ... \geq \beta_n \geq 0$. and order the eigenvalues of $\mathbf{Q}$ accordingly, i.e. $\alpha_1 \geq \alpha_2 \geq ... \geq \alpha_n \geq 0$. Then verify Schur's condition for fixed set of $\alpha_1,...,\alpha_n$

$$\begin{aligned}\frac{\partial g\left(\beta\right)}{\partial\beta_1} - \frac{\partial g\left(\beta\right)}{\partial\beta_2} &= 2\alpha_1\beta_1 - 2\alpha_2\beta_2 - 2\left(\sum_{k=1}^{n}\alpha_k\beta_k\right)\left(\alpha_1 - \alpha_2\right) \\ &= 2\alpha_1\left(\beta_1 - \sum_{k=1}^{n}\alpha_k\beta_k\right) - 2\alpha_2\left(\beta_2 - \sum_{k=1}^{n}\alpha_k\beta_k\right) \\ &\geq 2\left(\alpha_1 - \alpha_2\right)\left(\beta_2 - \sum_{k=1}^{n}\alpha_k\beta_k\right) \geq 0\end{aligned} \tag{56}$$

because $\left(\beta_1 - \sum_{k=1}^{n}\alpha_k\beta_k\right) \geq \left(\beta_2 - \sum_{k=1}^{n}\alpha_k\beta_k\right)$ and $\left(\beta_1 - \sum_{k=1}^{n}\alpha_k\beta_k\right) \geq 0$. This shows the Schur-convexity of g with respect to $\left[\beta_1,...,\beta_n\right]$ and establishes the lower bound

$$g\left(\alpha,\beta\right) = \sum_{k=1}^{n}\alpha_k\beta_k^2 - \left(\sum_{k=1}^{n}\alpha_k\beta_k\right)^2 \geq 0. \tag{57}$$

Therefore, the objective function is improved in every step, the objective function is concave with respect to $\mathbf{Q}$ and bounded and the iterative fixed point algorithm converges to the global optimum.

Discussion and Performance Examples

In this section, the algorithms for the two-phase bidirectional relay system developed in the last sections are illustrated and compared to the standard time-division approach outlined in section 1.9.1 as well as to the superposition coding and XOR approach described in (Hammerström, Kuh, Esli, Zaho, Wittneben, & Bauch, 2007).

Let us first describe the three suboptimal approaches in more detail.

1. In the time-division approach, the second phase is divided into two phases in which the relay forwards the message to node two and to node one, respectively. Only half of the time is used for transmission to each node. Therefore, in the BC phase the following pair of rates is achievable

$$
R_1^{TD} = \frac{1}{2} \max_{\substack{\mathbf{Q}_{r,1} \geq 0 \\ tr(\mathbf{Q}_{r,1}) \leq P_r}} \log \det \left(I + \frac{1}{\sigma^2} \mathbf{H}_1^H \mathbf{Q}_{r,1} \mathbf{H}_1 \right)
$$

$$
R_2^{TD} = \frac{1}{2} \max_{\substack{\mathbf{Q}_{r,2} \geq 0 \\ tr(\mathbf{Q}_{r,2}) \leq P_r}} \log \det \left(I + \frac{1}{\sigma^2} \mathbf{H}_2^H \mathbf{Q}_{r,2} \mathbf{H}_2 \right)
$$

(58)

The optimal transmit covariance matrix is given by SVD transmission and single-user waterfilling. The corresponding region is given by the downward positive comprehensive hull of this rate tuple, i.e.

$$
R_{TD} = dpch \left(\left\{ R_1^{TD}, R_2^{TD} \right\} \right).
$$

2. In the superposition coding scheme (SPC), the relay uses separate Gaussian codebooks for the two receiver nodes. The superposition of the codewords is sent to the receiver nodes. Assume that a portion β of the total transmit power P_r is allocated for the codeword to node one and $1 - \beta$ for codeword to node two, respectively. As outlined in (Oechtering & Boche, 2007), assuming perfect cancellation of the self-interference part, the achievable rates of node one and two in the broadcast phase are

$$
R_1^{SPC} = \max_{\substack{\mathbf{Q}_{r,1} \geq 0 \\ tr(\mathbf{Q}_{r,1}) \leq \beta P_r}} \log \det \left(I + \frac{1}{\sigma^2} \mathbf{H}_1^H \mathbf{Q}_{r,1} \mathbf{H}_1 \right)
$$

$$
R_2^{SPC} = \max_{\substack{\mathbf{Q}_{r,2} \geq 0 \\ tr(\mathbf{Q}_{r,2}) \leq (1-\beta) P_r}} \log \det \left(I + \frac{1}{\sigma^2} \mathbf{H}_2^H \mathbf{Q}_{r,2} \mathbf{H}_2 \right)
$$

(59)

and the corresponding rate region is given by the union of all dpch rates over the scalar parameter $0 \leq \beta \leq 1$ as

$$
R_{SPC} = \bigcup_{0 \leq \beta \leq 1} dpch \left(\left\{ R_1^{SPC} \left(\beta \right), R_2^{SPC} \left(\beta \right) \right\} \right).
$$

3. XOR coding is introduced in (Hammerström, Kuh, Esli, Zaho, Wittneben, & Bauch, 2007). The two codewords are combined prior to encoding at bit-level by XOR operations. Since both nodes know their own message, they can subtract their message by bit-level XOR operations (as described above for the network coding approach) and the achievable rates are given by

$$
R_{1,2}^{XOR} = \max_{\substack{\mathbf{Q}_r \geq 0 \\ tr(\mathbf{Q}_r) \leq P_r}} \min \left(\log \det \left(I + \frac{1}{\sigma^2} \mathbf{H}_1^H \mathbf{Q}_r \mathbf{H}_1 \right), \log \det \left(I + \frac{1}{\sigma^2} \mathbf{H}_2^H \mathbf{Q}_r \mathbf{H}_2 \right) \right)
$$

(60)

All achievable rate tuples lie in the dpch of $\left\{ R_1^{XOR}, R_2^{XOR} \right\}$.

In Figure 14, we illustrate the achievable rate region for the BC phase for the various suboptimal and the optimal scheme for a system in which the nodes as well as the relay have two antennas each. The individual transmit power constraints are given by $P_1 = P_2 = P_r = 1$ and the noise power is set to $\sigma^2 = 1$.

In the figure, it can be observed that the gain by using network coding (XOR encoding) is large compared to the standard TD approach as explained in subsection 1.9.1. Superposition coding achieves a large portion of the capacity region. The XOR-operating point is the maxmin point on the Pareto boundary of the region.

Another channel realization is shown in the next figure. Also shown is the capacity region of the MAC phase. In this example, the BC region lies within the MAC region. Depending on the concrete MIMO channel realization, also the opposite behavior can occur, i.e. the MAC region lies within the BC region. (See Figure 15)

The advantages and disadvantages of network coding are well known and discussed e.g. in (Fragouli & Soljanin, 2007). The main drawback of the precoding schemes for the MIMO MAC and BC phase is that we need perfect CSI at the transmitter side. For the BC phase this CSI might be obtained during the MAC phase transmissions if the same frequency is used and channel reciprocity applies. In the superposition, and optimal BC phase processing, the receiver nodes have to perform SIC whereas in XOR-precoding bit-level XOR is performed. In standard TD single-user receivers are optimal.

In this section, we have studied the capacity achieving signal processing strategy in a bidirectional two-phase Decode-and-forward MIMO relay channel. Furthermore, efficient algorithms were derived which solve the weighted sum rate maximization problem to achieve any point on the Pareto boundary of the capacity or rate region. Cooperation at the relay and at the nodes by self-interference subtraction provide significantly larger achievable rate regions.

Relay Deployment in Multihop Aystems

Introduction to the Problem

In this chapter, so far we only discussed single-hop scenarios, where only one relay is exploited in order to forward the message from the source to the intended receiver. In the following section, we consider a scenario in which multiple relays are available and exploited in a multihop fashion in order to transmit the source data to the destination. Depending on the signal-to-noise ratio (SNR), it might be helpful for the source to use the relays for routing its data to the destination. On the other hand, it might be sometimes of advantage to send the data directly to the destination. In the following, we determine a bound on the SNR up to which multihopping is the more favorable strategy.

System Model

The system under consideration is depicted in Figure 16. The source wants to communicate with the destination by exploiting the N-1 relays. Each relays simply decode-and-forwards the information received from the previous node to the following node.

As in the previous sections, we assume that the relays are not able to transmit and receive data simultaneously, thus, a half duplex-scheme is adopted. Further, we assume that there is only a link between adjacent nodes. Obviously, should there be also a signal from nonadjacent nodes, this signal could be used in order to improve the performance of detecting the signal correctly.

The system equation at hop n is given by

$$y_{n+1} = \left(\sqrt{\frac{L_n}{\left(\frac{D}{N} d_n \right)^{\alpha}}} h_n \right) s_n + w_{n+1},$$

where y_{n+1} is the received signal at node $n+1$, s_n is the transmitted signal (assuming a Gaussian codebook) from node n with an average power constraint of $P_n = \lambda_n P$. We assume that we have a fixed finite total power supply of $P = \sum_{n=1}^{N} \lambda_n P$ for the entire network, where $\sum_{n=1}^{N} \lambda_n = 1$. The parameter λ_n corresponds to the amount of time the hop n is used by the network. w_{n+1} is the complex white Gaussian noise with zero mean and variance 1, α is the path loss exponent (for simplicity, we assume equal path loss exponents at each hop). The parameter d_n, with $0 \le d_n \le N$, is the normalized Euclidean distance between node n and $n+1$, D is the overall distance between the source and the destination, N is the number of hops, i.e. number of relays plus destination). Note that $d_n=0$ corresponds to the case, where the node n and $n+1$ are on top of each other, i.e. effectively they merge to one node. In the remainder of the section, we assume that

$$\sum_{n=1}^{N} d_n = N,$$

Figure 14. Illustration of different achievable rate regions for the BC phase in bidirectional two-phase relaying: comparison between traditional TD, superposition coding, XOR encoding, and the proposed optimal scheme.

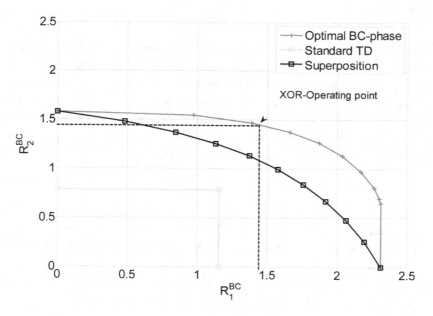

Figure 15. Illustration of different achievable rate regions for the BC phase in bidirectional two-phase relaying: comparison between traditional TD, superposition coding, XOR encoding, and the proposed optimal scheme. Also shown is the capacity region of the MAC phase.

i.e. the overall normalized distance equals N and thus the overall Euclidean distance is kept constant for all scenarios.

The parameter L_n arises due to the shadowing and is a lognormal distributed random variable. The microscopic fading is modeled by h_n, which is complex Gaussian random variable with zero mean and unit variance. It is assumed that perfect CSI is available at the transmitting nodes, such that adaptive rate allocation can be applied.

Capacity Analysis

In (Oyman & Sandhu, 2006), it was shown that the instantaneous end-to-end capacity of the multihop network is given by

$$C\left(SNR\right) = \max_{\sum_{n=1}^{N} \lambda = 1} \min\left\{\lambda_n C_n\left(SNR\right)\right\}$$

(61)

where the $C_n\left(SNR\right)$ is the instantaneous capacity of hop n given by

$$C_n\left(SNR\right) = \ln\left(1 + \frac{L_n|h|^2}{\left(\frac{D}{N}d_n\right)^p} SNR\right)$$

The optimization in equation (61) with respect to the λ_n results in (Oyman & Sandhu, 2006)

$$C\left(SNR\right) = \frac{1}{\sum_{n=1}^{N} \frac{1}{C_n\left(SNR\right)}},$$

(62)

i.e. the overall capacity is proportional to the harmonic mean of the individual capacities.

We are interested in understanding the impact of the inter-node distances, i.e. path loss, on the capacity in (1.10.2) averaged over the shadowing and the microscopic fading realizations. For the following analysis, however, we simply neglect the impact of shadowing and microscopic fading and focus only on the path loss effects. Since the inverse is a non-negative convex-function, it is sufficient to analyze the denominator in equation (62).

Figure16. Multihop Network comprising N-1 relays and a source and destination pair

In the following, we show that multihopping should be preferred to direct communication as long as one operates below a certain threshold *SNR* value. This is accomplished by proving the Schur-condition for equation (62). For this, we need the derivatives of

$$C^{-1}\left(SNR\right) = \sum_{n=1}^{N} \frac{1}{C_n\left(SNR\right)}, \tag{63}$$

with respect to the d_n given by

$$\frac{\partial C^{-1}}{\partial d_n} = \frac{SNRp}{\ln\left(1 + \frac{SNR}{\left(\frac{D}{N}d_n\right)^\alpha}\right)^2 d_n\left(\left(\frac{D}{N}d_n\right)^\alpha + SNR\right)} \tag{64}$$

From the above expression, we realize that there is a term

$$\ln\left(1 + \frac{SNR}{\left(\frac{D}{N}d_n\right)^\alpha}\right)^2,$$

which is decreasing with d_n and a term

$$d_n\left(\left(\frac{D}{N}d_n\right)^\alpha + SNR\right),$$

which is increasing with d_n. Thus, there is a *SNR* region, for which

$$\frac{\partial C^{-1}}{\partial d_1} - \frac{\partial C^{-1}}{\partial d_2} \geq 0, \tag{65}$$

and thus multihopping is not the optimal strategy. Using equation (64) in equation (65), it follows that equation (65) does not hold if

$$
\ln\left(1+\frac{SNR}{\left(\frac{D}{N}d_1\right)^{\alpha}}\right)^2 d_1\left(\left(\frac{D}{N}d_1\right)^{\alpha}+SNR\right) \geq \ln\left(1+\frac{SNR}{\left(\frac{D}{N}d_2\right)^{\alpha}}\right)^2 d_2\left(\left(\frac{D}{N}d_2\right)^{\alpha}+SNR\right) \tag{66}
$$

holds. At low SNR, $(D/N d_n)^{\alpha}$, with $n=1,2$, is dominating the *SNR* and thus the expression in equation (66) reduces to

$$
\frac{SNR}{2}d_1 \geq \frac{SNR}{2}d_2,
$$

where the approximation $log(1+x)^2 \approx 1/2x$, $x<1$, was used. Thus, as long as

$$
SNR \leq \left(\frac{D}{N}d_1\right)^{\alpha}, \tag{67}
$$

the expression in equation (65) is not fulfilled and the function in equation (63) is not a Schur-concave function. This in turn means that the capacity in equation (62) is not Schur-convex as long as equation (67) holds, i.e. a direct communication between source and destination is suboptimal. Thus the multihop strategy provides a higher capacity than a direct communication between source and destination. Note that equation (67) serves only as a lower bound, i.e. it only guarantees that direct communication is suboptimal as long as equation (67) holds. It might, however, still be favorable to apply the multihop strategy beyond equation (67). The condition is necessary for suboptimality of direct communication but not sufficient.

Simulation

In Figure 17, the capacity of a multihop system with up to N=4 hops is depicted. In the simulation, only the path-loss effect is considered. The overall distance from the source to the destination node was normalized $D=5$. The path loss is set to $\alpha=2$. From the figure, we observe that for low and average SNR it is optimal to use a multihop strategy using all *N-1* relays. From *SNR=11dB* on it turns out to be better using *N-2* relays for multihopping. The higher the SNR, the fewer the number of relays, which should be used in order to maximize the capacity. The direct communication from the source and to the destination is providing the best output for SNR values higher than *SNR=16dB*. The lower bound is indicating that up to *SNR=14dB* multihop is definitely superior to direct communication, which can also be observed from the simulation. However, since it is only a lower bound it might in general be still optimal to multihop beyond this point.

The effect of microscopic fading can be observed in Figure 18. All parameters are equivalent to the parameters in Figure 17. The SNR value up to which applying a multihop strategy with *N=4* is optimal

Figure17. Multi-hop capacity due to path loss effects, D=5, α=2, N=4.

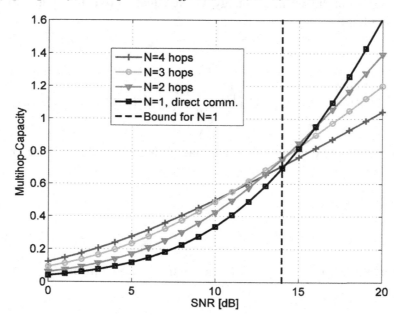

Figure 18. Multihop capacity due to path loss and microscopic facing effects, D=5, α=2, N=4

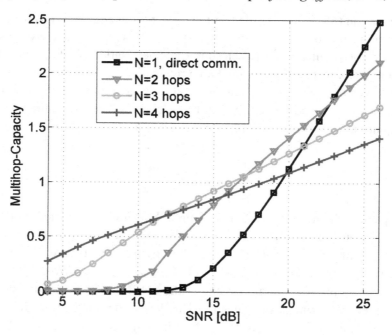

has only changed slightly from *SNR=11dB* to *SNR=12dB*. However, there is a significant change of the SNR value from which one on direct communication is preferable. This SNR value has moved from *SNR=16dB* to *SNR=23dB*. Thus, fading increases the SNR range of multihopping.

CONCLUSION

In this chapter, we have shown that majorization theory is a very useful tool in order to derive the optimal transmit strategy and to characterize the impact of different systems parameters on the performance of a cooperative wireless network. It provides some nice insights into the behavior of the system, even when the performance function is rather difficult to analyze, as e.g. for non-convex functions. Non-convex functions are very common in cooperative communication systems, i.e. analyzing the performance of such networks appears be much more difficult than single point-to-point links. Thus, although several important results in cooperative communications have been obtained, many open questions remain in both the theory as well as the practice.

This chapter discussed important aspects in cooperative communications using Majorization theory, spanning both theoretical foundations and practical issues. Majorization theory provides a large amount of tools and techniques which can be used in order to accelerate the pace of developments in this fascinating research area of cooperative communications. The aim of the chapter was to build good intuition and insight into this important field of cooperative communications and how Majorization theory can be used in order to solve quite complex problems in a very efficient and elegant way. Although we focused on some specific applications, the tools can be also applied to other setups and processing techniques.

REFERENCES

Boche, H., & Wiczanowski, M. (2007). Optimization-theoretic analysis of stability-optimal transmission policy for multiple antenna multiple access channel. *IEEE Transactions on Signal Processing, 55*(6), 2688–2702. doi:10.1109/TSP.2006.890926

Charafeddine, M. A., Sezgin, A., & Paulraj, A. (2007, September). Rate region frontiers for n-user interference channel with interference as noise. *45th Annual Allerton Conference on Communication, Control, and Computing,* Monticello, IL.

Cover, T. M., & Gamal, A. E. (1979). Capacity theorems for the relay channel. *IEEE Transactions on Information Theory, 25*(5), 572–584. doi:10.1109/TIT.1979.1056084

Cover, T. M., & Thomas, J. A. (1991). *Elements of information theory.* Wiley & Sons.

Dana, A. F., & Hassibi, B. (2006). On the power efficiency of sensory and ad hoc wireless networks. *IEEE Transactions on Information Theory, 52*(7), 2890–2914. doi:10.1109/TIT.2006.876245

Foschini, G. J., & Gans, M. J. (1998). On limits of wireless communications in a fading environment when using multiple antennas. *Wireless Personal Communications, 6*(3), 311–335. doi:10.1023/A:1008889222784

Fragouli, C., & Soljanin, E. (2007). Network coding applications. *Foundations and Trends in Networking, 2*(2), 135–269. doi:10.1561/1300000013

Hammerström, I., Kuhn, M., Esli, C., Zhao, J., Wittneben, A., & Bauch, G. (2007, June). MIMO two-way relaying with transmit CSI at the relay. In *Proc. SPAWC.*

Host-Madsen, A. (2002). On the capacity of wireless relaying. In *Proc. 56th-IEEE Conference on Vehicular Technology (VTC-Fall)* (pp. 1333-1337).

Jorswieck, E., & Boche, H. (2007). Majorization and matrix monotone functions in wireless communications. *Foundations and Trends in Communications and Information Theory, 3*(6), 553–701. doi:10.1561/0100000026

Jorswieck, E., Sezgin, A., Ottersten, B., & Paulraj, A. (2008). Feedback reduction in uplink MIMO OFDM systems by chunk optimization. *Journal on Advances in Signal Processing, 2008*, 597072.

Khajehnouri, N., & Sayed, A. H. (2007). Distributed MMSE relay strategies for wireless sensor networks. *IEEE Transactions on Signal Processing, 55*(7), 3336–3348. doi:10.1109/TSP.2007.894410

Knopp, R., & Humblet, P. (1995). Information capacity and power control in single-cell multiuser communications. In *Proc. IEEE International Conference on Communications, 1*(6), 331-335

Kramer, G., Berry, R., El Gamal, A., El Gamal, H., Franceschetti, M., Gastpar, M., & Laneman, J. N. (2007). Introduction to the special issue on models, theory, and codes for relaying and cooperation in communication networks. Special Issue in *IEEE Trans. Info. Theory, 53*(10).

Kramer, G., Gastpar, M., & Gupta, P. (2003). Capacity theorems for wireless relay channels. In *Proc. 41st Allerton Conf. on Comm., Control, and Comp.*

Marshall, A. W., & Olkin, I. (1979). Inequalities: Theory of majorization and its application. *Mathematics in Science and Engineering, 143.* Academic Press, Inc. (London) Ltd.

Morgenshtern, V. I., & Bölcskei, H. (2007). Crystallization in large wireless networks. *IEEE Transactions on Information Theory, 53*(10), 3319–3349. doi:10.1109/TIT.2007.904789

Nosratinia, A., & Hunter, T. E. (2007). Grouping and partner selection in cooperative wireless networks. *IEEE JSAC, 25*(2), 369–378.

Oechtering, T., & Boche, H. (2007). Optimal transmit strategies in multi-antenna bidirectional relaying. In *Proc. ICASSP,* (pp. III-145 – III-148).

Oechtering, T., Schnurr, C., Bjelakovic, I., & Boche, H. (2008). Broadcast capacity region of two-phase bidirectional relaying. *IEEE Transactions on Information Theory, 54*(1), 454–458. doi:10.1109/TIT.2007.911158

Oechtering, T., & Sezgin, A. (2004). A new cooperative transmission scheme using the space-time delay code. *ITG Workshop on Smart Antennas* (pp. 41-48).

Oechtering, T., Wyrembelski, R., & Boche, H. (2008). Optimal transmit strategy for the 2x1 MISO bidirectional broadcast channel. *SPAWC,* will appear.

Oyman, Ö., & Sandhu, S. (2006). A Shannon-theoretic perspective on fading multihop networks. *IEEE Conf. on Information Sciences and Systems (CISS)* (pp. 525-530).

Palomar, D. P., & Jiang, Y. (2006). MIMO transceiver design via majorization theory. *Foundations and Trends in Comm. and Info. Theory, 3*(4-5), 331–551.

Pereira, S., Sezgin, A., Paulraj, A., & Papanicolaou, G. C. (2008, July 6-11). Interference limited broadcast: Role of interferer geometry. *ISIT 2008,* Toronto, Canada.

Sendonaris, A., Erkip, E., & Aazhang, B. (1998). Increasing uplink capacity via user cooperation diversity. In *Proc. IEEE Intern. Symposium on Info. Theory (ISIT)* (p. 156).

Sendonaris, A., Erkip, E., & Aazhang, B. (2003). User cooperation diversity-part II: Implementation aspects and performance analysis. *IEEE Transactions on Communications, 51*(11), 1939–1948. doi:10.1109/TCOMM.2003.819238

Sendonaris, A., Erkip, E., & Aazhang, B.(n.d.). User cooperation diversity-part I: System description. *IEEE Transactions on Communications, 51*(11), 1927–1938. doi:10.1109/TCOMM.2003.818096

Sezgin, A., Jorswieck, E. A., & Charafeddine, M. (2008). Interaction between scheduling and user locations in an OSTBC coded downlink system. *7th International ITG Conference on Source and Channel Coding.*

Sezgin, A., Jorswieck, E. A., & Paulraj, A. (2008, May 7-9). Where to place interferers in a wireless network, *ITW 2008,* Porto, Portugal.

Sezgin, A., & Oechtering, T. (2004). A new resource efficient transmission scheme for cooperative systems. *IEEE SPAWC* (pp. 298-302).

Tarokh, V., Seshadri, N., & Calderbank, A. R. (1998). Space-time codes for high data rate wireless communication: Performance criterion and code construction. *IEEE Transactions on Information Theory, 44*(2), 744–765. doi:10.1109/18.661517

Telatar, E. (1999). Capacity of multi-antenna Gaussian channels. *European Trans. on Telecomm. ETT, 10*(6), 585–596.

Tse, D., & Viswanath, P. (2005). *Fundamentals of wireless communication.* Cambridge University Press.

Van der Meulen, E. C. (1971). Three-terminal communication channels. *Advances in Applied Probability, 3,* 120–154. doi:10.2307/1426331

Wyrembelski, R., Oechtering, T., Bjelakovic, I., Schnurr, C., & Boche, H. (2008). Capacity of Gaussian MIMO bidirectional broadcast channels. In *Proc. IEEE ISIT.*

Yeung, R. W., Li, S.-Y. R., Cai, N., & Zhang, Z. (2005). Network coding theory part I and II. *Foundation and Trends in Communications and Information Theory, 2*(4).

Yu, W., Rhee, W., Boyd, S., & Cioffi, J. M. (2004). Iterative water-filling for Gaussian vector multiple-access channels. *IEEE Transactions on Information Theory, 50*(1), 145–151. doi:10.1109/TIT.2003.821988

ENDNOTES

[1]   A function $f : \Re_+^n \to \Re_+$ is called convex if for any $\mathbf{a}, \mathbf{b} \in \Re_+^n$ and real $0 \le \lambda \le 1$ it holds $f\left(\lambda \mathbf{a} + \left(1 - \lambda\right) \mathbf{b}\right) \le \lambda f\left(\mathbf{a}\right) + \left(1 - \lambda\right) f\left(\mathbf{b}\right).$

2   A function $f : \Re_+^n \to \Re_+$ is called symmetric if for any permutation π and $\mathbf{a} \in \Re_+^n$ it holds $f(\pi \circ \mathbf{a}) = f(\mathbf{a})$.

3   The broadcast channel with common information is sometimes called multicast channel.

4   For other relaying strategies refer to the exposition in the last section.

Chapter 17
Data Gathering in Correlated Wireless Sensor Networks with Cooperative Transmission

Laxminarayana S. Pillutla
The University of British Columbia, Canada

Vikram Krishnamurthy
The University of British Columbia, and Canada

ABSTRACT

This chapter considers the problem of data gathering in correlated wireless sensor networks with distributed source coding (DSC), and virtual multiple input and multiple output (MIMO) based cooperative transmission. Using the concepts of super and sub modularity on a lattice, we analytically quantify as how the optimal constellation size, and optimal number of cooperating nodes, vary with respect to the correlation coefficient. In particular, we show that the optimal constellation size is an increasing function of the correlation coefficient. For the virtual MIMO transmission case, the optimal number of cooperating nodes is a decreasing function of the correlation coefficient. We also prove that in a virtual MIMO based transmission scheme, the optimal constellation size adopted by each cooperating node is a decreasing function of number of cooperating nodes. Also it is shown that, the optimal number of cooperating nodes is a decreasing function of the constellation size adopted by each cooperating node. We also study numerically that for short distance ranges, SISO transmission achieves better energy-mutual information (MI) tradeoff. However, for medium and large distance ranges, the virtual MIMO transmission achieves better energy-MI tradeoff.

DOI: 10.4018/978-1-60566-665-5.ch017

INTRODUCTION

A wireless sensor network (WSN) consists of large number of spatially distributed devices called *nodes,* which cooperate to accomplish various high level tasks. These nodes find applications in a wide range of remote sensing and environmental applications. Each of these nodes are equipped with a sensor and wireless transceiver, to perform sensing and communication tasks respectively. Typically the nodes are equipped with small batteries, and therefore are subject to severe energy constraints. Consequently, the main objective of sensor network research is to design energy-efficient devices and algorithms to support various aspects of network operation. The $^\mu$AMP project at MIT (Chandrakasan, Min, Bharadwaj, Cho & Wang, 2002), and the PicoRadio project at Berkeley (Rabaey, Ammer, dasilva, Patel & Roundy, 2000) are efforts in this direction. Another unique feature of WSNs is, the high degree of spatial correlation in the data sensed by the nodes, owing to their close proximity. We refer the readers to (Akyildiz, Su, Sankarasubramaniam & Cayirci, 2002), for a detailed discussion on WSNs and their applications.

Typically in a WSN, the nodes cooperate to accomplish a common task (unlike a conventional wireless network), therefore the idea of cooperation is central to the design of a WSN. The cooperation among WSNs can be of two types: *implicit* and *explicit*. In implicit cooperation, the nodes of a WSN cooperate implicitly based on the information obtained apriori. The concept of distributed source coding (DSC) where the sensor nodes determine their encoding rates, based on the prior information obtained on spatial correlation, is an instance of implicit cooperation. On the other hand, in the explicit cooperation, the nodes cooperate explicitly by sharing information. The concept of virtual multiple input and multiple output (MIMO) (or cooperative) transmission, where the sensor nodes share information and transmit cooperatively, is an instance of explicit cooperation. In this chapter, we consider an architecture based on implicit and explicit cooperation for WSNs. Specifically, we consider an architecture that employs, virtual MIMO to reduce the energy consumption of each sensor node and DSC to exploit the spatial correlation.

The concept of virtual MIMO aims to exploit the benefits of popular MIMO antenna techniques for WSNs. Direct application of the popular MIMO techniques to WSNs is not possible, owing to the relatively small size of the sensor nodes. However, by allowing the nodes to exchange their information, one can create a virtual MIMO system. We focus on *orthogonal* space-time block codes (STBC) (Tarokh, Jafarkhani & Calderbank, 1999) as a way to implement virtual MIMO in WSNs. The orthogonal STBCs are attractive because they facilitate simple decoding at the receiver. However, the orthogonal STBCs are bandwidth inefficient, especially if the number of cooperating nodes is greater than two. The reason for this bandwidth inefficiency can be attributed to the fact that, it may not be possible to design orthogonal STBCs with spatial rate equal to one, if the number of sensor nodes is greater than two (Liang, 2003). Thus it is of interest to know the effect of using an STBC with spatial rate less than one, on the energy efficiency of a WSN. To this end, we formulate the problem of minimum energy data gathering in correlated WSNs. We also study the tradeoff between energy-Mutual Information (MI) and energy-mean square error (MSE), for single input and single output (SISO) and virtual MIMO based transmission schemes in correlated WSNs.

An important aspect of our work is, we analytically study the effect of data correlation on physical layer design variables such as, constellation size and number of cooperating nodes, which was not considered before. In general, the energy expended by a sensor node is a function of constellation size. *Constellation size* represents the number of bits that can be transmitted in a given modulation symbol.

Further in cases where the nodes cooperate to transmit data, the energy expended is also a function of number of cooperating nodes. We accomplish this using *monotone comparative statics,* that are used extensively in economics literature, to study as how the optimal solution of a parametrized optimization problem, varies monotonically with respect to the parameter.

The main results of this chapter are summarized as follows:

Main Results:

1. We prove that the optimal constellation size adopted by a sensor, to minimize total energy consumption is, an increasing function of the correlation coefficient. This implies that when data is correlated, the sensor node can adopt higher constellation sizes, compared to the case when data is uncorrelated.
2. For the case of virtual MIMO based transmission, we prove that the optimal number of cooperating nodes, to minimize total energy consumption is, a decreasing function of the correlation coefficient. This implies that when data is correlated, the cooperation among nodes can be less than the case when data is uncorrelated.
3. In a virtual MIMO transmission scheme, we prove that the optimal constellation size adopted by each cooperating node, to minimize energy consumption, decreases with an increase in number of cooperating nodes. The reverse implication is also true, i.e., the optimal number of cooperating nodes, to minimize energy consumption, also decreases with an increase in constellation size.
4. Finally through our numerical results, we observe that for low bit error probability, virtual MIMO transmission is more energy efficient than the SISO transmission. Also in virtual MIMO transmission, there exists an optimal cluster size that can lead to better energy efficiency. Further for large values of intercluster spacing, virtual MIMO achieves better energy-MI (or MSE) tradeoff than the SISO transmission. However, for small values of intercluster spacing, SISO achieves better energy-MI (or MSE) tradeoff, see Table 1.

BACKGROUND

The design of WSNs differs significantly from the design of conventional wireless networks, owing to the severe energy constraint of nodes. To reduce energy consumption of each node and hence prolong the lifetime of network, a cooperation based architecture has been proposed in (Cui, Goldsmith & Bahai, 2004). Specifically, the authors assume that the sensor nodes exchange information and transmit cooperatively, using the well known multiple input and multiple output (MIMO) encoding technique due to Alamouti (Alamouti, 1998). Such cooperative based transmission schemes are attractive for applications viz., environmental, habitat and avalanche monitoring. In such applications, the nodes are expected to last for long time without battery replacement. Adopting virtual MIMO based transmission scheme for such applications can prolong lifetime of the network, since each node would spend a fraction of total energy required in non-cooperative based transmission.

The virtual MIMO scheme based on STBCs used in (Cui, Goldsmith & Bahai, 2004) is just one of the MIMO antenna based approaches to improve the energy efficiency of WSNs. For example in (Jayaweera, 2007), the author considers node cooperation based on the popular vertical Bell Laboratories layered space-time (V-BLAST) architecture (Foschini, 1996). Other related works on virtual MIMO include (Jayaweera, Chebolu & Donapati, 2007), where the authors demonstrate as how signal processing can

Table 1. Notation

Symbol	Description
$x_k, y_k, u_k, n_k, y_k^\Delta$	Denote respectively the state, observation, process noise, measurement noise and quantized observation values of the stage k.
P_0, σ^2, Γ	Denote signal power, observation noise variance and observation SNR respectively.
a_{k-1}	Denotes correlation between stages k and $k-1$.
A	Denotes field diffusion coefficient.
$d_{ij}, d_{max,k}$	Denote the length of the link (i,j) and the maximum radius of th cluster k.
$E_{SISO}^{ij}, E_{MIMO}^{ij}$	Denote the per bit energy consumption on the link (i,j) with SISO and MIMO transmission schemes.
L_{ij}, b_{ij}	Denote the number of bits transmitted and the constellation size on link (i,j)
p_b, N_0, G_0, M_l, B, r	Denote respectively, probability of error, channel noise variance, attenuation factor, link margin, symbol bandwidth and spatial rate of STBC.
α	Denotes a power amplifier constant that depends on its efficiency.
P_T, P_R	Denote the transmission and receiver power respectively.
$P_{ADC}, P_{DAC}, P_{filtx}, P_{filrx}, P_{IFA}, P_{LNA}, P_{mix}$	Denote respectively the power consumption of analog to digital converter (ADC), digital to analog converter (DAC), transmitter filter, receiver filter, intermediate frequency amplifier (IFA), low noise amplifier (LNA) and mixer.
R_k	Denotes encoding rate of stage k.
$\Sigma_{k\|k-1}$	Denotes the predictor covariance obtained via Kalman recursion.
n_k	Denotes the number of cooperating nodes in stage k

be used for data compression, thereby further enhancing the energy efficiency of virtual MIMO based transmission in WSNs

Unlike single-hop transmission considered in (Cui, Goldsmith & Bahai, 2004), the paper (Cui & Goldsmith, 2006) considers cooperative MIMO techniques for multi-hop routing. However, the issue of correlation in the data sensed by nodes was not considered in (Cui & Goldsmith, 2006). But in practice owing to the close proximity of nodes, the data sensed by neighboring nodes can be correlated, therefore, we consider the issue of data correlation in this work. Architecturally our system requires local information circulation to exploit the correlation, which is not the case in (Cui & Goldsmith, 2006). An important aspect of our work is, we obtain important structural results on, constellation size and number of cooperating nodes, using concepts in *monotone comparative statics* (Topkis, 1998), which to the best of our knowledge are not found elsewhere. These structural results provide important design guidelines for data gathering in correlated WSNs.

There exists other works in literature which consider correlated data gathering in WSNs. For example in (Cristescu, Beferull-Lozano & Vetterli, 2005), the authors consider distributed source coding (DSC) and routing in correlated sensor networks. In this approach, the joint problem of optimizing the transmission structure and source coding becomes decoupled. The optimal transmission structure can be determined in a simple manner, however, the coding becomes complex owing to the need for global network knowledge and correlation structure (Cristescu, Beferull-Lozano & Vetterli, 2005).

Contrary to the approach in (Cristescu, Beferull-Lozano & Vetterli, 2005), in (Cristescu, Beferull-Lozano,Vetterli & Wattenhofer, 2006), the authors consider the case where correlation structure is exploited through explicit communication. In this case, the rate allocation for source coding becomes simple, however, determining the optimal transmission structure becomes difficult. In fact, the authors prove the problem of determining optimal transmission structure to be NP-complete, and therefore propose heuristic approximation algorithms, that provide good solutions for this difficult problem.

Similar works where the interplay of data compression and routing was studied can be found in (Pattem, Krishnamachari & Govindan, 2004) and (Scaglione & Servetto, 2002). In (Pattem, Krishnamachari & Govindan, 2004), using an empirical correlation model, the authors evaluate three qualitatively different schemes namely: distributed source coding, routing driven compression and compression driven routing. Further, the authors prove cluster-based tree structures to have good performance, depending on the correlation level.

Wireless Sensor Network (WSN) Model and Assumptions

We consider data gathering using a WSN. For the purpose of our analysis, we assume the signal along the route to the destination forms, a one dimensional stationary Gauss-Markov process. Such a model was used in (Sung, Tong & Ephremides, 2005). This is valid in the case of acoustic sensor networks, where multiple acoustic sensor nodes are deployed to record sound field created by an acoustic source. In such a scenario, each of the sensor nodes deployed receives time delayed version of the same signal, and therefore it is reasonable to assume, the signal sensed by successive stages of the network is correlated. We use $k = 1, 2, \cdots, I$ to denote discrete time. Equivalently, k also represents stage number, since stage $(k + 1)$ receives observation at time $(k + 1)$ due to delay. The Gauss-Markov assumption on the underlying signal process leads us to the following state-space representation for the state and observation model (See (Sung, Tong & Ephremides, 2005)):

$$x_k = a_{k-1}x_{k-1} + u_{k-1}, \text{for } k = 1, 2, \cdots, I,$$
$$u_{k-1} \blacklozenge\blacklozenge\left(0, P_0(1 - a_{k-1}^2)\right),$$
(1)

where a_{k-1} denotes the correlation between $(k-1)^{\text{th}}$ and k^{th} stages of the network. In general, the models used for correlation are classified into four standard groups: *spherical, power Exponential, rational quadratic* and *Matérn* (Berger, Oliveira & Sanso, 2001).

The key properties such as, non-negativity of correlation and decrease in correlation with distance, are satisfied by all the four models. However, for our numerical study on energy-MI tradeoff, we model correlation a_{k-1} using the popular *power Exponential* model as $a_{k-1} = e^{-Ad_{k-1}}$, where A is the field diffusion coefficient and d_{k-1} denotes the distance between $(k-1)^{\text{th}}$ and k^{th} stage of the network. We also remark that the structural results on, constellation size and number of cooperating nodes, derived in this chapter are independent of the correlation model.

Each stage of the network on the way to the destination observes a noisy version of the process, i.e.,

$$y_k = x_k + n_k \;,\; n_k \blacklozenge\blacklozenge \left(0, \sigma^2\right), \tag{2}$$

where σ^2 denotes the variance of observation noise.

Due to stationarity of the random field, the observation SNR at each stage of the network is same, and is defined as

$$\Gamma \equiv \frac{P_0}{\sigma^2}. \tag{3}$$

The observations of each sensor are transmitted to the destination through multi-hop routing, which in turn aims at extracting relevant features of the original correlated random field using judicious signal processing.

We abstract the WSN at the network layer as a directed graph $G = \left(V, E\right)$, where V denotes the set of stages and E denotes the set of directed links. We consider the case of single sink data gathering only. We assume there are $\left(I + 1\right)$ stages in the network, where $1, 2, \cdots, \quad I$ denote the data gathering stages, and the $\left(I + 1\right)^{\text{th}}$ stage denotes the sink node. Every directed link $e_{ij} \in E$ that connects two stages $\left(i, j\right) \in V$ has a cost E_{ij} associated with it. In general the link cost E_{ij} depends on constellation size b_{ij} adopted on the link, square of length of the link d_{ij}, and circuit power consumption.

An $\left(I + 1\right)$ stage (a stage in a network represents single node for the case of SISO transmission or group of nodes for the case of virtual MIMO transmission. Henceforth, we use the word *stage* to denote either a single node (or) group of nodes.) network depicting the scenario considered in this work is shown in Fig. 1. Each stage can contain one node (for SISO transmission case) or more than one node (for MIMO transmission case). In the SISO case, the only node present in a given stage, senses, quantizes and transmits along with the data received from previous stages to the next stage. The next stage is determined by the solution of the energy optimization problem formulated in next section.

In the virtual MIMO case, a node called the *sensing node* makes an observation of the underlying process, quantizes and broadcasts locally to other nodes in that stage for cooperative transmission. Upon local information circulation within each stage, the nodes transmit cooperatively, the data originated at its own stage along with the data received from previous stages to the next stage. As in the SISO case, the next stage is determined by the solution of the energy optimization problem formulated in next section. The cooperative transmission between various stages of the network is accomplished based on the well-known STBCs (Tarokh, Jafarkhani & Calderbank, 1999).

For short range communications circuit energy can be significant, therefore, we consider circuit energy consumption along with the transmission energy consumption as in (Cui, Goldsmith & Bahai, 2004; Cui, Goldsmith & Bahai, 2005; Cui & Goldsmith, 2006). To model the power consumption blocks of a sensor node we use the model in (Cui, Goldsmith & Bahai, 2005). The antenna of a sensor node is typically at the ground level, therefore we model the channel between sensors to be of Rayleigh

Figure 1. The figure shows the scenario for data gathering in correlated random fields. Each stage of the network consists of a number of sensor nodes which cooperate during transmission of data. For each data collection period a node in each stage senses, quantizes and broadcasts to other nodes in that stage for cooperative transmission to the next stage. Note that apart from the data generated within each stage, the nodes in each stage also receive data from the previous stage that needs to be relayed to the next stage.

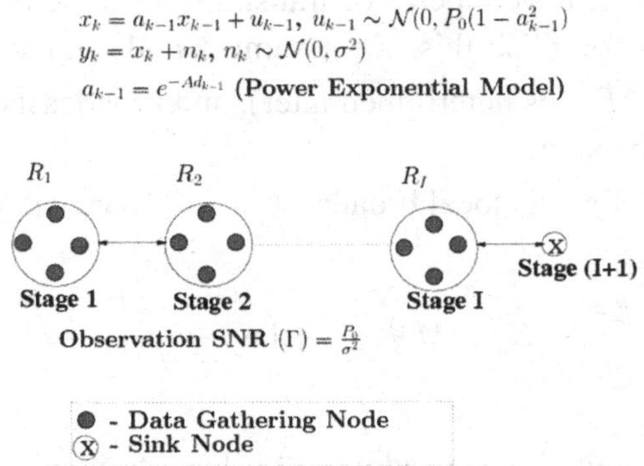

$$x_k = a_{k-1}x_{k-1} + u_{k-1},\ u_{k-1} \sim \mathcal{N}(0, P_0(1 - a_{k-1}^2))$$
$$y_k = x_k + n_k,\ n_k \sim \mathcal{N}(0, \sigma^2)$$
$$a_{k-1} = e^{-Ad_{k-1}}\ \textbf{(Power Exponential Model)}$$

Observation SNR $(\Gamma) = \frac{P_0}{\sigma^2}$

● - Data Gathering Node
Ⓧ - Sink Node

fading type. We assume each sensor node employs M-ary quadrature amplitude modulation (M-QAM) for digital modulation. We assume the cooperating nodes in stage k are distributed in a circle of radius $d_{\max,k}$. Since the cluster radius of stage k is much less than the inter cluster distance, it is reasonable to assume as in paper (Cui, Goldsmith & Bahai, 2004) that each node in a given stage is equidistant from the neighboring stage.

Minimum Energy Data Gathering (MEDG) in Correlated Wireless Sensor Networks

In this section, we formulate the problem of minimum energy data gathering in sensor networks, for the linear Gauss-Markov state-space model assumed in Eqs. (1) and (2).

Expression for Total Energy Consumption per bit using SISO Techniques

In SISO transmission since there is no local information circulation, the single node present in a given stage encodes at a rate R_k as determined later.

Thus the energy consumption per bit in transmitting over link (i, j) using SISO transmission can be written as

$$E_{SISO}^{ij} = \frac{2}{3}(1 + \alpha)\left(\frac{4}{p_b}\right)\left(\frac{2^{b_{ij}} - 1}{b_{ij}}\right)N_0 G_0 M_l d_{ij}^2 + \frac{(P_T + P_R)}{B}\frac{1}{b_{ij}}\left(\text{for } b_{ij} \geq 2\right). \tag{4}$$

where α is a constant defined by the power amplifier efficiency, p_b is the desired probability of bit error, b_{ij} denotes the constellation size adopted on link (i, j), N_0 is the noise spectral density, G_0 is the attenuation factor, M_l denotes the link margin, d_{ij} is length of the link (i, j), B is the symbol rate, and P_T, P_R denote the transmitter and receiver circuit power consumption.

Expression for Total Energy Consumption per bit using MIMO Techniques

As noted in the system model, for cooperative transmission to be possible there needs to be local information circulation. To accomplish this, we assume a node in each stage of the network makes an observation, encodes at rate R_k (as determined later), and broadcasts to the other $n_k - 1$ nodes in that stage for cooperative transmission.

The energy consumption for this local broadcast in k^{th} stage can be written as in Eq. (4) to be

$$E_{local}^k = \frac{2}{3}\left(1 + \alpha\right)\left(\frac{4}{p_b}\right)\left(\frac{2^{b_{local}} - 1}{b_{local}}\right)G_0 N_0 M_l d_{\max,k}^2 + \left(n_k - 1\right)\frac{\left(P_T + P_R\right)}{B\, b_{local}}, \tag{5}$$

where b_{local} denotes the constellation size adopted for local transmission and $d_{\max,k}$ denotes the cluster radius of stage k. The factor of $\left(n_k - 1\right)$ in the second term of Eq. (5) is due to the fact that during local broadcast there are always $\left(n_k - 1\right)$ circuits listening.

For the purpose of our analysis and simulation in this chapter we assume b_{local} as fixed. Consequently, the energy consumption for local broadcast at the kth stage E_{local}^k is a function of only the number of cooperating nodes n_k present in the kth stage.

If we assume n nodes in a cluster *cooperate* and transmit, then the energy consumption per bit $\left(E_b\right)$, for cooperative transmission can be upper bound as (Cui, Goldsmith & Bahai, 2004)

$$E_b \le \frac{2}{3}n\left(\frac{4}{p_b}\right)^{\frac{1}{n}}\left(\frac{2^b - 1}{b^{\frac{1}{n}+1}}\right)N_0 \left(\text{for } b \ge 2\right), \tag{6}$$

where b is the constellation size adopted for cooperative transmission, p_b is the desired probability of error and N_0 is the noise spectral density.

Approximating the energy consumption per bit in Eq. (6) as an equality, we can express the total power consumed for cooperative transmission over the link (i, j) as

$$P_{LH}^{ij} = \underbrace{\left(1 + \alpha\right)E_b b_{ij} G_0 M_l d_{ij}^2 B}_{\text{Transmission Power}} + \underbrace{n_i\left(P_T + P_R\right)}_{\text{Circuit Power}}, \tag{7}$$

where as before α is a constant defined by the power amplifier efficiency, E_b denotes the per bit energy conumption for cooperative transmission and is given by Eq. (6), b_{ij} denotes the constellation size adopted on the link (i, j), G_0 is the attenuation factor, M_l denotes the link margin, d_{ij} denotes the length of the link (i, j), B denotes the symbol rate, and P_T, P_R denotes the power consumption in transmitter and receiver circuits.

Thus, the total energy consumption per bit for transmission over the link (i, j) with virtual MIMO transmission can be written as

$$E_{MIMO}^{ij} = \underbrace{\left(\frac{P_{LH}^{ij}}{B_e}\right)\frac{1}{b_{ij}}}_{for\,long\,haul\,cooperative\,transmission} + \underbrace{E_{local}^{i}}_{for\,local\,information\,transmission}, \tag{8}$$

where B_e is the effective symbol rate and is equal to $(B\,r)$, r is the spatial rate of STBC employed, and E_{local}^{i} is given by Eq. (5).

In the case of SISO transmission B_e is equal to B. However, for the case of MIMO transmission, B_e depends on the *spatial rate r* of STBC employed for cooperative transmission of data. For the case of Alamouti code employed in (Cui & Goldsmith, 2006) (where number of cooperating nodes n is equal to two) r is equal to one.

For the general case of n cooperating nodes as observed in (Tarokh, Jafarkhani & Calderbank, 1999), it may not be possible to design $r = 1$ STBC for complex constellations such as QAM or phase shift keying (PSK) (this was formally proved in (Liang, 2003)). Consequently, r in our formulation depends on the number of cooperating nodes n.

Approximating Eq. (6) as an equality, we can express Eq. (8) as

$$E_{MIMO}^{ij} = \frac{2}{3}\left(1+\alpha\right)n_i\left(\frac{4}{p_b}\right)^{\frac{1}{n_i}}\left(\frac{2^{b_{ij}}-1}{b_{ij}^{\frac{1}{n_i}+1}}\right)N_0 G_0 M_l d_{ij}^2\left(\frac{1}{r}\right) + n_i\frac{\left(P_T+P_R\right)}{B\,r}\left(\frac{1}{b_{ij}}\right)$$

$$+E_{Tx-local} + \left(n_i-1\right)\frac{\left(P_T+P_R\right)}{B\,b_{local}}, \tag{9}$$

where $E_{Tx-local}$ is given by first term in Eq. (5).

A close look at Eq. (9) reveals that when a $r \le 1$ STBC is used, then it can lead to increased energy consumption compared to the case when $r = 1$.

Rate Allocations of Data Gathering Stages in Correlated WSNs

The rate allocations R_k of data gathering stages $k = 1, 2, \cdots, I$ are determined so that the *rate admissibility constraint* is satisfied. For a network with single sink the rate admissibility constraint can be written as below with y_k^{Δ} denoting the quantized value of the continuous observation y_k at stage k

$$\sum_{k=1}^{I} R_k \ge H\left(y_1^{\Delta}, y_2^{\Delta}, \cdots, y_I^{\Delta}\right), \tag{10}$$

where $H(z_1, z_2, \cdots, z_K)$ denotes joint entropy of $z_1, z_2, \cdots, z_K$.

In the presence of knowledge about correlation between various stages of the network, we can choose individual R_k as (Cristescu, Beferull-Lozano & Vetterli, 2005)

$$R_k = H\left(y_k^{\Delta} \mid y_{k+1}^{\Delta}, y_{k+2}^{\Delta}, \cdots, y_I^{\Delta}\right) \text{ for } k = 1, 2, \cdots, I-1, \tag{11}$$

$$R_I = H\left(y_I^{\Delta}\right), \tag{12}$$

where $H\left(z_k \mid z_{k+1}, z_{k+2}, \cdots, z_I\right)$ denotes the conditional entropy of z_k given $z_{k+1}, z_{k+2}, \cdots, z_I$ and $H\left(z_I\right)$ denotes the entropy of z_I.

To compute R_k for $k = 1, 2, \cdots, I$ in Eqs. (11) and (12) for the state and observation model in Eqs. (1) and (2), we approximate Eqs. (11) and (12) for sufficiently small quantization stepsize Δ as (Cover & Thomas, 1991)

$$R_k \approx h\left(y_k \mid y_{k+1}, y_{k+2}, \cdots, y_I\right) - \log_2 \Delta \text{ for } k = 1, 2, \cdots, I-1, \tag{13}$$

$$R_I \approx h\left(y_I\right) - \log_2 \Delta, \tag{14}$$

where $h\left(z_k \mid z_{k+1}, z_{k+2}, \cdots, z_I\right)$ denotes the conditional differential entropy of z_k given $z_{k+1}, z_{k+2}, \cdots, z_I$ and $h\left(z_I\right)$ denotes the differential entropy of z_I.

For the linear Gauss-Markov state-space model assumed in Eqs. (1) and (2), we can express $h\left(y_k \mid y_{k+1}, y_{k+2}, \cdots, y_I\right)$ in close form as equal to $\frac{1}{2}\log_2\left[2\pi e\sigma^2\left(1 + \frac{\Sigma_{k|k-1}}{\sigma^2}\right)\right]$, while $h(y_I)$ can be written as $\frac{1}{2}\log_2\left(2\pi e\sigma^2\left(\Gamma + 1\right)\right)$ (see Appendix for detailed derivation). Thus Eqs. (13) and (14) can be re-written as

$$R_k = \frac{1}{2}\log_2\left[\frac{2\pi e\sigma^2}{\Delta^2}\left(1 + \frac{\Sigma_{k|k-1}}{\sigma^2}\right)\right], \text{ for } k = 1, 2, \cdots, I-1, \tag{15}$$

$$R_I = \frac{1}{2}\log_2\left[\frac{2\pi e\sigma^2}{\Delta^2}\left(\Gamma + 1\right)\right], \tag{16}$$

where $\Sigma_{k|k-1}$ is the predictor covariance, Δ is the quantization stepsize, Γ is the observation SNR defined as in Eq. (3), and σ^2 is the observation noise variance.

For the case of stage 1, we choose $\Sigma_{1|0} = P_0$, therefore the encoding rate of stage 1 is equal to

$$R_1 = \frac{1}{2}\log_2\left[\frac{2\pi e\sigma^2}{\Delta^2}\left(\Gamma + 1\right)\right], \tag{17}$$

To simplify R_k for $k = 2, 3, \cdots, I-1$, we can write the predictor covariance $\Sigma_{k|k-1}$ using Kalman recursion (Anderson & Moore, 1979) as

$$\frac{\Sigma_{k|k-1}}{\sigma^2} = a_{k-1}^2 \gamma_{k-1} + \Gamma\left(1 - a_{k-1}^2\right),$$ (18)

where $\gamma_{k-1} = \dfrac{\Sigma_{k-1|k-2}}{\Sigma_{k-1|k-2} + \sigma^2}.$ (19)

By noting $\gamma_{k-1} \leq 1$, we can upper bound R_k for $k = 2, 3, \cdots, I-1$ in Eq. (15) as

$$R_k \leq \frac{1}{2} \log_2\left[\frac{2\pi e \sigma^2}{\Delta^2}\left((\Gamma + 1) - a_{k-1}^2\left(\Gamma - 1\right)\right)\right] \text{ for } k = 2, 3, \cdots, I-1.$$ (20)

The advantage of approximation in Eq. (20) over Eq. (15) is that to compute rate (R_k), the kth stage requires only the position of $(k-1)^{\text{th}}$ stage, which can be provided at the time of initial network setup.

For the rest of our analysis in this chapter, we assume that the sensing node in stage 1 *encodes at rate* R_1 given by Eq. (17),while stages $k = 2, 3, \cdots, I-1$ encode at a rate equal to the upper bound in Eq. (20), and the sensing node in stage I encodes at rate R_I given by Eq. (16).

Formulation of the Minimum Energy Data Gathering (MEDG) in Correlated WSNs

In WSNs since energy consumption needs to be minimized, to enhance lifetime of the network, therefore the optimization problem to minimize energy consumption can be stated as follows:

$$\underset{[b_{ij}\, L_{ij}]}{\text{minimize}} \sum_{(i,j) \in E} E_{SISO}^{ij}\left(b_{ij}\right) L_{ij} \text{ (for SISO case)}$$

$$\text{(OR) } \underset{[b_{ij}\, n\, L_{ij}]}{\text{minimize}} \sum_{(i,j) \in E} E_{MIMO}^{ij}\left(b_{ij}, n\right) L_{ij} \text{ (for MIMO case)}$$ (21)

Subject to

$$\sum_{j:(i,j) \in E} L_{ij} - \sum_{j:(j,i) \in E} L_{ji} = R_i \text{ for } i = 1, 2, 3, \cdots, I+1$$ (22)

$$2 \leq b_{ij} \leq b_{max}^{ij}$$ (23)

$$1 \leq n \leq N$$ (24)

where $E_{SISO}^{ij}\left(.\right)$ and $E_{MIMO}^{ij}\left(.,.\right)$ denote the cost in terms of energy associated with the link $\left(i, j\right)$ for the SISO and MIMO cases which are given by Eq. (4) and Eq. (9) respectively, L_{ij} denotes the *flow* (in terms of bits) associated with the link $\left(i, j\right)$. The constraint in Eq. (22) is the *flow conservation constraint,* which ensures at any stage the difference between outgoing flow and the incoming flow equals, the data generated at the corresponding stage. In Eq. (23) b_{max}^{ij} is determined so that the sensor node does

Table 2. System Parameters

$f_c = 2.45\,\text{GHz}$ $\lambda(\text{wavelength of radiation}) = 0.12$ m	$\eta = 0.35$
$N_0 = -174$ dBm/Hz	$B = 10$ KHz
$G_t(\text{transmit antenna gain}) = G_r(\text{receive antenna gain}) = 0$ dB	$P_{\text{ADC}} = 0.0067$ w
$P_{\text{DAC}} = 0.0163$ W	$P_{\text{filtx}} = P_{\text{filrx}} = 2.5$ mW
$P_{\text{IFA}} = 3$ mW	$P_{\text{LNA}} = 20$ mW
$P_{\text{mix}} = 30.3$ mW	$P_T \blacklozenge P_{\text{DAC}} + P_{\text{mix}} + P_{\text{filtx}}$
$P_R \blacklozenge P_{\text{LNA}} + P_{\text{mix}} + P_{\text{IFA}} + P_{\text{filrx}} + P_{\text{ADC}}$	$G_0 \blacklozenge \dfrac{(4\pi)^2}{G_t G_r \lambda^2} \approx 40$ dB
$\alpha = 7.6$	$M_l = 40$ dB

not transmit beyond its peak power.

The rate allocations R_i for $i = 1, 2, \cdots, I$ in Eq. (22) are given by Eqs. (17), (16) and (20) respectively. Since the sink node (corresponding to the stage with index $(I + 1)$) *absorbs* the data from the data gathering stages, i.e., stages $1, 2, \cdots, I$, therefore

$$R_{I+1} = -\sum_{k=1}^{I} R_k \,. \tag{25}$$

For a fixed constellation size $\left(b_{ij}\right)$ on a given link $\left(i, j\right)$ and fixed number of cooperating nodes n, the optimization problem in Eqs. (21) - (24) is a minimum cost network flow (MCNF) problem (Ahuja, Magnanti & Orlin, 1993), and can be efficiently solved using the Matlog toolbox (Matlog, 2006). By solving the optimization problem in Eqs. (21) - (24), we can determine the number of bits L_{ij} to be transmitted on a link $\left(i, j\right)$.

To give a numerical example, we consider a ten stage network (i.e., $(I + 1) = 10$), with each stage separated by a distance of thirty metres (i.e., $d = 30$ m). The first 9 stages of the network denote data gathering stages, while the 10th stage denotes the sink node. The system parameter details of 2.4 GHz radio in the Industrial-Scientific-Medical (ISM) band is shown in Table 2 (Cui, Goldsmith & Bahai, 2005).

We choose the probability of error $p_b = 10^{-3}$. For the virtual MIMO transmission case, we assume the number of cooperating nodes in each stage is equal to two (i.e., $n_i = 2$ for $i = 1, 2, \cdots, I$), therefore the spatial rate is equal to one (i.e., $r = 1$). We assume the cluster radius of stage k as equal to one metre i.e., $d_{\max,k} = 1$ m. We assume the observation SNR is equal to ten dB (i.e., $\Gamma = 10$ dB). We assume the observation noise variance $\sigma^2 = 0.01$ and the quantization stepsize $\Delta = 0.001$.

The plot of optimal energy consumption values versus the correlation coefficient is shown in Fig. 2. As expected, the energy consumption value decreases with an increase in correlation. It can be observed from Fig. 2 that when circuit energy is taken into account, then MIMO transmission may not be efficient.

However, when transmission energy alone is taken into account, then MIMO transmission is efficient. Also, when transmission energy alone is taken into account, then the optimal energy consumption value obtained by solving MCNF problem coincides with the energy consumption value obtained via shortest path routing. But when circuit energy consumption is taken into account, then the energy consumption value obtained via shortest path routing is higher. From this observation we can conclude that, when transmission energy alone is considered, then shortest path routing is optimal.

The plot of optimal energy consumption values versus the correlation coefficient, for the case when probability of error $p_b = 10^{-6}$ is shown in Fig. 3. For the case when number of cooperating nodes is equal to four, spatial rate r is equal to $3 / 4$, since it is possible to construct an orthogonal STBC with this spatial rate as shown in (Tarokh, Jafarkhani & Calderbank, 1999). For the case when number of cooperating nodes is equal to eight, spatial rate r is equal to $5 / 8$ (Liang, 2005).

From Fig. 3 it can be observed that, when the desired probability of error is small, virtual MIMO transmission outperforms SISO transmission, in terms of energy efficiency. For the given system parameter values, it has been found that a cluster with 4 cooperating nodes can be more energy efficient, compared to the case when the cluster has 2 and 8 cooperating nodes. For the case when number of cooperating nodes are equal to 4, the diversity advantage provided by additional cooperating nodes is more than, the increase in energy consumption due to local communication, and use of a spatial rate less than one STBC. While for the case of 8 cooperating nodes, the increase in energy consumption due to local communication, and use of STBC with spatial rate less than one, can be more than the diversity advantage provided by additional cooperating nodes. Thus there exists an optimal number of cooperating nodes, that can lead to high energy efficiency, which in the current case is equal to 4.

In this section, we derived the encoding rates adopted by sensor nodes, for data gathering in correlated WSNs. We also formulated the problem of MEDG in correlated WSNs, and demonstrated the efficacy

*Figure 2. The figure shows the plot of total energy consumption in the network versus the correlation coefficient. We fix the observation SNR Γ **equal to** 10 dB and assume the desired probability of error p_b equal to 10^{-3}. Number of stages in the network are assumed equal to 10. Distance between the various stages of the network is equal to 30 metres.*

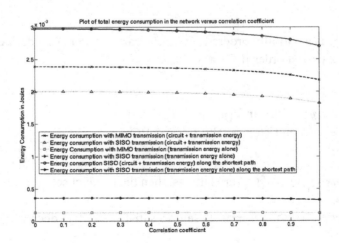

*Figure 3. The figure shows the plot of total energy consumption in the network versus the correlation coefficient. We fix the observation SNR Γ **equal to** 10 dB and assume the desired probability of error p_b equal to 10^{-6}. Number of stages in the network are assumed equal to 10. Distance between the various stages of the network is equal to 30 metres.*

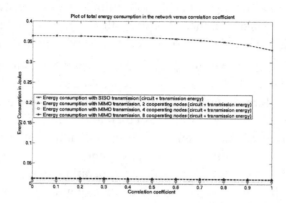

of exploiting correlation through a numerical example, for both virtual MIMO and SISO transmission schemes in WSNs.

However, as mentioned before, our aim in this chapter is to obtain structural results on physical layer design variables such as, constellation size and number of cooperating nodes. These structural results are derived next. In the next section, we present a brief survey on monotone comparative statics, that are useful in proving the structural results.

LATTICE, DECREASING AND INCREASING DIFFERENCES, SUPER AND SUB MODULAR FUNCTIONS AND PARAMETRIC MONOTONICITY

Definition 1.*A partially ordered set* Ω *is a lattice if for all* $x, y \in \Omega$, the componentwise maximum and minimum belong to Ω.

Definition 2.*A set* S *is a partially ordered set (i.e., a partial order (PO) defined over* $S \times S$ *), where a binary relation PO is a partial order if for any* $x, y, z \in S$

- Reflexive: $(x, x) \in \text{PO}$.
- Anti-Symmetric: $(x, y) \in \text{PO}$ and $(y, x) \in \text{PO}$ $\quad x = y$.
- Transitive: $(x, y) \in \text{PO}$ and $(y, z) \in \text{PO}$ $(x, z) \in \text{PO}$.

Theorem 1.*If the sets* $\Omega_1, \Omega_2, \cdots, \Omega_N$ *are lattices, then the product set* (Ω) *defined as* $\bigotimes_{i=1}^{N} \Omega_i$ *is also a lattice.*

Definition 3.*A function* $F : S \times A \rightarrow \blacklozenge$ *is said to satisfy (strictly) decreasing differences in* (s, a) *if*

$$F(s', a') - F(s', a)(<) \leq F(s, a') - F(s, a) \ \forall a' > a, s' > s, \tag{26}$$

and F is said to satisfy *(strictly) increasing differences* if

$$F(s', a') - F(s', a)(>) \geq F(s, a') - F(s, a) \ \forall a' > a, s' > s. \tag{27}$$

i.e., if the difference $F(:, a') - F(:, a)$ is a decreasing function for all $a' > a$, then the function F is said to satisfy decreasing differences. It is said to satisfy increasing differences if $F(:, a') - F(:, a)$ is an increasing function for all $a' > a$ (Amir, 2005).

Remark 1. *For differentiable functions, decreasing (increasing) differences is equivalent to submodularity (supermodularity) (Amir, 2005).*

Remark 2. *If the function F is continuously differentiable, then* $\dfrac{\partial^2 F(s, a)}{\partial a \partial s}(<) \leq 0$ *implies the function F satisfies (strict) decreasing differences and is submodular. On the other hand if* $\dfrac{\partial^2 F(s, a)}{\partial a \partial s}(>) \geq 0$, *then the function F is said to satisfy (strict) increasing differences and is supermodular (Amir, 2005).*

We now state the following simplified version of Topkis's monotonicity theorem in (Topkis, 1998). This simplified version provided in (Amir, 2005) is adequate to prove the structural results in this chapter.

Lemma 1. *For the optimization problem below:*

minimize $F(s, a)$ subjectto $s \in S$

Suppose that S forms a lattice and has at least one solution for each $a \in A$. Suppose also that F satisfies (strictly) increasing differences in (s, a). Then optimal s is always decreasing in a.

The above Lemma appears for the problem of maximizing the objective function in (Amir, 2005). To go from their framework to the current one, one needs to simply multiply the objective by a negative sign.

Remark 3. *In Lemma 1 if the function F satisfies decreasing differences, instead of increasing differences, with all the other assumptions in tact, then optimal s is increasing in a.*

Remark 4. *In Lemma 1 if we replace the objective function by a monotone transformation (such as logarithm) of the objective function, still the parametric monotonicity of Lemma 1 holds.*

The above remark might be useful in cases, where it is not possible to say conclusively, if the function is super or sub modular, without any additional assumptions. Please refer (Amir, 2005) for an interesting example.

STRUCTURAL RESULTS ON OPTIMAL CONSTELLATION SIZE AND NUMBER OF COOPERATING NODES

In this section, we present structural results on variation of optimal constellation size and optimal number of cooperating nodes, with respect to the correlation coefficient. The structural results stated in the form of theorems are provided without any proof, since they follow straightforwardly from the results in previous section.

To simplify analysis in this section, we assume for both SISO and virtual MIMO transmission, each stage adopts minimum distance routing, i.e., every stage of the network transmits to the immediate next stage (i^{th} stage to $(i+1)^{\text{th}}$ stage, $(i+1)^{\text{th}}$ stage to $(i+2)^{\text{th}}$ stage and so on). To simplify notation, we denote the constellation size adopted by i^{th} stage to transmit to $(i+1)^{\text{th}}$ stage as b_i. Note that the distance between i^{th} stage and $(i+1)^{\text{th}}$ stage is denoted as d_i.

Since each stage transmits the data generated locally, along with the data received from previous stages, the amount of data transmitted by k^{th} stage is given as

$$Q_k = \sum_{i=1}^{k} R_i \text{ for } k = 1, 2, \cdots, I, \tag{28}$$

where R_1 is given by Eq. (17), R_i for $i = 2, 3, \cdots, I-1$ is given by Eq. (20), and R_I is given by Eq. (16).

As described in system model description, in the SISO transmission scheme, there exists only one node per stage, that senses, quantizes and transmits to the next stage. Thus the total energy consumption in this case, using the expression for energy consumption per bit in Eq. (4) is given by

$$E_{SISO} = \sum_{k=1}^{I} \left[\varepsilon_{SISO} \left(\frac{2^{b_k} - 1}{b_k} \right) d_k^2 + \left(\frac{P_T + P_R}{B b_k} \right) \right] Q_k, \tag{29}$$

where we define $\varepsilon_{SISO} \equiv \frac{2}{3}(1+\alpha) N_0 G_0 M_l \left(\frac{4}{p_b} \right)$, b_k is constellation size adopted by stage k, d_k denotes the distance between k^{th} and $(k+1)^{\text{th}}$ stage, and Q_k is defined as in Eq. (28).

For the virtual MIMO transmission case, in a given stage k, one node would sense, encode at a rate R_k, and transmit to the remaining $(n_k - 1)$ nodes for cooperative transmission to the next stage of network. Using the expression for energy consumption per bit for the vitual MIMO transmission in Eq. (9), we can write the expression for total energy consumption using virtual MIMO as

$$E_{MIMO} = \sum_{k=1}^{I} \left[\varepsilon_{MIMO} n_k \left(\frac{2^{b_k} - 1}{b_k^{\frac{1}{n_k}+1}} \right) d_k^2 + \frac{n_k(P_T + P_R)}{B b_k r} + E_{Tx-local} + (n_k - 1)E_{C-local} \right] Q_k, \tag{30}$$

where we define $\varepsilon_{MIMO} \equiv \frac{2}{3}\left(1+\alpha\right)N_0 G_0 M_l\left(\frac{1}{r}\right)$, b_k is constellation size adopted for cooperative transmission by stage k, n_k is number of cooperating nodes in stage k, $E_{Tx-local}$ is defined as under Eq. (9) and $E_{C-local} \equiv \frac{\left(P_T + P_R\right)}{B b_{local}}$.

Expression for Energy Consumption for the case of Simple Homogeneous Network

For the purpose of proving structural results in this section, we assume a homogeneous network, where each stage of the network is separated by a distance of d metres as in (Cui & Goldsmith, 2006), i.e., $d_k = d$. Further, we assume each stage of the network adopts, the same constellation size of b for transmission, i.e., $b_k = b$. Also for virtual MIMO transmission case, we assume the number of cooperating nodes in each stage to be same, and equal to n, i.e., $n_k = n$.

Since we assume the spacing between two successive stages of the network to be same, therefore, the correlation between signal samples sensed by successive stages of the network is equal, i.e., $a_{k-1} = a$.

Thus the encoding rate R_k adopted by stages $k = 2, 3, \cdots, I-1$ given in Eq. (20) can be expressed in terms of common correlation coefficient (a) as

$$R = \frac{1}{2}\log_2\left(\frac{2\pi e \sigma^2}{\Delta^2}\left(\left(\Gamma + 1\right) - a^2\left(\Gamma - 1\right)\right)\right). \tag{31}$$

With these assumptions $Q_1 = R_1$, $Q_k = (k-1)R + R_1$ for $k = 2, 3, \cdots, I-1$, while $Q_I = (I-2)R + R_1 + R_I$, where R_1, R, and R_I are given by Eqs. (17), (31) and (16) respectively. Therefore the total energy consumption for SISO transmission can be written as

$$E_{Total}^{(1)} = \left\{\varepsilon_{SISO}\left(\frac{2^b - 1}{b}\right)d^2 + \frac{\left(P_T + P_R\right)}{B b}\right\}\left[\frac{\left(I-2\right)\left(I+1\right)R}{2} + IR_1 + R_I\right]. \tag{32}$$

Similarly, the total energy consumption for MIMO transmission can be written as

$$E_{Total}^{(2)} = \left\{\varepsilon_{MIMO}n\left(K\right)^{\frac{1}{n}}\left(\frac{2^b - 1}{b^{\frac{1}{n}+1}}\right)d^2 + \frac{n\left(P_T + P_R\right)}{B b} + \left(n-1\right)E_{C-local} + E_{Tx-local}\right\} \times$$

$$\left[\frac{\left(I-2\right)\left(I+1\right)R}{2} + IR_1 + R_I\right]. \tag{33}$$

Structural Result on Optimal Constellation Size (b^*) vs. Correlation Coefficient (a)

For the purpose of analysis in this subsection, consider the following optimization problem:

$$\underset{b}{\text{minimize}}\ E^{(1)}_{Total}\left(b,a\right)$$

$$\text{subject to } 2 \leq b \leq b_{max} \tag{34}$$

where $E^{(1)}_{total}$ is given by Eq. (32) and b_{max} is determined by peak power of the node.

We now state our first structural result as the following theorem. The theorem states that, the optimal constellation size b^* is an increasing function of correlation coefficient a.

Theorem 2.*For a given number of stages* I *in the network,* spacing d, observation SNR Γ, the optimal constellation size b^* of the optimization problem in Eq. (34) is, an increasing function of the correlation coefficient (a), i.e.,

$$b^*\left(a\right) \equiv \arg\underset{b}{\min}\left[E^{(1)}_{Total}\left(b,a\right) \mid 2 \leq b \leq b_{max}\right]$$

is an increasing function of (a).

Interpretation of Theorem 2: In general $a = 0$ corresponds to the case when data is uncorrelated, while $a > 0$ corresponds to the case when data is correlated, therefore from Theorem 2 it can be concluded that, while transmitting correlated data the sensor nodes can employ higher constellation sizes.

Remark 5.*Theorem 1 can also be proved with* $E^{(2)}_{Total}$ *as objective function in Eq. (34).*

Structural Result on Optimal Number of Cooperating Nodes (n^*) vs. Correlation Coefficient (a)

For the purpose of analysis in this subsection consider the following optimization problem

$$\underset{n}{\text{minimize}}\ E^{(2)}_{Total}\left(n,a\right)$$

$$\text{subject to } 1 \leq n \leq N \tag{35}$$

where $E^{(2)}_{Total}$ is given by Eq. (33).

We now state our second structural result as the following theorem. The theorem proves that for arbitrarily low probability of error p_b, the optimal number of cooperating nodes n^* is, a decreasing function of the correlation coefficient (a).

Theorem 3.*For a given number of stages* I *in the network, spacing* d, *constellation size* b, *observation SNR* Γ, the optimal number of cooperating nodes n^* of the optimization problem in Eq. (35)

is, a decreasing function of the correlation coefficient (a) for arbitrarily low probability of error p_b, i.e., as $p_b \to 0^+$

$$n^*(a) \equiv \arg \min_n \left[E_{Total}^{(2)}(n,a) \mid 1 \leq n \leq N \right]$$

is a decreasing function of correlation coefficient (a).

Interpretation of Theorem 3: The conclusion of Theorem 3 is interesting. The theorem suggests that for arbitrarily low probability of error p_b, when the data is correlated, the number of nodes participating in cooperative transmission can be less than the case when data is uncorrelated.

Structural Result on Optimal rate b^* *vs.* Number of Cooperating Nodes n

For analysis in this subsection, we ignore circuit power consumption of the sensor node.

By ignoring the circuit energy consumption of the node, the expression for total energy consumption using virtual MIMO transmission in Eq. (32) can be re-written as

$$E_{Total}^{(3)} = \left[\varepsilon_{MIMO} \, n \left(K \right)^{\frac{1}{n}} \left(\frac{2^b - 1}{b^{\frac{1}{n}+1}} \right) d^2 + E_{Tx-local} \right] \left[\frac{(I-2)(I+1)R}{2} + IR_1 + R_I \right] \tag{36}$$

where $\varepsilon_{MIMO} \equiv \frac{2}{3}(1+\alpha) N_0 G_0 M_l \left(\frac{1}{r} \right)$ and $K \equiv \left(\frac{4}{p_b} \right)$.

For the analysis in this subsection, consider the following optimization problem:

$$\begin{aligned}
&\text{minimize } E_{Total}^{(3)}(b,n) \\
&\text{subject to } 2 \leq b \leq b_{max},
\end{aligned} \tag{37}$$

where $E_{Total}^{(3)}$ is given by Eq. (36).

We now state our third structural result as the following theorem. In this theorem, we study as how the optimal constellation size b^* varies with the number of cooperating nodes.

Theorem 4. *For a given number of stages* I *in the network, observation SNR* Γ, *correlation coefficient* a *and spacing* d, *the optimal constellation size* b^* *at which each sensor node transmits, to minimize the transmission energy consumption is, a decreasing function of the number of cooperating nodes* n, *i.e.,*

$$b^*(n) \equiv \arg \min_b \left[E_{Total}^{(3)}(b,n) \mid 2 \leq b \leq b_{max} \right] \tag{38}$$

is a decreasing function of n.

Remark 6.*Since supermodularity does not differentiate between the two variables, therefore from Theorem 4 it can be inferred that, the optimal number of sensors* n^* *participating in cooperative transmission to the sink is, a decreasing function of the rate* b .

Interpretation of Theorem 4 and Remark 6: Theorem 4 and Remark 6 are intuitively appealing. Theorem 4 implies that owing to the cooperation among sensor nodes, the constellation size adopted by each sensor node participating in cooperative transmission decreases, as the number of participating nodes increase.

While the Remark 6 implies that as the constellation size adopted by each cooperating node increases, the number of cooperating nodes to minimize energy consumption decreases.

Remark 7.*Theorem 4 is valid only when transmission energy alone is considered. In presence of circuit energy the objective function in Eq. (37) is not supermodular. However, Theorem 4 is interesting in its own right, since transmission energy dominates circuit energy, when probability of error* p_b is arbitrarily small.

Tradeoff Between Mutual Information and Energy in Correlated WSNs

In this section, we study the tradeoff between mutual information (MI) and energy consumption in correlated WSNs, employing SISO and virtual MIMO transmission schemes.

For the purpose of exposition assume stage 1 of the network is fixed. From our earlier notation stage $(I+1)$ is the sink, which receives data from data gathering stages $1, 2, \cdots, I$. Also, $d_1, d_2, \cdots, d_{I-1}$ denote the distance between data gathering stages 1 and 2, 2 and 3, $\cdots$, $I-1$ and I respectively.

The following theorem computes the MI between the signal and observation process of data gathering stages $2, 3, \cdots, I$ for the state and observation model assumed in Eqs. (1) and (2).

Theorem 5.*Define* $X \equiv \left(x_2, x_3, \cdots, x_I \right)$ *and* $Y \equiv \left(y_2, y_3, \cdots, y_I \right)$,*where* x_i *and* y_i *denote the instance of state and observation process respectively, then the mutual information* $I\left(X, Y \right)$ *between the signal sequence* X *and the observation sequence* Y , *for the linear Gauss-Markov state and obsrvation model in Eqs. (1) and (2) is*

$$I\left(X; Y \right) = \frac{1}{2} \sum_{k=2}^{I} \log_2 \left(1 + \frac{\Sigma_{k|k-1}}{\sigma^2} \right). \tag{39}$$

Proof. See Appendix.

Using the expression for predictor covariance in Eq. (18), we can further express RHS in Eq. (39) as

$$I(X; Y) = \frac{1}{2} \sum_{k=2}^{I} \log_2 \left\{ 1 + \Gamma - \left(\Gamma - \gamma_{k-1} \right) a_{k-1}^2 \right\},$$

$$\overset{\text{highSNR}}{\approx} \frac{1}{2}\sum_{k=2}^{I}\log_2\left\{1+\Gamma-\left(\Gamma-1\right)a_{k-1}^2\right\}, \tag{40}$$

where Γ denotes the observation SNR, γ_{k-1} is defined as in Eq. (19), and a_{k-1} denotes the correlation between $(k-1)^{\text{th}}$ and k^{th} stage respectively.

With the power Exponential model assumed to model the correlation, i.e., $a_{k-1}=e^{-Ad_{k-1}}$, we can express Eq. (40) in terms of interstage spacing $d_1, d_2, \cdots, d_{I-1}$ as

$$I(d_1,d_2,\cdots,d_{I-1};\Gamma)=\frac{1}{2}\sum_{k=2}^{I}\log_2\left\{1+\Gamma-\left(\Gamma-1\right)e^{-2Ad_{k-1}}\right\}, \tag{41}$$

where d_{k-1} is the distance between $(k-1)^{\text{th}}$ and k^{th} stages of the network.

It is well known that traditional MIMO transmission achieves better MI and energy tradeoff than SISO transmission. However, in a correlated WSN employing virtual MIMO transmission scheme, it is not clear if virtual MIMO transmission achieves better MI and energy tradeoff, compared to the SISO transmission, because of the additional overhead in terms of local communication involved in virtual MIMO transmission (even if virtual MIMO transmissions are energy efficient, it is of interest to know over what distance ranges, they achieve better tradeoff compared to the SISO transmission).

To study this tradeoff between MI and total energy consumption in the network for virtual MIMO and SISO transmission schemes, we formulate the optimization problem as follows:

$$\underset{[d_1,d_2,\cdots,d_{I-1}]}{\text{Minimize}}\ \lambda\sum_{k=2}^{I}E_{SISO}(d_{k-1})-I(d_1,d_2,\cdots,d_{I-1};\Gamma)\ \text{(for SISO)}$$

$$\text{(OR) }\underset{[d_1,d_2,\cdots,d_{I-1}]}{\text{Minimize}}\ \lambda\sum_{k=2}^{I}E_{MIMO}(d_{k-1})-I(d_1,d_2,\cdots,d_{I-1};\Gamma)\ \text{(for MIMO)}$$

$$\text{Subject to: } 0\le d_{k-1}\le D \text{ for } k=2,3,\cdots,I \tag{42}$$

where $I(.)$ is given by Eq. (41), E_{SISO} and E_{MIMO} are energy consumption per bit values for SISO and virtual MIMO transmission schemes, given by Eqs. (4) and (9) respectively, $\lambda\,(>0)$ is the scanning parameter, and D is the maximum distance range that the sensor nodes can transmit, which is typically determined by peak power of the sensor nodes.

By varying the scanning parameter λ, we obtain the tradeoff curve between total energy consumption and MI. For a given λ, the optimization problem in Eq. (42) is convex (since the energy consumption is quadratic in d_{k-1} and the negative of MI can be verified to be convex), and therefore can be solved efficiently (Boyd & Vandenberghe, 2003).

To give a numerical example we consider the same ten stage network as before, i.e., $(I+1)=10$, where stages $1, 2, \cdots, 9$ correspond to data gathering stages and 10th stage corresponds to the sink. We choose the system parameter values as in Table I. We assume the constellation size for SISO and virtual

MIMO transmission scheme as equal to two, i.e., $b = 2$. We set the probability of error $p_b = 10^{-6}$. For the virtual MIMO transmission scheme, we assume the number of cooperating nodes as equal to two i.e., $n_i = 2$. Our optimization variables in this case are $d_1, d_2, \cdots, d_8$, which correspond to the distance between successive data gathering stages of the network. We set the maximum distance that the nodes can transmit D to be 30 metres. Finally, we set observation SNR $\Gamma = 10$ dB, and the field diffusion coefficient $A = 0.1$.

The optimal energy-MI tradeoff values and the optimal distance values for virtual MIMO and SISO transmission schemes are shown in Table 3. *From the values in the table it can be concluded that, for small distance ranges i.e., upto 6.5 metres, SISO transmission scheme requires less energy compared to the virtual MIMO scheme, to achieve the same amount of MI. But for large distances (roughly greater than 7.5 metres), virtual MIMO transmission scheme achieves better MI and energy tradeoff.*

Table 3 also shows the value of mean square error (MSE) achieved at the sink, for the corresponding optimal distance values. The MSE values are obtained by using the well-known Kalman filter recursion. *FromTable 3it can be observed that, in general, as the distance increases, the MSE increases. This is to be expected since an increase in distance reduces the correlation in the data sensed by nodes, which leads to performance degradation. Further it can be observed that for small distances, SISO transmission scheme consumes less energy than the virtual MIMO transmission scheme, to achieve the same MSE. However, for large distances, virtual MIMO transmission scheme achieves the same MSE, at a lower energy compared to the SISO transmission scheme.*

Thus from this simple numerical example, we can conclude that for small distances, SISO based transmission achieves better energy-MI (or MSE) tradeoff than the virtual MIMO based transmission scheme. However, for moderate and large distances, virtual MIMO based transmission scheme achieves better energy-MI (or MSE) tradeoff.

CONCLUSION

We considered cooperation based architecture for data gathering in correlated sensor networks. Specifically, we consider a combination of virtual MIMO and distributed source coding to improve energy efficiency of WSNs. Under a Gauss-Markov assumption on the signal process, we formulated minimum energy correlated data gathering in sensor networks. From our numerical results, we observed that there exists, an optimal number of cooperating nodes, that would improve the overall energy efficiency of a WSN. On the analytical front using the concepts in monotone comparative statics, we proved that the optimal constellation size adopted for transmission, is in general, an increasing function of the correlation coefficient. For cooperative transmission case, we proved that the optimal number of cooperating nodes, is in general, a decreasing function of the correlation coefficient. Further, the optimal constellation size adopted for cooperative transmission, decreases with an increase in number of cooperating nodes. Also, the optimal number of cooperating nodes, decreases with an increase in constellation size adopted for cooperative transmission. Finally from our numerical results, we observed that SISO transmission scheme achieves better energy-MI (or MSE) tradeoff than the virrtual MIMO transmission scheme for short distance ranges. However, for long distance ranges, virtual MIMO achieves better energy-MI (or MSE) tradeoff compared to the SISO transmission scheme.

Table 3. MI-Energy tradeoff for MIMO and SISO data gathering schemes

MI in bits	Optimal Spacing (d^*) in m	MSE	E_{SISO} in Joules	E_{MIMO} in Joules
11.90	3.20	0.0756	6.7730×10^{-5}	1.5239×10^{-4}
14.00	6.70	0.0874	1.2670×10^{-4}	1.5249×10^{-4}
14.35	7.85	0.0876	1.5486×10^{-4}	1.5254×10^{-4}
14.77	9.85	0.0883	2.1509×10^{-4}	1.5264×10^{-4}
14.94	10.96	0.0889	2.5466×10^{-4}	1.5271×10^{-4}
15.07	12.10	0.0893	2.9904×10^{-4}	1.5278×10^{-4}
15.31	15.30	0.0901	4.4759×10^{-4}	1.5304×10^{-4}
15.55	30	0.0909	16×10^{-4}	1.5496×10^{-4}

REFERENCES

Ahuja, R. K., Magnanti, T. L., & Orlin, J. B. (1993). *Network flows: Theory, algorithms, and applications.* Prentice Hall.

Akyildiz, I. F., Su, W., Sankarasubramaniam, Y., & Cayirci, E. (2002). Wireless sensor networks: A survey. *Elsevier Computer Networks, 38*(4), 393–422. doi:10.1016/S1389-1286(01)00302-4

Alamouti, S. M. (1998). A simple transmit diversity technique for wireless communications. *IEEE Journal on Selected Areas in Communications, 16*(8), 1451–1458. doi:10.1109/49.730453

Amir, R. (2005). Supermodularity and complementarity in economics: An elementary survey. *Southern Economic Journal, 71*(3), 636–660.

Anderson, B. D. O., & Moore, J. B. (1979). *Optimal filtering.* New Jersey: Prentice Hall.

Berger, J. O., Oliveira, V. D., & Sanso, B. (2001). Objective Bayesian analysis of spatially correlated data. *Journal of the American Statistical Association, 96*(456), 1361–1374. doi:10.1198/016214501753382282

Boyd, S., & Vandenberghe, L. (2003). *Convex optimization.* Cambridge, UK: Cambridge Univ. Press.

Chandrakasan, A. P., Min, R., Bharadwaj, M., Cho, S., & Wang, A. (2002, September). *Power aware wireless microsensor systems.* Paper presented at ESSCIRC.

Cover, T. M., & Thomas, J. A. (1991). *Elements of information theory.* New York: John Wiley.

Cristescu, R., Beferull-Lozano, B., & Vetterli, M. (2005). Networked Slepian-Wolf: Theory, algorithms, and scaling laws. *IEEE Transactions on Information Theory, 51*(12), 4057–4073. doi:10.1109/TIT.2005.858980

Cristescu, R., Beferull-Lozano, B., Vetterli, M., & Wattenhofer, R. (2006). Network correlated data gathering with explicit communication: NP-completeness and algorithms. *IEEE/ACM Transactions on Networking, 14*(1), 41–54.

Cui, S., & Goldsmith, A. J. (2006). Cross-layer design of energy-constrained networks using cooperative MIMO techniques. *EURASIP/Elsevier Signal Processing Journal, 86*(8), 1804–1814.

Cui, S., Goldsmith, A. J., & Bahai, A. (2004). Energy-efficiency of MIMO and cooperative MIMO techniques in sensor networks. *IEEE Journal on Selected Areas in Communications, 22*(6), 1089–1098. doi:10.1109/JSAC.2004.830916

Cui, S., Goldsmith, A. J., & Bahai, A. (2005). Energy-constrained modulation optimization. *IEEE Transactions on Wireless Communications, 4*(5), 2349–2360. doi:10.1109/TWC.2005.853882

Foschini, G. J. (1996). Layered space-time architecture for wireless communication in a flat fading environment using multi-element antennas. *Bell Laboratories Technical Journal, 1*(2), 41–59. doi:10.1002/bltj.2015

Jayaweera, S. K. (2007). V-BLAST-based virtual MIMO for distributed wireless sensor networks. *IEEE Transactions on Communications, 55*(10), 1867–1872. doi:10.1109/TCOMM.2007.906389

Jayaweera, S. K., Chebolu, M. L., & Donapati, R. K. (2007). Signal-processing-aided distributed compression in virtual MIMO-based wireless sensor networks. *IEEE Transactions on Vehicular Technology, 56*(5), 2630–2640. doi:10.1109/TVT.2007.900361

Liang, X. B. (2003). On the nonexistence of rate-one generalized complex orthogonal designs. *IEEE Transactions on Information Theory, 49*(11), 2984–2989. doi:10.1109/TIT.2003.818396

Liang, X. B. (2005). A complex orthogonal space-time block code for 8 transmit antennas. *IEEE Communications Letters, 9*(2), 115–117. doi:10.1109/LCOMM.2005.02007

Matlog. (2006). *Matlog: Logistics engineering Matlab toolbox (version 9).* Available at http://www.ie.ncsu.edu/kay/matlog/MatlomgRef.htm

Pattem, S., Krishnamachari, B., & Govindan, R. (2004, April). *The impact of spatial correlation on routing with compression in sensor networks.* Paper presented at IPSN.

Rabaey, J. M., Ammer, M. J., da Silva, J. L., Patel, D., & Roundy, S. (2000). PicoRadio supports ad hoc ultra-lowpower wireless networking. *IEEE Computers, 33*(7), 42–48.

Scaglione, A., & Servetto, S. D. (2002). *On the interdependence of routing and data compression in multi-hop sensor networks.* Paper presented at Mobicom.

Schweppe, F. C. (1973). *Uncertain dynamic systems.* Englewood Cliffs, NJ: Prentice Hall.

Sung, Y., Tong, L., & Ephremides, A. (2005, March). *Route selection for detection of correlated random fields in large sensor networks.* Paper presented at CISS.

Sung, Y., Tong, L., & Ephremides, A. (2005, October). *A new metric for routing in multihop wireless sensor networks for detection of correlated random fields*. Paper presented at MILCOM.

Tarokh, V., Jafarkhani, H., & Calderbank, R. A. (1999). Space-time block codes from orthogonal designs. *IEEE Transactions on Information Theory, 45*(5), 1456–1467. doi:10.1109/18.771146

Topkis, D. M. (1998). *Supermodularity and complementarity.* NJ: Princeton University Press.

APPENDIX

Derivation of $h\left(y_k \mid y_{k+1}, y_{k+2}, \cdots, y_I\right)$ and $h\left(y_I\right)$:

Using the definition of conditional differential entropy we can write

$$h\left(y_k \mid y_{k+1}, y_{k+2}, \cdots, y_I\right) = h\left(y_k, y_{k+1}, \cdots, y_I\right) - h\left(y_{k+1}, y_{k+2}, \cdots, y_I\right). \tag{46}$$

$$f\left(y_k, y_{k+1}, \cdots, y_I\right) \blacklozenge \blacklozenge \left(¼_{k|k-1}, diag(S_k)\right), \tag{47}$$

$$h\left(\mathbf{z}\right) = \frac{1}{2} \log_2 \left(2\pi e\right)^n \det \Upsilon, \text{ where det denotes determinant of the matrix.} \tag{48}$$

$$h\left(y_k \mid y_{k+1}, \cdots, y_I\right) = \frac{1}{2} \log_2 \left(2\pi e S_k\right)$$

$$= \frac{1}{2} \log_2 \left(2\pi e \sigma^2 \left(1 + \frac{\Sigma_{k|k-1}}{\sigma^2}\right)\right) \tag{49}$$

Similarly, using the definition of differential entropy we obtain $h\left(y_I\right) = \frac{1}{2} \log_2 \left(2\pi e \sigma^2 \left(\Gamma + 1\right)\right)$.

Proof of Theorem 5:

Define $X \equiv (x_2, x_3, \cdots, x_I)$ and $Y \equiv (y_2, y_3, \cdots, y_I)$, then the Mutual information $I(X;Y)$ denotes the reduction in the uncertainty of X when Y is observed and vice-versa (Cover & Thomas, 1991), i.e.,

$$I(X;Y) = H(X) - H(X \mid Y) = H(Y) - H(Y \mid X). \tag{56}$$

Using the definition of mutual information in Eq. (56), the fact that pdf of Y is given by Eq. (47) and the fact that the pdf of $Y \mid X$ is conditionally independent multi-variate Gaussian, the desired result follows.

Chapter 18
Cooperative Broadcast in Large-Scale Wireless Networks

Birsen Sirkeci-Mergen
San Jose State University, USA

Anna Scaglione
University of California at Davis, USA

Michael Gastpar
University of California at Berkeley, USA

ABSTRACT

This chapter studies the cooperative broadcasting in wireless networks. We especially focus on multistage cooperative broadcasting in which the message from a source node is relayed by multiple groups of cooperating nodes. Interestingly, group transmissions become beneficial in the case of broadcasting as opposed to the case in traditional networks where receptions from different transmitters are considered as collision and disregarded. Different aspects of multistage cooperative broadcasting are analyzed in the chapter: (i) coverage behavior; (ii) power efficiency; (ii) error propagation; (iv) maximum communication rate. Whenever possible, performance is compared with multihop broadcasting where transmissions are relayed by a single node at each hop. We consider a large-scale network with many nodes distributed randomly in a given area. In order to analyze such networks, an important methodology, the continuum limit, is introduced. In the continuum limit, random networks are approximated by their dense limits under sum relay power constraint. This method allows us to obtain analytical results for the analysis of cooperative multistage broadcasting.

INTRODUCTION

In distributed ad hoc networks, most network protocols require multicast or broadcast of certain control messages. These messages generally constitute a significant portion of network traffic, and they may cause performance bottlenecks. Several authors have studied how to optimally transmit broadcast

DOI: 10.4018/978-1-60566-665-5.ch018

information to minimize the total number of transmissions or the energy consumption in large wireless networks (e.g., see Williams & Camp, 2002).

Under traditional multi-hop broadcasting, each node receives signals from its nearest neighbor, and messages propagate over hops formed by single nodes. In this case, receptions from different transmitters are considered as collision even if they correspond to the same source message. Hence, networks with high number of nodes suffer from multiple retransmissions and waste resources (Tseng, Ni, Chen & Sheu, 1999; Korkmaz, Ekici & Ozguner, 2007).

In wireless networks, transmitted packets are heard not only by their intended recipients but also by other neighboring nodes. This is known as the broadcast property of wireless medium and it is the main motivation for cooperative schemes. In this chapter, we study cooperative broadcasting which takes advantage of the broadcast nature of the wireless medium by including all receivers in the relaying process. In broadcasting, the goal is to send a source message to the entire network in the most efficient way. In contrast with multi-hop broadcasting, cooperative broadcasting brings in several advantages: (i) increased connectivity and coverage; (ii) improved power efficiency; (iii) bounded error propagation for uncoded transmission; (iv) improved communication rate.

The advantages of cooperation for broadcasting applications have been recognized in many works. Different cooperative strategies are proposed in order to improve energy efficiency (Hong & Scaglione, 2003; Maric & Yates, 2004; Hong & Scaglione, 2006; Sirkeci-Mergen & Scaglione, 2007; Sirkeci-Mergen & Scaglione, 2006), network lifetime (Maric & Yates, 2005), network coverage (Sirkeci-Mergen, Scaglione & Mergen, 2006; Sirkeci-Mergen & Scaglione, 2004; Sirkeci-Mergen & Scaglione, 2005), and communication rate (Sirkeci-Mergen & Gastpar, 2007; Khisti, Erez & Wornell, 2006);

In (Hong & Scaglione, 2003; Maric & Yates, 2004; Hong & Scaglione, 2006), the authors investigate the energy efficiency of cooperative transmissions over multihop networks for different setups. In the basic formulation, they assume that the receiving nodes combine the receptions from all nodes that transmitted previously to harvest energy and, in turn, benefit from transmit diversity. In addition, the nodes transmit based on a predetermined schedule and power allocation policy such that total power consumption of the network is minimized. In (Hong & Scaglione, 2003; Maric & Yates, 2004; Hong & Scaglione, 2006), it was shown that for a given transmission schedule, the optimal power allocation can be formulated as a constrained optimization problem which can be solved in polynomial time by utilizing linear programming tools. On the other hand, the authors also showed that finding the optimal scheduling that leads to the minimum total power consumption is an NP-complete problem and thus, it is not computationally tractable in general. Both works proposed heuristic methods to determine the optimal schedule.

In (Khisti, Erez & Wornell, 2006), authors characterize the broadcast capacity for slowly fading channels in wireless relay networks. The authors consider a model where an outage is declared if any of the receivers fails to decode the source message, and the broadcast capacity is defined as the maximum data rate at which the outage probability converges to zero as the number of nodes goes to infinity. They showed that the broadcast capacity converges to $C = log\ (1 + P/N_0)$, where P is the sum power constraint on the network and N_0 is the noise power. This result is obtained under the assumption that there is i.i.d. (independent and identically distributed) Rayleigh fading between nodes, but there is no signal attenuation with distance. The achievability result is based on a two-phase cooperative broadcasting scheme.

The multistage cooperative broadcasting, which is the main focus of this chapter, is introduced and analyzed by Sirkeci-Mergen & Scaglione in various publications. A closely related scheme is *opportunistic large arrays* proposed in (Scaglione & Hong, 2003) in which the nodes transmit based on their accumulated energy in a distributed fashion. This scheme has low complexity compared to the centralized cooperative schemes, and it eliminates the problem of scheduling transmissions; however, it cannot guarantee diversity gains since the transmitted signals can overlap in time, and it requires non-negligible bandwidth overhead.

In *multistage cooperative broadcasting protocol*, nodes sequentially transmit the same message in large groups to increase the received power. This provides considerable advantages over traditional multi-hop broadcasting. The proposed protocol is described in the next section and then several advantages are provided in the following sections. In order to derive analytical results for random networks, we utilize a new technique which is an approximation for large-scale networks. A continuum model is obtained from the random network by letting the number of nodes go to infinity while the total relay power is fixed. The continuum approach was previously used in different contexts in (Shakkottai, Srikant & Shroff, 2004; Jacquet, 2004).

The analysis is done for two different types of channel models: (i) non-orthogonal channel model; (ii) orthogonal channel model. The reason behind presenting results for both models is to capture the fundamental trade-offs in cooperative broadcasting. In the orthogonal model, the transmitted signals go through orthogonal fading channels, and the received signal power after maximal ratio combining is the sum of the powers of the incoming links. The use of orthogonal channels is more reliable in the sense that the receiver eliminates the effect of fading via maximal ratio combining. In the non-orthogonal model all cooperating nodes synchronize to the preamble of the packet they have received and transmit phase asynchronously. In this case, the equivalent channel is modeled as Rayleigh fading in dense networks.

In most of the literature, cooperative communication is focused on unicast transmission where the orthogonal channel model seems to better serve the objective of having a reliable reception at the destination. However, in broadcasting the goal is fast delivery to many nodes, and the non-orthogonal channel model is a better fit. This is because, in a dense network, there is always a fraction of nodes that experiences good fading realization. Once these lucky nodes receive and retransmit, under the non-orthogonal channel model, the nodes neighboring them see a boost of signal power, and the signal power gets amplified as groups of nodes forward signals to several peers. The analytical framework in this chapter provides evidence of the fact that on the average non-orthogonal channel model has a faster delivery or better end-to-end delay.

MULTISTAGE COOPERATIVE BROADCAST

We consider a simple cooperation protocol for broadcast over multiple stages of relays. In the considered setup, a source node initiates the broadcast by transmitting a packet. Every cooperative node who can hear the source with sufficient signal-to-noise ratio (SNR), decodes and retransmits the same packet. A training preamble in the message helps nodes to detect the packet's presence, estimate the received power and synchronize the retransmissions. The first group excites a second group of nodes and the retransmissions continue until every node who hears the others with sufficient SNR, retransmits once. The subsequent groups of nodes that are activated are referred to as levels.

The decision criterion of when to relay packets is a subtle issue. In practice, the packets are coded according to a certain channel code, and CRC (Cyclic Redundancy Check) bits are placed into each packet. A packet reception is considered successful if the CRC test passes after decoding the channel code. Here, the nodes use a simple SNR threshold criterion to decide if they are going to retransmit or not, i.e., every node monitors its received SNR and decodes and retransmits if and only if its SNR exceeds a certain pre-determined threshold. In this way, the network can operate in a distributed fashion, since the nodes only use the locally available received SNR information to make transmission decisions. We assume that appropriate channel coding is used so that the decoding and retransmissions are correct as long as the received SNR is above the threshold.

For tractability, we consider a specific deployment where N nodes are uniformly and randomly distributed in a disc with radius R and a single source is located at the center of a circular region. Let the source transmit with power P_s and each cooperative relay transmit with power P_r. Let S denote the set of locations of relay nodes. The relay nodes decode and retransmit if and only if their SNR exceeds a certain threshold τ. At every step, the set of cooperative nodes with reception power exceeding τ, which has not transmitted so far, joins the next level.

Let S_1, S_2, denote the sets of locations for each level. Determining these sets is a challenging issue since the node locations are random, and a random number of nodes is reached by the source in every realization of the network. We study different aspects of the multistage cooperative broadcasting for random networks:

- **Coverage Analysis:** The objective of cooperative broadcast is to deliver the source message to the whole network. However, this goal may or may not be achieved depending on certain network parameters such as the source/relay transmission powers and the decoding threshold. In this part of the chapter, we analyze the effect of these parameters on the number of nodes reached by cooperative broadcast. In particular, we show that there exists a phase transition in the network behavior: if the decoding threshold is below a critical value, the message is delivered to the whole network. Otherwise, only a fraction of the nodes is reached proportional to the source transmit power.

- **Power Expenditure Analysis:** In large-scale networks, nodes are constrained in their size and battery power, and hence, it is crucial to design energy efficient and low complexity transmission schemes. We compare the power efficiency of multistage cooperative broadcasting which utilizes uniform power allocation with the non-cooperative multi-hop broadcast. We analytically quantify the gains of cooperative broadcast, and conclude that dense cooperative networks can be considerably advantageous in terms of power efficiency relative to the commonly employed multi-hop architecture. The power analysis is done for only orthogonal channels.

- **Error Propagation Analysis:** In non-cooperative multi-hop systems if uncoded transmission is utilized, the probability of error degrades steadily with the number of hops and, thus, with the distance covered by the message. The obvious question to ask is if cooperative broadcasting has worse error propagation. Intuition seems to be of little help because in cooperative broadcasting erroneous signals are mixed with the correct ones at the physical layer. In this part of the chapter, we show that when transmissions are added up at the physical layer, erroneous transmissions do not necessarily lead to catastrophic error propagation and that as the network size grows, the average error can be controlled precisely by controlling the size of the cooperative groups and the power density used in relaying the message. Furthermore, a bandwidth expansion is not necessary to attain this advantage.

- **Capacity Analysis:** A fundamental problem in wireless networks is determining the broadcast capacity, i.e. the maximum data transfer rate from a given node to every other node in a relay network. In this part of the chapter, we study extended networks (i.e. the network area goes to infinity while the node density is fixed) under per-node power constraint. We show that the upper bound on the broadcast capacity is $\theta(1)$ for path-loss exponent $\alpha > 2$. We determine that the transmission rate of the cooperative multistage broadcasting protocol scales as $\theta(1)$ which leads to optimality for $\alpha > 2$. Here, similar to the physical model in (Gupta & Kumar, 2000), we assume an SINR (signal to interference and noise ratio) criterion for successful transmission and do not consider fading.

CHANNEL MODELS

Consider a group of relay nodes L that transmit the same message simultaneously and a hypothetical node H which receives these signals. We consider two different models for receptions. First, we derive and consider a non-orthogonal channel model applicable for narrowband communication. Here, the impulse response of the channel with multiple transmitters is modeled as a Gaussian random vector. This model takes into account the effects of channel fading, time differences between simultaneous transmissions and random phases. On the other hand, the second model assumes that the received power of simultaneously transmitted packets after maximal ratio combining is equal to the sum of individual powers. This model will be referred as orthogonal channel model.

We make the following assumptions for analysis of the random networks:

- The relay nodes belonging to a group retransmit the same message simultaneously.
- All nodes in the network use identical mappings of the same message, that is the same pulse shaping filter $p(t)$.
- The signal between any two nodes is assumed to be attenuated due to both path-loss $\ell(d)$ and random multipath fading (modeled as Rayleigh fading).
- The random channel coefficients α_l's are assumed to be independent and identically distributed (i.i.d.).
- The phase differences, θ_l's, between modulator and demodulator clocks of relays are modeled as i.i.d. uniform random variables.
- Propagation delays from a group to a destination are assumed to be approximately the same.
- The signal is assumed to be affected also by additive white Gaussian noise of unit power.

Non-Orthogonal Channel Model

In the non-orthogonal channel model, we assume that the broadcast signal is narrowband, i.e., the coherence bandwidth of equivalent link provided to the relays is much larger than the transmission bandwidth. In this model, we assume the nodes transmitting in a given group are time-synchronized using the preamble of the packet received. The non-orthogonal scheme is also simple to use in a distributed network setting. Under the provided assumptions, the discrete baseband received signal is

$$r[n] = h[n] * c[n] + w[n], \tag{1}$$

where $c[n]$ denotes the transmitted samples, $w[n]$ AWGN samples and

Figure 1. The reception models for random fading corresponding to orthogonal and non-orthogonal relay transmission.

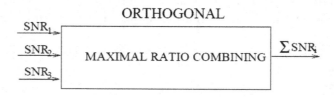

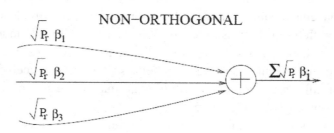

$$h[n] = \sum_l \sqrt{P_r}\, \beta_l\, p_l\,[n] \tag{2}$$

denotes the equivalent channel impulse response of the cooperative channel. Here $\beta_l = \sum_l \sqrt{l(d_l)}\, \alpha_l \exp(j\theta_l)$ denotes the *l*'th effective random channel coefficient and $p_l[n] = p(nT - t_l)$ denotes the corresponding pulse shaping filter sampled at $t = nT$ and delayed by the starting time of *l*'th node, t_l. In the following, we will also assume that the $h[n]$ is practically non-zero only for *2D + 1* terms, $h[-D], ..., h[D]$, and is zero elsewhere. In this case, the received SNR is

$$SNR = P_r \sum_n | \sum_l \beta_l\, p_l[n]|^2 \tag{3}$$

Orthogonal Channel Model

Here, we consider that the transmitted signals go through orthogonal fading channels, and the receiver does maximal ratio combining of channel outputs. In practice, maximal ratio combining achieves the sum power of incoming links under the following scenarios: (i) the nodes in a given level transmit in orthogonal channels (as in TDMA or FDMA); (ii) the nodes use orthogonal or pseudo-orthogonal spreading codes with desirable correlation properties; (iii) simultaneously transmitting nodes employ a distributed orthogonal space-time code (Laneman & Wornell, 2003; Sirkeci-Mergen & Scaglione, 2007) designed for a large number of nodes. The issue of allocating channels to many nodes seems daunting, but it can be solved with the randomized codes almost without any loss in performance (Sirkeci-Mergen & Scaglione, 2007).

In this scenario, for simplicity, we will assume that the receiver can perfectly recover the timing of each transmitted signal and sample it at time zero. Hence, the sampled received signal from the l'th node is

$r_l[n] = h_l[n] * p[n] + w_l[n]$, (4) where $w_l[n]$ is the noise in the l'th orthogonal channel. We assume that each $w_l[.]$ is white with unit power, and independent of one another. The effective channel from the l'th node is

$$h_l[n] = \sqrt{P_r} \, \beta_l p[n]. \tag{5}$$

If a Nyquist pulse $p(.)$ is used, the signal $r_l[.]$ has no intersymbol interference. Consequently, maximal ratio combining of $r_l[.], l \square L$ gives the highest SNR:

$$SNR = P_r \sum_l |\beta_l|^2 \, |p(0)|^2. \tag{6}$$

In the following, we present a special case of orthogonal model which we will refer to as deterministic channel model. This model greatly simplified the analysis of networks; in addition, we show that networks with orthogonal channels for certain scenarios can be analyzed using the deterministic channel model.

Under deterministic channel model, we assume that each transmission is affected by nonrandom path-loss attenuation. Suppose that every transmission with power P is received with power $P\ell(d)$ at distance d, where $\ell(.)$ is a path-loss attenuation function, which is assumed to be continuous and non-increasing (e.g., $\ell(d) = 1/d^2$).

Furthermore, it is assumed that if a set of relay nodes, say L, transmits simultaneously, then the destination node receives with SNR

$$SNR = P_r \sum_i \ell(d_i) \tag{7}$$

where d_i is the distance to the i'th transmitter. Note that our definition of deterministic channel is different from deterministic channel in the work (Avestimehr, Diggavi & Tse, 2007).

METHOD FOR ANALYSIS OF LARGE-SCALE NETWORKS: CONTINUUM LIMIT

In our analysis, we first consider a random network in which the node locations are randomly and uniformly distributed. In general, it is hard to obtain analytical results for random networks since each realization of the network will be a different arrangement for nodes. We obtain a continuum model from the random network by letting the number of nodes go to infinity while the total relay power is fixed. This will allow us to derive analytical results and intuitive conclusions about the behavior of random networks.

Let S denote the region containing the network. Let $\rho = N/Area(S)$ be the density [node/unit area] of relays within S. In the continuum model, we are interested in the behavior of high density networks with constant sum-power. That is, the number of relays N goes to infinity, while $P_r N$ is fixed. This implies that the relay power per unit area

$$\underline{P}\mathbf{r} = Pr\, N\, Area(S) = P_r\, \rho \tag{8}$$

is also fixed. Next, we derive the asymptotic channel distribution in the continuum limit.

Asymptotic Channel Distribution

The set L is viewed as the set of nodes belonging to a certain level of broadcast. Consider a hypothetical node that receives signals from the nodes in L. For tractability purposes, we make the following assumptions:

i)   The locations of the nodes in L are i.i.d., and therefore, the distances to the receiver, d_l's, are i.i.d. for different l.

ii)   The starting times t_l's, $l \square L$ are zero-mean i.i.d. distributed with a certain pdf $f(t)$. This models the situation that all the nodes in the same level transmit approximately around time zero, but there may be small variations due to differences between processing/relaying times at different nodes.

iii)   The fading coefficients α_l's are i.i.d. with unit variance for different $l \square L$. The phases θ_l's are i.i.d. Uniform$[0, 2\pi]$ (i.e., the modulator/demodulator clocks at different nodes are asynchronous).

iv)   The $d_l,\, t_l,\, \alpha_l,\, \theta_l,\, l \square L$ are independent.

The following theorems characterize the asymptotic distribution of the channel under both orthogonal and non-orthogonal channel models.

Theorem 1:*Non-orthogonal Channel Model:* Let L denote the number of nodes in L. Suppose that the relay power P_r varies with L, and the total relay power converges to PT as $L \to \infty$, i.e.,

$$LPr \to PT, \text{ as } L \to \infty. \tag{9}$$

Then, under assumptions i)-iv), the channel impulse response h, $(h[-D], \ldots, h[D])$, satisfies

$$h \xrightarrow{d} N_c(0, P_T\, E\{\ell(d_l)\}\Sigma) \text{ as } L \to \infty, \tag{10}$$

where $\xrightarrow{d}$ denotes convergence in distribution, and Σ is a matrix with entries

$$\Sigma_{mn} = \square\, f(t)p(nT - t)p^*(mT - t)dt, \text{ for } n, m \square \{-D, \ldots, D\}.$$

Remark 1: Theorem 1 indicates that the effective channel for the system with non-orthogonal channels converges to a Rayleigh multipath channel. Hence, the SNR in the limit is a random variable as opposed to the asymptotic channel behavior with orthogonal channels provided in Theorem 2.

Theorem 2: Orthogonal Channel Model: Under assumptions i)-iv) and (9), the SNR of the maximal-ratio-combined received signal converges to a deterministic limit as $L \to \infty$, i.e., $SNR \to P_T\, E\{\ell(d_l)\} |p(0)|^2$, as $L \to \infty$, (11) almost surely.

Remark 2: Theorem 2 indicates that the system with orthogonal channels, despite the existence of fading and randomness in the channel, has a deterministic SNR in the limit. As we will see in the next section, this property greatly simplifies the analysis in certain cases.

Remark 3: Notice that the bandwidth requirement of the orthogonal system is proportional to L if FDMA or CDMA type signaling is used. Hence, taking the limit $L \to \infty$ without sacrificing from the transmission rate requires potentially infinite bandwidth, i.e., the network operates in the wideband regime. On the other hand, by using distributed orthogonal space-time codes, it is possible to get to the above performance without sacrificing bandwidth.

COVERAGE ANALYSIS

In this section, we derive network behavior in the continuum limit using the asymptotic channel distribution obtained in the previous section for random networks. Consider a network with N relay nodes located in the disc S, where the source is located at the center. In the following, we will make the dependence of effective channel coefficient on the location of the receiving node *(x, y)* explicit wherever possible. For a subset of relay nodes $L \subset \{1, ..., N\}$, let $h_l(x, y)$ be the channel impulse response vector from level set L to a hypothetical node at *(x, y)*. See (Sirkeci-Mergen, Scaglione & Mergen, 2006) for details of the following results.

Network Behavior with Deterministic Channels

In this section, we consider the simple deterministic model for the received power of simultaneously transmitted signals. It is shown in (Sirkeci-Mergen, Scaglione & Mergen, 2006) that in the continuum limit, under the deterministic model, we can determine a rigorous approximation for high-density random networks ((Sirkeci-Mergen, Scaglione & Mergen, 2006, Theorem 1).

This theorem provides an approximation to the number of level-k nodes:

$$|S_k| \approx \rho \, Area(S_k), \text{ for large } N, \tag{12}$$

where ρ denotes the network density and S_k denotes the level sets obtained from the continuum analysis.

Example 1: In this example, we will give a more explicit characterization of the level sets $S_1, S_2,$ ... for $\ell(d) = 1/d^2$. For ease of presentation, we will assume an unbounded network, i.e., $R \to \infty$. The results for $R < \infty$ follow trivially. It can be shown that in this case each level is a disc shaped region with inner and outer radii given by r_{k-1} and r_k, respectively [12, Lemma 1]. In this case, the outer radius of k'th level is obtained as

$$r_k = \sqrt{Ps(\mu - 1) \, (1 - 1/ \, (\mu - 1)^k)/ \, \tau \, (\mu - 2)},$$

where $\mu = exp(\tau/\underline{P}r\pi)$. Depending on the value of μ, rk exhibits two different behaviors:

i)   Case 1: High Power ($\mu \leq 2$): The broadcast reaches to the whole network, i.e., l $\lim\limits_{k\to\infty} r_k = \infty$.

ii) Case 2: Low Power ($\mu > 2$): The total area reached by the broadcast is bounded, i.e.,

$$\lim\limits_{k\to\infty} r_k = \sqrt{Ps(\mu- 1)/\tau\ (\mu- 2)]}\ .\tag{13}$$

In other words, the network behavior goes through a phase transition depending on the value of μ. If

$$\mu \leq 2 \Leftrightarrow \tau \leq (\pi\ ln\ 2)\ \underline{P}r,\tag{14}$$

i.e., the detection threshold is low enough with respect to the relay power per unit area, then the signal propagates to the whole network (see Fig. 2). On the other hand, if $\mu > 2$, then a finite portion of the network is reached, and the total number of nodes the message is delivered to is approximately equal to

$$\left|\bigcup_k S_k\right| \approx \pi P_s \rho (\mu - 1) / [\tau (\mu - 2)]\ .\tag{15}$$

The right hand side of Eqn. (15) implies that the number of nodes reached by the broadcast is directly proportional to the source power.

Network Behavior with Random Channels

In the following, we derive the network behavior using the matched-filter upper bound $||h_L(x,\ y)||^2$ on received SNR, i.e., we assume that a reception is successful if $||h_L(x,\ y)||^2$ exceeds a certain threshold

Figure 2. (a) Transmissions propagate. (b) Transmissions die off. Notice that the scale of (a) and (b) are vastly different.

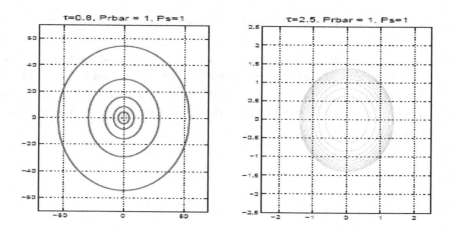

τ. A more elaborate model for receptions can be derived based on the notion of outage capacity (i.e., a reception is considered successful if the instantaneous mutual information of the equivalent channel exceeds a certain threshold).

When random channel fading is considered, we assume spatially independent fading for its simplicity that is effective channel coefficients are i.i.d. for different l and (x, y). For correlated fading scenarios, see (Sirkeci-Mergen, Scaglione & Mergen,, 2006).

Definition. Let $P_k(x, y)$ denote the probability that a node at location (x, y) joins level-k, and $\sigma^2_k(x, y)$ be the sum of signal powers from level-k at location (x, y). For $k = 1, 2, 3, \ldots$ the equations

$$P_{k+1}(x, y) = Pr\{||h_k(x, y)||2 \geq \tau\} \prod_{i=0}^{k-1} [1 - Pr\{||h_i(x, y)||^2 \geq \tau\}], \qquad (16)$$

$$\sigma^2_k(x, y) = \iint Pr\, P_k(x', y')\, \lambda(x - x', y - y')\, dx'dy', \qquad (17)$$

where $h_k(x, y) \sim N_c(0, \sigma^2_k(x, y))$, non-orthogonal $||h_k(x, y)||^2 = \sigma^2_k(x, y)\, |p(0)|^2$, orthogonal specify the continuum model for networks with random channels. See (Sirkeci-Mergen, Scaglione& Mergen, 2006) for details.

The functions P_k, σ^2_k define a non-linear dynamical system which evolves with k. Although the analytical solution of this system is hard, it can be usually evaluated numerically. Another property of the continuum model is that the $P_k(x, y)$ and $\sigma^2_k(x, y)$ are only functions of r where $r^2 = x^2 + y^2$. Therefore, the above dynamical system evolves only over 1-dimensional functions. For convenience, we will use the notations $P_k(x, y)$ and $P_k(r)$ interchangeably.

For our numerical evaluations in this section, we will use the following path-loss model

$$\ell(d) = \begin{cases} 1/d^2 & d \geq d_0 \\ 1/d_0^2 & d_0 \geq d \geq 0 \end{cases} \qquad (18)$$

with a small $d_0 > 0$ to avoid the singularity in the integral (17). The squared-distance attenuation model $\ell(d) = 1/d^2$ comes from the free-space attenuation of electromagnetic waves, and it does not hold when d is very small (Rappaport, 2003).

1) **Behavior of Continuum Network with Orthogonal Channels:** The above equations for continuum network greatly simplify in case of orthogonal channels. There are two possibilities:

i) If there is no fading from the source to the relays, then $P_1(x, y)$ is binary (i.e., $P_1(x, y) \in \{0, 1\}$). Furthermore, each $\sigma^2_k(x, y)$ for $k = 2, 3, \ldots$ is deterministic. Therefore, $Pr\{||h_k(x, y)||^2 \geq \tau\}$ is binary as well. If we define

$$S_k = \{(x, y) \in S: P_k(x, y) = 1\},$$

then

$\sigma^2_{k}(x, y) = \square\square\ \underline{Pr}\ \lambda(x'-x, y'-y)\ dx'\ dy'.$

Hence, for this scenario the continuum model reduces to the continuum model for deterministic channels.

ii) If the channels from the source to relays have fading, then $P_{1}(x, y)$ takes continuous values in [0, 1], but $Pr\{||h_{k}(x, y)||^2 \geq \tau\}$ is still binary. In our numerical evaluations we observed that the effect of $P_{1}(x, y)$ is transient, and the asymptotic behavior of the network is as in the deterministic model. In Figure 3, we plot $P_{k}(r)$ in both low and high threshold regimes.

2) **Behavior of Continuum Network with Non-orthogonal Channels:** Since $h_{k}(x, y)$, we can easily calculate $Pr\{||h_{k}(x, y)||^2 \geq \tau\}$ using standard probability distributions [12, Lemma 2]. The analytical solution of the continuum network in the case of non-orthogonal channels appears to be a non-trivial problem. In order to gain intuition, we evaluated (16) and (17) numerically for large R. Similar to the case of deterministic channels, it is observed that there exists a critical threshold τ_{c}. For $\tau > \tau_{c}$, the transmissions eventually die out, i.e.,

$sup_{(x,y)\ \square\ R}^2\ P_{k}(x, y) \to 0$ as $k \to \infty$.

Otherwise, the transmissions propagate to the whole network, while the level curves, $P_{k}(r)$, $r\ \square\ R$, become wider as k increases. See Figure 4 for these two regimes.

Comparison between Non-Orthogonal and Orthogonal Cooperative Broadcast

In this section, we compare the message propagation behavior in random orthogonal and random nonorthogonal channels. Figure 5 shows the $P_{k}(r)$ for both models. These plots are obtained for the parameters $\tau = 0.8$, $P_{s} = 4$, $\underline{Pr} = 1$, $M = 1$, $d_{0} = 1$.

Figure 3. (a) The parameters are $\tau = 1.2$, $P_{s} = 4$, $\underline{Pr} = 1$, $d_{0} = 1$, $\Sigma = 1$. The transmissions continue. (b) The parameters are $\tau = 2.58$, $P_{s} = 4$, $\underline{Pr} = 1$, $d_{0} = 0.5$, $\Sigma = 1$. The transmissions die out.

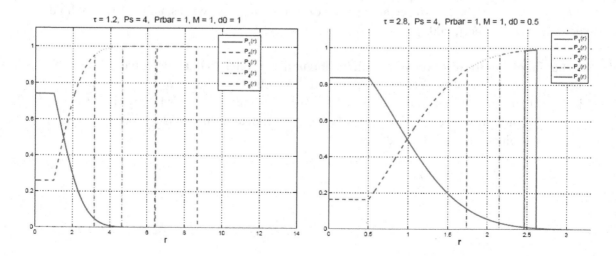

For both orthogonal and non-orthogonal models, $P_1(.)$ is the same; however, the level curves $P_k(.)$ differ significantly for large k. In particular, the level curves in the non-orthogonal case move faster. This is a rather counter-intuitive result, because the orthogonal system (with FDMA / CDMA) uses much more bandwidth than the non-orthogonal one, and the use of orthogonal channels is more reliable in the sense that the receiver eliminates/reduces the effects of fading via maximal ratio combining. Our result implies that the system with fading provides better broadcast behavior than the one without.

We believe that this fact can be explained as follows. The maximal ratio combining method reduces the probability that the combined signal experiences a deep fade at the cost of reducing the probability that the signal experiences a favorable fade. In a dense network, favorable fading realizations are very valuable, because when the node density is high, although there is a small probability of having a good fading realization, there is always a fraction of nodes that experiences them. Once these lucky nodes receive and retransmit, the nodes neighboring them see a boost of signal power because of the properties of $\ell(d)$. In conclusion, we believe that the nodes that enable faster level movement are the ones at the forefront of each level.

In the narrowband system, favorable fading realizations occur, when the phases of two or more simultaneously transmitting nodes happen to add coherently, or when one of the transmitting nodes experiences a very good channel with the receiver. Considering that non-orthogonal channel model do not require infinite bandwidth, we conclude that the non-orthogonal scheme is more advantageous also in terms of end-to-end delay.

POWER EFFICIENCY

In large-scale networks, nodes are constrained in their size and battery power, and hence, it is crucial to design energy efficient and low complexity transmission schemes. Furthermore, the energy consumption

Figure 4. The parameters are $P_s = 8$, $\underline{Pr} = 1, M = 2$, $d_0 = 1$, $S = 1/M\boldsymbol{I}$, where $\boldsymbol{I}$ is the identity matrix. (a) Transmissions continue. The decoding threshold $\tau = 0.8$. (b) Transmissions die out. The decoding threshold $\tau = 3.5$.

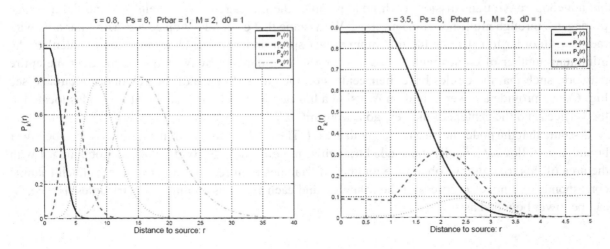

Figure 5. *Wideband orthogonal vs. narrowband non-orthogonal: the upper, rectangular shaped levels are the orthogonal, the lower wave-like levels are non-orthogonal. The parameters are* $\tau = 0.8$, $P_s = 4$, $\underline{P}r = 1$, $M = 1$, $d_0 = 1$, $\Sigma = 1$.

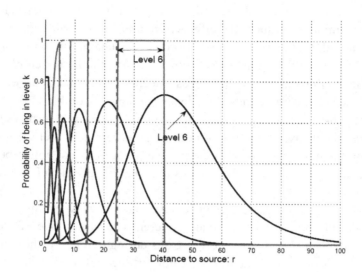

increases as the network size grows, and the power efficiency becomes more important. In the proposed scheme, the relays utilize a uniform power allocation. The main motivation behind the uniform power allocation scheme is the observation that the total power consumption of optimal cooperative broadcasting, i.e. multistage cooperative broadcasting with optimal power allocation, is $O(R^2/ \ln(R))$ for a circular network with radius R (see (Sirkeci-Mergen & Scaglione, 2007)). For large R, the effect of $\ln(R)$ is negligible when compared with R^2, hence for large networks, uniform power allocation is almost optimal.

In the case of uniform power allocation, the total relay power consumption is

$$P_{total} = \underline{P}r \, \pi \, R^2 = O(R^2). \qquad (19)$$

Next, we compare the cooperative multistage broadcasting with non-cooperative broadcasting. In the noncooperative transmission, each node receives the message from its neighbor and the message propagates through multi-hop transmissions. We assume that each relay hop covers a circular area with radius a. In order to calculate the minimum power spent under this scheme, we need to calculate the minimum number of circles with radius a, which will be denoted by N_a, required to cover the entire network, such that the circles have their centers on the circumference of the neighboring circles (see Fig. 6). We provide a lower bound on N_a using a hexagonal tessellation such that the nodes located at the vertices and the centers of the hexagons transmit.

Assume that the nodes lie on a region with area A. The total number of transmissions N_a can be lower bounded as follows. Consider the shaded triangular region in Fig. 6. By dividing the total area A with the area of this triangle, we obtain the number of triangles required to cover the area A. Each triangle corresponds to 3 nodes (vertices of the triangle), and each vertex is common to 6 triangles. Hence, N_a can be lower bounded as

$N_a \geq$ 0.5 A/ Triangular Area $= 2A / \sqrt{3}$ a², $\qquad$ (20)

where the shaded triangular region is shown in Fig. 6.

Let P_a denote the minimum power required by a transmitter so that the nodes within a radius a around the transmitting node, receive the message with at least SNR τ. For the path-loss attenuation function $\ell(a)$, $P_a = \tau/\ell(a)$. Note that by fixing a, we fix the total number of hops required to cover the area A. The best multi-hop scheme minimizes the total power consumption by optimizing over the number of hops, which is a function of a. Hence, the minimum total power spent by the multi-hop transmission is

$$P^{(noncoop)} = min_a \, N_a \, P_a \geq min_a \, 2A\tau / \sqrt{3} \, \ell(a) \, a^2 . \qquad (21)$$

Note that the derivation above is for the multi-hop transmission and does not include the direct transmission because of the way we calculate N_a (see Fig. 6). Under the path-loss attenuation model $\ell(a) = 1/a^2$, the total power spent by the multi-hop transmission is independent of the number of hops, and (21) simplifies further to

$$P^{(noncoop)} \geq 2A\tau / \sqrt{3} \,). \qquad (22)$$

The total power expenditure under the cooperative scheme is

$$P^{(coop)} = A \, \underline{Pr}_{,min} = A\tau/(\pi \, ln(2)). \qquad (23)$$

Figure 6. Noncooperative multi-hop broadcast: shaded area $= a^2 \sqrt{3} /4$

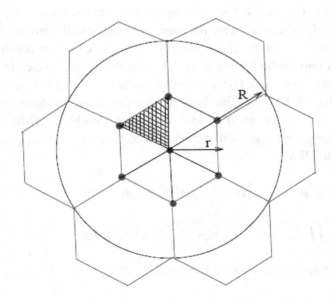

Using (22) and (23) the gain of cooperative transmission is lower bounded as,

$$Gain = [P^{(noncoop)} - P^{(coop)}] / P^{(noncoop)} \times 100 \geq 60\%. \tag{24}$$

This shows that the percentage gain attained with cooperation in dense networks is close to 60% under the path-loss model $\ell(a) = 1/a^2$.

Remark 4: Note that under pathloss model $\ell(r) = 1/r^2$, the power consumption of direct transmission is $P^{(direct)} = A\tau/\pi$, which is more power efficient than the multihop noncooperative broadcast and also cooperative broadcast. However, it is not reasonable to assume that a single node has too much battery power. Note that, for pathloss attenuation with exponents $\alpha > 2$, the performance of direct transmission gets substantially worse.

ERROR PROPAGATION

In this section, we analyze the error propagation behavior of the multistage cooperative broadcast. Different from previous sections, here an uncoded transmission is considered. We assume that N nodes are deployed with a random uniform distribution in this region of unit area. For simplicity, we will assume that the region is a strip of width W and length L *(WL = 1)* and that the source node located at one edge of the strip transmits with power Ps. We also assume that the source and relay data are BPSK modulated. The network region is divided into consecutive areas L_k, $k = 1, \ldots, K$ and $L_0 = (0, 0)$. All nodes that fall within L_k form a level k set S_k. Next we provide the analysis for both orthogonal and non-orthogonal transmissions. See (Scaglione, Kirti & Sirkeci-Mergen 2006) for details.

Orthogonal Channel Model

The network is pre-partitioned as mentioned and we assume that the source transmits a BPSK message with power P_s and the nodes use a coherent detector. Each relay will transmit their decoded bits over orthogonal channels. However, not all relay nodes will decode the bit they received correctly, therefore errors will be introduced into the bit retransmission flow at the physical layer. In this case, the received signal at any node will have two parts; (i) the contribution of nodes that have decoded the message correctly; (ii) the contribution of the nodes that have made an error during decoding. For ease of analysis we consider a suboptimal receiver that uses Maximal-Ratio Combining (MRC) receiver. Hence, our analysis of the error propagation is pessimistic. Similar to previous section, we analyze the error performance in the continuum limit.

Lemma 1: Assuming that all L_k cover an equal area, let:

$$\alpha = min_{(x,y)} \sqrt{2 \operatorname{Pr} / N_0 \iint\limits_{Lk} l(x - u, y - v) du dv} \quad \forall k = 1, \ldots, K. \tag{25}$$

For $\alpha > \pi/2$, the worst error performance is bounded by

$$Q(\alpha) \leq 1/2 - 1/2 \sqrt{-2/\alpha^2 \ln \sqrt{\pi / 2\alpha^2}} ,$$

where $Q(x) = \int\limits_{x}^{\infty} 1/\sqrt{2\pi} \ e^{-t^2}/2 \ dt$ denotes the Q-function.

Remark 5: The results obtained via continuum method in Lemma 1 and 2 show that is the error bounded by a function of the parameter α^2. Note that α can be interpreted as the cumulative SNR for each level. It is evident from above results that for the orthogonal channel model the lower bound on the BER, $Q(\alpha)$, becomes tight for $\alpha \geq 1$. This scheme nearly matches the performance of a cooperative transmission scheme over a deterministic channel i.e. an AWGN channel without fading.

Non-Orthogonal Channel Model

The problem setup is the same as that described for orthogonal channel model except that the nodes in the same level are no longer assigned orthogonal channels for their transmissions. Hence, for every transmitted binary symbol there will be only one received coefficient per level. We use the continuum limit in order to find a bound on the error performance.

Lemma 2: Let α be defined as in Lemma 1. For $\alpha^2/2 > 0.618$, the worst error performance is bounded by

$$\lambda = [1 + \sqrt{2 + (1 + \alpha^2/2)^{-1}}] / 2 [2 + \alpha^2/2] \tag{26}$$

For $\alpha^2/2 \leq 0.618$, the upper bound on λ is 0.5 rather than equation (26). For $\alpha \geq 1$,

$$\lambda \sim [1 + \sqrt{2}] / \alpha^2.$$

Remark 6: For the non-orthogonal scheme, the receiver can not average out the effects of fading since all cooperating nodes transmit asynchronously. Here, the bit error rate (BER) $\rightarrow 1/\alpha^2$ which is directly proportional to $1/SNR$ as one would intuitively guess. This result should not discourage us from using the non-orthogonal channel model because in terms of providing a controllable average BER performance the non-orthogonal scheme is as effective as the orthogonal scheme.

Comparison with Non-Cooperative Transmission

For certain ranges of α, we showed that the BER over each level of the cooperative network is bounded for both orthogonal and non-orthogonal channel models. Suppose that the cooperative physical layer is replaced by a single node at the edge of each level that can transmit at the total power of an entire level. Denoting the best point at each level to forward the message by (x_k, y_k), the average BER from level to level is

$$P_{k,k-1} = E\gamma_k \{ Q[2\underline{P}r \ |L_k| / N_0 \gamma_k l (x_k - x_{k-1}, y_k - y_{k-1})] \}$$

where γ_k is the square of the fading envelope of the 'hop.' Therefore, the BER for the multi-hop network is

$$\text{BER} = 1 - \prod_{k=1}^{K} (1 - P_{k,k-1}) \prod_{k=1}^{K} \approx (1 - P_{k,k-1}) \approx O(K) \tag{27}$$

because $\text{BER} \geq 1 - (1 - \min_k (P_{k,k-1}))^K = O(K)$ if $\min_k(P_{k,k-1}) \ll 1$. This indicates that eventually the error will grow to 0.5. The comparison made above is not entirely fair because the cooperative network we analyzed uses infinite bandwidth whereas the non-cooperative multihop network does not. However, we also showed that this asymptotic bandwidth expansion is unnecessary. The low achievable bounds on the error dynamics we have found above for the cooperative scheme using orthogonal channels are shared by non-cooperative broadcasting that utilizes an amount of bandwidth comparable to that of the multi-hop scheme.

Numerical Results

For both the orthogonal and non-orthogonal channel models we numerically solve the recursive equations that characterize the error propagation in each level. Setting the position of the source to be the origin, the plots below show the average BER of nodes located at coordinates *(x, 0)*.

For orthogonal channel model the BER for subsequent levels is determined and they are compared with the analytical results presented in Lemma 1. Fig. 7(a) shows that for values of $\underline{P}r$ and level size such that $\alpha > \sqrt{\pi / 2}$, the error propagation can be controlled and the worst error probability is close to $Q(\alpha)$. On the other hand, when $\alpha < \sqrt{\pi / 2}$ error propagation is catastrophic. This is shown in Fig. 7(b).

In Fig. 8, we plot the average BER for non-orthogonal channel model. If $\underline{P}r$ and level size are set to values such that $\alpha^2/2 > 0.618$ then the BER can be controlled. It is observed that upper bound predicted by Lemma 2 is larger than the fixed point obtained since Lemma 2 is obtained by upper bounding the worst error performance.

BROADCAST CAPACITY

In this section, we study the scenario where the source node transmits packets continuously and packets are long enough such that coded transmissions are errorless. Notice that in case of continuous source transmission, different levels transmit simultaneously, and this causes interference which did not exist in the single shot analysis utilized in previous sections.

Next, we analyze two-dimensional extended networks (i.e. the network area goes to infinity while the node density is fixed) under per-node power constraints. First, we provide an upper bound on the broadcast capacity that is valid for both deterministic path-loss and random fading models, and then we show that multistage cooperative scheme has broadcast rate that scales as $\theta(1)$. In the proof of achievability, we use an SINR criterion similar to physical model (Gupta & Kumar, 2000) and also we do not take fading into account.

Figure 7. Error propagation in Orthogonal Channel Model. Each tooth of the saw-tooth-like function represents the BER in a separate level. $P_s = 40$, $\underline{Pr} = 25$, Level size = 2.5. (a) $\alpha = 2.13 > \sqrt{\pi/2}$; (b) $\alpha = 0.60 < \sqrt{\pi/2}$.

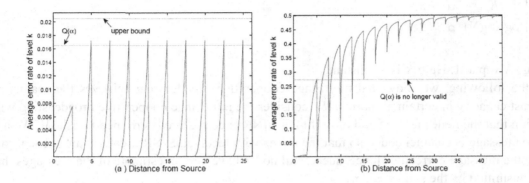

Figure 8. Error propagation in Non-Orthogonal Channel Model. Each tooth of the saw-tooth-like function represents the BER in a separate level. $P_s = 40$, $\underline{Pr} = 80$, Level size = 0.5, $\alpha = 7.85$.

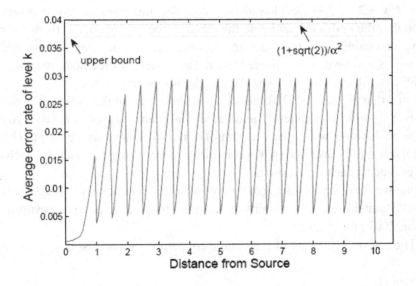

Suppose that N nodes are uniformly and randomly distributed within $S = \{x: \|x\|^2 \leq R^2\}$ for large R and the source node is located at the origin. Let α denote the path-loss exponent. Under per-node power constraint, it is assumed that

$$P_i \leq P_{ind} \quad i \in \{1, \ldots, N\} \tag{28}$$

where P_{ind} does not depend on N.

Theorem 3: Under per-node power constraint (28), the broadcast capacity for the extended random networks is upper bounded as

$$C_{ext} \leq \begin{cases} K & \alpha > 2 \\ \log(\log N) & \alpha = 2 \end{cases} \qquad (29)$$

for large N w.p.a. 1. Here K is a constant.

In the following, we show that multistage cooperative broadcasting achieves the scaling of the broadcast capacity in certain scenarios. We consider the multistage cooperative broadcasting with the condition that the nodes relay if and only if their SINR exceed a predetermined threshold. We assume that the message is channel coded so that the nodes with received SINR greater than or equal to τ can decode the message correctly. Hence, successful nodes have perfect estimate of the messages that has been transmitted by the source node.

In the considered scheme, transmissions occur as follows. The level-1 nodes retransmit the recently decoded block of data. In the meantime, the source node sends a new block of data. After the transmission of the level-1 nodes, the nodes that receive the data block with SINR greater and equal to τ will be called level-2 nodes. Here, we also assume that a node belonging to level-k can cancel-out the interference from level-m, $m \in \{k + 1, k + 2, . . .\}$, since it has already decoded the corresponding messages. This is based on the assumption that the nodes can estimate the channel coefficients perfectly at the receiver side.

For simplicity, we assume that the nodes are full-duplex, i.e., they can receive and transmit at the same time. Our results can be easily extended to the case where source node rests a while in between transmission of blocks, which will eliminate the full duplex assumption.

It should be noted that the above SINR model has an information theoretic basis. Namely, if a group of nodes (let's say S_{k-1}) use independently generated Gaussian codebooks (i.i.d. across time and user domains), then the achievable rate is same as that of a multiple-input single-output (MISO) link. Such transmissions are successfully decoded (with probability approaching 1 as packet length goes to infinity) as long as SINR exceeds a certain threshold.

In the following, we derive the achievable rate of multistage cooperative broadcast in the continuum asymptote (the node density goes to infinity under fixed total relay power constraint). See (Sirkeci-Mergen & Gastpar, 2007) for details.

Theorem 4: The transmission rate of multistage cooperative broadcast is

$$R_s = log(1 + \tau_c) \geq \theta(1),$$

where τ_c is a constant defined as $\tau_c = max_{L \in (0,1)} log(1+L) / [1/(\pi P_2) - log(1-L^2)]$ if $\alpha = 2$ and $\tau_c = \underline{P}r \, \pi / [2^{\alpha+1}(\Delta r)^{\alpha-2} + 16\pi \, \underline{P}r/(\alpha-2)]$ if $2 < \alpha < 4$. In addition, $\Delta r > 0$ denotes the width of first level. Note that $\tau_c > 0$ is obtained using the continuum model.

A legitimate question is if the performance is better when source node periodically transmits one block of new data and stays silent for $T_s - 1$ blocks where $T_s = 1$. We observe that the performance gets better as α increases and T_s decreases to 1.

Comparison with Non-Cooperative Broadcast

In the literature, most broadcast strategies are based on multi-hop transmissions, whose performance is limited by the worst link. In (Zheng, 2006), the authors show that the transmission rate of a multi-hopping scheme scales as

$$R_{multi-hop} = \theta((logN)^{-\alpha/d}), \text{ for } \alpha > d \tag{30}$$

for d-dimensional network with N relays, where α is the path-loss exponent. This result is obtained using the physical model (Gupta & Kumar, 2000) (a node is allowed to retransmit if its SINR exceeds a given threshold). In an extended multi-hop network, under the physical model, the performance is limited by the link between neighboring nodes that are furthest away. In (Zheng, 2006), the authors derive (Laneman & Wornell, 2003) using the scaling of the maximum distance between neighboring nodes, which is $\theta((logN)^{1/d})$ (Penrose, 1997).

Notice that multistage cooperative broadcasting has significant gains over multi-hop transmission whose transmission rate is $\theta((logN)^{-\alpha/2})$ decreases to zero as N increases.

CONCLUSION

In this chapter, we analyzed the behavior of wireless networks with cooperative broadcasting. We introduced multistage cooperative broadcasting and the analysis is based on the idea of continuum approximation, which models networks with high node density. We believe that the techniques used in this chapter can be useful in the analysis of other cooperative protocols. Different conclusions are drawn for multistage cooperative broadcasting:

- There exists a phase transition in the propagation of the message, which is a function of the node powers and the reception threshold.
- Multistage cooperative broadcasting has 60% power gain over traditional multi-hop broadcasting.
- The average BER of multistage cooperative broadcasting with uncoded transmission is bounded for both orthogonal and non-orthogonal channel model. It is also better than the BER for non-cooperative multi-hop communication.
- Multistage cooperative broadcasting achieves the upper bound, $\theta(1)$, on the broadcast capacity of extended networks when path-loss exponent $\alpha > 2$. This results quantify the gains obtained due to cooperation compared to multi-hop non-cooperative broadcasting, which has a maximum rate that scales $\theta(1/(logN)^{-\alpha/2})$.

REFERENCES

Avestimehr, A. S., Diggavi, S., & Tse, D. (2007). A deterministic approach to wireless relay networks. In *Proceedings of Allerton Conference*.

Boyer, J., Falconer, D. D., & Yanikomeroglu, H. (2004, October). Multihop diversity in wireless relaying channels. *IEEE Trans. Commun.*

Gupta, P., & Kumar, P. R. (2000, March). The capacity of wireless networks. *IEEE Transactions on Information Theory, 46*(2), 388–404. doi:10.1109/18.825799

Hong, Y.-W., & Scaglione, A. (2003, October). Energy-efficient broadcasting with cooperative transmission in wireless sensory ad hoc networks. In *Proc. of Allerton Conf. on Commun., Contr. and Comput. (ALLERTON)*.

Hong, Y.-W., & Scaglione, A. (2006, October). Energy-efficient broadcasting with cooperative transmission in wireless ad hoc networks. *IEEE Transactions on Wireless Communications, 5*(10), 2844–2855. doi:10.1109/TWC.2006.04608

Jacquet, P. (2004, May). Geometry of information propagation in massively dense ad hoc networks. *ACM MobiHoc, 04*, 157–162. doi:10.1145/989459.989479

Keshavarz-Haddad, A., Ribeiro, V., & Riedi, R. (2006, September 24-29). Broadcast capacity in multihop wireless networks. In *Proceedings of the 12th Annual International Conference on Mobile Computing and Networking, MobiCom'06* (pp. 239–250).

Khisti, A., Erez, U., & Wornell, G. (2006, June). Fundamental limits and scaling behavior of cooperative multicasting in wireless networks. *IEEE Transactions on Information Theory, 52*(6), 2762–2770. doi:10.1109/TIT.2006.874541

Korkmaz, G., Ekici, E., & Ozguner, F. (2007, September). Black-burst based multihop broadcast protocols for vehicular networks. *IEEE Trans. on Vehicular Tech., 56*(5).

Laneman, J. N., & Wornell, G. W. (2003, October). Distributed space-time coded protocols for exploiting cooperative diversity in wireless networks. *IEEE Transactions on Information Theory, 59*(10).

Maric, I., & Yates, R. D. (2004, August). Cooperative multihop broadcast for wireless networks. *IEEE J. Select. Areas Commun., 22*(6).

Maric, I., & Yates, R. D. (2005, January). Cooperative multicast for maximum network lifetime. *IEEE Journal on Selected Areas in Communications, 23*(1), 127–135. doi:10.1109/JSAC.2004.837343

Penrose, M. (1997). The longest edge of the random minimal spanning tree. *Annals of Applied Probability, 7*(2), 340–361. doi:10.1214/aoap/1034625335

Rappaport, T. S. (2003). *Wireless communications principles and practice*, 2nd ed. Prentice Hall.

Scaglione, A., & Hong, Y.-W. (2003, August). Opportunistic large arrays: Cooperative transmission in wireless multihop ad hoc networks to reach far distances. *IEEE Transactions on Signal Processing* m 8.

Scaglione, A., Kirti, S., & Sirkeci-Mergen, B. (2006, April 19-21). Error propagation in dense wireless networks with cooperation. In *Proc. of Information Processing in Sensor Networks (IPSN)*, Nashville, TN.

Shakkottai, S., Srikant, R., & Shroff, N. (2004). Unreliable sensor grids: Coverage, connectivity, and diameter. *AdHoc Networks*.

Sirkeci-Mergen, B., & Gastpar, M. (2007, January). On the cooperative broadcast capacity of high density wireless networks. *The 2007 Information Theory and Applications Workshop*, San Diego, CA (invited paper).

Sirkeci-Mergen, B., & Gastpar, M. (2007, June). (Manuscript submitted for publication). On the cooperative broadcast capacity of wireless networks with cooperative relays. *IEEE Transactions on Information Theory*.

Sirkeci-Mergen, B., & Gastpar, M. (2007, November 4-7). On the ccaling of the broadcast capacity of extended wireless networks with cooperative relays. In *Proc. Asilomar Conference on Signals, Systems, and Computers,* Pacific Grove, CA.

Sirkeci-Mergen, B., & Scaglione, A. (2004). Signal acquisition for cooperative transmissions in multihop ad hoc networks. In *Proc. of IEEE Inter. Conf. on Acoustics, Speech, and Signal Process. (ICASSP)*.

Sirkeci-Mergen, B., & Scaglione, A. (2004, July). Coverage analysis of cooperative broadcast in wireless networks. In *Proc. of IEEE Workshop on Signal Process. Advances in Wireless Commun. (SPAWC)*.

Sirkeci-Mergen, B., & Scaglione, A. (2005). A continuum approach to dense wireless networks with cooperation. In *Proc. of Annual Joint Conf. of the IEEE Computer and Commun. Societies (Infocom)*.

Sirkeci-Mergen, B., & Scaglione, A. (2005). Message propagation in a cooperative network with asynchronous receptions. In *Proc. of IEEE Inter. Conf. on Acoustics, Speech, and Signal Process. (ICASSP)*.

Sirkeci-Mergen, B., & Scaglione, A. (2006, July). On the optimal power allocation for broadcasting in dense cooperative networks. In *Proc. of IEEE International Symposium on Information Theory*, Seattle, WA.

Sirkeci-Mergen, B., & Scaglione, A. (2007, February). On the power efficiency of cooperative broadcast in dense wireless networks. [JSAC]. *IEEE Journal on Selected Areas in Communications, 25*(2), 497–507. doi:10.1109/JSAC.2007.070223

Sirkeci-Mergen, B., & Scaglione, A. (2007, October). Randomized space-time coding for distributed cooperative communication. *IEEE Transactions on Signal Processing, 55*(10). doi:10.1109/TSP.2007.896061

Sirkeci-Mergen, B., Scaglione, A., & Mergen, G. (2006, June). Asymptotic analysis of multistage cooperative broadcast in wireless networks. *IEEE Transactions on Information Theory, 52*(6), 2531–2550. doi:10.1109/TIT.2006.874514

Tavli, B. (2006, February). Broadcast capacity of wireless networks. *IEEE Communications Letters, 10*(2), 68–69. doi:10.1109/LCOMM.2006.02013.

Tseng, Y.-C., Ni, S.-Y., Chen, Y.-S., & Sheu, J.-P. (1999). The broadcast storm problem in a mobile ad hoc network. In *Proc. ACM/IEEE Int. Conf. on Mobile Computing and Networking (MOBICOM)* (pp. 151-162).

Williams, B., & Camp, T. (2002). Comparison of broadcasting techniques for mobile ad hoc networks. *ACM Proc. on MOBIHOC*.

Zheng, R. (2006). Information dissemination in power-constrained wireless networks. In *Proc. of the 25th Annual Joint Conference of the IEEE Computer and Communications Societies (INFOCOM)*.

Chapter

Cooperative Communication
System Architectures for

Section 6
An Industrial Perspective on Cooperative Communications

Chapter 19

Cooperative Communication System Architectures for Cellular Networks

Mischa Dohler
CTTC, Spain

Djamal-Eddine Meddour
Orange Labs, France

Sidi-Mohammed Senouci
Orange Labs, France

Hassnaa Moustafa
Orange Labs, France

ABSTRACT

An ever-growing demand for higher data-rates has facilitated the growth of wireless networks in the past decades. These networks, however, are known to exhibit capacity and coverage problems, hence jeopardizing the promised quality of service towards the end-user. To overcome these problems, prohibitive investment costs in terms of base station or access point rollouts would be required if traditional, non-scalable, cell-splitting, and micro-cell capacity dimension procedures were applied. The prime aim of current R&D initiatives is, hence, to develop innovative network solutions that decrease the cost per bit/s/Hz over the wireless link. To this end, cooperative networks have emerged as an efficient and promising solution. We discuss in this chapter some key research and deployment issues, with emphasis on cooperative architectures, networking, and security solutions. We expose some motivations to use such networks, as well as latest state-of-the-art developments, open research challenges, and business models.

DOI: 10.4018/978-1-60566-665-5.ch019

INTRODUCTORY NOTE ON COOPERATION

Background

Wireless networks have witnessed a tremendous upsurge in recent years; this is mainly attributed to a lasting demand of high data rates anywhere and at anytime, which has been partially realized by a variety of commercially viable voice and data oriented applications. Traditionally, a centralized network infrastructure, such as GSM, is deployed by service providers; this approach worked fine in the past but commences to exhibit drawbacks, such as high cost, high power consumption and limited throughput.

An extreme alternative are ad hoc networks, where packets are forwarded in a multihop fashion. In such networks, users cooperate to relay and process each other's information. Notwithstanding their low cost, rapid deployment and self-organization capabilities, ad hoc networks face QoS, security and scalability problems. Consequently, standalone ad hoc networks are not promising for service commercialization. Indeed, business models for real world deployments are fairly complicated, having prevented a commercially viable deployment of pure ad hoc networks to date.

A natural hybrid approach is to beneficially fuse both of the above wireless paradigms in order to construct a single network with high flexibility and improved network performance. In such a network, a centralized base station (BS) or access point (AP) communicates directly with some users or fixed low-cost relaying stations, which in turn cooperatively relay information in an ad hoc fashion to other users in connectivity range. In the cellular case, such networks are typically referred to as *multihop cellular networks (MCNs)* as introduced by Lin & Hsu (2000). Subsequently, we will partially focus on cooperative MCNs, bearing in mind that the majority of exposed techniques and architectures are equally applicable to non-cellular networks.

MCNs can reduce the required number of BSs/APs and/or improve the throughput performance, whilst limiting path vulnerabilities typically encountered in multihop networks. They are potentially opening new business opportunities for network operators and service providers, allowing commercial service provisioning with broader coverage. However, for wide-area deployments of MCNs, appropriate architectures are needed allowing for cooperative multihop communication between similar wireless technologies and cooperative communication between different operators and service providers, as well as different wireless technologies.

We hence focus on possible deployment architectures for cooperative MCNs and the major technical challenges that are currently being resolved in real deployment scenarios from cooperation perspectives, such as routing, appropriate QoS metrics, authentication, and authorization to services' access, etc. We will, however, precede the architectural description of such networks by some basics needed to understand cooperative communication systems.

Some Useful Definitions

A large body of recent publications has led to numerous independent terminologies, some of which we wish to harmonize below (Dohler & Aghvami, 2007). These definitions relate to the cooperative system, the cooperative information flow, the nodes' behavior and the actual method of relaying.

Often occurring in the exposures of cooperation is the concept of *infrastructure*. An infrastructure – be it physical or logical – can:

- be available prior to deployment (e.g. cellular networks or WLANs); or
- emerge after deployment or simply remain unavailable (e.g. ad hoc networks).

The former is also referred to as infrastructure-based, whereas the latter as infrastructure-less. The infrastructure can be managed in the following fashions:

- centralized (e.g. cellular network); or
- decentralized (e.g. WLAN mesh network).

Note that one may have a decentralized infrastructure-based system (e.g. systems with decentralized radio resource management) or a centralized infrastructure-less system (e.g. clustering). Subsequently, we will mainly deal with centralized but hybrid infrastructure-based/less systems.

Another key-concept is related to the *information flow* from source to destination/target, which can be:

- point-to-point (traditional);
- point-to-multipoint (broadcast/multicast);
- multipoint-to-point (multiple access);
- multipoint-to-multipoint (general).

Generally, such *information flows can be realized* by means of:

- direct links (no relays between source and target);
- relaying links (at least one relay between source and target);
- relaying stages (clusters where information passes approx. the same time).

Each of the involved nodes in the network can have the following behavior:

- egoistic (no help);
- supportive (unidirectional help);
- cooperative (mutual help).

The *relaying process* itself can be:

- transparent (retransmission of originally received analogue signal); or
- regenerative (retransmission of digitally modified received signal).

The former is seemingly simple as it usually only involves some form of frequency translation and amplification of the received signal, whereas the latter comprises some baseband processing but is generally known to outperform transparent techniques.

An operator would clearly be interested in designing an infrastructure-based system, for which service and QoS provisioning can be guaranteed and influenced. Also, both financial and performance gains – if any – due to relaying and cooperation ought to be quantified, as well as the optimum choice of relaying techniques and technologies established.

Academic Milestone Contributions

The method of relaying, i.e. a canonical form of cooperation, has been introduced by van der Meulen (1971). A first rigorous information theoretical analysis of the relay channel has been exposed by Cover & el Gamal (1979). In these contributions, a source mobile terminal (MT) communicates with a target MT directly and via a relaying MT. The maximum achievable communication rate has been derived in dependency of various communication scenarios, which include the cases with and without feedback to either source MT or relaying MT, or both. The capacity of such a relaying configuration was shown to exceed the capacity of a simple direct link. It should be noted that the analysis was performed for Gaussian communication channels only; therefore, neither the wireless fading channel has been considered, nor have the power gains due to shorter relaying communication distances been explicitly incorporated into the analysis.

Only in the middle of the 90s, the idea of utilizing relaying to boost the capacity of infrastructure-based wireless networks revived, thereby leading to the concept of opportunity driven multiple access or ODMA (3GPP, 1999). Here, the power gains due to the shorter relaying links have been the main incentive to investigate such systems to reach MTs out of base station (BS) coverage. The emphasis of the study was its applicability to cellular systems, as well as a suitable protocol designs.

Interesting milestones into the above-mentioned theoretical studies have been the contributions by Sendonaris, Erkip & Aazhang (1998). In their study, a very simple but effective user cooperation protocol has been suggested to boost the uplink capacity and lower the uplink outage probability for a given rate. Moreover, they showed that cooperation can reduce the MT's power consumption. The designed protocol stipulates a MT to broadcast its data frame to the BS and to a spatially adjacent MT, which then re-transmits the frame to the BS. Such a protocol certainly yields a higher degree of diversity because the channels from both MTs to the BS can be considered uncorrelated. The simple cooperative protocol has been extended by the same authors to more sophisticated schemes, which can be found in their subsequent excellent publications.

The contributions by Laneman & Wornell (2000) are a conceptual and mathematical extension to Sendonaris, Erkip & Aazhang (1998), where energy-efficient multiple access protocols are suggested based on decode-and forward and amplify-and-forward relaying technologies. It has been shown that significant diversity and outage gains are achieved by deploying the relaying protocols when compared to the direct link. The case of distributed space-time coding has also been analyzed by Laneman in his PhD dissertation. In his thesis, information theoretical results for distributed single-input-single-output (SISO) channels with possible feedback have been utilized to design simple communication protocols taking into account systems with and without temporal diversity, as well as various forms of cooperation. He has demonstrated that cooperation yields full spatial diversity, which allows drastic transmit power savings at the same level of outage probability for a given communication rate.

Gupta & Kumar (2000) were the first to statistically analyze the information theoretically offered throughput for large scale relaying networks. They showed that if the M terminals and associated traffic distributions are random, then the capacity per terminal decreases in the order of $1/\sqrt{(M \log M)}$. The analysis in (Gupta & Kumar, 2000) has been extended by the same authors to more general communication topologies, where the interested reader is referred to the landmark paper (Gupta & Kumar, 2003). Lately, a linear capacity scaling for a specific cooperative protocol has been exposed by Ozgur, Leveque & Tse (2007).

Whilst above milestone contributions concentrated on the simple relaying case, the concept of distributed cooperative relaying systems, also termed Virtual Antenna Arrays, with application to cellular networks has been introduced in February 2000 by Dohler (1999-2002). The generalization of the concept to distributed-MIMO multi-stage communication networks with application of distributed space-time codes has been introduced shortly after and consequently patented by M-VCE in June 2001 (Dohler, Said, Ghorashi & Aghvami, 2001).

Other excellent research in these areas has been performed thereafter, all of which led to the currently flourishing research area of cooperative wireless communication networks.

Industrial Motivation

The success of IP technologies jointly with the appearance of high data rate solutions at physical layer led to a rapid growth of telecommunications networks. For the wireless network, however, radio resources are limited (and expensive) and one cannot infinitely increase the network capacity. The availability of vacant bandwidth is not expected to increase significantly, and the gap will hence only widen.

To date, the only way to get around this is by controlling the transmission power and increasing the spatial reuse of frequencies by cell splitting/sectoring. These were the driving principles behind the design of cellular networks of the past (AMPS, GSM and also 3G). However, these methods incur huge installation and maintenance costs which explode in UMTS networks where micro-cells have a diameter of a few hundred meters. Further worsened by the high license fees, there is thus a burgeoning threat that high-quality wireless services may soon become an unaffordable luxury.

The need for a breakthrough in approaches to network dimensioning is hence evident. With the advance of recent academic developments as outlined above, cooperative relaying networks have proven to be a viable solution. Of particular commercial interest among operators are MCNs. Originally proposed by Lin & Hsu (2000), MCNs open the doors to a new paradigm of hybrid cellular and ad hoc networks. They rely on a set of BSs/APs connected to a backhaul network, as in conventional cellular networks, and on the mechanisms of multihop networks, in which the packets are relayed between peer wireless stations.

Numerous contributions have emerged ever since. To this end, aiming at fulfilling the requirement of IMT-Advanced, the WINNER project (https://www.ist-winner.org/) develops a new air interface that performs with scenarios ranging from Metropolitan Area to Local Area Networks. The WINNER interface inherently supports cooperative relaying features and provides cost-effective high data rate provisioning. Further, the economic evaluation of the solution showed that this can decrease the cost per bit transmitted by a factor of two to three (Esseling, Walke & Pabst, 2004).

Gunasekaran & Harmantzis (2005) present a comparative study of conventional point-to-multipoint (PMP) based IEEE 802.16 WiMAX with cooperative transmission-based mesh topology from an economic point of view. They show that – given reasonable assumptions on traffic, number of users, frequency allocation and number of hops – a mesh-based WiMAX solution is more affordable and advantageous. This is mainly due to lower wired backhaul costs and also the possibility of using lower antenna heights to serve as relays rather then conventional BSs.

From above, the expected coverage and throughput benefits of a cooperative relaying MCN approach with respect to conventional cellular networks are quantifiable and sufficiently large to attract industrial interest. From an economic view, the planning and optimization of BSs/APs together with the leasing costs of their locations could be reduced through MCNs. An unplanned deployment, however, gives justifiable gains only if the relay is about 10% of the BS cost and about 50-100% of the planning cost without relays (Timus, 2006). Furthermore, a potential increase in the wireless coverage due to cooperative relaying could be cost efficient for rural areas where the amount of users is lower and hence the income of the operator is limited.

Cooperative Relaying Techniques

From previously discussed building blocks, applied to the cooperative relaying case, we deem issues related to the wireless relaying channel, characterization of link and system capacity, as well as the various OSI layers of grand importance and hence briefly dwell on their state-of-the-art.

Characterization of Relaying Channels

Channel models are vital in the designing process of wireless systems, because they influence power budget dimensioning, transceiver design, performance behavior, etc. There are, however, only a few relaying channel measurements/models available and no explicit models, which cater for the distributed cooperative communication channel. We hence need to adapt known channel measurements and models to the distributed cooperative case, until explicit models will become available.

Channel models are composed in a multiplicative fashion of:

- pathloss (deterministic effect due to power loss over distance);
- shadowing (lognormally distributed random effect due to shadowed waves);
- fading (random effect due to phasor additions).

To characterize the above, we are particularly interested in the occurring pathloss coefficient, shadowing variance and shadowing correlation distance, fading statistics for each multipath component (MPC) and their correlation properties, and finally in the power delay profile with given delay spreads. Whilst some greater insights are given below for transparent and regenerative cooperative relaying channels, let us examine some general tendencies comparing narrowband/wideband cooperative/non-cooperative channel characteristics as observed by the destination terminal.

As exposed in Figure 1, compared to their narrowband counter-part, wideband communication systems manage to reduce the fading margin due to the additionally injected frequency diversity. Cooperative systems, in addition, have the advantage of reducing the shadowing margin due to a high spatial diversity. Such a reduction constitutes a serious advantage, as the performance of today's communication systems is dominated by the shadowing channel.

As for the *regenerative relaying channel*, the statistics of each individual cooperative relaying segment is of importance, thereby leading to point-to-point channel models. Since cooperative relaying systems are often composed of a cellular link from an elevated BS towards a relaying terminal as well as some non-elevated cooperative links among nodes, we will briefly summarize either characteristics (Konstantinou, Kang & Tzaras, 2007; Patel, Stüber & Pratt, 2006):

- pathloss coefficient n:
  - cellular links: $n = 2$ (LOS), $n = 2, \ldots, 4$ (nLOS);
  - cooperative links: $n = 2$ (LOS), $n = 4, \ldots, 6$ (nLOS);

Figure 1. Cooperative relaying channel power loss tendencies versus distance for narrow and wideband systems

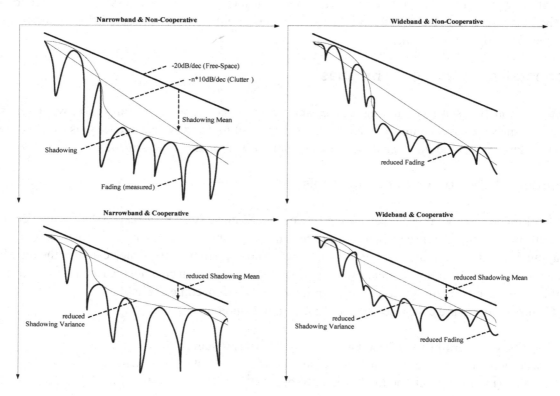

- shadowing variance:
  - cellular links: 2, . . ., 6dB (LOS), 6, . . ., 18dB (nLOS);
  - cooperative links: 0, . . ., 2dB (LOS), 2, . . ., 6dB (nLOS);
- shadowing coherence distance:
  - cellular links: >100m (LOS), tens of meters (nLOS)
  - cooperative links: 40-80m (LOS), 20-40m (nLOS);
- first MPC fading statistics (other MPCs are Rayleigh distributed):
  - cellular links: Ricean $K = 2, . . ., 10$ (LOS), Rayleigh (nLOS);
  - cooperative links: Ricean $K > 10$ (LOS), Rayleigh (nLOS);
- power delay profile:
  - cellular links: negative-exponential, clustered;
  - cooperative links: negative-exponential;
- root mean square (RMS) delay spread τ_{RMS}:
  - cellular links: depends on cell size, $\tau_{RMS} = 50ns, . . ., 4\mu s$;
  - cooperative links: $\tau_{RMS} = 10ns, . . ., 40ns$.

As for the *transparent relaying channel*, Laneman, Tse & Wornell (2004) have studied the statistical properties of a dual-hop amplify-and-forward cooperative relay channel. It has been shown that when the source-relay and relay-destination channels experience flat fading and their coefficients are independent complex Gaussian distributed, then the end-to-end channel (source-relay-destination) envelope is a modified Bessel function of zeroth order. It is interesting to point out that the temporal autocorrelation is a product of two first order Bessel functions. This leads to a faster decrease in correlation compared to the classic single relay channel, thereby complicated channel estimation procedures but aiding channel code performance.

Physical Layer Algorithms

At PHY layer, we distinguish three canonical relaying techniques, which can be used in conjunction with simple relaying or cooperative diversity relaying:

- amplify-and-forward;
- compress-and-forward; and
- decode-and-forward.

In the *amplify-and-forward* approach – being equivalent to *transparent relaying* – the cooperative relay down-converts the received analogue signal, amplifies it and up-converts it to another frequency band prior to re-transmitting it. The amplification requires some power constraints to be respected, where fixed or variable gain amplifications can be implemented. Note that this protocol suffers from severe performance losses at low signal-to-noise ratios (SNRs), because noise at the relay is also amplified. Furthermore, the analogue signal cannot be stored and hence requires immediate frequency translation; this implies two oscillators, two frequency bands and two fairly good filters – not necessarily making it a cheaper technology with respect to other relaying techniques. Apart from the below mentioned techniques, a feasible approach is to use quantization in order to store the analogue signal and then forward it in the same band in a TDMA fashion; see e.g. Djeumou, Lasaulce & Klein (2007).

The *compress-and-forward* approach is an extension of the amplify-and-forward method, where the analogue signal is sampled, quantized, compressed and re-transmitted. The advantage of doing so is to be able to temporarily store the signal or to relay it using a different communication standard. For instance, a 3G terminal could relay its received signal in compressed form via Bluetooth to adjacent terminals.

Finally, the *decode-and-forward* approach decodes the received signal and re-encodes it with a potentially different codebook prior to re-transmission. This clearly adds some complexity but at low SNR it exhibits a better performance than the amplify-and-forward approach. However, when the source-relay link is bad, this leads to a bottleneck for the transmission system since the relay is assumed to decode correctly the source message. Relay selection procedures are hence needed to overcome this problem and to increase the protocol's diversity order (Laneman & Wornell, 2003). Information theoretically, such a processing permits to adapt the relaying rate to the relay-destination capacity.

The requirement of two frequency bands and the inability to store the relayed signal makes, in our opinion, the amplify-and-forward a less likely deployment candidate when compared to the decode-and-forward protocol. We will hence concentrate on the latter, for which repetition based, channel code based, and space-time code based relaying methods are available. The first method repeats the received codeword (known to be sub-optimum from a code design point of view); the second method relays some parity information; and the third method constructs a space-time codeword between the source (s) and relaying (r) partners, thereby creating a distributed multiple-input multiple output (MIMO) antenna array with obvious performance gains (Dohler, 2003; Laneman & Wornell, 2003).

Repetition and channel code based methods require only a fairly loose synchronization at frame level between source and relay terminals, whereas the space-time code based relaying method requires a fairly tight synchronization at symbol level. This has lately been relaxed with the design of synchronization-robust space-time codes (Li & Xia, 2005).

Medium Access Control Mechanisms

Conflicts occur when more than one wireless link is active in a system. These conflicts are managed by the medium access control (MAC), which chooses:

- resources, i.e. which resources a link may use (e.g. specific time-slot);
- duplex method, i.e. whether the same frequency or different; and
- contention protocol, i.e. how each link gets access to the wireless medium.

Resources can usually be allocated using, e.g., time division multiple access (TDMA), frequency division multiple access (FDMA), code division multiple access (CDMA), or orthogonal frequency division multiple access (OFDMA). The available duplex methods are time division duplex (TDD) and frequency division duplex (FDD). Protocols resolving contention are reservation-based MACs for typically centralized applications – in conjunction with, e.g., TDMA; and contention-based MACs for distributed applications – e.g. carrier sensing multiple access (CSMA).

Whilst the MAC is traditionally informed by the network layer about the next-hop destination, it needs to *select one or several suitable relay partner(s)* to facilitate cooperation. Several such protocols, based on different underlying assumptions and design goals, have e.g. been proposed by Ahmed, Ibars, del Coso & Mohammed (2007); Chou, Yang, & Wang, (2007); Jakllari, Krishnamurthy, Faloutsos, Krishnamurthy & Ercetin (2006); Librino, Levorato & Zorzi (2007); Liu, Tao, Narayanan, Korakis & Panwar (2007); Beres & Adve (2008); Michalopoulos & Karagiannidis (2008).

Since the relay channel is an additional traffic channel, the choice of relaying mechanism will influence the *multiple access protocol*. For instance, amplify-and-forward approaches require FDMA to be implemented at the relay because it is difficult to temporally store the analog signal, whereas the other two approaches also allow for TDMA.

The TDMA mode is generally realized by means of two phases. In the first phase, the source broadcasts information to the destination and the relay(s). In the second one, the relay(s) transmit(s) the information towards the destination. At MAC, this can be implemented using an *orthogonal as well as non-orthogonal mode*. For the orthogonal mode, the source does either not transmit in the second phase which reduces interference at the receiver side (Laneman, Tse & Wornell, 2004) or uses entirely orthogonal space-time codes (Dohler, 2003). For the non-orthogonal mode, the source also transmits in the second phase, which is known to increase the rate (Azarian, el Gamal & Schniter, 2005). Several works studied different versions of the two orthogonal and non-orthogonal modes. Their performance is compared using the diversity-multiplexing trade-off (Zheng & Tse, 2003) introduced for MIMO systems. It is shown in general that the non-orthogonal mode outperforms the orthogonal one, because, for the same diversity order, they achieve higher rates (Zheng & Tse, 2003). This has also been extended to broadcast and multicast channels (Azarian, el Gamal & Schniter, 2005).

To facilitate a cooperative MAC from an implementation point of view, two cases need to be distinguished: the *homogenous MAC* cooperation, where one distinct MAC layer is present in the system; and the *heterogeneous MAC*, where MAC protocols from different systems are used for cooperation.

Cooperation using a *homogenous MAC* takes advantage of the inherent properties of the wireless medium, its shared nature as well as the broadcast support of wireless transmissions. In practice, the conventional wireless systems are designed such that any unicast communication involves the two concerned parties only, i.e. the sender and the receiver. Therefore, existing MAC protocols ignore any overheard information from neighboring nodes that are not involved in the transmission. In a cooperative scenario, this situation leads to a multitude of retransmission and therefore bandwidth waste.

In order to counteract this waste and also improve the system reliability, new wireless medium access control solutions enforce additional cooperative mechanisms at the neighboring nodes, which can act as relays to improve the transmission reliability. In such a case, we consider three entities: the source, the destination and the relay. The source transmits first its MAC packet data unit (PDU). If the destination successfully receives this PDU, it sends an acknowledgement which will be overheard by both the relay and the source. In the case where the destination does not receive the PDU correctly but the relay node does, the latter transmits the PDU to the destination. If both the destination and the relay fail, the packet gets retransmitted by the source node.

Several practical solutions based on the suggested scheme with some minor variations are proposed in the literature. For instance, Azarian, el Gamal & Schniter (2005) proposed a new MAC protocol called CoopMAC which is based on the IEEE 802.11 distributed coordination function (DCF). Another proposal has been put forward in the standardization group IEEE 802.15.

In the case of a *heterogeneous MAC*, we consider the co-existence of several MACs in the system, a configuration which will occur more often in future beyond 3rd generation (B3G) and 4th generation (4G) systems. The cooperative system must take profit of this diversity to improve the effectiveness of the network and shall enable the inter-working between the different solutions. It can work either in handover based mode such that it triggers the hand off between two MAC technologies using a predefined criterion like signal strength, or in a complementary fashion, i.e. the traffic is divided over all the existing links. Cooperative solution in this context did usually not imply any specific modification at the adjacent MACs; it is basically managed at L2.5 in order to ensure the backward compatibility with existing system. Many proposals have been made to handle cooperation issues in heterogeneous MAC environments; for instance, IEEE 802.21 or the unlicensed mobile access (UMA), as well as I-WLAN.

Research on cooperation mechanisms at MAC layer should also ensure that no user misbehaves. For example, in the IEEE 802.11 DCF, all participating nodes adhere to the backoff protocol to ensure – in the absence of hidden nodes – a fair share of the bandwidth for each node. A selfish node might want to obtain more than its fair share of the channel bandwidth by selecting smaller backoff values or using a different retransmission strategy, such as not to double the contention window value after a collision (Kyasanur & Vaidya, 2003). Such a selfish behavior seriously degrades the throughput of the fair/no-selfish nodes. To deal with this issue, protocols where changes to the backoff calculation are sought. In Kyasanur & Vaidya (2003), the authors propose some modifications to the IEEE 802.11 DCF with the supposition of the presence of a trusted base station that can identify sender misbehaviors.

Network Layer Protocols

Cooperation from a network viewpoint concerns the cooperation mechanisms between network elements for traffic forwarding. More specifically, it is about the design of an efficient routing protocol that enables effective network resource management. Interestingly, from the higher level perspectives, the wireless network is represented as a set of wireless nodes that attempt to increase the system's quality of service

(QoS) via cooperation. Therefore, the problem to alleviate at the routing level considering a multihop path is how to select the best cascaded cooperative relay set from a source towards the destination. It is worth noting that an effective cooperation at network level implies the usage of cooperative transmission at both MAC and PHY layers.

We find in the literature numerous protocols that deal with the proper selection of next hop relays as well as multihop paths in a wireless environment. However, only a few routing protocols exist that really consider the existence of cooperative terminals along the route.

Bletsas, Khisti, Reed & Lippman (2006) advocate the use of opportunistic relaying as a practical scheme for cooperative solutions. A distributed path selection mechanism is proposed where the best relay is selected by the source using instantaneous wireless channel conditions, i.e. signal-to-interference-and-noise ratio (SINR) measurements, and then used to realize the cooperation between the source and the destination. The simplicity of the solution facilitates the coordination between the cooperative entities and bounds the overall signaling overhead. Adam, Bettstetter & Senouci (2008) propose two refinements: (i) 'relay selection on demand' where relays are only selected if required by the destination and (ii) 'early retreat' where each potential relay assesses the channel state and decides whether to participate in the relay selection process or not.

Biswas & Morris (2005) combine both cross-layer optimization and spatial diversity by investigating the performance of a link/network layer diversity routing protocol. The process of packet delivery is as follows: iteratively, at each hop and for each packet, a 'candidate forwarder' is selected by the source node from its one hop neighbor nodes and prioritized based on their proximity, in terms of number of hops to the destination. Therefore, the node with the highest priority will relay the received packet, whereas the other candidate forwarders transmit only the unacknowledged packets. Such an approach was shown to outperform traditional routing, typically increasing the overall throughput by a factor of two.

Jerbi, Senouci, Ghamri & Beylot (2008) propose a self-organizing mechanism to emulate a geo-localized virtual infrastructure (GVI). This latter is emulated by a bounded-size subset of cooperative vehicles currently populating the geographic region solving by the way the infrastructure dependence problem of some existing dissemination protocols.

Finally, w.r.t. Kim & Bohacek (2005), the essence of this contribution led to the design and the implementation of the best-select protocol (BSP), which generalizes single-path routing with sets of nodes substituting the concept of a single node relay. Consequently, the data are transferred from a given relay-set towards another relay-set. The channel gain information obtained through message exchange between relay-sets is utilized to select the best node as the relay to transmit the data to the next relay-set. The process is reiterated until the destination is reached.

As discussed in the previous section, it is important to study the node behavior in the case of infrastructure-less ad hoc networks. In fact, in such networks, where no centralized entity exists, a malicious or self-interested user can misbehave and does not cooperate. A malicious user could inject false routing messages into the network in order to break the cooperative paradigm. However, a self-interested user does not intend to directly damage the overall functioning, but to save its own resources. A user's selfishness is comprehensible as it is often requires to forward packets for the benefit of others, consuming precious resources that they want to save for their own communication. The basic network functions subject to selfishness are broadcasting and routing.

Current approaches to counteract such behavior and enforce cooperation at network layer can be broadly classified into two categories:

- pricing or credit based schemes; and
- reputation based schemes (Conti, Gregori & Maselli, 2004) .

Credit-based schemes consider packet forwarding as a market model where nodes providing a service are remunerated, whilst nodes receiving a service are charged. Hence, if a node wants to send its own packets, it must forward packets for the benefit of others. However, these schemes require tamper-resistant hardware (Buttyan & Hubaux, 2003) or infrastructure-dependent credit clearance systems (Zhong, Chen & Yang, 2003) that other nodes can trust.

Reputation-based schemes discourage misbehavior by estimating the nodes reputation and punishing nodes with bad behavior (Buchegger & le Boudec, 2002; Michiardi & Molva, 2002). The scheme requires each node to rate every other node with which it communicates based on the service received or on observing the behavior of neighbors by listening to communications in the same transmission range. According to the collected information, the reputation system maintains a value for each observed node that represents a reputation of its behavior. The reputation mechanism allows avoiding sending packets through misbehaving nodes.

COOPERATIVE RELAYING ARCHITECTURES

In this section, we attempt to show to which systems and architectural designs above outlined cooperative techniques and protocols are currently applied. We will briefly discuss the need to distinguish homogenous and heterogeneous approaches, after which we will discuss each in some details.

Homogeneous vs. Heterogeneous Architectures

After having described the limitations encountered with conventional networks and motivating the use of cooperative systems as well as having dwelled on related state-of-the-art developments, we will review several proposals that describe possible realizations of cooperative multihop cellular networks.

Similar to the MAC layer, a classification ought also to be made on system architecture level based on the relaying technology used to facilitate the multihop communication. If the BSs/APs employ the same technology as the relay stations, e.g. both WLAN, we will refer to *homogeneous MCNs*. If they adopt a different technology, e.g. WiMax at the BS and WiFi and/or UWB for relaying (multi-mode relays are possible), we will refer to *heterogeneous MCNs*. This has been exemplified in Figure 2. Note that the BS shall handle all communication technologies used in the MCN.

Such a deployment would involve a repartition of intelligence and functionalities between the BSs and fixed relay stations, where the latter ensure the following functionalities:

- authentication, authorization and registration closer to the BS;
- topology discovery and update of routing table and traffic forwarding;
- resource scheduling QoS parameters establishment;
- managing user mobility and handover.

Figure 2. Deployment architectures of heterogeneous MCNs

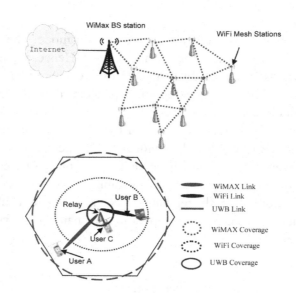

Homogeneous Cooperative Architectures

By advocating a homogeneous design, MCNs extend seamlessly the connectivity provided by the underlying system without further modifications at the user terminal's side. A mobile node can hence benefit from and access to network services independently from the presence of the relaying system. We shall subsequently discuss solutions based on IEEE 802.16j and 802.11s. It is worth mentioning that such solutions have been developed within the IST ROMANTIK project (www.ist-romantik.org) for UMTS networks. The personal network standardization efforts at the IEEE are also currently studying such approaches within TG IEEE 802.15.5.

Relaying in Wireless Metropolitan Area Networks (WMAN, IEEE 802.16j). The IEEE 802.16j aims at defining a multihop solution for WMANs. To this end, they propose to expand the standard IEEE 802.16 model which currently only allows direct communication between the mobile station (MS) and BS; the multihop relaying is advocated using relaying stations (RS). Considerable complexity reduction is expected at the relay station compared to legacy IEEE 802.16 BSs. Cooperative transmission and relaying is one of the important features provided by the IEEE 802.16j standard.

The standard defines two types of relays: fixed and nomadic; these are expected to provide expandable connectivity in buildings and for special events. The nomadic relay will be carried by mobile vehicles, such as a buses, cars or trains.

The initial version of the IEEE 802.16 standard is expected at the end of 2008, addressing routing, resource allocation, security and handover issues. Multihop relaying will be central to the upcoming IEEE 802.16m standard to achieve the expected throughput, i.e. 1Gbit/s for nomadic and 100 Mbit/s for mobile users.

Relaying in Wireless Local Area Networks (WLAN, IEEE 802.11s). IEEE 802.11s aims at defining an extended service set (ESS) multihop mesh networking as an extension of the IEEE 802.11 MAC. The

Figure 3. Cooperative wireless community networks in 4G

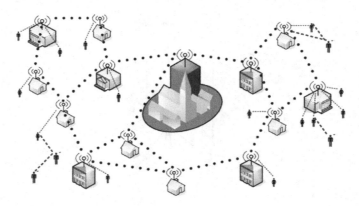

objective is to define architecture and protocols that enable broadcast/multicast and unicast data transmission and delivery modes over multihop mesh topologies, as well as facilitating auto-configuration and radio-aware routing.

Wireless community networks (commercial, public and non-profit), as shown in Figure 3, are another example of cooperation at the network level.

Heterogeneous Cooperative Architectures

In heterogeneous MCNs, mobile devices need to be multi-mode supporting multiple air interfaces (cellular, Bluetooth, IEEE 802.11, IEEE 802.16, etc.) and different data rates. This, however, is an extremely complex task and leads to numerous challenges at all layers of the protocol stack.

Factors that influence the design of such architecture include multi-interface MSs, transmission power and co-channel interference management, topology and routing, mobility and handoff, load balance, interoperability, and QoS provisioning. However, the network layer is the most challenging since MSs can have various physical and MAC layer protocols that need to be considered in an integrated routing process. The selection of the end-to-end route for any connection may be based on the user's service level agreement (SLA) and depends on several metrics (number of hops, delay, throughput, signal strength, etc.). Furthermore, the network layer has to handle horizontal handoffs between BSs/APs of the same technology (cellular IP) and vertical handoffs between different technologies (Mobile IP) in a seamless manner.

In such environments, different types of connections can be established between any two MSs. When we consider that MSs could have two interfaces (e.g. WLAN/cellular), three different heterogeneous scenarios are possible as per Figure 4:

1. Source *A* uses the WLAN interface to connect to *B*, which can establish a connection to destination *C* through a cellular BS in infrastructure mode.
2. Source *B* and destination *D* use cellular and WLAN interfaces, respectively, and the corresponding BS and AP are connected through the CN.

Figure 4. Connection alternatives between two dual-mode MSs in heterogeneous MCNs

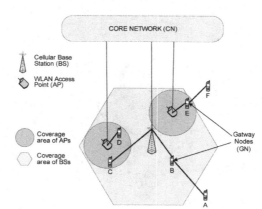

3. Source *A* uses its WLAN interface to connect to *B* that is connected to the BS, and this BS is connected through the CN to the AP that provides connectivity to the destination terminal *F* through *E*.

To facilitate communication, a number of architectures and hybrid routing protocols have been proposed in the literature; see, e.g., UCAN (Luo, Ramjee, Sinha, Li & Lu, 2003) or iCAR (Hu, Qiao, De & Tonguz, 2001). A detailed comparison of these integrated architectures is provided in (Cavalcanti, Cordeiro, Agrawal, Xie & Kumar, 2005).

Standardization

Ubiquitous cooperative protocols will likely find their way into standards and deployment with the advent of 4ᵗʰ generation (4G) systems. 4G is likely going to be composed of a heterogeneous plethora of seamlessly interconnected technologies (Fitzek & Katz, 2006). However, whilst cooperation at various layers between different systems has been in part discussed above, there are no finalized state-of-the-art standards available. Work only commenced; see, e.g., recent efforts of IEEE 802.16j, IEEE 802.21 and IEEE TG 802.15.5 as well as IEEE P1900 (www.ieeep1900.org) on reconfigurable networks facilitating cooperation.

CHALLENGES AND FUTURE TRENDS

Helping out other users in a cooperative fashion has its price – mainly, many unsolved problems in the area of routing, mobility management, authentication, incentive schemes and thus business modeling still prevail. In this section, we will dwell on their challenges and future trends.

Cooperative Relaying & Routing

Routing in cooperative MCNs utilizes additional resources available at the relays (which can be the user terminals). The objective is clearly to increase network radio coverage and optimize the utilization of shared network resources, i.e. it shifts a part of the processing and traffic load from the BSs to the relays. Therefore, routing needs to discover the integrated topology and find the best possible route.

Routing in cooperative MCNs may be simplified when the centralized part of the network has some control on the forwarding operations and is hence able to enforce cooperation policies. The centralized control in MCNs enhances the scalability of the entire system and dramatically improves self-x abilities. However, all these advantages come with signaling overheads; therefore, the trade-off between the overhead and route optimality needs to be considered.

A quantification of the increased network coverage by means of cooperative routing techniques remains an open question. Furthermore, the effect of egoistic relays on the limitation of the multihop route length between source and destination has not yet been considered.

Within a traditional network, networking operations are under the control of the operator. This is no longer the case with MCNs. Relay user nodes can easily disrupt forwarding operations for numerous reasons, such as:

- selfishness;
- temporary resource constraints;
- malicious purpose (intentional packet drops);
- mobility (radio link breaks).

The more hops a node is located from the BS/AP, the greater the probability of experiencing such disruptions. This gives an upper limit on the route length and also the relay infrastructure size, above which the proportion of data packets correctly received at the destination falls under an acceptable performance level.

We believe that in dynamic environments, such as MCNs, a route metric based solely on hop-count is not sufficient to maintain a good multihop connectivity between nodes and the BS/AP. To aid routing, there is a need to evaluate intermediary node behaviors and link qualities along the path, in order to quantify the expected cooperation level of the routes when nodes are located several hops away from the BS/AP. Several QoS metrics that reflect link quality can be used, such as ETX/WCETT (Draves, Padhye & Zill, 2004) or MIC (Yang, Wang & Kravets, 2005).

Mobility and Location Management

One of the important issues in providing ubiquitous communications is mobility management. In general, mobility management is a control plane that enables the network to locate a MS for call delivery and to maintain connectivity as the MS is moving to new service regions (mobility management also supports service discovery and vertical handoff; Xie, Kumar, Cavalcanti & Agrawal, 2006). The problem is acute in MCNs as the mobility of nodes affects the connectivity of not only the node that is moving but also of all other nodes maintaining links via it. The main objectives of mobility management architectures for homogeneous and heterogeneous wireless environments are to reduce the intra- and inter domain signaling load and handoff delay. Mobility support for heterogeneous networks has been addressed from different levels of the TCP/IP protocol stack. These include network level solutions, hiding the underlying wireless access technologies; link layer solutions, providing mobility-related features in the underlying radio systems; and cross-layer solutions for handoff management. Two general components of mobility management are handoff management and location management.

Several mobility management schemes attempt to reduce the packet loss incurred due to the node mobility between service regions. These include tunnel-based micro-mobility schemes like Mobile IP Regional registration (MIP-RR), Hierarchical Mobile-IP (HMIP), IDMP, etc., and routing-based micro-mobility schemes like Cellular IP (CIP), Hand-off Aware Wireless Access Internet Infrastructure (Hawaii), etc. (Akyildiz, Xie & Mohanty, 2004). While the micro-mobility solutions were proposed at network level for mobility of MSs between subnets of same domains for enabling transparent mobility for higher layers, the macro-mobility handled by Mobile IP tackle the mobility of users between different network domains. Mobile IP provides an effective solution for macro and global mobility management across homogeneous and heterogeneous systems.

Several cross-layer mobility management solutions were proposed in the context of heterogeneous wireless networks, and particularly MCNs with application to handoff management techniques. These include methods for low-latency MIP handoff and low-latency WLAN handoff. Seamless handoff techniques, including S-MIP, were also proposed for intelligent handoff management which provide a unique method of combining location tracking schemes and hierarchical MIP handoff schemes. The IEEE 802.21 standardization group is developing a MIH (media independent handover)-type cross-layer (layer-2 and layer-3) to enable mobility across heterogeneous networks by providing link layer intelligence and other related network information to upper layers to optimize handovers between heterogeneous media.

To transition MCN from a vision to reality requires the presence of efficient mobility management techniques which provides seamless connectivity with near-zero latency for the best possible user experience.

Security and Cooperation

As MCNs continue to grow and as their access is available for any wirelessly enabled device, cooperation between nodes should be guaranteed in order to assure the correct service provision. Hence, it should be ensured that only authorized users are granted network's access. We mainly notice two types of attacks in MCN environments: i) external attacks, where the attackers do not participate in the network, however they could carry out some attacks and malicious acts impacting the network and services performance, and ii) internal attacks, where the attackers participate in the network and have legitimate service access, however they penalize the network performance through malicious and non cooperative acts.

Prevention Against External Attacks. Indeed, authentication and access control are important counter-attack measures in MCN deployments, allowing only authorized clients to be connected and preventing external attackers to sneak into the network disrupting the normal cooperative operation or service provisioning. A simple solution to carry out authentication in MCNs is to employ an authentication key shared by all nodes in the network. Although this mechanism is considered as a *plug and play* solution and does not require the communication with centralized network entities, it is limited to closed scenarios of small number of participants in limited environments and belonging to the same provider. In addition, this shared secret authentication has two main pitfalls. Firstly, an attacker only needs to compromise one node to break the security of the system. Secondly, mobile nodes do not usually belong to the same community, which leads to a difficulty in installing/pre-configuring the shared keys. A challenge for wide scale commercial deployment of MCNs is to design authentication mechanisms for the more vulnerable yet more resource-constrained environment of MCNs. In most commercial deployments of WLANs, authentication and access control is mostly provided through employing IEEE 802.11i (IEEE 802.11i, 2004) authentication in which a centralized server is in place. In the context of MCNs, the challenge for applying the 802.11i approach mainly concerns the multihop characteristics and the hybrid infrastructure-based/less architecture. Hence, the 802.11i authentication model should be adapted to such environment through mainly considering two issues: i) introducing distributed authentication mechanisms, and ii) ensuring cooperation between nodes to support the hybrid architecture.

A possible approach for distributed authentication is the continuous discovery and mutual authentication between neighbors, whether they are mobile clients or fixed APs/BSs. Nevertheless, if mobile nodes move back to the range of previous authenticated neighbors or fixed nodes, it is necessary to perform re-authentication in order to prevent an adversary from taking advantage of the gap between the last security association and the current security association with the old neighbor. An approach adapting the 802.11i authentication model to multihop communication environments is presented by Moustafa, Bourdon & Gourhant (2006), proposing an extended forwarding capability to 802.11i and allowing mobile node authentication with the authentication server in a multihop fashion. The notion of friend nodes is introduced allowing each mobile node to initiate the authentication process through a selected node in its proximity, which plays the role of an auxiliary authenticator and forwards securely the authentication requests to the authentication server. Friend nodes are chosen to be trusted and cooperating nodes. This approach is suitable to the hybrid infrastructure-based/less architecture in MCNs, allowing mobile nodes beyond the APs/BSs coverage zone to get authenticated in a cooperative manner, through communicating with the authentication server at the infrastructure while passing by cooperative nodes (friend nodes). In addition, this approach allows authentication keys storage among intermediate (friend) nodes which optimizes the re-authentication process in case of roaming. Another possibility to facilitate multihop authentication is to employ a Protocol for carrying Authentication and Network Access or PANA (Forsberg, Ohba, Patil, Tschofenig, & Yegin, 2007). PANA allows the encapsulation of the used authentication protocol messages and their routing to the authentication server. The advantage of PANA mainly lies in its independence of the wireless media, and thus it is suitable for future cooperative MCNs having heterogeneous deployments and operator co-existence. However, PANA necessitates the existence of a routing infrastructure, which is a technical challenge for MCNs as previously outlined.

Prevention Against Internal Attacks. Although authentication and access control can reinforce cooperation through prevention against external attackers, internal attackers could always exist even in the presence of effective authentication and access control mechanisms. Internal attackers are nodes that are authenticated and authorized to participate in the network; however, they can be harmful nodes causing network and service performance degradation mainly through non cooperative behaviors (selfishness, greediness, and Denial-of-Services or DoS). Hence, there is a need for complementary mechanisms to authentication and access control. Nodes may behave selfishly by not forwarding packets for others in order to save power, bandwidth or just because of security and privacy concerns. Watchdog (Marti, Giuli, Lai & Baker, 2000), CONFIDANT (Buchegger & le Boudec, 2002) and Catch (Mahajan, Rodrig, Wetherall & Zahorjan, 2005) are three approaches developed to detect selfishness and enforce distributed cooperation and are suitable for MCNs multihop environment. Watchdog is based on monitoring neighbors to identify a misbehaving node that does not cooperate during data transmission. However, CONFIDANT and Catch incorporate an additional punishment mechanism making misbehavior unattractive through isolating misbehaving nodes. On the other hand, nodes may behave greedily in consuming channel and bandwidth for its own benefits at the expense of the other users. The DOMINO mechanism (Raya, Hubaux & Aad, 2004) solves the greedy sender problem in 802.11 WLANs with a possible extension to multihop wireless networks and MCNs. Internal attackers may also cause DoS through either faked messages injection or messages replay. DoS is a challenging problem greatly impacting cooperation, however it could be partially resolved through effective authentication of messages and messages' sources.

Business Models

MCNs require special accounting mechanisms and tailored billing systems, where appropriate business models should exist while considering the benefits of mobile users, network operators, and service providers. Business models should take into account the cooperation between different operators and service providers, which is a liable fact in future MCN deployment. Consequently, inter-domain accounting is required to assure services' access continuity. Billing in this context is expected to use gathered accounting information for each client, provided that a trust relationship exists between different operators and service providers.

Moreover, business models should take into account the cooperation between MCNs nodes. Mobile nodes should obtain credits for services' relaying, depending on their participation in the communication process, where nodes can be compensated (rewarded) according to their participation. In this context, payment mechanisms should be proposed for encouraging the cooperation between mobile nodes, where a sort of remuneration can be done for each participant according to his contribution. Another alternative is to let each node pay *cash traffic* for its own transfer and in turn gain *cash traffic* in order to forward packets for other nodes. The notion of *cash* could also be real money.

As a matter of opening a new business opportunity, business models should be rentable for telecom operators on one hand and affordable for wireless users on the other hand aiming to promoting services and attracting clients. Consequently, *paying access* models permitting different users privileges, according to the types of subscriptions, are the most appropriate ones. Privileged access to services in this context could include: secure communication, quality of connectivity, and different access rights to these services. In order to ensure the proper operation of the *paying access* models, the fulfillment of authentication and authorization to services' access is necessary. There are a number of possibilities to realize the *paying access* models; for example:

- Clients can go through the payment process each time they access the services. An o*n-use package* model can be applied in this case, through using pre-paid cards for instance. The advantage in this model is that payments are made according to the services' utilization. However, there is a need for complementary mechanisms allowing efficient management for clients' accounting information.
- A *pure package* (pay before use) business model, which could be associated to the clients' Internet subscription or telephone subscription and where the billing is monthly fixed whether the client uses the service or not. This model offers different privileges for clients according to their type of subscription, allowing the network operator/service provider to master the clients' access.
- A virtual operator model, in which the clients' access to different services is assured by service providers who do not own the deployed access network. On the other hand, the service providers are responsible of managing accounting and billing of mobile clients. This model permits the

integration between different service providers and the operators owning the access networks and can simplify managing the accounting information of mobile clients.

We believe that at least one of above models will yield non-negligible gains for operators and service providers.

CONCLUSION

As we had outlined in Dohler, Meddour, Senouci & Saadani (2008, p. 14), *"cooperation is not a natural characteristics attributed to human beings. The typical human horizon is focused on short-term gains, which might be due to our instinct-driven subconscious occupying a grander importance than we dare to admit (Gray, 2002). Cooperating with other individuals or entities, however, usually means that short-term losses may translate into long-term gains. Any cooperative technology depending solely on human decisions is hence a priori doomed to fail; history has shown this on numerous occasions. By contrast, if machines only have access to some decision making engines, cooperative schemes become viable communication techniques and are likely to occupy an important place in the technological landscape of the 21st century."*

Above statement is corroborated by numerous previous attempts to commercialize networks based on cooperative techniques. The most prominent example is ad hoc networks, which have already been researched for some decades without having produced a single viable commercial product. The main reason in our opinion is twofold (Dohler, Meddour, Senouci & Saadani, 2008): First, the design degrees of freedom have turned out to be too large to reach commercialization; i.e., a psychological barrier prevailed at the manufacturer and service provider side, which prevented the deployment of such technology that had not even been fully mastered for much simpler cellular systems. Second, the data relaying process required users to give away battery power and bandwidth, and possibly jeopardize the security of their own data, with no obvious instantaneous gains; i.e., a psychological barrier prevailed at the user side, which has turned out to be hardest to break. Other examples of cooperative relaying technologies to have failed are the UMTS Concept Group Epsilon, which proposed ODMA as a potential 3rd generation (3G) candidate solution (3GPP, 1999), and Ricochet® (www.ricochet.net), a US company which was well ahead of its time by rolling-out a broadband wireless network throughout major US cities more than 10 years ago without this technology ever having really taken off.

Very few technologies, no matter how compelling, were successful – mainly because they appeared at the right time, at the right place, at the right pace, and supported by the right team.

For offered services, where the end-user had the last word, many failed technologies were simply either far ahead of time (i.e. the user was psychologically not prepared to accept the new technology and got around it by using another – possibly worse – technology) or lagged behind time (i.e. the user was already saturated with similar technologies and saw no reason – particularly for incremental gains – to change technologies).

In our opinion, cooperative techniques will likely survive in scenarios which are independent of users but only depend on machines or operator-programmed decision engines. Examples of the former are machine-to-machine applications, such as wireless sensor networks, where cooperation benefits data reliability, energy savings, network longevity, etc (Dohler, Gkelias & Aghvami, 2006). An example of the latter are the architectures exposed in this chapter, i.e. the cooperative multihop cellular network architecture.

These cooperative MCNs are clearly emerging as a promising new technology, benefiting from both cellular and ad hoc networks technologies whilst alleviating some critical problems in these networks. We have shown that MCNs seem attractive in opening new business opportunity for network operators and service providers, enabling commercial service provisioning with broad coverage on one hand and seamless mobility for mobile clients with improved overall QoS on the other hand. We exposed the motivation and importance of deploying cooperative MCNs, highlighting some appropriate deployment architectures. In addition, we reviewed some important challenges for a real-world deployment of such networks. These challenges are interesting for operators and providers, enabling the feasibility study of cooperative deployment taking into consideration the design-cost principle.

We have shown that homogeneous and heterogeneous solutions MCN architectures prevail. These are currently being standardized in various standardization groups of the IEEE. We have explained why, in contrast to pure ad hoc routing protocols, routing in MCNs may be simplified due to the centralized part of the network having some control on routing operations. This enhances network scalability and significantly improves the self-organization abilities of the system.

We presented mobility and smart roaming solutions, which are applicable to MCNs. It has been emphasized that cross-layer solutions are vital in ensuring reliable mobility solutions. We have also alluded to the gamut of existing routing metrics, such as expected transmission count, expected transmission time, etc, and discussed their merits as well as short-comings.

Security and cooperation are two important issues that need to be resolved. Authentication and access control mechanisms should be in place, taking into consideration the dynamic and not fully centralized nature of cooperative networks. It is important to be mindful of the authentication overhead as wireless mobile users are often thin-clients with limited resources. Also, unacceptable authentication delay can impact the services' continuity. Complementary mechanisms to authentication and access control should exist as well, in order to assure the cooperative behavior and prevent against internal attackers.

From a commercial deployment perspective, business models should take into account the fact that cooperative MCNs can be managed by more than one operator/provider and hence allow for such co-existence. Special accounting mechanisms and tailored billing systems are needed. Business models should have mutual benefit in the sense of being rentable for operators and attractive for mobile clients (nodes), where they can integrate some mechanisms inciting the nodes cooperation.

The momentum of research into cooperative technologies in the context of incumbent and emerging cellular and local area networks is very large and it commences finding its input into various standardization bodies. It is hence safe to assume that this technology – in one way or another – will be part of our future wireless arena.

REFERENCES

Adam, H., Bettstetter, C., & Senouci, S. M. (2008). (Manuscript submitted for publication). Adaptive relay selection in cooperative wireless networks. *IEEE PIMRC*.

Ahmed, S., Ibars, C., del Coso, A., & Mohammed, A. (2007). Performance of multichannel MAC incorporating opportunistic cooperative diversity. *IEEE VTC-Spring, 2007*, 1297–1301.

Akyildiz, I. F., Xie, J., & Mohanty, S. (2004). A survey of mobility management in next-generation all-IP-based wireless systems. *IEEE Wireless Communications, 11*(4), 16–28. doi:10.1109/MWC.2004.1325888

Azarian, K., el Gamal, H., & Schniter, P. (2005). On the achievable diversity-multiplexing tradeoff in half-duplex cooperative channels. *IEEE Transactions on Information Theory, 51*(12), 4152–4172. doi:10.1109/TIT.2005.858920

Beres, E., & Adve, R. (2008). Selection cooperation in multisource cooperative networks. *IEEE Transactions on Wireless Communications, 7*(1), 118–127. doi:10.1109/TWC.2008.060184

Biswas, S., & Morris, R. (2005). ExOR: Opportunistic multihop routing for wireless networks. *ACM SIGCOMM, 25*(4), 133–144. doi:10.1145/1090191.1080108

Bletsas, A., Khisti, A., Reed, D. P., & Lippman, A. (2006). A simple cooperative diversity method based on network path selection. *IEEE JSAC, 24*(9), 659–672.

Buchegger, S., & le Boudec, J. Y. (2002). Performance analysis of the CONFIDANT protocol. *ACM MobiHoc* (pp. 226-236).

Buttyan, L., & Hubaux, J. (2003). Stimulating cooperation in self-organizing mobile ad hoc networks. *ACM/Kluwer Mobile Networks and Applications (MONET), 8*(5), 579-592.

Cavalcanti, D., Cordeiro, C. M., Agrawal, D. P., Xie, B., & Kumar, A. (2005). Issues in integrating cellular networks, WLANs, and MANETs: A futuristic heterogeneous wireless network. *IEEE Wireless Communications Magazine . Special Issue on Toward Seamless Internetworking of Wireless LAN and Cellular Networks, 12*(3), 30–34.

Chou, C. T., Yang, J., & Wang, D. (2007). Cooperative MAC protocol with automatic relay selection in distributed wireless networks. *IEEE PerCom Workshops* (pp. 526-531).

Conti, M., Gregori, E., & Maselli, G. (2004). Cooperation issues in mobile ad hoc networks. *ICDCSW* (pp. 803-808).

Cover, T., & el Gamal, A. (1979). Capacity theorems for the relay channel. *IEEE Transactions on Information Theory, IT-25*(5), 572–584. doi:10.1109/TIT.1979.1056084

Djeumou, B., Lasaulce, S., & Klein, A. G. (2007). Practical quantize-and-forward schemes for the frequency division relay channel. *EURASIP Journal on Wireless Communications and*

Dohler, M. (1999-2002). *VAA for hot-spots with applied STC.* ([III and IV). M-VCE, UK.]. *Internal Reports, I,* II.

Dohler, M. (2003). *Virtual antenna arrays.* Unpublished doctoral dissertation, King's College London, UK.

Dohler, M., & Aghvami, A. H. (2007). *A crash-course on cooperative wireless networks.* Tutorial presented at IEEE ICC, Glasgow, Scotland, UK.

Dohler, M., Gkelias, A., & Aghvami, A. H. (2006). Capacity of distributed PHY-layer sensor networks. *IEEE Transactions on Vehicular Technology, 55*(2), 622–639. doi:10.1109/TVT.2005.863470

Dohler, M., Meddour, D. E., Senouci, S. M., & Saadani, A. (2008). Cooperation in 4G-hype or ripe? *IEEE Technology and Society Magazine, 27*(1), 13–17. doi:10.1109/MTS.2008.918035

Dohler, M., Said, F., Ghorashi, S. A., & Aghvami, A. H. (2001). Improvements in or relating to electronic data communication systems. *Patent Publication No. WO 03/003672*, priority date 28 June 2001.

Draves, R., Padhye, J., & Zill, B. (2004). Routing in multiradio, multihop wireless mesh networks. *ACM MobiCom* (pp. 114-128).

Esseling, N., Walke, B., & Pabst, R. (2004). Performance evaluation of a fixed relay concept for next generation wireless systems. In R. Ganesh, S. Kota, K. Pahlavan & R. Agusti (Eds), *Emerging location aware broadband wireless ad hoc networks*. Boston, MA: Kluwer Academic Publishers.

Fitzek, F. H. P., & Katz, M. D. (2006). *Cooperation in wireless networks: Principles and applications.* New York: Springer-Verlag New York, Inc.

Forsberg, D., Ohba, O., Patil, B., Tschofenig, H., & Yegin, A. (2007). *Protocol for carrying authentication and network access (PANA)*. IETF, draft-ietf-pana-pana-18, work in progress.

3GPP. (1999). Opportunity driven multiple access. *3G TR 25.924 V1.0.0 (1999-12) - Technical Specification Group Radio Access Network.*

Gray, J. (2002). *Straw dogs: Thoughts on humans and other animals.* London, UK: Granta.

Gunasekaran, V., & Harmantzis, F. (2005). Affordable infrastructure for deploying WiMAX systems: Mesh vs. non mesh. *IEEE VTC-Spring, 5*, 2979–2983.

Gupta, P., & Kumar, P. R. (2000). The capacity of wireless networks. *IEEE Transactions on Information Theory, 46*(2), 388–404. doi:10.1109/18.825799

Gupta, P., & Kumar, P. R. (2003). Towards an information theory of large networks: An achievable rate region. *IEEE Transactions on Information Theory, 49*(8), 1877–1894. doi:10.1109/TIT.2003.814480

Hu, W., Qiao, C., De, S., & Tonguz, O. (2001). Integrated cellular and ad hoc relaying systems: iCAR. *IEEE JSAC, 19*(10), 2105–2115.

IEEE 802.11i. (2004). Medium access control security enhancements. *Standardization Document IEEE Std 802.11i.*

Jakllari, G., Krishnamurthy, S. V., Faloutsos, M., Krishnamurthy, P. V., & Ercetin, O. (2006). A framework for distributed spatio-temporal communications in mobile ad hoc networks. *IEEE INFOCOM* (pp. 1-13).

Jerbi, M., Senouci, S. M., Ghamri, Y., & Beylot, A. L. (2008). (Manuscript submitted for publication). Geo-localized virtual infrastructure for urban vehicular networks. *IEEE PIMRC.*

Kim, J., & Bohacek, S. (2005). Selection metrics for multihop cooperative relaying. *Annual Mediterranean Ad Hoc Networking Workshop.*

Konstantinou, K., Kang, S., & Tzaras, C. (2007). A measurement-based model for mobile-to-mobile UMTS links. *IEEE VTC-Spring, 2007*, 529–533.

Kyasanur, P., & Vaidya, N. H. (2003). Detection and handling of MAC layer misbehavior in wireless networks. *International Conference DSN'03* (pp. 173-182).

Laneman, J. N., Tse, D. N. C., & Wornell, G. W. (2004). Cooperative diversity in wireless networks: Efficient protocols and outage behavior. *IEEE Transactions on Information Theory, 50*(12), 3062–3080. doi:10.1109/TIT.2004.838089

Laneman, J. N., & Wornell, G. W. (2000). Energy-efficient antenna sharing and relaying for wireless networks. *IEEE WCNC, 1,* 7–12.

Laneman, J. N., & Wornell, G. W. (2003). Distributed space-time-coded protocols for exploiting cooperative diversity in wireless networks. *IEEE Transactions on Information Theory, 49*(10), 2415–2425. doi:10.1109/TIT.2003.817829

Li, Y., & Xia, X. G. (2005). Full diversity distributed space-time trellis codes for asynchronous cooperative communications. *IEEE ISIT, 2005,* 911–915.

Librino, F., Levorato, M., & Zorzi, M. (2007). Distributed cooperative routing and hybrid ARQ in MIMO-BLAST ad hoc networks. *IEEE Globecom, 2007,* 657–662.

Lin, Y.-D., & Hsu, Y.-C. (2000). Multihop cellular: A new architecture for wireless communications. *IEEE Infocom, 3,* 1273–1282.

Liu, P., Tao, Z., Narayanan, S., Korakis, T., & Panwar, S. S. (2007). CoopMAC: A cooperative MAC for wireless LANs. *IEEE JSAC, 25*(2), 340–354.

Luo, H., Ramjee, R., Sinha, P., Li, L., & Lu, S. (2003). UCAN: A unified cellular and ad hoc network architecture. *ACM Mobicom* (pp. 353-367).

Mahajan, R., Rodrig, M., Wetherall, D., & Zahorjan, J. (2005). Sustaining cooperation in multihop wireless networks. *ACM NSDI* (pp. 231-244).

Marti, S., Giuli, T. J., Lai, K., & Baker, M. (2000). Mitigating routing misbehavior in mobile ad hoc networks. *ACM MobiCom* (pp. 255-265).

Michalopoulos, D. S., & Karagiannidis, G. K. (2008). PHY-layer fairness in amplify and forward cooperative diversity systems. *IEEE Transactions on Wireless Communications, 7*(3), 1073–1082. doi:10.1109/TWC.2008.060825

Michiardi, P., & Molva, R. (2002). CORE: A collaborative reputation mechanism to enforce node cooperation in mobile ad hoc networks. *Communication and Multimedia Security Conference* (pp. 107-121).

Moustafa, H., Bourdon, G., & Gourhant, Y. (2006). Authentication, authorization, and accounting (AAA) in hybrid ad hoc hotspots' environments. *ACM WMASH* (pp. 37-46).

Networking, 4, Article ID 20258.

Ozgur, A., Leveque, O., & Tse, D. N. C. (2007). Hierarchical cooperation achieves optimal capacity scaling in ad hoc networks. *IEEE Transactions on Information Theory, 53*(10), 3549–3572. doi:10.1109/TIT.2007.905002

Patel, C. S., Stüber, G. L., & Pratt, T. G. (2006). Statistic properties of amplify and forward relay fading channels. *IEEE Transactions on Vehicular Technology*, *55*(1), 1–9. doi:10.1109/TVT.2005.861170

Raya, M., Hubaux, J. P., & Aad, I. (2004). Domino: A system to detect greedy behavior in IEEE 802.11 hotspots. *ACM MobiSys* (pp. 84-97).

Sendonaris, A., Erkip, E., & Aazhang, B. (1998). Increasing uplink capacity via user cooperation diversity. In *Proc. IEEE ISIT* (p. 196).

Timus, B. (2006). *Deployment cost efficiency in broadband delivery with fixed wireless relays*. Licentiate thesis, KTH, Sweden.

van der Meulen, E. (1971). Three-terminal communication channels. *Advances in Applied Probability*, *3*, 120–154. doi:10.2307/1426331

Xie, B., Kumar, A., Cavalcanti, D., & Agrawal, D. P. (2006). Multihop cellular IP: A new approach to heterogeneous wireless networks. *International Journal of Pervasive Computing and Communications*, *2*(4), 370–383.

Yang, Y., Wang, J., & Kravets, R. (2005). *Interference-aware load balancing for multihop wireless networks* (Tech. Rep.). University of Illinois at Urbana-Champaign.

Zheng, L., & Tse, D. N. C. (2003). Diversity and multiplexing: A fundamental tradeoff in multiple antenna channels. *IEEE Transactions on Information Theory*, *49*(5), 1073–1096. doi:10.1109/TIT.2003.810646

Zhong, S., Chen, J., & Yang, Y. R. (2003). Sprite: A simple, cheat-proof, credit-based system for mobile ad-hoc networks. *IEEE Infocom*, *3*, 1987–1997.

GLOSSARY

3G: 3rd Generation (Mobile System)
3GPP: 3rd Generation Partnership Project
4G: 4th Generation (Mobile System)
APL: Application (Layer)
BS: Base Station
BSPL: Best-Select Protocol
CAPEX: Capital Expenditure
CDMA: Code Division Multiple Access (Protocol)
CSMA: Carrier Sensing Multiple Access (Protocol)
DCF: Distributed Coordination Function
FDD: Frequency Division Duplex
FDMA: Frequency Division Multiple Access (Protocol)
GSM: Global System for Mobile Communications
ITS: Intelligent Transportation System
ITU: International Telecommunications Union

MAC: Medium Access Control
MHz: Mega Hertz
MIMO: Multiple-Input-Multiple-Output
MT: Mobile Terminal
M-VCE: Mobile Virtual Centre of Excellence
NTW: Network (Layer)
ODMA: Opportunity Driven Multiple Access
OFDMA: Orthogonal Frequency Division Multiple Access (Protocol)
OPEX: Operational Expenditure
OSI: Open Systems Interconnection (Reference Model)
PDU: Packet Data Unit
PHY: Physical (Layer)
QoS: Quality-of-Service
R&D: Research and Development
RRM: Radio Resource Management
SISO: Single-Input-Single-Output
SINR: Signal-to-Interference-and-Noise Ratio
SNR: Signal-to-Noise Ratio
TDD: Time Division Duplex
TDMA: Time Division Multiple Access (Protocol)
THz: Terra Hertz
TV: Television
UMA: Unlicensed Mobile Access
UMTS: Universal Mobile Telecommunications System
WLAN: Wireless Local Area Network
WRC: World Radio Conference
WSN: Wireless Sensor Networks

Compilation of References

3GPP. (1999). Opportunity driven multiple access. *3G TR 25.924 V1.0.0 (1999-12) - Technical Specification Group Radio Access Network*.

Aaron, A., & Girod, B. (2002). Compression with side information using turbo codes. *Proceedings of IEEE Data Compression Conference*, Snowbird, UT (pp. 252–261).

Abramowitz, M., & Stegun, I. A. (1970). *Handbook of mathematical functions with formulas, graphs, and mathematical tables*, 9th ed. New York: Dover.

Abramowitz, M., & Stegun, I. A. (1972). *Handbook of mathematical runctions*. New York: Dover Publications, Inc.

Abualhaol, I., & Matalgah, M. (2008). Subchannel-division adaptive resource allocation technique for cooperative relay-based MIMO-OFDM wireless communication systems. In *Proceedings IEEE Wireless Communications and Networking Conference*, Las Vegas, NV (pp. 1002-1007).

Adam, H., Bettstetter, C., & Senouci, S. M. (2008). (Manuscript submitted for publication). Adaptive relay selection in cooperative wireless networks. *IEEE PIMRC*.

Agrawal, D., Tarokh, V., Naguib, A., & Seshadri, N. (1998). Space-time coded OFDM for high data-rate wireless communication over wideband channels. In *Proceedings of IEEE Vehicular Technology Conference*, Ottawa, Canada (pp. 2232-2236).

Ahlswede, R. (1971). Multiway communication channels. In *Proceedings of IEEE International Symposium on Information Theory, Tsahkadsor, Armenian S.S.R.*, (pp. 23–52).

Ahlswede, R., & Han, T. (1983). On source coding with side information via a multiple-access channel and related problems in multi-user information theory. *IEEE Transactions on Information Theory, 29*, 396–412. doi:10.1109/TIT.1983.1056669

Ahlswede, R., Cai, N., Li, S.-Y. R., & Yeung, R. W. (2000, July). Network information flow. *IEEE Trans. Info. Theory*, 1204-1216.

Ahmed, M., & Vorobyov, S. (2008). Performance characteristics of collaborative beamforming for wireless sensor networks with Gaussian distributed sensor nodes. In *Proceedings of 2008 IEEE International Conference of Acoustic Speech Signal Processing*, Las Vegas, NV (pp. 3249-3252).

Ahmed, N., Khojastepour, M. A., Sabharwal, A., & Aazhang, B. (2006). Outage minimization with limited feedback for the fading relay channel. *IEEE Transactions on Communications, 54*(4), 659–669. doi:10.1109/TCOMM.2006.873074

Ahmed, N., Khojastepour, M., & Aazhang, B. (2004a). Outage minimization and optimal power control for the fading relay channel. In *IEEE Information Theory Workshop*, Houston, TX (pp. 458–462).

Ahmed, N., Khojastepour, M., Sabharwal, A., & Aazhang, B. (2004b). On power control with finite rate feedback for cooperative relay networks. In *International Symposium on Information Theory and Its Applications*.

Ahmed, S., Ibars, C., del Coso, A., & Mohammed, A. (2007). Performance of multichannel MAC incorporating opportunistic cooperative diversity. *IEEE VTC-Spring, 2007*, 1297–1301.

Ahuja, R. K., Magnanti, T. L., & Orlin, J. B. (1993). *Network flows: Theory, algorithms, and applications.* Prentice Hall.

Airy, M., Forenza, A., Heath, R. W. Jr, & Shakkottai, S. (2004). Practical Costa precoding for the multiple antenna broadcast channel. In . *Proceedings of IEEE Global Communications Conference, 6,* 3942–3946.

Aktas, E., Evans, J., & Hanly, S. (2008). Distributed decoding in a cellular multiple-access channel. *IEEE Transactions on Wireless Communications, 7*(1), 241–250. doi:10.1109/TWC.2008.060469

Akyildiz, I. F., Su, W., Sankarasubramaniam, Y., & Cayirci, E. (2002). Wireless sensor networks: A survey. *Elsevier Computer Networks, 38*(4), 393–422. doi:10.1016/S1389-1286(01)00302-4

Akyildiz, I. F., Su, W., Sankarasubramaniam, Y., & Cayirci, E. (2002). A survey on sensor networks. *IEEE Communications Magazine, 40*(8), 102–114. doi:10.1109/MCOM.2002.1024422

Akyildiz, I. F., Xie, J., & Mohanty, S. (2004). A survey of mobility management in next-generation all-IP-based wireless systems. *IEEE Wireless Communications, 11*(4), 16–28. doi:10.1109/MWC.2004.1325888

Alamouti, S. (1998). A simple transmit diversity technique for wireless communications. *IEEE Journal on Selected Areas in Communications, 16*(8), 1451–1458. doi:10.1109/49.730453

Al-Dhahir, N. (2001). Single-carrier frequency-domain equalization for space-time block-coded transmissions over frequency-selective fading channels. *IEEE Communications Letters, 5*(7), 304–306. doi:10.1109/4234.935750

Alexander, P., & Grant, A. (2000). Iterative channel and information sequence estimation in CDMA. In *IEEE Sixth International Symposium on Spread Spectrum Techniques and Applications,* (Vol. 2, pp. 593–597).

Al-Karaki, J. N., & Kamal, A. E. (2004). Routing techniques in wireless sensor networks: A survey. *IEEE Personal Communications, 11,* 6–28.

Alouini, M.-S., Tang, X., & Goldsmith, A. J. (1999). An adaptive modulation scheme for simultaneous voice and data transmission over fading channels. *IEEE Journal on Selected Areas in Communications, 5*(17), 837–850. doi:10.1109/49.768199

Amir, R. (2005). Supermodularity and complementarity in economics: An elementary survey. *Southern Economic Journal, 71*(3), 636–660.

Anderson, B. D. O., & Moore, J. B. (1979). *Optimal filtering.* New Jersey: Prentice Hall.

Andrews, J. (2005). Interference cancellation for cellular systems: A contemporary overview. *IEEE Transactions on Wireless Communications, 12*(2), 19–29. doi:10.1109/MWC.2005.1421925

Anghel, P. A., & Kaveh, M. (2004). Exact symbol error probability of a cooperative network in a Rayleigh-fading environment. *IEEE Transactions on Wireless Communications, 3*(9), 1416–1421. doi:10.1109/TWC.2004.833431

Anghel, P. A., & Kaveh, M. (2006). On the performance of distributed space-time coding systems with one and two nongenerative relays. *IEEE Transactions on Wireless Communications, 5,* 682–692.

Anghel, P., Leus, G., & Kaveh, M. (2003). Multi-user space-time coding in cooperative networks. In . *Proceedings of IEEE International Conference on Acoustic, Speech, and Signal Processing, 4,* 73–76.

Appuswamy, R., Franceschetti, M., & Zeger, K. (2006, July). Optimality of linear codes for broadcast-mode multicast networks. In *Proc. IEEE Intl. Symp. on Info. Theory,* Seattle, WA (pp. 50-53).

Atay Onat, F., Adinoyi, A., Yanikomeroglu, H., Thompson, J. S., & Marsland, I. D. (Accepted for publication). Threshold selection for SNR-based selective digital relaying in cooperative wireless networks. [retrieved from http://www.sce.carleton.ca/faculty/yanikomeroglu/cv/publications.shtml]. *IEEE Transactions on Wireless Communications.*

Atay Onat, F., Fan, Y., Yanikomeroglu, H., & Thompson, J. S. (Accepted for publication). Asymptotic BER analysis of threshold digital relaying schemes in cooperative wireless systems. [retrieved from http://www.sce.carleton.ca/faculty/yanikomeroglu/cv/publications.shtml]. *IEEE Transactions on Wireless Communications.*

Avestimehr, A. S., Diggavi, S. N., & Tse, D. N. C. (2007). A deterministic approach to wireless relay networks. In *Proceedings of Allerton Conference on Communications, Control, and Computing*, Allerton, IL.

Avestimehr, S., Diggavi, S., & Tse, D. (2008). Approximate capacity of Gaussian relay networks. In *Proceedings of IEEE International Symposium on Information Theory*.

Azarian, K., el Gamal, H., & Schniter, P. (2005). On the achievable diversity-multiplexing tradeoff in half-duplex cooperative channels. *IEEE Transactions on Information Theory*, 51(12), 4152–4172. doi:10.1109/TIT.2005.858920

Bacha, M., Evans, J., & Hanly, S. (2006). On the capacity of MIMO cellular networks with macrodiversity. In *7th Australian Communications Theory Workshop Proceedings*, (pp. 105–109).

Bahl, L. R., Cocke, J., Jelinek, F., & Raviv, J. (1974). Optimal decoding of linear codes for minimizing symbol error rates. *IEEE Transactions on Information Theory*, 20(2), 284–287. doi:10.1109/TIT.1974.1055186

Bajcsy, J., & Mitran, P. (2001). Coding for the Slepian-Wolf problem using turbo codes. In *Proceedings of IEEE Global Communications Conference*, San Antonio, TX (pp. 1400–1404).

Baker, G., & Graves-Morris, P. A. (1996). *Padé approximants*. Cambridge University Press.

Bao, X., & Li, J. (2005). Decode-amplify-forward (DAF): A new class of forwarding strategy for wireless relay channels. In *Proceedings of IEEE Workshop on Signal Processing Advances Wireless Communications* (pp. 816–820).

Bao, X., & Li, J. (2005, September). Matching code-on-graph with network-on-graph: Adaptive network coding for wireless relay networks. In *Proc. the 43rd Annual Allerton Conf. Communication, Control, Computing*, Champaign, IL.

Bao, X., & Li, J. (2006, July). A unified channel-network coding treatment for user cooperation in wireless ad-hoc networks. In *Proc. IEEE Int. Symp. Info. Theory*, Seattle, WA.

Bao, X., & Li, J. (2008). Adaptive network coded cooperation (ANCC) for wireless relay networks: Matching code-on-graph with network-on-graph. *IEEE Transactions on Wireless Communications*, 7(2), 574–583. doi:10.1109/TWC.2008.060439

Bao, X., & Li, J. (2008, January). Adaptive network coded cooperation (ANCC) for wireless relay networks: Matching code-on-graph with network-on-graph. *IEEE Transaction on Wireless Communications*, 574-583.

Barbarossa, S., & Scutari, G. (2004). Distributed space-time coding for multi-hop networks. In . *Proceedings of IEEE International Conference on Communications, ICC*, 2004.

Bauch, G. (2003). Space-time block codes versus space-frequency block codes. In *Proceedings of IEEE Vehicular Technology Conference*, Jeju, South Korea (pp. 567-571).

Belfiore, J. C., & Cipriano, A. M. (2006). Space-time coding for non-coherent channels. In H. Bolcskei, G. Gesbert, C. B. Papadias & V. Veen (Eds.), *Space time wireless systems: From array processing to MIMO communication* (pp. 198-217). Cambridge University Press.

Benedetto, S., Divsalar, D., Montorsi, G., & Pollara, F. (1998). Serial concatenation of interleaved codes: Performance analysis, design and iterative decoding. *IEEE Transactions on Information Theory*, 44, 909–926. doi:10.1109/18.669119

Beres, E., & Adve, R. (2008). Selection cooperation in multisource cooperative networks. *IEEE Transactions on Wireless Communications*, 7(1), 118–127. doi:10.1109/TWC.2008.060184

Berger, J. O., Oliveira, V. D., & Sanso, B. (2001). Objective Bayesian analysis of spatially correlated data. *Journal of the American Statistical Association*, 96(456), 1361–1374. doi:10.1198/016214501753382282

Berger, T. (1977, July). Multiterminal source coding. *Lectures Presented at CISM Summer School on the Information Theory Approach to Communication*.

Bergmans, P. P. (1973). Random coding theorem for broadcast channels with degraded components. *IEEE Transactions on Information Theory*, 19(2), 197–207. doi:10.1109/TIT.1973.1054980

Bergmans, P. P. (1974). A simple converse for broadcast channels with additive white Gaussian noise. *IEEE Transactions on Information Theory, 20*(2), 279–280. doi:10.1109/TIT.1974.1055184

Berrou, C., Glavieux, A., & Thitimajshima, P. (1993). Near Shannon limit error-correcting coding and decoding. In *Proceedings of IEEE International Conference Communications, 2*, Geneva, Switzerland (pp. 1064–1070).

Bertsekas, D. P. (1995). *Nonlinear programming.* Belmont, MA: Athena Scientific.

Bhatnagar, M. R., & Hjørungnes, A. (2007a). Downlink cooperative communication using differential relaying. In *Proc. 4th International Symposium on Wireless Communication Systems (ISWCS)*, Trondheim, Norway (pp. 1–5).

Bhatnagar, M. R., & Hjørungnes, A. (2007b). SER expressions for double-differential modulation. In *Proc. IEEE Inform. Theory Workshop*, Bergen, Norway (pp. 203–207).

Bhatnagar, M. R., & Hjørungnes, A. (2007c). Double-differential modulation for decode-and-forward cooperative communications. In *Proc. IEEE International Conference on Signal Processing and Communication (ICSPC)*, Dubai, United Arab Emirates (pp. 1–4).

Bhatnagar, M. R., & Hjørungnes, A. (2008b). Differential relay strategies over downlink channel in two-user cooperative communication system. *Wireless Personal Communications* (accepted for publication 18.09.2008).

Bhatnagar, M. R., & Hjørungnes, A. (2008f). Distributed double-differential orthogonal space-time code for cooperative networks. In *Proc. IEEE GLOBECOM*, New Orleans, LA.

Bhatnagar, M. R., Hjørungnes, A., & Bose, R. (2008h). Precoded DDOSTBC with non-unitary constellations over correlated Rayleigh channels with carrier offsets. In *Proc. IEEE International Symposium on Information Theory, (ISIT'08)*, Toronto, Canada.

Bhatnagar, M. R., Hjørungnes, A., & Song, L. (2007d). Double-differential orthogonal space-time block codes for arbitrarily correlated Rayleigh channels with carrier offsets. *IEEE Trans. Wireless Comm.* (conditionally accepted for publication subject to the reviewers' and editor's final decision after third round of review of the revised manuscript submitted 14.09.2008).

Bhatnagar, M. R., Hjørungnes, A., & Song, L. (2008a). Amplify-and-forward cooperative communications with double-differential modulation over Nakagami-*m* channels. In *Proc. IEEE Wireless Communications and Networking Conference (WCNC)*, Las Vegas, NV (pp. 1–5).

Bhatnagar, M. R., Hjørungnes, A., & Song, L. (2008b). Cooperative communications over flat fading channels with carrier offsets: A double-differential modulation approach. *EURASIP Journal on Advances in Signal, 2008*(531786), 1-11.

Bhatnagar, M. R., Hjørungnes, A., & Song, L. (2008e). Amplify based double-differential modulation for cooperative communications. In *Proc. for IEEE International Conference on Communication Systems and Networks (COMSNETS'09)*, Bangalore, India.

Bhatnagar, M. R., Hjørungnes, A., & Song, L. (2008g). Double differential orthogonal space-time modulation. In *Proc. IEEE Conference on Acoustics, Speech, and Signal Processing (ICASSP'08)*, Las Vegas, Nevada.

Bhatnagar, M. R., Hjørungnes, A., Song, L., & Bose, R. (2008c). Double-differential decode-and-forward cooperative communications over Nakagami-*m* channels with carrier offsets. In *Proc. IEEE Sarnoff Symposium*, Princeton, NJ.

Bhatnagar, M., Hjorunges, A., & Song, L. (2007). Non orthogonal differential space time block code with non-unitary constellations. In *Proc of IEEE international Workshop on Signal Processing Advances in Wireless Communications*.

Bhatnagar, M., Hjorunges, A., & Song, L. (2007). Precoded DOSTBCs over Rayleigh channels. *EURASIP Research Letters in Communications, 2007*, 16438.

Bhatnagar, M., Hjorunges, A., & Song, L. (2008). Precoded differential orthogonal space-time modulation over correlated Ricean MIMO channels. *IEEE Journal of Selected Topics in Signal Processing, 2*(2), 124–134.

Biglieri, E., Proakis, J., & Shamai (Shitz), S. (1998). Fading channels: Information-theoretic and communications aspects. *IEEE Transactions on Information Theory,*

44(6), 2619–2692. doi:10.1109/18.720551

Biswas, S., & Morris, R. (2005). ExOR: Opportunistic multihop routing for wireless networks. *ACM SIGCOMM, 25*(4), 133–144. doi:10.1145/1090191.1080108

Bletsas, A., Khisti, A., Reed, D. P., & Lippman, A. (2006). A simple cooperative diversity method based on network path selection. *IEEE JSAC, 24*(9), 659–672.

Bletsas, A., Shin, H., & Win, M. Z. (2007). Cooperative communications with outage-optimal opportunistic relay ins. *IEEE Transactions on Wireless Communications, 6*(9), 3450–3460. doi:10.1109/TWC.2007.06020050

Boche, H., & Wiczanowski, M. (2007). Optimization-theoretic analysis of stability-optimal transmission policy for multiple antenna multiple access channel. *IEEE Transactions on Signal Processing, 55*(6), 2688–2702. doi:10.1109/TSP.2006.890926

Boyd, S., & Vandenberghe, L. (2003). *Convex optimization.* Cambridge, UK: Cambridge Univ. Press.

Boyer, J., Falconer, D. D., & Yanikomeroglu, H. (2004, October). Multihop diversity in wireless relaying channels. *IEEE Trans. Commun.*

Boyer, J., Falconer, D., & Yanikomeroglu, H. (2004). Multihop diversity in wireless relaying channels. *IEEE Transactions on Communications, 52*(10), 1820–1830. doi:10.1109/TCOMM.2004.836447

Brennan, D. G. (2003). Linear diversity combining techniques. *Proceedings of the IEEE, 91*(2), 331–356. doi:10.1109/JPROC.2002.808163

Brown, D. R. III, & Poor, H. V. (2008). Time-slotted round-trip carrier synchronization for distributed beam-forming. *IEEE Transactions on Signal Processing, 56*(11), 5630–5643. doi:10.1109/TSP.2008.927073

Buchegger, S., & le Boudec, J. Y. (2002). Performance analysis of the CONFIDANT protocol. *ACM MobiHoc* (pp. 226-236).

Buttyan, L., & Hubaux, J. (2003). Stimulating cooperation in self-organizing mobile ad hoc networks. *ACM/Kluwer Mobile Networks and Applications (MONET), 8*(5), 579-592.

Cano, A., Morgado, E., Caamano, A., & Ramos, J. (2007). Distributed double-differential modulation for cooperative communications under CFO. In *Proc. IEEE GLOBECOM*, Washington, D.C. (pp. 3447–3441).

Cao, Y., & Vojcic, B. (2005). Cooperative coding using serial concatenated convolutional codes. In . *Proceedings of IEEE Wireless Communications and Networking Conference, 2*, 1001–1006. doi:10.1109/WCNC.2005.1424645

Carleial, A. B. (1982). Multiple access channels with different generalized feedback signals. *IEEE Transactions on Information Theory, 28*(6), 841–850. doi:10.1109/TIT.1982.1056587

Cavalcanti, D., Cordeiro, C. M., Agrawal, D. P., Xie, B., & Kumar, A. (2005). Issues in integrating cellular networks, WLANs, and MANETs: A futuristic heterogeneous wireless network. *IEEE Wireless Communications Magazine . Special Issue on Toward Seamless Internetworking of Wireless LAN and Cellular Networks, 12*(3), 30–34.

Chakrabarti, A., de Baynast, A., Sabharwal, A., & Aazhang, B. (2006). Half-duplex estimate-and-forward relaying: Bounds and code design. In *Proceedings of IEEE International Symposium on Information Theory* (pp. 1239–1243).

Chandrakasan, A. P., Min, R., Bharadwaj, M., Cho, S., & Wang, A. (2002, September). *Power aware wireless microsensor systems.* Paper presented at ESSCIRC.

Chang, Y., & Hua, Y. (2004). Diversity analysis of orthogonal space-time modulation for distributed wireless relays. *Proceedings of the IEEE, ICASSP2004*, 561–564.

Chang, Y.-C., Lin, Z.-S., & Chen, J.-L. (2006). Cluster based self-organization management protocols for wireless sensor networks. *IEEE Transactions on Consumer Electronics, 52*(1), 75–80. doi:10.1109/TCE.2006.1605028

Charafeddine, M. A., Sezgin, A., & Paulraj, A. (2007, September). Rate region frontiers for n-user interference channel with interference as noise. *45th Annual Allerton Conference on Communication, Control, and Computing*, Monticello, IL.

Chatterjee, D., & Wong, T. F. (in press). Active user cooperation in fading multiple-access channels. In . *Proceedings of IEEE Military Communications Conference.*

Chatterjee, D., Wong, T. F., & Lok, T. M. (2008). Cooperative transmission in a wireless cluster based on flow management. In *Proceedings of IEEE Wireless Communications Networking Conference*, Las Vegas, NV (pp. 30–35).

Chatzinotas, S., Imran, M. A., & Tzaras, C. (2008b). On the capacity of variable density cellular systems under multicell decoding. *IEEE Communications Letters, 12*(7), 496–498. doi:10.1109/LCOMM.2008.080439

Chatzinotas, S., Imran, M. A., & Tzaras, C. (2008c). Spectral efficiency of variable density cellular systems. In *IEEE International Symposium on Personal, Indoor and Mobile Radio Communications, International Workshop on Efficiency*, Cannes, France.

Chatzinotas, S., Imran, M., & Tzaras, C. (2008a). Optimal information theoretic capacity of planar cellular uplink channel. In *9th IEEE International Workshop on Signal Processing Advances in Wireless Communications*, Recife, Brazil (pp. 196-200).

Chatzinotas, S., Imran, M., & Tzaras, C. (2008d). On the multicell processing rates of the cellular downlink fading channel. Submitted to IET Communications.

Chatzinotas, S., Imran, M., & Tzaras, C. (2008e). Uplink capacity of MIMO cellular systems with multicell processing. In *IEEE International Symposium on Wireless Communication Systems*, Reykjavik, Iceland.

Chen, D., & Laneman, J. N. (2004a). Cooperative diversity for wireless fading channels without channel state information. In *Proc. 48th Asilomar Conf. Signals, Systems, and Computers*, Monterey, CA (pp. 1307–1312).

Chen, D., & Laneman, J. N. (2004b). Noncoherent demodulation for cooperative diversity in wireless systems. In *Proc. of Globecom*, Dallas, TX (pp. 32–35).

Chen, D., & Laneman, J. N. (2006). Modulation and demodulation for cooperative diversity in wireless systems. *IEEE Transactions on Wireless Communications, 5*, 1785–1794. doi:10.1109/TWC.2006.1673090

Chen, D., & Laneman, N. (2004). Noncoherent demodulation for cooperative diversity in wireless systems. In *Proceedings of IEEE GLOBECOM* (pp. 31-35).

Chen, D., & Laneman, N. (2006). Modulation and demodulation for cooperative diversity in wireless systems. *IEEE Transactions on Wireless Communications, 5*(07), 1785–1794.

Chen, Y., & Zhao, Q. (2005). On the lifetime of wireless sensor networks. *IEEE Communications Letters, 9*(11), 976–978. doi:10.1109/LCOMM.2005.11010

Chiani, M., Win, M., & Zanella, A. (2003). On the capacity of spatially correlated MIMO Rayleigh-fading channels. *IEEE Transactions on Information Theory, 49*(10), 2363–2371. doi:10.1109/TIT.2003.817437

Chizhik, D., Foschini, G., & Valenzuela, R. (2000). Capacities of multi-element transmit and receive antennas: Correlations and keyholes. *Electronics Letters, 36*(13), 1099–1100. doi:10.1049/el:20000828

Chizhik, D., Foschini, G., Gans, M., & Valenzuela, R. (2002). Keyholes, correlations, and capacities of multielement transmit and receive antennas. *IEEE Transactions on Wireless Communications, 1*(2), 361–368. doi:10.1109/7693.994830

Choi, C. H., Lim, J. B., & Im, G. H. (2007). Unique-word-based single carrier system with decision feedback equalization for space-time block coded transmissions. *IEEE Communications Letters, 11*(1), 28–30. doi:10.1109/LCOMM.2007.060875

Choi, W., & Andrews, J. G. (2008). The capacity gain from intercell scheduling in multi-antenna systems. *IEEE Transactions on Wireless Communications, 7*(2), 714–725. doi:10.1109/TWC.2008.060615

Chou, C. T., Yang, J., & Wang, D. (2007). Cooperative MAC protocol with automatic relay selection in distributed wireless networks. *IEEE PerCom Workshops* (pp. 526-531).

Chou, J., Pradhan, S., & Ramchandran, K. (2003). Turbo and trellis-based constructions for source coding with side information. In *Proceedings of IEEE Data Compression Conference*, Snowbird, UT (pp. 33–42).

Chou, P. A., & Wu, Y. (2007). Network coding for the Internet and wireless networks. *IEEE Signal Processing Magazine, 24*(5), 77–85. doi:10.1109/MSP.2007.904818

Chow, P. (1995). Practical discrete multitone transceiver loading algorithm for data transmission over spectrally shaped channels. *IEEE Transactions on Communications, 43*(2), 773–775. doi:10.1109/26.380108

Chu, D. (1972). Polyphase codes with good periodic correlation properties. *IEEE Transactions on Information*

Theory, 18(7), 531–532. doi:10.1109/TIT.1972.1054840

Chu, J., Adve, R. S., & Eckford, A. (2007). Relay selection for low-complexity coded cooperation. In *IEEE Globecom 2007*.

Chuah, C.-N., Tse, D., Kahn, J., & Valenzuela, R. (2002). Capacity scaling in MIMO wireless systems under correlated fading. *IEEE Transactions on Information Theory, 48*(3), 637–650. doi:10.1109/18.985982

Chung, S., Forney, G. D. Jr, Richardson, T. J., & Urbanke, R. (2001). On the design of low-density parity-check codes within 0.0045 dB of the Shannon limit. *IEEE Communications Letters, 5*(2), 58–60. doi:10.1109/4234.905935

Chung, S., Richardson, T., & Urbanke, R. (February 2001). Analysis of sum-product decoding of lowdensity parity-check codes using a Gaussian approximation. *IEEE Trans. Info. Theory*.

Coleman, T., Effros, M., Martinian, E., & Medard, M. (2005). Rate-splitting for the deterministic broadcast channel. In *Proceedings of IEEE International Symposium on Information Theory* (pp. 2189–2192).

Coleman, T., Martinian, E., Effros, M., & Medard, M. (2005). Interference management via capacity-achieving codes for the deterministic broadcast channel. In *Proceeding of IEEE Information Theory Workshop, 5*.

Compton, R. T. (1988). *Adaptive antennas-concepts and performance*. Englewood Cliffs, NJ: Prentice Hall.

Conti, M., Gregori, E., & Maselli, G. (2004). Cooperation issues in mobile ad hoc networks. *ICDCSW* (pp. 803-808).

COST 207 TD(86)51-REV 3(WG1). (1998). Proposal on channel transfer functions to be used in GSM tests late 1986.

Costa, M. (1972). On the Gaussian interference channel. *IEEE Transactions on Information Theory, 31*(5), 607–615. doi:10.1109/TIT.1985.1057085

Costa, M. (1983). Writing on dirty paper. *IEEE Transactions on Information Theory, IT-29*, 439–441. doi:10.1109/TIT.1983.1056659

Cover, T. M., & El Gamal, A. A. (1979). Capacity theorems for the relay channel. *IEEE Transactions on Information Theory, 25*(5), 572–584. doi:10.1109/TIT.1979.1056084

Cover, T. M., & Leung, C. (1981). An achievable rate region for the multiple-access channel with feedback. *IEEE Transactions on Information Theory, 27*(3), 292–298. doi:10.1109/TIT.1981.1056357

Cover, T. M., & Thomas, J. A. (1991). *Elements of information theory*. New York: John Wiley.

Cover, T. M., El Gamal, A., & Salehi, M. (1980). Multiple access channels with arbitrarily correlated sources. *IEEE Transactions on Information Theory, 26*(6), 648–657. doi:10.1109/TIT.1980.1056273

Cover, T., & el Gamal, A. (1979). Capacity theorems for the relay channel. *IEEE Transactions on Information Theory, IT-25*(5), 572–584. doi:10.1109/TIT.1979.1056084

Cover, T., & Thomas, J. A. (1991). *Elements of information theory*. NewYork: Wiley.

Cristescu, R., Beferull-Lozano, B., & Vetterli, M. (2005). Networked Slepian-Wolf: Theory, algorithms, and scaling laws. *IEEE Transactions on Information Theory, 51*(12), 4057–4073. doi:10.1109/TIT.2005.858980

Cristescu, R., Beferull-Lozano, B., Vetterli, M., & Wattenhofer, R. (2006). Network correlated data gathering with explicit communication: NP-completeness and algorithms. *IEEE/ACM Transactions on Networking, 14*(1), 41–54.

Cui, S., & Goldsmith, A. (2006). Cross-layer design in energy constrained networks using cooperative MIMO techniques. *EURASIP Signal Processing Journal, 86*(8), 1804–1814.

Cui, S., & Goldsmith, A. J. (2006). Cross-layer design of energy-constrained networks using cooperative MIMO techniques. *EURASIP/Elsevier Signal Processing Journal, 86*(8), 1804–1814.

Cui, S., Goldsmith, A. J., & Bahai, A. (2004). Energy efficiency of MIMO and cooperative MIMO in sensor networks. *IEEE Journal on Selected Areas in Communications, 22*(6), 1089–1098. doi:10.1109/JSAC.2004.830916

Cui, S., Goldsmith, A. J., & Bahai, A. (2005). Energy constrained modulation optimization. *IEEE Transactions on Wireless Communications, 4*(5), 2349–2360. doi:10.1109/TWC.2005.853882

Dallal, Y. E., & Shamai, S. (1992). Time diversity in DPSK noisy phase channels. *IEEE Transactions on Communications, 40*, 1703–1715. doi:10.1109/26.179934

Damen, M. O., & Hammons, A. R. (2007). Delay-tolerant distributed TAST codes for cooperative diversity. *IEEE Transactions on Information Theory, 53*, 3755–3773. doi:10.1109/TIT.2007.904773

Dana, A. F., & Hassibi, B. (2006). On the power efficiency of sensory and ad hoc wireless networks. *IEEE Transactions on Information Theory, 52*(7), 2890–2914. doi:10.1109/TIT.2006.876245

Dayal, P., Brehler, M., & Varanasi, M. K. (2004). Leveraging coherent space time codes for noncoherent communication via training. *IEEE Transactions on Information Theory, 50*(09), 2058–2080.

De Souza, J. C. (1999). *Discrete bit loading for multicarrier modulation systems.* Unpublished doctoral dissertation, Stanford University, CA.

Deb, R., & Medard, M. (2004). Algebraic gossip: A network coding approach to optimal multiple rumor mongering. In *Proc. the 42nd Annual Allerton Conf. on Commun. Control and Computing.*

Deng, X., & Haimovich, A. M. (2005). Power allocation for cooperative relaying in wireless networks. *IEEE Communications Letters, 11*(9), 994–996. doi:10.1109/LCOMM.2005.11012

Dimitris, P., & Karystinos, G. (2008). Efficient ML noncoherent orthogonal STBC detection. In *Proceedings of Allerton Conference.*

Ding, Y., Zhang, J.-K., & Wong, K. M. (2007). The amplify-and-forward half-duplex cooperative system: Pairwise error probability and precoder design. *IEEE Transactions on Signal Processing, 55*, 605–617. doi:10.1109/TSP.2006.885761

Ding, Z., Ratnarajah, T., & Cowan, C. C. F. (2007). On the diversity-multiplexing tradeoff for wireless cooperative multiple access systems. *IEEE Transactions on Signal Processing, 55*(9), 4627–4638. doi:10.1109/TSP.2007.896276

Divsalar, D., & Pollara, F. (1995, November). *On the design of turbo codes* (Tech. Rep.). *NASA Jet Propulsion Laboratory. TDA Progress Report 42-123.*

Divsalar, D., Dolinar, S., & Pollara, F. (2001). Iterative turbo decoder analysis based on density evolution. *IEEE Journal on Selected Areas in Communications, 19*(5), 891–907. doi:10.1109/49.924873

Divsalar, D., Jin, H., & McEliece, R. (1998). Coding theorems for 'turbo-like' codes. In *Proceedings of Allerton Conference on Communications, Control, and Computing* (pp. 201–210).

Djeumou, B., Lasaulce, S., & Klein, A. G. (2007). Practical quantize-and-forward schemes for the frequency division relay channel. *EURASIP Journal on Wireless Communications and*

Dohler, M. (1999-2002). *VAA for hot-spots with applied STC.* ([III and IV). M-VCE, UK.]. *Internal Reports, I,* II.

Dohler, M. (2003). *Virtual antenna arrays.* Unpublished doctoral dissertation, King's College London, UK.

Dohler, M., & Aghvami, A. H. (2007). *A crash-course on cooperative wireless networks.* Tutorial presented at IEEE ICC, Glasgow, Scotland, UK.

Dohler, M., Gkelias, A., & Aghvami, A. H. (2006). Capacity of distributed PHY-layer sensor networks. *IEEE Transactions on Vehicular Technology, 55*(2), 622–639. doi:10.1109/TVT.2005.863470

Dohler, M., Li, Y., Vucetic, B., Aghvami, A. H., Arndt, M., & Barthel, D. (2006). Performance analysis of distributed space-time block encoded sensor networks. *IEEE Transactions on Vehicular Technology, 55*(7), 1776–1789. doi:10.1109/TVT.2006.878729

Dohler, M., Meddour, D. E., Senouci, S. M., & Saadani, A. (2008). Cooperation in 4G-hype or ripe? *IEEE Technology and Society Magazine, 27*(1), 13–17. doi:10.1109/MTS.2008.918035

Dohler, M., Said, F., Ghorashi, S. A., & Aghvami, A. H. (2001). Improvements in or relating to electronic data communication systems. *Patent Publication No. WO 03/003672*, priority date 28 June 2001.

Dong, L., Petropulu, A. P., & Poor, H. V. (2008a). A cross-layer approach to collaborative beamforming for wireless ad hoc networks. *IEEE Transactions on Signal Processing, 56*(7), 2981–2993. doi:10.1109/TSP.2008.917352

Dong, L., Petropulu, A. P., & Poor, H. V. (2008b). Weighted cooperative beamforming for wireless networks. *IEEE Transactions on Signal Processing*, in revision.

Donvito, M. B., & Kassam, S. A. (1979). Characterization of the random array peak sidelobe. *IEEE Transactions on Antennas and Propagation, 27*(3), 379–385. doi:10.1109/TAP.1979.1142097

Douillard, C., Jezequel, M., Berrou, C., Picart, A., Didier, P., & Glavieux, A. (1995). Iterative correction of intersymbol interference: Turbo-equalization. *European Transactions on Telecommunications, 6*(5), 507–511. doi:10.1002/ett.4460060506

Draves, R., Padhye, J., & Zill, B. (2004). Routing in multiradio, multihop wireless mesh networks. *ACM MobiCom* (pp. 114-128).

El Gamal, H., Caire, G., & Damen, M. O. (2004). Lattice coding and decoding achieve the optimal diversity-multiplexing-delay tradeoff of mimo channels. *IEEE Transactions on Information Theory, 50*(6), 968–985. doi:10.1109/TIT.2004.828067

Elia, P., Kumar, K. R., Pawar, S. A., Kumar, P. V., & Lu, H. F. (2006). Explicit, minimum delay space time codes achieving the diversity multiplexing gain trade off. *IEEE Transactions on Information Theory, 52*(09), 896–900.

Elia, P., Kumar, K. R., Pawar, S. A., Kumar, P. V., & Lu, H. F. (2006). Explicit space-time codes achieving the diversity-multiplexing gain tradeoff. *IEEE Transactions on Information Theory, 52*(9), 3869–3884. doi:10.1109/TIT.2006.880037

Elia, P., Oggier, F., & Kumar, P. V. (2007). Asymptotically optimal cooperative wireless networks with reduced signaling complexity. *IEEE Journal on Selected Areas in Communications, 25*, 258–267. doi:10.1109/JSAC.2007.070203

El-Keyi, A., & Champagne, B. (2008). Cooperative MIMO-beamforming for multiuser relay network. In *Proceedings of 2008 IEEE International Conference of Acoustic Speech Signal Processing*, Las Vegas, NV (pp. 2749-2752).

Erez, U., & Brink, S. (2005). A close-to-capacity dirty paper coding scheme. *IEEE Transactions on Information Theory, 51*(10), 3417–3432. doi:10.1109/TIT.2005.855586

Erez, U., Shamai, S., & Zamir, R. (2000). Capacity and lattice strategies for canceling known interference. In *Proceedings of International Symposium on Information Theory and Its Application,* Honolulu, HI (pp. 681–684).

Erez, U., Shamai, S., & Zamir, R. (2005). Capacity & lattice strategies for canceling known interference. *IEEE Transactions on Information Theory, 51*(11), 3820–3833. doi:10.1109/TIT.2005.856935

Erkip, E. (2000). Capacity and power control for spatial diversity. In *Proceedings of the Conference on Information Sciences and Systems,* Princeton, NJ (pp. WA4-28/WA4-31).

Esseling, N., Walke, B., & Pabst, R. (2004). Performance evaluation of a fixed relay concept for next generation wireless systems. In R. Ganesh, S. Kota, K. Pahlavan & R. Agusti (Eds), *Emerging location aware broadband wireless ad hoc networks*. Boston, MA: Kluwer Academic Publishers.

Etkin, R., Tse, D. N. C., & Wang, H. (2007). Gaussian interference channel capacity to within one bit. Retrieved in 2007, from http://www.citebase.org/abstract?id=oai:arXiv.org:cs/0702045

ETSI TR 125 996 V7.0.0. (2007). Universal mobile telecommunications system (UMTS): Spatial channel model for multiple input multiple output (MIMO) simulations (3GPP tr 25.996 version 7.0.0 release 7).

Falconer, D., Ariyavisitakul, S. L., Benyamin-Seeyar, A., & Eidson, B. (2002). White paper: Frequency domain equalization for single-carrier broadband wireless systems. Retrieved from http://www.sce.carleton.ca/bbw/papers/whitepaper2.pdf

Fan, Y., Wang, C., Thompson, J., & Poor, H. V. (2007). Recovering multiplexing loss through successive relaying using repetition coding. *IEEE Transactions on Wireless Communications, 6*(12), 4484–4493. doi:10.1109/TWC.2007.060339

Fischer, R., Stierstorfer, C., & Windpassinger, C. (2003). Precoding and signal shaping for transmission over MIMO channels. In *Proceedings of Canadian Workshop Information Theory,* Waterloo, ON, Canada (pp. 83–87).

Fischer, R., Windpassinger, C., Lampe, A., & Huber, J. (2002a). MIMO precoding for decentralized receivers. In *Proceedings of IEEE International Symposium on Information Theory* (p. 496).

Fischer, R., Windpassinger, C., Lampe, A., & Huber, J. (2002b). Space-time transmission using Tomlinson-Harashima precoding. In *Proceedings of ITG Conference on Source and Channel Coding* (pp. 139–147).

Fitzek, F. H. P., & Katz, M. D. (2006). *Cooperation in wireless networks: Principles and applications.* New York: Springer-Verlag New York, Inc.

Forney, G. D. Jr. (1992). Trellis shaping. *IEEE Transactions on Information Theory*, *38*(2), 281–300. doi:10.1109/18.119687

Forsberg, D., Ohba, O., Patil, B., Tschofenig, H., & Yegin, A. (2007). *Protocol for carrying authentication and network access (PANA).* IETF, draft-ietf-pana-pana-18, work in progress.

Foschini, G. J. (1996). Layered space time architecture for wireless communication in a fading environment when using multielement antennas. *Bell Labs Technical Journal*, *01*(2).

Foschini, G. J., & Gans, M. J. (1998). On limits of wireless communications in a fading environment when using multiple antennas. *Wireless Personal Communications*, *6*(3), 311–335. doi:10.1023/A:1008889222784

Fragouli, C., & Soljanin, E. (2007). Network coding applications. *Foundations and Trends in Networking*, *2*(2), 135–269. doi:10.1561/1300000013

Furuskar, A., Mazur, S., Muller, F., & Olofsson, H. (1999). EDGE: Enhanced data rates of GSM and TDMA/136 evolution. *IEEE Personal Communications Magazine*, *6*(3), 56–66. doi:10.1109/98.772978

Gaarder, N., & Wolf, J. (1975). The capacity region of a multiple-access discrete memoryless channel can increase with feedback. *IEEE Transactions on Information Theory*, *21*(1), 100–102. doi:10.1109/TIT.1975.1055312

Gallager, R. G. (1962). Low-density parity-check codes. *IEEE Transactions on Information Theory*, *8*(1), 21–28. doi:10.1109/TIT.1962.1057683

Gallager, R. G. (1963). *Low-density parity-check codes.* Cambridge, MA: MIT Press.

Gallager, R. G. (1974). Capacity and coding for degraded broadcast channels. *Problemv Peredaci Informotsii*, *10*(3), 3–14.

Gallager, R. G. (1994). An inequality on the capacity region of multiaccess multipath channels. In Blahut, Costello, Maurer & Mittleholzer (Eds.), *Communications & cryptography, two sides of one tapestry* (pp. 129-140). Norwell, MA: Kluwer Academic Publishers.

Garcia-Frias, J. (2001a). Compression of correlated binary sources using turbo codes. *IEEE Communications Letters*, *5*(10), 417–419. doi:10.1109/4234.957380

Garcia-Frias, J. (2001b). Joint source-channel decoding of correlated sources over noisy channels. In *Proceedings of IEEE Data Compression Conference,* Snowbird, UT (pp. 283–292).

Garcia-Frias, J., & Xiong, Z. (2005). Distributed source and joint source-channel coding: From theory to practice. In . *Proceedings of IEEE International Conference on Acoustics, Speech, and Signal Processing*, *5*, 1093–1096. doi:10.1109/ICASSP.2005.1416498

Garcia-Frias, J., & Zhong, W. (2003). LDPC codes for compression of multi-terminal sources with hidden Markov correlation. *IEEE Communications Letters*, *7*(3), 115–117. doi:10.1109/LCOMM.2003.810001

Garcia-Frias, J., Zhong, W., & Zhao, Y. (2002). Iterative decoding schemes for source and joint source-channel coding of correlated sources. In *Proceedings of Asilomar Conference on Signals, Systems and Computers* (Vol. 1, pp. 250–256).

Gastpar, M. (2004). The Wyner-Ziv problem with multiple sources. *IEEE Transactions on Information Theory*, *50*(11), 2762–2768. doi:10.1109/TIT.2004.836707

Gastpar, M., & Vetterli, M. (2002). On the asymptotic capacity of Gaussian relay networks. In *Proceedings of the IEEE International Symposium on Information Theory*, Lausanne, Switzerland (p. 195).

Gastpar, M., & Vetterli, M. (2005). On the capacity of large Gaussian relay networks. *IEEE Transactions on Information Theory*, *51*(3), 765–779. doi:10.1109/TIT.2004.842566

Gesbert, D., Bolcskei, H., Gore, D., & Paulraj, A. (2002). Outdoor MIMO wireless channels: Models and performance prediction. *IEEE Transactions on Communications*, *50*(12), 1926–1934. doi:10.1109/TCOMM.2002.806555

Giallorenzi, T., & Wilson, S. (1996). Suboptimum multiuser receivers for convolutionally coded asynchronous DS-CDMA systems. *IEEE Transactions on Communications*, *44*(9), 1183–1196. doi:10.1109/26.536924

Ginis, G., & Cioffi, J. M. (2000). A multi-user precoding scheme achieving crosstalk cancellation with application to DSL systems. In *Proceedings of Asilomar Conference on Signals, Systems and Computers* (Vol. 2, pp. 1627–1631).

Godara, L. C. (1997). Application of antenna arrays to mobile communications, II: Beamforming and direction-of-arrival considerations. *Proceedings of the IEEE*, *85*, 1195–1245. doi:10.1109/5.622504

Goldsmith, A. J., & Varaiya, P. P. (1997). Capacity of fading channels with channel side information. *IEEE Transactions on Information Theory*, *43*(6), 1986–1992. doi:10.1109/18.641562

Golub, G. H., & Van Loan, C. F. (1996). *Matrix computations*. Baltimore, MD: Johns Hopkins University Press.

Grandshteyn, I. S., & Ryzhik, I. M. (2000). *Table of integrals, series, and products, 6th edition*. San Diego, CA: Academic Press.

Grant, A., Hanly, S., Evans, J., & Muller, R. (2004). Distributed decoding for Wyner cellular systems. In *5th Australian Communications Theory Workshop* (pp. 77–81).

Gray, J. (2002). *Straw dogs: Thoughts on humans and other animals*. London, UK: Granta.

Gunasekaran, V., & Harmantzis, F. (2005). Affordable infrastructure for deploying WiMAX systems: Mesh vs. non mesh. *IEEE VTC-Spring*, *5*, 2979–2983.

Guo, X., & Xia, X.-G. (2008). A distributed space-time coding in asynchronous wireless relay networks. *IEEE Transactions on Wireless Communications*, *7*, 1812–1816. doi:10.1109/TWC.2008.070042

Gupta, P., & Kumar, P. R. (2000). The capacity of wireless networks. *IEEE Transactions on Information Theory*, *46*(2), 388–404. doi:10.1109/18.825799

Gupta, P., & Kumar, P. R. (2000, March). The capacity of wireless networks. *IEEE Transactions on Information Theory*, *46*(2), 388–404. doi:10.1109/18.825799

Gupta, P., & Kumar, P. R. (2003). Towards an information theory of large networks: An achievable rate region. *IEEE Transactions on Information Theory*, *49*(8), 1877–1894. doi:10.1109/TIT.2003.814480

Hachem, W., Khorunzhiy, O., Loubaton, P., Najim, J., & Pastur, L. (2006). A new approach for capacity analysis of large dimensional multi-antenna channels. Submitted to *IEEE Transactions on Information Theory*.

Hagenauer, J. (1996). Forward error correcting for CDMA systems. In . *Proceedings of IEEE International Symposium on Spread Spectrum Techniques and Applications*, *2*, 566–569. doi:10.1109/ISSSTA.1996.563190

Hagenauer, J., Offer, E., & Papke, L. (1996). Iterative decoding of binary block and convolutional codes. *IEEE Transactions on Information Theory*, *42*(2), 429–445. doi:10.1109/18.485714

Haim, E., & Zamir, R. (2005). Quantization with variable resolution and coding for deterministic broadcast channels. In *Proceedings of Allerton Conference on Communications, Control, and Computing*.

Hammerström, I., Kuhn, M., & Wittneben, A. (2004). Impact of relay gain allocation on the performance of cooperative diversity networks. In *Proc. IEEE VTC'04* (pp. 1815–1819).

Hammerström, I., Kuhn, M., & Wittneben, A. (2004, September). Channel adaptive scheduling for cooperative relay networks. *Proc. IEEE Vehic. Tech. Conf.*, Los Angeles, CA (pp. 2784-2788).

Hammerström, I., Kuhn, M., Esli, C., Zhao, J., Wittneben, A., & Bauch, G. (2007, June). MIMO two-way relaying with transmit CSI at the relay. In *Proc. SPAWC*.

Han, T., & Kobayashi, K. (1981). A new achievable rate region for the interference channel. *IEEE Transactions on Information Theory*, *27*(1), 49–60. doi:10.1109/TIT.1981.1056307

Hanley, S., & Whiting, P. (1993). Information-theoretic capacity of multi-receiver networks. *Telecommunication Systems*, *1*, 1–42. doi:10.1007/BF02136153

Hanly, S., & Tse, D. N. C. (1998). Multiaccess fading channels-part I: Polymatroid structure, optimal resource allocation and throughput capacities. *IEEE Transactions on Information Theory*, *44*(7), 2796–2815. doi:10.1109/18.737514

Harashima, H., & Miyakawa, H. (1972). Matched-transmission technique for channels with intersymbol interference. *IEEE Transactions on Communications*, *20*, 774–780. doi:10.1109/TCOM.1972.1091221

Harshan, J., & Rajan, B. S. (2007). Coordinate interleaved distributed space time coding for two antenna relay networks. In *Proceedings of IEEE GLOBECOM* 2007, Washington D.C. (pp. 1719-1723).

Harshan, J., & Rajan, B. S. (2008). A nondifferential distributed space-time coding for partially-coherent cooperative communication. *IEEE Trans. Wireless Communication Letters*, *07*(11), 4076–4081.

Harshan, J., & Rajan, B. S. (2008, May 19-23). Nondifferential DSTBCs for partially-coherent cooperative communication. In *Proceedings of ICC 2008*, Beijing, China (pp.996-1000).

Hasna, M. O., & Alouini, M.-S. (2003). End-to-end performance of transmission systems with relays over Rayleigh-fading channels. *IEEE Transactions on Wireless Communications*, *2*(6), 1126–1131. doi:10.1109/TWC.2003.819030

Hasna, M. O., & Alouini, M.-S. (2004). A performance study of dual-hop transmissions with fixed gain relays. *IEEE Transactions on Wireless Communications*, *3*(6), 1963–1968. doi:10.1109/TWC.2004.837470

Hasna, M. O., & Alouini, M.-S. (2004). Optimal power allocation for relayed transmissions over rayleigh-fading channels. *IEEE Transactions on Wireless Communications*, *6*(3), 1999–2004. doi:10.1109/TWC.2004.833447

Hassibi, B., & Hochwald, B. (2002). Cayley differential unitary space time codes. *IEEE Transactions on Information Theory*, *48*(06), 1485–1503.

Hassibi, B., & Hochwald, B. M. (2002). High-rate codes that are linear in space and time. *IEEE Transactions on Information Theory*, *48*(7), 1804–1824.

Hassibi, B., & Marzetta, T. L. (2002). Multiple antennas and isotropically random inputs: The received signal density in closed form. *IEEE Transactions on Information Theory*, *48*(06), 1473–1484.

Hausl, C., & Dupraz, P. (2006). Joint network-channel coding for the multiple-access relay channel. In *Proc. of Int. Workshop Wireless Ad Hoc and Sensor Networks*.

Herhold, P., Zimmermann, E., & Fettweis, G. (2004). A simple cooperative extension to wireless relaying. *International Zurich Seminar on Communications* (pp. 36-39).

Herhold, P., Zimmermann, E., & Fettweis, G. (2005). Cooperative multi-hop transmission in wireless networks. *Computer Networks Journal*, *49*(3), 229–324.

Himsoon, T., Su, W., & Liu, K. J. R. (2006). Differential transmission for amplify-and-forward cooperative communications. *IEEE Signal Processing Letters*, *12*(9), 597–600. doi:10.1109/LSP.2005.853067

Hochwald, B., & Marzetta, T. L. (1999). Capacity of a mobile multiple antenna communication links in Rayleigh flat fading. *IEEE Transactions on Information Theory*, *45*(01), 139–157.

Hochwald, B., & Marzetta, T. L. (2000). Unitary space time modulation for multiple antenna communication in Rayleigh flat fading. *IEEE Transactions on Information Theory*, *46*(02), 543–564.

Hochwald, B., & Sweldens, W. (2000). Differential unitary space-time modulation. *IEEE Transactions on Communications*, *48*(12), 2041–2052.

Hochwald, B., Peel, C., & Swindlehurst, A. (2005). A vector-perturbation technique for near-capacity multi-antenna multiuser communication-part II: Perturbation. *IEEE Transactions on Communications*, *53*, 537–544. doi:10.1109/TCOMM.2004.841997

Hong, Y.-W., & Scaglione, A. (2003, October). Energy-efficient broadcasting with cooperative transmission in wireless sensory ad hoc networks. In *Proc. of Allerton Conf. on Commun., Contr. and Comput. (ALLERTON)*.

Hong, Y.-W., Huang, W.-J., Chiu, F.-H., & Kuo, C.-C. J. (2007). Cooperative communications in resource-constrained wireless networks. *IEEE Signal Processing Magazine*, *24*, 47–57. doi:10.1109/MSP.2007.361601

Horadam, K. J. (2005). A generalized Hadamard transform. *ISIT, Adelaide* (pp. 1006-1008).

Host-Madsen, A. (2002). On the capacity of wireless relaying. In *Proc. 56ᵗʰ-IEEE Conference on Vehicular Technology (VTC-Fall)* (pp. 1333-1337).

Høst-Madsen, A. (2002). On the capacity of wireless relaying. In . *Proceedings of IEEE Vehicular Technology Conference, 3*, 1333–1337.

Host-Madsen, A., & Zhang, J. (2005). Capacity bounds and power allocation for wireless relay channels. *IEEE Transactions on Information Theory, 51*(6), 2020–2040. doi:10.1109/TIT.2005.847703

Høst-Madsen, A., & Zhang, J. (2005). Capacity bounds and power allocation for wireless relay channels. *IEEE Transactions on Information Theory, 51*(6), 2020–2040. doi:10.1109/TIT.2005.847703

Ho-Van, K., & Le-Ngoc, T. (2007). A bandwidth-efficient coded user-cooperation scheme for flat block fading channels. In *Proceedings of International Symposium on Wireless Communications Systems* (pp. 421–425).

Hu, J., & Beaulieu, N. C. (2006). A closed-form expression for the outage probability of decode-and-forward relaying in dissimilar Rayleigh fading channels. *IEEE Communications Letters, 10*(12), 813–815. doi:10.1109/LCOMM.2006.061048

Hu, J., & Duman, T. (2007). Low density parity check codes over wireless relay channels. *IEEE Transactions on Wireless Communications, 6*(9), 3384–3394. doi:10.1109/TWC.2007.06083

Hu, R., & Li, J. (2005). Exploiting Slepian-Wolf codes in wireless user cooperation. In *Proceedings of IEEE Workshop on Signal Processing Advances in Wireless Communications* (pp. 275–279).

Hu, R., & Li, J. (2006). Practical compress-forward in user cooperation: Wyner-Ziv cooperation. In *Proceedings of IEEE International Symposium on Information Theory* (pp. 489–493).

Hu, W., Qiao, C., De, S., & Tonguz, O. (2001). Integrated cellular and ad hoc relaying systems: iCAR. *IEEE JSAC, 19*(10), 2105–2115.

Hua, Y., Mei, Y., & Cheng, Y. (2003). Wireless antennas-making wireless communications perform like wireline communications. In *Proc. IEEE AP-S Topical Conference on Wireless Communication Technology* (pp. 47–73).

Hughes, B. L. (2000). Differential space-time modulation. *IEEE Transactions on Information Theory, 46*(07), 2567–2578.

Hunter, T. E., & Nosratinia, A. (2006). Diversity through coded cooperation. *IEEE Transactions on Wireless Communications, 5*(2), 283–289. doi:10.1109/TWC.2006.1611050

Hunter, T., & Nosratinia, A. (2002). Cooperation diversity through coding. In *Proceedings of IEEE International Symposium on Information Theory* (p. 220).

Hwang, C. S., Nam, S. H., Chung, J., & Tarokh, V. (2003). Differential space time block codes using non-constant modulus constellations. *IEEE Transactions on Signal Processing, 51*(11), 2955–2964. doi:10.1109/TSP.2003.818157

IEEE 802.11i. (2004). Medium access control security enhancements. *Standardization Document IEEE Std 802.11i.*

Ikki, S., & Ahmed, M. H. (2007). Performance analysis of cooperative diversity wireless networks over Nakagami-m fading channel. *IEEE Communications Letters, 11*(4), 334–336. doi:10.1109/LCOM.2007.348292

Ishii, K. (2007). Coded cooperation protocol utilizing superposition modulation for half-duplex scenario. In *Proceedings of IEEE Personal, Indoor, and Mobile Radio Communications*, Athens, Greece.

ITs. (2007). *IEEE Transactions on Information Theory.* Special Issue on Models, Theory, and Codes for Relaying and Cooperation in Communication Networks. Avestimehr, S., Diggavi, S., & Tse, D. (2007). A deterministic approach to wireless relay networks. In *Proceedings of Allerton Conference on Communication, Control, and Computing.*

Jacquet, P. (2004, May). Geometry of information propagation in massively dense ad hoc networks. *ACM MobiHoc, 04*, 157–162. doi:10.1145/989459.989479

Jafarkhani, H. (2001). A quasi-orthogonal space-time block code. *IEEE Transactions on Communications, 49*(1), 1–4.

Jagadeesh, H., & Rajan, B. S. (2008). Single-symbol ML decodable precoded DSTBCs for cooperative networks. In *Proc. IEEE ICC'08* (pp. 991—995).

Jakllari, G., Krishnamurthy, S. V., Faloutsos, M., Krishnamurthy, P. V., & Ercetin, O. (2006). A framework for distributed spatio-temporal communications in mobile ad hoc networks. *IEEE INFOCOM* (pp. 1-13).

Janani, M., Hedayat, A., Hunter, T., & Nosratinia, A. (2004). Coded cooperation in wireless communications: Space-time transmission and iterative decoding. *IEEE Transactions on Signal Processing, 52*(2), 362–371. doi:10.1109/TSP.2003.821100

Jang, J. H., Won, H. C., & Im, G. H. (2006). Cyclic prefixed single carrier transmission with SFBC over mobile wireless channels. *IEEE Signal Processing Letters, 13*(5), 261–264. doi:10.1109/LSP.2006.870374

Jayaweera, S. K. (2007). V-BLAST-based virtual MIMO for distributed wireless sensor networks. *IEEE Transactions on Communications, 55*(10), 1867–1872. doi:10.1109/TCOMM.2007.906389

Jayaweera, S. K., Chebolu, M. L., & Donapati, R. K. (2007). Signal-processing-aided distributed compression in virtual MIMO-based wireless sensor networks. *IEEE Transactions on Vehicular Technology, 56*(5), 2630–2640. doi:10.1109/TVT.2007.900361

Jerbi, M., Senouci, S. M., Ghamri, Y., & Beylot, A. L. (2008). (Manuscript submitted for publication). Geo-localized virtual infrastructure for urban vehicular networks. *IEEE PIMRC*.

Jin, H., Khandekar, A., & McEliece, R. (2000, September). Irregular repeat-accumulate codes. In *Proceedings of International Symposium on Turbo Codes Related Topics,* Brest, France (pp. 1–8).

Jindal, N., Rhee, W., Vishwanath, S., Jafar, S., & Goldsmith, A. (2005). Sum power iterative water-filling for multi-antenna Gaussian broadcast channels. *IEEE Transactions on Information Theory, 51*(4), 1570–1580. doi:10.1109/TIT.2005.844082

Jing, Y., & Hassibi, B. (2003). Unitary space-time modulation via Cayley transforms. *IEEE Transactions on Signal Processing, 51*(11), 2891–2904.

Jing, Y., & Hassibi, B. (2006). Distributed space-time coding in wireless relay networks. *IEEE Transactions on Wireless Communications, 5*(12), 3524–3536.

Jing, Y., & Hassibi, B. (2008). Diversity analyses of distributed space time codes in relay networks with multiple transmit/receive antennas. *Eurosip Journal on Advances in Signal Processing, 2008*.

Jing, Y., & Jafarkhani, H. (2007). Network beamforming using relays with perfect channel information. In *Proc. IEEE ICASSP'07* (pp. 15—20).

Jing, Y., & Jafarkhani, H. (2007). Using orthogonal and quasi-orthogonal designs in wireless relay networks. *IEEE Transactions on Information Theory, 53*, 4106–4118. doi:10.1109/TIT.2007.907516

Jing, Y., & Jafarkhani, H. (2008). Distributed differential space-time coding for wireless relay networks. *IEEE Transactions on Communications, 56*(7), 1092–1100.

Jing, Y., & Jafarkhani, H. (2008). Network beamforming with channel mean and covariance at relays. In *Proc. IEEE ICC'08* (pp. 19—23).

Jing, Y., & Jafarkhani, H. (2008). Single and multiple relay selection schemes and their diversity orders. In *Proc. IEEE ICC'08* (pp. 349—353).

Joham, M., Brehmer, J., & Utschick, W. (2002). MMSE approaches to multiuser spatio-temporal Tomlinson-Harashima precoding. In *Proceedings of ITG Conference on Source and Channel Coding* (pp. 387–394).

Johnson, M., Ishwar, P., Prabhakaran, V., Schonberg, D., & Ramchandran, K. (2004). On compressing encrypted data. *IEEE Transactions on Signal Processing, 52*(10), 2992–3006. doi:10.1109/TSP.2004.833860

Jorswieck, E., & Boche, H. (2006a). Majorization and matrix-monotone functions in wireless communications. *Foundations and Trends in Communications and Information Theory, 3*(6), 553–701. doi:10.1561/0100000026

Jorswieck, E., & Boche, H. (2006b). Performance analysis of MIMO systems in spatially correlated fading using matrix-monotone functions. *IEICE Transactions, 89-A*(5), 1454–1472.

Jorswieck, E., & Boche, H. (2007). Majorization and matrix monotone functions in wireless communications.

Foundations and Trends in Communications and Information Theory, 3(6), 553–701. doi:10.1561/0100000026

Jorswieck, E., Sezgin, A., Ottersten, B., & Paulraj, A. (2008). Feedback reduction in uplink MIMO OFDM systems by chunk optimization. *Journal on Advances in Signal Processing, 2008*, 597072.

Ju, M.-C., Song, H.-K., & Kim, I.-M. (2008). Exact BER analysis of distributed Alamouti's code for cooperative diversity networks. *IEEE Trans. Commun.*, in second round of revision.

Kaewgrapha, P., Puttarak, N., Wang, H., & Li, J. (2008, November). Receiver-cooperation: Network coding and distributed scheduling. In *Proc. IEEE Global Communications Conference*, New Orleans.

Kaleh, G. K. (1996). Frequency-diversity spread-spectrum communication system to counter bandlimited gaussian interference. *IEEE Transactions on Communications, 44*, 886–893. doi:10.1109/26.508308

Kammoun, I., & Belfiore, J. C. (2003). A new family of Grassmann space time codes for non-coherent MIMO systems. *IEEE Trans. Communication Letters, 07*(11), 528–530.

Karagiannidis, G. K. (2006). Performance bounds of multihop wireless communications with blind relays over generalized fading channels. *IEEE Transactions on Wireless Communications, 5*(3), 498–503. doi:10.1109/TWC.2006.1611077

Karagiannidis, G. K., Tsiftsis, T. A., & Mallik, R. K. (2006). Bounds for multihop relayed communications in Nakagami-*m* fading. *IEEE Transactions on Communications, 54*(1), 18–22. doi:10.1109/TCOMM.2005.861679

Karaki, J. N. A., & Kamal, A. E. (2004). Routing techniques in wireless sensor networks: A survey. *IEEE Wirel. Commun., 6*(11), 6–28. doi:10.1109/MWC.2004.1368893

Karmakar, S., & Rajan, B. S. (2006, July 9-14). Minimum-decoding complexity, maximum-rate space-time block codes from Clifford algebras. In *Proc. IEEE International Symposium on Information Theory*, Seattle, WA (pp. 788-792).

Karp, R., Schindelhauer, C., Shenker, S., & Vocking, B. (2000). Randomized rumor spreading. In *Proc. of*

the *41ˢᵗ Annual IEEE Symp. on Foundations of Comp. Sci.* (pp. 565-574).

Kaspi, A., & Berger, T. (1982). Rate-distortion for correlated sources with partially separated encoders. *IEEE Transactions on Information Theory, 28*(6), 828–840. doi:10.1109/TIT.1982.1056586

Katti, S., Rahul, H., Hu, W., Katabi, D., Medard, M., & Crowcroft, J. (2006, September). XORs in the air: Practical wireless network coding. In *Proc. ACM SIGCOMM* (pp. 243–254).

Kay, S. M. (1993). *Fundamentals of statistical signal processing: Estimation theory*. Englewood Cliffs, NJ: Prentice Hall.

Kaya, O. (2006). Window and backwards decoding achieve the same sum rate for the fading cooperative Gaussian multiple access channel. In *Proceedings of IEEE Global Communications Conference,* San Francisco, CA (pp. 1–5).

Kaya, O., & Ulukus, S. (2006). Achieving the capacity region boundary of fading CDMA channels via generalized iterative waterfilling. *IEEE Transactions on Wireless Communications, 5*(11), 3215–3223. doi:10.1109/TWC.2006.04785

Kaya, O., & Ulukus, S. (2007). Power control for fading cooperative multiple access channels. *IEEE Transactions on Communications, 6*, 2915–2923. doi:10.1109/TWC.2007.05858

Keller, T., & Hanzo, L. (2000). Adaptive modulation techniques for duplex OFDM transmission. *IEEE Transactions on Vehicular Technology, 49*(5), 1893–1906. doi:10.1109/25.892592

Kermoal, J., Schumacher, L., Pedersen, K., Mogensen, P., & Frederiksen, F. (2002). A stochastic MIMO radio channel model with experimental validation. *IEEE Journal on Selected Areas in Communications, 20*(6), 1211–1226. doi:10.1109/JSAC.2002.801223

Keshavarz-Haddad, A., Ribeiro, V., & Riedi, R. (2006, September 24-29). Broadcast capacity in multihop wireless networks. In *Proceedings of the 12ᵗʰ Annual International Conference on Mobile Computing and Networking, MobiCom'06* (pp. 239– 250).

Khajehnouri, N., & Sayed, A. H. (2007). Distributed MMSE relay strategies for wireless sensor networks. *IEEE Transactions on Signal Processing, 55*(7), 3336–3348. doi:10.1109/TSP.2007.894410

Khan, M. Z., & Rajan, B. S. (2006). Single-symbol maximum likelihood decodable linear STBCs. *IEEE Transactions on Information Theory, 52*, 2062–2091. doi:10.1109/TIT.2006.872970

Khisti, A., Erez, U., & Wornell, G. (2006, June). Fundamental limits and scaling behavior of cooperative multicasting in wireless networks. *IEEE Transactions on Information Theory, 52*(6), 2762–2770. doi:10.1109/TIT.2006.874541

Khojastepour, M., Ahmed, N., & Aazhang, B. (2004). Code design for the relay channel and factor graph decoding. In *Proceedings of Asilomar Conference on Signals, Systems and Computers* (Vol. 2, pp. 2000–2004).

Khojastepour, M., Sabharwal, A., & Aazhang, B. (2003). On capacity of Gaussian 'cheap' relay channel. In . *Proceedings of IEEE Global Communications Conference, 3*, 1776–1780.

Kim, D., Kwon, U. K., & Im, G. H. (2008). Pilot position selection and detection for channel estimation of SC-FDE. *IEEE Communications Letters, 12*(5), 350–352. doi:10.1109/LCOMM.2008.080071

Kim, J., & Bohacek, S. (2005). Selection metrics for multihop cooperative relaying. *Annual Mediterranean Ad Hoc Networking Workshop.*

Kiran, T., & Rajan, B. S. (2005). STBC schemes with nonvanishing determinant for certain number of transmit antennas. *IEEE Transactions on Information Theory, 51*(08), 2984–2992.

Kiran, T., & Rajan, B. S. (2006). Distributed space-time codes with reduced decoding complexity. In *Proc. IEEE ISIT'06* (pp. 542—546).

Kiran, T., & Rajan, B. S. (2006). Partially-coherent distributed space-time codes with differential encoder and decoder. *IEEE Journal on Selected Areas in Communications, 25*(2), 426–433.

Knopp, R., & Humblet, P. (1995). Information capacity and power control in single-cell multiuser communications. In *Proc. IEEE International Conference on Communications, 1*(6), 331-335

Kohno, R., Imai, H., Hatori, M., & Pasupathy, S. (1990). Combinations of an adaptive array antenna and a canceller of interference for direct-sequence spread-spectrum multiple-access system. *IEEE Journal on Selected Areas in Communications, 8*(4), 675–682. doi:10.1109/49.54463

Konstantinou, K., Kang, S., & Tzaras, C. (2007). A measurement-based model for mobile-to-mobile UMTS links. *IEEE VTC-Spring, 2007*, 529–533.

Korkmaz, G., Ekici, E., & Ozguner, F. (2007, September). Black-burst based multihop broadcast protocols for vehicular networks. *IEEE Trans. on Vehicular Tech., 56*(5).

Koyuncu, E., Jing, Y., & Jafarkhani, H. (2008). Distributed beamforming in wireless relay networks with quantized feedback. *IEEE Journal on Selected Areas in Communications, 26*, 1429–1439. doi:10.1109/JSAC.2008.081009

Kramer, G. (2004). Models and theory for relay channels with receive constraints. In *Proceedings of 42nd Allerton Conference on Communication, Control and Computing.*

Kramer, G., & van Wijngaarden, A. J. (2000). On the white Gaussian multiple-access relay channel. In *Proceedings of the IEEE Inernational Symposium on Information Theory (ISIT)*, Sorrento, Italy (p. 40).

Kramer, G., Berry, R., El Gamal, A., El Gamal, H., Franceschetti, M., Gastpar, M., & Laneman, J. N. (2007). Introduction to the special issue on models, theory, and codes for relaying and cooperation in communication networks. Special Issue in *IEEE Trans. Info. Theory, 53*(10).

Kramer, G., Gastpar, M., & Gupta, P. (2003). Capacity theorems for wireless relay channels. In *Proc. 41st Allerton Conf. on Comm., Control, and Comp.*

Kramer, G., Gastpar, M., & Gupta, P. (2003). Capacity theorems for wireless relay channels. In *Proceedings of the Allerton Conference*, Monticello, IL (pp. 1074-1083).

Kramer, G., Gastpar, M., & Gupta, P. (2005). Cooperative strategies and capacity theorems for relay networks. *IEEE Transactions on Information Theory, 51*(9), 3037–3063. doi:10.1109/TIT.2005.853304

Kramer, G., Maric, I., & Yates, R. D. (2006). Cooperative communications. *Foundations and Trends in Networking, 1*(3-4). NOW Publishers Inc.

Krikidis, I., & Belfiore, J.-C. (2007a). Three scheduling schemes for amplify-and-forward relay environments. *IEEE Communications Letters, 5*(11), 414–416. doi:10.1109/LCOMM.2007.070077

Krikidis, I., & Belfiore, J.-C. (2007b). Scheduling for amplify-and-forward cooperative networks. *IEEE Trans. Vehic. Tech., 6*(56), 3780–379. doi:10.1109/TVT.2007.901062

Krikidis, I., Thompson, J., & Goertz, N. (2008a). A cross-layer approach for cooperative networks. *IEEE Trans. Vehic. Tech., 57*(5), 3257–3263. doi:10.1109/TVT.2008.915508

Krikidis, I., Thompson, J., McLaughlin, S., & Goertz, N. (2008). Amplify-and-forward with partial relay selection. *IEEE Communications Letters, 12*(4), 235–237. doi:10.1109/LCOMM.2008.071987

Krikidis, I., Thompson, J., McLaughlin, S., & Goertz, N. (2008c). Optimization issues for cooperative amplify-and-forward systems over block-fading channels. *IEEE Trans. Vehic. Tech., 57*(5), 2868–2884. doi:10.1109/TVT.2008.917235

Krikidis, I., Thompson, J., McLaughlin, S., & Goertz, N. (2008d). Relay selection in interference-limited amplify-and-forward systems. *IEEE Trans. Wirel. Commun.*, submitted for publication.

Krim, H., & Viberg, M. (1996). Two decades of array signal processing research. *IEEE Signal Processing Magazine, 13*(4), 67–94. doi:10.1109/79.526899

Kschischang, F., Frey, B., & Loeliger, H.-A. (2001). Factor graphs and the sum-product algorithm. *IEEE Transactions on Information Theory, 47*(2), 498–519. doi:10.1109/18.910572

Kusume, K., Joham, M., Utschick, W., & Bauch, G. (2005). Efficient Tomlinson-Harashima precoding for spatial multiplexing on flat MIMO channel. In . *Proceedings of IEEE International Conference Communications, 3,* 2021–2025.

Kwon, U. K., Kim, D., & Im, G. H. (in press). Amplitude clipping and iterative reconstruction of MIMO-OFDM signals with optimum equalization. *IEEE Transactions on Wireless Communications.*

Kyasanur, P., & Vaidya, N. H. (2003). Detection and handling of MAC layer misbehavior in wireless networks. *International Conference DSN'03* (pp. 173-182).

Lai, L., Liu, K., & El Gamal, H. (2006). The three-node wireless network: Achievable rates and cooperation strategies. *IEEE Transactions on Information Theory, 52*(3), 805–828. doi:10.1109/TIT.2005.864421

Lai, L., Liu, K., & Gamal, H. E. (2006). The three node wireless network: Achievable rates and cooperation strategies. *IEEE Transactions on Information Theory, 52*(3), 805. doi:10.1109/TIT.2005.864421

Lam, C.-T., Falconer, D., Danilo-Lemoine, F., & Dinis, R. (2006). Channel estimation for SC-FDE systems using frequency domain multiplexed pilots. In *Proceedings of IEEE Vehicular Technology Conference*, Montreal, Canada.

Laneman, J. N. (2002). *Cooperative diversity in wireless networks: Algorithms and architectures.* Unpublished doctoral thesis, Massachusetts Institute of Technology, Cambridge, MA.

Laneman, J. N., & Kramer, G. (2004). Window decoding for the multiple access channel with generalized feedback. In *Proceedings of the IEEE International Symposium on Information Theory*, Chicago, IL (p. 279).

Laneman, J. N., & Wornell, G. W. (2000). Energy-efficient antenna sharing and relaying for wireless networks. *IEEE WCNC, 1*, 7–12.

Laneman, J. N., & Wornell, G. W. (2003). Distributed space-time coded protocols for exploiting cooperative diversity in wireless networks. *IEEE Transactions on Information Theory, 49*(10), 2415. doi:10.1109/TIT.2003.817829

Laneman, J. N., Tse, D. N. C., & Wornell, G. (2004). Cooperative diversity in wireless networks: Efficient protocols and outage behavior. *IEEE Transactions on Information Theory, 50*(12), 3062–3080. doi:10.1109/TIT.2004.838089

Laneman, J. N., Tse, D. N. C., & Wornell, G. W. (2004). Cooperative diversity in wireless networks: Efficient protocols and outage behavior. *IEEE Transactions on*

Information Theory, *50*(12), 3062–3080. doi:10.1109/TIT.2004.838089

Laneman, J. N., Wornell, G. W., & Tse, D. N. C. (2001). An efficient protocol for realizing cooperative diversity in wireless networks. In *Proceedings of IEEE International Symposium on Information Theory* (p. 294).

Laneman, N., & Wornell, G. W. (2003). Distributed space-time-coded protocols for exploiting cooperative diversity in wireless networks. *IEEE Transactions on Information Theory*, *49*(10), 2415–2425.

Lange, K. (2004). *Optimization (Springer texts in statistics)*. Springer-Verlag.

Larsson, E. G., & Stoica, P. (2003). *Space-time block coding for wireless communications*. Cambridge, UK: Cambridge University Press.

Larsson, E. G., & Vojcic, B. R. (2005). Cooperative transmit diversity based on superposition coding. *IEEE Communications Letters*, *9*(9), 778–780. doi:10.1109/LCOMM.2005.1506700

Larsson, E. G., & Vojcic, B. R. (2005). Cooperative transmit diversity based on superposition modulation. *IEEE Communications Letters*, *9*(9), 778–780. doi:10.1109/LCOMM.2005.1506700

Lee, I.-H., & Kim, D. (2007). BER analysis for decode-and-forward relaying in dissimilar Rayleigh fading channels. *IEEE Communications Letters*, *11*(1), 52–54. doi:10.1109/LCOMM.2007.061375

Lee, W. (1990). *Mobile cellular telecommunications systems*. New York: McGraw-Hill, Inc.

Letzepis, N. (2005). *Gaussian cellular multiple access channels*. Unpublished doctoral dissertation, Institute for Telecommunications Research, University of South Australia.

Li, C., & Wang, X. (2008). Cooperative multibeamforming in ad hoc networks. *EURASIP Journal on Advances in Signal Processing*, Article ID 310247.

Li, J., & Hu, R. (2005). Slepian-Wolf cooperation: A practical and efficient compress-and-forward relay scheme. In *Proceedings of Allerton Conference on Communications, Control, and Computing*, Champaign, IL.

Li, J., Tu, Z., & Blum, R. (2004). Slepian-Wolf coding for nonuniform sources using turbo codes. In *Proceedings of IEEE Data Compression Conference*, Snowbird, UT (pp. 312–321).

Li, S.-Y. R., Yeung, R. W., & Cai, N. (2003, February). Linear network coding. *IEEE Trans. Info. Theory*, 371-381.

Li, X. (2004). Space-time coded multitransmission among distributed transmitters without perfect synchronization. *IEEE Signal Processing Letters*, *11*(12), 948–952. doi:10.1109/LSP.2004.838213

Li, X., & Ritcey, J. A. (1999a). Bit-interleaved coded modulation with iterative decoding. In *Proceedings of IEEE International Conference on Communications* (Vol. 2, pp. 858–863).

Li, X., & Ritcey, J. A. (1999b). Trellis-coded modulation with bit interleaving and iterative decoding. *IEEE Journal on Selected Areas in Communications*, *17*(4), 715–724. doi:10.1109/49.761047

Li, Y., & Kishore, S. (2007). Asymptotic analysis of amplify-and-forward relaying in Nakagami-fading environments. *IEEE Transactions on Wireless Communications*, *6*(12), 4256–4262. doi:10.1109/TWC.2007.060368

Li, Y., & Xia, X. G. (2005). Full diversity distributed space-time trellis codes for asynchronous cooperative communications. *IEEE ISIT*, *2005*, 911–915.

Li, Y., & Xia, X.-G. (2007). A family of distributed space-time trellis codes with asynchronous cooperative diversity. *IEEE Transactions on Communications*, *55*, 790–800. doi:10.1109/TCOMM.2007.894117

Li, Y., Murata, H., & Yoshida, S. (1998). Coding for multi-user detection in interference channel. In *Proceedings of IEEE Global Communications Conference* (Vol. 6, pp. 3596–3601).

Li, Z., & Xia, X.-G. (2007). A simple Alamouti space-time transmission scheme for asynchronous cooperative systems. *IEEE Signal Processing Letters*, *14*, 804–807. doi:10.1109/LSP.2007.900224

Liang, K.-C., Wang, X., & Berenguer, I. (2007). Minimum error rate linear dispersion codes for cooperative relays. *IEEE Transactions on Vehicular Technology*, *6*, 2143–2157. doi:10.1109/TVT.2007.897636

Liang, P.-P., & Stark, W. (2001). Iterative multiuser detection for turbo-coded FHMA communications. *IEEE Journal on Selected Areas in Communications, 19*(9), 1810–1819. doi:10.1109/49.947045

Liang, X. (2003). Orthogonal designs with maximal rates. *IEEE Transactions on Information Theory, 49*(10), 2468–2503.

Liang, X. B. (2003). On the nonexistence of rate-one generalized complex orthogonal designs. *IEEE Transactions on Information Theory, 49*(11), 2984–2989. doi:10.1109/TIT.2003.818396

Liang, X. B. (2005). A complex orthogonal space-time block code for 8 transmit antennas. *IEEE Communications Letters, 9*(2), 115–117. doi:10.1109/LCOMM.2005.02007

Liang, Y., Yoo, T., & Goldsmith, A. (2006). Coverage spectral efficiency of cellular systems with cooperative base stations. In *IEEE Global Telecommunications Conference*, San Francisco, CA.

Liang, Y.-C., & Zhang, R. (2008). Optimal analogue relaying with multiple-antennas for physical layer network coding. In *Proceedings of IEEE International Conference on Communications*, Beijing, China.

Liao, H. (1972). A coding theorem for multiple access communications. In *Proceedings of IEEE International Symposium on Information Theory*, Asilomar, CA.

Librino, F., Levorato, M., & Zorzi, M. (2007). Distributed cooperative routing and hybrid ARQ in MIMO-BLAST ad hoc networks. *IEEE Globecom, 2007*, 657–662.

Lim, J. B., Choi, C. H., Yune, T. W., & Im, G. H. (2007). Iterative multiuser detection for single-carrier frequency-domain equalization. *IEEE Communications Letters, 11*(6), 471–473. doi:10.1109/LCOMM.2007.070063

Lin, S., & Costello, D. J. (1983). *Error control coding: Fundamentals and applications*. NJ: Prentice-Hall.

Lin, Y.-D., & Hsu, Y.-C. (2000). Multihop cellular: A new architecture for wireless communications. *IEEE Infocom, 3*, 1273–1282.

Litva, J., & Lo, T. K. Y. (1996). *Digital beamforming in wireless communications*. Boston, MA: Artech House.

Liu, P., Tao, Z., Lin, Z., Erkip, E., & Panwar, S. (2006). Cooperative wireless communications: A cross-layer approach. *IEEE Wireless Communications, 13*(4), 84–92. doi:10.1109/MWC.2006.1678169

Liu, P., Tao, Z., Narayanan, S., Korakis, T., & Panwar, S. S. (2007). CoopMAC: A cooperative MAC for wireless LANs. *IEEE JSAC, 25*(2), 340–354.

Liu, Q., Zhou, S., & Giannakis, G. B. (2004). Cross-layer combining of adaptive modulation and coding with truncated ARQ over wireless links. *IEEE Transactions on Wireless Communications, 22*(5), 1746–1755. doi:10.1109/TWC.2004.833474

Liu, R., Spasojevic, P., & Soljanin, E. (2003). User cooperation with punctured turbo codes. In *Proceedings of Allerton Conference on Communications, Control, and Computing*, Monticello, IL (pp. 201–210).

Liu, Z., Cheng, S., Liveris, A., & Xiong, Z. (2004). Slepian-Wolf coded nested quantization (SWC-NQ) for Wyner-Ziv coding: Performance analysis and code design. In *Proceedings of IEEE Data Compression Conference*, Snowbird, UT (pp. 322–331).

Liu, Z., Cheng, S., Liveris, A., & Xiong, Z. (2006). Slepian-Wolf coded nested lattice quantization for Wyner-Ziv coding: High-rate performance analysis and code design. *IEEE Transactions on Information Theory, 52*(10), 4358–4379. doi:10.1109/TIT.2006.881708

Liu, Z., Giannakis, G. B., & Hughes, B. L. (2001). Double-differential space-time block coding for time-selective fading channels. *IEEE Transactions on Communications, 49*(9), 1529–1539. doi:10.1109/26.950340

Liu, Z., Stankovic, V., & Xiong, Z. (2005). Wyner-Ziv coding for the half-duplex relay channel. In *Proceedings of IEEE International Conference on Acoustic, Speech, and Signal Processing* (Vol. 5, pp. 1113–1116).

Liveris, A., Xiong, Z., & Georghiades, C. (2002a). Compression of binary sources with side information at the decoder using LDPC codes. *IEEE Communications Letters, 6*(10), 440–442. doi:10.1109/LCOMM.2002.804244

Liveris, A., Xiong, Z., & Georghiades, C. (2002b). A distributed source coding technique for correlated images using turbo-codes. *IEEE Communications Letters, 6*(9), 379–381. doi:10.1109/LCOMM.2002.803479

Liveris, A., Xiong, Z., & Georghiades, C. (2002c). Joint source-channel coding of binary sources with side information at the decoder using IRA codes. In *Proceedings of IEEE Multimedia Signal Processing Workshop* (pp. 53–56).

LMSC. (2008). *Draft amendment to IEEE standard for local and metropolitan area networks-part 16: Air interface for fixed and mobile broadband wireless access systems-multihop relay specification*. Retrieved from http://www.ieee802.org/16/

Lo, C., Vishwanath, S., & Heath, R. W., Jr. (2005). Rate bounds for MIMO relay channels using precoding. In *Proceedings of IEEE Global Communications Conference* (Vol. *3*, pp. 1172–1176).

Lo, Y. T. (1964). A mathematical theory of antenna arrays with randomly spaced elements. *IEEE Transactions on Antennas and Propagation, 12*(3), 257–268. doi:10.1109/TAP.1964.1138220

Loyka, S. (2001). Channel capacity of MIMO architecture using the exponential correlation matrix. *IEEE Communications Letters, 5*(9), 369–371. doi:10.1109/4234.951380

Luby, M. (2002). LT codes. In *Proc. 43rd Annual IEEE Symposium on Foundations of Computer Science (FOCS)* (pp. 271-282).

Luby, M. (2002). LT codes. In *Proc. IEEE Symp. on Foundations of Computer Sci.* (pp. 271-280).

Luby, M. G., Mitzenmacher, M., Shokrollahi, M. A., & Spielman, D. A. (1998). Analysis of low density codes and improved designs using irregular graphs. In *Proceedings of ACM Symposium on Theory Computing,* Dallas, TX (pp. 249–258).

Luby, M. G., Mitzenmacher, M., Shokrollahi, M. A., & Spielman, D. A. (2001). Efficient erasure correcting codes. *IEEE Transactions on Information Theory, 47*(2), 569–584. doi:10.1109/18.910575

Luby, M. G., Mitzenmacher, M., Shokrollahi, M. A., Spielman, D. A., & Stemann, V. (1997). Practical loss-resilient codes. In *Proceedings of ACM Symposium on Theory Computing,* El Paso, TX (pp. 50–159).

Luby, M., Mitzenmacher, M., Shokrollahi, A., & Spielman, D. (2001, February). Efficient erasure correction codes. *IEEE Trans. Info. Theory, 47*.

Lucky, R. W. (1965). Automatic equalization for digital communication. *The Bell System Technical Journal, 44*(4), 547–588.

Luo, H., Ramjee, R., Sinha, P., Li, L., & Lu, S. (2003). UCAN: A unified cellular and ad hoc network architecture. *ACM Mobicom* (pp. 353-367).

MacKay, D. (1999). Good error-correcting codes based on very sparse matrices. *IEEE Trans. Info. Theory*.

MacKay, D., & Neal, R. (1997). Near Shannon limit performance of low density parity check codes. *IEE Electronics Letters, 33*(6), 457–458. doi:10.1049/el:19970362

Madan, R., Metha, N., Molisch, A., & Zhang, J. (2006). *Energy-efficient cooperative relaying over fading channels with simple relay selection* (Tech. Rep. No. 2006.075). Mitsubishi Electrical Research Lab.

Mahajan, R., Rodrig, M., Wetherall, D., & Zahorjan, J. (2005). Sustaining cooperation in multihop wireless networks. *ACM NSDI* (pp. 231-244).

Maham, B., & Hjørungnes, A. (2007). Distributed GABBA space-time codes in amplify-and-forward cooperation. In *Proc. IEEE ITW'07* (pp. 1-5).

Mahinthan V., Cai L., Mark J. W., & Shen X. (S.) (2008). Partner selection based on optimal power allocation in cooperative-diversity systems. *IEEE Trans. Vehic. Tech., 1*(57), 511-520.

Marcellin, M., & Fischer, T. (1990). Trellis coded quantization of memoryless and Gauss-Markov sources. *IEEE Transactions on Communications, 38*(1), 82–93. doi:10.1109/26.46532

Marcenko, V., & Pastur, L. (1967). Distributions of eigenvalues of some sets of random matrices. *Math. USSR-Sb., 1*, 507–536.

Maric, I., & Yates, R. D. (2004, August). Cooperative multihop broadcast for wireless networks. *IEEE J. Select. Areas Commun., 22*(6).

Maric, I., & Yates, R. D. (2005, January). Cooperative multicast for maximum network lifetime. *IEEE Journal on Selected Areas in Communications, 23*(1), 127–135. doi:10.1109/JSAC.2004.837343

Marina, M. K., & Das, S. R. (2001). On-demand multipath distance vector routing in ad hoc networks. In *Proc. IEEE Int. Conf. Netw. Protoc.*, Mission Inn, CA (pp. 14-23).

Marques, P., & Abrantes, S. (2008). On the derivation of the exact, closed-form capacity formulas for receiver-sided correlated MIMO channels. *IEEE Transactions on Information Theory, 54*(3), 1139–1161. doi:10.1109/TIT.2007.915692

Marsch, P., & Fettweis, G. (2008). On multicell cooperative transmission in backhaul-constrained cellular systems. *Annales des Télécommunications, 63*(5-6). doi:10.1007/s12243-008-0028-3

Marshall, A. W., & Olkin, I. (1979). Inequalities: Theory of majorization and its application. *Mathematics in Science and Engineering, 143*. Academic Press, Inc. (London) Ltd.

Marti, S., Giuli, T. J., Lai, K., & Baker, M. (2000). Mitigating routing misbehavior in mobile ad hoc networks. *ACM MobiCom* (pp. 255-265).

Martin, C., & Ottersten, B. (2004). Asymptotic eigenvalue distributions and capacity for MIMO channels under correlated fading. *IEEE Transactions on Wireless Communications, 3*(4), 1350–1359. doi:10.1109/TWC.2004.830856

Marton, K. (1979, May). A coding theorem for the discrete memoryless broadcast channel. *IEEE Transactions on Information Theory, 25*(3), 306–311. doi:10.1109/TIT.1979.1056046

Matlog. (2006). *Matlog: Logistics engineering Matlab toolbox (version 9)*. Available at http://www.ie.ncsu.edu/kay/matlog/MatlomgRef.htm

Maure, P., & Tarokh, V. (2001). Transmit diversity when receiver does not know the number of transmit antennas. In *Proceedings of International symposium on Wireless Personal Multimedia Communications*.

Merkey, P., & Posner, E. C. (1984). Optimum cyclic redundancy codes for noisy channels. *IEEE Transactions on Information Theory, 30*(6), 865–867. doi:10.1109/TIT.1984.1056971

Mesbah, W., & Davidson, T. N. (2006a). Optimal power allocation for full-duplex cooperative multiple access. In *Proceedings of IEEE International Conference on Acoustic, Speech, and Signal Processing,* Toulouse, France (Vol. *4*, pp. 689–692).

Mesbah, W., & Davidson, T. N. (2006b). Optimal power and resource allocation for half-duplex cooperative multiple access. In *Proceedings of IEEE International Conference Communications,* Istanbul, Turkey (Vol. *10*, pp. 4469–4473).

Mestre, X., Fonollosa, J., & Pages-Zamora, A. (2003). Capacity of MIMO channels: Asymptotic evaluation under correlated fading. *IEEE Journal on Selected Areas in Communications, 21*(5), 829–838. doi:10.1109/JSAC.2003.810352

Mheidat, H., & Uysal, M. (2007). Non-coherent and mismatched-coherent receivers for distributed STBCs with amplify-and-forward relaying. *IEEE Transactions on Wireless Communications, 6*, 4060–4070. doi:10.1109/TWC.2007.060180

Mheidat, H., Uysal, M., & Al-Dhahir, N. (2007). Equalization techniques for distributed space-time block codes with amplify-and-forward relaying. *IEEE Transactions on Signal Processing, 55*(5), 1839–1852. doi:10.1109/TSP.2006.889974

Michalopoulos, D. S., & Karagiannidis, G. K. (2007). Distributed switch and stay combining (DSSC) with a single decode and forward relay. *IEEE Communications Letters, 11*(5), 408–410. doi:10.1109/LCOMM.2007.070018

Michalopoulos, D. S., & Karagiannidis, G. K. (2008). (Accepted for publication). Performance analysis of single relay selection in Rayleigh fading. [retrieved from http://users.auth.gr/~dmixalo/]. *IEEE Transactions on Wireless Communications.*

Michalopoulos, D. S., & Karagiannidis, G. K. (2008). -{1}). PHY-layer fairness in amplify and forward cooperative diversity systems. *IEEE Transactions on Wireless Communications, 7*(3), 1073–1082. doi:10.1109/TWC.2008.060825

Michalopoulos, D. S., & Karagiannidis, G. K. (2008). -{2}). Performance analysis of single relay selection in Rayleigh fading. *IEEE Transactions on Wireless Communications, 7*(10), 3718–3724. doi:10.1109/TWC.2008.070492

Michalopoulos, D. S., & Karagiannidis, G. K. (2008). -{3}). Two-relay distributed switch and stay combining. *IEEE Transactions on Communications, 56*(11), 1790–1794. doi:10.1109/TCOMM.2008.070017

Michalopoulos, D. S., & Karagiannidis, G. K. (2008). PHY-layer fairness in amplify and forward cooperative diversity systems. *IEEE Transactions on Wireless Communications*, 7(3), 1073–1082. doi:10.1109/TWC.2008.060825

Michalopoulos, D. S., & Karagiannidis, G. K. (2008-{4}). Spectral efficient cooperative communications via spatial signal separation. *International Conference on Communications (ICC'08), IEEE* (pp. 1039-1043).

Michiardi, P., & Molva, R. (2002). CORE: A collaborative reputation mechanism to enforce node cooperation in mobile ad hoc networks. *Communication and Multimedia Security Conference* (pp. 107-121).

Mitran, P., & Bajcsy, J. (2002). Coding for the Wyner-Ziv problem with turbo-like codes. In *Proceedings of IEEE International Symposium on Information Theory*, Sorrento, Italy (p. 91).

Mitran, P., & Bajcsy, J. (2002). Turbo source coding: A noise-robust approach to data compression. In *Proceedings of IEEE Data Compression Conference*, Snowbird, UT (p. 465).

Miyakawa, M., & Harashima, H. (1969). A method of code conversion for a digital communication channel with intersymbol interference. *Transactions of the Institute of Electronics and Communications Engineers of Japan*, 52-A, 272–273.

Moher, M., & Guinand, P. (1998). An iterative algorithm for asynchronous coded multi-user detection. *IEEE Communications Letters*, 2(8), 229–231. doi:10.1109/4234.709440

Molisch, A. F. (2005). *Wireless communications*. New York: Wiley-IEEE Press.

Morgenshtern, V. I., & Bölcskei, H. (2007). Crystallization in large wireless networks. *IEEE Transactions on Information Theory*, 53(10), 3319–3349. doi:10.1109/TIT.2007.904789

Mosk-Aoyama, D., & Shah, D. (2006, July). Information dissemination via network coding. In *Proc. IEEE Int. Symp. Info. Theory*, Seattle, WA (pp. 1748-1752).

Moustafa, H., Bourdon, G., & Gourhant, Y. (2006). Authentication, authorization, and accounting (AAA) in hybrid ad hoc hotspots' environments. *ACM WMASH* (pp. 37-46).

Moustakas, A., & Simon, S. (2007). On the outage capacity of correlated multiple-path MIMO channels. *IEEE Transactions on Information Theory*, 53(11), 3887–3903. doi:10.1109/TIT.2007.907468

Mudumbai, R., Barriac, G., & Madhow, U. (2007). On the feasibility of distributed beamforming in wireless networks. *IEEE Transactions on Wireless Communications*, 6(5), 1754–1763. doi:10.1109/TWC.2007.360377

Mudumbai, R., Wild, B., Madhow, U., & Ramchandran, K. (2006). Distributed beamforming using 1 bit feedback: From concept to realization. In *Proceedings of 44th Allerton Conference on Communication, Control and Computing*, Monticello, IL.

Muller, R. (2002). A random matrix model of communication via antenna arrays. *IEEE Transactions on Information Theory*, 48(9), 2495–2506. doi:10.1109/TIT.2002.801467

Murugan, A., Azarian, K., & El Gamal, H. (2007). Cooperative lattice coding and decoding in half-duplex channels. *IEEE Journal on Selected Areas in Communications*, 25(2), 268–279. doi:10.1109/JSAC.2007.070204

Murugan, A., El Gamal, H., Damen, M. O., & Caire, G. (2006). A unified framework for tree search decoding: Rediscovering the sequential decoder. *IEEE Transactions on Information Theory*, 52(3), 933–953. doi:10.1109/TIT.2005.864418

Murugan, A., Gopala, P., & El Gamal, H. (2004). Correlated sources over wireless channels: Cooperative source-channel coding. *IEEE Journal on Selected Areas in Communications*, 22(6), 988–998. doi:10.1109/JSAC.2004.830889

Myung, H. G., Lim, J., & Goodman, D. J. (2006). Single carrier FDMA for uplink wireless transmission. *IEEE Vehicular Technology Magazine*, 1(3), 30–38. doi:10.1109/MVT.2006.307304

Nabar, R. U., Bölcskei, H., & Kneubühler, F. W. (2004). Fading relay channels: Performance limits and space-time signal designs. *IEEE Journal on Selected Areas in Communications*, 22, 1099–1109. doi:10.1109/JSAC.2004.830922

Nabar, R., & Bolcskei, H. (2003). Space-time signal design for fading relay channels. In *Proceedings of IEEE Global Communications Conference* (Vol. *4*, pp. 1952–1956).

Nadarajah, S., & Kotz, S. (2006). On the product and ratio of gamma and weibull random variables. *Econometric theory-Cambridge University Press*, 22, 338-344.

Nam, Y.-H., & El Gamal, H. (in press). Joint lattice decoding achieves the optimal diversity-multiplexing tradeoff of multiple access channels. Retrieved from http://www.ece.osu.edu/ helgamal/MAC_TITR1_H.pdf

Networking, 4, Article ID 20258.

Ng, B., Evans, J., Hanly, S., & Grant, A. (2004). Distributed linear multiuser detection in cellular networks. In *5th Australian Communications Theory Workshop* (pp. 127–132).

Nosratinia, A., & Hunter, T. E. (2007). Grouping and partner selection in cooperative wireless networks. *IEEE JSAC, 25*(2), 369–378.

Nosratinia, A., Hunter, T. E., & Hedayat, A. (2004). Cooperative communication in wireless networks. *IEEE Communications Magazine, 42*, 74–80. doi:10.1109/MCOM.2004.1341264

Ochiai, H., Mitran, P., Poor, H. V., & Tarokh, V. (2005). Collaborative beamforming for distributed wireless ad hoc sensor networks. *IEEE Transactions on Signal Processing, 53*(11), 4110–4124. doi:10.1109/TSP.2005.857028

Oechtering, T., & Boche, H. (2007). Optimal transmit strategies in multi-antenna bidirectional relaying. In *Proc. ICASSP,* (pp. III-145 – III-148).

Oechtering, T., & Sezgin, A. (2004). A new cooperative transmission scheme using the space-time delay code. *ITG Workshop on Smart Antennas* (pp. 41-48).

Oechtering, T., Schnurr, C., Bjelakovic, I., & Boche, H. (2008). Broadcast capacity region of two-phase bidirectional relaying. *IEEE Transactions on Information Theory, 54*(1), 454–458. doi:10.1109/TIT.2007.911158

Oechtering, T., Wyrembelski, R., & Boche, H. (2008). Optimal transmit strategy for the 2x1 MISO bidirectional broadcast channel. *SPAWC,* will appear.

Oggier, F. (2005). First application of cyclic algebras to non-coherent MIMO channels. In *Allerton Conference.*

Oggier, F., & Hassibi, B. (2006). *A coding strategy for wireless networks with no channel information.* Presented at the Forty-Fourth Annual Allerton Conference on Communication, Control, and Computing.

Oggier, F., & Hassibi, B. (2007). Algebraic Cayley differential space time codes. *IEEE Transactions on Information Theory, 53*(05).

Oggier, F., & Lequeu, E. (2008, July 6-11). Differential distributed space-time coding based on Cayley codes. In *Proceedings of IEEE International Symposium on Information Theory (ISIT) 2008*, Toronto, Canada (pp. 2548-2552).

Oggier, F., Sloane, A., Diggavi, S., & Calderbank, A. R. (2005). Nonintersecting subspaces based on finite alphabets. *IEEE Transactions on Information Theory, 51*(12), 4320–4325.

Onat, F. A., Adinoyi, A., Fan, Y., Yanikomeroglu, H., & Thompson, J. S. (2007). Optimum threshold for SNR-based selective digital relaying schemes in cooperative wireless networks. *Wireless Communications and Networking Conference, 2007,WCNC 2007, IEEE* (pp. 969-974).

Ong, L., & Motani, M. (2007). On the capacity of the single source multiple relay single destination mesh network. *Ad Hoc Networks, 5*(6), 786–800. doi:10.1016/j.adhoc.2006.12.006

Oohama, Y. (1997). Gaussian multiterminal source coding. *IEEE Transactions on Information Theory, 43*(6), 1912–1923. doi:10.1109/18.641555

Oohama, Y. (1998). The rate-distortion function for the quadratic Gaussian CEO problem. *IEEE Transactions on Information Theory, 44*(3), 1057–1070. doi:10.1109/18.669162

Oohama, Y. (2005). Rate-distortion theory for Gaussian multiterminal source coding systems with several side informations at the decoder. *IEEE Transactions on Information Theory, 51*(7), 2577–2593. doi:10.1109/TIT.2005.850110

Oppenheim, A., & Schafer, R. (1989). *Discrete-time signal processing*. Englewood Cliffs, NJ: Prentice-Hall.

Ordano, L., & Tallone, F. (1997). Dual polarized propagation channel: Theoretical model and experimental results. In *Proceeding 10th Int. Conf. Antennas and Propagation* (Vol. 2, pp. 363–366).

Oyman, Ö., & Sandhu, S. (2006). A Shannon-theoretic perspective on fading multihop networks. *IEEE Conf. on Information Sciences and Systems (CISS)* (pp. 525-530).

Ozarow, L. (1984). The capacity of the white Gaussian multiple access channel with feedback. *IEEE Transactions on Information Theory, 30*(4), 623–629. doi:10.1109/TIT.1984.1056935

Ozgur, A., Leveque, O., & Tse, D. N. C. (2007). Hierarchical cooperation achieves optimal capacity scaling in ad hoc networks. *IEEE Transactions on Information Theory, 53*(10), 3549–3572. doi:10.1109/TIT.2007.905002

Pabst, R., Walke, B. H., Schultz, D. C., Herhold, P., Yanikomeroglu, H., & Mukherjeee, S. (2004). Relay-based deployment concepts for wireless and mobile broadband radio. *IEEE Communications Magazine, 42*(9), 80–89. doi:10.1109/MCOM.2004.1336724

Palomar, D. P., & Jiang, Y. (2006). MIMO transceiver design via majorization theory. *Foundations and Trends in Comm. and Info. Theory, 3*(4-5), 331–551.

Papoulis, A. (1991). *Probability, random variables, and stochastic processes, 3rd edition*. Singapore: McGraw-Hill Book Company.

Patel, C. S., Stüber, G. L., & Pratt, T. G. (2006). Statistic properties of amplify and forward relay fading channels. *IEEE Transactions on Vehicular Technology, 55*(1), 1–9. doi:10.1109/TVT.2005.861170

Pattem, S., Krishnamachari, B., & Govindan, R. (2004, April). *The impact of spatial correlation on routing with compression in sensor networks*. Paper presented at IPSN.

Peel, C. (2003). On "dirty-paper coding." . *IEEE Signal Processing Magazine, 20*(3), 112–113. doi:10.1109/MSP.2003.1203214

Peng, C., Zhang, Q., Zhao, M., & Yao, Y. (2007). SNCC: A selective network-coded cooperation scheme in wireless networks. In *Proc. of IEEE Int. Conf. Commun.*

Penrose, M. (1997). The longest edge of the random minimal spanning tree. *Annals of Applied Probability, 7*(2), 340–361. doi:10.1214/aoap/1034625335

Pereira, S., Sezgin, A., Paulraj, A., & Papanicolaou, G. C. (2008, July 6-11). Interference limited broadcast: Role of interferer geometry. *ISIT 2008*, Toronto, Canada.

Perez, L. C., Seghers, J., & Costello, D. J. Jr. (1996). A distance spectrum interpretation of turbo codes. *IEEE Transactions on Information Theory, 42*(6), 1698–1709. doi:10.1109/18.556666

Ping, L., Leung, W. K., & Wu, K. Y. (2003). Low-rate turbo-Hadamard codes. *IEEE Transactions on Information Theory, 49*(12), 3213–3224. doi:10.1109/TIT.2003.820018

Ping, L., Liu, L., & Leung, W. K. (2003). A simple approach to near-optimal multiuser detection: Interleave-division multiple-access. In *Proceedings of IEEE Wireless Communications and Networking Conference* (pp. 391–396).

Ping, L., Liu, L., Wu, K. Y., & Leung, W. K. (2004). Approaching the capacity of multiple access channels using interleaved low-rate codes. *IEEE Communications Letters, 8*(1), 4–6. doi:10.1109/LCOMM.2003.822534

Pradhan, S. S., & Ramchandran, K. (1999). Distributed source coding using syndromes (DISCUS): Design and construction. In *Proceedings of IEEE Data Compression Conference*, Snowbird, UT (pp. 158–167).

Pradhan, S. S., & Ramchandran, K. (2000a). A constructive approach to distributed source coding with symmetric rates. In *Proceedings of IEEE International Symposium on Information Theory* (p. 178).

Pradhan, S. S., & Ramchandran, K. (2000b). Distributed source coding: Symmetric rates and applications to sensor networks. In *Proceedings of IEEE Data Compression Conference*, Snowbird, UT (pp. 363–372).

Prasad, N., & Varanasi, M. K. (2004). Diversity and multiplexing tradeoff bounds for cooperative diversity protocols. In *Proceedings of IEEE International Symposium on Information Theory* (p. 268).

Prasad, R. (2004). *OFDM for wireless communications systems.* London: Artech House.

Proakis, J. G. (2000). *Digital Communications,* fourth ed. New York: McGraw-Hill, Inc.

Proakis, J. G. (2001). *Digital Communications.* New York: McGraw-Hill.

Qien, G. E., Holm, H., & Hole, K. J. (2004). Impact of channel prediction on adaptive coded modulation performance in Rayleigh fading. *IEEE Transactions on Vehicular Technology, 53*(3), 758–769. doi:10.1109/TVT.2004.827156

Rabaey, J. M., Ammer, M. J., da Silva, J. L., Patel, D., & Roundy, S. (2000). PicoRadio supports ad hoc ultra-lowpower wireless networking. *IEEE Computers, 33*(7), 42–48.

Rabiner, W., & Heizelman, A. Chandrakasan, & Hari, B. (2000). Energy-efficient communication protocol for wireless microsensor networks. In *Proceedings of the Hawaii International Conference on System Science.*

Raghavan, V., Kotecha, J. H., & Sayeed, A. M. (2008). Why does a Kronecker model result in misleading capacity estimates? Submitted to *IEEE Transactions on Information Theory.*

Raj Rao, N., & Edelman, A. (2007). The polynomial method for random matrices. In *Foundations of Computational Mathematics.*

Rajan, G. S., & Rajan, B. S. (2006). A non-orthogonal distributed space-time coded protocol, part-II: Code construction and DM-G tradeoff. In *Proc. IEEE ITW'06* (pp. 488—492).

Rajan, G. S., & Rajan, B. S. (2007). Algebraic distributed space-time codes with low ML decoding complexity. In *Proc. IEEE ISIT'07* (pp. 1516—1520).

Rajan, G. S., & Rajan, B. S. (2007). Distributed space-time codes for cooperative networks with partial CSI. In *Proc. IEEE WCNC'07* (pp. 902–906).

Rajan, G. S., & Rajan, B. S. (2007). Noncoherent low-decoding-complexity space-time codes for wireless relay networks. In *Proc. IEEE International Symposium on Information Theory (ISIT'07),* Nice, France (pp.1521–1525).

Rajan, G. S., & Rajan, B. S. (2007). Signal set design for full-diversity low-decoding-complexity differential scaled-unitary STBCs. In *Proceedings of IEEE International Symposium on Information Theory, (ISIT 2007),* Nice, France (pp. 1616-1620).

Rajan, G. S., & Rajan, B. S. (2008). Algebraic distributed differential space time codes with low decoding complexity. *IEEE Transactions on Wireless Communications, 7*(10), 3962–3971.

Rajan, G. S., & Rajan, B. S. (2008). Leveraging coherent distributed space-time codes for noncoherent communication in relay networks via training. To appear in *IEEE Trans. Wireless Communication.* Also available at arXiv: 0804.3155v2 [cs.IT] 3 Sep 2008.

Rajan, G. S., & Rajan, B. S. (2008). OFDM based distributed space time coding for asynchronous relay networks. In *Proc. IEEE ICC'08* (pp. 1118—1122).

Rajan, G. S., Tandon, A., & Rajan, B. S. (2007). On four-group ML decodable distributed space time codes for cooperative communication. In *Proc. IEEE WCNC'07* (pp. 11—15).

Rankov, B., & Wittneben, A. (2007). Spectral efficient protocols for half-duplex fading relay channels. *IEEE Journal on Selected Areas in Communications, 25*(2), 379–389. doi:10.1109/JSAC.2007.070213

Rankov, B., & Wittneben, A. (2007). Spectral efficient protocols for half-duplex fading relay channels. *IEEE Journal on Selected Areas in Communications, 25*(2), 379–389. doi:10.1109/JSAC.2007.070213

Rappaport, T. S. (2003). *Wireless communications principles and practice,* 2nd ed. Prentice Hall.

Raya, M., Hubaux, J. P., & Aad, I. (2004). Domino: A system to detect greedy behavior in IEEE 802.11 hotspots. *ACM MobiSys* (pp. 84-97).

Reed, M., Schlegel, C., Alexander, P., & Asenstorfer, J. (1998). Iterative multiuser detection for CDMA with FEC: Near-single-user performance. *IEEE Transactions on Communications, 46*(12), 1693–1699. doi:10.1109/26.737408

Reznik, A., Kulkarni, S., & Verdu, S. (2004). Degraded Gaussian multirelay channel: Capacity and optimal power allocation. *IEEE Transactions on Information Theory*, *50*(12), 3037–3046. doi:10.1109/TIT.2004.838373

Ribeiro, A., Cai, X., & Giannakis, G. B. (2005). Symbol error probabilities for general cooperative links. *IEEE Transactions on Wireless Communications*, *4*(3), 1264–1273. doi:10.1109/TWC.2005.846989

Richardson, T., & Urbanke, R. (2001). The capacity of low-density parity-check codes under message-passing decoding. *IEEE Transactions on Information Theory*, *47*(2), 599–618. doi:10.1109/18.910577

Richardson, T., Shokrollahi, M., & Urbanke, R. (2001). Design of capacity-approaching irregular low-density parity-check codes. *IEEE Transactions on Information Theory*, *47*(2), 619–637. doi:10.1109/18.910578

Roumy, A., Guemghar, S., Caire, G., & Verdu, S. (2004). Design methods for irregular repeat-accumulate codes. *IEEE Transactions on Information Theory*, *50*(8), 1711–1727. doi:10.1109/TIT.2004.831778

Sadek, M., Tarighat, A., & Sayed, A. H. (2007). A leakage-based precoding scheme for downlink multi-user MIMO channels. *IEEE Transactions on Wireless Communications*, *6*(5), 1711–1721. doi:10.1109/TWC.2007.360373

Sampath, H., Stoica, P., & Paulraj, A. (2001). Generalized linear precoder and decoder design for MIMO channels using the weighted MMSE criterion. *IEEE Transactions on Communications*, *49*(12), 2198–2206. doi:10.1109/26.974266

Sankaranarayanan, L., Kramer, G., & Mandayam, N. B. (2004). Capacity theorems for the multiple-access relay channel. In *Proceedings of the 42nd Annual Allerton Conference on Communication, Control, and Computing*, Montocello, IL (pp. 1782-1791).

Savazzi, S., & Spagnolini, U. (2007). Distributed orthogonal space time coding: Design and outage analysis for randomized cooperation. *IEEE Transactions on Signal Processing*, *6*, 4546–4557.

Scaglione, A., & Hong, Y.-W. (2003, August). Opportunistic large arrays: Cooperative transmission in wireless multihop ad hoc networks to reach far distances. *IEEE Transactions on Signal Processing*m 8.

Scaglione, A., & Servetto, S. D. (2002). *On the interdependence of routing and data compression in multi-hop sensor networks*. Paper presented at Mobicom.

Scaglione, A., Geockel, D. L., & Laneman, J. N. (2006). Cooperative communications in mobile ad hoc networks: Rethinking the link abstraction. *IEEE Signal Processing Magazine, Special Issues on Signal Processing for Wireless Ad hoc Communication Networks*, *23*(5), 18–29.

Scaglione, A., Kirti, S., & Sirkeci-Mergen, B. (2006, April 19-21). Error propagation in dense wireless networks with cooperation. In *Proc. of Information Processing in Sensor Networks (IPSN)*, Nashville, TN.

Scaglione, A., Stoica, P., Barbarossa, S., Giannakis, G. B., & Sampath, H. (2002). Optimal designs for space-time linear precoders and decoders. *IEEE Transactions on Signal Processing*, *50*(5), 1051–1064. doi:10.1109/78.995062

Schein, B., & Gallager, R. (2000). The Gaussian parallel relay network. In *Proceedings IEEE International Symposium on Information Theory*, Sorrento, Italy (p. 22).

Schein, B., & Gallager, R. (2000). The parallel Gaussian relay network. In *Proceedings of the IEEE International Symposium on Information Theory*, Sorrento, Italy (p. 22).

Schweppe, F. C. (1973). *Uncertain dynamic systems*. Englewood Cliffs, NJ: Prentice Hall.

Scutari, G., & Barbarossa, S. (2005). Distributed space-time coding for regenerative relay newtroks. *IEEE Transactions on Wireless Communications*, *5*(4), 2387–2399. doi:10.1109/TWC.2005.853883

Sendonaris, A., Erkip, E., & Aazhang, B. (1998). Increasing uplink capacity via user cooperation diversity. In *Proc. IEEE Intern. Symposium on Info. Theory (ISIT)* (p. 156).

Sendonaris, A., Erkip, E., & Aazhang, B. (2003). User cooperation diversity-part II: Implementation aspects and performance analysis. *IEEE Transactions on Communications*, *51*(11), 1939–1948. doi:10.1109/TCOMM.2003.819238

Sendonaris, A., Erkip, E., & Aazhang, B. (2003). User cooperation diversity. Part I. System description. *IEEE Transactions on Communications*, *51*(11), 1927–1938. doi:10.1109/TCOMM.2003.818096

Sendonaris, A., Erkip, E., & Aazhang, B. (2003). User cooperation diversity—part I: System description. *IEEE Transactions on Communications*, *51*(11), 1927–1938.

Sendonaris, A., Erkip, E., & Aazhang, B. (2003). User cooperation diversity—part II: Implementation aspects and performance analysis. *IEEE Transactions on Communications, 51*(11), 1939–1948.

Sendonaris, A., Erkip, E., & Aazhang, B. (2003). User cooperation diversity-part I: System description. *IEEE Transactions on Communications, 51*(11), 1927–1938. doi:10.1109/TCOMM.2003.818096

Sengupta, A., & Mitra, P. (1999). Distributions of singular values for some random matrices. *Physical Review E: Statistical Physics, Plasmas, Fluids, and Related Interdisciplinary Topics, 60*(3), 3389–3392. doi:10.1103/PhysRevE.60.3389

Sengupta, A., & Mitra, P. (2006). Capacity of multivariate channels with multiplicative noise: Random matrix techniques and large-n expansions (2). *Journal of Statistical Physics, 125*(5), 1223–1242. doi:10.1007/s10955-006-9076-0

Seol, D. Y., Kwon, U. K., & Im, G. H. (in press). Performance of single carrier transmission with cooperative diversity over fast fading channels. *IEEE Transactions on Communications*.

Servetto, S. D. (2000). Lattice quantization with side information. In *Proceedings of IEEE Data Compression Conference,* Snowbird, UT (pp. 510–519).

Servetto, S. D. (2007). Lattice quantization with side information: Codes, asymptotics, and applications in sensor networks. *IEEE Transactions on Information Theory, 53*(2), 714–731. doi:10.1109/TIT.2006.889697

Sezgin, A., & Oechtering, T. (2004). A new resource efficient transmission scheme for cooperative systems. *IEEE SPAWC* (pp. 298-302).

Sezgin, A., Jorswieck, E. A., & Charafeddine, M. (2008). Interaction between scheduling and user locations in an OSTBC coded downlink system. *7th International ITG Conference on Source and Channel Coding*.

Sezgin, A., Jorswieck, E. A., & Paulraj, A. (2008, May 7-9). Where to place interferers in a wireless network, *ITW 2008,* Porto, Portugal.

Shakkottai, S., Srikant, R., & Shroff, N. (2004). Unreliable sensor grids: Coverage, connectivity, and diameter. *AdHoc Networks*.

Shamai (Shitz), S., & Wyner, A. (1997). Information theoretic considerations for symmetric, cellular multiple access fading channels part I and part II. *IEEE Transactions on Information Theory, 43*, 1877–1911. doi:10.1109/18.641553

Shamai, S., Somekh, O., & Zaidel, M. (2004). Multicell communications: An information theoretic perspective. In *Joint Workshop on Communications and Coding,* Donnini (Florence), Italy.

Shamai, S., Somekh, O., Simeone, O., Sanderovich, A., Zaidel, B., & Poor, H. (2007). Cooperative multicell networks: Impact of limited-capacity backhaul and interusers links. In *Joint Workshop on Coding and Communications,* Austria.

Shang, Y., & Xia, X.-G. (2006). Shift-full-rank matrices and applications in space-time trellis codes for relay networks with asynchronous cooperative diversity. *IEEE Transactions on Information Theory, 52*, 3153–3167. doi:10.1109/TIT.2006.876222

Shannon, C. E. (1961). Two-way communication channels. In *Proceedings of the 4th Berkeley Symposium on Mathematical Statistics and Probability* (pp. 611-614).

Shi, Z., & Schlegel, C. (2001). Joint iterative decoding of serially concatenated error control coded CDMA. *IEEE Journal on Selected Areas in Communications, 19*(8), 1646–1653. doi:10.1109/49.942525

Shin, H., Win, M., & Chiani, M. (2008). Asymptotic statistics of mutual information for doubly correlated MIMO channels. *IEEE Transactions on Wireless Communications, 7*(2), 562–573. doi:10.1109/TWC.2008.060271

Shiu, D.-S., Foschini, G., & Kahn, M. G. J. (2000). Fading correlation and its effect on the capacity of multielement antenna systems. *IEEE Transactions on Communications, 48*(3), 502–513. doi:10.1109/26.837052

Shor, N. Z. (1985). *Minimization methods for non-differentiable functions.* Berlin, Germany: Springer-Verlag.

Simeone, O., Somekh, O., Bar-Ness, Y., & Spagnolini, U. (2006). Low-SNR analysis of cellular systems with cooperative base stations and mobiles. In *Fortieth Asilomar Conference on Signals, Systems, and Computers* (pp. 626–630).

Simeone, O., Somekh, O., Bar-Ness, Y., & Spagnolini, U. (2007). Uplink throughput of TDMA cellular systems with multicell processing and amplify-and-forward cooperation between mobiles. *IEEE Transactions on Wireless Communications, 6*(8), 2942–2951. doi:10.1109/TWC.2007.051026

Simon, M. K., & Alouini, M.-S. (2000). *Digital communication over fading channels: A unified approach to performance analysis.* New York: Wiley-Interscience.

Simon, M. K., & Alouini, M.-S. (2005). *Digital communication over fading channels: A unified approach to performance analysis.* New York: John Wiley & Son Inc.

Simon, M. K., & Alouini, M.-S. (2005). *Digital communications over fading channels, 2nd edition.* New York: Wiley

Simon, M. K., & Divsalar, D. (1992). On the implementation and performance of single and double-differential detection schemes. *IEEE Transactions on Communications, 40*(2), 278–291. doi:10.1109/26.129190

Simon, M. K., Hinedi, S. M., & Lindsey, W. C. (1994). *Digital communication techniques: Signal design and detection.* NJ: Prentice-Hall.

Siriwongpairat, W. P., Himsoon, T., Su, W., & Liu, K. J. R. (2006). Optimum threshold-selection relaying for decode-and-forward cooperation protocol. *Wireless Communications and Networking Conference, 2006, WCNC 2006, IEEE* (pp. 1015-1020).

Sirkeci-Mergen, B., & Gastpar, M. (2007, January). On the cooperative broadcast capacity of high density wireless networks. *The 2007 Information Theory and Applications Workshop*, San Diego, CA (invited paper).

Sirkeci-Mergen, B., & Gastpar, M. (2007, June). (Manuscript submitted for publication). On the cooperative broadcast capacity of wireless networks with cooperative relays. *IEEE Transactions on Information Theory.*

Sirkeci-Mergen, B., & Gastpar, M. (2007, November 4-7). On the ccaling of the broadcast capacity of extended wireless networks with cooperative relays. In *Proc. Asilomar Conference on Signals, Systems, and Computers*, Pacific Grove, CA.

Sirkeci-Mergen, B., & Scaglione, A. (2004). Signal acquisition for cooperative transmissions in multihop ad hoc networks. In *Proc. of IEEE Inter. Conf. on Acoustics, Speech, and Signal Process. (ICASSP).*

Sirkeci-Mergen, B., & Scaglione, A. (2004, July). Coverage analysis of cooperative broadcast in wireless networks. In *Proc. of IEEE Workshop on Signal Process. Advances in Wireless Commun. (SPAWC).*

Sirkeci-Mergen, B., & Scaglione, A. (2005). A continuum approach to dense wireless networks with cooperation. In *Proc. of Annual Joint Conf. of the IEEE Computer and Commun. Societies (Infocom).*

Sirkeci-Mergen, B., & Scaglione, A. (2005). Message propagation in a cooperative network with asynchronous receptions. In *Proc. of IEEE Inter. Conf. on Acoustics, Speech, and Signal Process. (ICASSP).*

Sirkeci-Mergen, B., & Scaglione, A. (2005). Randomized distributed space-time coding for cooperative communication in self organized networks. In *IEEE Proceedings of SPAWC* (pp. 500–504).

Sirkeci-Mergen, B., & Scaglione, A. (2006, July). On the optimal power allocation for broadcasting in dense cooperative networks. In *Proc. of IEEE International Symposium on Information Theory*, Seattle, WA.

Sirkeci-Mergen, B., & Scaglione, A. (2007). Randomized space-time coding for distributed cooperative communication. *IEEE Transactions on Signal Processing, 55*, 5003–5017. doi:10.1109/TSP.2007.896061

Sirkeci-Mergen, B., & Scaglione, A. (2007, February). On the power efficiency of cooperative broadcast in dense wireless networks. [JSAC]. *IEEE Journal on Selected Areas in Communications, 25*(2), 497–507. doi:10.1109/JSAC.2007.070223

Sirkeci-Mergen, B., & Scaglione, A. (2007, October). Randomized space-time coding for distributed cooperative communication. *IEEE Transactions on Signal Processing, 55*(10). doi:10.1109/TSP.2007.896061

Sirkeci-Mergen, B., Scaglione, A., & Mergen, G. (2006, June). Asymptotic analysis of multistage cooperative broadcast in wireless networks. *IEEE Transactions on Information Theory, 52*(6), 2531–2550. doi:10.1109/TIT.2006.874514

Skupch, A., Seethaler, D., & Hlawatsch, F. (2005). Free probability based capacity calculation for MIMO channels with transmit or receive correlation. In *International Conference on Wireless Networks, Communications and Mobile Computing* (Vol.2, pp. 1041–1046).

Slepian, D., & Wolf, J. (1973a). A coding theorem for multiple access channels with correlated sources. *The Bell System Technical Journal, 52,* 1037–1076.

Slepian, D., & Wolf, J. (1973b). Noiseless coding of correlated information sources. *IEEE Transactions on Information Theory, 19*(4), 471–480. doi:10.1109/TIT.1973.1055037

Somekh, O., & Shamai, S. (1998). A Shannon-theoretic view of Wyner's multiple-access cellular channel model in the presence of fading. In *IEEE International Symposium on Information Theory* (p. 393).

Somekh, O., & Shamai, S. (2000). Shannon-theoretic approach to a Gaussian cellular multiple-access channel with fading. *IEEE Transactions on Information Theory, 46*(4), 1401–1425. doi:10.1109/18.850679

Somekh, O., Simeone, O., Poor, H. V., & Shamai, S. (2007). Cellular systems with full-duplex amplify-and-forward relaying and cooperative base-stations. In *IEEE International Symposium on Information Theory,* Nice, France.

Somekh, O., Zaidel, B., & Shamai, S. (2007). Sum rate characterization of joint multiple cell-site processing. *IEEE Transactions on Information Theory, 53*(12), 4473–4497. doi:10.1109/TIT.2007.909170

Sreedhar, D., Chockalingam, A., & Rajan, B. S. (2008). Single-symbol ML decodable distributed STBCs for partially-coherent cooperative networks. *IEEE ICC, 08,* 1029–1033.

Srinivasan, V., Nuggehalli, P., Chiasserini, C. F., & Rao, R. R. (2003, March). Cooperation in wireless ad hoc networks. In *Proc. IEEE INFOCOM,* San Francisco, CA (pp. 808-817).

Stankovic, V., Liveris, A., Xiong, Z., & Georghiades, C. (2004). Design of Slepian-Wolf codes by channel code partitioning. In *Proceedings of IEEE Data Compression Conference,* Snowbird, UT (pp. 302–311).

Stefanov, A., & Erkip, E. (2004). Cooperative coding for wireless networks. *IEEE Transactions on Communications, 52*(9), 2415–2425. doi:10.1109/TCOMM.2004.833070

Steinberg, B. D. (1972). The peak sidelobe of the phased array having randomly located elements. *IEEE Transactions on Antennas and Propagation, 20*(2), 129–136. doi:10.1109/TAP.1972.1140162

Stoica, P., Liu, J., & Li, J. (2004). Maximum-likelihood double-differential detection clarified. *IEEE Transactions on Information Theory, 50*(3), 572–576. doi:10.1109/TIT.2004.825501

Stüber, G. L. (2000). *Principle of mobile communication,* 2nd ed. Kluwer Academic Publishers.

Su, W., & Sadek, K. A., & Liu, K. J. R. (2005). SER performance analysis and optimum power allocation for decode-and-forward cooperation protocol in wireless networks. In *Proceedings of IEEE Wireless Communication and Networking Conference* (Vol. 2, pp. 984–989).

Su, W., Sadek, A. K., & Liu, K. J. R. (2005). SER performance analysis and optimal power allocation for decode-and-forward cooperation protocol in wireless networks. In *Proc. IEEE Wireless Communications and Networking Conf. (WCNC),* New Orleans, LA (pp. 984–989).

Sun, Y., Liveris, A., Stankovic, V., & Xiong, Z. (2005). Near-capacity dirty-paper code designs based on TCQ and IRA codes. In *Proceedings of IEEE International Symposium on Information Theory* (pp. 184–188).

Sung, Y., Tong, L., & Ephremides, A. (2005, March). *Route selection for detection of correlated random fields in large sensor networks.* Paper presented at CISS.

Sung, Y., Tong, L., & Ephremides, A. (2005, October). *A new metric for routing in multihop wireless sensor networks for detection of correlated random fields.* Paper presented at MILCOM.

Suraweera, H. A., & Karagiannidis, G. K. (2008). Closed-form error analysis of the non-identical Nakagami-m relay fading channel. *IEEE Communications Letters, 12*(4), 259–261. doi:10.1109/LCOMM.2008.071922

Suraweera, H. A., Smith, P. J., & Armstrong, J. (2006). Outage probability of cooperative relay networks in Nakagami-m fading channels. *IEEE Communications Letters,*

10(12), 834–836. doi:10.1109/LCOMM.2006.060834

Tam, W. P., Lok, T. M., & Wong, T. F. (in press). Flow optimization in parallel relay networks with cooperative relaying. *IEEE Transactions on Wireless Communications*.

Tang, T., Chae, C. B., Heath, R. W., Jr., & Cho, S. (2006). On achievable sum rates of a multiuser MIMO relay channel. In *Proceedings of IEEE International Symposium on Information Theory* (pp. 1026–1030).

Tanner, R. (1981). A recursive approach to low complexity codes. *IEEE Transactions on Information Theory, 27*(5), 533–547. doi:10.1109/TIT.1981.1056404

Tarasak, P., Minn, H., & Bhargava, V. K. (2004). Differential modulation for two-user cooperative diversity systems. In *Proc. of Globecom*, Dallas, TX (pp. 3389–3393).

Tarasak, P., Minn, H., & Bhargava, V. K. (2005). Differential modulation for two-user cooperative diversity systems. *IEEE Journal on Selected Areas in Communications, 23*(9), 1891–1900. doi:10.1109/JSAC.2005.853792

Taricco, G. (2008). Asymptotic mutual information statistics of separately-correlated Rician fading MIMO channels. *IEEE Transactions on Information Theory, 54*(8), 3490–3504. doi:10.1109/TIT.2008.926415

Tarokh, V., & Kim, I. M. (2002). Existence and construction of noncoherent unitary space-time codes. *IEEE Transactions on Information Theory, 48*(12), 3112–3117.

Tarokh, V., Jafarkhani, H., & Calderbank, A. R. (1999). Space-time block codes from orthogonal designs. *IEEE Transactions on Information Theory, 45*(5), 1456–1467.

Tarokh, V., Seshadri, N., & Calderbank, A. R. (1998). Space time codes for high data rate wireless communication: Performance criterion and code construction. *IEEE Transactions on Information Theory, 44*(02), 744–765.

Tavli, B. (2006, February). Broadcast capacity of wireless networks. *IEEE Communications Letters, 10*(2), 68–69. doi:10.1109/LCOMM.2006.02013.

Telatar, E. (1999). Capacity of multi-antenna Gaussian channels. *European Trans. on Telecomm. ETT, 10*(6), 585–596.

Tellambura, C., Annamalai, A., & Bhargava, V. K. (2001). Unified analysis of switched diversity systems in independent and correlated fading channels. *IEEE Transactions on Communications, 49*(11), 1955–1965. doi:10.1109/26.966072

ten Brink, S. (1999). Convergence of iterative decoding. *IEE Electronics Letters, 35*(13), 1117–1119.

ten Brink, S. (2001). Convergence behavior of iteratively decoded parallel concatenated codes. *IEEE Transactions on Communications, 49*(10), 1727–1737. doi:10.1109/26.957394

Timus, B. (2006). *Deployment cost efficiency in broadband delivery with fixed wireless relays*. Licentiate thesis, KTH, Sweden.

Tomlinson, M. (1971). New automatic equaliser employing modulo arithmetic. *IEE Electronics Letters, 7*(5), 138–139. doi:10.1049/el:19710089

Topkis, D. M. (1998). *Supermodularity and complementarity*. NJ: Princeton University Press.

Tsatsanis, M. K., Zhang, R., & Banerjee, S. (2000). Network-assisted diversity for random access wireless networks. *IEEE Transactions on Signal Processing, 48*(3), 702–711. doi:10.1109/78.824666

Tse, D. N. C., Viswanath, P., & Zheng, L. (2004). Diversity-multiplexing tradeoff in multiple-access channels. *IEEE Transactions on Information Theory, 50*(9), 1859–1874. doi:10.1109/TIT.2004.833347

Tse, D., & Viswanath, P. (2004). *Fundamentals of wireless communications*. Cambridge University Press.

Tse, D., & Viswanath, P. (2005). *Fundamentals of wireless communication*. Cambridge University Press.

Tseng, Y.-C., Ni, S.-Y., Chen, Y.-S., & Sheu, J.-P. (1999). The broadcast storm problem in a mobile ad hoc network. In *Proc. ACM/IEEE Int. Conf. on Mobile Computing and Networking (MOBICOM)* (pp. 151-162).

Tsiftsis, T. A., Karagiannidis, G. K., Kotsopoulos, S. A., & Pavlidou, F.-N. (2004). BER analysis of collaborative dual-hop wireless transmissions. *IET Electronics Letters, 40*(11), 679–681. doi:10.1049/el:20040393

Tsiftsis, T. A., Karagiannidis, G. K., Mathiopoulos, P. T., & Kotsopoulos, S. A. (2006). Nonregenerative dual-hop cooperative links with selection diversity. *EURASIP Journal on Wireless Comm. and Net., 2006*(17862), 1–8.

Tsiftsis, T. A., Karagiannidis, G. K., Mathiopoulos, T., & Kotsopoulos, S. A. (2006). Nonregenerative dual-hop cooperative links with selection diversity. *EURASIP Journal on Wireless Communications and Networking, 2006 . Article ID, 17862*, 1–8.

Tuchler, M., Koetter, R., & Singer, A. (2002). Turbo equalization: Principles and new results. *IEEE Transactions on Communications, 50*(5), 754–767. doi:10.1109/TCOMM.2002.1006557

Tulino, A. M., & Verdu, S. (2004). Random matrix theory and wireless communications. *Communications and Information Theory, 1*(1), 1–182. doi:10.1561/0100000001

Tulino, A. M., Lozano, A., & Verdu, S. (2006). Capacity-achieving input covariance for single-user multi-antenna channels. *IEEE Transactions on Wireless Communications, 5*(3), 662–671. doi:10.1109/TWC.2006.1611096

Tulino, A., Lozano, A., & Verdu, S. (2005). Impact of antenna correlation on the capacity of multiantenna channels. *IEEE Transactions on Information Theory, 51*(7), 2491–2509. doi:10.1109/TIT.2005.850094

Tung, S.-Y. (1978). *Multiterminal source coding.* Unpublished doctoral dissertation, Cornell University, Ithaca, NY.

Uysal, M., & Canpolat, O. (2005). On the distributed space-time signal design for a large number of relay terminals. In *Proceedings of IEEE Wireless Communications and Networking Conference, 2005.*

van der Meulen, E. (1971). Three-terminal communication channels. *Advances in Applied Probability, 3*, 120–154. doi:10.2307/1426331

Van der Meulen, E. C. (1968). *Transmission of information in a t-terminal discrete memoryless channel.* Unpublished doctoral dissertation, Univ. of California, Berkeley, CA.

Van der Meulen, E. C. (1971). Three-terminal communication channels. *Advances in Applied Probability, 3*, 120–154. doi:10.2307/1426331

Varanasi, M. (1995). Group detection for synchronous Gaussian code-division multiple-access channels. *IEEE Transactions on Information Theory, 41*(4), 1083–1096. doi:10.1109/18.391251

Varanasi, M. K., & Guess, T. (1997). Optimum decision feedback multiuser equalization with successive decoding achieves the total capacity of the Gaussian multiple-access channel. In *Thirty-First Asilomar Conference on Signals, Systems, & Computers,* (Vol.2, pp.1405-1409).

Vardhe, K., & Reynolds, D. (2007, July). Fast communication: User cooperation in an asynchronous cellular uplink. *Signal Processing, 87*(7), 1799–1807. doi:10.1016/j.sigpro.2007.01.002

Verdu, S. (1986). Minimum probability of error for asynchronous Gaussian multiple-access channels. *IEEE Transactions on Information Theory, 32*(1), 85–96. doi:10.1109/TIT.1986.1057121

Verdu, S. (1998). *Multiuser detection.* Cambridge: University Press.

Vishwanath, S., Jindal, N., & Goldsmith, A. (2003). Duality, achievable rates, and sum-rate capacity of Gaussian MIMO broadcast channels. *IEEE Transactions on Information Theory, 49*(10), 2658–2668. doi:10.1109/TIT.2003.817421

Viswanath, P., & Tse, D. (2003). Sum capacity of the vector Gaussian broadcast channel and uplink-downlink duality. *IEEE Transactions on Information Theory, 49*(8), 1912–1921. doi:10.1109/TIT.2003.814483

Viterbi, A. J. (1995). *CDMA: Principles of spread spectrum communication.* MA: Addison-Wesley.

Viterbi, A., Viterbi, A., Gilhousen, K., & Zehavi, E. (1994). Soft handoff extends CDMA cell coverage and increases reverse link capacity. *IEEE Journal on Selected Areas in Communications, 12*(8), 1281–1288. doi:10.1109/49.329346

Voiculescu, D. (1983). Asymptotically commuting finite rank unitary operators without commuting approximants. *Acta Sci. Math., 45*, 429–431.

Wagner, A., & Anantharam, V. (2008). An improved outer bound for multiterminal source coding. *IEEE Transactions on Information Theory, 54*(5), 1919–1937. doi:10.1109/TIT.2008.920249

Wang, B., Zhang, J., & Høst-Madsen, A. (2005). On the capacity of MIMO relay channels. *IEEE Transactions on Information Theory, 51*(1), 29–43. doi:10.1109/TIT.2004.839487

Wang, G., Zhang, Y., & Amin, M. (2006). Differential distributed space-time modulation for cooperative networks. *IEEE Transactions on Communications, 05*(11).

Wang, H., & Xia, X.-G. (2003). Upper bounds of rates of complex orthogonal space-time block codes. *IEEE Transactions on Information Theory, 49*, 2788–2796. doi:10.1109/TIT.2003.817830

Wang, L., Kwok, Y.-K., Lau, W.-C., & Lau, V. K. N. (2004). Efficient packet scheduling using channel adaptive fair queueing in distributed mobile computing systems. *ACM Mobile Netw. & Appl., 9*, 297–309. doi:10.1023/B:MONE.0000031589.32967.0f

Wang, T., & Giannakis, G. B. (2007). High-throughput cooperative communications with complex field network coding. In *Proc. Conf. Info. Sci. and Systems.*

Wang, T., & Giannakis, G. B. (2008). Complex field network coding for multiuser cooperative communications. *IEEE Journal on Selected Areas in Communications, 26*(3), 561–571. doi:10.1109/JSAC.2008.4481380

Wang, T., Cano, A., Giannakis, G. B., & Laneman, J. N. (2007). High-performance cooperative demodulation with decode-and-forward relays. *IEEE Transactions on Communications, 55*(7), 1427–1438. doi:10.1109/TCOMM.2007.900631

Wang, T., Wang, A., & Giannakis, G. B. (2006). Smart regenerative relays for link-adaptive cooperative communications. *40th Conference on Information Sciences and Systems 2006, CISS 2006, IEEE* (pp. 1038-1043).

Wang, T., Wang, R., & Giannakis, G. B. (2006). Smart regenerative relays for link-adaptive cooperative communications. In *Proc. 40th Conference on Information Sciences and Systems 2006 (CISS'06)* (pp. 1038-1043).

Wang, X., & Host-Madsen, A. (1999). Group-blind multiuser detection for uplink CDMA. *IEEE Journal on Selected Areas in Communications, 17*(11), 1971–1984. doi:10.1109/49.806826

Wang, X., & Orchard, M. (2001). Design of trellis codes for source coding with side information at the decoder. In *Proceedings of IEEE Data Compression Conference,* Snowbird, UT (pp. 361–370).

Wang, X., & Poor, H. (1999). Iterative (turbo) soft interference cancellation and decoding for coded CDMA. *IEEE Transactions on Communications, 47*(7), 1046–1061. doi:10.1109/26.774855

Wang, Z., & Giannakis, G. B. (2003). A simple and general parameterization quantifying performance in fading channels. *IEEE Transactions on Communications, 51*, 1389–1398. doi:10.1109/TCOMM.2003.815053

Wei, N., Zhang, Z., & Li, S. (2006). An adaptive space-time coded cooperation scheme in wireless communication. *IEICE Transactions on Communications . E (Norwalk, Conn.), 89-B*(11), 2973–2981.

Weingarten, H., Steinberg, Y., & Shamai, S. (2006). The capacity region of the Gaussian multiple-input multiple-output broadcast channel. *IEEE Transactions on Information Theory, 52*(9), 3936–3964. doi:10.1109/TIT.2006.880064

Weingarten, H., Steinberg, Y., & Shamai, S. (2006). The capacity region of the Gaussian multiple-input multiple-output broadcast channel. *IEEE Transactions on Information Theory, 52*(9), 3936–3964. doi:10.1109/TIT.2006.880064

Wijk, F., Janssen, G., & Prasad, R. (1995). Groupwise successive interference cancellation in a DS/CDMA system. In *IEEE International Symposium on Personal, Indoor and Mobile Radio Communications,* (Vol. 2, pp. 742–746), Toronto, Canada.

Willems, F. (1982). The feedback capacity region of a class of discrete memoryless multiple access channels. *IEEE Transactions on Information Theory, 28*(1), 93–95. doi:10.1109/TIT.1982.1056437

Willems, F. M. J. (1982). *Informationtheoretical results for the discrete memoryless multiple access channel.* Unpublished doctoral dissertation, Katholieke Universiteit Leuven, Leuven, Belgium.

Willems, F. M. J. (1983). The discrete memoryless multiple access channel with partially cooperating encoders. *IEEE Transactions on Information Theory, 29*(3), 441–445. doi:10.1109/TIT.1983.1056660

Willems, F. M. J., & van der Meulen, E. C. (1983). Partial feedback for the discrete memoryless multiple access channel. *IEEE Transactions on Information Theory, 29*(2), 287–290. doi:10.1109/TIT.1983.1056646

Willems, F. M. J., & van der Meulen, E. C. (1985). The discrete memoryless multiple access channel with cribbing encoders. *IEEE Transactions on Information Theory, 31*(3), 313–327. doi:10.1109/TIT.1985.1057042

Willems, F. M. J., & van der Meulen, E. C. (1985). The discrete memoryless multiple-access channel with cribbing encoders. *IEEE Transactions on Information Theory, 31*, 313. doi:10.1109/TIT.1985.1057042

Willems, F. M. J., van der Meulen, E. C., & Schalkwijk, J. P. M. (1983). An achievable rate region for the multiple access channel with generalized feedback. In *Proceedings of the Allerton Conference,* Monticello, IL (pp. 284-292).

Willems, F. M. J., van der Meulen, E. C., & Schalkwijk, J. P. M. (1983). Achievable rate region for the multiple access channel with generalized feedback. In *Proc. Annu. Allerton Conf. Communication, Control, Computing* (p. 284).

Williams, B., & Camp, T. (2002). Comparison of broadcasting techniques for mobile ad hoc networks. *ACM Proc. on MOBIHOC.*

Windpassinger, C., Fischer, R., Vencel, T., & Huber, J. (2004). Precoding in multiantenna and multiuser communications. *IEEE Transactions on Wireless Communications, 3*(4), 1305–1316. doi:10.1109/TWC.2004.830852

Winters, J. H. (1998). Smart antennas for wireless systems. *IEEE Personal Commun., 5*, 23–27. doi:10.1109/98.656155

Winters, J., Salz, J., & Gitlin, R. D. (1994). The impact of antenna diversity on the capacity of wireless communication systems. *IEEE Transactions on Communications, 42*(1), 1740–1751. doi:10.1109/TCOMM.1994.582882

Wong, D., & Lim, T. (1997). Soft handoffs in CDMA mobile systems. *IEEE Wireless Communications, 4*(6), 6–17.

Wong, T. F., Lok, T. M., & Shea, J. M. (in press). Flow-optimized cooperative transmission for the relay channel. *IEEE Transactions on Information Theory,* Retrieved from http://arxiv.org/PScache/cs/pdf/0701/0701019v3.pdf

Wu, Y. (2006, July). A trellis connectivity analysis of random linear network coding with buffering. In *Proc. Intl. Symp. Info. Theory,* Seattle, MA (pp. 768-772).

Wyner, A. (1974). Recent results in the Shannon theory. *IEEE Transactions on Information Theory, 20*(1), 2–10. doi:10.1109/TIT.1974.1055171

Wyner, A. (1994). Shannon-theoretic approach to a Gaussian cellular multiple-access channel. *IEEE Transactions on Information Theory, 40*(6), 1713–1727. doi:10.1109/18.340450

Wyner, A., & Ziv, J. (1976). The rate-distortion function for source coding with side information at the decoder. *IEEE Transactions on Information Theory, 22*(1), 1–10. doi:10.1109/TIT.1976.1055508

Wyrembelski, R., Oechtering, T., Bjelakovic, I., Schnurr, C., & Boche, H. (2008). Capacity of Gaussian MIMO bidirectional broadcast channels. In *Proc. IEEE ISIT.*

Xie, B., Kumar, A., Cavalcanti, D., & Agrawal, D. P. (2006). Multihop cellular IP: A new approach to heterogeneous wireless networks. *International Journal of Pervasive Computing and Communications, 2*(4), 370–383.

Xie, L., & Kumar, P. R. (2004). A network information theory for wireless communication: Scaling laws and optimal operation. *IEEE Transactions on Information Theory, 50*(5), 748. doi:10.1109/TIT.2004.826631

Xie, L.-L., & Kumar, P. (2005). An achievable rate for the multiple-level relay channel. *IEEE Transactions on Information Theory, 51*(4), 1348–1358. doi:10.1109/TIT.2005.844066

Yan, Y., Chen, M., & Kwon, T. (2006). A novel cluster-based cooperative MIMO scheme for multi-hop wireless sensor networks. *EURASIP Journal on Wireless Communications and Networking, 2*(4), 1–9. doi:10.1155/WCN/2006/72493

Yang, S., & Belfiore, J.-C. (2007). Optimal space-time codes for the MIMO amplify-and-forward cooperative channel. *IEEE Transactions on Information Theory, 53*(2), 647–663. doi:10.1109/TIT.2006.888998

Yang, S., & Belfiore, J.-C. (2007). Towards the optimal amplify-and-forward cooperative diversity scheme. *IEEE Transactions on Information Theory, 9*(53), 3114–3126. doi:10.1109/TIT.2007.903133

Yang, S., & Belfiore, J.-C. (2008). Diversity of MIMO multi-hop relay channels. *IEEE Trans. Inf. Theory.* Submitted for publication.

Yang, S., & Belfiore, J-C. (2006). Diversity-multiplexing tradeoff of double scattering MIMO channels. Submitted to *IEEE Transactions on Information Theory.*

Yang, Y., Cheng, S., Xiong, Z., & Zhao, W. (2003). Wyner-Ziv coding based on TCQ and LDPC codes. In *Proceedings of Asilomar Conference on Signals, Systems and Computers* (Vol. *1*, pp. 825–829).

Yang, Y., Wang, J., & Kravets, R. (2005). *Interference-aware load balancing for multihop wireless networks* (Tech. Rep.). University of Illinois at Urbana-Champaign.

Yang, Z., & Høst-Madsen, A. (2007). Cooperation through interference amplification. *IEEE Communications Letters, 11*(5), 369–371. doi:10.1109/LCOMM.2007.070028

Yao, Y., Cai, X., & Giannakis, G. B. (2005). On energy efficiency and optimum resource allocation of relay transmissions in the low-power regime. *IEEE Transactions on Wireless Communications, 6*(4), 2917–2927.

Ye, W., Heidemann, J., & Estrin, D. (2004). Medium access control with coordinated adaptive sleeping for wireless sensor networks. *IEEE/ACM Transactions on Networking, 12*(3), 493–506.

Yeung, R. W., Li, S.-Y. R., Cai, N., & Zhang, Z. (2005). Network coding theory part I and II. *Foundation and Trends in Communications and Information Theory, 2*(4).

Yi, Z., & Kim, I.-M. (2007). Joint optimization of relay-precoders and decoders with partial channel side information in cooperative networks. *IEEE Journal on Selected Areas in Communications, 25*, 447–458. doi:10.1109/JSAC.2007.070219

Yi, Z., & Kim, I.-M. (2007). Single-symbol ML decodable distributed STBCs for cooperative networks. *IEEE Transactions on Information Theory, 53*, 2977–2985. doi:10.1109/TIT.2007.901177

Yi, Z., & Kim, I.-M. (2008). Diversity order analysis of the decode-and-forward cooperative networks with relay selection. *IEEE Transactions on Wireless Communications, 7*, 1792–1799. doi:10.1109/TWC.2008.061041

Yi, Z., & Kim, I.-M. (2008). Row-monomial distributed orthogonal space-time block codes with channel phase information. In *IEEE ICC'08.*

Yi, Z., & Kim, I.-M. (2008). The impact of noise correlation on the single-symbol ML decodable distributed STBCs. In *IEEE ICC'08.*

Yiu, S., Schober, R., & Lampe, L. (2005). Noncoherent distributed space-time block coding. *Vehicular Technology Conference* (pp. 947-951).

Yiu, S., Schober, R., & Lampe, L. (2006). Distributed space-time block coding. *IEEE Transactions on Communications, 54*(07), 1195–1206.

Yu, M., & Li, J. (2005). Is amplify-and-forward practically better than decode-and-forward or vice versa? In *Proceedings of IEEE International Conference on Acoustic, Speech, and Signal Processing* (Vol. *3*, pp. 365–368).

Yu, W. (2006a). Uplink-downlink duality via minimax duality. *IEEE Transactions on Information Theory, 52*(2), 361–374. doi:10.1109/TIT.2005.862102

Yu, W. (2006b). Sum-capacity computation for the Gaussian vector broadcast channel via dual decomposition. *IEEE Transactions on Information Theory, 52*(2), 754–759. doi:10.1109/TIT.2005.862106

Yu, W., & Aleksic, M. (2005). Coding for the Blackwell channel: A survey propagation approach. In *Proceedings of IEEE International Symposium on Information Theory* (pp. 1583–1587).

Yu, W., & Cioffi, J. (2001). Trellis precoding for the broadcast channel. In *Proceedings of IEEE Global Communications Conference* (Vol. *2*, pp. 1344–1348).

Yu, W., & Cioffi, J. (2002). Sum capacity of a Gaussian vector broadcast channel. In *Proceedings of IEEE International Symposium on Information Theory* (p. 498).

Yu, W., & Lan, T. (2007). Transmitter optimization for the multi-antenna downlink with per-antenna power constraints. *IEEE Transactions on Signal Processing, 55*(6), 2646–2660. doi:10.1109/TSP.2006.890905

Yu, W., Rhee, W., Boyd, S., & Cioffi, J. M. (2004). Iterative water-filling for Gaussian vector multiple-access channels. *IEEE Transactions on Information Theory, 50*(1), 145–151. doi:10.1109/TIT.2003.821988

Yu, W., Varodayan, D., & Cioffi, J. (2005). Trellis and convolutional precoding for transmitter-based interference presubtraction. *IEEE Transactions on Communications, 53*(7), 1220–1230. doi:10.1109/TCOMM.2005.851605

Yuen, C., Guan, Y. L., & Tjhung, T. T. (2005). Quasi-orthogonal STBC with minimum decoding complexity. *IEEE Transactions on Wireless Communications, 4*(5), 2089–2094.

Yuen, C., Guan, Y. L., & Tjhung, T. T. (2006). Single-symbol decodable differential space-time modulation based on QO-STBCs. *IEEE Transactions on Wireless Communications, 05*, 3329–3335.

Yuksel, M., & Erkip, E. (2007). Multiple-antenna cooperative wireless systems: A diversity-multiplexing tradeoff perspective. *IEEE Transactions on Information Theory, Special Issue on Models, Theory, and Codes for Relaying and Cooperation in Communication Networks, 53*(10), 3371.

Zamir, R., & Berger, T. (1999). Multiterminal source coding with high resolution. *IEEE Transactions on Information Theory, 45*(1), 106–117. doi:10.1109/18.746775

Zamir, R., & Shamai, S. (1998). Nested linear/lattice codes for Wyner-Ziv encoding. In *Proceedings of Information Theory Workshop* (pp. 92–93).

Zeng, C., Kuhlmann, F., & Buzo, A. (1989). Achievability proof of some multiuser channel coding theorems using backward decoding. *IEEE Transactions on Information Theory, 35*(6), 1160. doi:10.1109/18.45272

Zhang, H., & Dai, H. (2004). Cochannel interference mitigation and cooperative processing in downlink multicell multiuser MIMO networks. *EURASIP Journal on Wireless Communications and Networking, 2004*(2), 222-235.

Zhang, J., & Zhang, Q. (2008). Cooperative routing in multi-source multi-destination multi-hop wireless networks. In *Proceeding of 27ᵗʰ IEEE International Conference on Computer Communications* (mini symposium).

Zhang, J., Zhang, Q., Shao, C., Wang, Y., Zhang, P., & Zhang, Z. (2004). Adaptive optimal transmit power allocation for two-hop non-regenerative wireless relay system. In *Proc. IEEE Vehicular Technology Conf.,* Milan, Italy (pp. 1213-1217).

Zhang, R., Sidiropoulos, N. D., & Tsatsanis, M. K. (2002). Collision resolution in packet radio networks using rotational invariance techniques. *IEEE Transactions on Communications, 59*(1), 146–155. doi:10.1109/26.975780

Zhang, S., Liew, S. C., & Lam, P. P. (2006). Hot topic: Physical-layer network coding. In *Proceedings of 12ᵗʰ Annual International Conference on Mobile Computing and Networking,* Los Angeles, CA (pp. 23-26).

Zhang, S., Liew, S., & Lam, P. (2006, September). Physical layer network coding. In *Proc. 12ᵗʰ Annual International Conference on Mobile Computing and Networking.*

Zhang, Y. (2005). Differential modulation schemes for decode-and-forward cooperative diversity. In *Proc. IEEE Int. Conf. Acoust., Speech, SignalProc. (ICASSP'05),* Philadelphia, PA (pp. 918–920).

Zhang, Y. J., & Ben Lataief, K. B. (2005). An efficient resource-allocation scheme for spatial multiuser access in MIMO-OFDM systems. *IEEE Transactions on Communications, 53*(1), 107–116. doi:10.1109/TCOMM.2004.840666

Zhang, Y. J., & Letaief, K. B. (2003). Optimizing power and resource management for multiuser MIMO-OFDM systems. *IEEE Global Telecommunications Conference,* San Francisco, CA (Vol. 1, pp. 179-183).

Zhang, Z., & Duman, T. (2002). Capacity approaching turbo coding for half duplex relaying. In *Proceedings of IEEE International Symposium on Information Theory* (pp. 1888–1892).

Zhao, B., & Valenti, M. C. (2003). Distributed turbo coded diversity for relay channel. *IEEE Electronics Letters, 39*(10), 786–787. doi:10.1049/el:20030526

Zhao, B., & Valenti, M. C. (2005). Practical relay networks: A generalization of hybrid ARQ. *IEEE Journal on Selected Areas in Communications, 1*(23), 7–18. doi:10.1109/JSAC.2004.837352

Zhao, Q., & Li, H. (2005a). Performance analysis of an amplify-based differential modulation for wireless relay networks under Nakagami-m fading channels. In *Proc. IEEE Int. Workshop on Signal Process. Advances in Wireless Communications,* New York (pp. 211–215).

Zhao, Q., & Li, H. (2005b). Performance of differential modulation with wireless relays in Rayleigh fading channels. *IEEE Communications Letters, 9*(4), 343–345. doi:10.1109/LCOMM.2005.1413628

Zhao, Q., & Li, H. (2006). Performance of decode-based differential modulation for wireless relay networks in Nakagami-m channels. In *Proc. 2006 IEEE Int. Conf. Acoustics, Speech, Signal Processing (ICASSP)*, Toulouse, France (pp. 673–675).

Zhao, Q., & Li, H. (2007). Differential modulation for cooperative wireless systems. *IEEE Transactions on Signal Processing, 55*(5), 2273–2283. doi:10.1109/TSP.2006.890922

Zhao, Y. (2008). Cooperative networks with multiple relays: Selection, repetition, and beamforming. Unpublished doctoral dissertation, University of Toronto, Toronto, ON.

Zhao, Y., & Garcia-Frias, J. (2005). Joint estimation and compression of correlated nonbinary sources using punctured turbo codes. *IEEE Transactions on Communications, 53*(3), 385–390. doi:10.1109/TCOMM.2005.843414

Zhao, Y., Adve, R. S., & Lim, T. J. (2007). Improving amplify-and-forward relay networks: Optimal power allocation versus selection. *IEEE Transactions on Wireless Communications, 6*(8), 3114–3123.

Zhao, Y., Adve, R. S., & Lim, T. J. (2007b). Beamforming with limited feedback in amplify-and-forward cooperative networks. In *IEEE Globecom 2007*.

Zhao, Y., Adve, R., & Lim, T. J. (2006). Symbol error rate of selection amplify-and-forward relay systems. *IEEE Communications Letters, 10*, 757–759. doi:10.1109/LCOMM.2006.060774

Zhao, Y., Adve, R., & Lim, T. J. (2007). Improving amplify-and-forward relay networks: Optimal power allocation versus selection. *IEEE Transactions on Wireless Communications, 6*(8), 3114–3123.

Zheng, H., Zhu, Y., Shen, C., & Wang, X. (2005). On the effectiveness of cooperative diversity in ad hoc networks: A MAC layer study. In *IEEE Proceedings of ICASSP* (pp. 509–512).

Zheng, L., & Tse, D. (2003). Diversity and multiplexing: A fundamental tradeoff in multiple antenna channels. *IEEE*

Transactions on Information Theory, 49(5), 1073–1096. doi:10.1109/TIT.2003.810646

Zheng, L., & Tse, D. N. (2003). Diversity and multiplexing: A fundamental tradeoff in multiple-antenna channels. *IEEE Transactions on Information Theory, 5*(49), 1073–1096. doi:10.1109/TIT.2003.810646

Zheng, L., & Tse, D. N. C. (2002). Communication on the Grassmann manifold: A geometric approach to the noncoherent multiple antenna channels. *IEEE Transactions on Information Theory, 48*(2), 359–383.

Zheng, L., & Tse, D. N. C. (2003). Diversity and multiplexing: A fundamental tradeoff in multiple antenna channels. *IEEE Transactions on Information Theory, 49*(5), 1073–1096. doi:10.1109/TIT.2003.810646

Zheng, L., & Tse, D. N. C. (2003). Diversity and multiplexing: A fundamental tradeoff in multiple-antenna channels. *IEEE Transactions on Information Theory, 49*(5), 1073–1096. doi:10.1109/TIT.2003.810646

Zheng, R. (2006). Information dissemination in power-constrained wireless networks. In *Proc. of the 25th Annual Joint Conference of the IEEE Computer and Communications Societies (INFOCOM)*.

Zhong, S., Chen, J., & Yang, Y. R. (2003). Sprite: A simple, cheat-proof, credit-based system for mobile ad-hoc networks. *IEEE Infocom, 3*, 1987–1997.

Zhong, W., Zhao, Y., & Garcia-Frias, J. (2003). Turbo-like codes for distributed joint source-channel coding of correlated senders in multiple access channels. In *Proceedings of Asilomar Conference on Signals, Systems, and Computers* (Vol. *1*, pp. 840–844).

Zhou, Z., Zhou, S., Cui, J.-H., & Cui, S. (2007), Energy-efficient cooperative communication based on power control and selective relay in wireless sensor networks. In *IEEE Proceedings of Military Communications*.

Zhou, Z., Zhou, S., Cui, S., & Cui, J.-H. (2006). Energy efficient cooperative communication in clustered wireless sensor networks. In *IEEE Proceedings of Military Communications*.

Zorzi, M., & Rao, R. R. (2003). Geographic random forwarding for ad hoc and sensor networks: Energy and latency performance. *IEEE Transactions on Mobile Computing, 2*(4), 349–365. doi:10.1109/TMC.2003.1255650

About the Contributors

Murat Uysal was born in Istanbul, Turkey in 1973. He received the B.Sc. and the M.Sc. degree in electronics and communication engineering from Istanbul Technical University, Istanbul, Turkey, in 1995 and 1998, respectively, and the Ph.D. degree in electrical engineering from Texas A&M University, College Station, Texas, in 2001. From 1995 to 1998, he worked as a Research and Teaching Assistant in the Communication Theory Group at Istanbul Technical University. From 1998 to 2002, he was affiliated with the Wireless Communication Laboratory, Texas A&M University. During the Fall of 2000, he worked as a Research Intern at AT&T Labs-Research, Florham Park, New Jersey. Since 2002, he has been with the Department of Electrical and Computer Engineering, University of Waterloo, Canada, where he is now an Associate Professor. Dr. Uysal is an Associate Editor for IEEE Transactions on Wireless Communications and IEEE Communications Letters. He served a Guest Co-Editor for Journal on Wireless Communications and Mobile Computing's Special Issue on "MIMO Communications" published in 2004 and IEEE Journal on Selected Areas in Communications' Special Issue on "Optical Wireless Communications" to be published in 2010. Over the years, he has served on the technical program committee of more than 50 international conferences in the communications area. He recently co-chaired IEEE ICC'07 Communication Theory Symposium and chaired CCECE'08 Communications and Networking Symposium. Dr. Uysal is a Senior IEEE member. His general research interests are in the area of communications theory with particular emphasis on wireless applications. Specific areas include multi-input multi-output (MIMO) communications, space-time coding, diversity techniques and coding for fading channels, cooperative transmission, performance analysis over fading channels, channel estimation and equalization, orthogonal frequency division multiplexing (OFDM), and free-space optical communication.

* * *

Ibrahim Y. Abualhaol received his Master and Bachelor degree in Electrical Engineering from Jordan University of Science and Technology in 2000 and 2004, respectively. He received his Ph.D. degree in Electrical Engineering from the University of Mississippi in 2008. After that he joined Qualcomm Incorporation then Broadcom Corporation where he is working now as wireless system engineer. He is a member of Honor Society of Phi Kappa Phi, Institute of Electrical and Electronics Engineers, and Jordan Engineers Association. His areas of research include digital signal processing for communication systems, cooperative networks, diversity techniques, resource allocation, and practical system design that involves MIMO, OFDM, OFDMA, and MC-CDMA systems.

Ravi Adve received his B. Tech. degree in Electrical Engineering from IIT, Bombay in 1990 and his Ph.D. from Syracuse University in 1996, where his dissertation received a Doctoral Prize. Between 1997 and August 2000, he worked for Research Associates for Defense Conversion Inc. on contract with the Air Force Research Laboratory at Rome, NY. He joined the faculty at the University of Toronto in August 2000. Dr. Adve's research interests include practical signal processing algorithms for smart antennas with applications in wireless communications and radar. He is currently focused on linear precoding in wireless communications, cooperative communications and augmenting space-time adaptive processing with a waveform dimension.

Muhammad Ali Imran received his B.Sc. degree in Electrical Engineering from University of Engineering and Technology Lahore, Pakistan, in 1999, and the M.Sc. and Ph.D. degrees from Imperial College London, UK, in 2002 and 2007, respectively. He is currently a research fellow at the Centre for Communication Systems Research (CCSR) at the University of Surrey, UK. He has significant contributions to the project on "Fundamental Limits of Wireless Network Capacity": an Elective Research Programme of the Mobile VCE, www.mobilevce.com, funded by BBC, BT, Huawei, Nokia Siemens Networks, Nortel and Vodafone. He is also actively involved in the project on "Reconfigurable OFDMA-based Cooperative Networks Enabled by Agile Spectrum Use (ROCKET)" funded by the European Union's 7th Framework Programme. His main research interests include the analysis and modelling of the physical layer, resource allocation in multiuser communication networks and evaluation of the fundamental capacity limits of wireless networks.

Elzbieta Beres was born in Wroclaw, Poland in 1978. She received the B.Eng. degree in electrical engineering in 2002 from Carleton University, Ottawa, ON, Canada, and the M.A.Sc. degree in 2004 from the University of Toronto, Toronto, ON, Canada. From 1999 to 2000 she was an exchange student at the Institut National Polytechnique de Grenoble, in Grenoble, France. She is currently working toward the Ph.D. degree at the University of Toronto. Elzbieta Beres is the recipient of the postgraduate National Sciences and Engineering Research of Canada scholarship in 2002-2004 and 2006-2008. Her current research interests are in wireless communication systems, with focus on cooperative and multi-hop communications in mesh networks.

Manav R. Bhatnagar (IEEE Member) received his PhD from the University of Oslo, Norway in 2008 and M. Tech. in communications engineering from the Indian Institute of Technology Delhi, India. Currently, he is working as Post Doctoral Research Fellow at the UNIK-University Graduate Center, University of Oslo, Norway. He visited the Wireless Research Group at the Indian Institute of Technology Delhi, India from November 2007 to January 2008 for two months, the SPINCOM Group of the Department of Electrical and Computer Engineering, University of Minnesota Twin Cities from January 2008 to September 2008 for nine months, and the Alcatel-Lucent Chair at SUPÉLEC in France in November 2008 for ten days. His research interests include multiple-input multiple-output (MIMO) system, space-time coding, and cooperative communications.

Symeon Chatzinotas received his B.Sc. degree in Electrical and Computer Engineering from Aristotle University of Thessaloniki, Greece, in 2003, and the M.Sc. in Electronic Engineering from University of Surrey, UK, in 2006. He is currently a final-year PhD researcher at the Centre for Communication Systems Research (CCSR) at the University of Surrey, UK. He is actively involved in the project "Fundamental Limits of Wireless Network Capacity": an Elective Research Programme of the Mobile VCE, www.mobilevce.com, funded by BBC, BT, Huawei, Nokia Siemens Networks, Nortel and Vodafone. In the past, he has worked as a researcher in the Centre for Research & Technology, Hellas and in the National Centre of Scientific Research "DEMOKRITOS" in Greece. His main research interests include information-theoretic capacity limits, cooperative multiuser wireless systems, random matrix theory, free probability theory and optimization theory.

Byonghyok Choi received the B.S. in electronics engineering and the M.S. in electronic communication engineering from Hanyang University, Republic of Korea in 1997 and 1999, respectively. From 1999 to 2002, he has worked in LG electronics Co., LTD., Seoul, South Korea, as a research engineer in developing user interface software for CDMA(IS-95, J-STD-008, CDMA2000) mobile handsets. Since 2005, he has been working toward the Ph.D. degree with department of electrical and computer engineering, University of Florida, Gainesville, FL. His current research interests are in the area of communication theory applied to cooperative communication including broadcast and relay channel.

Jun-Hong Cui received her B.S. degree in Computer Science from Jilin University, China in 1995, her M.S. degree in Computer Engineering from Chinese Academy of Sciences in 1998, and her Ph.D. degree in Computer Science from UCLA in 2003. Currently, she is an Associate Professor in the Computer Science and Engineering Department at University of Connecticut. Her research interests cover the design, modelling, and performance evaluation of networks and distributed systems. Recently, her research mainly focuses on exploiting the spatial properties in the modeling of network topology, network mobility, and group membership, scalable and efficient communication support in overlay and peer-to-peer networks, algorithm and protocol design in underwater sensor networks. At UConn, she leads UbiNet (Ubiquitous Networking) Lab and UWSN (UnderWater Sensor Network) Lab. Please see http://www.cse.uconn.edu/~jcui/ for her recent projects and publications. Jun-Hong is actively involved in the community as an organizer, a TPC member, and a reviewer for many conferences and journals. She was a guest editor for ACM MCCR (Mobile Computing and Communications Review) and Elsevier Ad Hoc Networks. She now serves as an Associate Editor for Elsevier Ad Hoc Networks. She co-founded the first ACM International Workshop on UnderWater Networks (WUWNet'06), and she is now serving as the WUWNet steering committee chair. Jun-Hong received 2007 NSF CAREER Award and 2008 ONR YIP Award. She also received the United Technologies Corporation (UTC) Professorship in Engineering Innovation award at UConn in 2008.

Shuguang Cui received the B.Eng. degree (with the highest distinction) in radio engineering from Beijing University of Posts and Telecommunications, Beijing, China, in 1997, the M. Eng. degree in electrical engineering from McMaster University, Hamilton, ON, Canada, in 2000, and the Ph.D. degree in electrical engineering from Stanford University, Stanford, CA, in 2005. From 1997 to 1998, he was with Hewlett-Packard, Beijing, as a System Engineer. In the summer of 2003, he was with National Semiconductor, Santa Clara, CA, where he worked on the ZigBee project. From 2005 to 2007, he was an Assistant Professor with the Department of Electrical and Computer Engineering, University of Arizona, Tucson. He is currently an Assistant Professor of electrical and computer engineering with Texas A&M University, College Station. His current research interests include cross-layer optimization for resource-constrained networks, hardware and system synergies for high performance wireless radios, statistical signal processing, and general communication theories. Dr. Cui was a recipient of the NSERC graduate fellowship from the National Science and Engineering Research Council of Canada and the Canadian Wireless Telecommunications Association graduate scholarship. He has been serving as the TPC co-chair for the 2007 IEEE Communication Theory Workshop and the ICC'08 Communication Theory Symposium. He is currently serving as an Associate Editor for the IEEE Communications Letters and the IEEE Transactions on Vehicular Technology.

Mischa Dohler is now Senior Researcher with CTTC in Barcelona. Prior to this, he has been Senior Research Expert in the R&D division of France Telecom. He obtained his PhD in Telecommunications from King's College London, UK, in 2003. He has pioneered research on distributed cooperative space-time encoded communication systems, dating back to December 1999. He has published more than 100 technical journal and conference papers at a citation h-index of 15 and citation g-index of 29, holds several patents, co-edited and contributed to several books, has given numerous international short-courses, and participated in standardisation activities. He has been TPC member and co-chair of various conferences, is editor for the IEEE Communications Letters, the IEEE Transactions on Vehicular Technology, the IEEE Communications Magazine, the IEEE Wireless Communications, the IET Communications, the Elsevier Physical Communications journal, the EURASIP JWCN journal, among others. He is a Senior Member of the IEEE.

Lun Dong received the B.S. and M.S. degrees in 2001 and 2004 from Tsinghua University, Beijing, China, and the Ph.D. degree in 2008 from Drexel University, Philadelphia, PA, all in electrical engineering. From 2004 to 2008, he was a research assistant with Electrical & Computer Engineering department of Drexel University, working on the area of cooperative communications, cross-layer design and wireless security. From June 2007 to September 2007, he was a research intern at Mitsubishi Electric Research Laboratories (MERL), Cambridge, MA. From June 2008 to August 2008, he was a research intern at NTT DOCOMO Communications Laboratories USA, Palo Alto, CA. He is a recipient of the 2007 George Hill Jr. Endowed Fellowship of Drexel University.

Elza Erkip received the Ph.D. and M.S. degrees in Electrical Engineering from Stanford University, and the B.S. degree in Electrical and Electronic Engineering from Middle East Technical University, Turkey. She is now an Associate Professor of Electrical and Computer Engineering at Polytechnic Institute of NYU. Dr. Erkip received NSF CAREER award in 2001 and IBM Faculty Partnership Award in 2000. Her papers won *2004 Communications Society Stephen O. Rice Paper Prize in the Field of Communications Theory*, the *Best Paper Award of the Communication Theory Symposium, ICC 2007* and the *Student Paper Award, ISIT 2007*. Dr. Erkip is a Senior Member of IEEE. She is an Associate Editor of *IEEE Transactions on Information Theory*, and *IEEE Transactions on Communications* and a Guest Editor of *IEEE Signal Processing Magazine*. She is the Technical Program Co-Chair of *Communication Theory Symposium, GLOBECOM 2009*. She organized *Workshop on Cooperative Communications*, sponsored by WICAT, an NSF Industry/University Cooperative Research Center at Polytechnic Institute of NYU. She was the Technical Area Chair for the *MIMO Communications and Signal Processing* track of 41st Annual Asilomar Conference on Signals, Systems, and Computers, and the Technical Program Co-Chair of *2006 Communication Theory Workshop*. She has also served on Technical Program Committees of numerous IEEE Conferences. Her general research interests are in wireless communications, information theory and communication theory.

Michael Gastpar (Ph.D. EPFL, 2002, M.S. UIUC, 1999, Dipl. El-Ing, ETH, 1997) is currently an Associate Professor in the Department of Electrical Engineering and Computer Sciences at the University of California, Berkeley. He was also a student in electrical engineering and philosophy at the Universities of Edinburgh and Lausanne, and a summer researcher in the Mathematics of Communications Department at Bell Labs, Lucent Technologies. He won the 2002 EPFL Best Thesis Award, an NSF CAREER award in 2004, and an Okawa Foundation Research Grant in 2008.

J. Harshan received Bachelor of Engineering degree in Electronics and Communication from the National Institute of Engineering, Mysore, India in 2004. From October 2004 to December 2005, he was with Robert Bosch (India) limited, Bangalore, India. Currently, he is a working towards his Ph.D. in the Department of Electrical Communication Engineering, Indian Institute of Science, Bangalore, India. His primary research interests include wireless communication, space-time coding, coding for wireless relay networks and coding for multiple access channels.

Are Hjørungnes (IEEE Senior Member) works as an Associate Professor at UNIK - University Graduate Center, at the University of Oslo, Norway. He obtained his Sivilingeniør (M.Sc.) degree (with honors) in 1995 from the Department of Telecommunications at the Norwegian Institute of Technology in Trondheim, Norway, and his Doktor ingeniør (Ph.D.) degree in 2000 from the Signal Processing Group at the Norwegian University of Science and Technology. His current main research areas are signal processing, communications, and wireless networks. From August 2000 to December 2000, he worked as a researcher at Tampere University of Technology, in Finland, within the Tampere International Center for Signal Processing. From March 2001 to July 2002, he worked as a postdoctoral fellow at the Federal University of Rio de Janeiro in Brazil, within the Signal Processing Laboratory. From September 2002 to August 2003, he worked as a postdoctoral fellow at the Helsinki University of Technology in Finland, within the Signal Processing Laboratory. From September 2003 to August 2004, he was working as a postdoctoral fellow at the University of Oslo in Norway, at the Department of Informatics, within the Digital Signal Processing and Image Analysis Group. He has held visiting appointments at the Image and Signal Processing Laboratory at the University of California, Santa Barbara, the Signal Processing Laboratory of the Federal University of Rio de Janeiro, the Mobile Communications Department at Eurecom Institute in France, the University of Manitoba in Canada, the Alcatel-Lucent Chair at SUPÉLEC in France, and the Electrical and Computer Engineering at the University of Houston in USA. From March 2007, he has been serving as an Editor for IEEE Transactions on Wireless Communications. He co-authored the paper winning the 2007 Best Paper Award for the IEEE International Conference on Wireless Communications, Networking and Mobile Computing (WiCOM 2007).

Gi-Hong Im was with AT&T Bell Laboratories, Holmdel, NJ, where he was responsible of design and implementation of high-speed digital transmission systems for loop plant, local area network and broadband access applications (1990-1996). He has authored or co-authored more than twenty standards contributions to standards organizations such as ANSI T1E1.4, ETSI, IEEE 802.9, ANSI X3T9.5, and the ATM Forum. These contributions have led to the adoption of three AT&T proposals for new standards for high-speed LANs and broadband access. In 1995, he was appointed as Distinguished Member of Technical Staff at AT&T Bell Laboratories. Since 1996, he has been with POSTECH as a professor. From 1996 to 2000, he was a Bell Laboratories Technical Consultant. From 2002 to 2003, he was a visiting vice president of Samsung Electronics, where he worked on 4G wireless communication systems. Dr.

Im received the 1996 Leonard G. Abraham Prize Paper Award from the IEEE Communications Society, and the 2000 LG Award from LG Electronics, and the 2005 National Scientist Award from Korea government. He serves as an editor for the IEEE Transactions on Communications and an associate editor for the IEEE Communications Letters, and also serves on Technical Program Committee of Globecom, ICC and WCNC. He holds ten U.S. patents with seven more patents pending.

Reza Hoshyar received the B.S. degree in communications Eng. and the M.S. and Ph.D. degrees both in mobile communications from Tehran University, Tehran, Iran, in 1991, 1996, and 2001, respectively. He received the top and second rank awards for his B.S. and M.S. degrees from Tehran University. He is currently a senior research fellow in the Centre for Communication Systems Research (CCSR) of University of Surrey (UniS). His current interest is on iterative detection, signal processing and scheduling for MIMO and cooperative transmission techniques. He has been actively involved in several European projects. He is currently involved in ROCKET ICT project and is principal investigator of Mobile VCE elective program on fundamental capacity limits. Besides his academic activities, Dr Hoshyar was involved in many research companies and continues collaboration with main industrial players inside and outside United Kingdom.

Eduard A. Jorswieck received his Diplom-Ingenieur degree and Doktor-Ingenieur (Ph.D.) degree, both in electrical engineering and computer science from the Berlin University of Technology (TUB), Germany, in 2000 and 2004, respectively. He was with the Fraunhofer Institute for Telecommunications, Heinrich-Hertz-Institute (HHI) Berlin, from 2001 to 2006. Since 2005 he has been a lecturer at the TUB. In 2006, he joined the Signal Processing Department at the Royal Institute of Technology (KTH) as a post-doc and became a Assistant Professor in 2007. Since February 2008, he has been the head of the Chair of Communications Theory and Full Professor at Dresden University of Technology (TUD), Germany. Dr. Jorswieck is senior member of IEEE and member of the IEEE SPCOM Technical Committee. From 2008-2010 he serves as an Associate Editor for IEEE Signal Processing Letters. In 2006, he was co-recipient of the IEEE Signal Processing Society Best Paper Award.

Dr. **George K. Karagiannidis** was born in Pithagorion Town, Samos Island , Greece. He received his university degree in 1987 and his PhD degree in 1999, both in Electrical Engineering, from the University of Patras, Greece. From 2000 to 2004 he was Researcher at the Institute for Space Applications and Remote Sensing, National Observatory of Athens, Greece. In June 2004, he joined the faculty of Aristotle University of Thessaloniki, Greece where he is currently an Assistant Professor at the Electrical and Computer Engineering Department. His major research interests include Wireless Communications Theory, Digital Communications over Fading Channels, Cooperative Diversity Systems, Satellite Communications, Cognitive Radio and Free-Space Optical Communications. Dr. Karagiannidis has published and presented more than 80 technical papers in scientific journals and international conferences, he is co-author in three chapters in books and also co-author in a Greek Edition Book on Mobile Communications. He is co-recipient of the Best Paper Award of the Wireless Communications Symposium (WCS) in 2007 IEEE International Conference on Communications (ICC'07). He is member of the Editorial Boards of IEEE Transactions on Communications, IEEE Communications Letters and EURASIP Journal on Wireless Communications and Networking.

Onur Kaya received the B.S. degree in electrical and electronics engineering from Bilkent University, Ankara, Turkey, in 2000, and the M.S. and Ph.D. degrees in electrical and computer engineering from University of Maryland, College Park, in 2002 and 2005 respectively. Currently he is with the Electronics Engineering Department at Işık University in Istanbul, Turkey as an Assistant Professor. His research interests are in the fields of communication theory and information theory, with particular focus on wireless communications and network information theory. Onur Kaya is a co-recipient of one of the "IEEE Wireless Communications and Networking Conference 2008 Best Paper Awards", for his paper titled "Achievable Rates for the Three User Cooperative Multiple Access Channel."

Dongsik Kim received the B.S. degree in electronic and electrical engineering from Pohang University of Science and Technology (POSTECH), Kyungbuk, Korea, in 2005. He is currently working toward the Ph.D. degree at Communications Research Laboratory, POSTECH. His current research interests are adaptive signal processing, cooperative communications, and MIMO systems.

Il-Min Kim received the B.S. degree in electronics engineering from Yonsei University, Seoul, Korea, in 1996, and the M.S. and Ph.D. degrees in electrical engineering from the Korea Advanced Institute of Science and Technology (KAIST), Taejon, Korea, in 1998 and 2001, respectively. From July 1997 to Aug. 2001, he worked as a Member of Technical Staff at the Electronics and Telecommunications Research Institute (ETRI). From October 2001 to August 2002 he was with the Dept. of Electrical Engineering and Computer Sciences at MIT, Cambridge, USA, and from September 2002 to June 2003 he was with the Dept. of Electrical Engineering at Harvard, Cambridge, USA, as a Postdoctoral Research Fellow. In July 2003, he joined the Dept. of Electrical and Computer Engineering at Queen's University, Kingston, Canada, where he is currently an Assistant professor. His research interests include distributed space-time codes, cooperative diversity network, cognitive radio, wireless network coding, multiple-input multiple-output (MIMO) systems, resource management for wireless communications, and wireless video. He received a number of awards including the Best Young Alumnus award from KAIST, Gold prize of 2001 Samsung International Paper Contest and Bronze Prize of 1999 IEEE-Korea Session Paper Contest. He also received three teaching awards from the ECE department at Queen's University in 2005, 2006, and 2008 in recognition of outstanding effort and enthusiasm in undergraduate teaching. He is currently serving as Editor for IEEE Transactions on Wireless Communications and Journal of Communications and Networks (JCN) . He is a Senior Member of IEEE and a registered Professional Engineer with Professional Engineers Ontario (PEO). He is Vice Chair, Programme, of IEEE Kingston Section.

Ioannis Krikidis was born in Athens, Greece, in 1977. He received the B.S. degree in computer engineering from the Computer Engineering and Informatics Department, University of Patras, Patras, Greece, in 2000 and the M.Sc. and Ph.D. degrees in electrical engineering from Ecole Nationale Supérieure des Télécommunications (ENST), Paris, France, in 2001 and 2005, respectively. From 2001 to 2002, he was a Research Associate with the National Capodistrean, University of Athens. From 2006 to 2007, he was a Postdoctoral Researcher with ENST. He is currently a Research Fellow with the School of Engineering and Electronics, University of Edinburgh, Edinburgh, U.K. His research interests include reconfigurable architectures, wireless communication systems, and cooperative ad hoc networks. Dr. Krikidis is a member of the Technical Chamber of Greece.

Vikram Krishnamurthy was born in 1966. He received the Bachelor's degree from the University of Auckland, New Zealand, in 1988 and the Ph.D. degree from the Australian National University, Canberra, in 1992. Prior to 2002, he was a Chaired Professor at the Department of Electrical and Electronic Engineering, University of Melbourne, Australia, where he also served as deputy head of department. He is currently a Professor and holds the Canada Research Chair at the Department of Electrical Engineering, University of British Columbia, Vancouver, Canada. His current research interests include computational game theory, stochastic dynamical systems for modeling of biological ion channels, and stochastic optimization and scheduling. Dr. Krishnamurthy is a Fellow of the IEEE and has served as Associate Editor for several journals, including the IEEE Transactions on Automatic Control, the IEEE Transactions on Signal Processing, the IEEE Transactions on Aerospace and Electronic Systems, the IEEE Transactions on Circuits and Systems B, the IEEE Transactions on Nanobioscience, and Systems and Control Letters. In 2009 and 2010, he serves as IEEE distringushed lecturer for the signal processing society.

Jing Li received the B.S. degree in computer science from Beijing (Peking) University in 1997, and the masters and the Ph.D. degrees in electrical engineering from Texas A&M University, College Station, TX, in 1999 and 2002. She spent the summer of 2000 and 2001 with Seagate Research, Pittsburgh, PA, and TycoCommunications Laboratory, Eatontown, NJ, respectively. After obtaining her Ph.D. degree, she joined Lehigh University where she is currently an associate professor. Her research interests fall in the general area of wireless communications, and data storage systems, with a focus on coding theory and practice. She served as a symposium co-chair in *IEEE GLOBECOM* 2006, *ChinaCom* 2006, and *IEEE ICC* 2008, an associate editor for *IEEE Communications Letters* from 2004 to 2008, and is currently a member of editorial board for *IEEE Communication Surveys and Tutorials*.

Mustafa M. Matalgah received his Ph.D. in Electrical and Computer Engineering in 1996 from The University of Missouri, Columbia. From 1996 to 2002, he was with Sprint, Kansas City, MO, USA, where he led various projects dealing with optical transmission systems and the evaluation and assessment of 3G wireless communication emerging technologies. In 2000, he was a Visiting Assistant Professor at The University of Missouri, Kansas City, MO, USA. Since August 2002, he has been with The University of Mississippi in Oxford where he is now an Associate Professor in the Electrical Engineering Department. His current technical and research experience is in the fields of emerging wireless communications systems and communication networks. He also previously published in the fields of signal processing and optical binary matched filters. He has published more than 70 refereed journal and conference proceeding papers, five book chapters, and more than 20 technical industrial applied research reports in these areas. Dr. Matalgah received several certificates of recognition for his work accomplishments in the industry and academia. He is the recipient of the Best Paper Award of the IEEE ISCC 2005, La Manga del Mar Menor, Spain. He is the recipient of the 2006 School of Engineering Junior Faculty Research Award at The University of Mississippi. He served on several international conferences technical program and organization committees.

Djamal-Eddine Meddour received his computer engineering degree (Honours) from I.N.I (Institut National d'Informatique), Algiers, Algeria in 2000 and Master's degree in computer science from University of Versailles in 2001 and his PhD in computer science from University of Paris VI in 2004. He is currently a research engineer with Orange Labs, Lannion, France. His main research activities concern

Resource Management for wireless mesh networking, quality of service for multimedia peer to peer networks, management and interoperability in new generation wireless networks. Dr D-E. Meddour is very active in research communities as reviewer and TPC member for ICC 08, WCNC 08, Med hoc Net 08, PIMRC 08, Globecom 07 Med Hoc Net '06, IWWAN 06, MobileSummit '06.He was a guest editor for the "Wireless Personal Communications" journal. Dr Meddour is serving as vice-chair for WWRF SIG4 on Home and Enterprise networks, he is an active member in 802.16j and 802.16m group.

Diomidis S. Michalopoulos was born in Thessaloniki, Greece, in 1983. He received the Diploma of Electrical and Computer Engineering from the Aristotle University of Thessaloniki in 2005, and since then he is pursuing a Ph.D degree in the ECE department. His research areas include digital communications over fading channels, with particular emphasis on the physical layer of Cooperative and Multihop Wireless Communications. Mr. Michalopoulos is co-recipient of the Best Paper Award of the Wireless Communications Symposium (WCS) in 2007 IEEE International Conference on Communications (ICC'07).

Hassnaa Moustafa obtained her PhD from the Computer and Network Department at the Ecole Nationale Superieure des Telecommunications (ENST)-Paris in the area of Unicast and Multicast Routing in Mobile Ad hoc Networks in 2004. Since then Hassnaa is a Research Engineer at France Telecom R&D (Orange Labs). Her current research activities concern Security, Authentication, Authorization, and Accounting (AAA), Services Access Control and IP Autoconfiguration. She is the scientific responsible of an integrated French national project with academic and industrial participants and is involved in some standardization activities, mainly within the context of the IETF. Hassnaa is an IEEE member and a member of the Engineering and Scientific Research Groups (ESRGroups). She also served as a TPC member of a number of International conferences and a as a reviewer for several international conferences and IEEE Transactions.

Athina P. Petropulu is currently a Professor at the Department of Electrical and Computer Engineering at Drexel University. She has held visiting appointments at SUPELEC in France and at Princeton University. Her research interests span the area of statistical signal processing, wireless communications and networking. She is the recipient of the 1995 Presidential Faculty Fellow Award in Electrical Engineering given by NSF and the White House. She is the co-author (with C.L. Nikias) of the textbook entitled, "Higher-Order Spectra Analysis: A Nonlinear Signal Processing Framework," (Englewood Cliffs, NJ: Prentice-Hall, Inc., 1993). She served as Editor-In-Chief of the IEEE Transactions on Signal Processing (2009-2011). She was IEEE SPS Vice President-Conferences and member of the IEEE Signal Processing Board of Governors. She was the General Chair of the 2005 International Conference on Acoustics Speech and Signal Processing (ICASSP-05), Philadelphia PA. She is co-recipient of the 2005 IEEE Signal Processing Magazine Best Paper Award.

Laxminarayana S. Pillutla recently (December, 2008) obtained his Ph.D. in electrical engineering from the University of British Columbia, Vancouver, Canada. Prior to his Ph.D., he obtained his masters (in 2004) and bachelors degree (in 2000) in electrical engineering from Wichita State University, Wichita, KS, USA and University of Madras, India respectively. His research interests are in physical and MAC layer aspects of ad hoc, sensor and cognitive radio networks, computational game theory and statistical signal processing.

H. Vincent Poor is the Dean of Engineering and Applied Science at Princeton University, where he is also the Michael Henry Strater University Professor of Electrical Engineering. His interests lie in the area of statistical signal processing, with applications in wireless networks and related fields. Among his publications are the recent books MIMO Wireless Communications (Cambridge, 2007) and Quickest Detection (Cambridge, 2009). Dr. Poor is a member of the National Academy of Engineering, a Fellow of the American Academy of Arts and Sciences, and a former Guggenheim Fellow. He is also a Fellow of the IEEE, the Institute of Mathematical Statistics, and other scientific and technical organizations. In 2005, he received the IEEE Education Medal. Recent recognition of his work includes the 2007 Marconi Prize Paper Award, the 2007 Technical Achievement Award of the IEEE Signal Processing Society, and the 2008 Aaron Wyner Award of the IEEE Information Theory Society.

Dr. **B. Sundar Rajan** received the B.Sc. degree in mathematics from Madras University, Madras, India, the B.Tech degree in electronics from Madras Institute of Technology, Madras, and the M.Tech and Ph.D. degrees in electrical engineering from the Indian Institute of Technology, Kanpur, India. He was a faculty member with the Department of Electrical Engineering at the Indian Institute of Technology in Delhi, India, from 1990 to 1997. Since 1998, he has been a Professor in the Department of Electrical Communication Engineering at the Indian Institute of Science, Bangalore, India. His primary research interests include space-time coding for MIMO channels, distributed space-time coding and cooperative communication, coding for multiple-access and relay channels, with emphasis on algebraic techniques. Dr. Rajan is an Associate Editor of the IEEE Transactions on Information Theory, an Editor of the IEEE Transactions on Wireless Communications, and an Editorial Board Member of International Journal of Information and Coding Theory. He served as Technical Program Co-Chair of the IEEE Information Theory Workshop (ITW'02), held in Bangalore, in 2002. He is a Fellow of Indian National Academy of Engineering and recipient of the IETE Pune Center's S.V.C Aiya Award for Telecom Education in 2004. Also, Dr. Rajan is a Member of the American Mathematical Society.

G. Susinder Rajan received his Bachelor of Engineering degree in Electronics and Communications from the College of Engineering Guindy, Anna University, India in the year 2005. He completed Ph.D. in Electrical Communication Engineering from the Indian Institute of Science in 2008. He is currently with Atheros Communications Inc., Chennai, India. His research interests include space-time coding, algebraic code constructions for MIMO systems and distributed space-time coding.

Anna Scaglione joined UC Davis in July 2008 as Associate Professor. Prior to this she was on the faculty at Cornell University where she joined in 2001 and was promoted to Associate Professor in 2006. Her first academic appointment as assistant professor was in 2000, at the University of New Mexico (Albuquerque, NM). She was awarded and is co-recipient of some awards: the 2000 IEEE Signal Processing Transactions Best Paper Award; the NSF Career Award in 2002, the Ellersick Best Paper Award (MILCOM 2005), the 2005 Best paper for Young Authors of the Taiwan IEEE Comsoc/ Information theory section. Her research interests are in the broad area of signal processing applied to communication networks and sensor systems.

Sidi-Mohammed Senouci received the M.Sc. degree from the University of Paris 13, France, and the Ph.D. degree from the University of Paris 6, France in 1999 and 2003 respectively. From 2002 to 2004, he was an associate lecturer in the University of Cergy-Pontoise, France. From December 2004, he joined France Telecom R&D Lannion where he is project leader of different European and French projects dealing with infrastructure-less networks. He acted as general chair of Ubiroads2007 workshop (collocated with IEEE GIIS2007) and editor of a special issue at UBICC journal. He is founding co-editor of the IEEE ComSoc Ad Hoc and Sensor Network Technical Committee (AHSN TC) Newsletter. He has been TPC member of various conferences. He acted as TPC Co-chair of VehiCom2009 Workshop associated with IEEE IWCMC'2009. He is a Member of IEEE and is a Senior Member of the French engineering society SEE.

Dae-Young Seol received the B.S. degree in electronic and electrical engineering from Pohang University of Science and Technology (POSTECH), Kyungbuk, Korea, in 2005. He is currently working toward the Ph.D. degree at Communications Research Laboratory, POSTECH. His current research interests are cognitive and cooperative communications for homo-/heterogeneous networks.

Aydin Sezgin received the Dipl.-Ing. (M.S.) degree in communications engineering and the Dr.-Ing. (Ph.D.) degree in electrical engineering (both with distinction) from the University of Applied Sciences, Berlin, Germany, in 2000 and the University of Technology Berlin, in 2005, respectively. From 2001 to 2006, he was with the Department of Broadband Mobile Communication Networks, Fraunhofer Institute for Telecommunications, Heinrich-Hertz-Institut (HHI), Berlin. Since 2006, he has been lecturer at the Department of Mobile Communications, University of Technology Berlin. Since 2006 and 2007, he has been a Postdoctoral Research Associate and Lecturer, respectively, with the Information Systems Laboratory, Department of Electrical Engineering, Stanford University, Stanford, CA. His current research interests are in the area of information theory, signal processing and multiple antenna systems (e.g., MIMO transceiver design and interference channels).

John M. Shea received the BS degree (with highest honors) in computer engineering from Clemson University in 1993 and the MS and PhD degrees in electrical engineering from Clemson University in 1995 and 1998, respectively. He is currently an Associate Professor in the Department of Electrical and Computer Engineering, University of Florida, Gainesville. Prior to that, he was an Assistant Professor at the University of Florida from July 1999 to August 2005 and a Postdoctoral Research Fellow at Clemson University from January 1999 to August 1999. Dr. Shea is an Editor for the IEEE Transactions on Wireless Communications. He was an Associate Editor for the IEEE Transactions on Vehicular Technology from 2002 to 2007. He was selected as a finalist for the 2004 Eta Kappa Nu Outstanding Young Electrical Engineer Award and was a National Science Foundation fellow from 1994 to 1998.

Birsen Sirkeci-Mergen is an assistant professor at San Jose State University (SJSU) in the department of electrical engineering. Prior to joining SJSU, she was a postdoctoral researcher at UC Berkeley, CA. She received her Ph.D. from Cornell University, Ithaca, NY in 2006 and her M.S. from Northeastern University, Boston, MA in 2001, both in Electrical and Computer Engineering. She received her B.Sc. degrees in Electrical and Electronics Engineering and Mathematics in 1998 from Middle East Technical University (METU), Ankara, Turkey. Her research lies in the areas of wireless communications, sensor networks and statistical signal processing. She received the Fred Ellersick Award for the best unclassified paper in MILCOM 2005 with her co-authors.

John S. Thompson received the B.Eng. and Ph.D. degrees from the University of Edinburgh, Edinburgh, U.K., in 1992 and 1996, respectively. From July 1995 to August 1999, he was a Postdoctoral Researcher with the University of Edinburgh, which was funded by the U.K. Engineering and Physical Sciences Research Council and Nortel Networks. In September 1999, he was a Lecturer with the School of Engineering and Electronics, University of Edinburgh, where, since October 2005, he has been a Reader. He has authored approximately 150 papers to date, including a number of invited papers, book chapters, and tutorial talks; he is currently coauthoring an undergraduate textbook on digital signal processing. He is currently the Editor-in-Chief of the *IET Signal Processing Journal*. His research interests include signal-processing algorithms for wireless systems, antenna array techniques, and multihop wireless communications. Dr. Thompson was a Technical Program Cochair for the IEEE International Conference on Communications, which was held in Glasgow, U.K., in June 2007.

Sennur Ulukus received the B.S. and M.S. degrees in electrical and electronics engineering from Bilkent University, Ankara, Turkey, in 1991 and 1993, respectively, and the Ph.D. degree in electrical and computer engineering from Rutgers University, NJ, in 1998. During her Ph.D. studies, she was with the Wireless Information Network Laboratory (WINLAB), Rutgers University. From 1998 to 2001, she was a Senior Technical Staff Member at AT&T Labs-Research in NJ. In 2001, she joined the University of Maryland at College Park, where she is currently an Associate Professor in the Department of Electrical and Computer Engineering, with a joint appointment at the Institute for Systems Research (ISR). Her research interests are in wireless communication theory and networking, network information theory for wireless networks, signal processing for wireless communications and security for multi-user wireless communications. Sennur Ulukus is a recipient of the 2005 NSF CAREER Award, and a co-recipient of the 2003 IEEE Marconi Prize Paper Award in Wireless Communications. She serves/served as an Associate Editor for the IEEE Transactions on Information Theory since 2007, as an Associate Editor for the IEEE Transactions on Communications between 2003-2007, as a Guest Editor for the IEEE Journal on Selected Areas in Communications in 2006-2008, as the co-chair of the Communication Theory Symposium at the 2007 IEEE Global Telecommunications Conference, as the co-chair of the Medium Access Control (MAC) Track at the 2008 IEEE Wireless Communications and Networking Conference, as the co-chair of the Wireless Communications Symposium at the 2010 IEEE International Conference on Communications, and as the Secretary of the IEEE Communication Theory Technical Committee (CTTC) since 2007.

Haidong Wang is a staff systems engineer at Thales Communications, Inc. Prior to this he was a principal member of technical staff at Armillaire Technologies, developing ATM and IP multiservice switch products. He also worked at Hughes Network Systems on various cellular and satellite communication projects. He received his M.S.E.E. from Georgia Institute of Technology, Atlanta, Georgia, in 1995 and his B.Sc from Tsinghua University, Beijing, China, in 1993.

Chan Wong Wong received the B.S. and the M.S. in electrical engineering from National Taiwan University, Taipei, Taiwan in 2002 and 2004, respectively. Chan Wong is currently pursuing his PhD degree at University of Florida, Gainesville, FL. From 2002 to 2004 he was a teaching assistant at Graduate Institute of Communications Engineering, National Taiwan University, Taipei, Taiwan. During 2003 to 2006 he was with Afa Technologies, Inc., Taipei, Taiwan, as a DSP system engineer in developing demodulators for various digital video broadcasting standards. His current research interests lie in the area of communication theory applied to equalization and coding for wireless communication. Chan Wong is the member of The Phi Tau Phi Scholastic Honor Society of the Republic of China.

Tan F. Wong received the B.Sc. degree (1st class honors) in electronic engineering from the Chinese University of Hong Kong in 1991, and the M.S.E.E. and Ph. D. degrees in electrical engineering from Purdue University in 1992 and 1997, respectively. He was a research engineer working on the high speed wireless networks project in the Department of Electronics at Macquarie University, Sydney, Australia. He also served as a post-doctoral research associate in the School of Electrical and Computer Engineering at Purdue University. Since August 1998 he has been with the University of Florida, where he is currently an associate professor of electrical and computer engineering. He was Editor for Wideband and Multiple Access Wireless Systems for the *IEEE Transactions on Communications* and was the Editor-in-Chief for the *IEEE Transactions on Vehicular Technology* from 2003 to 2006.

Zhihang Yi received his B.Eng degree from Information Science and Electrical Engineering Department, Zhejiang University, Hangzhou, China PR, in July 2003. In Jan 2004, he joined the Wireless Information Transmission Lab (WITL) of Electrical and Computer Engineering Department, Queen's University, Kingston, Canada, where he received his M.Sc (Eng) degree in Aug 2005. Now he is a Ph.D. candidate and working as a Research Assistant. Mr. Yi's research interests include cooperative networks, multiuser systems, OFDM systems, and MIMO systems. He has published five papers in IEEE journals and six other journal papers have been submitted recently.

Meng Yu received the B.S. and M.S degrees in electrical engineering from Southeast University, Nanjing, China in 1999 and 2002 respectively. He received the Ph.D degree from the Department of Electrical and Computer Engineering, Lehigh University, Bethlehem, PA, USA in 2008, where he worked as a research assistant. He is currently a member of technical staff in Sandbridge Technologies, Inc, White Plains, NY. His research interests are in the areas of communication theory, signal processing, relay and user cooperation. Prior to attending graduate school at Lehigh University, he was a hardware engineer with Motorola Electronics Ltd. where he was engaged in the development of GSM smart phones.

Melda Yuksel received the Ph.D. degree in electrical engineering at Polytechnic University, Brooklyn, NY, in August 2007, and the B.S. degree in electrical and electronics engineering from Middle East Technical University, Ankara, Turkey, in 2001. She joined TOBB University of Economics and Technology, Ankara, Turkey, in Fall 2007. In 2004, she was a summer researcher in Mathematical Sciences Research Center, Bell-Labs, Lucent Technologies, Murray Hill, NJ. Her research interests include communication theory and information theory and more specifically cooperative communications, network information theory, and information-theoretic security over communication channels. Melda Yuksel is the recipient of the best paper award at the Communication Theory Symposium of the 2007 IEEE International Conference on Communications.

Tae-Won Yune received the B.S. degree in electronic and electrical engineering from Hanyang University, Seoul, Korea, in 2004. He is currently working toward the Ph.D. degree at Communications Research Laboratory, Pohang University of Science and Technology (POSTECH), Kyungbuk, Korea. His current research interests are adaptive signal processing, cooperative communications, and MIMO systems.

Shengli Zhou received the B.S. and M.Sc. degrees in electrical engineering and information science from the University of Science and Technology of China, Hefei, China, in 1995 and 1998, respectively, and the Ph.D. degree in electrical engineering from the University of Minnesota, Minneapolis, in 2002. Since 2003, he has been an Assistant Professor with the Department of Electrical and Computer Engineering, University of Connecticut, Storrs. He now holds a United Technologies Corporation (UTC) Professorship in Engineering Innovation. His research interests include wireless communications and signal processing. His current research interests are in underwater acoustic communications and networking. Dr. Zhou served as an Associate Editor for the IEEE Transactions on Wireless Communications from February 2005 to January 2007. 2007, and is now an associate editor for IEEE Transactions on Signal Processing. He was the recipient of the 2007 Office of Naval Research Young Investigator Award, and the 2007 Presidential Early Career Award for Scientists and Engineers.

Zhong Zhou received the B.Eng. degree in telecommunication engineering and the M. Eng degree in computer science from Beijing University of Posts and Telecommunications, Beijing, China, in 2000 and 2003, respectively. He is currently working toward the Ph.D. degree in computer science and engineering with the University of Connecticut (UCONN), Storrs. Since January 2006, he has been with the Underwater Sensor Network Laboratory and the Ubiquitous Networking Research Laboratory, Department of Electrical and Computer Engineering, UCONN, as a Graduate Research Assistant. His current research interests are underwater acoustic communication and networking, localization, and cross-layer design for wireless networks.

Index